Algebraic, Number Theoretic, and Topological Aspects of Ring Theory

Jean-Luc Chabert • Marco Fontana •
Sophie Frisch • Sarah Glaz • Keith Johnson
Editors

Algebraic, Number Theoretic, and Topological Aspects of Ring Theory

Springer

Editors

Jean-Luc Chabert
Département de Mathématiques
Université de Picardie
Amiens, France

Marco Fontana
Università degli Studi Roma Tre
Dipartimento di Matematica e Fisica
Rome, Italy

Sophie Frisch
Institut für Analysis und Zahlentheorie
Technische Universität Graz
Graz, Austria

Sarah Glaz
Department of Mathematics
University of Connecticut
Storrs, CT, USA

Keith Johnson
Department of Mathematics and Statistics
Dalhousie University
Halifax, NS, Canada

ISBN 978-3-031-28846-3 ISBN 978-3-031-28847-0 (eBook)
https://doi.org/10.1007/978-3-031-28847-0

Mathematics Subject Classification: 13-06, 16-06, 16D, 16L, 16N, 16P, 16S

This Springer imprint is published by the registered company Springer Nature Switzerland AG
The registered company address is: Gewerbestrasse 11, 6330 Cham, Switzerland

In memory
of our friend and colleague, Paul-Jean Cahen

Preface

This volume was originally planned to be the joint proceedings of two conferences: The Conference on Rings and Polynomials which was to be held in July 2020 at the Technische Universität, in Graz, Austria, and The Fourth International Meeting on Integer-Valued Polynomials and Related Topics which was to be held in December 2020 at the CIRM in Luminy, France. The COVID-19 pandemic and the associated cancelation of most international travel forced the cancellation of the second of these and the rescheduling of the first to July 19–24, 2021, and its conversion to a hybrid format in which many of the speakers and attendees participated remotely by video link. This volume is, therefore, a combination of papers intended for the second conference and papers presented, either remotely or in person, at the first. Their high quality is a testament to how effectively the participants met and overcame a considerable challenge.

The papers in the collection range widely over the algebraic, number theoretic and topological aspects of rings, algebras and polynomials. There are, however, two areas of particular note: topological methods in ring theory and integer-valued polynomials. These were two areas of concentration at the Graz conference, and the second was, of course, to be the focus of the Luminy conference. An additional note is that the volume as a whole is dedicated to the memory of Paul-Jean Cahen, a co-author or research collaborator with some of the conference participants and a friend to many, many of the others. This collection contains a memorial article about Paul-Jean, written by his long-time research collaborator and co-author Jean-Luc Chabert with a translation of it from French by Sophie Frisch. Sadly, Paul-Jean is not the only loss that the commutative algebra community has suffered recently. Robert Gilmer, Wolmer Vasconcelos, Muhammad Zafrullah, Nicholas Baeth and Dan Anderson have all recently died. The last two of these, Nick Baeth and Dan Anderson, each have a paper that is appearing posthumously in this collection, and for each, there is a short obituary appended.

Among the people who helped to make this volume possible are Sabine Evrard, who did most of the organizational work for the Luminy conference, and the local organizing committee for the Graz conference consisting of Amr Al-Maktry, Chimere Anabanti, Victor Fadinger, Sarah Nakato and Daniel Windisch.

We would like to thank the following sponsors of the Graz conference: Technische Universität Graz (Graz University of Technology), FWF (the Austrian Science Foundation), Land Steiermark (the state of Styria) and Österreichische Mathematische Gesellschaft (the Austrian Mathematical society).

Last but not least, we thank the editorial staff at Springer, in particular Elizabeth Loew and Saveetha Balasundaram, for their care, patience and encouragement during the preparation of this volume.

Amiens, France — Jean-Luc Chabert
Rome, Italy — Marco Fontana
Graz, Austria — Sophie Frisch
Storrs, USA — Sarah Glaz
Halifax, Canada — Keith Johnson

Contents

Paul-Jean Cahen (1946–2019)

Jean-Luc Chabert

Paul-Jean Cahen, born on February 10, 1946, in La Flèche (France), died on June 14, 2019, in Aix en Provence. He was the fifth of seven siblings. Growing up in La Flèche, he received his secondary education at the military academy Le Prytanée. In 1963, in his senior year of high school (which, at the time, was called the class of elementary mathematics), Paul-Jean's mathematics professor happened to be the future Abel prize laureate Yves Meyer. He instilled in Paul-Jean a taste for mathematics (and still remembers him in 2019 in an interview for the journal of the CNRS [1]). In 1965, Paul-Jean Cahen entered the École Polytechnique.

J.-L. Chabert (✉)
Université de Picardie, Amiens, France
e-mail: jlchabert@me.com

J.-L. Chabert et al. (eds.), *Algebraic, Number Theoretic, and Topological Aspects of Ring Theory*, https://doi.org/10.1007/978-3-031-28847-0_1

In the summer of 1967, he obtained his engineering diploma from the École Polytechnique and married Martine Taton. A son was born: Alexandre. At that time, Paul-Jean was employed as a researcher at the Centre National de la Recherche Scientifique (*CNRS*) in Paris. In 1970, he defended his doctoral thesis (*thèse de 3ème cycle*) directed by Pierre Samuel.

The following year, he went as a postdoc to Canada and spent 3 years there: first at Queen's university in Kingston, where he defended a doctoral thesis directed by Paulo Ribenboim, then at McGill in Montreal, and finally as a visiting assistant professor at the University of British Columbia. In 1973, back in France, he earned his *Doctorat d'État* (the advanced degree corresponding to the current *habilitation à diriger des recherches*) at Orsay (Université Paris XI), again supervised by Pierre Samuel.

In 1974, he was appointed professor at the University of Tunis, where he stayed for 14 years. During his time in Tunisia, Paul-Jean and Martine divorced. In a discussion group on "mathematics and philosophy," Paul-Jean met Thérèse Baduel. In 1981, they married. With Thérèse, Paul-Jean had two children, Raphaël and Anne-Sophie. Also, Paul-Jean cared for Thérèse's two children from a previous marriage, Talel and Amina, as for his own children.

In 1988, Paul-Jean returned to France, to accept a professorship at the faculty of sciences and technology at the Université Aix-Marseille III, where he remained until his retirement.

He took a very active role at his university, even beyond the mathematics department. For a period of 5 years, he was director of the faculty of sciences and technology (UFR Sciences et Techniques) of the university, and collaborated closely with the university's board of administration.

Later, he was vice-president of the Mathematical Society of France (SMF) [2] for 6 years and, during this time, engaged himself a lot for the CIRM, the center of research in mathematics at Luminy, Marseilles.

Paul-Jean has eight mathematical descendants, whose doctoral theses he guided: Othman Echi (1990), Ahmed Ayache (1991), Mohamad El-Hajjar (1995), Nathalie Gonzales (1997), Francesca Tartarone (1998), Richard Robert (2001), Sébastien Pellerin (2002), and Julie Yeramian (2004).

In addition to attending and organizing conferences, Paul-Jean often visited universities abroad, for instance, the University la Sapienza in 1992 and later at least four times the University Roma Tre (invited by Marco Fontana), the University of North Carolina at Charlotte (invited by Evan Houston) for the Fall semester of 1996, Florida State University in Tallahassee (invited by Robert Gilmer) for the Spring semester of 1997, Technische Universität Graz (invited by Sophie Frisch) in June 2001, the University of Connecticut in Storrs (invited by Sarah Glaz) in Fall 2005, the University of North Carolina again in Spring 2006, and Princeton (invited by Manjul Bhargava) for a month in 2006.

Paul-Jean's research specialty was commutative algebra, with one recurring theme—integer-valued polynomials—that he investigated throughout his academic life, from his two theses until his last papers in 2018.

According to Math Reviews (MathSciNet), Paul-Jean Cahen published 66 papers, 40% of which do not concern polynomials. They are often the fruit of international collaborations (in particular, with Tunisia, Morocco, Italy, and the USA) and encompass many and varied topics:

Henselian fields (his first paper), torsion theory and associated prime ideals (subject of his doctoral thesis at Queen's university), pairs of rings sharing an ideal, Krull dimension and valuative dimension, the "dimension formula," saturated chains of rings, minimal extensions of rings, Jaffard rings, Nagata rings, Mori rings, Hilbert rings, pseudo-valuation rings, and quasi-Prüfer rings, among others.

The profusion of collaborations is explained by Paul-Jean's ease of entering into discussions and his capacity to give intense thought to any question proposed by others (or by himself).

The remaining 60% of his papers treat various questions concerning integer-valued polynomials. The topic had been introduced by Pólya et Ostrowski in 1919, in the context of number fields, and then forgotten (except by Skolem, who wrote a beautiful paper on integer-valued polynomials in 1936) until Paul-Jean initiated the study of integer-valued polynomials in the context of commutative algebra in his theses. The subject is situated between algebra and number theory, but also touches p-adic analysis, and later also combinatorics, topology, dynamic systems, probability theory, and, in recent years, even non-commutative algebra. Paul-Jean entered into many international collaborations, but, it turns out that his principal collaborator was myself.

I met Paul-Jean when we were both preparing our doctoral theses in Paris under Pierre Samuel. We regularly met as a small group to work on problems posed by Benzaghou concerning the so-called Fatou property. To show that this property passes from a ring D to its polynomial ring $D[X]$, Paul-Jean had the idea to use integer-valued polynomials on D :

$$\mathrm{Int}(D) = \{f \in K[X] \mid f(D) \subseteq D\},$$

where K denotes the quotient field of the integral domain D. (Actually, the notation $\mathrm{Int}(D)$ became standard only in the 1990s under the influence of Robert Gilmer.) The integer-valued polynomials thus introduced as a tool became our thesis topic. We defended our theses, the first part of which was a joint work, in 1970 in Paris before the same jury.

Three years later, we each defended our thesis for the *Doctorat d'État* at Orsay, again both before the same jury. Our theses bore the same title, but had very different content. Paul-Jean had used his work on torsion theory to study polynomial torsion. He generalized Pólya's and Ostrowski's results from rings of integers in number fields to arbitrary Dedekind rings, and also to the case of polynomials in several indeterminates.

He had the idea to define and study the so-called Pólya-Ostrowski group of a Dedekind ring D, meaning the subgroup of the class group generated by the classes of products of all prime ideals of the same finite norm. This group measures how far

Int(D) is from having a regular basis. Paul-Jean showed that this group is trivial in the case of cyclotomic fields.

Using the density of the ring Int(V), where V is a discrete valuation domain with finite residue field in the ring of continuous functions $\mathcal{C}(\widehat{V}, \widehat{V})$, he gave a new proof of the characterization of the spectrum of Int(V). By the same principle, he was able to describe the spectrum of Int(D) for more general domains than Dedekind rings, for example, analytically irreducible Noetherian domains.

During the first International Meeting on Integer-Valued Polynomials that we had organized at the CIRM in Marseilles in 1990, the idea was born to write a comprehensive monograph on the subject. This was intensive work, but working with Paul-Jean was very stimulating. We rewrote each chapter several times, because there was always a better way to present our stuff. Our book Integer-Valued Polynomials was published in 1997 by the AMS in their Mathematical Surveys and Monographs series. The subject "rings of integer-valued polynomials" first appeared in the mathematics subject classification of the AMS in 2000, under (13F20).

It is impossible to list exhaustively all of Paul-Jean's work on the subject. Let us just mention polynomials that are integer-valued as well as their derivatives or their divided differences, integer-valued rational functions, Newton and Schinzel type sequences in number fields, etc. Also, he had the idea to study polynomials that are integer-valued on an arbitrary subset of a domain, and the polynomial closure of subsets. This enumeration does not do justice to Paul-Jean, because he always brought new ideas to each subject that he took up.

The second International Meeting on *Integer-Valued Polynomials* took place in 2000, again at the CIRM, and we were happy to welcome Manjul Bhargava, whose first papers treated integer-valued polynomials. He presented some of his remarkably simple, but efficient, ideas, in particular the concept of P-ordering, which permits one to obtain in an algorithmic fashion the characteristic ideals of a ring of polynomials integer-valued on a subset, thus adding additional depth to a subject introduced by Paul-Jean. Paul-Jean later wrote a joint paper with Bhargava.

There was a third International Meeting on Integer-Valued Polynomials at the CIRM in 2010, and a fourth, planned for 2020, was intended to be a tribute to the memory of Paul-Jean, but, alas, it could not take place.

When Paul-Jean retired, one might have expected, and he himself might have, too, that he would give up mathematics in favor of his hobby, singing. (He could, however, not resist and published several more mathematical papers 2017–2018.)

Paul-Jean's musical career began in Tunis, where he used to accompany a singer on the piano at her recitals in the cultural center in the Medina.

But mainly, Paul-Jean himself sang, in a tenor voice, wherever he went. In particular, in university choirs: in Tunis, in Storrs, in Charlotte with Randy Haldeman, at the university of St. Charles in Marseilles with Jérôme Conttenceau, and at the conservatory Darius Milhaud in Aix en Provence with Michel Piquemal. All who met him at a conference knew well his passion for singing, and especially, for Schumann's Lieder.

But he also played chess and bridge, loved poetry, wordplay, literature, theater, philosophy, etc.

Paul-Jean died of cancer. He fought it with courage, uncomplainingly, always thinking of others, of those who would remain. Also, in spite of chemotherapy and radiotherapy, he continued to sing in the choir of the region of Marseilles.

With Paul-Jean we lose a gifted mathematician, but also a colleague and friend of remarkable qualities. Those who knew him will certainly recognize him in the following two testimonies. The first is by a colleague from his university, a lawyer: "We will always remember our colleague's great elegance of soul and mind, his infallible sense for public service, joyful intellectual curiosity, and immense culture."

The second testimony is by Thérèse: "He steadily rested a luminous, loving, patient, attentive, benevolent, admiring gaze on five children, seven grand-children, and many of those who were so lucky as to meet him."

Jean-Luc Chabert
translated into English
by Sophie Frisch

References

1. Y. Meyer, https://images.math.cnrs.fr/Comment-enseigner-les-mathematiques-selon-Yves-Meyer.html
2. S.M.F., https://smf.emath.fr/actualites-smf/deces-de-paul-jean-cahen

Paul-Jean Cahen's Mathematical Publications

1. Cahen, Paul-Jean and Picavet, Gabriel and Picavet-L'Hermitte, Martine, Pointwise minimal extensions, *Arabian Journal of Mathematics* **7** (2018), 249–271.
2. Cahen, Paul-Jean and Chabert, Jean-Luc, Test sets for polynomials: n-universal subsets and Newton sequences, *Journal of Algebra* **502** (2018), 277–314.
3. Cahen, Paul-Jean, Chabert, Jean-Luc and Kedlaya, Kiran S., Bhargava's early work: the genesis of P-orderings, *American Mathematical Monthly* **124** (2017), 773–790.
4. Cahen, Paul-Jean and Chabert, Jean-Luc, What you should know about integer-valued polynomials, *American Mathematical Monthly* **123** (2016), 311–337.
5. Cahen, Paul-Jean, Dobbs, David E. and Lucas, Thomas G., Valuative pairs of commutative rings, *Houston Journal of Mathematics* **41** (2015), 397–431.
6. Cahen, Paul-Jean and Rissner, Roswitha, Finiteness and Skolem closure of ideals for non-unibranched domains, *Communications in Algebra* **43** (2015), 2231–2239.
7. Cahen, Paul-Jean, Fontana, Marco, Frisch, Sophie and Glaz, Sarah, Open problems in commutative ring theory, in *Commutative algebra*, 353–375, Springer, New York, 2014.
8. Cahen, Paul-Jean, Dobbs, David E., and Lucas, Thomas G., Finitely valuative domains, *Journal of Algebra and its Applications* **11** (2012), 1250112,-139.
9. Cahen, Paul-Jean, Dobbs, David E., and Lucas, Thomas G., Characterizing minimal ring extensions, *The Rocky Mountain Journal of Mathematics* **41** (2011), 1081–1125.
10. Adam, David and Cahen, Paul-Jean, Newtonian and Schinzel quadratic fields, *Journal of Pure and Applied Algebra*, **215** (2011), 1902–1918.

11. Cahen, Paul-Jean, Dobbs, David E., and Lucas, Thomas G., Valuative domains, *Journal of Algebra and its Applications* **9** (2010), 43–72.
12. Bhargava, Manjul, Cahen, Paul-Jean and Yeramian, Julie, Finite generation properties for various rings of integer-valued polynomials, *Journal of Algebra* **322** (2009), 1129–1150.
13. Cahen, Paul-Jean, Newtonian and Schinzel sequences in a domain, *Journal of Pure and Applied Algebra* **213** (2009), 2117–2133.
14. Chabert, Jean-Luc and Cahen, Paul-Jean, Old problems and new questions around integer-valued polynomials and factorial sequences, in *Multiplicative ideal theory in commutative algebra*, 89–108, Springer, New York, 2006,
15. Cahen, Paul-Jean and Chabert, Jean-Luc, On the ultrametric Stone-Weierstrass theorem and Mahler's expansion, *Journal de Théorie des Nombres de Bordeaux* **14** (2002), 43–57.
16. Cahen, Paul-Jean, Chabert, Jean-Luc and Loper, K. Alan, High dimension Prüfer domains of integer-valued polynomials, *Journal of the Korean Mathematical Society* **38** (2001), 915–935.
17. Cahen, Paul-Jean, Houston, Evan and Tartarone, Francesca, The t- and v-spectra of the ring of integer-valued polynomials over a valuation domain, in *Ideal theoretic methods in commutative algebra* ,Lecture Notes in Pure and Appl. Math. **220**, 109–117, Dekker, New York, 2001.
18. Cahen, Paul-Jean and Chabert, Jean-Luc, What's new about integer-valued polynomials on a subset?, in *Non-Noetherian commutative ring theory*, Math. Appl., **520**, 75–96, Kluwer Acad. Publ., Dordrecht, 2000,
19. Cahen, Paul-Jean, Gabelli, Stefania and Houston, Evan, Mori domains of integer-valued polynomials, *Journal of Pure and Applied Algebra* **153** (2000), 1–15.
20. Cahen, Paul-Jean, Loper, Alan and Tartarone, Francesca, Integer-valued polynomials and Prüfer v-multiplication domains, *Journal of Algebra* **226** (2000), 765–787.
21. Cahen, Paul-Jean, Chabert, Jean-Luc and Frisch, Sophie, Interpolation domains, *Journal of Algebra* **225**, (2000), 794–803.
22. Cahen, Paul-Jean, Chabert, Jean-Luc, Dobbs, David E. and Tartarone, Francesca, On locally divided domains of the form $\text{Int}(D)$, *Archiv der Mathematik* **74** (2000), 183–191.
23. Cahen, Paul-Jean and Chabert, Jean-Luc, Skolem properties and integer-valued polynomials: a survey, in *Advances in commutative ring theory*, Lecture Notes in Pure and Appl. Math. **205**, 175–195, Dekker, New York, 1999.
24. Ayache, Ahmed and Cahen, Paul-Jean, Anneaux noethériens de Hilbert, *Communications in Algebra* **27** (1999), 3833–3850.
25. Cahen, Paul-Jean, Chabert, Jean-Luc, Houston, Evan and Lucas, Thomas G., Skolem properties, value-functions, and divisorial ideals, *Journal of Pure and Applied Algebra* **135** (1999), 207–223.
26. Loper, Alan and Cahen, Paul-Jean, Rings of integer-valued rational functions, *Journal of Pure and Applied Algebra* **131** (1998), 179–193.
27. Cahen, Paul-Jean and Ayache, Ahmed, Radical valuatif et sous-extensions, *Communications in Algebra* **26** (1998), 2767–2787.
28. Cahen, Paul-Jean and Elkhayyari, Zahra and Kabbaj, Salah, Krull and valuative dimension of the Serre conjecture ring $R\langle n\rangle$, in *Commutative ring theory*, Lecture Notes in Pure and Appl. Math. **185**, 173–180, Dekker, New York, 1997.
29. Cahen, Paul-Jean and Lucas, Thomas G., The special trace property, in *Commutative ring theory*, Lecture Notes in Pure and Appl. Math. **185**, 161–172, Dekker, New York,1997.
30. Cahen, Paul-Jean and Chabert, Jean-Luc, *Integer-valued polynomials*, Mathematical Surveys and Monographs **48**,American Mathematical Society, Providence, RI, 1997, 322p.
31. Cahen, Paul-Jean and Haouat, Youssef, Spectre des anneaux de polynômes à valeurs entières à plusieurs indéterminées, in *Anneaux et modules*, Travaux en cours **51**, 27–54, Hermann, Paris, 1996.
32. Cahen, P-J., Chapman, S. T., Roegner, K., and Smith, W. W., Two classes of ideals determined by integer-valued polynomials, *Rendiconti di Matematica e delle sue Applicazioni* **16** (1996), 625–636
33. Cahen, Paul-Jean, Polynomial closure, *Journal of Number Theory* **61** (1996), 226–247.

34. Ayache, Ahmed, Cahen, Paul-Jean, and Echi, Othman, Anneaux quasi-prüfériens et P-anneaux, *Unione Matematica Italiana. Bollettino* **10** (1996), 1–24.
35. Cahen, Paul-Jean, Houston, Evan G., and Lucas, Thomas G., Discrete valuation overrings of Noetherian domains, *Proceedings of the American Mathematical Society* **124** (1996), 1719–1721.
36. Cahen, Paul-Jean and Chabert, Jean-Luc, Elasticity for integral-valued polynomials, *Journal of Pure and Applied Algebra* **103** (1995), 303–311.
37. Cahen, Paul-Jean, Houston, Evan G., and Lucas, Thomas G., Answer to a question on the principal ideal theorem, in *Zero-dimensional commutative rings*, Lecture Notes in Pure and Appl. Math. **171**, 163–166, Dekker, New York, 1995.
38. Anderson, David F., Cahen, Paul-Jean, Chapman, Scott T., and Smith, William W., Some factorization properties of the ring of integer-valued polynomials, in *Zero-dimensional commutative rings*, Lecture Notes in Pure and Appl. Math. **171**, 125–142, Dekker, New York, 1995.
39. Ayache, Ahmed, Cahen, Paul-Jean, and Echi, Othman, Valuative heights and infinite Nagata rings, *Communications in Algebra* **23** (1995), 1913–1926.
40. Ayache, Ahmed, Cahen, Paul-Jean, and Echi, Othman, Intersection de produits fibres et formule de la dimension, *Communications in Algebra* **22** (1994), 3495–3509.
41. Cahen, Paul-Jean, Ascending chain conditions and associated primes, in *Commutative ring theory*, Lecture Notes in Pure and Appl. Math. **153**, 41–46, Dekker, New York, 1994.
42. Cahen, Paul-Jean, Parties pleines d'un anneau noethérien, *Journal of Algebra* **157** (1993), 199–212
43. Cahen, Paul-Jean, Integer-valued polynomials on a subset, *Proceedings of the American Mathematical Society* **117** (1993), 919–929.
44. Ayache, Ahmed and Cahen, Paul-Jean, Anneaux vérifiant absolument l'inégalité ou la formule de la dimension, *Unione Matematica Italiana. Bollettino* **6** (1992), 39–65.
45. Cahen, Paul-Jean, Polynômes à valeurs entières sur un anneau non analytiquement irréductible, *Journal für die Reine und Angewandte Mathematik* **418** (1991),131–137.
46. Cahen, Paul-Jean, Dimension de l'anneau des polynômes à valeurs entières, *Manuscripta Mathematica* **67** (1990), 333–343.
47. Cahen, Paul-Jean, Construction B, I, D et anneaux localement ou résiduellement de Jaffard, *Archiv der Mathematik* **54** (1990), 125–141.
48. Cahen, Paul-Jean, Dimension des couples d'anneaux partageant un idéal, in *Publications du Département de Mathématiques.* Nouvelle Série. B **88**, 69–75, Univ. Claude-Bernard, Lyon, 1988.
49. Cahen, Paul-Jean, Couples d'anneaux partageant un idéal, *Archiv der Mathematik* **51** (1988), 505–514.
50. Cahen, Paul-Jean and Haouat, Youssef, Polynômes, dérivées et différences finies divisées à valeurs entières sur un anneau de pseudo-valuation, *Comptes Rendus des Séances de l'Académie des Sciences.* Série I. Mathématique **306** (1988), 581–584.
51. Cahen, Paul-Jean and Haouat, Youssef, Polynômes à valeurs entières sur un anneau de pseudo-valuation, *Manuscripta Mathematica* **61** (1988), 23–31.
52. Cahen, Paul-Jean, Anneaux presque integralement clos, *Annales Scientifiques de l'Université de Clermont-Ferrand II. Mathématiques* **24** (1987), 61–64.
53. Cahen, Paul-Jean and Haouat, Youssef, Spectre d'anneaux de polynômes sur une suite croissante d'anneaux, *Archiv der Mathematik* **49** (1987), 281–285.
54. Cahen, Paul-Jean, Grazzini, Fulvio, and Haouat, Youssef, Intégrité du complété et théorème de Stone-Weierstrass, *Annales Scientifiques de l'Université de Clermont-Ferrand II. Mathématiques* **21** (1982), 47–58.
55. Cahen, P. J., Fractions rationnelles à valeurs entières, *Annales Scientifiques de l'Université de Clermont. Série Mathématique* **16** (1978), 85–100.
56. Cahen, Paul-Jean, Polynômes et dérivées à valeurs entières, *Annals of Science* **54** (1975), 25–43.

57. Cahen, Paul-Jean and Chabert, Jean-Luc, Éléments quasi-entiers et extensions de Fatou, *Journal of Algebra* **36** (1975), 185–192.
58. Cahen, Paul-Jean, Premiers de Goldman au-dessus des idéaux premiers d'un anneau commutatif, *Comptes Rendus Hebdomadaires des Séances de l'Académie des Sciences. Séries A et B* **277** (1973), A305–A306.
59. Cahen, Paul-Jean, Premiers et copremiers sur un anneau noethérien, *Comptes Rendus Hebdomadaires des Séances de l'Académie des Sciences. Séries A et B* **277** (1973), A277–A280.
60. Cahen, Paul-Jean, Torsion theory and associated primes, *Proceedings of the American Mathematical Society* **38** (1973), 471–476.
61. Cahen, Paul-Jean, Premiers, copremiers et fibres, *Publications du Département de Mathématiques. Faculté des Sciences de Lyon* **10** (1973), 9–24.
62. Cahen, Paul-Jean, Commutative torsion theory, *Transactions of the American Mathematical Society* **184** (1973), 73–85.
63. Cahen, Paul-Jean, Polynômes à valeurs entières, *Canadian Journal of Mathematics* **24**, (1972), 747–754.
64. Cahen, Paul-Jean and Chabert, Jean-Luc, Coefficients et valeurs d'un polynôme, *Bulletin des Sciences Mathématiques. 2e Série* **95** (1971), 295–304.
65. Cahen, Paul-Jean, Transfert de la propriété de Fatou aux anneaux de polynômes, *Bulletin des Sciences Mathématiques. 2e Série* **94** (1970), 81–83.
66. Cahen, Paul-Jean, Corps henséliens et corps munis d'une seule valuation discrète, *Bulletin des Sciences Mathématiques. 2e Série* **93** (1969), 81–88.

Bhargava's Exponential Functions and Bernoulli Numbers Associated to the Set of Prime Numbers

David Adam and Jean-Luc Chabert

1 Introduction

In [3], Bhargava associates to each infinite subset E of $\mathbb{Z}$ a sequence of positive integers $\{n!_E\}_{n\geq 0}$ called the *factorial sequence* of E. Such sequences have many properties of classical factorials. These integers $n!_E$ are defined locally with combinatorial calculations which are then globalized. They may also be defined by means of the polynomials that are integer-valued on E. Recall that the ring of *integer-valued polynomials* on a subset E of $\mathbb{Z}$ is

$$\text{Int}(E, \mathbb{Z}) = \{f \in \mathbb{Q}[X] \mid f(E) \subseteq \mathbb{Z}\}.$$

Definition 1 The *nth factorial* of a subset E of $\mathbb{Z}$ is the positive integer, denoted by $n!_E$, such that $\frac{1}{n!_E}\mathbb{Z}$ is the fractional ideal formed by the leading coefficients of the polynomials of $\text{Int}(E, \mathbb{Z})$ with degree $\leq n$.

At the end of his paper, Bhargava asked whether the generalization of some classical numbers and functions associated to the factorials could have interesting properties. For instance, the *generalized binomial coefficients* defined, for $0 \leq k \leq n$, by

$$\binom{n}{k}_E = \frac{n!_E}{k!_E\ (n-k)!_E}$$

D. Adam (✉)
GAATI, Université de la Polynésie Française, Tahiti, Polynésie française, France
e-mail: david.adam@upf.pf

J.-L. Chabert
LAMFA, Université de Picardie Jules Verne, Amiens, France
e-mail: jean-luc.chabert@u-picardie.fr

J.-L. Chabert et al. (eds.), *Algebraic, Number Theoretic, and Topological Aspects of Ring Theory*, https://doi.org/10.1007/978-3-031-28847-0_2

are integers. This is an obvious consequence of Definition 1. Does there exist a combinatorial interpretation of these integers? One does not know such an interpretation except in the case of the set $\{q^n \mid n \geq 0\}$ where q is an integer ≥ 2 (cf. [11, §3]). We can also consider a generalization of the exponential function.

Definition 2 The *generalized exponential function* associated to an infinite subset E of $\mathbb{Z}$ is defined by

$$\exp_E(x) = \sum_{n=0}^{+\infty} \frac{1}{n!_E} x^n .$$

One can wonder if the values of the function are rational, irrational, algebraic, or transcendental numbers. We are also interested in a natural generalization of the Bernoulli numbers.

Definition 3 The *generalized Bernoulli numbers* associated to an infinite subset E of $\mathbb{Z}$ are defined by

$$\frac{x}{\exp_E(x) - 1} = \sum_{n \geq 0} \frac{B_{E,n}}{n!_E} x^n .$$

We will consider in particular the case where E is the set $\mathbb{P}$ of prime numbers. We obtain in particular that $\exp_{\mathbb{P}}(r)$ is irrational for the rationals $r \neq -2$ of the form $\frac{\pm 1}{d}$ or $\frac{\pm 2}{d}$ where $d \in \mathbb{N}$ and that the Bernoulli polynomials without constant term $B_{\mathbb{P},n}(X) - B_{\mathbb{P},n}(0) = \sum_{k=0}^{n-1} \binom{n}{k}_{\mathbb{P}} B_{\mathbb{P},k} X^{n-k}$ have integral coefficients.

2 Bhargava's Exponential Functions

The power series

$$\exp_E(x) = \sum_{n=0}^{+\infty} \frac{1}{n!_E} x^n \tag{1}$$

converges everywhere since, for every $n \geq 0$, $n!$ divides $n!_E$. Obviously, $\exp_E(0) = 1$ and $|\exp_E(x)| \leq \exp_E(|x|)$. The *generalized Euler number* associated to E is

$$e_E = \exp_E(1) = \sum_{n=0}^{+\infty} \frac{1}{n!_E} .$$

Clearly, $1 < e_E \leq e$. Mingarelli [13] proved that e_E is always irrational (cf. Proposition 1 below). Thanks to known examples where the factorials are easy to compute (for instance, see [1]), we may compute some generalized exponential functions.

Examples 1

(a) Let $aE + b = \{ax + b \mid x \in E\}$ where $a \in \mathbb{N}^*$ and $b \in \mathbb{Z}$. One knows that $n!_{aE+b} = a^n\, n!_E$. Consequently,

$$\exp_{aE+b}(x) = \sum_{n=0}^{+\infty} \frac{x^n}{a^n n!_E} = \exp_E\left(\frac{x}{a}\right)$$

In particular,

$$e_{a\mathbb{Z}+b} = e^{\frac{1}{a}}\,.$$

(b) For $E = \mathbb{N}^{(2)} = \{n^2 \mid n \in \mathbb{N}\}$, one has $n!_{\mathbb{N}^{(2)}} = \frac{1}{2}(2n)!$ $(n \geq 1)$. Consequently,

$$\exp_{\mathbb{N}^{(2)}}(x) = 1 + 2\sum_{n=1}^{+\infty} \frac{x^n}{(2n)!} = \begin{cases} 2\cosh\sqrt{x} - 1 & \text{if } x \geq 0 \\ 2\cos\sqrt{-x} - 1 & \text{if } x \leq 0\,. \end{cases}$$

(c) For $E = \left\{\frac{n(n+1)}{2} \mid n \geq 0\right\}$, one has

$$\exp_{\{n(n+1)/2\,|\,n\geq 0\}}(x) = \sum_{n=0}^{+\infty} \frac{1}{\frac{(2n)!}{2^n}}\, x^n = \sum_{n=0}^{+\infty} \frac{(2x)^n}{(2n)!} = \begin{cases} \cosh\sqrt{2x} & \text{if } x \geq 0 \\ \cos\sqrt{-2x} & \text{if } x \leq 0\,. \end{cases}$$

Proposition 1 ([13, Theorem 53]) *For every infinite subset E, the associated Euler number $e_E = \exp_E(1) = \sum_{n\geq 0} \frac{1}{n!_E}$ is irrational.*

Here is a short proof of this fact.

Proof Assume that e_E is rational and write $e_E = \frac{a}{b}$ where $a, b \in \mathbb{N}$. For every $k \geq b$, let

$$a_k = k!_E \left(\frac{a}{b} - \sum_{n=0}^{k} \frac{1}{n!_E}\right) = k!_E \left(\sum_{n=k+1}^{+\infty} \frac{1}{n!_E}\right). \tag{2}$$

On the one hand, $b \leq k$ implies that b divides $k!_E$, and on the other hand, $n \leq k$ implies that $n!_E$ divides $k!_E$; thus a_k is a positive integer. Since for $n \geq k+1$ the binomial coefficient $\binom{n}{k+1}_E$ is an integer, $\frac{1}{n!_E} \leq \frac{1}{(k+1)!_E} \times \frac{1}{(n-(k+1))!_E}$. Thus,

$$0 < a_k \leq \frac{k!_E}{(k+1)!_E} \left(\sum_{n=k+1}^{+\infty} \frac{1}{(n-(k+1))!_E}\right) = \frac{k!_E}{(k+1)!_E}\, e_E\,.$$

Assume that there exists a constant A such that $\frac{k_E}{(k+1)!_E} < \frac{1}{A}$ for at most finitely many k. Then, there would exist a positive constant c such that, for every n, $\frac{1}{n!_E} = \prod_{k=0}^{n-1} \frac{k!_E}{(k+1)!_E} \geq c \times \frac{1}{A^n}$. But the inequality $\frac{A^n}{n!_E} \geq c$ would contradict the fact that the power series $\exp_E A$ converges. Consequently, there exist infinitely many k such that $\frac{k!_E}{(k+1)!_E} < 3$, and hence, such that $0 < a_k < 1$ which leads again to a contradiction. □

Corollary 1 *For all $d \in \mathbb{N}^*$, $\exp_E\left(\frac{1}{d}\right) \notin \mathbb{Q}$.*

Proof It follows from Example 1 (a) that for every positive integer d

$$\exp_E\left(\frac{1}{d}\right) = \sum_{n=0}^{+\infty} \frac{1}{d^n \times n!_E} = \sum_{n=0}^{+\infty} \frac{1}{n!_{dE}} = \exp_{dE}(1).$$

□

Analogously, we have:

Proposition 2 *For every infinite subset E and for every integer $d \geq 2$, $\exp_E\left(\frac{-1}{d}\right) \notin \mathbb{Q}$.*

Proof Assume that $\exp_E\left(\frac{-1}{d}\right)$ is rational and write $\exp_E\left(\frac{-1}{d}\right) = \frac{a}{b}$ where $a, b \in \mathbb{N}$. For every $k \geq b$, the following number a_k is an integer:

$$a_k = d^k k!_E \left(\frac{a}{b} - \sum_{n=0}^{k} \frac{(-1)^n}{d^n n!_E}\right) = d^k k!_E \sum_{n=k+1}^{\infty} \frac{(-1)^n}{d^n n!_E}.$$

The fact that $\frac{d^n n!_E}{d^{n+1}(n+1)!_E} \leq \frac{1}{2}$ implies that

$$0 < (-1)^{k+1} a_k < \frac{1}{d} \times \frac{k!_E}{(k+1)!_E} < 1.$$

This is a contradiction. □

In the case where $d = 1$, the previous proof does not work because the inequalities $0 \leq |a_k| \leq \frac{k!_E}{(k+1)!_E} \leq 1$ are not necessarily strict. In order to prove that $\exp_E(-1)$ is irrational, we have to introduce a hypothesis about the behavior of the sequence $\{\frac{n!_E}{(n+1)!_E}\}_{n\geq 0}$. As the sequence $\{n!_E\}_{n\geq 0}$ is increasing, we always have $0 \leq \overline{\lim}_{n\to+\infty} \frac{n!_E}{(n+1)!_E} \leq 1$, but as $n!_E$ divides $(n+1)!_E$, we have more precisely

$$\text{either } \overline{\lim}_{n\to+\infty} \frac{n!_E}{(n+1)!_E} = 1 \quad \text{or} \quad 0 \leq \overline{\lim}_{n\to+\infty} \frac{n!_E}{(n+1)!_E} \leq \frac{1}{2}. \tag{3}$$

Note that the bounds of the second case are sharp: the limit is clearly 0 for $E = \mathbb{N}$ while, by (11), it is $\frac{1}{2}$ for $E = \mathbb{P}$. The first case corresponds to subsets E such that $k!_E = (k+1)!_E$ for infinitely many k. Until now it was not known whether there exists a subset E satisfying this condition. Hence,

Mingarelli's Question ([13]) *Does there exist an infinite subset E of $\mathbb{Z}$ such that, for infinitely many k, $k!_E = (k+1)!_E$?*

In Sect. 7, we prove that the subset $E = \mathbb{P} \cup 2\mathbb{P}$ is such an example.

Proposition 3 *If E is an infinite subset such that $\overline{\lim}_{n\to+\infty} \frac{n!_E}{(n+1)!_E} < 1$, then the number* $\exp_E(-1)$ *is irrational.*

Proof Assume that exp(-1) is rational. Then, for k large enough, we have $k!_E \neq (k+1)!_E$ and the following number a_k is an integer:

$$a_k = k!_E \left(\exp_E(-1) - \sum_{n=0}^{k} \frac{(-1)^n}{n!_E} \right) = k!_E \, R_k.$$

Clearly, $|R_k| \leq \frac{1}{(k+1)!_E}$. Thus, $|a_k| \leq \frac{k!_E}{(k+1)!_E} < 1$. As $R_k = \frac{(-1)^{k+1}}{(k+1)!_E} + R_{k+1}$ and $|R_{k+1}| \leq \frac{1}{(k+2)!_E} < \frac{1}{(k+1)!_E}$, we have $R_k \neq 0$. Finally, $0 < |a_k| < 1$. This is impossible. □

Remarks 1

1. Mingarelli states Proposition 3 (see [13, Theorem 53]) without any assumption. However, it is not clear from his proof that R_k is not zero.
2. In the case $\overline{\lim}_{n\to+\infty} \frac{n!_E}{(n+1)!_E} = 1$, the conclusion of Proposition 3 still holds under the weaker condition that there exist infinitely many n such that $n!_E < (n+1)!_E < (n+2)!_E$. Perhaps, this condition is always satisfied. Thus, the authors propose the following question.

Open Problem Does there exist a subset E of $\mathbb{Z}$ such that, for n large enough, either $n!_E = (n+1)!_E$ or $(n+1)!_E = (n+2)!_E$?

Here is another condition on the behavior of the sequence $\{\frac{n!_E}{(n+1)!_E}\}_{n\geq 0}$.

Proposition 4 *Let $\tau_E^+ = \overline{\lim}_{n\to+\infty} \sqrt[n]{\frac{(n+1)!_E}{n!_E}}$. For every positive rational number $\alpha = \frac{c}{d}$ where $c, d \in \mathbb{N}$ and $c < \tau_E^+$, $\exp_E(\alpha)$ is irrational.*

Proof Assume that $\exp_E\left(\frac{c}{d}\right) = \frac{a}{b}$ where $a, b \in \mathbb{N}$. Similarly to the proof of Proposition 1, for $k \geq b$, the following number is a positive integer:

$$a_k = d^k \times k!_E \left(\exp_E(\alpha) - \sum_{n=0}^{k} \frac{\alpha^n}{n!_E} \right) = d^k \times k!_E \left(\sum_{n=k+1}^{+\infty} \frac{\alpha^n}{n!_E} \right)$$

and

$$a_k < \frac{k!_E}{(k+1)!_E} \times \frac{c^{k+1}}{d} \times \sum_{n=0}^{+\infty} \frac{\alpha^n}{n!_E} = \frac{k!_E}{(k+1)!_E} \times \frac{c^{k+1}}{d} \times \exp_E(\alpha) .$$

As $c < \tau_E^+$, we may write $c = \tau_E^+(1-2\delta)$ where $\delta < \frac{1}{2}$. By definition of τ_E^+, there exist infinitely many indices k such that $\sqrt[k]{\frac{(k+1)!_E}{k!_E}} > \tau_E^+(1-\delta)$. For such a k, we have

$$\frac{k!_E}{(k+1)!_E}\, c^k < \left(\frac{1-2\delta}{1-\delta}\right)^k$$

and, for k large enough, the right hand side of this inequality is less than $\frac{d}{c}/\exp_E(\alpha)$. This leads to a contradiction since the nonzero integer a_k would be less than 1. □

The constant τ_E^+ is not always equal to 1 (cf. Example 2); it can even be infinite (cf. Proposition 6).

Example 2 Fix an integer q such that $|q| \geq 2$ and let $E_q = \{q^n \mid n \geq 0\}$. Then $n!_{E_q} = |\prod_{j=0}^{n-1}(q^n - q^j)|$ (cf. [3]) and $\tau_{E_q}^+ = q^2$. Consequently, for $0 < c < q^2$, $\exp_{E_q}\left(\frac{c}{d}\right) \notin \mathbb{Q}$. In fact, a more particular study of the function $\exp_{E_q}$ allows to conclude that $\exp_{E_q}(c/d)$ is irrational for any nonzero rational c/d (see [16]).

For negative rational numbers, we have the counterpart.

Proposition 5 *Let $\tau_E^- = \underline{\lim}_{n\to+\infty} \sqrt[n]{\frac{(n+1)!_E}{n!_E}}$. For every nonzero rational number $\alpha = -\frac{c}{d}$ where $c, d \in \mathbb{N}$ and $c < \tau_E^-$, $\exp_E(\alpha)$ is irrational.*

Proof Without loss of generality, one can assume $\tau_E^- > 1$. Let $\varepsilon > 0$ be such that $1 \leq c < \tau_E^- - \varepsilon < \tau_E^-$. For n large enough, $\frac{(n+1)!_E}{n!_E} \geq (\tau_E^- - \varepsilon)^n$. Thus

$$\left|\frac{\frac{\alpha^{n+1}}{(n+1)!_E}}{\frac{\alpha^n}{n!_E}}\right| = \frac{|\alpha| n!_E}{(n+1)!_E} \leq \frac{|\alpha|}{(\tau_E^- - \varepsilon)^n} \leq \frac{c}{\tau_E^- - \varepsilon} < 1.$$

Hence the sequence $\left(\frac{|\alpha|^n}{n!_E}\right)_{n\geq k}$ is strictly decreasing. Assume that $\exp_E(\alpha)$ is rational, then for k large enough the following number a_k is an integer:

$$a_k = d^k \times k!_E \left(\exp_E(\alpha) - \sum_{n=0}^{k} \frac{\alpha^n}{n!_E}\right) = d^k \times k!_E \times R_k,$$

where $R_k = \sum_{n=k+1}^{+\infty} \frac{\alpha^n}{n!_E}$. As $R_k = \frac{\alpha^{k+1}}{(k+1)!_E} + R_{k+1}$, $|R_{k+1}| \leq \frac{|\alpha|^{k+2}}{(k+2)!_E} < \frac{|\alpha|^{k+1}}{(k+1)!_E}$ implies that $R_k \neq 0$, and hence, $a_k \neq 0$. Moreover,

$$|a_k| \leq d^k k!_E \frac{|\alpha|^{k+1}}{(k+1)!_E} \leq \frac{c^{k+1} k!_E}{d(k+1)_E} \leq |\alpha| \left(\frac{c}{\tau_E^- - \varepsilon} \right)^k .$$

Since $\frac{c}{\tau_E^- - \varepsilon} < 1$, $\lim_{k \to +\infty} \left(\frac{c}{\tau_E^- - \varepsilon} \right)^k = 0$, and again, for k large enough, $0 < |a_k| < 1$ □

Corollary 2 *Let E be a subset of $\mathbb{Z}$ for which there exist two positive constants C and ε such that, for n large enough, $(n+1)!_E \geq C\,(n!_E)^{1+\varepsilon}$. Then, for every nonzero rational number α, $\exp_E(\alpha)$ is irrational.*

Proof By Stirling's formula, for n large enough, we have

$$n!_E \geq n! \geq \left(\frac{n}{e} \right)^n . \tag{4}$$

As a consequence, for n large enough

$$\sqrt[n]{\frac{(n+1)!_E}{n!_E}} \geq \sqrt[n]{(n!_E)^\varepsilon} \geq \left(\frac{n}{e} \right)^\varepsilon .$$

This implies that $\tau_E^- = \tau_E^+ = +\infty$ and, by Propositions 4 and 5, $\exp_E(\alpha)$ is irrational. □

The following example shows that the hypothesis of Corollary 2 may be realized.

Proposition 6 *Let $f(X) = a_d X^d + \cdots + a_1 X + a_0 \in \mathbb{Z}[X]$ be a polynomial of degree $d \geq 2$, and let a be an integer that is not preperiodic for f. Denote by $f^{[n]}$ the nth iterate of f. Then, for the set $E_{f,a} = \left\{ f^{[n]}(a) \mid n \geq 0 \right\}$, one has $(n+1)!_{E_{f,a}} \geq (n!_{E_{f,a}})^{d-\frac{1}{4}}$ for n large enough.*

Proof One knows (see [1]) that the sequence $\{f^{[n]}(a)\}_{n \geq 0}$ is self-simultaneously ordered, and hence, the n-th factorial of the set $E_{f,a} = \left\{ f^{[n]}(a) \mid n \geq 0 \right\}$ is equal to

$$n!_{E_{f,a}} = \prod_{j=0}^{n-1} \left| f^{[n]}(a) - f^{[j]}(a) \right| .$$

Let $u_n = f^{[n]}(a)$ and $v_n = |u_n|$. As a is not preperiodic, $\lim_{n \to +\infty} v_n = +\infty$ and $v_{n+1} \underset{n \to +\infty}{\sim} |a_d| v_n^d$. Let $n_0 \in \mathbb{N}$ be such that $n \geq n_0$ implies $v_n \geq 2^{4d+7} \geq 4$ and $v_{n+1} \geq \frac{1}{2} |a_d| v_n^d$. Now, let $n_1 \geq n_0$ be such that $n \geq n_1$ implies $v_n > 2 \max(v_0, \cdots, v_{n_0})$. Then

$$(n \geq n_1 \text{ and } 0 \leq k < n) \Rightarrow \frac{v_k}{v_n} < \frac{1}{2} .$$

Indeed, if $k \leq n_0$, the inequality follows from the fact that $n \geq n_1$ and, if $k > n_0$, it follows from the inequalities:

$$\frac{v_k}{v_{k+1}} \leq \frac{1}{\frac{1}{2}|a_d|v_k^{d-1}} < \frac{1}{4^{d-1} \times \frac{1}{2}} \leq \frac{1}{2}.$$

Consequently,

$$n!_{E_{f,a}} = v_n^n \times \prod_{k=0}^{n-1}(1 - \frac{u_k}{u_n}) \leq v_n^n \times \prod_{k=0}^{n-1}(1 + \frac{v_k}{v_n}) \leq 2^n \times v_n^n$$

while

$$(n+1)!_{E_{f,a}} \geq v_{n+1}^{n+1} \times \prod_{k=0}^{n}(1 - \frac{v_k}{v_{n+1}}) \geq \frac{1}{2^{n+1}} \times v_{n+1}^{n+1}.$$

Finally, for all $n \geq n_1$, one has

$$\frac{(n+1)!_{E_{f,a}}}{(n!_{E_{f,a}})^{d-1/4}} \geq \frac{1}{2^{(d+3/4)n}} \cdot \frac{v_{n+1}}{2} \left(\frac{v_{n+1}}{v_n^{d-1/4}} \right)^n \geq \frac{1}{2^{(d+3/4)n}} \cdot \frac{v_{n+1}}{2} \left(\frac{1}{2}|a_d| v_n^{1/4} \right)^n$$

$$\geq \frac{v_{n+1}}{2} \geq 1.$$

□

For instance, let $\mathbb{F} = \{2^{2^n} + 1 \mid n \geq 0\}$ be the set formed by Fermat numbers. In fact, $\mathbb{F}$ is the orbit of 3 under the iteration of the polynomial $X^2 - 2X + 2$. Then, for every $\alpha \in \mathbb{Q}^*$, $\exp_{\mathbb{F}}(\alpha) \notin \mathbb{Q}$.

Hermite's theorem says that, if α is a nonzero rational number, then $\exp(\alpha)$ is transcendental. The following theorem shows that such a result holds for the exponential function of some subsets E of $\mathbb{Z}$.

Theorem 1 *Let E be a subset of $\mathbb{Z}$ for which there are two positive constants C and $\delta > 2$ such that, for n large enough, $(n+1)!_E \geq C\,(n!_E)^\delta$. Then, for every nonzero rational number α, $\exp_E(\alpha)$ is transcendental.*

Proof By Corollary 2, $\exp_E(\alpha)$ is irrational. Write $\alpha = \frac{c}{d}$ with $c \in \mathbb{Z}$ and $d \in \mathbb{N}$. Analogously to the proof of Proposition 4, we have

$$\left| \exp_E(\alpha) - \sum_{n=0}^{k} \frac{\alpha^n}{n!_E} \right| \leq \frac{|\alpha|^{k+1}}{(k+1)!_E} \exp_E(|\alpha|) \leq \frac{|\alpha|^{k+1}}{C(k!_E)^\delta} \exp_E(|\alpha|).$$

Equivalently,

$$\left|\exp_E(\alpha) - \frac{\sum_{n=0}^{k} c^n\, d^{k-n}\, \frac{k!_E}{n!_E}}{d^k\, k!_E}\right| \leq \frac{|c|^{k+1}}{C\, d^{k+1}\, (k!_E)^{\delta}} \exp_E(|\alpha|).$$

Let ε be such that $0 < \delta - \varepsilon < \delta$. By inequality (4), for k large enough, we have

$$C\, d^{k+1}\, (k!_E)^{\delta} \geq \left(d^k\, k!_E\right)^{\delta-\varepsilon} \times |c|^{k+1} \exp_E(|\alpha|),$$

and hence,

$$\left|\exp_E(\alpha) - \frac{\sum_{n=0}^{k} c^n\, d^{k-n}\, \frac{k!_E}{n!_E}}{d^k\, k!_E}\right| \leq \frac{1}{\left(d^k\, k!_E\right)^{\delta-\varepsilon}}. \tag{5}$$

Clearly, $\sum_{n=0}^{k} c^n\, d^{k-n} \frac{k!_E}{n!_E}$ is an integer. Choosing ε such that $\delta - \varepsilon > 2$, by Roth theorem [17], $\exp_E(\alpha)$ is transcendental. □

One cannot assert the transcendence of $\exp_{\mathbb{F}}(1)$ by Theorem 1, since for every $\varepsilon > 0$, one may check that $\lim_{n\to+\infty} \frac{(n+1)!_{\mathbb{F}}}{(n!_{\mathbb{F}})^{2+\varepsilon}} = 0$. However, for this very specific set, for all nonzero rational α, the transcendence of $\exp_{\mathbb{F}}(\alpha)$ is a consequence of a Lang's conjecture. For the reader's convenience, we announce it.

Lang's Conjecture ([12, Chapter IX]) Let α be a real non-rational algebraic number. Then, for every $\varepsilon > 0$, the following inequality

$$|\alpha - \frac{p}{q}| \leq \frac{1}{q^2 \ln^{1+\varepsilon} |q|}$$

has only a finite number of solutions in rationals p/q.

Proposition 7 *Suppose Lang's conjecture is true. Let $a \in \mathbb{Z} \setminus \{-1, 0, 1\}$. Then for every nonzero rational α, $\exp_{E_{X^2,a}}(\alpha)$ is transcendental.*

Proof Put $E = E_{X^2,a}$. Obviously, a is not a preperiodic point of the map $z \mapsto z^2$. Let k be a positive integer. Clearly,

$$k!_E = \prod_{j=0}^{k-1} (a^{2^k} - a^{2^j}).$$

It follows that this inequality holds:

$$k!_E \leq 2a^{k2^k}. \tag{6}$$

Moreover, one has

$$(k+1)!_E = (a^{2^{k+1}} - a)\prod_{l=1}^{k}(a^{2^{k+1}} - a^{2^l}) = (a^{2^{k+1}} - a)\prod_{l=0}^{k-1}(a^{2^{k+1}} - a^{2^{l+1}})$$

$$= (a^{2^{k+1}} - a)\prod_{l=0}^{k-1}(a^{2^k} - a^{2^l})(a^{2^k} + a^{2^l}) = (a^{2^{k+1}} - a)k!_E\prod_{l=0}^{k-1}(a^{2^k} + a^{2^l}).$$

Hence, one obtains for all $k \geq 3$

$$(k+1)!_E \geq \frac{1}{2}(a^{2^{k+1}} - a)(k!_E)^2 \geq \frac{1}{4}a^{2^{k+1}}(k!_E)^2. \tag{7}$$

Writing $\alpha = \frac{c}{d}$ as in the proof of Theorem 1, inequalities (6) and (7) imply that $\lim_{k\to+\infty} \frac{(k+1)!_E}{(k!_E d^k)^2(\ln(k!_E d^k))^{3/2}|\alpha|^{k+1}\exp_E(|\alpha|)} = +\infty$. In particular, one obtains that for k large enough

$$(k+1)!_E \geq (k!_E d^k)^2\,(\ln(k!_E d^k))^{3/2}\,|\alpha|^{k+1}\exp_E(|\alpha|).$$

On another side, the usual estimate leads to

$$\left|\exp_E(\alpha) - \frac{\sum_{n=0}^{k} c^n d^{k-n}\frac{k!_E}{n!_E}}{d^k k!_E}\right| \leq \frac{|\alpha|^{k+1}}{(k+1)!_E}\exp_E(|\alpha|) \leq \frac{1}{(k!_E d^k)^2(\ln(k!_E d^k))^{3/2}}.$$

By Lang's conjecture, $\exp_E(\alpha)$ is transcendental. □

Corollary 3 *Under Lang's conjecture, for every $\alpha \in \mathbb{Q}^*$, $\exp_{\mathbb{F}}(\alpha)$ is transcendental.*

Proof Remark that the sets $\mathbb{F}$ and $E_{X^2,2}$ have the same sequence of factorials. □

Conjecture The authors dare to conjecture that, if E is a subset of $\mathbb{Z}$ for which there are two positive constants C and $\delta > 1$ such that, for every $n \in \mathbb{N}$, one has $(n+1)!_E \geq C(n!_E)^\delta$ then, for all $\alpha \in \overline{\mathbb{Q}}^*$, $\exp_E(\alpha)$ is transcendental.

Hermite's theorem can be rephrased also as: for any nonzero distinct rationals $r_1, \cdots, r_s$, the numbers $1, e^{r_1}, \cdots, e^{r_s}$ are $\mathbb{Q}$-linearly independent. Now, we prove the analogous result for the exponential associated to a subset of $\mathbb{Z}$ satisfying the conditions of Corollary 2.

Theorem 2 *Let E be a subset of $\mathbb{Z}$ for which there are two positive constants C and ε such that, for n large enough, $(n+1)!_E \geq C\,(n!_E)^{1+\varepsilon}$. The numbers $1, \exp_E(r_1), \cdots, \exp_E(r_s)$ are $\mathbb{Q}$-linearly independent for any nonzero distinct rational numbers $r_1, \cdots, r_s$.*

Proof Let $A_0, A_1, \cdots, A_s$ be integers such that

$$A_0 + \sum_{i=1}^{s} A_i \exp_E(r_i) = 0.$$

Write $r_i = p_i/q_i$ with $(p_i, q_i) \in \mathbb{Z} \times \mathbb{N}^*$, let $d = \mathrm{LCM}(q_i)_{1 \le i \le s}$ and, for every n, consider the equality

$$d^n n!_E \left(A_0 + \sum_{i=1}^{s} A_i \sum_{k=0}^{n} \frac{r_i^k}{k!_E} \right) + R_n = 0$$

where

$$R_n := d^n n!_E \sum_{i=1}^{s} A_i \sum_{k \ge n+1} \frac{r_i^k}{k!_E}.$$

Clearly, $d^n n!_E \left(A_0 + \sum_{i=1}^{s} A_i \sum_{k=0}^{n} \frac{r_i^k}{k!_E} \right)$ is an integer and R_n is too. For every $i \in [1, s]$, one has

$$\sum_{k \ge n+1} \frac{r_i^k}{k!_E} = \frac{r_i^{n+1}}{(n+1)!_E} + r_i^{n+2} \sum_{k \ge 0} \frac{r_i^k}{(k+n+2)!_E} \underset{n \to +\infty}{\sim} \frac{r_i^{n+1}}{(n+1)!_E}. \tag{8}$$

The equivalence follows from the inequality $\left| \sum_{k \ge 0} \frac{r_i^k}{(k+n+2)!_E} \right| \le \frac{\exp_E(|r_i|)}{(n+2)!_E}$ and the preponderance $n!_E = o((n+1)!_E)$. Consequently, for n large enough, say $n \ge N$,

$$|R_n| < 2d^n \frac{n!_E}{(n+1)!_E} \sum_{i=1}^{s} |A_i||r_i|^{n+1} < 1$$

because of (4). As R_n is an integer, necessarily, for $n \ge N$, $R_n = 0$. Finally, $\frac{R_{n+1}}{d^{n+1}} - \frac{R_n}{d^n} \frac{(n+1)!_E}{n!_E} = 0$ implies that, for $n \ge N$, $\sum_{i=1}^{s} A_i r_i^{n+1} = 0$, which implies that all the A_i's are equal to 0. □

Corollary 4 *Let E be a subset of $\mathbb{Z}$ for which there are two positive constants C and ε such that, for n large enough, $(n+1)!_E \ge C\,(n!_E)^{1+\varepsilon}$. There are at most $\lfloor 1/\varepsilon \rfloor$ nonzero rationals r such that $\exp_E(r)$ is algebraic.*

Proof Suppose that there are m nonzero rationals $r_1, \cdots, r_m$ such that $\exp_E(r_i) \in \overline{\mathbb{Q}}$ for every $i \in [1, m]$ and that $m > \frac{1}{\varepsilon}$. Fix some ε' such that $0 < \varepsilon' < \varepsilon - \frac{1}{m}$. By the previous theorem, $\exp_E(r_1), \cdots, \exp_E(r_m)$ are $\mathbb{Q}$-linearly independent. Write $r_i = p_i/q$ with $p_i \in \mathbb{Z}$ and $q \in \mathbb{N}$.

Following the now classical way of proof, for all integers n large enough

$$\left|\exp_E(r_i) - \frac{\sum_{k=0}^{n} \frac{n!_E q^n p_i^k}{k!_E q^k}}{n!_E q^n}\right| \leq \frac{|r_i|^{n+1} \exp_E(|r_i|)}{(n+1)!_E} \leq \frac{1}{(n!_E q^n)^{1+\frac{1}{m}+\varepsilon'}}.$$

Since for every $i \in [1, m]$, $\sum_{k=0}^{n} \frac{n!_E q^n p_i^k}{k!_E q^k}$ is an integer, by Schmidt's extension of Roth's theorem [20, Corollary of Theorem 1], at least one of the $\exp_E(r_i)$'s is transcendental. □

An immediate consequence of this corollary is

Corollary 5 *Let $P \in \mathbb{Z}[X]$ of degree 2 and a be an integer that is not a preperiodic point of P. Then, there is at most one nonzero rational such that $\exp_{E_{P,a}}(r)$ is algebraic.*

Proof By Proposition 6, there is a positive constant C such that for every $n \in \mathbb{N}$, $(n+1)!_{E_{P,a}} \geq C(n!_{E_{P,a}})^{7/4}$. By the previous corollary, there are at most $\lfloor 4/3 \rfloor = 1$ nonzero rational such that $\exp_{E_{P,a}}(r)$ is algebraic. □

Remark 2 Corollary 4 allows to prove Theorem 1 too.

3 The Factorial Sequence of the Set of Prime Numbers

Thanks to Definition 1, Bhargava's factorial sequence $\{n!_{\mathbb{P}}\}_{n\geq 0}$ of the set $\mathbb{P}$ of prime numbers was known before its formal introduction (see [4] or [6]). Its elements can be easily computed with the formula

$$n!_{\mathbb{P}} = \prod_{p\in\mathbb{P}} p^{w_{\mathbb{P},p}(n)} \quad \text{where} \quad w_{\mathbb{P},p} = \sum_{k\geq 0} \left\lfloor \frac{n-1}{(p-1)p^k} \right\rfloor \tag{9}$$

and we have the following equivalence:

$$n!_{\mathbb{P}} \;=\; n!\, e^{Cn+o(n)} \quad \text{where } C = \sum_{p\in\mathbb{P}} \frac{\ln p}{(p-1)^2} \quad \text{(Diaz [8])}. \tag{10}$$

Remark 3 Sequence (9) is Sequence A053657 of *The On-Line Encyclopedia of Integer Sequences* [19] which begins with 1, 1, 2, 24, 48, 5760, 11520, ... It turns out that this sequence appears in different contexts, especially:

- in group theory: $(n+1)!_{\mathbb{P}}$ is equal to the n-th *Minkowski number* M_n, that is, the least common multiple of the orders of all finite subgroups of $GL_n(\mathbb{Q})$ (cf. Minkowski [14] and Schur [18]).

- in algebraic topology: the $n!_{\mathbb{P}}$'s are the denominators of the Laurent polynomials forming a regular basis for the Hopf algebroid of stable cooperations for complex K-theory (cf. Johnson [9]).
- in power series expansions: for instance in

$$\left(-\frac{\ln(1-x)}{x}\right)^m = \left(\sum_{k=1}^{+\infty} \frac{x^k}{k+1}\right)^m$$

$$= 1 + \frac{m}{2}x + \frac{m(3m+5)}{24}x^2 + \cdots = \sum_{n=0}^{+\infty} \frac{C_n(m)}{(n+1)!_{\mathbb{P}}} x^n$$

where $C_n(X) \in \mathbb{Z}[X]$ is primitive with degree n (cf. [5]).

Clearly,

$$w_{\mathbb{P},2}(2k) = 1 + w_{\mathbb{P},2}(2k-1) \text{ and, for } p \neq 2,\ w_{\mathbb{P},p}(2k) = w_{\mathbb{P},p}(2k-1).$$

So that, for $k \geq 1$:

$$(2k)!_{\mathbb{P}} = 2 \times (2k-1)!_{\mathbb{P}}\,. \tag{11}$$

The following lemma gives the value of all the quotients $\frac{(n+1)!_{\mathbb{P}}}{n!_{\mathbb{P}}}$ of two consecutive factorials, that is the value of $\binom{n+1}{n}_{\mathbb{P}}$.

Lemma 1 *For every* $n \geq 1$,

$$\frac{(2n+1)!_{\mathbb{P}}}{(2n)!_{\mathbb{P}}} = \prod_{p \in \mathbb{P},\ p-1 \mid 2n} p^{1+v_p(2n)} \quad \text{and} \quad \frac{(2n)!_{\mathbb{P}}}{(2n-1)!_{\mathbb{P}}} = 2\,. \tag{12}$$

Proof

$$v_p\left(\binom{2n+1}{2n}_{\mathbb{P}}\right) = \sum_{k \geq 0} \left\lfloor \frac{2n}{(p-1)p^k} \right\rfloor - \sum_{k \geq 0} \left\lfloor \frac{2n-1}{(p-1)p^k} \right\rfloor.$$

If $p-1 \nmid 2n$, this difference is zero and, if $p-1 \mid 2n$, it is equal to

$$\left\lfloor \frac{2n}{p-1} \right\rfloor - \left\lfloor \frac{2n-1}{p-1} \right\rfloor + \sum_{k \geq 1} \left\lfloor \frac{2n}{(p-1)p^k} \right\rfloor - \sum_{k \geq 1} \left\lfloor \frac{2n-1}{(p-1)p^k} \right\rfloor$$

$$= 1 + v_p\left(\left(\frac{2n}{p-1}\right)!\right) - v_p\left(\left(\frac{2n}{p-1}-1\right)!\right) = 1 + v_p\left(\frac{2n}{p-1}\right) = 1 + v_p(2n).$$

The second equality is already known; this is Formula (11). □

This lemma shows in particular that

$$\overline{\lim}_{n\to+\infty} \frac{n!_{\mathbb{P}}}{(n+1)!_{\mathbb{P}}} = \frac{1}{2}.$$

We said in the introduction that, whatever the infinite subset E of $\mathbb{Z}$, for $0 \le k \le n$, the binomial coefficient

$$\binom{n}{k}_E = \frac{n!_E}{k!_E(n-k)!_E}$$

is an integer. This is obvious because the product of an integer-valued polynomial on E of degree k by another of degree $n-k$ is itself integer-valued with degree n. But, in the case of the subset $\mathbb{P}$, we have a stronger result.

Lemma 2 *Whatever the integers $k_1, \cdots, k_r \in \mathbb{N}$ such that $k_1 + \cdots + k_r = n$*

$$\frac{(n+1)!_{\mathbb{P}}}{(k_1+1)!_{\mathbb{P}} \cdots (k_r+1)!_{\mathbb{P}}} \in \mathbb{N}. \tag{13}$$

Proof For every $p \in \mathbb{P}$, we have:

$$v_p\left(\frac{(n+1)!_{\mathbb{P}}}{(k_1+1)!_{\mathbb{P}} \cdots (k_r+1)!_{\mathbb{P}}}\right) = \sum_{s\ge 0}\left\{\left[\frac{n}{(p-1)p^s}\right] - \sum_{j=1}^{r}\left[\frac{k_j}{(p-1)p^s}\right]\right\} \ge 0$$

since $n = \sum_{j=1}^{r} k_j$. □

For instance, for every $n \ge 1$, $3!_{\mathbb{P}} = 24$ divides $\frac{(n+2)!_{\mathbb{P}}}{n!_{\mathbb{P}}}$.

4 The Exponential Function Associated to $\mathbb{P}$

It follows from Formula (9) that

$$\exp_{\mathbb{P}}(x) = \sum_{n=0}^{+\infty} \frac{1}{n!_{\mathbb{P}}} x^n = 1 + x + \frac{1}{2}x^2 + \frac{1}{24}x^3 + \frac{1}{48}x^4 + \cdots \tag{14}$$

In particular,

$$\frac{5}{2} < e_{\mathbb{P}} < e. \tag{15}$$

Formula (11) leads to

$$\exp_{\mathbb{P}}(x) = 1 \ + \ (2+x) \sum_{k=1}^{+\infty} \frac{1}{(2k)!_{\mathbb{P}}} \, x^{2k-1} \, . \tag{16}$$

Formula (16) shows that

$$\exp_{\mathbb{P}}(x) \begin{cases} > 1 \text{ for } x < -2 \text{ or } x > 0 \\ = 1 \text{ for } x = -2 \text{ or } x = 0 \\ < 1 \text{ for } \quad -2 < x < 0. \end{cases} \tag{17}$$

Formula (16) shows also that the function $\frac{\exp_{\mathbb{P}}(x)-1}{x+2}$ is odd. Consequently, we have the following functional equation

$$(x-2)(\exp_{\mathbb{P}}(x)-1) = (x+2)(\exp_{\mathbb{P}}(-x)-1)\, . \tag{18}$$

By Corollary 1 and Propositions 2 and 3, we know that, for each integer $d \neq 0$, $\exp_{\mathbb{P}}\left(\frac{1}{d}\right) \notin \mathbb{Q}$. Can we say more? Here, Proposition 4 is useless since $\tau_{\mathbb{P}} = 1$ as easily shown using (10). Nevertheless we can say a little more.

Proposition 8 *If a rational $r \neq -2$ is of the form $\frac{\pm 1}{d}$ or $\frac{\pm 2}{d}$ where $d \in \mathbb{N}^*$, then $\exp_{\mathbb{P}}(r)$ is irrational.*

Proof The assertion is already known for $r = \frac{\pm 1}{d}$. The case $r = -\frac{2}{d}$ will follow from the case $r = \frac{2}{d}$ thanks to the functional equation (16). Let us assume now that $r = \frac{2}{d}$ where $d \in \mathbb{N}^*$. Then, by Formula (16), we have:

$$\exp_{\mathbb{P}}(r) = 1 + \left(1 + \frac{1}{d}\right) \sum_{n=0}^{+\infty} \frac{1}{(2n+1)!_{\mathbb{P}}} \left(\frac{2}{d}\right)^{2n+1} .$$

Assume that $\exp_{\mathbb{P}}(r) \in \mathbb{Q}$, and hence, that there exist a and $b \in \mathbb{N}^*$ such that

$$\frac{a}{b} = (\exp_{\mathbb{P}}(r) - 1) \times \frac{d}{d+1} = \sum_{n=0}^{+\infty} \frac{1}{(2n+1)!_{\mathbb{P}}} \left(\frac{2}{d}\right)^{2n+1} .$$

For every positive integer m, let

$$m!_{\mathbb{P}^*} = \prod_{p \in \mathbb{P},\ p \neq 2} p^{w_{\mathbb{P},p}(m)} .$$

Note that $w_{\mathbb{P},2}(2n+1) = 2n + v_2((2n)!) = 3n + v_2(n!) = 4n - \sigma_2(n)$ where $\sigma_2(n)$ denotes the sum of the digits of n in base 2. The last equality follows from the classical formula

$$v_p(n!) = \frac{n - \sigma_p(n)}{p-1}.$$

Thus, we have

$$(2n+1)!_{\mathbb{P}} = 2^{4n-\sigma_2(n)} \times (2n+1)!_{\mathbb{P}^*}.$$

Consequently, the following number is a positive integer

$$a_k = b \times (2d)^{2k+1} \times (2k+1)!_{\mathbb{P}^*} \left(\frac{a}{b} - \sum_{n=0}^{k} \frac{1}{(2n+1)!_{\mathbb{P}}} \left(\frac{2}{d}\right)^{2n+1} \right).$$

Equivalently,

$$a_k = b \times (2d)^{2k+1} \times (2k+1)!_{\mathbb{P}^*} \left(\sum_{n=k+1}^{+\infty} \frac{1}{(2n+1)!_{\mathbb{P}}} \left(\frac{2}{d}\right)^{2n+1} \right).$$

Thus,

$$a_k \leq 4b \times (2k+1)!_{\mathbb{P}^*} \sum_{n=k+1}^{+\infty} \frac{1}{2^{2(n-k)-\sigma_2(n)}} \times \frac{1}{(2n+1)!_{\mathbb{P}^*}}.$$

The inequality $(2n+1)!_{\mathbb{P}^*} \geq (2k+3)!_{\mathbb{P}^*}$ implies that

$$a_k \leq b \times \frac{(2k+1)!_{\mathbb{P}^*}}{(2k+3)!_{\mathbb{P}^*}} \sum_{m=0}^{+\infty} \frac{1}{2^{2m-\sigma_2(m+k+1)}}.$$

As $\sigma_2(m+k+1) \leq \sigma_2(m) + \sigma_2(k) + 1$ and $\sigma_2(m) \leq m$,

$$a_k \leq b \times 2^{2+\sigma_2(k)} \frac{(2k+1)!_{\mathbb{P}^*}}{(2k+3)!_{\mathbb{P}^*}}.$$

To be able to control the number $\sigma_2(k)$, we consider integers k of the form 2^{s-1} and, by Formula (12), $v_3\left(\frac{(2^s+3)!_{\mathbb{P}^*}}{(2^s+1)!_{\mathbb{P}^*}}\right) = 1 + v_3(2^s+2)$. For every $n \in \mathbb{N}^*$, 2 spans the multiplicative group $(\mathbb{Z}/3^n\mathbb{Z})^\times$ (see [15, Lemma 1.10]); thus there exists a positive integer s_n such that $2^{s_n} = -2 \pmod{3^n}$. Therefore $\lim_{n\to+\infty} v_3\left(\frac{(2^{s_n}+3)!_{\mathbb{P}^*}}{(2^{s_n}+1)!_{\mathbb{P}^*}}\right) = +\infty$ and the sequence $(a_{2^{s_n}})_{n\geq 1}$ converges to 0. This is a contradiction. □

5 Bernoulli Numbers Associated to $\mathbb{P}$

For every infinite subset E of $\mathbb{Z}$, we define the Bernoulli numbers $B_{E,n}$ by

$$\frac{x}{\exp_E(x)-1}=\sum_{n=0}^{+\infty}\frac{B_{E,n}}{n!_E}\,x^n\,. \tag{19}$$

Consequently,

$$\left(\sum_{n=0}^{+\infty}\frac{1}{(n+1)!_E}\,x^n\right)\left(\sum_{n=0}^{+\infty}\frac{B_{E,n}}{n!_E}\,x^n\right)=1\,.$$

Thus,

$$B_{E,0}=1 \text{ and, for } n\geq 1,\ \sum_{r+s=n}\frac{1}{(r+1)!_E}\,\frac{B_{E,s}}{s!_E}=0\,.$$

Equivalently, for $n \geq 1$

$$\sum_{k=0}^{n}\binom{n+1}{k}_E B_{E,k}=0 \quad\text{where}\quad \binom{n}{k}_E=\frac{n!_E}{k!_E(n-k)!_E}\,. \tag{20}$$

It is the analogue of the induction formula for the classical Bernoulli numbers. From now on, we only consider the Bernoulli numbers associated to the set of prime numbers. They are defined either by the equation:

$$\frac{x}{\exp_{\mathbb{P}}(x)-1}=\sum_{n=0}^{+\infty}\frac{B_{\mathbb{P},n}}{n!_{\mathbb{P}}}\,x^n \tag{21}$$

or by the induction formula:

$$B_{\mathbb{P},0}=1 \ \text{ et } \ \forall n\geq 1 \qquad \sum_{k=0}^{n}\binom{n+1}{k}_{\mathbb{P}} B_{\mathbb{P},k}=0\,. \tag{22}$$

The first values are the following:

$$B_{\mathbb{P},0}=1\,,\ B_{\mathbb{P},1}=-\frac{1}{2}\,,\ B_{\mathbb{P},2}=\frac{5}{12}\,,\ B_{\mathbb{P},3}=-\frac{5}{2}\,,\ B_{\mathbb{P},4}=\frac{103}{40}\,,\cdots$$

From Eq. (21) and the functional equation (18) satisfied by $\exp_{\mathbb{P}}(x)$, we deduce:

Proposition 9

$$\forall n \geq 0 \quad 2\, B_{\mathbb{P},2n+1} = -\,\frac{(2n+1)!_{\mathbb{P}}}{(2n)!_{\mathbb{P}}}\, B_{\mathbb{P},2n}\,. \tag{23}$$

Proof The functional equation may be written:

$$(2+x)\times\frac{x}{\exp_{\mathbb{P}}(x)-1} = (2-x)\times\frac{x}{\exp_{\mathbb{P}}(-x)-1}.$$

Thus,

$$(2+x)\sum_{n\geq 0}\frac{B_{\mathbb{P},\,n}}{n!_{\mathbb{P}}}\,x^n = (2-x)\sum_{n\geq 0}\frac{B_{\mathbb{P},n}}{n!_{\mathbb{P}}}\,(-x)^n\,,$$

and by identification:

$$\forall n \geq 1 \qquad 2\,\frac{B_{\mathbb{P},n}}{n!_{\mathbb{P}}}\,(1-(-1)^n) = -\,\frac{B_{\mathbb{P},n-1}}{(n-1)!_{\mathbb{P}}}\,(1+(-1)^{n-1})\,.$$

This leads to Formula (23) which holds also for $n = 0$. □

On the other hand, the induction formula (22) leads to the following "explicit" formula:

Proposition 10 *For every $n \geq 1$, we have:*

$$B_{\mathbb{P},n} = n!_{\mathbb{P}}\times\sum_{k=1}^{n}(-1)^k \sum_{\substack{i_1,\ldots,i_k\geq 1\\ i_1+\cdots+i_k=n}}\frac{1}{(i_1+1)!_{\mathbb{P}}\cdots(i_k+1)!_{\mathbb{P}}}\,. \tag{24}$$

Proof The formula holds for $n \leq 2$: $B_{\mathbb{P},1} = -\frac{1}{2}$ and $B_{\mathbb{P},2} = \frac{5}{12}$. Assume that the formula is true up to rank $n-1$. Formulas (22) and (24) up to rank $n-1$ together lead to

$$\begin{aligned} B_{\mathbb{P},n} = -\tfrac{n!_{\mathbb{P}}}{(n+1)!_{\mathbb{P}}}\Bigg(1 &+ \sum_{k=1}^{n-1}\tbinom{n+1}{k}_{\mathbb{P}} k!_{\mathbb{P}} \\ &\times \sum_{j=1}^{k}(-1)^j \sum_{i_1+\cdots+i_j=k}\tfrac{1}{(i_1+1)!_{\mathbb{P}}\cdots(i_j+1)!_{\mathbb{P}}}\Bigg). \end{aligned}$$

that may be written

$$-\frac{B_{\mathbb{P},n}}{n!_{\mathbb{P}}} = \frac{1}{(n+1)!_{\mathbb{P}}} + \sum_{k=1}^{n-1}\frac{1}{(n+1-k)!_{\mathbb{P}}}\sum_{j=1}^{k}(-1)^j \sum_{i_1+\cdots+i_j=k}\frac{1}{(i_1+1)!_{\mathbb{P}}\cdots(i_j+1)!_{\mathbb{P}}}$$

$$= \frac{1}{(n+1)!_{\mathbb{P}}} + \sum_{k=1}^{n-1}\sum_{j=1}^{k}(-1)^j \sum_{i_1+\cdots+i_j=k} \frac{1}{(i_1+1)!_{\mathbb{P}}\cdots(i_j+1)!_{\mathbb{P}}(n+1-k)!_{\mathbb{P}}}$$

$$= -\frac{1}{(n+1)!_{\mathbb{P}}} + \sum_{j=1}^{n-1}(-1)^j \sum_{k=j}^{n-1}\sum_{i_1+\cdots+i_j=k} \frac{1}{(i_1+1)!_{\mathbb{P}}\cdots(i_j+1)!_{\mathbb{P}}(n-k+1)!_{\mathbb{P}}}$$

$$= \frac{1}{(n+1)!_{\mathbb{P}}} + \sum_{j=1}^{n-1}(-1)^j \sum_{i_1+\cdots+i_j+i_{j+1}=n} \frac{1}{(i_1+1)!_{\mathbb{P}}\cdots(i_j+1)!_{\mathbb{P}}(i_{j+1}+1)!_{\mathbb{P}}}$$

$$= \frac{1}{(n+1)!_{\mathbb{P}}} + \sum_{j=2}^{n}(-1)^{j+1} \sum_{i_1+\cdots+i_j=n} \frac{1}{(i_1+1)!_{\mathbb{P}}\cdots(i_j+1)!_{\mathbb{P}}}\,.$$

This is Formula (24). □

Proposition 11

$$\forall\, n \geq 0 \qquad \frac{(n+1)!_{\mathbb{P}}}{n!_{\mathbb{P}}}\, B_{\mathbb{P},n} \in \mathbb{Z}\,. \tag{25}$$

Proof Formula (24) may also be written:

$$\frac{(n+1)!_{\mathbb{P}}}{n!_{\mathbb{P}}}\, B_{\mathbb{P},n} = -1 + \sum_{k=2}^{n-1}(-1)^{k+1} \sum_{\substack{i_1,\ldots,i_k\geq 1\\ i_1+\cdots+i_k=n}} \frac{(n+1)!_{\mathbb{P}}}{(i_1+1)!_{\mathbb{P}}\cdots(i_k+1)!_{\mathbb{P}}} \tag{26}$$

We can conclude with Lemma 2 which shows that all the fractions of the right hand side are integers. □

Since $\binom{2n+2}{2n+1} = 2$, a first obvious consequence of assertion (25) is that $2\, B_{\mathbb{P},2n+1} \in \mathbb{Z}$. In fact, one can say more starting with Formula (26).

Proposition 12

$$\forall n \geq 0 \quad B_{\mathbb{P},2n+1} \in \frac{1}{2} + \mathbb{Z}\,. \tag{27}$$

Proof We know that $2B_{\mathbb{P},2n+1} \in \mathbb{Z}$, we have to show that this integer is odd. It follows from (11) that

$$B_{\mathbb{P},2n+1} = -\frac{1}{2} + \sum_{k=2}^{n-1}\frac{(-1)^{k+1}}{2} \sum_{\substack{i_1,\ldots,i_k\geq 1\\ i_1+\cdots+i_k=2n+1}} \frac{(2n+2)!_{\mathbb{P}}}{(i_1+1)!_{\mathbb{P}}\cdots(i_k+1)!_{\mathbb{P}}}\,.$$

Thus, we just have to prove that $\sum_{\substack{i_1,\ldots,i_k\geq 1\\ i_1+\cdots+i_k=2n+1}} \frac{(2n+2)!_{\mathbb{P}}}{(i_1+1)!_{\mathbb{P}}\cdots(i_k+1)!_{\mathbb{P}}} \in 2\mathbb{Z}_{(2)}$.

If at least two of the integers i_j are equal, then the value of the corresponding fraction is even. Indeed, assume, for instance, that $i_1 = i_2$. Then

$$\frac{(2n+2)!_{\mathbb{P}}}{(i_1+1)!_{\mathbb{P}}\cdots(i_k+1)!_{\mathbb{P}}} = \frac{(2n+2)!_{\mathbb{P}}}{(2i_1+1)!_{\mathbb{P}}(i_3+1)!_{\mathbb{P}}\cdots(i_k+1)!_{\mathbb{P}}} \times \frac{(2i_1+1)!_{\mathbb{P}}}{(i_1+1)!_{\mathbb{P}}^2}.$$

By (13), we know that both terms of the product are integers. Moreover, the second one is even since $v_2\left(\frac{(2i_1+1)!_{\mathbb{P}}}{(i_1+1)!_{\mathbb{P}}^2}\right) = v_2((2i_1)!) - 2v_2(i_1!) > 0$.

As to the remaining fractions, we have

$$\sum_{\substack{i_1,\ldots,i_k \text{ distinct}\\ i_1+\cdots+i_k=2n+1}} \frac{(2n+2)!_{\mathbb{P}}}{(i_1+1)!_{\mathbb{P}}\cdots(i_k+1)!_{\mathbb{P}}} = k! \sum_{\substack{1\leq i_1<i_2<\cdots<i_k\\ i_1+\cdots+i_k=2n+1}} \frac{(2n+2)!_{\mathbb{P}}}{(i_1+1)!_{\mathbb{P}}\cdots(i_k+1)!_{\mathbb{P}}}$$

where $2|k!$ since $k \geq 2$. □

Noticing that, by (27), the left hand side of Formula (23) is the numerator of $B_{\mathbb{P},2n+1}$, we have the following relations between numerators and denominators of the Bernoulli numbers.

Proposition 13 *For every $n \geq 0$,*

$$\mathrm{num}(B_{\mathbb{P},2n+1}) \times \mathrm{den}(B_{\mathbb{P},2n}) = -\frac{(2n+1)!_{\mathbb{P}}}{(2n)!_{\mathbb{P}}} \times \mathrm{num}(B_{\mathbb{P},2n}).$$

In particular,

- $\mathrm{num}(B_{\mathbb{P},2n}) \mid \mathrm{num}(B_{\mathbb{P},2n+1})$
- *if $p-1|2n$ then, either $p \mid \mathrm{den}(B_{\mathbb{P},2n})$, or $p \mid \mathrm{num}(B_{\mathbb{P},2n+1})$,*
- *if $p \mid \mathrm{den}(B_{\mathbb{P},2n})$, then $p-1 \mid 2n$ and $v_p(\mathrm{den}(B_{\mathbb{P},2n})) \leq 1 + v_p(2n)$,*
- *for every $m \geq 1$, $v_2(\mathrm{den}(B_{\mathbb{P},m})) = 1 + v_2(m)$.*

Proof It suffices to recall that, by (12), p divides $\frac{(2n+1)!_{\mathbb{P}}}{(2n)!_{\mathbb{P}}}$ if and only if $p-1$ divides $2n$, and then, $v_p\left(\frac{(2n+1)!_{\mathbb{P}}}{(2n)!_{\mathbb{P}}}\right) = 1 + v_p(2n)$. For the last assertion, note that the numerators of all the Bernoulli numbers are odd. □

Examples 3 For $p = 3$ and $2n = 4$, $p-1|4$, $p \nmid \mathrm{den}(B_4)$, but $p|\mathrm{num}(B_5)$.
For $p = 5$ and $2n = 2$, $p|\mathrm{num}(B_3)$, but $p-1 \nmid 2$,

Thanks to Proposition 13, Formula (27) is more precise for prime indices.

Corollary 6 *If p is an odd prime number, then*

$$2\,B_{\mathbb{P},p} \equiv 1 \pmod{p}. \tag{28}$$

Proof Consider Formula (20) for $n = p$:

$$2\, B_{\mathbb{P},p} + 1 + \sum_{k=1}^{p-1} \binom{p+1}{k}_{\mathbb{P}} B_{\mathbb{P},k} = 0\,.$$

For $2 \leq k \leq p-1$, we have

$$v_p\left(\binom{p+1}{k}_{\mathbb{P}}\right) = v_p((p+1)!_{\mathbb{P}}) - v_p(k!_{\mathbb{P}}) - v_p((p+1-k)!_{\mathbb{P}}) = 1,$$

while, for $1 \leq k \leq p-2$, following Formula (27) and Proposition 13, $v_p(B_{\mathbb{P},k}) \geq 0$. Thus,

$$2\, B_{\mathbb{P},p} + 1 + \binom{p+1}{p}_{\mathbb{P}} B_{\mathbb{P},1} + \binom{p+1}{p-1}_{\mathbb{P}} B_{\mathbb{P},p-1} \equiv 0 \pmod{p}\,.$$

Moreover,

$$\binom{p+1}{p}_{\mathbb{P}} B_{\mathbb{P},1} = 2 \times (-\frac{1}{2}) = -1 \text{ et } \binom{p+1}{p-1}_{\mathbb{P}} = \frac{p!_{\mathbb{P}}}{(p-1)!_{\mathbb{P}}}\,.$$

Consequently,

$$2\, B_{\mathbb{P},p} \equiv -\frac{p!_{\mathbb{P}}}{(p-1)!_{\mathbb{P}}} B_{\mathbb{P},p-1} \pmod{p}\,.$$

By Formula (20) again, we have:

$$\frac{p!_{\mathbb{P}}}{(p-1)!_{\mathbb{P}}} B_{\mathbb{P},p-1} + 1 + \sum_{k=1}^{p-2} \binom{p}{k}_{\mathbb{P}} B_{\mathbb{P},k} = 0\,.$$

For $1 \leq k \leq p-2$, $v_p(B_{\mathbb{P},k}) \geq 0$ and $v_p(\binom{p}{k}_{\mathbb{P}}) = 1$. Thus,

$$\frac{p!_{\mathbb{P}}}{(p-1)!_{\mathbb{P}}} B_{\mathbb{P},p-1} + 1 \equiv 0 \text{ and } 2\, B_{\mathbb{P},p} \equiv 1 \pmod{p}\,.$$

□

The classical Bernoulli numbers have the following property:

$$\forall n \geq 1 \ \ \forall a \in \mathbb{Z} \qquad a^n(a^n - 1)\frac{B_n}{n} \in \mathbb{Z} \quad \text{(Sylvester [22])}\,.$$

We have a similar result for the $B_{\mathbb{P},n}$'s if we replace the integer $n = \frac{n!}{(n-1)!}$ by the integer $\frac{n!_{\mathbb{P}}}{(n-1)!_{\mathbb{P}}}$.

Proposition 14

$$\forall n \neq 2 \ \forall a \in \mathbb{Z} \qquad a^n(a^{2\lfloor n/2 \rfloor} - 1)\,\frac{(n-1)!_{\mathbb{P}}}{n!_{\mathbb{P}}}\,B_{\mathbb{P},n} \in \mathbb{Z}\,.$$

Proof Fix $n \neq 2$, $a \in \mathbb{Z}$, and $p \in \mathbb{P}$. Write $A_n = a^n(a^{2\lfloor n/2 \rfloor} - 1)$

- Assume that $n = 2m$ $(m \neq 1)$.
  As $\frac{(2m-1)!_{\mathbb{P}}}{2m!_{\mathbb{P}}} = \frac{1}{2}$, we have to show that $\frac{1}{2}\,A_{2m} \times B_{\mathbb{P},2m} \in \mathbb{Z}_{(p)}$.
  - If $p - 1 \nmid 2m$, then $v_p(B_{\mathbb{P},2m}) \geq 0$, and hence, $\frac{1}{2}\,A_{2m} \times B_{\mathbb{P},2m} \in \mathbb{Z}_{(p)}$.
  - If $p - 1 \mid 2m$, then $v_p(B_{\mathbb{P},2m}) \geq -(1 + v_p(2m))$ and it is enough to prove

$$(*) \quad v_2(A_{2m}) \geq 2 + v_2(2m) \text{ and, for } p \neq 2,\ v_p(A_{2m}) \geq 1 + v_p(2m).$$

- Assume that $n = 2m + 1$.
  As $B_{\mathbb{P},2m+1} \in \frac{1}{2} + \mathbb{Z}$, we have to show that $\frac{1}{2}\,A_{2m+1}\,\frac{(2m)!_{\mathbb{P}}}{(2m+1)!_{\mathbb{P}}} \in \mathbb{Z}_{(p)}$.
  - If $p - 1 \nmid 2m$, then $v_p(\frac{(2m)!_{\mathbb{P}}}{(2m+1)!_{\mathbb{P}}}) \geq 0$, and hence, $\frac{1}{2}\,A_{2m+1}\,\frac{(2m)!_{\mathbb{P}}}{(2m+1)!_{\mathbb{P}}} \in \mathbb{Z}_{(p)}$.
  - If $p - 1 | 2m$, then $v_p(\frac{(2m)!_{\mathbb{P}}}{(2m+1)!_{\mathbb{P}}}) = -(1 + v_p(2m))$ and, as in the even case, it is enough to prove that

$$(*) \quad v_2(A_{2m+1}) \geq 2 + v_2(2m) \text{ and, for } p \neq 2,\ v_p(A_{2m+1}) \geq 1 + v_p(2m).$$

- It remains to prove that inequalities (*) are satisfied when $p - 1 | 2m$. Let us write $2m = (p-1)p^{\alpha}r$ where $\alpha = v_p(2m)$.
  - If $p \mid a$, $v_p(A_n) \geq n \geq 2m \geq 2 + v_p(2m)$.
  - If $p \nmid a$, then $v_p(a^{p-1} - 1) \geq 1$ and, by Lemma 3 below,
  - if $p \neq 2$, $v_p(a^{2m} - 1) = v_p(a^{((p-1)q) \times p^{\alpha}} - 1) \geq 1 + \alpha = 1 + v_p(2m)$,
  - if $p = 2$, $v_2(a^{2m} - 1) \geq 2 + v_2(2m)$. □

Lemma 3 ([7]) *For every integer a and every prime p,*

$$v_p(a - 1) \geq \alpha \ \Rightarrow \ v_p(a^p - 1) \geq 1 + \alpha\,.$$

Moreover,

$$v_2(a - 1) \geq 1 \ \Rightarrow \ v_2(a^{2m} - 1) \geq 2 + v_2(2m)\,.$$

Proof One has

$$v_p(a^p - 1) = v_p(((a-1)+1)^p - 1) = v_p\left(\sum_{k=1}^{p}\binom{p}{k}(a-1)^k\right) \geq 1 + v_p(a-1).$$

The inequality $v_2(a-1) \geq 1$ implies that $v_2(a^2-1) \geq 3$ and $v_2(a^{2m}-1) \geq 3 + v_2(m) = 2 + v_2(2m)$. □

6 Generalized Bernoulli Polynomials

Similarly, we may also consider *generalized Bernoulli polynomials* associated to a subset E of $\mathbb{Z}$ as defined by the power series

$$\frac{x\exp_E(tx)}{\exp_E(x)-1} = \sum_{n\geq 0} B_{E,n}(t)\,\frac{x^n}{n!_E} \tag{29}$$

From (1) and (21), we easily deduce that the nth Bernoulli polynomial associated to E is equal to

$$B_{E,n}(X) = \sum_{k=0}^{n}\binom{n}{k}_E B_{E,k}X^{n-k}\,.$$

As in the classical case, we have

$$B_{E,n}(0) = B_{E,n}\ (n\geq 0)\ ,\ B_{E,0}(X) = 1 \ \text{ and } \ B_{E,n}(1) = B_{E,n}\ (n \neq 1).$$

Following Alkmvist and Meurman [2], the classical Bernoulli polynomials satisfy: $B_n(X) - B_n(0) \in \mathrm{Int}(\mathbb{Z})$. In fact, for all positive integers a and b, $b^n\left(B_n\left(\frac{a}{b}\right) - B_n(0)\right) \in \mathbb{Z}$ (see also [21]). We have a similar property, in fact a much stronger property, for $B_{\mathbb{P},n}(X)$.

Proposition 15 *For every n,*

$$B_{\mathbb{P},n}(X) - B_{\mathbb{P},n}(0) = \sum_{k=0}^{n-1}\binom{n}{k}_{\mathbb{P}} B_{\mathbb{P},k}\,X^{n-k} \in \mathbb{Z}[X]\,.$$

Proof We have to prove that, for $0 \leq k \leq n-1$, $\binom{n}{k}_{\mathbb{P}}\,B_{\mathbb{P},k} \in \mathbb{Z}$. By Proposition 11, we know that $\binom{k+1}{k}_{\mathbb{P}}\,B_{\mathbb{P},k} \in \mathbb{Z}$. It is then easy to conclude with Lemma 4 below. □

Lemma 4 *For* $0 \leq k \leq n$, $\frac{(k+1)!_{\mathbb{P}}}{k!_{\mathbb{P}}}$ *divides* $\binom{n+1}{k}_{\mathbb{P}}$.

Proof The assertion is obvious for $k = 0$ or n. Thus, we assume now that $0 < k < n$. Let us prove first that

$$\forall\, 1 \leq k \leq n \qquad 2 \,\Big|\, \binom{n+1}{k}_{\mathbb{P}} \tag{30}$$

We know that $v_2((n+1)!_{\mathbb{P}}) = n + v_2(n!)$. Consequently,
$v_2(\binom{n+1}{k}_{\mathbb{P}}) = v_2((n+1)!_{\mathbb{P}}) - v_2(k!_{\mathbb{P}}) - v_2((n+1-k)!_{\mathbb{P}})$
$= 1 + v_2(n!) - v_2((k-1)!) - v_2((n-k)!)$
$\geq 1 + v_2(n!) - v_2((n-1)!) = 1 + v_2(n)$.
Consider now the following equalities of fractions:

$$\binom{n+1}{k}_{\mathbb{P}} \Big/ \frac{(k+1)!_{\mathbb{P}}}{k!_{\mathbb{P}}} = \binom{n+1}{k+1}_{\mathbb{P}} \Big/ \frac{(n+1-k)!_{\mathbb{P}}}{(n-k)!_{\mathbb{P}}} = \binom{n+2}{k+1}_{\mathbb{P}} \Big/ \frac{(n+2)!_{\mathbb{P}}}{(n+1)!_{\mathbb{P}}}$$

We know that, if n is odd, then $(n+1)!_{\mathbb{P}} = 2 \times n!_{\mathbb{P}}$. Thus, the denominators of the fractions are respectively equal to 2 when $k+1$, $n+1-k$ and n are respectively even. All the possible cases having been considered, we can conclude with (30). □

The first Bernoulli polynomials associated to $\mathbb{P}$ are as follows:

$$B_{\mathbb{P},1}(X) = X - \frac{1}{2}\,,\ B_{\mathbb{P},2}(X) = X^2 - X + \frac{5}{12}\,,\ B_{\mathbb{P},3}(X) = X^3 - 6X^2 + 5X - \frac{5}{2}\,,$$

$$B_{\mathbb{P},4}(X) = X^4 - X^3 + 5X^2 - 5X + \frac{103}{40}\,, \ldots$$

7 A Subset E Such That $k!_E = (k+1)!_E$ for Infinitely Many k

For each subset E of $\mathbb{Z}$ and each $p \in \mathbb{P}$, denote by E_p the topological closure for the p-adic topology of E in $\mathbb{Z}_{(p)}$. It is clear from Definition 1 that $n!_E\,\mathbb{Z}_{(p)} = n!_{E_p}\,\mathbb{Z}_{(p)}$, or equivalently, writing $n!_E = \prod_{p\in\mathbb{P}} p^{w_{E,\,p}(n)}$, that we have $w_{E,\,p}(n) = w_{E_p,\,p}(n)$.
From now on, we let $E = \mathbb{P} \cup 2\mathbb{P}$.

Lemma 5 *For $E = \mathbb{P} \cup 2\mathbb{P}$, we have:*

(a) $E_p = \{p, 2p\} \cup (\mathbb{Z}_{(p)} \setminus p\mathbb{Z}_{(p)}) \quad (p \neq 2)$,
(b) $E_2 = \{4\} \cup (\mathbb{Z}_{(2)} \setminus 4\mathbb{Z}_{(2)})$.

Proof Thanks to Dirichlet's theorem about arithmetical sequences, it is known that $\mathbb{P}_p = \{p\} \cup (\mathbb{Z}_{(p)} \setminus p\mathbb{Z}_{(p)})$ (see [4] or [6]). Assertion (a) follows from the fact that $2\mathbb{P} \subset \{2p\} \cup (\mathbb{Z} \setminus p\mathbb{Z})$ and Assertion (b) from $\mathbb{P}_2 = \{2\} \cup (\mathbb{Z}_{(2)} \setminus 2\mathbb{Z}_{(2)})$ and $2\mathbb{P}_2 = \{4\} \cup (2\mathbb{Z}_{(2)} \setminus 4\mathbb{Z}_{(2)})$. □

Lemma 6 *For every odd prime p, we have*

$$w_{E,p}(n) = \begin{cases} 0 & \text{if } 0 \le n < p \\ 1 & \text{if } p \le n < 2p \\ h + v_p(h!) & \text{if } 2 + (p-1)h \le n \le p + (p-1)h \text{ for } h \ge 2. \end{cases}$$

In particular, if $n \neq p$ is odd, $w_{E,p}(n) = w_{E,p}(n-1)$.

Proof Let $E_p = \cup_{j=0}^{p-1} E_j$ where $E_0 = \{p, 2p\}$ and $E_j = j + p\mathbb{Z}_{(p)}$ for $1 \le j < p$. Then $w_{E_0,p}(0) = 0$, $w_{E_0,p}(1) = 1$ and, for $n \ge 2$, $w_{E_0,p}(n) = +\infty$ and, for $1 \le j \le p-1$, $w_{E_j,\,p}(n) = w_{p\mathbb{Z}_{(p)},\,p}(n) = n + v_p(n!)$. As for $0 \le i < j \le p-1$, $x \in E_i$ and $y \in E_j$, we have $v_p(x-y) = 0$; the sequence $\{w_{E,p}(n)\}_{n\ge 0}$ is the increasing shuffle of the p previous sequences, that is, the disjoint union of these sequences sorted into a nondecreasing order [10, §2]. Noticing that all the p sequences begin with 0 and 1, we obtain the announced formula.

The last assertion is an obvious consequence since an odd integer n distinct from p is strictly greater than the lower bound of the interval to which it belongs, and hence, n and $n-1$ belong to the same interval. □

Lemma 7 *The sequence $\{w_{E,2}(n)\}_{n\ge 0}$ is the increasing shuffle of the sequences $\{n + v_2(n!)\}_{n\ge 0}$ and $\{0\} \cup \{2m - 1 + v_2((m-1)!)\}_{m\ge 1}$.*

In particular, if n and m are two integers such that

$$2n - \sigma_2(n) = 3m - 2 - \sigma_2(m-1)$$

where $\sigma_2(k)$ denotes the sum of the digits of k in base 2, then

$$w_{E,2}(n+m+1) = w_{E,2}(n+m).$$

Proof Let us write $E_2 = \{4\} \cup (1 + 4\mathbb{Z}_{(2)}) \cup (2 + 4\mathbb{Z}_{(2)}) \cup (3 + 4\mathbb{Z}_{(2)})$. As $(1 + 4\mathbb{Z}_{(2)}) \cup (3 + 4\mathbb{Z}_{(2)}) = 1 + 2\mathbb{Z}_{(2)}$, we have

$$w_{(1+4\mathbb{Z}_{(2)})\cup(3+4\mathbb{Z}_{(2)}),\,2}(n) = w_{2\mathbb{Z}_{(2)},\,2}(n) = n + v_2(n!) = 2n - \sigma_2(n).$$

On the other hand, as $\{4\} \cup (2 + 4\mathbb{Z}_{(2)}) = 2 \times \left(\{2\} \cup (1 + 2\mathbb{Z}_{(2)})\right)$, we have

$$w_{\{4\}\cup(2+4\mathbb{Z}_{(2)}),\,2}(m) = m + w_{\{2\}\cup(1+2\mathbb{Z}_{(2)}),\,2}(m) = m + w_{2\mathbb{Z}_{(2)},\,2}(m-1)$$

$$= 2m - 1 + w_{\mathbb{Z}_{(2)},\,2}(m-1) = 2m - 1 + v_2((m-1)!) = 3m - 2 - \sigma_2(m-1).$$

The last assertion of the proposition follows from the fact that, in the shuffle of the two sequences, $w_{E,\,2}(n+m)$ belongs to the constant subsequence formed by the values equal to $w_{E,\,2}(n+m+1)$. □

Proposition 16 *Let* $E = \mathbb{P} \cup 2\mathbb{P}$. *There are infinitely many integers* n *such that* $n!_E = (n-1)!_E$, *namely, for instance, the integers* $5F_k^2 + 6$ *where* F_k *denotes the* k*th Fermat number and* k *is divisible by* 4.

Proof Let $n = 3\,(4^{2^k} + 2 \cdot 2^{2^k} + 2)$ and $m = 2\,(4^{2^k} + 2 \cdot 2^{2^k} + 2)$. For $k \geq 2$, $\sigma_2(n) = 6$ and $\sigma_2(m-1) = 4$. Thus, $3m - 2n = 0 = 2 + \sigma_2(m-1) - \sigma_2(n)$. Consequently, by Lemma 7, $w_2(n+m+1) = w_2(n+m)$.

Noticing that $n + m + 1 = 5F_k^2 + 6$ is odd and, for $k \equiv 0 \pmod 4$, is also divisible by 11, we thus have by Lemma 6 that $w_p(5F_k^2 + 6) = w_p(5F_k^2 + 5)$ for every odd prime p. □

References

1. D. Adam, J.-L. Chabert, Y. Fares, Subsets of $\mathbb{Z}$ with simultaneous orderings, *Integers* **10** (2010), 435–451.
2. G. Alkmvist and A. Meurman, Values of Bernoulli polynomials and Hurwitz's Zeta function at rational points, *C. R. Math. Rep. Acad. Sci. Canada* **13** (1991), 104–109.
3. M. Bhargava, The factorial function and generalizations, *Amer. Math. Monthly* **107** (2000), 783–799.
4. J.-L. Chabert, Une caractérisation des polynômes prenant des valeurs entières sur tous les nombres premiers, *Canad. Math. Bull.* **39** (1996), 402–407.
5. J.-L. Chabert, Integer-valued polynomials on prime numbers and logarithm power expansion, *European J. Combin.* **28** (2007), 754–761.
6. J.-L. Chabert, S. Chapman, W. Smith, A basis for the ring of polynomials integer-valued on prime numbers, in *Factorization in Integral Domains*. Edited by D. D. Anderson. Lecture Notes in Pure and Appl. Math., **189**, Dekker, New York, 1997, 271–284.
7. J.-L. Chabert, A.-H. Fan, and Y. Fares, Minimal dynamical systems on a discrete valuation domain, *Discrete Contin. Dyn. Syst.* **35** (2009), 777–795.
8. B. Diaz, Asymptotics on a class of Legendre formulas, arXiv: 2010.13645v3 [math.NT] 4 Apr 2021.
9. K. Johnson, The invariant subalgebra and anti-invariant submodule of $K_*K_{(p)}$, *J. K-theory* **2** (2008), 123–145.
10. K. Johnson, Limits of characteristic sequences of integer-valued polynomials on homogeneous sets, *J. Number Theory* **129** (2009), 2933–2942.
11. J. Konvalina, A Unified Interpretation of the Binomial Coefficients, the Stirling Numbers, and the Gaussian Coefficients, *Amer. Math. Monthly* **107** (2000), 901–910.
12. S. Lang, *Number Theory III*, Encyclopedia of Mathematical Sciences, **60**, Springer-Verlag, New York, 1991
13. A. Mingarelli, Abstract factorials, *Notes on Number Theory and Discrete Mathematics* **19.3** (2013), 43–76.
14. H. Minkowski, Zur Theorie der quadratischen Formen, *J. Reine Angew. Math.* **101** (1897), 196–202.
15. W. Narkiewicz, The development of prime number theory, *Monographs in Mathematics*, Springer.
16. I. Rochev, On linear independence of values of certain q-series, *Izv. Math.* **75** (2011), 177–221.
17. K. F. Roth, Rational approximations to algebraic numbers, *Mathematika* **2.1** (1955), 1–20.
18. I. Schur, Über eine Klasse von endlichen Gruppen linearer Substitutionen, *Sitzungsber. Preuss. Akad. Wiss.* (1905), 77–91.
19. N. Sloane, *The On-Line Encyclopedia of Integer Sequences*, https://oeis.org.

20. W. M. Schmidt, Simultaneous approximation to algebraic numbers by rationals, *Acta Math.* **125** (1970), 189–201.
21. B. Sury, The Value of Bernoulli Polynomials at Rational Numbers, *Bull. London Math. Soc.* **25** (1993), 327–329.
22. J.J. Sylvester, Note on the Numbers of Bernoulli and Euler, and a new Theorem concerning Prime Numbers, *The London, Edinburgh, and Dublin Philosophical Magazine and Journal of Science* **21** (1861), 127–136.

Polynomial Root Extensions

D. D. Anderson and David F. Anderson

In memory of Paul-Jean Cahen

2020 Mathematics Subject Classification Primary: 13B02; Secondary: 13B21; 13B22; 13B99

1 Introduction

All rings are commutative with $1 \neq 0$, and all ring extensions $R \subseteq S$ are unitary (i.e., R and S have the same identity element). In this paper, we study several generalizations of the integral closure of R in S by replacing monic polynomials in $R[X]$ with other subsets of $R[X]$. Examples include the algebraic closure, root closure, (2, 3)-closure, and complete integral closure of R in S.

In Sect. 2, we define the $\mathscr{S}$*-closure* of R in S for $\mathscr{S} \subseteq R[X]$ as $\overline{R}^{\mathscr{S}} = \{s \in S \mid f(s) = 0 \text{ for some } f(X) \in \mathscr{S}\}$ and give many results and examples. In Sect. 3, we give a more general definition of closure which includes the complete integral closure and (2, 3)-closure. For $\mathcal{S} \subseteq \mathcal{P}(\prod_{\alpha \in \Lambda} R[X])$, we define the $\mathcal{S}$*-closure* of R in S as $\overline{R}^{\mathcal{S}} = \{s \in S \mid \text{for some } A \in \mathcal{S} \text{ and } (f_\alpha(X)) \in A,\ f_\alpha(s) = 0 \text{ for every } \alpha \in \Lambda\}$. For $\mathscr{S} \subseteq R[X]$, we have $\overline{R}^{\mathcal{S}} = \overline{R}^{\mathscr{S}}$, where $\mathcal{S} = \{\mathscr{S}\} \subseteq \mathcal{P}(R[X])$.

Let $\mathbb{N}$, $\mathbb{Z}$, $\mathbb{Q}$, $\mathbb{R}$, and $\mathbb{F}_{p^n}$ denote the set of positive integers, the ring of integers, the field of rational numbers, the field of real numbers, and the finite field with p^n elements, respectively. For a commutative ring R, let $Z(R)$ be the set of zero-divisors of R, $Reg(R) = R \setminus Z(R)$ the set of regular elements of R, $T(R) = R_{Reg(R)}$

D. D. Anderson (✉)
Department of Mathematics, University of Iowa, Iowa City, IA, USA
e-mail: dan-anderson@uiowa.edu

D. F. Anderson
Department of Mathematics, The University of Tennessee, Knoxville, TN, USA
e-mail: danders5@utk.edu

J.-L. Chabert et al. (eds.), *Algebraic, Number Theoretic, and Topological Aspects of Ring Theory*, https://doi.org/10.1007/978-3-031-28847-0_3

the total quotient ring of R, and $A^* = A \setminus \{0\}$ for $A \subseteq R$. The power set of a set A will be denoted by $\mathcal{P}(A)$, and X, Y will be indeterminates. If $R \subseteq S$ is an extension of commutative rings and $R = R_1 \times \cdots \times R_n$, then also $S = S_1 \times \cdots \times S_n$ with each $R_i \subseteq S_i$ a unitary ring extension since $R \subseteq S$ is unitary. Let $\{R_\alpha\}_{\alpha \in \Lambda}$ be a family of commutative rings and $R = \prod_{\alpha\in\Lambda} R_\alpha$. We will often identify $R[X]$ as a subring of $\prod_{\alpha\in\Lambda}(R_\alpha[X])$; it is a proper subring precisely when Λ is infinite. For an extension $R \subseteq S$ of commutative rings and $s \in S$, let $deg_R(s) = min\{degf(X) \mid 0 \neq f(X) \in R[X]$ and $f(s) = 0\}$; if no such $f(X)$ exist, we let $deg_R(s) = \infty$. In a similar manner, we define $mdeg_R(s)$ (resp., $rdeg_R(s)$) by restricting to monic polynomials (resp., polynomials whose leading coefficient is in $Reg(R)$) in $R[X]$.

General references for ring theory are [4] and [5]. For a detailed study of root closure, see the survey article [2].

2 The $\mathscr{S}$-Closure

Let $R \subseteq S$ be a (unitary) extension of commutative rings. As usual, an element $s \in S$ is *integral over* R if $f(s) = 0$ for some monic polynomial $f(X) \in R[X]$. The set $\overline{R} = \{s \in S \mid s$ is integral over $R\}$, called the *integral closure of* R *in* S, is a subring of S containing R, and the extension $R \subseteq \overline{R}$ enjoys many nice properties concerning prime ideals of the two rings (see [4, Sections 11 and 12] and [5, Section 1.6]). If $S = T(R)$, we just call $\overline{R}$ the *integral closure of* R. More generally, an element $s \in S$ is *algebraic over* R if $f(s) = 0$ for some $0 \neq f(X) \in R[X]$. Let $\overline{R}^{\text{alg}} = \{s \in S \mid s$ is algebraic over $R\}$ be the *algebraic closure of* R *in* S. Then $\overline{R}^{\text{alg}}$ contains R, but unlike the integral closure, $\overline{R}^{\text{alg}}$ need not be a subring of S. Indeed, we can have $a, b \in \overline{R}^{\text{alg}}$, but $a \pm b, ab \notin \overline{R}^{\text{alg}}$ when R is not an integral domain (see Example 2.12(a) and Corollary 2.6).

Let $\mathscr{S} \subseteq R[X]$ (we usually assume that $\emptyset \neq \mathscr{S} \subseteq R[X]^*$). We define the *$\mathscr{S}$-closure of* R *in* S as $\overline{R}^{\mathscr{S}} = \{s \in S \mid f(s) = 0$ for some $f(X) \in \mathscr{S}\}$. We have $\overline{R}^{\mathscr{S}} \subseteq S$, but $\overline{R}^{\mathscr{S}}$ need not be a subring of S, need not contain R, and may even be empty (e.g., $\overline{R}^{\mathscr{S}} = \emptyset$ when $R = \mathbb{Z} \subseteq \mathbb{Q} = S$ and $\mathscr{S} = \{X^2 + 1\}$). We say that R is *$\mathscr{S}$-closed in* S if $\overline{R}^{\mathscr{S}} \subseteq R$, and R is *$\mathscr{S}$-closed* if $\overline{R}^{\mathscr{S}} \subseteq R$ when $S = T(R)$. Thus $\overline{R} = \overline{R}^{\mathscr{S}}$ for $\mathscr{S} = \{f(X) \in R[X] \mid f(X)$ monic$\}$, while $\overline{R}^{\text{alg}} = \overline{R}^{\mathscr{S}}$ for $\mathscr{S} = R[X]^*$.

Note that we may have $\emptyset \neq \overline{R}^{\mathscr{S}} \subsetneq R$ (e.g., if $R = \mathbb{Z} \subseteq \mathbb{Q} = S$ and $\mathscr{S} = \{X^2 - n \mid n \in \mathbb{N}\}$, then $\overline{R}^{\mathscr{S}} = R^* \subsetneq R$). One way around this would be to define $\mathscr{S}_* = \mathscr{S} \cup \{X - r \mid r \in R\}$. Then $R \subseteq \overline{R}^{\mathscr{S}_*} = R \cup \overline{R}^{\mathscr{S}}$ (see Example 2.22(b)). Another way would be to consider the ring $R[\overline{R}^{\mathscr{S}}]$, the subring of S generated by $R \cup R^{\mathscr{S}}$, i.e., the smallest subring of S containing $R \cup R^{\mathscr{S}}$.

For $\emptyset \neq T \subseteq R^*$, let $\mathscr{S}_T = \{f(X) \in R[X] \mid$ the leading coefficient of $f(X)$ is in $T\}$ and $\overline{R}^T = \overline{R}^{\mathscr{S}_T}$. Then $\overline{R}^T = \overline{R}$ for $\{1\} \subseteq T \subseteq U(R)$, while $\overline{R}^T = \overline{R}^{\text{alg}}$

for $T = R^*$. For $T = Reg(R)$, we define the *regular closure of R in S* as $\overline{R}^{\text{reg}} = \overline{R}^{\text{Reg}(R)}$. If R is an integral domain, then $\mathscr{S}_{Reg(R)} = R[X]^*$. Thus $\overline{R}^{\text{reg}} = \overline{R}^{\text{alg}}$ when R is an integral domain. Clearly, $R \subseteq \overline{R}^T$, but $\overline{R}^T$ need not be a subring of S (see Theorems 2.4, 2.5, and Example 2.7).

As another example of $\mathscr{S}$-closure, let $\mathscr{S} = \{X^n - r \mid n \in \mathbb{N}, r \in R\}$. Then $\overline{R}^{\text{rt}} = \overline{R}^{\mathscr{S}} = \{s \in S \mid s^n \in R \text{ for some } n \in \mathbb{N}\}$ is the *root closure of R in S*. More generally, for a fixed $n \in \mathbb{N}$, let $\mathscr{S}_n = \{X^n - r \mid r \in R\}$ and $\overline{R}^{\text{rt}_\text{n}} = \overline{R}^{\mathscr{S}_n} = \{s \in S \mid s^n \in R\}$ be the *n-root closure of R in S*. Thus $R \subseteq \overline{R}^{\text{rt}_\text{n}} \subseteq \overline{R}^{\text{rt}} \subseteq \overline{R} \subseteq \overline{R}^{\text{reg}} \subseteq \overline{R}^{\text{alg}}$ for every $n \in \mathbb{N}$. Also, note that $\overline{R}^{\text{rt}_1} = R$, $\overline{R}^{\text{rt}_\text{n}} \subseteq \overline{R}^{\text{rt}_\text{kn}}$ for every $k, n \in \mathbb{N}$, and $\overline{R}^{\text{rt}} = \bigcup_{n \in \mathbb{N}} \overline{R}^{\text{rt}_\text{n}} = \overline{R}^{\mathscr{S}}$, where $\mathscr{S} = \bigcup_{n \in \mathbb{N}} \mathscr{S}_n$ (cf. Theorem 2.21(e)). Even more generally, for $\emptyset \neq T \subseteq \mathbb{N}$, let $\overline{R}^{\text{rt}_\text{T}} = \{s \in S \mid s^n \in R \text{ for some } n \in T\}$. Then $\overline{R}^{\text{rt}_\text{T}} = \bigcup_{n \in T} \overline{R}^{\text{rt}_\text{n}} = \overline{R}^{\bigcup_{n \in T} \mathscr{S}_n}$ (cf. Theorem 2.21(e) again).

More generally, suppose that A is only a subset of S. Then we define $A[X] = \{f(X) \in S[X] \mid f(x) = a_n X^n + \cdots + a_0 \text{ with every } a_i \in A\}$ and $\overline{A}^{\text{alg}} = \{s \in S \mid f(s) = 0, \text{ where } 0 \neq f(X) \in A[X]\}$. For example, $\overline{\overline{R}^{\text{alg}}}^{\text{alg}} = \{s \in S \mid f(s) = 0$ for some $0 \neq f(X) = a_n X^n + \cdots + a_0$ with every $a_i \in \overline{R}^{\text{alg}}\}$. In a similar manner, we can define $\overline{A}^{\text{reg}} = \overline{A}^{\text{Reg}(A)}$, $\overline{A}^{\text{rt}} = \{s \in S \mid s^n \in A \text{ for some } n \in \mathbb{N}\}$, and $\overline{A}^{\text{rt}_\text{n}} = \{s \in S \mid s^n \in A\}$ for any $A \subseteq S$.

We next state some well-known results concerning integral closure.

Theorem 2.1 *Let $R \subseteq S$ be an extension of commutative rings and $\overline{R}$ the integral closure of R in S.*

(a) *$\overline{R}$ is a subring of S containing R and $\overline{\overline{R}} = \overline{R}$.*
(b) *Let T be a multiplicatively closed subset of R. Then $\overline{R}_T = \overline{R_T}$, the integral closure of R_T in S_T.*
(c) *$\overline{R}[X] = \overline{R[X]}$, the integral closure of $R[X]$ in $S[X]$.*
(d) *Let I be an ideal of S. Then $\overline{R}/(I \cap \overline{R}) \subseteq \overline{R/(I \cap R)}$, the integral closure of $R/(I \cap R)$ in S/I.*
(e) *Let $\{R_\alpha \subseteq S_\alpha \mid \alpha \in \Lambda\}$ be a family of extensions of commutative rings, $R = \prod_{\alpha \in \Lambda} R_\alpha$, and $S = \prod_{\alpha \in \Lambda} S_\alpha$. Then $\overline{R} \subseteq \prod_{\alpha \in \Lambda} \overline{R_\alpha}$. Moreover, $\overline{R} = \prod_{\alpha \in \Lambda} \overline{R_\alpha}$ when Λ is finite.*

Proof (a)–(c) are well known and may be found in almost any book on commutative algebra; see, for example, [4, Proposition 10.2 and Theorem 10.7, pp. 86-87].

(d) We have $R/(I \cap R) \subseteq \overline{R}/(I \cap \overline{R}) \subseteq S/I$. Let $x \in \overline{R}$ and $\overline{x} = x + (I \cap \overline{R}) \in \overline{R}/(I \cap \overline{R})$. Then $x^n + r_{n-1}x^{n-1} + \cdots + r_0 = 0$ for some $r_i \in R$ with every $\overline{r}_i \in R/(I \cap R)$; so $\overline{x}^n + \overline{r}_{n-1}\overline{x}^{n-1} + \cdots + \overline{r}_0 = \overline{0}$, that is, $\overline{x}$ is integral over $R/(I \cap R)$. Thus $\overline{x} \in \overline{R/(I \cap R)}$; so $\overline{R}/(I \cap \overline{R}) \subseteq \overline{R/(I \cap R)}$.

(e) Let $s = (s_\alpha) \in \overline{R} \subseteq \prod_{\alpha \in \Lambda} S$. Then $f(s) = 0$, where $f(X) \in R[X] \subseteq \prod_{\alpha \in \Lambda}(R_\alpha[X])$ is monic of degree n. Thus $f(X) = (f_\alpha(X))$, where every $f_\alpha(X) \in R_\alpha[X]$ is monic of degree n. Then $f_\alpha(s_\alpha) = 0$ for every $\alpha \in \Lambda$; so $s_\alpha \in \overline{R_\alpha}$ for every $\alpha \in \Lambda$. Hence $s = (s_\alpha) \in \prod_{\alpha \in \Lambda} \overline{R_\alpha}$; so $\overline{R} \subseteq \prod_{\alpha \in \Lambda} \overline{R_\alpha}$.

Suppose that Λ is finite, say $R_i \subseteq S_i$, $i = 1, \ldots, n$. Then $R = R_1 \times \cdots \times R_n \subseteq S_1 \times \cdots \times S_n = S$. For $s_i \in \overline{R_i}$, let $f_i(s_i) = 0$, where $f_i(X) \in R_i[X]$ is monic of degree m. For $j \neq i$, $1 \leq j \leq n$, let $f_j(X) = X^m$. Then $f(X) = (f_1(X), \ldots, f_n(X)) \in R[X]$ is monic of degree m and $f(0, \ldots, 0, s_i, 0, \ldots, 0) = 0$; so $(0, \ldots, 0, s_i, 0, \ldots, 0) \in \overline{R}$. If $s_i \in \overline{R_i}$ for $i = 1, \ldots, n$, then $(s_1, \ldots, s_n) = (s_1, 0, \ldots, 0) + (0, s_2, 0, \ldots, 0) + \cdots + (0, \ldots, 0, s_n) \in \overline{R}$ since $\overline{R}$ is a subring of S. Thus $\overline{R_1} \times \cdots \times \overline{R_n} \subseteq \overline{R}$; so $\overline{R} = \overline{R_1} \times \cdots \times \overline{R_n}$. □

The following example shows that the inclusions in parts (d) and (e) of Theorem 2.1 may be strict.

Example 2.2

(a) (We may have $\overline{R}/(I \cap \overline{R}) \subsetneq \overline{R/(I \cap R)}$ in Theorem 2.1(d).) Let $R = \mathbb{Z} \subseteq \mathbb{Z}[X] = S$, and $I = (X^2)$; so $R/(I \cap R) = \mathbb{Z} \subseteq \mathbb{Z}[X]/(X^2) = S/I$. Then $\overline{R} = \mathbb{Z}$; so $\overline{R}/(I \cap \overline{R}) = \mathbb{Z}/((X^2) \cap \mathbb{Z}) = \mathbb{Z}$, but $\overline{R/(I \cap R)} = S/I = \mathbb{Z}[X]/(X^2)$. Thus $\overline{R}/(I \cap \overline{R}) \subsetneq \overline{R/(I \cap R)}$.

(b) (For Λ infinite, we may have $\overline{\prod_{\alpha \in \Lambda} R_\alpha} \subsetneq \prod_{\alpha \in \Lambda} \overline{R_\alpha}$ in Theorem 2.1(e).) Let $R_n = \mathbb{F}_2 \subsetneq \mathbb{F}_{2^n} = S_n$ for every $n \in \mathbb{N}$, and $R = \prod_{n \in \mathbb{N}} \mathbb{F}_2 \subsetneq \prod_{n \in \mathbb{N}} \mathbb{F}_{2^n} = S$. Let $t_n \in \mathbb{F}_{2^n}$ with $\langle t_n \rangle = \mathbb{F}_{2^n}^*$; so the minimal polynomial for t_n (over $\mathbb{F}_2$) has degree n. Let $t = (t_n)_{n \in \mathbb{N}}$. Then every $t_n \in \mathbb{F}_{2^n} = \overline{\mathbb{F}_2}$; so $t \in \prod_{n \in \mathbb{N}} \mathbb{F}_{2^n} = \prod_{n \in \mathbb{N}} \overline{\mathbb{F}_2}$. But $t \notin \overline{R}$. Thus $\overline{\prod_{n \in \mathbb{N}} R_n} \subsetneq \prod_{n \in \mathbb{N}} \overline{R_n}$.

The next result identifies $\overline{\prod_{\alpha \in \Lambda} R_\alpha}$ as a subset of $\prod_{\alpha \in \Lambda} \overline{R_\alpha}$ and recovers the "moreover" statement in Theorem 2.1(e).

Theorem 2.3 *Let $\{R_\alpha \subseteq S_\alpha \mid \alpha \in \Lambda\}$ be a family of extensions of commutative rings, $R = \prod_{\alpha \in \Lambda} R_\alpha$, and $S = \prod_{\alpha \in \Lambda} S_\alpha$. Then $\overline{R} = \{(s_\alpha) \in S \mid \{mdeg_{R_\alpha}(s_\alpha)\}_{\alpha \in \Lambda}$ is bounded$\} \subseteq \prod_{\alpha \in \Lambda} \overline{R_\alpha}$.*

Proof Let $T = \{(s_\alpha) \in S \mid \{mdeg_{R_\alpha}(s_\alpha)\}_{\alpha \in \Lambda}$ is bounded$\}$. Then the proof of Theorem 2.1(e) shows that $\overline{R} \subseteq T$. For the reverse inclusion, let $s = (s_\alpha) \in T$, and for every $\alpha \in \Lambda$, let $f_\alpha(s_\alpha) = 0$ for a monic $f_\alpha(X) \in R_\alpha[X]$ with $deg(f_\alpha(X)) = n_\alpha \leq N$. Then $f(X) = (X^{N-n_\alpha} f_\alpha(X)) \in R[X]$ is monic of degree N and $f(s) = 0$. Thus $s \in \overline{R}$; so $T \subseteq \overline{R}$. Hence $\overline{R} = T$. □

We have already observed several of the inclusions in the following theorem.

Theorem 2.4 *Let $\emptyset \neq T_1 \subseteq T_2 \subseteq R^*$ and $n \in \mathbb{N}$. Then $R \subseteq \overline{R}^{\mathrm{rtn}} \subseteq \overline{R}^{\mathrm{rt}} \subseteq \overline{R} \subseteq \overline{R}^{T_1} \subseteq \overline{R}^{T_2} \subseteq \overline{R}^{\mathrm{alg}}$. In particular, $R \subseteq \overline{R}^{\mathrm{rt}} \subseteq \overline{R} \subseteq \overline{R}^{\mathrm{reg}} \subseteq \overline{R}^{\mathrm{alg}}$.*

Proof Clearly $R \subseteq \overline{R}^{\mathrm{rtn}} \subseteq \overline{R}^{\mathrm{rt}} \subseteq \overline{R}$ and $\overline{R}^{T_1} \subseteq \overline{R}^{T_2} \subseteq \overline{R}^{\mathrm{alg}}$. Suppose that $s \in \overline{R}$; so $f(s) = 0$, where $f(X) \in R[X]$ is monic. Then $(tf)(s) = 0$ for every $t \in T_1$, and $tf(X) \in \mathscr{S}_{T_1}$; so $s \in \overline{R}^{T_1}$. Thus $\overline{R} \subseteq \overline{R}^{T_1}$; so $R \subseteq \overline{R}^{\mathrm{rtn}} \subseteq \overline{R}^{\mathrm{rt}} \subseteq \overline{R} \subseteq \overline{R}^{T_1} \subseteq \overline{R}^{T_2} \subseteq \overline{R}^{\mathrm{alg}}$.

The "in particular" statement is clear. □

We next show that $\overline{R}^{\text{reg}}$, like $\overline{R}$, but unlike $\overline{R}^{\text{alg}}$, $\overline{R}^{\text{rt}}$, and $\overline{R}^{\text{rt}_n}$ is always a subring of S (see Examples 2.12(a) and 2.14). In the next several results, by the usual abuse of notation, we let $\overline{R_T} \cap S = \{s \in S \mid st/t \in \overline{R_T}$ for some $t \in T\}$.

Theorem 2.5 *Let $\emptyset \neq T \subseteq \text{Reg}(R)$ be multiplicatively closed. Then $\overline{R}^T = \overline{R_T} \cap S$; so $\overline{R}^T$ is a subring of S. In particular, $\overline{R}^{\text{reg}} = \overline{T(R)} \cap S$; so $\overline{R}^{\text{reg}}$ is a subring of S.*

Proof For notational convenience, we assume that $1 \in T$. Let $s \in \overline{R}^T$, say $f(s) = 0$, where $f(X) = tX^n + a_{n-1}X^{n-1} + \cdots + a_0 \in R[X]$ and $t \in T$. Then $f(X)/t$ is monic over R_T. Thus $s/1 \in S_T$ is integral over R_T; so $s \in \overline{R_T} \cap S$. Hence $\overline{R}^T \subseteq \overline{R_T} \cap S$.

For the reverse inclusion, let $s \in \overline{R_T} \cap S$. Then $s/1 \in S_T$ is integral over R_T; so $s/1$ satisfies a monic polynomial $f(X) \in R_T[X]$ of degree $n \geq 1$. Thus $(ts^n + a_{n-1}s^{n-1} + \cdots + a_0)/1 = 0/1$ for some $t \in T$ and $a_i \in R$, and hence $(tt')s^n + t'a_{n-1}s^{n-1} + \cdots + t'a_0 = 0$ for some $t' \in T$. Thus $s \in \overline{R}^T$ since $tt' \in T$; so $\overline{R_T} \cap S \subseteq \overline{R}^T$. Hence $\overline{R}^T = \overline{R_T} \cap S$.

The "in particular" statement is clear since $Reg(R)$ is always multiplicatively closed. □

Corollary 2.6 *Let D be an integral domain with quotient field K and $D \subseteq S$. Then $\overline{D}^{\text{alg}} = \overline{K} \cap S$ is a subring of S, where $\overline{K}$ is the integral closure of K in S_{D^*}.*

The following example shows that the "multiplicatively closed" hypothesis is needed in Theorem 2.5.

Example 2.7 Let $R = \mathbb{Z} \subseteq \mathbb{Q} = S$ and $T = \{2\} \subseteq Reg(\mathbb{Z}) = \mathbb{Z}^*$. Then $R \subseteq \overline{R}^T$ and $1/2 \in \overline{R}^T$, but $1/4 = 1/2 \cdot 1/2 \notin \overline{R}^T$. Thus $\overline{R}^T$ is not a subring of S.

However, we may have $\overline{R}^T$ a subring of S when T is not multiplicatively closed. For example, let $R = \mathbb{Z} \subseteq \mathbb{Q} = S$ and $T = \{-n \mid n \in \mathbb{N}\} \subseteq Reg(R)$. Then T is not multiplicatively closed, but $\overline{R}^T = \overline{R}^{\text{reg}} = S$.

We next show that $\overline{R}^{\text{reg}}$ is always $\mathscr{S}_{Reg(\overline{R}^{\text{reg}})}$-closed.

Theorem 2.8

(a) *Let $R \subseteq S$ be an extension of commutative rings. Then $\overline{\left(\overline{R}^{\text{reg}}\right)}^{\text{reg}} = \overline{R}^{\text{reg}}$.*
(b) *Let $R_i \subseteq S_i$, $i = 1, \ldots, n$, be extensions of commutative rings and $R = R_1 \times \cdots \times R_n \subseteq S_1 \times \cdots \times S_n = S$. Then $\overline{R}^{\text{reg}} = \overline{R_1}^{\text{reg}} \times \cdots \times \overline{R_n}^{\text{reg}}$.*

Proof

(a) By Theorem 2.5, $\overline{R}^{\text{reg}} = \overline{T(R)} \cap S$ and $\overline{R}^{\text{reg}} \subseteq \overline{\left(\overline{R}^{\text{reg}}\right)}^{\text{reg}} = \overline{T(\overline{R}^{\text{reg}})} \cap S$. Thus $\overline{\left(\overline{R}^{\text{reg}}\right)}^{\text{reg}} = \overline{T(\overline{R}^{\text{reg}})} \cap S = \overline{\overline{T(R)} \cap S} \cap S \subseteq \overline{\overline{T(R)}} \cap S = \overline{T(R)} \cap S = \overline{R}^{\text{reg}}$, and hence $\overline{\left(\overline{R}^{\text{reg}}\right)}^{\text{reg}} = \overline{R}^{\text{reg}}$.

(b) Observe that for $f(X) = (f_1(X), \ldots, f_n(X)) \in R[X] = R_1[X] \times \cdots \times R_n[X]$, $f(X) \in \mathscr{S}_{\mathrm{Reg}(R)} \Leftrightarrow \deg f_1(X) = \cdots = \deg f_n(X)$ and every $f_i(X) \in \mathscr{S}_{\mathrm{Reg}(R_i)}$. So $s = (s_1, \ldots, s_n) \in \overline{R}^{\mathrm{reg}} \Rightarrow$ every $s_i \in \overline{R_i}^{\mathrm{reg}}$. Thus $\overline{R}^{\mathrm{reg}} \subseteq \overline{R_1}^{\mathrm{reg}} \times \cdots \times \overline{R_n}^{\mathrm{reg}}$. Next, suppose that $s_i \in \overline{R_i}^{\mathrm{reg}}$, with $f_i(s_i) = 0$, where $f_i(X) \in \mathscr{S}_{\mathrm{Reg}(R_i)}$. For $j \neq i$, $1 \leq j \leq n$, let $f_j(X) = X^m$, where $m = \deg f_i(X)$. Then $f(X) = (f_1(X), \ldots, f_n(X)) \in \mathscr{S}_{\mathrm{Reg}(R)}$ has degree m and $f(0, \ldots, 0, s_i, 0, \ldots, 0) = 0$; so $(0, \ldots, 0, s_i, 0, \ldots, 0) \in \overline{R}^{\mathrm{reg}}$. Since $\overline{R}^{\mathrm{reg}}$ is a subring of S by Theorem 2.5, if $s_i \in \overline{R_i}^{\mathrm{reg}}$ for $i = 1 \leq i \leq n$, then $(s_1, \ldots, s_n) = (s_1, 0, \ldots, 0) + (0, s_2, 0, \ldots, 0) + \cdots + (0, \ldots, 0, s_n) \in \overline{R}^{\mathrm{reg}}$. Thus $\overline{R_1}^{\mathrm{reg}} \times \cdots \times \overline{R_n}^{\mathrm{reg}} \subseteq \overline{R}^{\mathrm{reg}}$, and hence $\overline{R}^{\mathrm{reg}} = \overline{R_1}^{\mathrm{reg}} \times \cdots \times \overline{R_n}^{\mathrm{reg}}$.

□

Example 2.2(b) also shows that we may have $\overline{\prod_{\alpha\in\Lambda} R_\alpha}^{reg} \subsetneq \prod_{\alpha\in\Lambda} \overline{R_\alpha}^{reg}$ when Λ is infinite. The next result identifies $\overline{\prod_{\alpha\in\Lambda} R_\alpha}^{reg}$ as a subset of $\prod_{\alpha\in\Lambda} \overline{R_\alpha}^{reg}$ and recovers Theorem 2.8(b).

Theorem 2.9 *Let $\{R_\alpha \subseteq S_\alpha \mid \alpha \in \Lambda\}$ be a family of extensions of commutative rings, $R = \prod_{\alpha\in\Lambda} R_\alpha$, and $S = \prod_{\alpha\in\Lambda} S_\alpha$. Then $\overline{R}^{reg} = \{(s_\alpha) \in S \mid \{rdeg_{R_\alpha}(s_\alpha)\}_{\alpha\in\Lambda}$ is bounded$\} \subseteq \prod_{\alpha\in\Lambda} \overline{R_\alpha}^{reg}$.*

Proof The proof is similar to that of Theorem 2.3, but replace $mdeg$ with $rdeg$. □

It is interesting to compare the following theorem with Theorem 2.1(e) and Example 2.2(b); being integral versus algebraic behaves very differently with respect to direct products.

Theorem 2.10 *Let $\{R_\alpha \subseteq S_\alpha \mid \alpha \in \Lambda\}$ be a family of commutative ring extensions, $R = \prod_{\alpha\in\Lambda} R_\alpha$, and $S = \prod_{\alpha\in\Lambda} S_\alpha$. Then $s = (s_\alpha) \in S$ is algebraic over R if and only if $s_\beta \in S_\beta$ is algebraic over R_β for some $\beta \in \Lambda$.*

Proof ($\Rightarrow$) Suppose that s is algebraic over R; so $f(s) = 0$, where $0 \neq f(X) \in R[X] \subseteq \prod_{\alpha\in\Lambda}(R_\alpha[X])$. Since $0 \neq f(X) = (f_\alpha(X))_{\alpha\in\Lambda}$ with every $f_\alpha(X) \in R_\alpha[X]$, we have $f_\beta(X) \neq 0$ for some $\beta \in \Lambda$. Thus $f_\beta(s_\beta) = 0$ gives that s_β is algebraic over R_β.

($\Leftarrow$) Suppose that $s = (s_\alpha) \in S$ and $s_\beta \in S_\beta$ is algebraic over R_β for some $\beta \in \Lambda$, say $f_\beta(s_\beta) = 0$, where $0 \neq f_\beta(X) \in R_\beta[X]$. For $\alpha \in \Lambda \setminus \{\beta\}$, let $f_\alpha(X) = 0$, and then let $f(X) = (f_\alpha(X))$. Thus $0 \neq f(X) \in R[X]$ and $f(s) = 0$; so s is algebraic over R. □

Corollary 2.11 *Let $\{R_\alpha \subseteq S_\alpha \mid \alpha \in \Lambda\}$ be a family of commutative ring extensions, $R = \prod_{\alpha\in\Lambda} R_\alpha$, and $S = \prod_{\alpha\in\Lambda} S_\alpha$. Then $\overline{R}^{\mathrm{alg}} = \bigcup_{\alpha\in\Lambda} A_\alpha$, where $A_\alpha = \prod_{\beta\in\Lambda} T_\beta$ with $T_\beta = S_\beta$ if $\beta \neq \alpha$ and $T_\alpha = \overline{R_\alpha}^{alg}$. In particular, if $R_i \subseteq S_i$, $i = 1, \ldots, n$, are extensions of commutative rings and $R = R_1 \times \cdots \times R_n$, then $\overline{R}^{\mathrm{alg}} = \overline{R_1}^{\mathrm{alg}} \times S_2 \times \cdots \times S_n \cup S_1 \times \overline{R_2}^{\mathrm{alg}} \times S_3 \times \cdots \times S_n \cup \cdots \cup S_1 \times \cdots \times S_{n-1} \times \overline{R_n}^{\mathrm{alg}}$.*

We can use Theorem 2.10 to provide the promised example where $\overline{R}^{\mathrm{alg}}$ is not a subring of S.

Example 2.12

(a) ($\overline{R}^{\mathrm{alg}}$ need not be a subring of S.) Let A be a commutative ring and $R = A \times A \subsetneq A[X] \times A[X] = R[X] = S$. Let $a = (X, 1)$ and $b = (1, X)$; so $a, b \in S$ are algebraic over R by Theorem 2.10. But $a \pm b = (X \pm 1, 1 \pm X)$ and $ab = (X, X)$ are not algebraic over R by Theorem 2.10. Thus $\overline{R}^{\mathrm{alg}}$ is not a subring of S.

(b) (We may have $\overline{R}^{\mathrm{alg}} \subsetneq \overline{\overline{R}^{\mathrm{alg}}}^{\mathrm{alg}}$.) Let $R = A \times A \subsetneq A[X] \times A[X] = R[X] = S$ as in part (a). Then $\overline{R}^{\mathrm{alg}} = (A \times A[X]) \cup (A[X] \times A)$ by Corollary 2.11. By Theorem 2.10, $(X, X) \in S$ is algebraic over $\overline{R}^{\mathrm{alg}}$; so $(X, X) \in \overline{\overline{R}^{\mathrm{alg}}}^{\mathrm{alg}} \setminus \overline{R}^{\mathrm{alg}}$. Thus $\overline{R}^{\mathrm{alg}} \subsetneq \overline{\overline{R}^{\mathrm{alg}}}^{\mathrm{alg}}$. In fact, $\overline{\overline{R}^{\mathrm{alg}}}^{\mathrm{alg}} = A[X] \times A[X]$ (see Example 2.19(a), Theorem 2.16, and Theorem 2.20).

We next consider $\overline{R}^{\mathrm{rt}}$ and $\overline{R}^{\mathrm{rt}_n}$ for $n \in \mathbb{N}$ (recall that $\overline{R}^{\mathrm{rt}_1} = R$).

Theorem 2.13 *Let $R \subseteq S$ be an extension of commutative rings and $m, n \in \mathbb{N}$. Then $\overline{R}^{\mathrm{rt}}$ and $\overline{R}^{\mathrm{rt}_n}$ are closed under multiplication, $\overline{\overline{R}^{\mathrm{rt}}}^{\mathrm{rt}} = \overline{R}^{\mathrm{rt}}$, and $\overline{\overline{R}^{\mathrm{rt}_m}}^{\mathrm{rt}_n} = \overline{R}^{\mathrm{rt}_{mn}}$ for every $m, n \in \mathbb{N}$.*

Proof Clearly, $\overline{R}^{\mathrm{rt}}$ and $\overline{R}^{\mathrm{rt}_n}$ are both closed under multiplication. To show that $\overline{\overline{R}^{\mathrm{rt}_m}}^{\mathrm{rt}_n} = \overline{R}^{\mathrm{rt}_{mn}}$, let $x \in \overline{R}^{\mathrm{rt}_{mn}}$. Then $(x^n)^m = x^{mn} \in R$; so $x^n \in \overline{R}^{\mathrm{rt}_m}$, and thus $x \in \overline{\overline{R}^{\mathrm{rt}_m}}^{\mathrm{rt}_n}$. Hence $\overline{R}^{\mathrm{rt}_{mn}} \subseteq \overline{\overline{R}^{\mathrm{rt}_m}}^{\mathrm{rt}_n}$. Conversely, let $x \in \overline{\overline{R}^{\mathrm{rt}_m}}^{\mathrm{rt}_n}$. Then $x^n \in \overline{R}^{\mathrm{rt}_m}$, and thus $x^{mn} = (x^n)^m \in R$. Hence $x \in \overline{R}^{\mathrm{rt}_{mn}}$; so $\overline{\overline{R}^{\mathrm{rt}_m}}^{\mathrm{rt}_n} \subseteq \overline{R}^{\mathrm{rt}_{mn}}$. Thus $\overline{\overline{R}^{\mathrm{rt}_m}}^{\mathrm{rt}_n} = \overline{R}^{\mathrm{t}_{mn}}$. The proof that $\overline{\overline{R}^{\mathrm{rt}}}^{\mathrm{rt}} = \overline{R}^{\mathrm{rt}}$ is similar. □

Although $\overline{R}^{\mathrm{rt}}$ and $\overline{R}^{\mathrm{rt}_n}$ are always closed under multiplication and contain R, the following example shows that neither need be closed under addition, and thus need not be a subring of S (for $n \geq 2$).

Example 2.14

(a) ($\overline{R}^{\mathrm{rt}}$ and $\overline{R}^{\mathrm{rt}_n}$ need not be subrings of S.) Although $\overline{R}^{\mathrm{t}}$ is always closed under multiplication, $\overline{R}^{\mathrm{rt}}$ need not be closed under addition. Let $R = \mathbb{Z} \subseteq \mathbb{R} = S$. Then $\sqrt{2}, 1 \in \overline{R}^{\mathrm{rt}}$, but $\sqrt{2} + 1 \notin \overline{R}^{\mathrm{rt}}$ since $(\sqrt{2} + 1)^n \notin \mathbb{Z}$ for every $n \in \mathbb{N}$. Thus $\overline{R}^{\mathrm{rt}}$ is not a subring of S. Similar examples show that $\overline{R}^{\mathrm{rt}_n}$ need not be a subring of S for $n \geq 2$. If we let $R = \mathbb{Z} + Y\mathbb{R}[Y] \subseteq \mathbb{R}[Y]$, then we can choose $S = T(R) = \mathbb{R}(Y)$, and $\overline{R}^{\mathrm{rt}_n}$ is not a subring of S for every $n \geq 2$.

(b) Let $R = K[Y^2, Y^3] \subseteq K(Y) = T(R) = S$, where K is a field. If $char(K) = p > 0$, then $\overline{R}^{\mathrm{rt}} = K[Y]$ is a subring of S, and $\overline{R}^{\mathrm{rt}_n}$, for $n \geq 2$, is a subring of S if and only if $p \mid n$. However, if $char(K) = 0$, then $\overline{R}^{\mathrm{rt}_n} \subsetneq K[Y]$ is not a subring of S for every integer $n \geq 2$; in fact, $\overline{R}^{\mathrm{rt}_n} = R \cup YK[Y] = K[Y] \setminus \{a_m Y^m + \cdots + a_1 Y + a_0 \in K[Y] \mid a_1, a_0 \neq 0\}$ for every integer $n \geq 2$. Thus $\overline{\overline{R}^{\mathrm{rt}_n}}^{\mathrm{rt}_n} = \overline{R}^{\mathrm{rt}_n}$.

(c) (We may have $\overline{R}^{\mathrm{rt_n}} \subsetneq \overline{\overline{R}^{\mathrm{rt_n}}}^{\mathrm{rt_n}}$). For $n \geq 2$, let $R = K[Y^{2n-1}, Y^{2n}, \ldots, Y^{4n-3}] \subseteq K(Y) = T(R) = S$, where K is a field. Then $Y \in \overline{\overline{R}^{\mathrm{rt_n}}}^{\mathrm{rt_n}} \setminus \overline{R}^{\mathrm{rt_n}}$ since $n^2 \geq 2n-1$; so $\overline{R}^{\mathrm{rt_n}} \subsetneq \overline{\overline{R}^{\mathrm{rt_n}}}^{\mathrm{rt_n}}$.

Example 2.2(b) again shows that we may have $\overline{\prod_{\alpha\in\Lambda} R_\alpha}^{\mathrm{rt}} \subsetneq \prod_{\alpha\in\Lambda} \overline{R_\alpha}^{\mathrm{rt}}$ when Λ is infinite. The next result identifies $\overline{\prod_{\alpha\in\Lambda} R_\alpha}^{\mathrm{rt}}$ as a subset of $\prod_{\alpha\in\Lambda} \overline{R_\alpha}^{\mathrm{rt}}$.

Theorem 2.15 *Let $\{R_\alpha \subseteq S_\alpha \mid \alpha \in \Lambda\}$ be a family of extensions of commutative rings, $R = \prod_{\alpha\in\Lambda} R_\alpha$, and $S = \prod_{\alpha\in\Lambda} S_\alpha$. Then $\overline{R}^{rt_n} = \prod_{\alpha\in\Lambda} \overline{R_\alpha}^{rt_n}$ for every $n \in \mathbb{N}$ and $\overline{R}^{rt} = \bigcup_{n\in\mathbb{N}} \overline{R}^{rt_n} = \bigcup_{n\in\mathbb{N}} \prod_{\alpha\in\Lambda} \overline{R_\alpha}^{rt_n}$.*

Proof This follows easily from the definitions. □

The next theorem clarifies when $\overline{R}^{\mathrm{alg}}$ is a subring of S when R is a direct product (also see Example 2.19 and Theorem 2.20).

Theorem 2.16 *Let $R \subseteq S$ be an extension of commutative rings, where R is <u>not</u> indecomposable. Then $\overline{R}^{\mathrm{alg}}$ is a subring of S if and only if $\overline{R}^{\mathrm{alg}} = S$.*

Proof If $\overline{R}^{\mathrm{alg}} = S$, then $\overline{R}^{\mathrm{alg}}$ is certainly a subring of S. Conversely, suppose that $\overline{R}^{\mathrm{alg}}$ is a subring of S and R is not indecomposable. Let $R = R_1 \times R_2 \subseteq S_1 \times S_2 = S$ and $(s_1, s_2) \in S$. Then $(s_1, 0), (0, s_2) \in \overline{R}^{\mathrm{alg}}$ by Theorem 2.10; so $(s_1, s_2) = (s_1, 0) + (0, s_2) \in \overline{R}^{\mathrm{alg}}$ since $\overline{R}^{\mathrm{alg}}$ is a subring of S. Thus $\overline{R}^{\mathrm{alg}} = S$. □

We have seen in Example 2.12(c) (resp., Example 2.14(b)) that $\overline{R}^{\mathrm{alg}}$ (resp., $\overline{R}^{\mathrm{rt}_n}$) need not be algebraically closed (resp., n-root closed) in S. However, by iterating the closure, we do obtain an algebraically closed (resp., n-root closed) set. Specifically, let $\overline{R}^{\mathrm{alg}_0} = R$, $\overline{R}^{\mathrm{alg}_{n+1}} = \overline{\left(\overline{R}^{\mathrm{alg}_n}\right)}^{\mathrm{alg}}$ for every integer $n \geq 0$, and $\overline{R}^{\mathrm{alg}_\infty} = \bigcup_{n=0}^{\infty} \overline{R}^{\mathrm{alg}_n}$. In a similar manner, we define $\overline{R}^{(\mathrm{rt_n})_0} = R$, $\overline{R}^{(\mathrm{rt_n})_{m+1}} = \overline{\left(\overline{R}^{(\mathrm{rt_n})_m}\right)}^{\mathrm{rt_n}}$ for every integer $m \geq 0$, and $\overline{R}^{(\mathrm{rt_n})_\infty} = \bigcup_{m=0}^{\infty} \overline{R}^{(\mathrm{rt_n})_m}$. We next show that $\overline{R}^{\mathrm{alg}_\infty}$ (resp., $\overline{R}^{(rt_n)_\infty}$) is indeed algebraically closed (resp., n-root closed) in S (also see Theorem 3.6).

Theorem 2.17 *Let $R \subseteq S$ be an extension of commutative rings. Then $\overline{\overline{R}^{\mathrm{alg}_\infty}}^{\mathrm{alg}} = \overline{R}^{\mathrm{alg}_\infty}$ and $\overline{\overline{R}^{(\mathrm{rt_n})_\infty}}^{\mathrm{rt_n}} = \overline{R}^{(\mathrm{rt}_n)_\infty}$.*

Proof Let $R' = \overline{R}^{\mathrm{alg}_\infty}$. Of course, $R' \subseteq \overline{R'}^{\mathrm{alg}}$. Let $s \in \overline{R'}^{\mathrm{alg}}$; so $f(s) = 0$ for some $0 \neq f(X) \in R'[X]$. Then $f(X) = a_nX^n + \cdots + a_0$, where every $a_i \in \overline{R}^{\mathrm{alg}_{m_i}}$ for some $m_i \in \mathbb{N}$. Thus $f(X) \in \overline{R}^{\mathrm{alg}_m}[X]$ for $m = max\{m_0, \ldots, m_n\}$; so $s \in \overline{R}^{\mathrm{alg}_{m+1}} \subseteq R'$. Hence $\overline{R'}^{\mathrm{alg}} \subseteq R'$, and thus $\overline{R'}^{\mathrm{alg}} = R'$. The proof for $\overline{R}^{(\mathrm{rt}_n)_\infty}$ is similar. □

Remark 2.18 Another approach would be to let R^{rt} be the smallest root closed subring of S containing R, and $R_0 = R$, $R_{m+1} = R_m[\overline{(R_m)}^{\mathrm{rt}}]$ for every integer $m \geq 0$, and $R_\infty = \bigcup_{m=0}^{\infty} R_m$. Then $R^{\mathrm{rt}} = R_\infty$. (Similarly, we can define R^{rt_n} for every $n \in \mathbb{N}$.) See [2, Section 5] and [3] for more details.

The following example shows that if R is a direct product, then $\overline{R}^{\mathrm{alg}_\infty} = \overline{R}^{\mathrm{alg}_2}$ is a subring of S.

Example 2.19 Let $R = R_1 \times R_2 \subseteq S_1 \times S_2 = S$. Then $\overline{R}^{\mathrm{alg}} = (\overline{R_1}^{\mathrm{alg}} \times S_2) \cup (S_1 \times \overline{R_2}^{\mathrm{alg}})$ by Corollary 2.11. Now $\overline{\overline{R_1}^{\mathrm{alg}} \times S_2}^{\mathrm{alg}} = S_1 \times S_2$ by Corollary 2.11 again; so $\overline{R}^{\mathrm{alg}_\infty} = \bigcup_{n=0}^{\infty} \overline{R}^{\mathrm{alg}_n} = \overline{R}^{\mathrm{alg}_2} = S_1 \times S_2 = S$.

We next give several additional criteria for $\overline{R}^{\mathrm{alg}}$ to be a subring of S when R is a direct product of rings.

Theorem 2.20 *Let $R \subseteq S$ be an extension of commutative rings, where R is* <u>*not*</u> *indecomposable. Then the following statements are equivalent.*

(1) $\overline{R}^{\mathrm{alg}}$ *is a subring of* S.
(2) $\overline{R}^{\mathrm{alg}}$ *is closed under addition.*
(3) $\overline{R}^{\mathrm{alg}}$ *is closed under subtraction.*
(4) $\overline{R}^{\mathrm{alg}}$ *is closed under multiplication.*
(5) $\overline{R}^{\mathrm{alg}}$ *is algebraically closed in* S.
(6) $\overline{R}^{\mathrm{alg}} = S$.

Proof Let $R = R_1 \times R_2$ and $S = S_1 \times S_2$, where each $R_i \subseteq S_i$. Let $(s_1, s_2) \in S$.

(1) $\Leftrightarrow$ (6): This is Theorem 2.16.

(1) $\Rightarrow$ (2),(3),(4): This is clear.

(2) $\Rightarrow$ (6): Let $(s_1, s_2) \in S$. Then $(s_1, 0), (0, s_2) \in \overline{R}^{\mathrm{alg}}$ by Theorem 2.10; so $(s_1, s_2) = (s_1, 0)+(0, s_2) \in \overline{R}^{\mathrm{alg}}$ since $\overline{R}^{\mathrm{alg}}$ is closed under addition. Thus $S \subseteq \overline{R}^{\mathrm{alg}}$, and hence $\overline{R}^{\mathrm{alg}} = S$.

(3) $\Rightarrow$ (6): Let$(s_1, s_2) \in S$. Then $(s_1, 0), (0, -s_2) \in \overline{R}^{\mathrm{alg}}$ by Theorem 2.10; so $(s_1, s_2) = (s_1, 0) - (0, -s_2) \in \overline{R}^{\mathrm{alg}}$ since $\overline{R}^{\mathrm{alg}}$ is closed under subtraction. Thus $S \subseteq \overline{R}^{\mathrm{alg}}$, and hence $\overline{R}^{\mathrm{alg}} = S$.

(4) $\Rightarrow$ (6): Let $(s_1, s_2) \in S$. Then $(s_1, 1), (1, s_2) \in \overline{R}^{\mathrm{alg}}$ by Theorem 2.10; so $(s_1, s_2) = (s_1, 1)(1, s_2) \in \overline{R}^{\mathrm{alg}}$ since $\overline{R}^{\mathrm{alg}}$ is closed under multiplication. Thus $S \subseteq \overline{R}^{\mathrm{alg}}$, and hence $\overline{R}^{\mathrm{alg}} = S$.

(6) $\Rightarrow$ (5): This is clear.

(5) $\Rightarrow$ (6): This follows from Example 2.19. □

The following theorem gives some general results about the $\mathscr{S}$-closure of R in S. Note that here we allow $0 \in \mathscr{S}$ and $\mathscr{S} = \emptyset$.

Theorem 2.21 *Let $R \subseteq S$ be an extension of commutative rings and $\mathscr{S}, \mathscr{S}_1, \mathscr{S}_2, \mathscr{S}_\alpha \subseteq R[X]$, $\alpha \in \Lambda$.*

(a) $\overline{R}^{R[X]} = \overline{R}^{0} = S$.
(b) $\overline{R}^{\emptyset} = \emptyset$.
(c) *If* $\mathscr{S} \subseteq R[X]^*$, *then* $\overline{R}^{\mathscr{S}} \subseteq \overline{R}^{\text{alg}} = \overline{R}^{R[X]^*}$.
(d) *If* $\mathscr{S}_1 \subseteq \mathscr{S}_2$, *then* $\overline{R}^{\mathscr{S}_1} \subseteq \overline{R}^{\mathscr{S}_2}$.
(e) $\overline{R}^{\bigcup_{\alpha\in\Lambda}\mathscr{S}_\alpha} = \bigcup_{\alpha\in\Lambda}\overline{R}^{\mathscr{S}_\alpha}$.
(f) $\overline{R}^{\bigcap_{\alpha\in\Lambda}\mathscr{S}_\alpha} \subseteq \bigcap_{\alpha\in\Lambda}\overline{R}^{\mathscr{S}_\alpha}$.

Proof (a)–(d) These are all clear.

(e) We have $\overline{R}^{\mathscr{S}_\alpha} \subseteq \overline{R}^{\bigcup_{\alpha\in\Lambda}\mathscr{S}_\alpha}$ for every $\alpha \in \Lambda$ by part (d); so $\bigcup_{\alpha\in\Lambda}\overline{R}^{\mathscr{S}_\alpha} \subseteq \overline{R}^{\bigcup_{\alpha\in\Lambda}\mathscr{S}_\alpha}$. Let $s \in \overline{R}^{\bigcup_{\alpha\in\Lambda}\mathscr{S}_\alpha}$; so $f(s) = 0$, where $f \in \mathscr{S}_\beta$ for some $\beta \in \Lambda$. Thus $s \in \overline{R}^{\mathscr{S}_\beta} \subseteq \bigcup_{\alpha\in\Lambda}\overline{R}^{\mathscr{S}_\alpha}$; so $\overline{R}^{\bigcup_{\alpha\in\Lambda}\mathscr{S}_\alpha} \subseteq \bigcup_{\alpha\in\Lambda}\overline{R}^{\mathscr{S}_\alpha}$. Hence $\overline{R}^{\bigcup_{\alpha\in\Lambda}\mathscr{S}_\alpha} = \bigcup_{\alpha\in\Lambda}\overline{R}^{\mathscr{S}_\alpha}$.

(f) Let $\beta \in \Lambda$. Then $\bigcap_{\alpha\in\Lambda}\mathscr{S}_\alpha \subseteq \mathscr{S}_\beta$ gives $\overline{R}^{\bigcap_{\alpha\in\Lambda}\mathscr{S}_\alpha} \subseteq \overline{R}^{\mathscr{S}_\beta}$ by part (d). Thus $\overline{R}^{\bigcap_{\alpha\in\Lambda}\mathscr{S}_\alpha} \subseteq \bigcap_{\alpha\in\Lambda}\overline{R}^{\mathscr{S}_\alpha}$. □

The next example shows that the inclusion in part (f) of Theorem 2.21 may be strict.

Example 2.22 (a) (We may have $\overline{R}^{\bigcap_{\alpha\in\Lambda}\mathscr{S}_\alpha} \subsetneq \bigcap_{\alpha\in\Lambda}\overline{R}^{\mathscr{S}_\alpha}$ in Theorem 2.21(f).) Let $\mathscr{S}_1 = \{X\}$ and $\mathscr{S}_2 = \{2X\}$. Then $0 \in \overline{R}^{\mathscr{S}_1} \cap \overline{R}^{\mathscr{S}_2}$, but $\overline{R}^{\mathscr{S}_1\cap\mathscr{S}_2} = \overline{R}^{\emptyset} = \emptyset$. Thus $\overline{R}^{\mathscr{S}_1\cap\mathscr{S}_2} \subsetneq \overline{R}^{\mathscr{S}_1} \cap \overline{R}^{\mathscr{S}_2}$.

For a less trivial example, let $R = \mathbb{Z} \subseteq \mathbb{R} = S$, $\mathscr{S}_1 = \{X-r, X^2-r \mid r \in R\}$, and $\mathscr{S}_2 = \{X-r, X^4-r \mid r \in R\}$. Then $\mathscr{S}_1 \cap \mathscr{S}_2 = \{X-r \mid r \in R\}$; so $\overline{R}^{\mathscr{S}_1\cap\mathscr{S}_2} = R = \mathbb{Z}$. However, $\sqrt{2} \in \overline{R}^{\mathscr{S}_1} \cap \overline{R}^{\mathscr{S}_2}$; so $\overline{R}^{\mathscr{S}_1\cap\mathscr{S}_2} \subsetneq \overline{R}^{\mathscr{S}_1} \cap \overline{R}^{\mathscr{S}_2}$.

(b) We have seen that we need not have $R \subseteq \overline{R}^{\mathscr{S}}$ for $\mathscr{S} \subseteq R[X]$. If we let $\mathscr{S}' = \{x - r \mid r \in R\}$, then $\overline{R}^{\mathscr{S}'} = R$. Thus, if we define $\mathscr{S}_* = \mathscr{S} \cup \mathscr{S}'$, then $R \subseteq \overline{R}^{\mathscr{S}_*} = \overline{R}^{\mathscr{S}'} \cup R$ by Theorem 2.21(e).

(c) Let $R \subseteq S$ be an extension of commutative rings and $\mathscr{S} = \{rX - b \mid r \in Reg(R), b \in R\} \subseteq R[X]$. Then $\overline{R}^{\mathscr{S}} = \{s \in S \mid rs \in R \text{ for some } r \in Reg(R)\}$ is a subring of S. If $Reg(R) \subseteq Reg(S)$, then $\overline{R}^{\mathscr{S}} = T(R) \cap S$, where $T(R)$ is the total quotient ring of R.

3 The $\mathcal{S}$-Closure

There are several other types of polynomial root extensions that do not fit into the $\mathscr{S}$-closure scheme of Sect. 2. In this section, we discuss two of them, the complete integral closure and (2, 3)-closure of R in S. We then introduce the $\mathcal{S}$-closure of R, which includes the $\mathscr{S}$-closure, complete integral closure, and (2, 3)-closure as special cases.

For an extension $R \subseteq S$ of commutative rings, recall that $s \in S$ is *almost integral over R* if there is an $a \in Reg(R)$ such that $as^n \in R$ for every $n \in \mathbb{N}$, and the *complete integral closure of R in S* is $\overline{R}^{\text{cic}} = \{s \in S \mid s \text{ is almost integral over } R\}$. We say that R is *completely integrally closed in S* (resp., *completely integrally closed*) if $\overline{R}^{\text{cic}} = R$ (resp., $\overline{R}^{\text{cic}} = R$ when $S = T(R)$). We have $R \subseteq \overline{R}^{\text{cic}} \subseteq \overline{R}^{\text{reg}}$ and $\overline{R}^{\text{cic}}$ is a subring of S (cf. [4, Theorem 13.1(1)]).

We define the *(2, 3)-closure of R in S* as $\overline{R}^{\text{sem}} = \{s \in S \mid s^2, s^3 \in R\}$ (here, we use "sem" since a (2, 3)-closed ring is often called seminormal). We say that R is *(2, 3)-closed in S* (resp., *(2, 3)-closed*) if $\overline{R}^{\text{sem}} = R$ (resp., $\overline{R}^{\text{sem}} = R$ when $S = T(R)$). We have $R \subseteq \overline{R}^{\text{sem}} \subseteq \overline{R}^{\text{rt}_n}$ for every integer $n \geq 2$, but the following examples show that $\overline{R}^{\text{sem}}$ need not be a subring of S and $\overline{R}^{\text{sem}}$ need not be (2, 3)-closed.

Example 3.1

(a) ($\overline{R}^{\text{sem}}$ need not be a subring of S.) Let $R = K[Y^2, Y^3] \subseteq K(Y) = T(R) = S$, where K is a field, as in Example 2.14(a). Then $\overline{R}^{\text{sem}} \subseteq K[Y]$ is not a subring of S since $Y, 1 \in \overline{R}^{\text{sem}}$, but $Y + 1 \notin \overline{R}^{\text{sem}}$. In fact, it may easily be shown that $\overline{R}^{\text{sem}} = R \cup YK[Y] = K[Y] \setminus \{a_m Y^m + \cdots + a_1 Y + a_0 \in K[Y] \mid a_1, a_0 \neq 0\}$, and thus $\overline{\overline{R}^{\text{sem}}}^{\text{sem}} = \overline{R}^{\text{sem}}$.

(b) (We may have $\overline{R}^{\text{sem}} \subsetneq \overline{\overline{R}^{\text{sem}}}^{\text{sem}}$.) Let $R = K[Y^3, Y^4, Y^5] \subseteq K(Y) = T(R) = S$, where K is a field. Then $\overline{R}^{\text{sem}}$ is not a subring of S since $Y^2, 1 \in \overline{R}^{\text{sem}}$, but $Y^2 + 1 \notin \overline{R}^{\text{sem}}$. Also, $\overline{R}^{\text{sem}} \subsetneq \overline{\overline{R}^{\text{sem}}}^{\text{sem}}$ since $Y \in \overline{\overline{R}^{\text{sem}}}^{\text{sem}} \setminus \overline{R}^{\text{sem}}$.

We next show that these two closures commute with direct products.

Theorem 3.2 *Let $\{R_\alpha \subseteq S_\alpha \mid \alpha \in \Lambda\}$ be a family of extensions of commutative rings, $R = \prod_{\alpha\in\Lambda} R_\alpha$, and $S = \prod_{\alpha\in\Lambda} S_\alpha$. Then $\overline{R}^{\text{cic}} = \prod_{\alpha\in\Lambda} \overline{R_\alpha}^{\text{cic}}$ and $\overline{R}^{\text{sem}} = \prod_{\alpha\in\Lambda} \overline{R_\alpha}^{\text{sem}}$.*

Proof These both follow easily from the definitions since $Reg(\prod_{\alpha\in\Lambda} R_\alpha) = \prod_{\alpha\in\Lambda} Reg(R_\alpha)$. □

Remark 3.3 More generally, for $\emptyset \neq T \subseteq \mathbb{N}$, we can define the *$T$-closure* of R in S as $\overline{R}^{\text{T}-\text{clos}} = \{s \in S \mid s^n \in R \text{ for every } n \in T\}$. We say that R is *T-closed in S* (resp., *T-closed*) if $\overline{R}^{\text{T}-\text{clos}} = R$ (resp., $\overline{R}^{\text{T}-\text{clos}} = R$ when $S = T(R)$). Thus $\overline{R}^{\{2,3\}-\text{clos}} = \overline{R}^{\text{sem}}$ and $\overline{R}^{\{n\}-\text{clos}} = \overline{R}^{\text{rt}_n}$. However, this generalization yields nothing new. Let $[T]$ be the additive subsemigroup of $\mathbb{N}$ generated by T; so $[T] = [n_1, \ldots, n_t]$ for some $n_1, \ldots n_t \in \mathbb{N}$. If $[T] \subsetneq \mathbb{N}$ and $d = gcd\{n_1, \ldots, n_t\}$, then $\overline{R}^{\text{T}-\text{clos}} = \overline{R}^{\text{sem}}$ if $d = 1$ and $\overline{R}^{\text{T}-\text{clos}} = \overline{R}^{\text{rt}_d}$ if $d > 1$ by [1, Theorem 3.2].

We now give the promised generalization of $\mathscr{S}$-closure.

Let $\mathcal{S} \subseteq \mathcal{P}(\prod_{\alpha\in\Lambda} R[X])$. We define the *$\mathcal{S}$-closure of R in S* as $\overline{R}^{\mathcal{S}} = \{s \in S \mid \text{for some } A \in \mathcal{S} \text{ and } (f_\alpha(X)) \in A,\ f_\alpha(s) = 0 \text{ for every } \alpha \in \Lambda\}$. We say that R is *$\mathcal{S}$-closed in S* if $\overline{R}^{\mathcal{S}} \subseteq R$, and R is *$\mathcal{S}$-closed* if $\overline{R}^{\mathcal{S}} \subseteq R$ when $S = T(R)$.

For $\mathscr{S} \subseteq R[X]$, let $\mathcal{S} = \{\mathscr{S}\} \subseteq \mathcal{P}(R[X])$. Then $\overline{R}^{\mathcal{S}} = \overline{R}^{\mathscr{S}}$. Thus the $\mathcal{S}$-closure generalizes the $\mathscr{S}$-closure from Sect. 2.

We next show that $\overline{R}^{\text{sem}}$ and $\overline{R}^{\text{cic}}$ are both special cases of the $\mathcal{S}$-closure.

Example 3.4

(a) Let $W_2 = \{X^2 - r \mid r \in R\}$, $W_3 = \{X^3 - r \mid r \in R\} \subseteq R[X]$; so $W_2 \times W_3 \subseteq R[X] \times R[X]$. For $\mathcal{S} = \{W_2 \times W_3\} \subseteq \mathcal{P}(R[X] \times R[X])$, we have $\overline{R}^{\mathcal{S}} = \overline{R}^{\text{sem}}$.
(b) For every $r \in Reg(R)$ and $n \in \mathbb{N}$, let $T_{n,r} = \{rX^n - s \mid s \in R\} \subseteq R[X]$ and $A_r = \prod_{n\in\mathbb{N}} T_{n,r} \subseteq \prod_{n\in\mathbb{N}} R[X]$; so $\mathcal{S} = \{A_r \mid r \in Reg(R)\} \subseteq \mathcal{P}(\prod_{n\in\mathbb{N}} R[X])$. For $\mathcal{S} = \{A_r \mid r \in Reg(R)\}$, we have $\overline{R}^{\mathcal{S}} = \overline{R}^{\text{cic}}$.

We next give some general results for the $\mathcal{S}$-closure which are analogous to Theorem 2.21. Example 2.22(a) shows that the inclusion in part (c) may be strict.

Theorem 3.5 *Let $R \subseteq S$ be an extension of commutative rings and $\mathcal{S}_1, \mathcal{S}_2, \mathcal{S}_\beta \subseteq \mathcal{P}(\prod_{\alpha\in\Lambda} R[X])$, $\beta \in \Delta$.*

(a) *If $\mathcal{S}_1 \subseteq \mathcal{S}_2$, then $\overline{R}^{\mathcal{S}_1} \subseteq \overline{R}^{\mathcal{S}_2}$.*
(b) $\overline{R}^{\bigcup_{\beta\in\Delta}\mathcal{S}_\beta} = \bigcup_{\beta\in\Delta} \overline{R}^{\mathcal{S}_\beta}$.
(c) $\overline{R}^{\bigcap_{\beta\in\Delta}\mathcal{S}_\beta} \subseteq \bigcap_{\beta\in\Delta} \overline{R}^{\mathcal{S}_\beta}$.

Proof (a) and (b) are clear, and (c) follows directly from (a). □

It is well known that $\overline{R}^{\text{cic}}$ need not be completely integrally closed. For $R \subseteq S$ an extension of commutative rings, we define $\overline{R}^{\text{cic}_0} = R$, $\overline{R}^{\text{cic}_{\alpha+1}} = \overline{\overline{R}^{\text{cic}_\alpha}}^{\text{cic}}$ for α any ordinal number, and $\overline{R}^{\text{cic}_\beta} = \bigcup_{\alpha<\beta} \overline{R}^{\text{cic}_\alpha}$ for β a limit ordinal. D. Lantz [6] has shown (1) for every countable ordinal number α, there is a Bézout domain D_α such that $\overline{D_\alpha}^{\text{cic}_\alpha} \subsetneq \overline{D_\alpha}^{\text{cic}_{\alpha+1}} = \overline{D_\alpha}^{\text{cic}_{\alpha+2}}$ and (2) for any integral domain D, $\overline{\overline{D}^{\text{cic}_\Omega}}^{\text{cic}} = \overline{D}^{\text{cic}_\Omega}$, where Ω is the first uncountable ordinal number. Also, in a similar manner, we define $\overline{R}^{\text{sem}_0} = R$, $\overline{R}^{\text{sem}_{n+1}} = \overline{(\overline{R}^{\text{sem}_n})}^{\text{sem}}$ for every integer $n \geq 0$, and $\overline{R}^{\text{sem}_\infty} = \bigcup_{n=0}^{\infty} \overline{R}^{\text{sem}_n}$. We next show that $\overline{R}^{\text{cic}_\Omega}$ (resp., $\overline{R}^{\text{sem}_\infty}$) is completely integrally closed (resp., (2, 3-closed) in S (also see Theorem 2.17).

Theorem 3.6 *Let $R \subseteq S$ be an extension of commutative rings. Then $\overline{\overline{R}^{\text{cic}_\Omega}}^{\text{cic}} = \overline{R}^{\text{cic}_\Omega}$, where Ω is the first uncountable ordinal number, and $\overline{\overline{R}^{\text{sem}_\infty}}^{\text{sem}} = \overline{R}^{\text{sem}_\infty}$.*

Proof We first show that $\overline{\overline{R}^{\text{cic}_\Omega}}^{\text{cic}} = \overline{R}^{\text{cic}_\Omega}$. Let $R' = \overline{R}^{\text{cic}_\Omega}$. Clearly $R' \subseteq \overline{R'}^{\text{cic}}$. Conversely, let $s \in \overline{R'}^{\text{cic}}$. Then $as^n \in R'$ for some $a \in Reg(R')$ and every $n \in \mathbb{N}$. Thus $as^n \in \overline{R}^{\text{cic}_\beta}$ for some $a \in Reg(\overline{R}^{\text{cic}_\beta})$ and every $n \in \mathbb{N}$; so $s \in \overline{R}^{\text{cic}_{\beta+1}} \subseteq R'$. Hence $\overline{R'}^{\text{cic}} \subseteq R'$; so $\overline{R'}^{\text{cic}} = R'$.

We now show that $\overline{\overline{R}^{\mathrm{sem}_\infty}}^{\mathrm{sem}} = \overline{R}^{\mathrm{sem}_\infty}$. Let $R' = \overline{R}^{\mathrm{sem}_\infty}$. Clearly $R' \subseteq \overline{R'}^{\mathrm{sem}}$. Conversely, let $s \in \overline{R'}^{\mathrm{sem}}$. Then $s^2, s^3 \in R'$; so $s^2, s^3 \in \overline{R}^{\mathrm{sem}_n}$ for some $n \in \mathbb{N}$. Thus $s \in \overline{R}^{\mathrm{sem}_{n+1}} \subseteq R'$; so $\overline{R'}^{\mathrm{sem}} = R'$.

□

Corollary 3.7 (Lantz [6]) *Let D be an integral domain. Then* $\overline{\overline{D}^{\mathrm{cic}_\Omega}}^{\mathrm{cic}} = \overline{D}^{\mathrm{cic}_\Omega}$, *where* Ω *is the first uncountable ordinal number.*

We can also define the *u-closure* and *t-closure* of R in S. Let $R \subseteq S$ be an extension of commutative rings. Then R is *u-closed in* S (resp., *t-closed in* S) if $t^2 - t, t^3 - t^2 \in R$ for $t \in S$ implies $t \in R$ (resp., $t^2 - rt, t^3 - rt^2 \in R$ for $t \in S$ and some $r \in R$ implies $t \in R$). We define $\overline{R}^{\mathrm{uclos}} = \{s \in S \mid s^2 - s, s^3 - s^2 \in R\}$ and $\overline{R}^{\mathrm{tclos}} = \{s \in S \mid s^2 - rs, s^3 - rs^2 \in R \text{ for some } r \in R\}$. We have $R \subseteq \overline{R}^{\mathrm{uclos}} \subseteq \overline{R}^{\mathrm{tclos}} \subseteq \overline{R}$. See [2, Section 7] for connections between these two concepts and seminormal and quasinormal rings and other related references.

The final example shows that $\overline{R}^{\mathrm{uclos}}$ and $\overline{R}^{\mathrm{tclos}}$ can both be realized as $\mathcal{S}$-closures.

Example 3.8

(a) Let $W = \{X^2 - X - r \mid r \in R\} \times \{X^3 - X^2 - r \mid r \in R\} \subseteq R[X] \times R[X]$ and $\mathcal{S}_u = \{W\} \subseteq \mathcal{P}(R[X] \times R[X])$. Then $\overline{R}^{\mathcal{S}_u} = \overline{R}^{\mathrm{uclos}}$.

(b) Let $A_r = \{X^2 - rX - s \mid s \in R\} \times \{X^3 - rX^2 - s \mid s \in R\} \subseteq R[X] \times R[X]$ for $r \in R$ and $\mathcal{S}_t = \{A_r \mid r \in R\} \subseteq \mathcal{P}(R[X] \times R[X])$. Then $\overline{R}^{\mathcal{S}_t} = \overline{R}^{\mathrm{tclos}}$.

Appendix

Dan Anderson (1948–2022) by David F. Anderson

After a long battle with esophageal cancer, my twin brother, Daniel D. Anderson (Dan), died on April 24, 2022. Dan has had a lasting effect on the mathematical community through his research, Ph.D. students, and service at the University of Iowa. This was recognized by being named an AMS Fellow in 2018.

Dan received his BA from the University of Iowa in 1971 and his Ph.D. from the University of Chicago in 1974 under Irving Kaplansky; his thesis title was *Multiplicative Lattices*. His first job was as a visiting assistant professor at the University of Iowa (1974–1975). After being an assistant professor at Virginia Tech (1975–1976) and then at the University of Missouri (1976–1977), he returned to Iowa in the fall of 1977 as an assistant professor. He quickly rose through the ranks to full professor in 1983, and then the Chairman of the Mathematics Department (2011–2017).

Dan published many influential papers covering a wide range of topics in commutative algebra and related areas, including factorization and divisibility, rings with zero-divisors, and lattice theory (abstract ideal theory). MathSciNet lists Dan's over 235 publications, with over 3000 citations and 80 coauthors. These publications include short notes, very long articles, survey/expository articles, lectures notes, and edited conference proceedings. Dan had 36 Ph.D. students and coauthored about 75 papers with 30 of these students. Dan was also a serious collector of US coins. He was a leading expert on (Iowa) trade tokens and has written several scholarly articles on tokens.

Dan is survived by his wife of 51 years, Kathy, his daughter, Caitlin, two grandsons, and one great grandson. For more on Dan's life and mathematics, see his obituary at www.gayandciha.com and my article *Dan Anderson and his Mathematics in Rings, Monoids and Module Theory*, Springer, 2021, edited by A. Badawi and J. Coykendall.

References

1. D. F. Anderson, *Root closure in integral domains*, J. Algebra 79 (1982), 51–59.
2. D. F. Anderson, *Root closure in commutative rings: A survey*, Advances in Commutative Ring Theory (Fez, 1997), 55-71, Lecture Notes in Pure and Appl. Math., vol. 205, Dekker, New York, 1999.
3. D. F. Anderson, D. E. Dobbs, and M. Roitman, *Root closure in commutative rings*, Ann. Sci. Univ. Clermont-Ferrand II Math., No. 26 (1990), 1–11.
4. R. Gilmer, *Multiplicative Ideal Theory*, Queen's Papers in Pure and Applied Mathematics, vol. 90, Queen's University, Kingston, Ontario, 1992.
5. I. Kaplansky, *Commutative Rings, Revised Edition*, The University of Chicago Press, Chicago, 1974.
6. D. C. Lantz, *Finite Krull dimension, complete integral closure and gcd-domains*, Comm. Algebra 3 (1975), 951–958.

Absorbing Ideals in Commutative Rings: A Survey

Ayman Badawi

In memory of Paul-Jean Cahen

1 Introduction

Let R be a commutative ring with $1 \neq 0$ and I be a proper ideal of R. Then I is called a 2-absorbing ideal of R as in [10] if whenever $abc \in I$ for some $a, b, c \in R$, then $ab \in I$ or $bc \in I$ or $ac \in I$. Over the past 15 years, there has been considerable attention in the literature to 2-absorbing ideals of commutative rings and their generalizations, for example, see [1–5, 9–11, 13–21, 23–26, 30–38, 40–56]. A more general concept than 2-absorbing ideals is the concept of n-absorbing ideals. Let $n \geq 1$ be a positive integer. A proper ideal I of R is called an *n-absorbing ideal* of R as in [2] if $a_1, a_2, \ldots, a_{n+1} \in R$ and $a_1 a_2 \cdots a_{n+1} \in I$, then there are n of the a_i's whose product is in I. In this article, we survey some recent developments on conjectures (see, [2, 9], and [23]) concerning n-absorbing ideals of commutative rings. We survey some classifications of factorization-commutative rings in terms of absorbing ideals. We survey some properties of n-absorbing ideals in ring extensions. We strongly recommend that the reader keeps the first survey article [9] in hand while reading this paper.

2 Conjectures on n-Absorbing Ideals of Commutative Rings

Let I be a proper ideal of a commutative ring R. Then $\sqrt{I}$ denoted the radical ideal of R. A proper ideal of R is called a *strongly n-absorbing ideal* of R as in [2] if whenever $I_1 \cdots I_{n+1} \subseteq I$ for ideals $I_1, \ldots, I_{n+1}$ of R, then the product of some

A. Badawi (✉)
Department of Mathematics & Statistics, The American University of Sharjah, Sharjah, UAE
e-mail: abadawi@aus.edu

J.-L. Chabert et al. (eds.), *Algebraic, Number Theoretic, and Topological Aspects of Ring Theory*, https://doi.org/10.1007/978-3-031-28847-0_4

n of the I_j's is contained in I. It is clear that a strongly-n-absorbing ideal of a commutative ring R is an n-absorbing ideal of R.

Anderson and Badawi in [2] made the following conjectures:

Conjecture I If I is an n-absorbing ideal of a commutative ring R, then $(\sqrt{I})^n \subseteq I$.

Conjecture II If I is an n-absorbing ideal of a commutative ring R, then I is a strongly n-absorbing ideal of R.

Conjecture III If I is an n-absorbing ideal of a commutative ring R, then $I[X]$ is an n-absorbing ideal of $R[X]$.

Choi and Walker in [28] gave an affirmative answer for Conjecture I for any positive integer n, and G. Donadze independently in [35] gave an alternative proof of Conjecture I. It was shown in [10] that Conjecture II is correct for $n = 2$. Conjectures II and III were verified in [2] for any positive integer n when R is a Prüfer domain. Also, Conjecture III was verified in [2] when $n = 2$. Laradji in [47] proved that Conjectures II and III are valid for any positive integer n when R is an arithmetical ring (e.g., if R is a Prufer domain). It was shown in [47] that if $I[X]$ is an n-absorbing ideal of $R[X]$, then I is a strongly n-absorbing ideal of R, and hence if Conjecture III is true, then Conjecture II is true.

We recall that a commutative ring R is said to be a *U-ring* provided R has the property that an ideal contained in a finite union of ideals must be contained in one of those ideals. Recall that a Prufer domain is a U-ring. The authors in [53] proved the following result.

Theorem 2.1 ([53, Theorem 2.4]) *If R is a U-ring, then Conjecture II holds.*

We recall from [39] and [6] that an integral domain R is called a *pseudo-valuation domain* (PVD) if R has exactly one maximal ideal M such that $(M : M)$ is a valuation domain. We recall that if $f(x) = a_n x^n + \cdots + a_0 \in R[x]$, then $C(f)$ is the ideal $(a_n, \ldots, a_0)R$. A ring R is called a *Gaussian ring* if $C(fg) = C(f)C(g)$ for every $f, g \in R[x]$. The authors in [53] proved the following result.

Theorem 2.2

(1) *[53, Theorem 2.6]. If R is a U-ring that is a Gaussian ring, then Conjecture III holds.*
(2) *[53, Theorem 2.7]. Let $n \geq 2$. Suppose that R is a PVD with maximal ideal M and I is a proper ideal of R such that $\sqrt{I} \neq M$. Then I is an n-absorbing ideal of R if and only if $I[x]$ is an n-absorbing ideal of $R[x]$.*

Since if Conjecture III holds, then Conjecture II holds by [47, Theorem 2.9(i)], in light of Theorem 2.2 we have the following result.

Corollary 2.3

(1) *If R is a U-ring that is a Gaussian ring, then Conjectures II and III hold.*
(2) *Let $n \geq 2$. Suppose that R is a PVD with maximal ideal M and I is a proper ideal of R such that $\sqrt{I} \neq M$. Then I is an n-absorbing ideal of R if and only if*

$I[x]$ is an n-absorbing ideal of $R[x]$, if and only if I is a strongly n-absorbing ideal of R.

We recall from [33] and [12] that a commutative ring R is called a *divided ring* if $Q \subset xR$ for every prime ideal Q of R, and $x \in R \setminus Q$ and it is called a *locally divided* ring as in [15] if R_P is a divided ring for every prime ideal P of R.

Recently, Choi in [27] proved the following result.

Theorem 2.4 ([27, Corollary 13]) *Let R be a locally divided ring. Then Conjectures II and III hold.*

Since a PVD is a divided ring (and hence locally divided), we conclude that Corollary 2.3(ii) is a particular case of Theorem 2.4.

We recall from [1] that the AF-dimension of a ring R, denoted by AF-dim(R), is the smallest positive integer n such that each proper ideal of R can be written as a finite product of n-absorbing ideals of R; if no such n exists, then AF-dim(R) = ∞. A ring R is an FAF-ring if AF-dim(R) $< \infty$.

The following are examples of FAF-rings.

Example 2.5

(1) [1, Corollary 3.9]. Let $d \in \mathbb{Z}-\{0, 1\}$ be a square-free integer such that $4 \mid (d-1)$ and $8 \mid (d-5)$. Then $R = \mathbb{Z}[\sqrt{d}]$ is an FAF-ring and $AF\text{-}dim(R) = 2$.
(2) [1, Corollary 4.4]. Let R be a finite direct product of fields. Then R and $R[X]$ are FAF-rings.

Choi in [27] proved the following result.

Theorem 2.6 ([27, Theorem 39 (4)]) *Assume that R is an FAF-ring. Then Conjectures II and III hold.*

3 2-AB-Rings and Factorization Rings

We recall from [21] that a commutative ring R is called a 2-AB-ring if every 2-absorbing ideal of R is prime.

The authors in [21] proved the following results.

Theorem 3.1 ([21, Theorem 2.3]) *Let R be a commutative ring with $1 \neq 0$. The following statements are equivalent.*

(1) *R is a 2-AB-ring.*
(2) *R has exactly one maximal ideal, say M, such that the prime ideals of R are linearly ordered (by inclusion) and $IM = P$ for every 2-absorbing ideal I of R and every minimal prime ideal P over I.*
(3) *R has exactly one maximal ideal, say M, such that the prime ideals of R are linearly ordered (by inclusion) and P is the only minimal 2-absorbing ideal over P^2 for every prime ideal P of R.*

Let $n \geq 2$ be a positive integer. The authors in [43] extended the concept of 2-AB-rings to n-AB-rings. We recall from [43] that a commutative ring R is called an n-AB-ring if every n-absorbing ideal of R is a prime ideal of R. They obtained similar results to those in Theorem 3.1.

Theorem 3.2 ([43, Theorem 2.13]) *Let R be a commutative ring with $1 \neq 0$. The following statements are equivalent.*

(1) *R is an n-AB-ring.*
(2) *R has exactly one maximal ideal, say M, such that the prime ideals of R are linearly ordered (by inclusion) and $IM = P$ for every n-absorbing ideal I of R and every minimal prime ideal P over I.*
(3) *R has exactly one maximal ideal, say M, such that the prime ideals of R are linearly ordered (by inclusion) and P is the only minimal n-absorbing ideal over P^n for every prime ideal P of R.*

4 Commutative Rings with 2-Absorbing Factorization

Let R be a commutative ring with $1 \neq 0$. Then R is called a *TAF*-ring if every ideal of R is a finite product of 2-absorbing ideals. The authors in [50] obtained the following results.

Theorem 4.1 ([50, Theorem 3.3]) *Any TAF-ring is a finite direct product of one-dimensional domains and zero-dimensional quasi-local rings having nilpotent maximal ideal. In particular, a TAF-ring of dimension one having a unique height-zero prime ideal is a domain.*

Theorem 4.2 ([50, Corollary 3.4]) *Let R be a commutative ring. The following are equivalent.*

(1) *$R[X]$ is a TAF-ring.*
(2) *R is a von Neumann regular TAF-ring.*
(3) *R is a finite direct product of fields.*

In view of Theorem 4.2, we have the following example.

Example 4.3 Let $R = \mathbb{Z}_5 \times \mathbb{Q} \times \mathbb{R} \times \mathbb{Z}_{11}$. Then R and $R[X]$ are TAF-rings by Theorem 4.2.

The authors in [22] proved the following result.

Theorem 4.4 ([22, Theorem 2.3]) *Let R be a commutative ring. Then $R[X]$ is a principal ideal ring if and only if R is ring-isomorphic to a finite direct product of fields.*

In view of Theorems 4.4 and 4.2, we have the following result.

Corollary 4.5 *Let R be a commutative ring. The following are equivalent.*

(1) *$R[X]$ is a TAF-ring.*
(2) *R is a von Neumann regular TAF-ring.*
(3) *R is a finite direct product of fields.*
(4) *$R[X]$ is a principal ideal ring.*

Let R be an integral domain. We recall the following definitions.

(1) We say R has *finite character* if every $x \in R - \{0\}$ belongs to only finitely many maximal ideals of R.
(2) R is called an *atomic* domain if every nonzero non-unit can be written in at least one way as a finite product of irreducible elements.
(3) R is a *discrete valuation ring (DVR)* if R is a principal ideal domain (PID) with exactly one nonzero maximal ideal.
(4) R is an ACCP-domain if there is no infinite strictly ascending chain of principal ideals.

We recall from [50] that a proper ideal I of R is called a TA-ideal if I is a finite product of 2-absorbing ideals.

Theorem 4.6 ([50, Theorem 4.3]) *Let R be an integral domain that is not a field with exactly one maximal ideal M. The following are equivalent.*

(1) *R is a TAF-domain.*
(2) *R is one-dimensional and every principal ideal of R is a TA-ideal.*
(3) *R is atomic, one-dimensional and every atom of R generates a TA-ideal.*
(4) *R is atomic and M^2 is universal (i.e. $M^2 \subseteq aR$ for each atom $a \in R$).*
(5) *R is an atomic PVD.*
(6) *R is a PVD which satisfies ACCP.*
(7) *$(M : M)$ is a DVR with maximal ideal M.*
Furthermore, if R is Noetherian, then the integral closure R' of R is a DVR with maximal ideal M.

Theorem 4.7 ([50, Theorem 4.4])
Let R be an integral domain. The following are equivalent.

(1) *R is a TAF-domain.*
(2) *R has finite character and R_M is a TAF-domain for each maximal ideal M of R.*
(3) *R has finite character and R_M is an atomic PVD for each maximal ideal M of R.*
(4) *R has finite character and R_M is an ACCP PVD for each maximal ideal M of R.*
(5) *R is a one-dimensional domain which has finite character and every principal ideal of R is a TA-ideal.*
(6) *R is a one-dimensional ACCP-domain that has finite character and every principal ideal generated by an atom is a TA-ideal.*

If R is a Noetherian domain, then we have the following result.

Theorem 4.8 ([50, Corollary 4.5]) *For a Noetherian domain R that is not a field, the following are equivalent.*

(1) *R is a TAF-domain.*
(2) *R_M is a TAF-domain for each maximal ideal M of R.*
(3) *R_M is a PVD for each maximal ideal M of R.*
(4) *R'_M is a DVR with maximal ideal MR_M for each maximal ideal M of R.*
(5) *R is one-dimensional and every principal ideal generated by an atom is a TA-ideal.*

Theorem 4.9

(1) *[50, Corollary 4.7]. Let R be a Noetherian domain. If R is a TAF-domain, then so is every overring of R.*
(2) *[50, Corollary 4.8]. Let $K \subseteq L$ be a field extension. Then $K + XL[X]$ is a TAF-domain.*
(3) *[50, Corollary 4.11]. Let $d \in \mathbb{Z} - \{0, 1\}$ be a square-free integer such that $4 \mid (d - 1)$. Then $\mathbb{Z}[\sqrt{d}]$ is a TAF-domain if and only if $8 \mid (d - 5)$.*

5 Commutative Rings with Absorbing Factorization

We recall from [1] that the AF-dimension of a ring R, denoted by AF-dim(R), is the smallest positive integer n such that each proper ideal of R can be written as a finite product of n-absorbing ideals of R; if no such n exists, then AF-dim(R) = ∞. A ring R is an FAF-ring if AF-dim(R) $< \infty$. Recall that a *ZPI-ring* is a ring whose proper ideals can be written as a product of prime ideals. Hence, $AF - dim(R)$ measures, in some sense, how far R is from being a ZPI-ring.

The following is a structure theorem for the FAF-rings.

Theorem 5.1 ([1, Theorem 4.2]) *Any FAF-ring is a finite direct product of one-dimensional domains and zero-dimensional local rings with nilpotent maximal ideal. In particular, an FAF-ring of Krull dimension one having unique height-zero prime ideal is a domain.*

Recall that a ring R is said to be *special primary* if R has exactly one maximal ideal M and every proper ideal of R is a power of M. Note that if R is a ZPI ring, then R is a special primary ring.

Recall that R is called a chained ring if $a \mid b$ or $b \mid a$ for every $a, b \in R$.

Theorem 5.2 ([1, Proposition 3.4]) *A chained ring R is an FAF-ring if and only if R is a special primary ring.*

The next result says that the AF-dimension of a factor (resp. fraction) ring is bounded above by the AF-dimension of the ring.

Theorem 5.3 ([1, Proposition 3.5]) *Let R be an FAF-ring and T a factor or a fraction ring of R. Then $AF - dim(T) \leq AF - dim(R)$.*

Theorem 5.4 ([1, Proposition 3.6]) *Let $R_1, \ldots, R_k$ be FAF-rings and $R = R_1 \times \cdots \times R_k$. Then $AF - dim(R) = max\{AF - dim(R_i) \mid 1 \leq i \leq k\}$.*

Denote by $Min(I)$ the set of minimal prime ideals over an ideal I.

Theorem 5.5 ([1, Proposition 3.7]) *Let R be an FAF-ring and I a proper ideal. Then $Min(I)$ is finite.*

Theorem 5.6 ([1, Proposition 3.8]) *Let R be a finite ring of order m such that $p^{n+2} \nmid m$ for each prime p. Then $AF - dim(R) \leq n$. Moreover, $AF - dim(\mathbb{Z}_{p^{n+1}}[X]/(X^2, pX)) = n + 1$.*

Recall that if R is a ring, then $Spec(R) = \{P \mid \text{P is a prime ideal of } R\}$.

Theorem 5.7 ([1, Theorem 5.4]) *Let R be a commutative Noetherian one-dimensional domain with nonzero conductor $(R : R')$, where R' is the integral closure of R. The following are equivalent.*

(1) *R is an FAF-domain.*
(2) *R_M is an FAF-domain for each maximal ideal M of R.*
(3) *The spectral map $Spec(R') \to Spec(R)$ is bijective.*

In view of Theorem 5.7, we have the following example.

Example 5.8 ([1, Example 5.5])

(1) $AF - dim(\mathbb{Z}[2i]) = 3$.
(2) $R = \mathbb{Z}[\sqrt[3]{4}]$ is an FAF-ring. Since $R' = \mathbb{Z}[\sqrt[3]{2}]$ and $R \subseteq R'$ is a root extension (i.e., $z^2 \in R$ for each $z \in R'$), the map $Spec(R') \to Spec(R)$ is bijective. Hence R is an FAF-domain by Theorem 5.7.
(3) $R = \mathbb{Z}[\sqrt[3]{10}]$ is not an FAF-ring. Note that $R' = \mathbb{Z}[t]$ with $t = \frac{1+\sqrt[3]{10}+\sqrt[3]{100}}{3}$. Furthermore, $(3, t)$ and $(3, t-1)$ are two distinct prime ideals lying over $(3, 1-\sqrt[3]{10})$ in $\mathbb{Z}[\sqrt[3]{10}]$. Thus R is not an FAF-ring by Theorem 5.7.
(4) Let K be a field. Consider the Noetherian one-dimensional domains $A = K + X(X-1)K[X]$ and $B = K + X^n K[X]$ for some $n \geq 2$. Their integral closure is $K[X]$. Consider the spectral maps $Spec(K[X]) \to Spec(A)$ and $Spec(K[X]) \to Spec(B)$. Since only the second one is bijective, we get that B is an FAF-domain while A is not.

Theorem 5.9 ([1, Corollary 4.4]) *Let R be a commutative ring. The following are equivalent.*

(1) *$R[X]$ is an FAF-ring.*
(2) *R is a von Neumann regular FAF-ring.*
(3) *R is a finite direct product of fields.*
(4) *$R[X]$ is a ZPI-ring.*

Since $R[X]$ is a TFT-ring if and only if R is a finite direct product of fields by Corollary 4.5 if and only if R is an FAF-ring by Theorem 5.9, we have the following result.

Corollary 5.10 *Let R be a commutative ring. The following are equivalent.*

(1) *$R[X]$ is a TAF-ring.*
(2) *R is a von Neumann regular TAF-ring.*
(3) *R is a finite direct product of fields.*
(4) *$R[X]$ is a principal ideal ring.*
(5) *$R[X]$ is an FAF-ring.*
(6) *R is a von Neumann regular FAF-ring.*
(7) *$R[X]$ is a ZPI-ring.*

For a one-dimensional domain R, we have the following result.

Theorem 5.11 ([1, Theorem 4.3]) *Let R be a one-dimensional domain. The following are equivalent.*

(1) *R is an FAF-domain.*
(2) *R has finite character and there is some positive integer d such that $AF - dim(R_M) \leq d$ for each maximal ideal M of R.*

Acknowledgments The author would like to thank the referee for a careful reading of the paper.

References

1. M.T. Ahmed, T. Dumitrescu and M. Azeem Khadam, *Commutative rings with absorbing factorization* , Comm. Algebra, **48**(12), 5067-5075 (2020).
2. D.F. Anderson and A. Badawi , *On n-absorbing ideals of commutative rings*, Comm. Algebra, **39**, 1646-1672 (2011).
3. D.F. Anderson and A. Badawi , *Von Neumann regular and related elements in commutative rings*, Algebra Colloq., **19**(Spec 1), 1017-1040 (2012).
4. D.F Anderson and A. Badawi , *On (m, n)-closed ideals of commutative rings* , J. Algebra Appl.,**16**(1), 2016.
5. D.F. Anderson, A. Badawi and B. Fahid, *Weakly (m, n)-closed ideals and (m, n)-Von Neumann regular rings*, J. Korean Math. Soc., **55**(5), 1031–1043 (2018).
6. D.F. Anderson and D.E. Dobbs, *Pairs of rings with the same prime ideals*, Canad. J. Math., **32**, 362–384 (1980).
7. D.D. Anderson and J.L. Mott, *Cohen-Kaplansky domains: Integral domains with a finite number of irreducible elements*, J. Algebra, **148**(1), 17–41 (1992).
8. D.D. Anderson and E. Smith, *Weakly prime ideals*, Houston J. Math., **29**(4), 831–840 (2008).
9. A. Badawi, *n-Absorbing Ideals of Commutative Rings and Recent Progress on Three Conjectures: A Survey*. In Rings, Polynomials, and Modules, edited by Marco Fontana, Sophie Frisch, Sarah Glaz, Francesca Tartarone, and Paolo Zanardo, Springer, 33-52 (2017).
10. A. Badawi, *On 2-absorbing ideals of commutative rings*, Bull. Austral. Math. Soc. **75**, 417-429 (2007).
11. A. Badawi, *On weakly semiprime ideals of commutative rings*, Beitr Algebra Geom., **57**, 589-597 (2016).
12. A. Badawi, *On commutative divided rings*, Comm. Algebra, **27**(3), 1465–1474 (1999).
13. A. Badawi and E.Y. Celikel, *On 1-absorbing primary ideals of commutative rings*, J. Algebra Appl., **19**(6), 2050111 (2020).

14. A. Badawi and A.Y. Darani , *On weakly 2-absorbing ideals of commutative rings*, Houston J. Math., **39**(2), 441–452 (2013).
15. A. Badawi and D. E. Dobbs, *On locally divided rings and going-down rings*, Comm. Algebra, **29**(7), 2805–2825 (2001).
16. A. Badawi , B. Fahid, *On weakly 2-absorbing δ-primary ideals of commutative rings*, Georgian. Math. J, **27**(4), 503–516 (2020).
17. A. Badawi, M. Issoual and N. Mahdou, *On n-absorbing ideals and (m, n)-closed ideals in trivial ring extensions of commutative rings*, J. Alg. Appl., **18**(7), 2019, 1950123, 19 pp (2019).
18. A. Badawi , U. Tekir and E.Yetkin, *On 2-absorbing primary ideals in commutative rings*, Bull. Korean. Math. Soc., **51**(4), 1163–1173 (2014).
19. A. Badawi, U. Tekir, and E. Yetkin, *On weakly 2-absorbing primary ideals of commutative rings*, J. Korean Math. Soc., **52**(1), 97-111 (2015).
20. A.Badawi, D. Sonmez, G.Yesilot, *On weakly δ-semiprimary ideals of commutative rings*, Algebra Colloq., **25**, 387–398 (2018).
21. D. Bennis and B. Fahid, *Rings in which every 2-absorbing ideal is prime*, Beitr Algebra Geom, **59**, 391–396 (2018)
22. H. Chimal-Dzul and C. A. Lopez-Andrade, *When is R[x] a principal ideal ring?*, Rev. Integr. Temas Mat., **35**(2), 143–148 (2017).
23. P. J. Cahen, M. Fontana, S. Frisch, and S. Glaz, *Open problems in commutative ring theory, Commutative Algebra*, Springer, 353-375 (2014).
24. F. Callialp, E. Yetkin and U. Tekir, *On 2-absorbing primary and weakly 2-absorbing elements in multiplicative lattices*, Ital. J. Pure Appl. Math., **34**, 263–276 (2015).
25. E. Y. Celikel, E. A. Ugurlu and G. Ulucak, *On ϕ-2-absorbing elements in multiplicative lattices*, Palest. J. Math., **5**, 127–135 (2016).
26. E. Y. Celikel, E. A. Ugurlu and G. Ulucak, *On ϕ-2-absorbing primary elements in multiplicative lattices*, Palest. J. Math., **5**, 136–146 (2016).
27. H. Choi, *On n-absorbing ideals of locally divided commutative rings*, J. Algebra, **594**, 483–518 (2022).
28. H. Choi and A. Walker, *The radical of an n-absorbing ideal*. J. Commut. Algebra, **12**(2), 171–177 (2020).
29. H. Choi and Andrew Walker, *n-Absorbing monomial ideals in polynomial ring*, Int. Electron. J. Algebra, **26**, 204–223 (2019).
30. A.Y. Darani , *On 2-absorbing and weakly 2-absorbing ideals of commutative semirings*, Kyungpook Math. J., **52**, 91–97 (2012).
31. A. Y. Darani and H. Mostafanasab, *On 2-absorbing preradicals*, J. Algebra Appl., **14**, 22 pages (2015).
32. A.Y. Darani and E.R. Puczylowski, *On 2-absorbing commutative semigroups and their applications to rings*, Semigroup Forum, **86**, 83–91 (2013).
33. D. E. Dobbs, *Divided rings and going-down*, Pacific J. Math., **67**(2), 353–363 (1976).
34. G. Donadze, *The Anderson-Badawi conjecture for commutative algebras over infinite fields*, Indian J. Pure Appl. Math., **47**, 691-696 (2016).
35. G.Donadze, *A proof of the Anderson-Badawi $\sqrt{I}^n \subseteq I$ formula for n-absorbing ideal*, Proc. Indian Acad. Sci. Math. Sci., **128**(1), paper no. 6, 6 pp (2018).
36. M. Ebrahimpour and R. Nekooei, *On generalizations of prime ideals*, Comm. Algebra, **40**, 1268–1279 (2012).
37. A. El Khalfi, M. Issoual, N. Mahdou and A. Reinhart, *Commutative rings with one-absorbing factorization*, Comm. Algebra, **49**(6), 2689–2703 (2021).
38. A. El Khalfi, N. Mahdou, U. Tekir and S. Koc, *On 1-absorbing δ-primary ideals*, An. Ştiinţ. Univ. "Ovidius" Constanţa Ser. Mat., Accepted for publication.
39. J. R. Hedstorm and E. G. Houston, *Pseudo-valuation domains*, Pacific J. Math., **75**, 137-147 (1978).

40. M. Issoual, *Rings in which every 2-absorbing primary ideal is primary ideal*, Beitr. Algebra Geom., **62**, 605–614 (2021).
41. M. Issoual and N. Mahdou, *Trivial Extensions defined by 2-absorbing-like conditions*, J. Algebra Appl., **17**(11), 1850208 (2018).
42. M. Issoual and N. Mahdou, *On n-AB ring and on $\Omega(R)$ where R is a commutative ring*, submitted for publication.
43. M. Issoual, N. Mahdou and A. M. Moutui, *New results about n-absorbing ideals of commutative rings*, preprint.
44. M. Issoual, N. Mahdou and A.M. Moutui, *On n-absorbing and strongly n-absorbing ideals in amalgamated algebras*, J. Algebra Appl., **19**(10), 2050199 (2019).
45. M. Issoual, N. Mahdou and A.M. Moutui, *On (m, n)-closed ideals in amalgamated algebras*, Int. Electron. J. Algebra **29**(2), 134–147 (2021).
46. C. Jayaram, U. Tekir and E. Yetkin, *2-absorbing and weakly 2-absorbing elements in multiplicative lattices*, Comm. Algebra. **42**, 2338–2353 (2014).
47. A. Laradji, *On n-absorbing rings and ideals*, Colloq. Math., **147**, 265-273 (2017).
48. A. Malek, A. Hamed and A. Benhissi, *2-absorbing ideals in formal power series rings*, Palest. J. Math., **6**(2), 502–506 (2017).
49. H.F.Moghimi and S.R.Naghani, *On n-absorbing ideals and the n-Krull dimension of a commutative ring*, J.Korean Math.Soc., **53**(6), 1225–1236 (2016).
50. M.Muzammil, A.M.Tusif and T.Dumitrescu, *Commutative rings with two absorbing factorization*, **46**(3), 970–978 (2018).
51. P. Nasehpour, *On the Anderson-Badawi $w_{R[X]}(I[X]) = w_R(I)$ Conjecture, Arch*, Math. Brno. **52**, 71–78 (2016).
52. S. Payrovi ans S. Babaei, *On the 2-absorbing ideals*, Int. Math. Forum, **7**, 265–271 (2012).
53. S. Smach and S. Hizem, *On Anderson-Badawi Conjectures*, Beitr. Algebra Geom., **58**, 775–785 (2017).
54. M. Tamekkante and E. M. Bouba, *$(2, n)$-ideals of commutative rings*, J. Algebra Appl., **18**(6), 1950103 (2019).
55. U. Tekir, S. Koc and K.H. Oral, *n-Ideals of commutative rings*, Filomat, **31**(10), 2933–2941 (2017).
56. A. Yassine, M.J. Nikmehr and R. Nikandish, *On 1-absorbing prime ideals of commutative rings*, J. Algebra Appl., **20**(10), 2150175 (2021).

Complement-Finite Ideals

N. Baeth

1 Introduction and Motivation

Let $F = \mathcal{F}(P)$ be a free abelian monoid with basis P, the set of primes in F. Then every element in F is uniquely a product of elements in P; that is, for $f \in F$ there are unique $\nu_p(f) \in \mathbb{N}_0$, all but finitely many of which are zero, so that $f = \prod_{p \in P} p^{\nu_p(f)}$. Let $S \neq F$ be a submonoid of F satisfying the following two properties.

(CF1) $|F \setminus S| < \infty$ and
(CF2) $fs \in S$ for all $f \in F$ and all $s \in S \setminus \{1\}$.

That is, S has finite complement in the free monoid F and $S \setminus \{1\}$ is an ideal of F. We call a proper submonoid S of a free monoid F with identity 1 and satisfying Properties **(CF1)** and **(CF2)** a *complement-finite ideal (of F)*. Properties **(CF1)** and **(CF2)** are quite strong and, from a set theoretic point of view, make S quite similar to the free monoid F. However, as we shall see in the subsequent sections, S is extremely far from being free, both algebraically and arithmetically. The purpose of this manuscript is to initiate a study of such monoids and to illustrate that while the two defining conditions appear to be rather strong, the resulting monoids are nontrivial, have an interesting arithmetic, and are related to several other well-studied commutative monoids.

I would like to thank Alfred Geroldinger and Pedro García-Sánchez for useful conversations about this topic.

N. Baeth (✉)
Department of Mathematics, Franklin and Marshall College, Lancaster, PA, USA
e-mail: nicholas.baeth@fandm.edu

J.-L. Chabert et al. (eds.), *Algebraic, Number Theoretic, and Topological Aspects of Ring Theory*, https://doi.org/10.1007/978-3-031-28847-0_5

We begin by fixing some notation. A more detailed structure will be outlined in Set-up 2.1. With $F = \mathcal{F}(P)$, we set $P \setminus S = \{q_1, \ldots, q_t\}$, by **(CF1)** a necessarily finite set of primes in P. It is clear from the defining properties of S that each of its elements is either divisible by some $p \in P \setminus \{q_1, \ldots, q_t\}$ or is a product of the form $q_1^{m_1} \cdots q_t^{m_t}$ for sufficiently large m_i. To set the stage, we now introduce several examples of complement-finite ideals related to more familiar algebraic structures, namely, numerical semigroups, affine semigroups, and monoids of zero-sum sequences.

Example 1.1

(1) Let N be a numerical monoid, that is, an additive submonoid of $\mathbb{N}_0$ with $|\mathbb{N}_0 \setminus N| < \infty$. The algebraic and arithmetic properties of numerical monoids have been studied extensively; see, for example, [2, 6, 7, 13, 20]. Clearly, each numerical monoid is also closed under multiplication. In fact, if $n \in N$ and $m \in \mathbb{N}_0$, then $mn \in N$ and so $(N \setminus \{0\}) \cup \{1\}$ is a complement-finite ideal of $(\mathbb{N}, \cdot)$. The study of multiplicative factorization in numerical monoids was introduced in [4]. As we will see, many of the results of that work can be extended to the more general setting of complement-finite ideals.

(2) Let $\mathbb{N}_0^p$ denote the free affine monoid with $p > 0$. Fix t incomparable (with respect to the standard component-wise ordering) elements $\alpha_1, \ldots, \alpha_t \in \mathbb{N}_0^p$ and set $I = \bigcup_{i=1}^t \alpha_i + \mathbb{N}_0^p$, an ideal of $\mathbb{N}_0^p$. With the assumption that for each $j \in \{1, \ldots, p\}$ there exists $k_j \in \mathbb{N}$ such that $k_j \mathbf{e}_j \in I$, take $S = I \cup \{0\}$. Then S is a complement-finite ideal of $\mathbb{N}_0^p$. Moreover, S belongs to the class of generalized numerical monoids studied in [9] and [8].

(3) Let G be an additive finite abelian group with identity 0, and let $\mathcal{F}(G)$ denote the free abelian monoid with basis G, that is, the set of all formal products of elements in G without regard to order. The classical monoid $\mathcal{B}(G)$ of zero-sum sequences is the submonoid of $\mathcal{F}(G)$ consisting of all sequences $g_1 \cdot \ldots \cdot g_t \in \mathcal{F}(G)$ such that $g_1 + \cdots + g_t = 0$ in G. These monoids play a central role in factorization theory and are also interesting in their own right. See [11] and [23] for surveys of work devoted to $\mathcal{B}(G)$. Here we introduce a similarly defined monoid $\mathcal{F}_{\mathcal{B}}(G)$ consisting of all sequences in $\mathcal{F}(G)$ that are not zero-sum free; that is,

$$\mathcal{F}_{\mathcal{B}}(G) = \left\{ g_1 \cdots g_t \in \mathcal{F}(G) \colon \exists\, I \subseteq [1, t] \text{ with } \sum_{i \in I} g_I = 0 \right\} \subseteq \mathcal{F}(G).$$

It is clear that $\mathcal{B}(G) \subseteq \mathcal{F}_{\mathcal{B}}(G) \subseteq \mathcal{F}(G)$ and that $|\mathcal{F}(G) \setminus \mathcal{F}_{\mathcal{B}}(G)| < \infty$. Consequently, $\mathcal{F}_{\mathcal{B}}(G)$ is a complement-finite ideal of $\mathcal{F}(G)$. To further motivate this particular example, which will be studied in Sect. 5, we note the following. Let H be a Krull monoid with finite class group G, and let $S \subset \mathcal{I}_v^*(H)$ be the subsemigroup of v-invertible v-ideals consisting of ideals that are divisible by a principle ideal. Then there is a transfer homomorphism from S to the monoid

$\mathcal{F}_{\mathcal{B}}(G)$. This parallels nicely the fact that there is a transfer homomorphism from H to $\mathcal{B}(G)$.

In Sect. 2 we provide the necessary background and set the notation for the subsequent sections. Section 3 provides an alternate way to view each complement-finite ideal and investigates the algebraic properties. In particular, it is shown that complement-finite ideals are never Krull but are always C-monoids. In fact, the class semigroup of each complement-finite ideal S has a nice structure that can be utilized to define a new monoid $\widetilde{B}(S)$, analogous to the monoid $\mathcal{B}(H)$ of zero-sum sequences for a Krull monoid H and whose arithmetic is related to the Erdös-Burgess constant. In Sect. 4 we classify the irreducible elements of complement-finite ideals and investigate their arithmetic. Finally, in Sect. 5, we investigate more thoroughly the monoid $\mathcal{F}_{\mathcal{B}}(G)$ introduced in Example 1.1(3).

2 Background and Terminology

Throughout, a *monoid* is a set S together with a commutative, associative binary operation. Usually we write this operation multiplicatively unless $S \subseteq \mathbb{N}_0^t$ in which case we use additive notation. We further assume that the operation is cancellative (if $x, y, z \in S$ with $xy = xz$, then $y = z$) and that S has an identity element 1_S ($1_S x = x$ for all $x \in S$). The monoid S is said to be *finitely generated* provided there is some finite subset $A \subset S$ so that every element in S can be written as a finite product of elements from A. We will be particularly interested in several families of monoids that we now describe. Additional information can be found in the following volumes: [15] and [22].

With P a (not necessarily finite) set, we denote by $F = \mathcal{F}(P)$ the *free abelian monoid with basis* P. Each element f in F is a unique product of the form $f = \prod_{p \in P} p^{v_p(f)}$ for some $v_p(f) \in \mathbb{N}_0$, all but finitely many zero. We put a partial ordering $\preccurlyeq$ on F by declaring $f \preccurlyeq f'$ whenever $f = \prod_{p \in P} p^{v_p(f)}$ and $f' = \prod_{p \in P} p^{v_p'(f')}$ if and only if $v_p(f) \leq v_p'(f')$ for all $p \in P$. For a subset X of F, we denote by $\min(X)$, the set of minimal elements in X with respect to this partial ordering. For convenience, if $1 \in X$, then $\min_+(X)$ denotes the set of minimal elements, again with respect to the partial ordering $\preccurlyeq$, of $X \setminus \{1\}$.

For a positive integer t, any finitely generated submonoid S of $\mathbb{N}_0^t$ is called an *affine monoid*. Such monoids are part of the larger class of monoids that can be embedded in a finite-rank free abelian monoid. Such monoids appear often in the literature, especially because of their intrinsic geometric and combinatorial properties. See [17] and the references therein.

The so-called *block monoid* $\mathcal{B}(G)$ of zero-sum sequences was introduced in Example 1.1(3). If G is a finite abelian group, $\mathcal{B}(G)$ is the subset of the free monoid $\mathcal{F}(G)$ consisting of formal products $g_1 \cdots g_t$ with $g_1 + \cdots + g_t = 0$ in G. With operation given by concatenation and with identity element the empty string, $\mathcal{B}(G)$

is a monoid. The largest t such that $g_1 \cdots g_t$ is in $\mathcal{B}(G)$ with no proper subsequence also in $\mathcal{B}(G)$ is called the *Davenport constant* and is denoted with $\mathsf{D}(G)$. This constant, though difficult to compute for most groups, appears often in formulas related to the arithmetic of certain Krull monoids. See, for example, [5]. It is known, for example, that $\mathsf{D}(G) \le |G|$ for each finite abelian group G and that $\mathsf{D}(C_n) = n$ when C_n is a cyclic group of order n.

Let H be a commutative cancellative monoid, written multiplicatively. Then H embeds naturally into $\mathsf{q}(H) = \left\{\frac{a}{b} : a, b \in H\right\}$, the group of formal quotients of elements in H. If X and Y are subsets of H, then $(X : Y) = \{z \in \mathsf{q}(H) : zY \subseteq X\}$. We then denote by $X^{-1} = (H : X)$ and $X_v = (X^{-1})^{-1}$. The subset X of $\mathsf{q}(H)$ is a *v-ideal* provided that $X = X_v$ and that there is $h \in H$ with $hX \subseteq H$. The *complete integral closure* of H is

$$\widehat{H} = \{x \in \mathsf{q}(H) : \text{ there is } c \in H \text{ so that } cx^n \in H \text{ for all } n \in \mathbb{N}\}$$

and H is said to be completely integrally closed if $H = \widehat{H}$. We also recall that a monoid H is *seminormal* provided that

$$H = \{x \in \mathsf{q}(H) : \exists N \in \mathbb{N} \text{ so that } x^n \in H \,\forall n \ge N\}$$

and note that all completely integrally closed monoids are seminormal. Although there are several equivalent ways to define a Krull monoid (see, e.g., [15]), here we recall that H is Krull provided that H is completely integrally closed and v-noetherian (satisfies the ascending chain condition on v-ideals). We now define the more general concept of a C-monoid (see [15, §2.8-2.9] for a more thorough treatment). With H a multiplicative subsemigroup of a free monoid F, we define the class relation $\sim$ on F by $y \sim y'$ if for all $x \in F$, $xy \in H \iff xy' \in H$. The set of equivalence classes together with the binary operation $[x][y] = [xy]$ gives the *class semigroup* $\mathcal{C}^*(H, F)$. When $\mathcal{C}^*(H, F)$ is finite, we say that H is a *C-Monoid.* If $x^{2\alpha}F \cap H = x^{\alpha}(x^{\alpha}F \cap H)$ for all $x \in F\backslash\{1\}$, then we say that H is a C-monoid with *exponent* α. We also say that a subsemigroup H of F is *simple* if whenever $\{p_1, \ldots, p_t\}$ is a finite subset of the primes in F, there exist $r_1, \ldots, r_t \in \mathbb{N}$ with $x = p_1^{r_1} \cdots p_t^{r_t} \in H$. A multiplicative (necessarily noncancellative) semigroup H is a *nulloid* provided H has a multiplicative identity 1 and a zero element 0 such that $x \cdot 0 = 0$ for all $x \in H$.

We now briefly introduce the factorization-theoretic terms that we will use in Sect. 4. Take H to be a commutative cancellative monoid. A *unit* in H is an element $u \in H$ so that there is $v \in H$ with $uv = 1$, the identity of H. We restrict now to *reduced* monoids where the identity element 1 is the only unit. In this case, an element $a \in H$ is an atom (or is irreducible) provided that one cannot write $a = bc$ in H with $b, c \neq 1$. We denote the set of atoms of H by $\mathcal{A}(H)$. The irreducible element a is *prime* if whenever $a \mid bc$ in H, either $a \mid b$ or $a \mid c$ and is *absolutely irreducible* if a is the only atom of H that divides any power a^n of a. The ω-value of an atom a, $\omega(a)$, is the smallest $t \in \mathbb{N}$ so that whenever $a \mid h_1 \cdots h_n$ in H, there is $I \subseteq [1, t]$ such that $a \mid \prod_{i \in I} h_i$. Clearly $\omega(a) = 1$ precisely when a is prime. In general, $\omega(a)$ quantifies how far the atom a is from being prime. Then $\omega(H) = \sup\{\omega(a) : a \in \mathcal{A}(H)\}$ gives a measure of how nonunique factorization can be in H.

A *factorization* z of an element $h \in H$ is a formal product $\mathsf{z} := a_1 \cdots a_l$ of atoms where the product $a_1 \cdots a_l = h$ in H. The *length* of the factorization $\mathsf{z} = a_1 \cdots a_l$ is l and the *set of lengths* of an element $h \in H$ is $\mathsf{L}(h) = \{l : h \text{ has a factorization of length } l\}$. If, for each $h \in H$, there are only finitely many atoms $a \in \mathcal{A}(H)$ with $a \mid h$, we say that H is a *finite factorization monoid* (FFM). If, for each $h \in H$, $\mathsf{L}(h)$ is finite, we say that H is a *bounded factorization monoid* (BFM). Of course $|\mathsf{L}(h)| = 1$ for all $h \in H$ precisely when H is a *half-factorial monoid* (HFM). Note that this includes the class of *unique factorization monoids* (UFMs). When this is not the case, the length sets and related invariants provide a quantification of how far from being a UFM the monoid is. In particular, we consider unions of sets of lengths, elasticities, and the delta set. The *elasticity* of a nonunit $h \in H$ is $\rho(h) = \sup \mathsf{L}(h)/\min \mathsf{L}(h)$. With $\rho(u) = 1$ for each unit u of H, we set $\rho(H) = \sup\{\rho(h) : h \in H\}$ to be the *elasticity of* H. The elasticity is *accepted* if there is $h \in H$ so that $\rho(H) = \rho(h)$ and is *full* if $\rho(H) < \infty$ and for each $q \in [1, \rho(H)]$ there is $h \in H$ so that $\rho(h) = q$. With k an integer at least two,

$$\mathcal{U}_k = \bigcup_{k \in \mathsf{L}(h)} \mathsf{L}(h)$$

is the *union of sets of lengths containing* k. Then the *refined elasticity* $\rho_k(H)$ is $\rho_k(H) = \sup \mathcal{U}_k(H)$. It is not hard to see that $\rho(H) = \lim_{k \to \infty} \rho_k(H)/k$. If $\mathsf{L}(h) = \{l_1 < l_2 < \cdots\}$, we set $\Delta(h) = \{l_{i+1} - l_i : i \geq 1\}$ to be the *set of distances of* a and $\Delta(H) = \bigcup_{h \in H} \Delta(h)$. The set $\Delta^*(H)$ denotes the set of all $d = \min \Delta(K)$ where K is a divisor-closed submonoid of H with $\Delta(K) \neq \emptyset$. Finer invariants such as the *catenary degree* $\mathsf{c}(H)$ and *tame degree* $\mathsf{t}(H)$ measure how distinct two factorizations of a given element can be, regardless of the lengths of these factorizations. We do not define these invariants here since we mention them only briefly in Proposition 4.9 and Theorem 5.5. See [12] for a recent survey on these invariants. We note that if G is a finite abelian group, then $\rho(\mathcal{B}(G)) = \frac{\mathsf{D}(G)}{2}$. We will present similar results for a similarly defined constant in Sect. 5.

Finally, we set the stage for all subsequent sections by introducing notation we will be using throughout.

Set-up 2.1 *Let $F = \mathcal{F}(P)$ be the free abelian monoid with basis P and let $\preccurlyeq$ be the standard partial ordering on F (see Sect. 2). Let S be a complement-finite ideal of F; that is, $S \neq F$ is a submonoid of F such that $|F \setminus S| < \infty$ where $S \setminus \{1\}$ is an ideal of F. Since $|F \setminus S| < \infty$, $P \setminus S = \{q_1, \ldots, q_t\}$ for some nonempty finite collection $\{q_1, \ldots, q_t\} \subseteq P$. Moreover, for each $i \in [1, t]$, there is a minimal $\mathsf{m}_i \in \mathbb{N}_{\geq 2}$ so $q_i^{\mathsf{m}_i} \in S$. We denote by $\mathsf{m}(S)$ and $\mathsf{M}(S)$ the minimal and maximal values in the set $\{m_1 + \cdots + m_t : q_1^{m_1} \cdots q_t^{m_t} \in \min_+(S)\}$.*

3 Additional Structure and Algebraic Properties

In this section we consider the algebraic properties of complement-finite ideals and show that complement-finite ideal is closely related to a certain kind of affine monoid. We begin by giving an alternative definition for complement-finite ideals that illustrates that such a monoid is completely determined by its minimal nonidentity elements.

Proposition 3.1 *Let X and Y be subsets of $F \setminus \{1\}$ with $F = \mathcal{F}(P)$ a free abelian monoid with basis P.*

(1) $S = XF \cup \{1\} = \min(X)F \cup \{1\}$ *is a complement-finite ideal of F if and only if the following two properties hold:*

  (a) *For all $p \in P$ there exists (a minimal) $\mu_p \in \mathbb{N}$ with $p^{\mu_p} \in X$.*
  (b) *For all but finitely many $p \in P$, $\mu_p = 1$.*

  In particular, if $S = XF \cup \{1\}$ and $T = YF \cup \{1\}$ are complement-finite ideals, then $S = T$ if and only if $\min(X) = \min(Y)$.

(2) *If S is a complement-finite ideal of a free abelian monoid F, then $S = XF \cup \{1\}$ whenever $\min(X) = \min(S \setminus \{1\})$.*

Proof First note that for any subset X of $F \setminus \{1\}$, $\min(X)F \cup \{1\} \subseteq XF \cup \{1\}$. Suppose $xf \in XF$. If $x \notin \min(X)$, then $\frac{x}{g} = y$ with $y \in \min(X)$ and $g \in F$. Then $xf = y(fg) \in \min(X)F$. Thus $\min(X)F \cup \{1\} = XF \cup \{1\}$. Set $S = \min(X)F \cup \{1\}$. If there is $p \in P$ such that $p^n \notin X$ for any $n \in \mathbb{N}$, then $\{p^n : n \in \mathbb{N}\} \subseteq F \setminus S$ and so $|F \setminus S| = \infty$ and S is not a complement-finite ideal. Suppose now that properties (a) and (b) hold and let $f \in F$. If $\nu_p(f) > 0$ for some $p \in P \cap X$ (and hence $p \in \min(X)$), then $f = pg$ for some $g \in F$ and so $f \in S$. Thus $F \setminus S \subseteq \{f \in F : \nu_p(f) = 0 \, \forall \, p \in P \cap X\}$. Let $q_1, \dots, q_t$ be the finitely many primes from P that are not in X (so that $\mu_{q_i} \neq 1$). Since for each i, $q_i^{\mathsf{m}_i} g \in S$ for all $g \in F$, $F \setminus S \subseteq \{q_1^{m_1} \cdots q_t^{m_t} : m_i < \mu_{q_i} \, \forall \, i\}$ is finite. Since $S = XF \cup \{1\}$, clearly $sf \in S$ for all $f \in F$ and $s \in S \setminus \{1\}$. Thus S is a complement-finite ideal of F. Conversely, suppose that S is a complement-finite ideal and set $X = S \setminus \{1\}$. Since $sf \in S$ for all $s \in X$ and $f \in F$, $S = XF \cup \{1\}$ and so from (1) $S = \min(X)F \cup \{1\}$. This proves (1). Claim (2) follows immediately. □

Proposition 3.1 gives rise to the following corollary which gives a simple and useful way to think about complement-finite ideals.

Corollary 3.2 *Let S be a complement-finite ideal of a free monoid $F = \mathcal{F}(P)$ endowed with the partial ordering $\preccurlyeq$ given in Set-up 2.1. Then $S = \{f \in F : x \preccurlyeq f$ for some $x \in \min_+(S)\}$. Moreover, $p \in \min_+(S)$ for all but finitely many $p \in P$ and for all $q_i \in P \setminus S$, $q_i^{\mathsf{m}_i} \in \min_+(S)$.*

Suppose that S is a complement-finite ideal of a free monoid $F = \mathcal{F}(P)$ with P finite. Identifying F with $\mathbb{N}_0^{|P|}$, S embeds in this free monoid. We now argue that S is finitely generated. Writing $P = \{p_1, \dots, p_t\}$, set $B = \{p_1^{n_1} \cdots p_t^{n_t} : n_i <$

m_i for some $i\}$ and, for each $i \in [1,t]$, $B_i = \{p_i^{n_i} : \mathsf{m}_i \le n_i \le 2\mathsf{m}_i - 1\}$. With $X = B \cup \bigcup_{i=1}^t B_i$, a finite set, we show that $S = \langle X \rangle$. Let $s = p_1^{k_1} \cdots p_t^{k_t}$ with each $k_i \in \mathbb{N}_0$. If $k_i < \mathsf{m}_i$ for some i, then $s \in B$. Otherwise $k_i \ge \mathsf{m}_i$ for all i. Now, for each $i \in [1,t]$, write $k_i = \mathsf{m}_i q_i + r_i$ where $q_i \in \mathbb{N}$ and $r_i \in \mathbb{N}_0$ with $r_i < \mathsf{m}_i$. Then $p_i^{k_i} = p_i^{\mathsf{m}_i(q_i-1)} p^{\mathsf{m}_i + r_i}$. Since $\mathsf{m}_i \le \mathsf{m}_i + r_i \le 2\mathsf{m}_i - 1$, $p_i^{k_i} \in B_i^{q_i}$. Thus $s \in \prod_{i=1}^t B_i^{q_i} \subseteq \langle X \rangle$. As a consequence, we have the following result.

Proposition 3.3 *If S is a complement-finite ideal of a free monoid $F = \mathcal{F}(P)$ with P finite, then S is isomorphic to an affine monoid. If P is not finite, then S is a product of an affine semigroup and a free monoid.*

As a consequence, if S is a complement-finite ideal of $F = \mathcal{F}(P)$ with P finite, S is a FFM, and is hence a BFM, satisfies the ACCP, and is atomic (see [1]). In fact, the finite generation criterion is not required to obtain these properties (see [17]). In fact, since every complement-finite ideal is a reduced submonoid of a free monoid, it is a FFM (see [15, Theorem 1.5.6(2)]).

Before we move on, we consider the following elementary lemma. We will use these basic facts throughout.

Lemma 3.4 *Let S be a complement-finite ideal of a free monoid $F = \mathcal{F}(P)$ with basis P.*

(1) *For every $f \in F$, there is $N \in \mathbb{N}$ such that $f^n \in S$ for all $n \ge N$.*
(2) *There is $c \in F \setminus S$ with $c^2 \in S$.*
(3) *If $q \in P \setminus S$, there exists $x \in F \setminus S$ with $qx \in S$.*

Proof We first prove (1). If $f \in S$, take $N = 1$. Otherwise, $f \notin S$, and so $f \neq 1$. Consider the set $U = \{f^i : i \in \mathbb{N}\}$. As F is reduced, $f^i \neq f^j$ unless $i = j$, and so U is infinite. By **(CF1)**, $U \cap S \neq \emptyset$ and so $f^N \in S$ for some N. Now by property **(CF2)** $f^n = f^N f^{n-N} \in S$ for all $n \ge N$, proving (1). Choose $d \in F \setminus S$. By (1) there must be some smallest $K \ge 2$ such that $d^k \in S$ for all $k \ge K$. With $c = d^{K-1}$, $c \notin S$, yet $c^2 \in S$. This proves (2). Suppose (3) is false. Then for all $x \in F \setminus S$, $qx \in F \setminus S$. Write $F \setminus S = \{x_1, \ldots, x_n\}$. Then $\{qx_1, \ldots, qx_n\} \subseteq \{x_1, \ldots, x_n\}$. If $qx_i = qx_j$, then by cancellation $x_i = x_j$. In fact, $\{qx_1, \ldots, qx_n\} = \{x_1, \ldots, x_n\}$ and $q^i x_1 \in F \setminus S$ for all i. If $q^i x_1 = q^j x_1$, then by cancellation $i = j$. Now $\{q^i x_1 : i \in \mathbb{N}\} \subseteq \{x_1, \ldots, x_n\}$, a contradiction to $|F \setminus S| < \infty$, proving (3). □

3.1 Restrictions of Primes

In this section we consider a certain divisor-closed submonoid of each complement-finite ideal that, itself, is also a complement-finite ideal. With notation as in Setup 2.1, let S be a complement-finite ideal of a free abelian monoid $F = \mathcal{F}(P)$. With $P \setminus S = \{q_1, \ldots, q_t\} =: P_{\mathsf{C}}$, set $F_{\mathsf{C}} = \mathcal{F}(P_{\mathsf{C}})$ and $\mathsf{C}(S) = S \cap F_{\mathsf{C}}$. We show now that $\mathsf{C}(S)$ is a divisor-closed submonoid of S and also a complement-finite ideal of F_{C}. As will be seen in the exposition after Proposition 4.1, $\mathsf{C}(S)$ is the submonoid of S generated by a core set of atoms. For this reason we call $\mathsf{C}(S)$ the *core* of S.

Proposition 3.5 *Let S be a complement-finite ideal of the free monoid $F = \mathcal{F}(P)$ with basis P.*

(1) *The core of S, $\mathsf{C}(S)$, is divisor-closed in S.*
(2) *$S \setminus \mathsf{C}(S)$ is an s-ideal of S.*
(3) *$\mathsf{C}(S)$ is a complement-finite ideal in $F_{\mathsf{C}} = \mathcal{F}(P_{\mathsf{C}})$.*
(4) *$\mathsf{C}(S)$ is isomorphic to an affine monoid.*

Proof Let $c \in \mathsf{C}(S)$ and write $c = q_1^{n_1} \cdots q_t^{n_t}$ with each $q_i \in P_{\mathsf{C}}$ and each $n_i \in \mathbb{N}_0$. If $s \in S$ with $s \mid c$ in S, then $s \mid c$ in F and so $s = q_1^{m_1} \cdots q_t^{m_t}$ with $m_i \leq n_i$ for each $i \in [1, t]$. Thus $s \in \mathsf{C}(S)$ as well, proving (1).

If $P \cap S = P$; that is, $P_{\mathsf{C}} = \emptyset$, then $S \setminus \mathsf{C}(S) = \emptyset$ is, by definition, an s-ideal of S. Otherwise, if $x \in S \setminus \mathsf{C}(S)$, then $x = py$ with $y \in F$ and $p \in P \cap S$. Then, if $s \in S$, $sx = p(sy) \in S \setminus \mathsf{C}(S)$, and so $S(S \setminus \mathsf{C}(S)) = S \setminus \mathsf{C}(S)$.

Clearly $\mathsf{C}(S)$ is a submonoid of $F_{\mathsf{C}} = \mathcal{F}(P_{\mathsf{C}})$. Observe that $F_{\mathsf{C}} \setminus \mathsf{C}(S) = F_{\mathsf{C}} \setminus (S \cap F_{\mathsf{C}}) = F_{\mathsf{C}} \setminus S \subseteq F \setminus S$ is finite. If $s = q_1^{m_1} \cdots q_t^{m_t} \in \mathsf{C}(S)$ and $f = q_1^{n_1} \cdots q_t^{n_t} \in F_{\mathsf{C}}$, then $fs \in F_{\mathsf{C}}$. Moreover, $fs \in S$, and so $fs \in F_{\mathsf{C}} \cap S = \mathsf{C}(S)$. That is, $\mathsf{C}(S)$ is a complement-finite ideal of F_{C} proving (3). Since $\mathsf{C}(S)$ is finitely generated, fact (4) follows immediately from Proposition 3.3. □

3.2 Algebraic Properties

We now consider some algebraic properties of complement-finite ideals. Proposition 3.7 generalizes the results from [4, Section 4] on the multiplicative structure of numerical monoids to more general complement-finite ideals. Before giving the result we introduce a new object.

Definition 3.6 Let C be a commutative semigroup, not necessarily cancellative or with identity. Let $\mathcal{F}(C)$ denote the free abelian monoid with basis C. Then we denote by $\widetilde{\mathcal{B}}(C)$ to be the submonoid

$$\widetilde{\mathcal{B}}(C) = \left\{ c_1 \cdots c_t : \prod_{i=1}^{t} c_i \text{ is idempotent in } C \right\} \subseteq \mathcal{F}(C).$$

With operation given by concatenation and with identity given by the empty string in $\mathcal{F}(C)$, it is clear (since the product of two idempotent elements is again idempotent) that $\widetilde{\mathcal{B}}(C)$ is a commutative cancellative monoid.

Proposition 3.7 *Let S be a complement-finite ideal of a free monoid $F = \mathcal{F}(P)$.*

(1) *S is not seminormal, root-closed, or completely integrally closed, and so is not Krull.*
(2) *S is a simple C-monoid with exponent $\mathsf{M}(S)$ and $\mathcal{C}^*(S, F)$ is a nulloid with $\{1\}$ and $S \setminus \{1\}$ the only idempotent elements of $\mathcal{C}^*(S, F)$.*
(3) *There is a transfer homomorphism from S to $\widetilde{\mathcal{B}}(\mathcal{C}^*(S, F))$.*

(4) *$\widetilde{\mathcal{B}}(\mathcal{C}^*(S, F))$ is a complement-finite ideal of $\mathcal{F}(\mathcal{C}^*(S, F))$.*

The proofs of (1) and (2) are similar to the arguments in [4], and we include them here for convenience. To prove (3) and (4) we require the following result from [15], restated within the current context.

Proposition 3.8 ([15], Theorem 3.3.4) *Let $F = \mathcal{F}(P)$ be a free monoid and $H \subset F$ a C-monoid with identity $1_H = 1_F$. Set $P_0 = \{p \in P : p^{-1}H \cap F = H \setminus \{1\}\}$, $\widetilde{P} = \{[p] : p \in P \setminus P_0\} \subseteq \mathcal{C}^*(H, F)$, and $\widetilde{F} = \mathcal{F}(\widetilde{P})$ be the free monoid with basis $\widetilde{P}$. Let $\widetilde{\beta} : F \to \widetilde{F}$ be the unique homomorphism satisfying $\widetilde{\beta}(p) = [p]$ for all $p \in P \setminus P_0$, $\widetilde{\beta}(p) = [1]$ for all $p \in P_0$, and $\widetilde{\beta}(1) = [1]$. With $\widetilde{H} = \widetilde{\beta}(H)$ and $\beta = \widetilde{\beta}|_H : H \to \widetilde{H}$, β is a transfer homomorphism.*

Proof of Proposition 3.7 Observe that if $x = \frac{a}{b} \in \mathsf{q}(F)$, then for any $y \in S$, $x = \frac{ay}{by} \in \mathsf{q}(S)$. In particular, $F \subseteq \mathsf{q}(F) = \mathsf{q}(S)$. Take $f \in F \setminus S$. By Lemma 3.4 there is $N \in \mathbb{N}$ such that $f^n \in S$ for all $n \geq N$ and so S is not seminormal. As a consequence, it is not root-closed, completely integrally closed, or Krull.

By definition, the equivalence classes in $\mathcal{C}^*(S, F)$ partition F. If $s \in S \setminus \{1\}$, then by one of the defining properties of complement-finite ideals (Property **(CF2)**), $sf \in S$ for all $f \in F$. As a consequence, for any $s \in S \setminus \{1\}$, $[s] = S \setminus \{1\}$ is an element of $\mathcal{C}^*(S, F)$. As F is reduced, $[1] = \{1\}$. Now, for any $f \in F$ with $[f] \in \mathcal{C}^*(S, F) \setminus \{[1], S \setminus \{1\}\}$, $f \in F \setminus S$, a finite set (by Property **(CF1)**). Thus $|\mathcal{C}^*(S, F)| \leq |F \setminus S| + 2$ is finite and so S is a C-monoid. In fact, S is a simple C-monoid since for any set $\{p_1, \ldots, p_t\} \subseteq P$, $p_1^{\mathsf{m}_1} \cdots p_t^{\mathsf{m}_t} \in S$.

We now show that S has exponent $\mathsf{M}(S)$. Let $f \in F$ and write $f = p_1^{m_1} \cdots p_s^{m_s} q_1^{n_1} \cdots q_t^{n_t}$ with primes $p_1, \ldots, p_s \in P \cap S$ and $q_1, \ldots q_t \in P \setminus S$ and nonnegative integers $m_1, \ldots, m_s, n_1, \ldots, n_t$. Then for $\epsilon \in \{1, 2\}$, $f^{\epsilon \mathsf{M}(S)} F \subseteq S$ and $f^{\epsilon \mathsf{M}(S)} F \cap S = f^{\epsilon \mathsf{M}(S)} F$. Thus

$$f^{\mathsf{M}(S)}(f^{\mathsf{M}(S)} F \cap S) = f^{\mathsf{M}(S)}(f^{\mathsf{M}(S)} F) = f^{2\mathsf{M}(S)} F = f^{\mathsf{M}(S)} F \cap S.$$

The multiplicative identity of $\mathcal{C}^*(S, F)$ is $[1]$ and the zero element is $S \setminus \{1\} = [s]$ for each $s \in S \setminus \{1\}$ since for any $[x] \in \mathcal{C}^*(S, F)$, $[x][s] = [xs] = [s]$ by Property **(CF2)**. Thus $\mathcal{C}^*(S, F)$ is a nulloid.

We now prove (3). Let $p \in P$. If $p \in S$, then for any $f \in F$, $pf \in S$ and $f = p^{-1}(pf) \in p^{-1}S \cap F$. If $p \notin S$, then there is some $x \in F \setminus S$ with $pf \in S$ (see Lemma 3.4(3)). Again, $f = p^{-1}(pf) \in p^{-1}S \cap F$. In either case, $p^{-1}S \cap F \neq S \setminus \{1\}$ and so, with notation as in Proposition 3.8, $P_0 = \emptyset$. Thus $\widetilde{P} = \{[p] : p \in P\}$ and, by (2) and Proposition 3.8, there is a transfer homomorphism to

$$\left\{ [p_1] \cdots [p_t] \in \mathcal{F}(\widetilde{P}) : \underbrace{p_1 \cdots p_t \in S}_{[p_1] \cdots [p_t] \text{ idempotent}} \right\} = \widetilde{\mathcal{B}}(\mathcal{C}^*(S, F)).$$

To prove (4), we note that for any $p \in P$, $p^m \in S$ for some $m \in \mathbb{N}$, for each $[p] \in \widetilde{P}$ there is $m \in \mathbb{N}$ so that $[p]^m \in \widetilde{\mathcal{B}}(\mathcal{C}^*(S, F))$. Since $|\widetilde{P}|$ is finite, Proposition 3.1 implies that $\widetilde{\mathcal{B}}(\mathcal{C}^*(S, F))$. is a complement-finite ideal. □

Continuing with the notation in this section and recalling Proposition 3.7, write

$$\mathcal{C}^*(S, F) = \{[1], [a_1], \ldots, [a_r], [z]\}$$

with $[1] = \{1\}$ the identity of $\mathcal{C}^*(S, F)$, $[z] = S \setminus \{1\}$ the zero element, and where $[a_1], \ldots, [a_r]$ are the r non-idempotent elements of $\mathcal{C} := \mathcal{C}^*(S, F)$. With $\widetilde{\mathcal{B}}(\mathcal{C})$, defined in Definition 3.6, the monoid of formal products in the free monoid with basis $\{[a_1], \ldots, [a_r], [z]\}$ whose actual product in $\mathcal{C}^*(S, F)$ is idempotent, the irreducible elements in $\widetilde{\mathcal{B}}(\mathcal{C})$ are the idempotent products containing no proper idempotent subsequence. In [16] (answering a question of Paul Erdős) it was shown that if S is a finite semigroup with r nonidempotent elements, then every product of at least $r + 1$ elements in S contains an idempotent subproduct. The maximal length of an idempotent sequence having no idempotent subproduct is referred to the Erdős-Burgess constant. It has been studied for various semigroups (e.g., [24]) and is related to the analogous Davenport constant for groups (cf. [18]). We propose a careful study of the arithmetic of the monoid $\widetilde{\mathcal{B}}(\mathcal{C})$ as has been done extensively with the monoid $\mathcal{B}(H)$ of zero-sum sequences (see, e.g., [23] and the references therein). Here we give one general result about the elasticity in semigroups of idempotent products, the corollary of which gives an upper bound on the elasticity for complement-finite ideals.

Proposition 3.9 *Let $T = \{1, a_1, \ldots, a_r, z\}$ be a commutative semigroup with identity 1, zero element z, and where $a_1, \ldots, a_r$ are nonidempotent elements of T. Let $\widetilde{\mathcal{B}}(T)$ be the submonoid of $\mathcal{F}(T)$ from Definition 3.6. Then $\rho(\widetilde{\mathcal{B}}(T)) \leq r + \frac{1}{2}$.*

Proof Suppose that $\alpha_1, \ldots, \alpha_s, \alpha'_1, \ldots, \alpha'_t, \beta_1, \ldots, \beta_u, \beta'_1, \ldots, \beta'_v$ are irreducible elements of $\widetilde{\mathcal{B}}(T)$ such that, in $\mathcal{F}(T)$, $z \mid \alpha_i$ and $z \mid \beta_j$ for each i and j, yet $z \nmid \alpha'_i$ and $z \nmid \beta'_j$ for each i and j. Observe that since z itself is an irreducible element of $\widetilde{\mathcal{B}}(T)$, $z^2 \nmid \alpha_i$ and $z^2 \nmid \beta_j$ for each i and j. Suppose that $\gamma = \alpha_1 \cdots \alpha_s \alpha'_1 \cdots \alpha'_t = \beta_1 \cdots \beta_u \beta'_1 \cdots \beta'_v$ in $\widetilde{\mathcal{B}}(T)$ with $\frac{s+t}{u+v}$ as large as possible; that is, $\rho(\gamma) = \frac{s+t}{u+v}$. Since each α_i and β_j is irreducible and as $z^2 \nmid \alpha_i$ and $z^2 \nmid \beta_j$ for each i and j, $s = u$. With $v \leq t$, $\frac{s+t}{s+v} \leq \frac{t}{v}$ and so to obtain an upper bound on the elasticity we may as well assume that $s = u = 0$. In $\mathcal{F}(T)$, write $\gamma = f_1 \cdots f_\ell$. Since no α'_i involves z, the length of each α'_i in $\mathcal{F}(T)$ is at least two and so $t \leq \frac{\ell}{2}$. Since T contains exactly r nonidempotent elements, by Gillam et al. [16], the length in $\mathcal{F}(T)$ of each β'_i is at most $2r + 1$. Thus $v \geq \frac{\ell}{2r+1}$. Now $\rho(\gamma) \leq \frac{t}{v} \leq \frac{\ell/2}{\ell/(2r+1)} = \frac{2r+1}{2} = r + \frac{1}{2}$. As γ was arbitrary, the result follows. □

In particular, if S is a complement-finite ideal, then we immediately obtain a bound on the elasticity of S, an improvement on the general bound on elasticities for C-monoids given in [15, Theorem 3.3.1]. In particular, if S is a complement-finite ideal of the free monoid F, we have that $\rho(S) \leq |F \setminus S| - 1 + \frac{1}{2}$. This bound will be greatly improved in the next section.

Corollary 3.10 *Let S be a complement-finite ideal of a free monoid F. Then*

$$\rho(S) \leq |\mathcal{C}^*(S, F)| + \frac{1}{2}.$$

Remark 3.11 With the same proof, the result of Proposition 3.9 can be easily strengthened by replacing $r + 1$ with $\mathsf{I}(T)$, the Erdös-Burgess constant of T. Then, this result nicely matches the analogous result for the monoid $\mathcal{B}(G)$ of zero-sum sequences over an abelian group G: $\rho(\mathcal{B}(G)) = \frac{\mathsf{D}(G)}{2}$, where $\mathsf{D}(G)$ is the Davenport constant of G.

Example 3.12 Let n be a positive integer and let $S = \langle n, \ldots, 2n - 1\rangle$ be the numerical monoid generated by the full interval $[n, 2n-1]$. Let S_1 be the associated multiplicative semigroup, a complement-finite ideal of $(\mathbb{N}, \cdot)$, so that $S = \{1\} \cup \mathbb{N}_{\geq n}$. Then

$$\mathcal{C}^*(S_1, \mathbb{N}) = \left\{[1], \left[\lceil\frac{n}{\lceil n/2\rceil}\rceil\right], \ldots, \left[\lceil\frac{n}{2}\rceil\right], [n] = S_1 \setminus \{1\}\right\}.$$

Take, for example, $n = 4$ so that $\mathcal{C}^*(S_1, \mathbb{N}) = \{[1], [2], [4]\}$. Now $r = 1$ and Corollary 3.10 gives that $\rho(\widetilde{\mathcal{B}}(\mathcal{C}^*(S, F))) \leq 1 + \frac{1}{2} = \frac{3}{2}$. This bound is obtained by the following two factorizations in $\widetilde{\mathcal{B}}(\mathcal{C}^*(S, F))$.

$$([2] \cdot [2] \cdot [2])^3 = ([2] \cdot [2])^3 .$$

Now if $n = 8$, then $\mathcal{C}^*(S_1, \mathbb{N}) = \{[1], [2], [3], [4], [8]\}$. In this case $r = 3$ and Corollary 3.10 gives that $\rho(\widetilde{\mathcal{B}}(\mathcal{C}^*(S, F))) \leq 3 + \frac{1}{2} = \frac{7}{2}$. We now argue that this elasticity cannot be obtained. As in the proof of Corollary 3.10, to obtain an upper bound on the elasticity, we need only consider elements involving [2], [3], and [4]. Thus we can instead consider the elasticity of the affine submonoid T of $\mathbb{N}_0^3$ (Proposition 3.3) wherein [2], [3], and [4] are mapped to $\mathbf{e}_1$, $\mathbf{e}_2$, and $\mathbf{e}_3$, respectively. Consider the function $\varphi : T \to \mathbb{N}_0$ given by $\varphi(a, b, c) = a + b + c$. It is not hard to see that this is a semilength function. A bit of computation shows that $\min\{\varphi(a, b, c) \colon (a, b, c) \text{ is an atom of } T\} = 2$ and that $\max\{\varphi(a, b, c) \colon (a, b, c) \text{ is an atom of } T\} = 5$. Thus $\rho(\widetilde{\mathcal{B}}(\mathcal{C}^*(S, F)) = \rho(T) \leq \frac{5}{2} < \frac{7}{2}$. Alternatively, one could note that the Erdös-Burgess constant here is 3, also giving the upper bound $\frac{5}{2}$ for the elasticity.

4 Atoms and Arithmetic

In this section we establish some results about the arithmetic of complement-finite ideals, some of which are straight-forward extensions of what was already done for multiplicative numerical semigroups in [4]. We begin by classifying the atoms of a given complement-finite ideal. The classification, and its proof, is very much like that for multiplicative numerical semigroups. But, for the general case we must use Lemma 3.4.

We are now able to give a classification of the atoms as well as state that there are no prime elements and that only certain types of atoms are absolutely irreducible.

Proposition 4.1 *Let S be a complement-finite ideal of a free monoid $F = \mathcal{F}(P)$ as in Step-up 2.1.*

(1) $\mathcal{A}(S) = \mathcal{A}_1 \cup \mathcal{A}_2 \cup \mathcal{A}_3$, *a disjoint union where*

  (I) $\mathcal{A}_1 = \{p\colon \in P \cap S\}$,
  (II) $\mathcal{A}_2 = \{pf\colon p \in P \cap S \text{ and } f \in F \setminus S\}$, *and*
  (III) $\mathcal{A}_3 = \mathcal{A}(\mathsf{C}(S))$.

(2) *No atom of S is prime.*
(3) *Only atoms of type (I) are absolutely irreducible.*

Proof Let $p \in P \cap S$ and let $f \in F \setminus S$. If $pf = s_1 s_2$ with $s_1, s_2 \in S$, then $p \mid s_1 s_2$ in F. Since p is prime in F, without loss of generality, $p \mid s_1$ in F. Then $s_1 = px$ with $x \in F$. But then $pf = pxs_1$ and by cancellation $f = xs_1 \in S$, a contradiction. If $s \in S$ and $p \nmid s$ for any $p \in P \cap S$, then $s \in \mathsf{C}(S)$. Since $\mathsf{C}(S)$ is divisor-closed in S, any atom of S of this form is an atom of the core $\mathsf{C}(S)$.

We now show that no atom of S is prime. Let $p \in P \cap S$, an atom of type (I), and choose, by Lemma 3.4, $c \in F \setminus S$ such that $c^2 \in S$. Then $p \mid p(pc^2) = (pc)^2$, yet $p \nmid pc$. Let pf be an atom of type (II) with $p \in P \cap S$ and $f \notin S$. Then $pf \mid (pf)^2 = p \cdot (pf^2)$, yet $pf \nmid p$ and $pf \nmid pf^2$. Finally, let $s = q_1^{m_1} \cdots q_t^{m_t}$ be an atom of type (III). Then $\sum_{i=1}^{t} m_i \geq 2$ and we can write $q_1^{m_1} \cdots q_t^{m_t} = \underbrace{(q_1^{l_1} \cdots q_t^{l_t})}_{r} \cdot \underbrace{(q_1^{m_1-l_1} \cdots q_t^{m_t-l_t})}_{r'}$ with $l_i \leq m_i$ for all i and with $l_j > 0$ for some j and $m_k - l_k > 0$ for some k. With $p \in P \cap S$, obviously $s \mid p^2 s$ in S. However, $p^2 s = (p \cdot r) \cdot (p \cdot r')$ and yet $s \nmid p \cdot r$ and $s \nmid p \cdot r'$. Thus S has no prime elements.

From the classification of atoms in S, and because of unique prime factorization in F, it is clear that each atom of type (I) is absolutely irreducible. If pf is an atom of type (II), then $(pf)^2 = p \cdot pf^2$ and so $p \mid (pf)^2$. Thus atoms of type (II) are not absolutely irreducible. Finally, let $s = q_1^{m_1} \cdots q_t^{m_t}$ be an atom of type (III). By Lemma 3.4, we can choose N so that $q_1^N \in S$. Then $\left(q_1^{m_1} \cdots q_t^{m_t}\right)^N = q_1^N \cdot \left(q_1^{Nm_1 - m_1} q_2^{Nm_2} \cdots q_t^{Nm_t}\right)$ in S and so $q_1^N \mid s^N$, whence s is not absolutely irreducible. □

Remark 4.2 Observe that the atoms of type (III) are precisely the elements of S the form $q_1^{m_1} \cdots q_t^{m_t}$ so that $(m_1, \ldots, m_t) \in \mathcal{A}(T)$ where T is the affine semigroup $T = \{(n_1, \ldots, n_t) \in \mathbb{N}_0^t\colon q_1^{n_1} \cdots q_t^{n_t} \in S\}$ (see the paragraph preceding Proposition 3.3). With the partial ordering $\preccurlyeq$ and $\min_+(S)$ as defined in Step-up 2.1, we can describe the atoms of type (III) as those elements in $\mathsf{C}(S)$ satisfying the following two conditions.

- $q_1^{m_1} \cdots q_t^{m_t} \succcurlyeq q_1^{n_1} \cdots q_t^{n_t}$ with $q_1^{n_1} \cdots q_t^{n_t} \in \min_+(S)$.
- $q_1^{m_1} \cdots q_t^{m_t} \not\succcurlyeq q_1^{k_1} \cdots q_t^{k_t} \cdot q_1^{l_1} \cdots q_t^{l_t}$ whenever $q_1^{k_1} \cdots q_t^{k_t}, q_1^{l_1} \cdots q_t^{l_t} \in \min_+(S)$.

That is, atoms of type (III) correspond to elements in $\mathsf{C}(S)$ "between" (via the partial ordering $\preccurlyeq$) $\min_+(S)$ and $\min_+(S)^2$.

Remark 4.3 Let $q \in P \setminus S$ and let m be as in Step-up 2.1 so that $\mathsf{m} = \min\{m : q^m \in S\}$. Now, if $q^k \in S$, then $k \geq \mathsf{m}$. On the other hand, if $k \geq 2\mathsf{m}$, then $q^k = (q^{\mathsf{m}})(q^{k-\mathsf{m}})$ and so q^k is not irreducible. Conversely, suppose that $m \leq k < 2m$. Then q^k is irreducible since the only way to factor q^k (in F) is as $q^m \cdot q^n$ with $1 < m, n < k$, an impossibility in S if $\mathsf{m} \leq k < 2\mathsf{m}$. Thus we observe that q^k is an atom in S if and only if $\mathsf{m} \leq k < 2\mathsf{m}$.

Because the atoms of S of type (III) are atoms in the core of S, we refer to them as *core* atoms. We refer to atoms of types (I) and (II) as *peripheral* atoms. While no atom in S is prime (see Proposition 4.1), as we will see in Lemma 4.7, the peripheral atoms do have some prime-like properties, namely, that the number of such atoms in a factorization of a fixed element of S is unchanged, regardless of the factorization. Moreover in Corollary 4.10, we will see that $\omega(p) = 2$ for atoms of type (I), while ω values are largest for the core atoms.

The following result shows how lengths sets in complement-finite ideals are, in general, much more well-structured than in more general C-monoids, or even general Krull monoids.

Theorem 4.4 *Let S be a complement-finite ideal in a free monoid $F = \mathcal{F}(P)$.*

(1) *S is half-factorial if and only if S is factorial.*
(2) *There is $M \in \mathbb{N}$ such that every $L \in \mathcal{L}(S)$ is an AAP with distance 1 and bound M.*

Proof If $S \neq F$, then there is $q \in P \setminus S$. With m as in Remark 4.3, $\mathsf{m} \geq 2$ and both q^{m} and $q^{\mathsf{m}+1}$ are atoms of S. Then $(q^{\mathsf{m}})^{\mathsf{m}+1} = (q^{\mathsf{m}+1})^{\mathsf{m}}$, showing that S is not half-factorial. Clearly S is half-factorial if $S = F$. This proves (1).

Item (2) is immediate if S is half-factorial, so assume otherwise. Since S is a C-monoid, by Geroldinger and Halter-Koch [15, Theorem 4.6.6] there is $M \in \mathbb{N}$ such that every $L \in \mathcal{L}(S)$ is an AAMP with some distance $d \in \Delta^*(S)$ and bound M. Thus we need only show that $\Delta^*(S) = \{1\}$. Let T be a divisor-closed submonoid of S with T not half-factorial. Then there is $q \in P \setminus T$, and we can again let $m = \mathsf{m}(q) \geq 2$ so that q^m and q^{m+1} are atoms of T. Then q^{m^2+m} factors both as $(q^{m+1})^m$ and as $(q^m)^{m+1}$, factorizations of lengths m and $m+1$. In particular, $\min \Delta(T) = 1$ and so $\Delta^*(S) = \{1\}$. □

In fact, if we take for T in the above proof, the submonoid $T = \{1, q^m, q^{m+1}, q^{m+2}, \ldots\}$ of S with $q \in P \setminus S$, then we can note that T is isomorphic to the additive numerical monoid $M = \langle m, m+1, \ldots, 2m-1 \rangle$. Then from [2] we immediately obtain that $\Delta(T) = \Delta(M) = \{1\}$. Not only is every length set in S an AAP with distance 1, but the system of sets of lengths contains length sets containing arbitrarily large intervals as is evidenced by the following proposition, adapted from [4].

Proposition 4.5 *Let S be a complement-finite ideal as in Step-up 2.1. For each $i \in [1, t]$, and each $k \in \mathbb{N}_0$, $\mathsf{L}(q_i^k) = \left[\left\lceil \frac{k}{2\mathsf{m}_i-1} \right\rceil, \left\lfloor \frac{k}{\mathsf{m}_i} \right\rfloor\right]$. Moreover, if $k_1, \ldots, k_t \in \mathbb{N}_0$ with $k_i \geq \mathsf{m}_i$ for each i,*

$$\mathsf{L}(q_1^{k_1} \cdots q_t^{k_t}) \supseteq \left[\sum_{i=1}^{t} \left\lceil \frac{k_i}{2\mathsf{m}_i - 1} \right\rceil, \sum_{i=1}^{t} \left\lfloor \frac{k}{\mathsf{m}_i} \right\rfloor\right].$$

Proof For each $i \in [1, t]$, the only atoms of S that divide q_i^k are the atoms $q_i^{\mathsf{m}_i}, \ldots, q_i^{2\mathsf{m}_i-1}$ of type (III). Thus multiplicative factorizations of q_i^k in S correspond exactly to the additive factorizations of k in the numerical monoid $\langle \mathsf{m}_i, \ldots, 2\mathsf{m}_i - 1 \rangle$. By Amos et al. [2, Theorem 2.2], $\mathsf{L}(q_i^k) = \left[\left\lceil \frac{k}{2\mathsf{m}_i-1} \right\rceil, \left\lfloor \frac{k}{\mathsf{m}_i} \right\rfloor\right]$. The moreover statement follows by subadditivity of length sets. □

We now consider an example illustrating Theorem 4.4. In particular we show that because the monoid $\mathcal{F}_{\mathcal{B}}(G)$, introduced in Example 1.1(3), has so many more atoms than the corresponding monoid $\mathcal{B}(G)$, even if $B \in \mathcal{B}(G) \cap \mathcal{F}_{\mathcal{B}}(G)$, $\mathsf{L}_{\mathcal{F}_{\mathcal{B}}(C_n)}(B)$ is much larger than $\mathsf{L}_{\mathcal{B}(C_n)}(B)$.

Example 4.6 Let $C_n = \langle g \rangle$ be a cyclic group of order n and consider $B = g^n(-g)^n$ in the monoid of zero-sum sequences $\mathcal{B}(C_n)$. The only atoms of $\mathcal{B}(C_n)$ that divide B are $g(-g)$, g^n, and $(-g)^n$ and so the only factorizations in $\mathcal{B}(C_n)$ of B are $B = g^n \cdot (-g)^n$ and $B = (g(-g))^n$. Consequently, $\mathsf{L}_{\mathcal{B}(C_n)}(B) = \{2, n\}$.

We now consider the same element B in $\mathcal{F}_{\mathcal{B}}(C_n)$. In this monoid there are many more atoms. In particular, the atoms of $\mathcal{F}_{\mathcal{B}}(C_n)$ that divide B are: g^k and $(-g)^k$ with $k \in [n, 2n-1]$ as well as $g^l(-g)$ and $g(-g)^l$ for $l \in [1, n]$. As in $\mathcal{B}(C_n)$, $\min \mathsf{L}_{\mathcal{F}_{\mathcal{B}}(C_n)}(B) = 2$ since $B = g^n \cdot g^{-n}$. Similarly, $\max \mathsf{L}_{\mathcal{F}_{\mathcal{B}}(C_n)}(B) = n$ as evidenced by the factorization $(g(-g))^n$. But, in $\mathcal{F}_{\mathcal{B}}(C_n)$, for all $l \in [2, n]$ there is a factorization z_l of B with $|\mathsf{z}_l| = l$, namely,

$$\mathsf{z}_l := (g(-g))^{l-2} \cdot g^{n-l+1}(-g) \cdot g(-g)^{n-l+1}.$$

Thus $\mathsf{L}_{\mathcal{F}_{\mathcal{B}}(C_n)}(B) = [2, n]$.

Lemma 4.7 *Let S be a complement-finite ideal of $F = \mathcal{F}(P)$ as in Step-up 2.1, and let $\mathsf{C}(S)$ denote the submonoid of S generated by the core atoms.*

(1) *Let $w_1 \cdots w_a x_1 \cdots x_b = y_1 \cdots y_c z_1 \cdots z_d$ be two factorizations of an element in S with each $w_i, x_j, y_k, z_l \in \mathcal{A}(S)$. With the notation as in Proposition 4.1, if $w_i, y_k \in \mathcal{A}_1 \cup \mathcal{A}_2$ for all $i \in [1, a]$ and $k \in [1, c]$ and $x_j, z_l \in \mathcal{A}_3$ for all $i \in [1, b]$ and $l \in [1, d]$, then $a = c$.*

(2) $\rho(S) = \rho(\mathsf{C}(S))$.
(3) $\mathsf{C}(S)$*, and hence* S*, is fully elastic.*
(4) *If* $y = p_1 \cdots p_s x \in S$ *with* $p_1, \ldots, p_s \in P \cap S$ *and* $x \in \mathsf{C}(S)$, $s \le \min \mathsf{L}(y) \le s + \min \mathsf{L}(x)$ *and* $\max \mathsf{L}(y) = s + \max \mathsf{L}(x)$.

Proof (1) follows from unique factorization in F and since elements in $\mathcal{A}_1 \cup \mathcal{A}_2$ are precisely those atoms of S divisible, in F, by some prime $p \in P \cap S$.

Since $\mathsf{C}(S)$ is a divisor-closed submonoid of S (Proposition 3.5), $\rho(\mathsf{C}(S)) \le \rho(S)$. Let $s \in S \setminus \mathsf{C}(S)$. Let $w_1 \cdots w_a x_1 \cdots x_b = y_1 \cdots y_c z_1 \cdots z_d$ be two factorizations of s. By (1), $a = c$. With $d \ge b$, $\frac{d+c}{b+a} \le \frac{d}{b}$, and so $\rho(S) \le \rho(\mathsf{C}(S))$. This proves (2).

Now, since $\mathsf{C}(S)$ is (locally) finitely generated, so is $S' = \{r^n x \colon n \in \mathbb{N}_0, x \in \mathsf{C}(S)\} \subseteq F' = \mathcal{F}(\{q_1, \ldots, q_t, r\})$. Now, by Zhong [25, Theorem 1.1], S' is fully elastic if there is $s \in S'$ with $\overline{\rho}(s) = \lim_{n\to\infty} \rho(s^n) = 1$. Taking $s = r$ and noting that r is absolutely irreducible in S' (see Lemma 3.4) completes the proof of (3).

By (1), every factorization of y involves exactly s atoms of type (I) and (II) and so $s \le \min \mathsf{L}(y)$. On the other hand, for each factorization $\mathsf{z} := a_1 \cdots a_l$ of x, $p_1 \cdots p_s \mathsf{z}$ is a factorization of y. Thus $\min \mathsf{L}(y) \le s + \min \mathsf{L}(x)$ and $\max \mathsf{L}(x) \ge s + \max \mathsf{L}(x)$. If $\max \mathsf{L}(y) > s + \max \mathsf{L}(x)$, then there is a factorization $\mathsf{z} := p_1 \cdots p_r x_{r+1} \cdots x_s x'_1 \cdots x'_t$ of y with each x_i and x'_j irreducible and with $p_i \mid x_i$ (in F) for each $i \in [r+1, s]$ and $p_i \nmid x'_j$ for any i or j. That is, each x_i is an atom of type (II) and each x'_j is in $\mathsf{C}(S)$. But then $\max \mathsf{L}(x'_1 \cdots x'_t) > \max \mathsf{L}(x)$ which is impossible since $x'_1 \cdots x'_t \mid x$ in F. □

Remark 4.8 Observe that if, for example, $x = q_1^{r_1} \cdots q_t^{r_t}$ in Lemma 4.7(4) with $\sum_{i=1}^t r_i \le s$, y has a factorization as a product involving only atoms of types (I) and (II) and so $\min \mathsf{L}(y) = s$, the stated lower bound. But, if $\sum_{i=1}^t r_i$ is considerably larger than s, $\min \mathsf{L}(y)$ may be much larger than s.

As a consequence of Lemma 4.7, it makes sense to restrict our attention to complement-finite ideals S with $S = \mathsf{C}(S)$. We now briefly consider the catenary degree, tame degree, and the ω invariant of complement-finite ideals.

Proposition 4.9 *Let* S *be a complement-finite ideal of* $F = \mathcal{F}(P)$ *as in Step-up 2.1.*

(1) $\omega(S) \in \{\mathsf{M}(S), \mathsf{M}(S) + 1\}$.
(2) $\mathsf{c}(S) \le \mathsf{M}(S) + 1$.
(3) $\mathsf{M}(S) \le \mathsf{t}(S) \le (\mathsf{M}(S) + 1)^2$.

Proof Let $a \in \mathcal{A}(S)$ and suppose that $a \mid x_1 \cdots x_k$ for some $x_1, \ldots, x_k \in S$. Write $a = p_1 \cdots p_{|a|}$ with $p_1, \ldots, p_{|a|} \in F$. After reordering the x_is as necessary, for each $i \in [1, |a|]$, $x_i = p_i x'_i$ in F. Then clearly $a \mid x_1 \cdots x_{|a|}$ in F and so $ay = x_1 \cdots x_{|a|}$ for some $y \in F$. If $y \in S$, then $a \mid x_1 \cdots x_{|a|}$ in S. Otherwise, writing $a(yx_{|a|+1}) = x_1 \cdots x_{|a|+1}$ and noting that $yx_{|a|+1} \in S$, $a \mid x_1 \cdots x_{|a|+1}$ in S. Thus $\omega(a) \in \{|a|, |a|+1\}$ and $\omega(S) \in \{\mathsf{M}(S), \mathsf{M}(S) + 1\}$, proving (1).

(2) follows directly from the general result [15, Chapter 1] that if H is an atomic monoid $\mathsf{c}(H) \le \omega(H)$. By Geroldinger and Halter-Koch [15, Chapter 8] we have that $\omega(H) \le \mathsf{t}(H) \le \omega^2(H)$ whenever H is an atomic monoid. This proves (3). □

Putting together Propositions 4.1 and 4.9, we have the following corollary.

Corollary 4.10 *Let S be a complement-finite ideal of $F = \mathcal{F}(P)$ as in Step-up 2.1 and let $a \in \mathcal{A}(S)$ as characterized in Proposition 4.1.*

(1) $\omega(a) = 2$ *if* $a \in \mathcal{A}_1$.
(2) $2 \le \omega(a) \le \sum_{i=1}^t \mathsf{m}_i - t$ *if* $a \in \mathcal{A}_2$.
(3) $\mathsf{m}(S) \le \omega(a) \le \mathsf{M}(S) + 1$ *if* $a \in \mathcal{A}_3$.

Proof By (1) of Proposition 4.9 $\omega(p) \in \{1, 2\}$ for each atom p of type (I). But since no element in S is prime (Proposition 4.1), $\omega(p) = 2$ for each $p \in \mathcal{A}_1$. If $a \in \mathcal{A}_2$, then $a = pq_1^{m_1} \cdots q_t^{m_t}$ with $p \in P \cap S$ and $q_1^{m_1} \cdots q_t^{m_t} \notin S$. The lower bound follows since $\sum_{i=1}^t m_i \ge 1$. The upper bound follows since $m_i \le \mathsf{m}_i - 1$ for each i. Finally, if $a \in \mathcal{A}_3$, then $a = q_1^{m_1} \cdots q_t^{m_t}$ with $\mathsf{m}(S) \le \sum_{i=1}^t m_i \le \mathsf{M}(S)$. The result then follows from Proposition 4.9(1). □

We now consider the elasticity of complement-finite ideals.

Lemma 4.11 *Let S be a complement-finite ideal of $F = \mathcal{F}(P)$ as in Step-up 2.1. With $\mathsf{m}_1 \le \cdots \le \mathsf{m}_t$,*

$$\rho(S) \le \frac{(t+1)\prod_{i=1}^t \mathsf{m}_i - \sum_{i=1}^t \prod_{\substack{j=1\\ j\ne i}}^t \mathsf{m}_j}{\min_{\mathbf{x}\in S\setminus\{1\}} \sum_{i=1}^t \prod_{\substack{j=1\\ j\ne i}}^t \mathsf{m}_j x_i}.$$

Proof By Lemma 4.7 we may suppose that $S = \mathsf{C}(S)$. Consider the function $\varphi : F \to \mathbb{N}_0$ defined by

$$\varphi(q_1^{x_1}, \ldots, q_t^{x_t}) = \sum_{i=1}^t \prod_{\substack{j=1\\ j\ne i}}^t \mathsf{m}_j x_i.$$

Restricted to S, $\varphi|_S : S \to \mathbb{N}_0$ is a semilength function. As a consequence, $\rho(S) \le \frac{\max\{\varphi(\mathbf{x}) \colon \mathbf{x}\in\mathcal{A}(S)\}}{\min\{\varphi(\mathbf{x}) \colon \mathbf{x}\in\mathcal{A}(S)\}}$. It is clear that $\max\{\varphi(\mathbf{x}) \colon \mathbf{x} \in \mathcal{A}(S)\} \le (t+1)\prod_{i=1}^t \mathsf{m}_i - \sum_{i=1}^t \prod_{\substack{j=1\\ j\ne i}}^t \mathsf{m}_j$ as evidenced by the element $q_1^{\mathsf{m}_1 - 1} \cdots q_{t-1}^{\mathsf{m}_{t-1}-1} \cdot q_t^{2\mathsf{m}_t - 1}$. □

With the following example, generalized in Proposition 4.15, we illustrate that the bound in Lemma 4.11 is sometimes obtained.

Example 4.12 Consider the affine complement-finite ideal

$$S = \{(0,0)\} \cup \{(x, y) : x \geq 3 \text{ or } y \geq 4\} \subseteq \mathbb{N}_0^2.$$

Then $\mathsf{m}_1 = 3$, $\mathsf{m}_2 = 4$, and the set of atoms of S is:

$$\{(x, y) : 3 \leq x \leq 5 \text{ and } 0 \leq y \leq 3\} \cup \{(x, y) : 0 \leq x \leq 2 \text{ and } 4 \leq y \leq 7\}.$$

The upper bound given in Lemma 4.11 gives

$$\rho(S) \leq \frac{(t+1)\prod_{i=1}^{t} \mathsf{m}_i - \sum_{i=1}^{t} \prod_{\substack{j=1 \\ j \neq i}}^{t} \mathsf{m}_j}{\min_{\mathbf{x} \in S \setminus \{1\}} \sum_{i=1}^{t} \prod_{\substack{j=1 \\ j \neq i}}^{t} \mathsf{m}_j x_i} \leq 29/12.$$

This upper bound is obtained since

$$12 \cdot (2, 7) = 8 \cdot (3, 0) + 21 \cdot (0, 4).$$

We now illustrate that the upper bound in Lemma 4.11 may not be obtained if not all minimal elements of S lie on extremal rays.

Example 4.13 Consider the affine complement-finite ideal

$$S = \{(0,0)\} \cup \{(x, y) : x \geq 3,\ y \geq 4, \text{ or } xy \geq 1\} \subseteq \mathbb{N}_0^2.$$

Then $\mathsf{m}_1 = 3$, $\mathsf{m}_2 = 4$, and the atoms of S are:

$$(3, 0), (4, 0), (5, 0), (0, 4), (0, 5), (0, 6), (0, 7), (1, 1), (1, 2), (1, 3), (1, 4),$$
$$(2, 1), \text{ and } (3, 1).$$

The upper bound given in Lemma 4.11 gives

$$\rho(S) \leq \frac{(t+1)\prod_{i=1}^{t} \mathsf{m}_i - \sum_{i=1}^{t} \prod_{\substack{j=1 \\ j \neq i}}^{t} \mathsf{m}_j}{\min_{\mathbf{x} \in S \setminus \{1\}} \sum_{i=1}^{t} \prod_{\substack{j=1 \\ j \neq i}}^{t} \mathsf{m}_j x_i} \leq 29/7.$$

Using the more basic fact from the proof, we have that

$$\rho(S) \leq \frac{\max\{\varphi(\mathbf{x}) : \mathbf{x} \in \mathcal{A}(S)\}}{\min\{\varphi(\mathbf{x}) : \mathbf{x} \in \mathcal{A}(S)\}} = 21/7 = 3.$$

However, with a bit of computation [10, 14] we find that $\rho(S) = 35/12$, less than either of these approximations.

4.1 Complement-Finite Ideals Whose Minimal Elements Lie on Extremal Rays

Observe that if S is a complement-finite ideal of $\mathbb{N}_0^t$, say with $\mathsf{C}(S) = S$ for simplification, then there are $m_1, \ldots, m_t \in \mathbb{N}_{\geq 2}$ such that $m_i\mathbf{e}_1, \ldots, m_t\mathbf{e}_t$ are the minimal nonzero elements of S. In this section we study semigroups S so that these are the *only* minimal nonzero elements of S; that is, all minimal elements lie on the extremal rays of the affine monoid. This work naturally generalizes what happens in an additive numerical semigroup $S = \langle m, m_1, \ldots, 2m - 1\rangle$ generated by a maximal length interval and give some credence to the forthcoming Lemma 4.16. By the classification of atoms of $\mathsf{C}(S)$ and by Lemma 4.2, we see that the irreducible elements are those elements $\alpha \in S$ such that $\alpha \succcurlyeq \beta$ for some minimal element β of S but yet $\alpha \not\succcurlyeq \beta + \gamma$ for any two minimal elements β and γ of S. Consequently, we have the following.

Proposition 4.14 *Let S be a complement-finite ideal of $\mathbb{N}_0^t$ (with $\mathsf{C}(S) = S$) so that the only minimal nonzero elements of S are $m_1\mathbf{e}_1, \ldots, m_t\mathbf{e}_t$ for some $m_1, \ldots, m_t \in \mathbb{N}_{\geq 2}$. Then the atoms of S are*

$$\mathcal{A}(S) = \{(a_1, \ldots, a_t) \colon m_i \leq a_i < 2m_i \text{ for some } i \text{ and } 0 \leq a_j < m_j \text{ for all } j \neq i\}.$$

In this special situation, we can describe precisely the length sets of each element in S. To do so, note that for each $i \in [1, t]$, $(0, \ldots, 0, x_i, 0, \ldots, 0) \in S$ if and only if $x_i \in \langle m_i, \ldots, 2m_i - 1\rangle$, an additive numerical semigroup generated by the full interval $[m_i, 2m_i - 1]$. Moreover, $\{(0, \ldots, 0, x_i, 0, \ldots, 0) \colon x_i = 0 \text{ or } x_i \geq m_i\}$ is a divisor-closed subsemigroup of S, isomorphic to $\langle m_i, \ldots, 2m_i - 1\rangle$. Thus the results of [2] can be used to study some sets of lengths in S.

Proposition 4.15 *Let S be a complement-finite ideal of $F = \mathcal{F}(P)$ as in Lemma 2.1 so that $q_1^{\mathsf{m}_1}, \ldots, q_t^{\mathsf{m}_t}$ are the only minimal elements of $S \setminus \{1\}$. Then every length set is an interval (so that $\Delta(S) = \{1\}$) and*

$$\rho(S) = t + 1 - \sum_{i=1}^{t} \frac{1}{\mathsf{m}_i}.$$

Proof We first show that for all $s \in S$, $\mathsf{L}(s)$ is an interval. For each $k \in [1, t]$ let

$$I_k = \{q_1^{n_1} \cdots q_t^{n_t} : \mathsf{m}_k \leq n_k \leq 2\mathsf{m}_k - 1, 0 \leq n_i \leq \mathsf{m}_i - 1 \,\forall i \neq k\}.$$

Then $\mathcal{A}(S) = \coprod_{k=1}^{t} I_k$. Moreover, for any $s \in S$,

$$\mathsf{L}(s) = \left\{\sum_{k=1}^{t} a_k : s \in \prod_{k=1}^{t} I_k^{a_k}\right\}.$$

Let $s = q_1^{x_1} \cdots q_t^{x_t}$ with $l, l' \in \mathsf{L}(s)$ so that $l' \geq l + 2$. Then $l = \sum_{k=1}^{t} a_k$ and $l' = \sum_{k=1}^{t} b_k$ where $s \in \left(\prod_{k=1}^{t} I_k^{a_k}\right) \cap \left(\prod_{k=1}^{t} I_k^{b_k}\right)$. Since $\sum_{k=1}^{t} b_k \leq \sum_{k=1}^{t} a_k + 2$, without loss of generality $b_1 \geq a_1 + 1$. Since $s \in \prod_{k=1}^{t} I_k^{a_k}$,

$$a_i \mathsf{m}_i \leq x_i \leq a_i(2\mathsf{m}_i - 1) + \sum_{\substack{k=1 \\ k \neq i}}^{t} a_k(\mathsf{m}_i - 1) = \sum_{k=1}^{t} a_k(\mathsf{m}_i - 1) + a_i \mathsf{m}_i .$$

Similarly, since $s \in \prod_{k=1}^{t} I_k^{b_k}$,

$$b_i \mathsf{m}_i \leq x_i \leq b_i(2\mathsf{m}_i - 1) + \sum_{\substack{k=1 \\ k \neq i}}^{t} b_k(\mathsf{m}_i - 1) = \sum_{k=1}^{t} b_k(\mathsf{m}_i - 1) + b_i \mathsf{m}_i .$$

Because $a_1 + 1 \leq b_1$, replacing a_1 with $a_1 + 1$ we have that

$$(a_1 + 1)\mathsf{m}_1 \leq b_1 \mathsf{m}_1 \leq x_1 \leq \sum_{k=1}^{t} a_k(\mathsf{m}_1 - 1) + a_1 \mathsf{m}_1 + (2\mathsf{m}_1 - 1) .$$

Consequently, $s \in I_1^{a_1+1} \prod_{k=2}^{t} I_k^{a_k}$ and so $l + 1 = (a_1 + 1) + \sum_{k=2}^{t} a_k \in \mathsf{L}(s)$. Now, since whenever $l, l' \in \mathsf{L}(s)$ with $l' \geq l + 2$, $l + 1 \in \mathsf{L}(s)$, $\mathsf{L}(s)$ is an interval.

Lemma 4.11 immediately gives the upper bound of

$$\rho(S) \leq \frac{(t+1)\prod_{i=1}^{t} \mathsf{m}_i - \sum_{i=1}^{t} \prod_{\substack{j=1 \\ j \neq i}}^{t} \mathsf{m}_j}{\prod_{i=1}^{t} \mathsf{m}_i} = t + 1 - \sum_{i=1}^{t} \frac{1}{\mathsf{m}_i} .$$

Equality follows as we can obtain the upper bound through the following two factorizations of an element:

$$\left(q_1^{\mathsf{m}_1 - 1} \cdots q_{t-1}^{\mathsf{m}_{t-1} - 1} \cdot q_t^{2\mathsf{m}_t - 1}\right)^{\prod_{i=1}^{t} \mathsf{m}_i}$$
$$= \prod_{i=1}^{t-1} \left(q_i^{\mathsf{m}_i}\right)^{(\mathsf{m}_i - 1)\prod_{\substack{j=1 \\ j \neq i}}^{t-1} \mathsf{m}_j} \left(q_t^{\mathsf{m}_t}\right)^{(2\mathsf{m}_t - 1)\prod_{i=1}^{t-1} \mathsf{m}_i} .$$

□

Now, based on Theorem 4.4, Proposition 4.5, Example 4.6, and Proposition 4.15, and many computations using [14], we make the following conjecture.

Conjecture 4.16 *Let S be a complement-finite ideal of a free monoid F. For every $x \in S$, $\mathsf{L}(x)$ is an interval.*

5 Sequences Over a Finite Abelian Group That Are Not Zero-sum Free

In this final section we return to the monoid $\mathcal{F}_{\mathcal{B}}(G)$ as introduced in Example 1.1(3). After examining more closely how $\mathcal{F}_{\mathcal{B}}(G)$ sits between $\mathcal{B}(G)$ and $\mathcal{F}(G)$, we introduce a new constant related to the Davenport constant of the finite abelian group G. We then restate general results about the arithmetic of complement-finite ideals in the context of sequences that are not zero-sum free, and use the additional structure of $\mathcal{F}_{\mathcal{B}}(G)$ to obtain further arithmetical results. We begin by recalling the definition of $\mathcal{F}_{\mathcal{B}}(G)$.

Let G be an additive finite abelian group with identity 0, and let $\mathcal{F}(G)$ denote the free abelian monoid with basis G, that is, the set of all formal products of elements in G without regard to order. The classical *block monoid* $\mathcal{B}(G)$ of zero-sum sequences is the submonoid of $\mathcal{F}(G)$ consisting of all sequences $g_1 \cdot \ldots \cdot g_t \in \mathcal{F}(G)$ such that $g_1 + \cdots + g_t = 0$ in G. Here we consider a similarly defined monoid $\mathcal{F}_{\mathcal{B}}(G)$ consisting of all sequences in $\mathcal{F}(G)$ that are not zero-sum free; that is,

$$\mathcal{F}_{\mathcal{B}}(G) = \left\{ g_1 \cdots g_t \in \mathcal{F}(G) \colon \exists\, I \subseteq [1,t] \text{ with } \sum_{i \in I} g_I = 0 \right\} \subseteq \mathcal{F}(G).$$

It is clear that

$$\mathcal{B}(G) \subseteq \mathcal{F}_{\mathcal{B}}(G) = \mathcal{F}(G)\,(\mathcal{B}(G) \setminus \{\{\}\}) \cup \{\} \subseteq \mathcal{F}(G)$$

and that $|\mathcal{F}(G) \setminus \mathcal{F}_{\mathcal{B}}(G)| < \infty$. Consequently, $\mathcal{F}_{\mathcal{B}}(G)$ is a complement-finite ideal of $\mathcal{F}(G)$. Proposition 5.1 sheds more light on how $\mathcal{F}_{\mathcal{B}}(G)$ sits between $\mathcal{B}(G)$ and $\mathcal{F}(G)$. However, we first define, as with $\mathcal{B}(G)$, natural submonoids $\mathcal{F}_{\mathcal{B}}(G_0)$ of $\mathcal{F}_{\mathcal{B}}(G)$ for any subset G_0 of G.

Let G_0 be a nonempty subset of G. Then

$$\mathcal{F}_{\mathcal{B}}(G_0) = \{g_1 \cdots g_t \in \mathcal{F}_{\mathcal{B}}(G) \colon g_1, \ldots, g_t \in G_0\}.$$

Observe that $\mathsf{C}(\mathcal{F}_{\mathcal{B}}(G)) = \mathcal{F}_{\mathcal{B}}(G \setminus \{0\})$ and that each $\mathcal{F}_{\mathcal{B}}(G_0)$ is a divisor-closed submonoid of $\mathcal{F}_{\mathcal{B}}(G)$. In fact, we see in Proposition 5.1 (4) that the converse holds. Also observe that $\mathcal{F}_{\mathcal{B}}(G_0)$ is a complement-finite ideal of the free abelian monoid $\mathcal{F}(G_0)$. Before stating the next result, recall that if $A = g_1 \cdots g_t \in \mathcal{F}(G)$, $\sigma(A) = g_1 + \cdots + g_t \in G$ and that $v_g(A) = |\{i \in [1,t] \colon g_i = g\}|$.

Proposition 5.1 *Let G be an additive finite abelian group.*

(1) *If $|G| = 1$, then $\mathcal{B}(G) = \mathcal{F}_{\mathcal{B}}(G) = \mathcal{F}(G)$.*
(2) *If $|G| \geq 2$, then $\mathcal{B}(G)$ is not divisor-closed in $\mathcal{F}_{\mathcal{B}}(G)$, but is saturated.*
(3) *$\mathsf{q}(\mathcal{F}_{\mathcal{B}}(G)) = \mathsf{q}(\mathcal{F}(G))$ and $\mathcal{C}(\mathcal{B}(G), \mathcal{F}_{\mathcal{B}}(G)) \cong G$.*
(4) *If $S \subseteq \mathcal{F}_{\mathcal{B}}(G)$ is divisor-closed, then $S = \mathcal{F}_{\mathcal{B}}(G_0)$ for some subset $G_0 \subseteq G$.*

Proof The proof of (1) is clear since if $G = \{0\}$, $\mathcal{F}(G) = \{0^n : n \in \mathbb{N}_0\}$.

If $|G| \geq 2$, then there is an element $g \in G$ with order $n \geq 2$. Then $g^{2n-1} \cdot g^{n+1} = g^{3n}$ in $\mathcal{F}_{\mathcal{B}}(G)$. However, neither g^{2n-1} nor g^{n+1} are in $\mathcal{B}(G)$, and so $\mathcal{B}(G) \subseteq \mathcal{F}_{\mathcal{B}}(G)$ is not divisor closed. However, if $A, B \in \mathcal{B}(G)$ with $A|_{\mathcal{F}_{\mathcal{B}}(G)} B$, then $B = AC$ in $\mathcal{F}(G)$. Since $\sigma(B) = \sigma(A) + \sigma(C) = 0$ and $\sigma(B) = \sigma(A) = 0$, $\sigma(C) = 0$ as well and $C \in \mathcal{B}(G)$. Thus $\mathcal{B}(G) \subseteq \mathcal{F}_{\mathcal{B}}(G)$ is saturated. This proves (2).

That $\mathsf{q}(\mathcal{F}_{\mathcal{B}}(G)) = \mathsf{q}(\mathcal{F}(G))$ follows from the first part of the proof of Proposition 3.7. Now $\mathcal{C}(\mathcal{B}(G), \mathcal{F}_{\mathcal{B}}(G)) \cong \mathcal{C}(\mathcal{B}(G), \mathcal{F}(G)) \cong G$.

Finally, we prove (4). Let S be a divisor-closed submonoid of $\mathcal{F}_{\mathcal{B}}(G)$, and let G_0 denote the set of all $g \in G$ with $v_g(B) > 0$ for some $B \in S$. Then clearly $S \subseteq \mathcal{F}_{\mathcal{B}}(G_0)$. To prove equality, note that if $B = g_1 \cdots g_t \in \mathcal{F}_{\mathcal{B}}(G_0)$, then $g_i \in G_0$ for each i and thus there are $B_i \in S$ with $v_{g_i}(B_i) > 0$. But then $B_1 \cdots B_t \in S$ and $B \mid_{\mathcal{F}_{\mathcal{B}}(G)} B_1 \cdots B_t$. As S is divisor-closed in $\mathcal{F}_{\mathcal{B}}(G)$, $B \in S$ as well. □

The next proposition is a restatement of Proposition 4.1 in the context of $\mathcal{F}_{\mathcal{B}}(G)$.

Proposition 5.2 *Let G be a finite abelian group. The atoms of $\mathcal{F}_{\mathcal{B}}(G)$ are:*

(I) 0,
(II) $0g_1 \cdots g_t$ *where* $g_1 \cdots g_t \in \mathcal{F}(G) \setminus \mathcal{F}_{\mathcal{B}}(G)$, *and*
(III) $\mathcal{A}(\mathcal{B}(G))\mathcal{F}(G) \setminus \mathcal{A}(\mathcal{B}(G))^2\mathcal{F}(G)$.

That is, the atoms of $\mathcal{F}_{\mathcal{B}}(G)$ are precisely the sequences in $\mathcal{F}(G)$ containing some subsequence in $\mathcal{B}(G)$ but not any two disjoint subsequences in $\mathcal{B}(G)$.

We now provide an example to illustrate how many more atoms $\mathcal{F}_{\mathcal{B}}(G)$ has than $\mathcal{B}(G)$. Like the monoid in Proposition 4.6 this is a modification of a standard example in zero-sum theory.

Example 5.3 Let G be a finite abelian group and let $e_1, \ldots, e_r$ be a set of independent elements of G, each of order n. Set $e_0 = -(e_1 + \cdots e_r)$ and $G_0 = \{e_0, e_1, \ldots, e_r\}$. We now consider $\mathcal{B}(G_0)$ and $\mathcal{F}_{\mathcal{B}}(G_0)$, submonoids of $\mathcal{F}(G_0)$. The monoid $\mathcal{B}(G_0)$ is considered in [15, Proposition 4.1.2]. There it is shown that there are $r+1$ atoms: $U_i = e_i^n$ for each $i \in [1, r]$ and $W = e_0 e_1 \cdots e_r$. It is also computed that if

$$A = \prod_{i=0}^{r} e_i^{nk_i+s}$$

with $s \in [0, n-1]$, $k_0, \ldots, k_r \in \mathbb{N}_0$, and $k^* = \min\{k_0, \ldots, k_r\}$, then

$$\mathsf{L}_{\mathcal{B}(G_0)} = \left\{s + k_0 + \cdots + k_r + \nu(n - r - 1) \colon \nu \in [0, k^*]\right\}.$$

However,

$$A = \prod_{i=0}^{r} e_i^{nk_i+s}$$

is in $\mathcal{F}_{\mathcal{B}}(G_0)$ if and only if either $k_i > 0$ for some i or $k_i = 0$ for all i and $s_i > 0$ for all i. Thus A is an atom of $\mathcal{F}_{\mathcal{B}}(G_0)$ precisely when either $k_i = 1$ for some i and $k_j = 0$ when $j \neq i$ or $k_i = 0$ for all i and $s_i \in \{1, 2\}$ for all i with $s_j = 1$ for some j. Because there are many more atoms in $\mathcal{F}_{\mathcal{B}}(G_0)$ than in $\mathcal{B}(G_0)$, length sets are also larger. In particular, $\mathsf{L}(A)$ is again an interval.

Before discussing the arithmetic of $\mathcal{F}_{\mathcal{B}}(G)$, we introduce a new combinatorial invariant. Recall that if G is a finite abelian group, the *Davenport constant* $\mathsf{D}(G)$ is the length of the longest atom (measured as the length of a sequence in $\mathcal{F}(G)$) in $\mathcal{B}(G)$. We now define the constant $\mathsf{D}_{\mathcal{B}}(G)$ to be the length of the longest atom in $\mathcal{F}_{\mathcal{B}}(G)$. We now give a few minor results about $\mathsf{D}_{\mathcal{B}}(G)$ and propose a further study of this constant.

Proposition 5.4 *Let G be a finite abelian group with $|G| \geq 2$.*

(1) $\mathsf{D}(G) + 1 \leq \mathsf{D}_{\mathcal{B}}(G) \leq 2\mathsf{D}(G) - 1$.
(2) *If G is cyclic of order n, then* $\mathsf{D}_{\mathcal{B}}(G) = 2\mathsf{D}(G) - 1 = 2n - 1$.

Proof Let $B = g_1 \cdots g_{\mathsf{D}_{\mathcal{B}}(G)} \in \mathcal{A}(\mathcal{F}_{\mathcal{B}}(G))$. If $\mathsf{D}_{\mathcal{B}}(G) \geq 2\mathsf{D}(G)$, then $g_1 \cdots g_{\mathsf{D}(G)}, g_{\mathsf{D}(G)+1} \cdots g_{\mathsf{D}_{\mathcal{B}}(G)} \in \mathcal{B}(G) \subseteq \mathcal{F}_{\mathcal{B}}(G)$, contradicting the fact that B is irreducible. Thus $\mathsf{D}_{\mathcal{B}}(G) \leq 2\mathsf{D}(G) - 1$. Since $|G| \geq 2$, there is $g \in G \setminus \{0\}$. Suppose that $g_1 \cdots g_t$ is an atom of $\mathcal{B}(G)$ with no $g_i = 0$, so that $t \geq 2$. If $t = 2$, then $g_2 = -g_1$. In this case, the only possible ways to nontrivially factor $g_1 g_2 g$ in $\mathcal{F}(G)$ are as $(g_1 g_2) g$, $(g_1 g) g_2$, and $(g_2 g) g$, neither of which is a factorization in $\mathcal{F}_{\mathcal{B}}(G)$ since $g_1, g_2, g \neq 0$. Suppose now that $t \geq 3$. We claim that for any $g \in G \setminus \{0\}$, $g_1 \cdots g_t \cdot g$ is an atom of $\mathcal{F}_{\mathcal{B}}(G)$. Suppose this is not the case. Then $g_1 \cdots g_t \cdot g$ factors in $\mathcal{F}_{\mathcal{B}}(G)$ and so by the description of atoms of type (III) in Proposition 5.2 there are positive integers k and l with $1 < k < l \leq t$ so that after reordering the terms as necessary, either $g_1 \cdots g_k$ and $g_{k+1} \cdots g_l$ are atoms of $\mathcal{B}(G)$ or $g_1 \cdots g_k$ and $g_{k+1} \cdots g_l \cdot g$ are atoms of $\mathcal{B}(G)$. The former is impossible since $g_1 \cdots g_t$ is an atom of $\mathcal{B}(G)$ and is divisible (in $\mathcal{B}(G)$) by the product $g_1 \cdots g_k \cdot g_{k+1} \cdots g_l$. In the later case, since $\sigma(g_1 \cdots g_k) = \sigma(g_{k+1} \cdots g_l \cdot g) = 0$ and $\sigma(g_1 \cdots g_t \cdot g) = g$, $\sigma(g_{l+1} \cdots g_t) = g$ and $\sigma(g_{k+1} \cdots g_l) = -g$. Then $g_{k+1} \cdots g_t \in \mathcal{B}(G)$. Now $g_1 \cdots g_k, g_{k+1} \cdots g_t \in \mathcal{B}(G)$, contradicting the fact that $g_1 \cdots g_t$ is an atom of $\mathcal{B}(G)$. Taking $t = \mathsf{D}(G)$ gives the lower bound in (1).

From (1), $\mathsf{D}_{\mathcal{B}}(G) \leq 2n - 1$ if G is cyclic of order n. If G is generated by g, then g^{2n-1} is an atom of $\mathcal{F}_{\mathcal{B}}(G)$, giving the equality in (2). □

In the following theorem we summarize results from previous sections, but in the context of $\mathcal{F}_{\mathcal{B}}(G)$. In addition, we give more refined results about elasticities, results about unions of sets of lengths and the delta set, analogous to results about $\mathcal{B}(G)$. Perhaps unsurprising, many of these arguments are similar to the standard arguments used in the case of $\mathcal{B}(G)$. We note that (1) and (2) would be trivially obtained should Conjecture 4.16 be true.

Theorem 5.5 *Let G be an additive finite abelian group. The following arithmetical finiteness conditions hold for $\mathcal{F}_{\mathcal{B}}(G)$.*

(1) *There is $M \in \mathbb{N}$ such that every $L \in \mathcal{L}(\mathcal{F}_{\mathcal{B}}(G))$ is an AAP with distance 1 and with bound M.*
(2) *For every $k \in \mathbb{N}$, $\mathcal{U}_k(\mathcal{F}_{\mathcal{B}}(G))$ is an interval.*
(3) *$\rho(\mathcal{F}_{\mathcal{B}}(G)) = \frac{\mathsf{D}_{\mathcal{B}}(G)}{2}$ and, for each $k \in \mathbb{N}$,*

$$\rho_{2k}(\mathcal{F}_{\mathcal{B}}(G)) = k\mathsf{D}_{\mathcal{B}}(G) \text{ and } 1 + k\mathsf{D}_{\mathcal{B}}(G) \le \rho_{2k+1}(\mathcal{F}_{\mathcal{B}}(G)) \le k\mathsf{D}_{\mathcal{B}}(G) + \lfloor \frac{\mathsf{D}_{\mathcal{B}}(G)}{2} \rfloor.$$

(4) *$\mathcal{F}_{\mathcal{B}}(G)$ has full elasticity.*
(5) *$\omega(\mathcal{F}_{\mathcal{B}}(G)) \in \{\mathsf{D}_{\mathcal{B}}(G), \mathsf{D}_{\mathcal{B}}(G) + 1\}$.*
(6) *$\mathsf{c}(\mathcal{F}_{\mathcal{B}}(G)) \le \mathsf{D}_{\mathcal{B}}(G) + 1$.*
(7) *$\mathsf{D}_{\mathcal{B}}(G) \le \mathsf{t}(\mathcal{F}_{\mathcal{B}}(G)) \le (\mathsf{D}_{\mathcal{B}}(G) + 1)^2$.*

Proof The statements (1) and (4)–(7) follow from Theorem 4.4, Proposition 4.9, and the fact that $\mathsf{M}(\mathcal{F}_{\mathcal{B}}(G)) = \mathsf{D}_{\mathcal{B}}(G)$. We now prove (2) and (3).

Observe that if $[k, \rho_k(\mathcal{F}_{\mathcal{B}}(G))] \subseteq \mathcal{U}_k(\mathcal{F}_{\mathcal{B}}(G))$ and $l \in [\min \mathcal{U}_k(\mathcal{F}_{\mathcal{B}}(G)), k]$, then $l \le k \le \rho_l(\mathcal{F}_{\mathcal{B}}(G))$, and so $k \in \mathcal{U}_l(\mathcal{F}_{\mathcal{B}}(G))$. Therefore $l \in \mathcal{U}_k(\mathcal{F}_{\mathcal{B}}(G))$ and we need only prove that $[k, \rho_k(\mathcal{F}_{\mathcal{B}}(G))] \subseteq \mathcal{U}_k(\mathcal{F}_{\mathcal{B}}(G))$.

Let l be the smallest element of $[k, \rho_k(\mathcal{F}_{\mathcal{B}}(G))]$ with $[l, \rho_k(\mathcal{F}_{\mathcal{B}}(G))]$ and assume, for the sake of contradiction, that $l > k$. Choose $B \in \mathcal{F}_{\mathcal{B}}(G)$ with $|B|$ (the length of B in $\mathcal{F}(G)$) minimal such that $\{j, k\} \subseteq \mathsf{L}(B)$ for some $j \ge l$. Since 0 is the only absolutely irreducible element in $\mathcal{F}_{\mathcal{B}}(G)$, $B \ne 0^k$, and we can write $B = U_1 \cdots U_k = V_1 \cdots V_j$ for atoms $U_1, \ldots, U_k, V_1, \ldots, V_j$. After reordering the terms as needed, $U_k = g_1 g_2 U'$ and $V_{j-1} V_j = g_1 g_2 V'$ in $\mathcal{F}(G)$, possibly with $U' = \emptyset$ and with $g_1, g_2 \in G$ not necessarily distinct. Observe that $U_k' = (g_1 + g_2)U'$ is also an atom of $\mathcal{F}_{\mathcal{B}}(G)$. Set $V'' = (g_1 + g_2)V'$ and factor V'' as a product of atoms in $\mathcal{F}_{\mathcal{B}}(G)$ as $V'' = W_1 \cdots W_t$. Now set $B' = U_1 \cdots U_{k-1} U_k'$ and observe that $|B'| = |B| - 1 < |B|$. Moreover, $B' = V_1 \cdots V_{j-2} W_1 \cdots W_t$. Because of the minimality condition on $|B|$, we know that $j - 2 + t < l$. This can only happen if $t = 1$ and $j = l$, whence $l - 1 \in \mathcal{U}_k(\mathcal{F}_{\mathcal{B}}(G))$, contradicting the definition of l. Thus $[k \rho_k(\mathcal{F}_{\mathcal{B}}(G))] \subseteq \mathcal{U}_k(\mathcal{F}_{\mathcal{B}}(G))$ proving (2).

Since $\mathsf{C}(\mathcal{F}_{\mathcal{B}}(G)) = \mathcal{F}_{\mathcal{B}}(G \setminus \{0\})$, by Lemma 4.7 we know that $\rho(\mathcal{F}_{\mathcal{B}}(G)) \le \rho(\mathcal{F}_{\mathcal{B}}(G \setminus \{0\}))$. Suppose $A_1 \cdots A_m = B_1 \cdots B_n$ are two factorizations of an element C in $\mathcal{F}_{\mathcal{B}}(G \setminus \{0\})$ with $n \le m$. Write $C = g_1 \cdots g_t$ in $\mathcal{F}(G)$. In $\mathcal{F}(G)$, $|A_i| \ge 2$ for each i and $|B_j| \le \mathsf{D}_{\mathcal{B}}(G)$ for each j. Thus $2m \le t \le n\mathsf{D}_{\mathcal{B}}(G)$. Now

$$\frac{m}{n} \le \frac{t/2}{t/\mathsf{D}_{\mathcal{B}}(G)} = \frac{\mathsf{D}_{\mathcal{B}}(G)}{2}.$$

With $U = g_1 \cdots g_{\mathsf{D}_{\mathcal{B}}(G)}$ an atom in $\mathcal{F}_{\mathcal{B}}(G)$ of maximal length, note that $-U = (-g_1) \cdots (-g_{\mathsf{D}_{\mathcal{B}}(G)})$ is an atom in $\mathcal{F}_{\mathcal{B}}(G)$ as well. Set, for each $i \in [1, \mathsf{D}_{\mathcal{B}}(G)]$,

$V_i = (g_i)(-g_i) \in \mathcal{F}_{\mathcal{B}}(G)$ $\mathsf{D}_{\mathcal{B}}(G)$ additional atoms. The equality $U(-U) = V_1 \cdots V_{\mathsf{D}_{\mathcal{B}}(G)}$ illustrates that $\rho(\mathcal{F}_{\mathcal{B}}(G)) = \frac{\mathsf{D}_{\mathcal{B}}(G)}{2}$.

Because S is an atomic monoid, by Geroldinger and Halter-Koch [15, Proposition 1.4.2] we now have that

$$\rho_k(\mathcal{F}_{\mathcal{B}}(G)) \leq k\rho(\mathcal{F}_{\mathcal{B}}(G)) \leq k\frac{\mathsf{D}_{\mathcal{B}}(G)}{2}.$$

With $U = g_1 \cdots g_{\mathsf{D}_{\mathcal{B}}(G)}$ an atom of maximal length, $U^k(-U)^k = (g_1(-g_1))^k \cdots (g_{\mathsf{D}_{\mathcal{B}}(G)}(-g_{\mathsf{D}_{\mathcal{B}}(G)}))^k$ shows that $k\mathsf{D}_{\mathcal{B}}(G) \leq \rho_{2k}(\mathcal{F}_{\mathcal{B}}(G))$. Therefore $\rho_{2k}(\mathcal{F}_{\mathcal{B}}(G)) = k\mathsf{D}_{\mathcal{B}}(G)$. Finally,

$$1 + k\mathsf{D}_{\mathcal{B}}(G) = \rho_1(\mathcal{F}_{\mathcal{B}}(G)) + \rho_{2k}(\mathcal{F}_{\mathcal{B}}(G)) \leq \rho_{2k+1}(\mathcal{F}_{\mathcal{B}}(G)) \leq \frac{(2k+1)\mathsf{D}_{\mathcal{B}}(G)}{2}.$$

This proves (3). □

Remark 5.6 Recall that for a finite nonabelian group G, $\mathcal{B}(G)$ denotes the set of *product-one sequences*; that is, $\mathcal{B}(G)$ is the submonoid of the free abelian monoid $\mathcal{F}(G)$ consisting of sequences (formal products) $g_1 \cdots g_t$ so that there is some permutation σ of $[1, t]$ so that in G (the actual product) $g_{\sigma(1)} \cdots g_{\sigma(t)} = 1$. See, for example, [21], for more on product-one sequences. As we defined $\mathcal{F}_{\mathcal{B}}(G)$ as the submonoid of $\mathcal{F}(G)$ consisting of sequences that are not zero-sum free whenever G is a finite abelian group, we can define, for any finite nonabelian group G, $\mathcal{F}_{\mathcal{B}}(G)$, the monoid of sequences that are not product-one free. It is easy to see that such monoids are always complement-finite ideals of $\mathcal{F}(G)$ and hence (1) and (4)–(7) of Theorem 5.5 hold. Using [21, Lemma 5.1], modifications of the proofs of [21, Theorem 5.5, Proposition 5.6] can be made to show that (2) and (3) of Theorem 5.5 also hold for the monoid $\mathcal{F}_{\mathcal{B}}(G)$ of sequences that are not product-one free.

Appendix

Nick Baeth (1978–2021) by Scott Chapman and James B. Coykendall

After a brief battle with pancreatic cancer, Nick Baeth died on December 11, 2021. Nick completed his Ph.D. degree at the University of Nebraska writing a dissertation in module theory under the direction of Roger Wiegand. During his 16-year academic career, he spent 13 years at Central Missouri State University in Warrensburg, Missouri, and slightly more than 3 years at Franklin and Marshall College in Lancaster, Pennsylvania.

Nick's work in mathematics made him highly visible both nationally and internationally. He published in a number of subfields of commutative algebra including factorization theory, general monoid theory, and module theory. His research papers have been highly cited; since his 2005 Ph.D. thesis, MathSciNet records that his 31 publications have been cited 181 times by 100 authors (Google Scholar credits him with 372 citations of which 238 are since 2016). Additionally, he was a highly sought colleague and collaborator—in a field where collaboration is often difficult, he wrote papers with 32 distinct individuals.

Nick's recent work was heavily influenced by his 2013 NAWI-Graz Fulbright Visiting Professorship at Karl Franzens Universität—Graz, Austria. His work with the research group in Graz gave rise to several very influential and pioneering papers in factorization theory. Nick's contribution to the current volume, "Complement-finite ideals" is a great example of Nick's later work and the kind of mathematics that he truly loved.

Nick is survived by his wife Katherine and son George. He is already greatly missed by his numerous friends and colleagues.

References

1. Anderson, D. D.; Anderson, D. F.; and Zafrullah, M., *Factorizations in integral domains*, J. Pure Appl. Algebra **69**, (1990), 1–19.
2. Amos, J.; Chapman, S.T.; Hine, N.; Paixão, J., *Sets of lengths do not characterize numerical monoids*, Integers, **7**, #A50 (2007), 8pp.
3. Anderson, D.D; Anderson, D.F., *Elasticity of factorization in integral domains*, J. Pure Appl. Algebra **80** (1992), 217–235.
4. Baeth, N. R.; Enlow, M., *Multiplicative factorization in numerical semigroups*, Internat. J. Algebra Comput. **30**, no. 1 (2020) 419–430.
5. Chapman, S. T. *On the Davenport constant, the cross number, and their application in factorization theory*, Lecture Notes in Pure and Applied Mathematics (1995) 167–167.
6. Chapman, S. T.; Corrales, M.; Miller, A.; Miller, C.; Phatel D., *The catenary degrees of elements in numerical monoids generated by arithmetic sequences,* Comm. Algebra. **45** no. 12, (2017), 5443–5452.
7. Chapman, S. T.; García-Sánchez, P.A.; Tripp, Z.; Viola, C., *Measuring primality in numerical semigroups with embedding dimension three*, J. Algebra Appl. **15** no. 1, (2016), 16 pp.
8. Cisto, C.; Delgado, M.; García-Sánchez, P. A., *Algorithms for generalized numerical semigroups*, J. Algebra Appl. **20** no. 5, (2021), 24 pp.
9. Cisto, C.; Failla, G.; Utano, R., *On the generators of a generalized numerical semigroup*, Analele Univ. "Ovidius", **27** no. 1, (2019), 49–59.
10. Delgado, M.; García-Sánchez, P; Morais, J., *NumericalSgps, A package for numerical semigroups*, Version 1.1.0 (2017) (GAP package) https://gap-packages.github.io/numericalsgps/
11. Gao, W.; Geroldinger, A., *Zero-sum problems in finite abelian groups: A survey*, Expositiones Mathematicae, **24**, no. 4, (2006), 337–369.
12. Geroldinger, A.; Zhong, Q., *Factorization theory in commutative monoids*, Semigroup Forum, **100**, no. 1, (2020), 22–51.
13. Rosales, J. C.; García-Sánchez, P. A., *Numerical Semigroups*, Developments in Mathematics, **20**, Springer, New York, 2009.
14. The GAP Group, *GAP—Groups, Algorithms, and Programming, Version 4.11.1*; 2021, https://www.gap-system.org.
15. Geroldinger, A. and Halter-Koch, F., *Non-unique factorizations*, Pure and Applied Mathematics (Boca Raton), **278**, Chapman & Hall/CRC, Boca Raton, FL, 2006.

16. Gillam, D.W.H; Hall, T.E; Williams, N.H., *On finite semigroups and idempotents*, Bull. London Math. Soc. **4** (1972) 143-144.
17. Gotti, F., *Geometric and combinatorial aspects of submonoids of a finite-rank free commutative monoids*, Linear Alg. and Appl, **604**, (2020), 146–186.
18. Kravitz, N.; Sah, A., *A stronger connection between the Erdős-Burgess and Davenport constants*, preprint, arXiv:1808.06031.
19. Jenssen, M.; Montealegre, D.; Ponomarenko, V., *Irreducible factorization lengths and the elasticity problem within* $\mathbb{N}$, Amer. Math. Monthly. **120** no. 4, (2013), 322–328.
20. Kiers, C.; O'Neill, C.; Ponomarenko, V., *Numerical semigroups on compound sequences*, Comm. Algebra **44** no. 9, (2016), 3842–3852.
21. Oh, J. S., *On the algebraic and arithmetic structure of the monoid of product-one sequences*, J. Commut. Algebra **12**, no. 3, (2020), 409–433.
22. Grillet, P. A., *Commutative Semigroups*, Advances in Mathematics, Kluwer Academic Publishers, Dordrecht, The Netherlands, 2001.
23. Schmid, W., *Arithmetic of block monoids*, Mathematica Slovaca **54**, no. 5, (2004), 503–526.
24. Wang, G., *Erdős-Burgess constant of the direct product of cyclic semigroups*, preprint, arXiv:1802.08791.
25. Zhong, Q., *On elasticities of locally finitely generated monoids*, J. Algebra, **534**, (2019), 145–167.

When Is a Group Algebra Antimatter?

Mohamed Benelmekki and Said El Baghdadi

1 Introduction

Let D be an integral domain. By an irreducible element or atom of D, we mean a nonzero nonunit $x \in D$ such that $x = ab, a, b \in D$, implies a or b is a unit. The domain D is atomic if each nonzero nonunit element can be written as a finite product of atoms. A domain with no atoms is called an antimatter domain. The class of antimatter domains was introduced by Coykendall, Dobbs, and Mullins in [3]. The study of antimatter domains is very useful in factorization theory in non-atomic setting.

Let D be an integral domain, and let S be a commutative cancellative torsion-free monoid, written additively. Denote by $\langle S \rangle$ the quotient group of S and $(S, <)$ a total order on S. The monoid domain of S over D is defined by $D[S] = \{\sum_i a_i X^{r_i} | a_i \in D \text{ and } r_i \in S\}$. Note that $D[S]$ is an integral domain and each nonzero element $f \in D[S]$ has a unique representation in the form

$$f = a_1 X^{r_1} + \cdots + a_n X^{r_n},$$

where $n \in \mathbb{Z}_+, 0 \neq a_i \in D$ and $r_i \in S$ $(i = 1, \ldots, n)$ such that $r_1 < r_2 < \cdots < r_n$. The subset $\text{Supp}(f) = \{r_1, \ldots, r_n\}$ of S is called the support of f. Notice that the units of $D[S]$ are the monomials aX^s, where $a \in D$ and $s \in S$ are both units. In the case where S is a group, $D[S]$ is the group ring of S over D. If K is a field and G is a torsion-free abelian group, the group algebra $K[G]$ is a GCD-domain. An excellent reference for monoid domains is [8].

M. Benelmekki · S. El Baghdadi (✉)
Department of Mathematics, Faculté des Sciences et Techniques, Beni Mellal University, Beni Mellal, Morocco
e-mail: s.elbaghdadi@usms.ma

J.-L. Chabert et al. (eds.), *Algebraic, Number Theoretic, and Topological Aspects of Ring Theory*, https://doi.org/10.1007/978-3-031-28847-0_6

The goal of this paper is to study the following question. Given a commutative field K and a torsion-free abelian group G, under which conditions the group algebra $K[G]$ is an antimatter domain? Note that, if G contains an element g such that the subgroup $\langle g\rangle$ of G generated by g is pure in G (e.g., g of height $(0, 0, \ldots)$; see [6, p. 108] for definition), then it follows from [9, Corollary 7.7] that $1 - X^g$ is irreducible in $K[G]$. In particular, $K[G]$ is not an antimatter domain. As a consequence, the condition that G contains no element of height $(0, 0, \ldots)$ is necessary for $K[G]$ to be an antimatter domain. This latter observation suggests that more attention should be devoted to antimatter group algebras with rational exponents, and this will be the subject of Sect. 2. Given a subgroup G of the additive group of rationals $\mathbb{Q}$, we investigate some irreducibility criteria in $K[G]$, and then we derive a characterization of the antimatter property for these group algebras. This extends some well-known results. Note that the case of the group algebra $K[\mathbb{Q}]$ was investigated by the authors in [2]. As an application of Sect. 2, in Sect. 3, we give necessary and sufficient conditions under which the algebra $K[G]$ is an antimatter domain, in the case where K is a field of characteristic zero and G an arbitrary torsion-free abelian group. Examples in either characteristic are provided to illustrate our results.

Throughout this paper, we let $\mathbb{N}$, $\mathbb{Z}$, and $\mathbb{Q}$ denote the positive integers, the integers, and the rational numbers, respectively. The set of non-negative integers (resp., rationals) will be denoted by $\mathbb{Z}_+$ (resp., $\mathbb{Q}_+$). All monoids (resp., groups) considered in this paper will be assumed to be commutative.

For $r = \frac{n}{d} \in \mathbb{Q}$, $(n, d) \in \mathbb{Z} \times \mathbb{N}$, reduced (i.e., $\gcd(n, d) = 1$), we call the unique pair of positive integers n and d, the numerator and denominator of r, and denote them by $n(r)$ and $d(r)$, respectively. All elements of $\mathbb{Q}$ considered in the following will be assumed to be reduced. For a nonempty subset A of $\mathbb{Q}$, we set $d(A) = \{d(r) \mid r \in A\}$, that is, the set of the denominators of all elements of A.

General references for any undefined terminology or notation are [7, 8, 12].

2 Antimatter Group Algebras with Rational Exponents

Throughout this section, K is a commutative field, T is an indeterminate over K, and G is a subgroup of the additive group of rationals $\mathbb{Q}$. In this section, we investigate some irreducibility criteria in $K[G]$, and as an application, we derive a characterization of the antimatter property for $K[G]$.

We start with the following lemma which gives a nice form of the elements of $K[G]$.

Lemma 1 *Let K be a field, $K[T]$ the polynomial ring in the indeterminate T, and G a subgroup of the additive group $\mathbb{Q}$. Let $f \in K[G] \setminus \{0\}$, then $f = X^s h(X^g)$, where $g, s \in G$, $g \geq 0$, and $h(T) \in K[T]$ such that $h(0) \neq 0$.*

Proof Let $f = \sum_{i=1}^{n} a_i X^{r_i} \in K[G]$ with $r_1 < \cdots < r_n$. Let H be the subgroup of G generated by $\mathrm{Supp}(f)$. Since H is a finitely generated subgroup of the additive group $\mathbb{Q}$, H is cyclic. Hence $H = \langle g\rangle$ for some $g \in G$, $g \geq 0$. Thus, $f = \sum_{i=1}^{n} a_i X^{\lambda_i g}$ for some integers $\lambda_1, \ldots, \lambda_n$. Set $\lambda = \min_i \lambda_i$. Then $f = X^{\lambda g} \sum_{i=1}^{n} a_i X^{(\lambda_i - \lambda)g}$. Consequently, $f = X^s h(X^g)$ where $s = \lambda g \in G$ and $h(T) = \sum_{i=1}^{n} a_i T^{(\lambda_i - \lambda)} \in K[T]$ such that $h(0) \neq 0$. □

To investigate the irreducibility in $K[G]$, the previous lemma leads to consider only those elements of the form $f(X^g)$ for some $g \in G$ and polynomial $f(T) \in K[T]$ with $f(0) \neq 0$.

Let S be a monoid and $a \in S$. Given an integer m, we say that m divides a in S and denote $m|a$ if $a = m\alpha$ for some $\alpha \in S$. The following proposition is a version of [10, Proposition 5.7] for the group algebra $K[G]$.

Proposition 1 *Let K be a field and G a subgroup of the additive group $\mathbb{Q}$. Let $f(T)$ be an irreducible polynomial in $K[T]$ such that $f(0) \neq 0$, and let $0 \neq g \in G$. Then the following conditions are equivalent.*

(1) *$f(X^g)$ is irreducible in $K[G]$.*
(2) *For every positive integer m dividing g in G, the polynomial $f(T^m)$ is irreducible in $K[T]$.*

Proof (1)⇒(2) Suppose that $f(X^g)$ is irreducible in $K[G]$. Let m be a positive integer such that $g = m\alpha$ for some $\alpha \in G$, and suppose that $f(T^m) = \varphi_1(T)\varphi_2(T)$, where $\varphi_1, \varphi_2 \in K[T]$. Clearly, $\varphi_i(0) \neq 0, i = 1, 2$. Then, $f(X^g) = \varphi_1(X^\alpha)\varphi_2(X^\alpha)$. Since $f(X^g)$ is irreducible in $K[G]$ and $\varphi_i(0) \neq 0$ $(i = 1, 2)$, then either $\varphi_1(T)$ or $\varphi_2(T)$ is in K. Consequently, the polynomial $f(T^m)$ is irreducible in $K[T]$.

(2)⇒(1) Suppose that $f(X^g) = h_1 h_2$, where $h_1, h_2 \in K[G]$. Let H be the subgroup of G generated by $\mathrm{Supp}(h_1) \cup \mathrm{Supp}(h_2) \cup \{g\}$. Since H is a finitely generated subgroup of the additive group $\mathbb{Q}$, H is cyclic. Then, there exists an element $0 \neq \alpha \in G$ such that $H = \langle\alpha\rangle$ and $g = m\alpha$ for some positive integer m. Moreover, there exists $f_i \in K[T, T^{-1}]$ such that $h_i = f_i(X^\alpha)$, $i = 1, 2$. Thus $f(X^{m\alpha}) = f_1(X^\alpha)f_2(X^\alpha)$, and hence $f(T^m) = f_1(T)f_2(T)$. Now, since $f(0) \neq 0$, condition (2) implies that the polynomial $f(T^m)$ is irreducible in $K[T, T^{-1}]$, and then either f_1 or f_2 is a unit in $K[T, T^{-1}]$. Consequently, either h_1 or h_2 is a unit in $K[G]$. Therefore, $f(X^g)$ is irreducible in $K[G]$. □

Example 1 Let $K = \mathbb{F}_2$ be the finite field of 2 elements, $f(T) = T^2 + T + 1$, and G the subgroup of the additive group of rational numbers $\mathbb{Q}$ generated by the subset $\{\frac{1}{3^k} \mid k \in \mathbb{N}\}$. Fix an integer $n \geq 1$, then $f(X^{\frac{1}{3^n}}) = X^{\frac{2}{3^n}} + X^{\frac{1}{3^n}} + 1$ is irreducible in $K[G]$. Indeed, if m is a positive integer such that $\frac{1}{3^n} = m\alpha$ for some $\alpha \in G$, then $m = 3^k$ for some non-negative integer k. By Coykendall and Gotti [4, Lemma 5.3], it follows that the polynomial $f(T^m) = T^{2.3^k} + T^{3^k} + 1$ is irreducible in $K[T]$, and hence Proposition 1 implies that $f(X^{\frac{1}{3^n}})$ is irreducible in $K[G]$. In particular, the group algebra $K[G]$ is not an antimatter domain.

The next result offers a Capelli's type criterion for the irreducibility of binomials. For a field K and a positive integer n, set $K^n = \{x^n, x \in K\}$.

Proposition 2 *Let K be a field and G a subgroup of the additive group $\mathbb{Q}$. Let $0 \neq g \in G$ and $a \in K$. Then $X^g - a$ is irreducible in $K[G]$ if and only if $a \notin K^p$ for every prime number p dividing g in G, and $a \notin -4K^4$ if $4|g$ in G.*

Proof We may assume that $a \neq 0$. From Proposition 1, $X^g - a$ is irreducible in $K[G]$ if and only if the binomial $T^m - a$ is irreducible in $K[T]$ for every positive integer m dividing g in G. Then, the result follows immediately from Capelli's Theorem [12, Section 2.1; Theorem 19]. □

Corollary 1 *Let K be a field and G a subgroup of the additive group $\mathbb{Q}$. Let $0 \neq g \in G$, $f \in K[T]$ an irreducible polynomial, and α an element in an extension field of K such that $f(\alpha) = 0$. Then the following conditions are equivalent.*

1. *$f(X^g)$ is irreducible in $K[G]$.*
2. *$X^g - \alpha$ is irreducible in $K(\alpha)[G]$.*
3. *$\alpha \notin [K(\alpha)]^p$ for every prime number p dividing g in G, and $\alpha \notin -4[K(\alpha)]^4$ if $4|g$ in G.*

Proof (1) $\Leftrightarrow$ (2) By Proposition 1, $f(X^g)$ is irreducible in $K[G]$ if and only if the polynomial $f(T^m)$ is irreducible in $K[T]$ for every positive integer m dividing g in G. By Schinzel [12, Section 2.1; Theorem 22], this is equivalent to that $T^m - \alpha$ is irreducible in $K(\alpha)[T]$ for every positive integer m dividing g in G. Thus, the result follows, again, from Proposition 1.

(2) $\Leftrightarrow$ (3) Follows from Proposition 2. □

As an application of the above results, the next theorem characterizes the antimatter property for the group algebra $K[G]$.

Theorem 1 *Let K be a field and G a subgroup of the additive group $\mathbb{Q}$. Then the following conditions are equivalent.*

1. *$K[G]$ is an antimatter domain.*
2. *$X^g - \alpha$ is reducible in $K(\alpha)[G]$ for every $0 \neq g \in G$ and every α in an algebraic extension field of K.*
3. *For every $0 \neq g \in G$ and every α in an algebraic extension field of K, either $4|g$ in G and $\alpha \in -4[K(\alpha)]^4$, or $\alpha \in [K(\alpha)]^p$ for some prime number p dividing g in G.*

Proof (1) $\Leftrightarrow$ (2) Suppose that $K[G]$ is an antimatter domain. Let $0 \neq g \in G$, α an algebraic element over K, and $f \in K[T]$ the minimal polynomial of α over K. Assume that $f(X^g)$ is a nonunit of $K[G]$. Since $K[G]$ is antimatter, $f(X^g)$ is reducible in $K[G]$. Hence by Corollary 1, $X^g - \alpha$ is reducible in $K(\alpha)[G]$. Conversely, let f be a nonzero nonunit element of $K[G]$. By Lemma 1, there exist $s, g \in G$, $g > 0$, and $h \in K[T]$, with $h(0) \neq 0$, such that $f = X^s h(X^g)$. We need only show that $h(X^g)$ is reducible in $K[G]$. If h is reducible in $K[T]$, then $h(X^g)$ is reducible in $K[G]$ since $h(0) \neq 0$, and we are done. Otherwise, assume that h is irreducible in $K[T]$, and let $h(\alpha) = 0$. By the assumption $X^g - \alpha$ is reducible in $K(\alpha)[G]$. Then, it follows from Corollary 1 that $h(X^g)$ is reducible in $K[G]$. Hence $f = X^s h(X^g)$ is reducible in $K[G]$. Therefore, $K[G]$ is an antimatter domain.

(2) $\Leftrightarrow$ (3) Follows from Corollary 1. □

Before giving another application, we recall some definitions about factorization in monoids. Let S be a commutative cancellative torsion-free monoid, written additively. For $s, t \in S$, we say that s divides t, and write $s \mid t$, if there exists $r \in S$ such that $t = r + s$. If $t = u + s$ for some unit u of S, we say that s and t are associate. A nonzero nonunit element $s \in S$ is said to be irreducible (or atom) of S if $s = t + r, t, r \in S$, implies that t or r is a unit of S. The monoid S is antimatter if it contains no atoms. An element $x \in \langle S \rangle$ is said to be integral over S, if $nx \in S$ for some integer $n \geq 1$, then $x \in S$. The monoid S is integrally closed if every element of $\langle S \rangle$ which is integral over S is in S. For more details, see [8].

Recently, the additive submonoids of $\mathbb{Q}_+$ have received a lot of attention in nonunique factorization theory. F. Gotti called such monoids Puiseux monoids. Antimatter Puiseux monoid algebras (called Puiseux algebras) were considered in [1, 11]. Let G be a subgroup of the additive group $\mathbb{Q}$. Set $G_+ = G \cap \mathbb{Q}_+$, the monoid of non-negative elements in G. Note that the monoid G_+ is an integrally closed Puiseux monoid. We next characterize the antimatter property for the Puiseux algebra $K[G_+]$. We need the following lemma.

Lemma 2 *Let K be a field, and let G be a subgroup of the additive group $\mathbb{Q}$. Then the following hold.*

1. *Let $f \in K[G_+]$ such that $f(0) \neq 0$. Then f is irreducible in $K[G_+]$ if and only if it is irreducible in $K[G]$.*
2. *The monoid G_+ is not antimatter if and only if it is cyclic. In this case, $K[G_+] = K[X]$ and $K[G] = K[X, X^{-1}]$.*

Proof

(1) Let $f \in K[G_+]$ such that $f(0) \neq 0$. Then, a nontrivial factorization of f in $K[G_+]$ is also a nontrivial factorization of f in $K[G]$. Conversely, assume that $f = h_1h_2$ is a nontrivial factorization in $K[G]$. By Lemma 1, there exist nonzero elements $g_1, g_2 \in G_+$ such that, for $i = 1, 2$, $h_i = X^{s_i} f_i(X^{g_i})$ for some $s_i \in G$ and a polynomial $f_i \in K[T] \setminus K$ with $f_i(0) \neq 0$. Hence $s_1 + s_2 = 0$ and $f = f_1(X^{g_1}) f_2(X^{g_2})$ is a nontrivial factorization of f in $K[G_+]$. Therefore, f is reducible in $K[G_+]$.

(2) We need only to show that if the monoid G_+ is not antimatter, it is cyclic. Let $a, b \in G_+$ be atoms in G_+. The subgroup $\langle a, b \rangle$ of G is cyclic, so $\langle a, b \rangle = \langle c \rangle$ for some $c \in G$, $c > 0$. But c will be a common divisor of a and b in G_+; this forces $a = b$. Denote by a the unique atom of G_+. We next show that $G_+ = a\mathbb{Z}_+$. Let $0 \neq x \in G_+$. Since a is an atom, $a \leq x$. Let n be a positive integer such that $na \leq x < (n + 1)a$. Then $x - na \in G_+$ and $x - na < a$. So $x - na = 0$, that is $x = na$. Thus, if G_+ is not antimatter, G_+ is a cyclic monoid generated by its unique atom.

□

Theorem 2 *Let K be a field and G a subgroup of the additive group $\mathbb{Q}$. Then the following conditions are equivalent.*

1. *$K[G_+]$ is an antimatter domain.*
2. *$K[G]$ is an antimatter domain.*
3. *$X^g - \alpha$ is reducible in $K(\alpha)[G_+]$ for every $0 \neq g \in G_+$ and every α in an algebraic extension field of K.*
4. *For every $0 \neq g \in G_+$ and every α in an algebraic extension field of K, either $4|g$ in G_+ and $\alpha \in -4[K(\alpha)]^4$, or $\alpha \in [K(\alpha)]^p$ for some prime number p dividing g in G_+.*

Proof (1) $\Leftrightarrow$ (2) Suppose that $K[G_+]$ is an antimatter domain. Let f be a nonzero nonunit element of $K[G]$. Then by Lemma 1, $f = X^s h(X^g)$ for some $s \in G$, $0 \neq g \in G_+$, and $h \in K[T]$ with $h(0) \neq 0$. Thus, $h(X^g) \in K[G_+]$ is reducible in $K[G_+]$. Hence by Lemma 2(1), $h(X^g)$ is reducible in $K[G]$. Therefore, f is reducible in $K[G]$, and hence $K[G]$ is antimatter. Conversely, suppose that $K[G]$ is an antimatter domain, and let f be a nonzero nonunit element of $K[G_+]$. Again, by Lemma 1, $f = X^s h(X^g)$ for some $s \in G_+$, $g \in G_+$, and $h \in K[T]$ with $h(0) \neq 0$. If $h(X^g) \notin K$, Lemma 2(1) implies that $h(X^g)$ is reducible in $K[G_+]$, since $K[G]$ is antimatter. Hence f is reducible in $K[G_+]$. Otherwise, we need to show that X^s is reducible in $K[G_+]$. Since $K[G]$ is antimatter, by Lemma 2(2), the monoid G_+ is antimatter. Hence X^s is reducible in $K[G_+]$. Therefore, $K[G_+]$ is antimatter.

The remaining equivalences follow from Theorem 1 and Lemmas 1 and 2. □

Monoid algebras $K[M]$ such that M is an integrally closed Puiseux monoid containing 1 are an important class of Puiseux algebras (cf. [8, Theorem 13.5]). The antimatter property for these algebras was studied in [11, Section 5]. The following corollary gives a full characterization of the antimatter property for these Puiseux algebras.

Corollary 2 *Let K be a field and M an integrally closed Puiseux monoid containing 1. Then the following conditions are equivalent.*

1. *$K[M]$ is an antimatter domain.*
2. *$X^g - \alpha$ is reducible in $K(\alpha)[M]$ for every $0 \neq g \in M$ and every α in an algebraic extension field of K.*
3. *For every $0 \neq g \in M$ and every α in an algebraic extension field of K, either $4|g$ in M and $\alpha \in -4[K(\alpha)]^4$, or $\alpha \in [K(\alpha)]^p$ for some prime number p dividing g in M.*

Proof This follows from Theorem 2, since $\langle M\rangle_+ = \langle M\rangle \cap \mathbb{Q}_+ = M$ [8, Theorem 13.5]. □

Corollary 3 *Let K be a field that is either algebraically closed or real closed, and let G be a subgroup of the additive group $\mathbb{Q}$. If G is not cyclic, then $K[G]$ (resp., $K[G_+]$) is an antimatter domain.*

Proof This follows from Theorem 2 (4). Indeed, let $\alpha \notin K$ be an algebraic element over K and p a prime number. Then, if K is either algebraically closed (in this case, $K(\alpha) = K$) or real closed (in this case, $K(\alpha)$ is algebraically closed), then $\alpha \in [K(\alpha)]^p$. If $\alpha \in K$, we have $\alpha \in [K(\alpha)]^p = K^p$ in either case for every prime $p \geq 3$. □

3 Antimatter Group Algebras

As an application of Sect. 2, we next give a complete characterization of when a group algebra $K[G]$ is antimatter in characteristic 0.

We start our results of this section by studying the group algebra of the additive group $\mathbb{Q} \oplus \mathbb{Q}$ over a field K defined by

$$K[\mathbb{Q} \oplus \mathbb{Q}] = \{\sum_{i,j} a_{i,j} X^{r_i} Y^{s_j} | a_{i,j} \in K \text{ and } r_i, s_j \in \mathbb{Q}\},$$

where X, Y are two indeterminates over K.

For a nonzero element $F(X, Y) = \sum_{i,j} a_{i,j} X^{r_i} Y^{s_j}$ of $K[\mathbb{Q} \oplus \mathbb{Q}]$ with $a_{i,j} \neq 0$ and $(r_i, s_j) \neq (r_k, s_l)$ for $(i, j) \neq (k, l)$, let $\mathrm{Supp}_X(F)$ (resp., $\mathrm{Supp}_Y(F)$) denote the support of F with respect to the indeterminate X (resp., Y).

The following lemma is straightforward.

Lemma 3 *Let K be a field, and let $r, s \in \mathbb{Q} \setminus \{0\}$. The following map*

$$\begin{aligned} \phi_{r,s} : K[\mathbb{Q} \oplus \mathbb{Q}] &\longrightarrow K[\mathbb{Q} \oplus \mathbb{Q}] \\ F(X, Y) = \textstyle\sum_{i,j} a_{i,j} X^{r_i} Y^{s_j} &\longmapsto F(X^r, Y^s) = \textstyle\sum_{i,j} a_{i,j} X^{rr_i} Y^{ss_j} \end{aligned}$$

is an automorphism of $K[\mathbb{Q} \oplus \mathbb{Q}]$.

Let $r, s \in \mathbb{Q} \setminus \{0\}$, by the substitution $(X, Y) \to (X^r, Y^s)$ we mean the automorphism $\phi_{r,s}$ of $K[\mathbb{Q} \oplus \mathbb{Q}]$ defined in Lemma 3 such that $\phi_{r,s}(F(X, Y)) = F(X^r, Y^s)$. The next result gives a connection between the irreducibility in $K[\mathbb{Q} \oplus \mathbb{Q}]$ and that in the polynomial ring $K[X, Y]$.

Following the notation in [12], for a nonzero rational function $F(X, Y) \in K[X^{\pm 1}, Y^{\pm 1}]$, we define

$$JF(X, Y) = X^{u_1} Y^{u_2} F(X, Y),$$

where each u_i $(i = 1, 2)$ is an integer chosen as small as possible so that JF is a polynomial in X and Y (i.e., $JF \in K[X, Y]$). We set $J0 = 0$. Note that the operation J is distributive with respect to multiplication, and for every $F \in K[X^{\pm 1}, Y^{\pm 1}]$, $F \neq 0$, the polynomial JF is prime to XY. Also, note that $JF \in K \setminus \{0\}$ if and only if F is a unit of $K[X^{\pm 1}, Y^{\pm 1}]$ (i.e., $F = aX^u Y^v$, for some $0 \neq a \in K$ and $u, v \in \mathbb{Z}$).

Proposition 3 *Let K be a field, and let $F \in K[X, Y]$ irreducible in $K[X, Y]$ such that F is not associate to either X or Y. The following conditions are equivalent.*

1. *F is irreducible in $K[\mathbb{Q} \oplus \mathbb{Q}]$.*
2. *For every positive integers m and n, the polynomial $F(X^m, Y^n)$ is irreducible in $K[X, Y]$.*

Proof (1)$\Rightarrow$(2) Suppose that F is irreducible in $K[\mathbb{Q} \oplus \mathbb{Q}]$ and $F(X^m, Y^n)$ is reducible in $K[X, Y]$ for some positive integers m, n. Then there exist $F_1, F_2 \in K[X, Y] \setminus K$ such that

$$F(X^m, Y^n) = F_1(X, Y)F_2(X, Y).$$

Since $gcd(F(X, Y), XY) = 1$, then $gcd(F(X^m, Y^n), X^mY^n) = 1$ [12, Section 2.3; Lemma 6], and hence $gcd(F(X^m, Y^n), XY) = 1$. It follows that F_1 and F_2 are not associate to monomials in $K[X, Y]$. Now, substituting $(X^{\frac{1}{m}}, Y^{\frac{1}{n}})$ for (X, Y), we obtain

$$F(X, Y) = F_1(X^{\frac{1}{m}}, Y^{\frac{1}{n}})F_2(X^{\frac{1}{m}}, Y^{\frac{1}{n}}),$$

with $F_1(X^{\frac{1}{m}}, Y^{\frac{1}{n}})$ and $F_2(X^{\frac{1}{m}}, Y^{\frac{1}{n}})$ are nonunits of $K[\mathbb{Q}\oplus\mathbb{Q}]$. Hence F is reducible in $K[\mathbb{Q} \oplus \mathbb{Q}]$, which is a contradiction.

(2)$\Rightarrow$(1) Suppose that $F(X, Y) = F_1(X, Y)F_2(X, Y)$, where $F_1, F_2 \in K[\mathbb{Q} \oplus \mathbb{Q}]$. Let $A_X = \mathrm{Supp}_X(F_1) \cup \mathrm{Supp}_X(F_2)$, $A_Y = \mathrm{Supp}_Y(F_1) \cup \mathrm{Supp}_Y(F_2)$, $m = \mathrm{lcm}\{d(A_X)\}$, and $n = \mathrm{lcm}\{d(A_Y)\}$, where $d(A_X)$ and $d(A_Y)$ are the sets of denominators of all (reduced) elements of A_X and A_Y, respectively. Substituting (X^m, Y^n) for (X, Y), we get

$$F(X^m, Y^n) = F_1(X^m, Y^n)F_2(X^m, Y^n).$$

Note that $F_1(X^m, Y^n)$ and $F_2(X^m, Y^n)$ are elements of $K[X^{\pm 1}, Y^{\pm 1}]$.

As above, we have $gcd(F(X^m, Y^n), XY) = 1$, hence

$$F(X^m, Y^n) = JF_1(X^m, Y^n)JF_2(X^m, Y^n).$$

By condition (2), either $JF_1(X^m, Y^n)$ or $JF_2(X^m, Y^n)$ is in K, which implies that either $F_1(X^m, Y^n)$ or $F_2(X^m, Y^n)$ is a unit of $K[X^{\pm 1}, Y^{\pm 1}]$. Consequently, either $F_1(X, Y)$ or $F_2(X, Y)$ is a unit of $K[\mathbb{Q} \oplus \mathbb{Q}]$. Therefore, F is irreducible in $K[\mathbb{Q} \oplus \mathbb{Q}]$. □

Let K be a field. Since $K[X, Y]$ is a UFD, each nonzero nonunit polynomial $F \in K[X, Y]$ is uniquely expressed as product of a constant and coprime powers of irreducible polynomials in $K[X, Y]$, i.e.,

$$F(X, Y) = const \prod_{i=1}^{t} F_i^{e_i}(X, Y),$$

where F_i is an irreducible polynomial over K and $t, e_i \in \mathbb{Z}_+$, $i = 1, \ldots, t$. This factorization will be called the standard form of F over K.

The following result, Proposition 4, shows that there are some special elements of $K[\mathbb{Q} \oplus \mathbb{Q}]$ that have a prime factorization in $K[\mathbb{Q} \oplus \mathbb{Q}]$, in the case where K is a field of characteristic 0.

Proposition 4 *Let K be a field of characteristic 0, and let $F \in K[X, Y]$ be an irreducible polynomial, different (up to associate) from X and Y and from $J F_0(X^{\delta_1} Y^{\delta_2})$ for any $F_0(T) \in K[T]$ and any $\delta = (\delta_1, \delta_2) \in \mathbb{Z}^2 \setminus \{0\}$. Then F can be expressed as a finite product of prime elements of $K[\mathbb{Q} \oplus \mathbb{Q}]$. Moreover, F has no multiple divisor in $K[\mathbb{Q} \oplus \mathbb{Q}]$.*

Proof Let m and n be two positive integers. By Schinzel [12, Section 2.3; Corollary 2], there exists $\upsilon = (\upsilon_1, \upsilon_2) \in \mathbb{N}^2$ such that

(i) $\operatorname{lcm}\{\upsilon_1, \upsilon_2\} \leq (\deg F)^2$,
(ii) $m = \upsilon_1 u_1$ and $n = \upsilon_2 u_2$, where $u_1, u_2 \in \mathbb{N}$,
(iii) If $F(X^{\upsilon_1}, Y^{\upsilon_2}) = const \prod_{\tau=1}^{t} F_\tau(X, Y)^{e_\tau}$, in the standard form, then $e_\tau = 1$ $(\tau = 1, \ldots, t)$ and $F(X^m, Y^n) = const \prod_{\tau=1}^{t} F_\tau(X^{u_1}, Y^{u_2})$, in the standard form.

Since $\operatorname{lcm}\{\upsilon_1, \upsilon_2\} \leq (\deg F)^2$, there exists only a finite number of such $\upsilon \in \mathbb{N}^2$, for every positive integers m, n. Let Λ be the subset of all υ that satisfy the conditions (i)–(iii). Then

$$t \leq \left(\max_{\upsilon \in \Lambda} \{\upsilon_1 + \upsilon_2\} \right) . \deg F \leq 2 (\deg F)^3 .$$

Thus, for every $(m, n) \in \mathbb{N}^2$, the polynomial $F(X^m, Y^n)$ has, say $t_{m,n}$, non-associate irreducible divisors (with no multiple factor) in $K[X, Y]$ such that $t_{m,n} \leq 2 (\deg F)^3$. Let $t_0 = \max_{m,n \in \mathbb{N}} \{t_{m,n}\}$, and let (m_0, n_0) be any pair of positive integers such that $t_0 = t_{m_0,n_0}$. Then the standard form of $F(X^{m_0}, Y^{n_0})$ over K is as follows:

$$F(X^{m_0}, Y^{n_0}) = const \prod_{i=1}^{t_0} G_i(X, Y),$$

where each $G_i \in K[X, Y]$ is irreducible in $K[X, Y]$. On substituting $(X^{\frac{1}{m_0}}, Y^{\frac{1}{n_0}})$ for (X, Y), we obtain

$$(*) \qquad F(X, Y) = const \prod_{i=1}^{t_0} G_i(X^{\frac{1}{m_0}}, Y^{\frac{1}{n_0}}).$$

Now, we proceed to prove that the expression $(*)$ is a prime factorization of F in $K[\mathbb{Q} \oplus \mathbb{Q}]$ by showing that each G_i is irreducible in $K[\mathbb{Q} \oplus \mathbb{Q}]$ $(i = 1, \ldots, t_0)$. Let k and l be two positive integers, then

$$F(X^{km_0}, Y^{ln_0}) = const \prod_{i=1}^{t_0} G_i(X^k, Y^l).$$

Since the G_i's are non-associate irreducible elements in $K[X, Y]$, $\gcd(G_i, G_j) = 1$, for $i \neq j$. Hence by Schinzel [12, Section 2.3; Lemma 6], $\gcd(G_i(X^k, Y^l), G_j(X^k, Y^l)) = 1$ whenever $i \neq j$. By the maximality of t_0, each polynomial $G_i(X^k, Y^l)$ must be irreducible in $K[X, Y]$ $(i = 1, \ldots, t_0)$. On the other hand, as above, we have $gcd(G_i(X, Y), XY) = 1$ for every i; hence by Proposition 3, G_i is irreducible in $K[\mathbb{Q} \oplus \mathbb{Q}]$ for every i. In view of the automorphism of $K[\mathbb{Q} \oplus \mathbb{Q}]$ defined in Lemma 3, each $G_i(X^{\frac{1}{m_0}}, Y^{\frac{1}{n_0}})$ is irreducible in $K[\mathbb{Q} \oplus \mathbb{Q}]$. Thus, $(*)$ is the prime factorization of F in the GCD-domain $K[\mathbb{Q} \oplus \mathbb{Q}]$. □

Corollary 4 *Let K be a field of characteristic 0; the group algebra $K[\mathbb{Q} \oplus \mathbb{Q}]$ is not an antimatter domain.*

Proof Consider any irreducible polynomial $F \in K[X, Y]$ that is not a scalar multiple of any of the indeterminates X, Y, and different from $JF_0(X^{\delta_1} Y^{\delta_2})$ for any $F_0(T) \in K[T]$ and any $\delta = (\delta_1, \delta_2) \in \mathbb{Z}^2 \setminus \{0\}$. For instance, $F(X, Y) = X^n + Y + 1 \in K[X, Y]$ where n is an arbitrary positive integer. Then, Proposition 4 implies that F has a prime factorization in $K[\mathbb{Q} \oplus \mathbb{Q}]$. Hence $K[\mathbb{Q} \oplus \mathbb{Q}]$ is not an antimatter domain. □

The condition in Corollary 4 that the field K is of characteristic 0 cannot be omitted as the following example shows:

Example 2 Let K be a perfect field of characteristic $\pi > 0$. Let F be a nonzero nonunit element of $K[\mathbb{Q} \oplus \mathbb{Q}]$; say $F(X, Y) = \sum_{i,j} a_{i,j} X^{r_i} Y^{s_j}$, where $a_{i,j} \in K$, and $r_i, s_j \in \mathbb{Q}$. Then

$$F(X, Y) = \left(\sum_{i,j} a_{i,j}^{\frac{1}{\pi}} X^{\frac{r_i}{\pi}} Y^{\frac{s_j}{\pi}} \right)^{\pi}.$$

Hence F is not irreducible in $K[\mathbb{Q} \oplus \mathbb{Q}]$, and therefore $K[\mathbb{Q} \oplus \mathbb{Q}]$ is an antimatter domain.

We next recall some results about the divisible hull of an abelian group. Let G be an abelian additive group. Recall that a positive integer n divides an element $g \in G$, in G, if the equation $nx = g$ is solvable in G (i.e., $n\alpha = g$ for some $\alpha \in G$). The group G is n-divisible if the equation $nx = g$ is solvable in G for every $g \in G$ (i.e., $G = nG$). The group G is said to be divisible if it is n-divisible for every positive integer n. Also, every abelian group G can be embedded in a minimal divisible abelian group G^*, called the divisible hull of G [5, Theorem 24.1]. For a torsion-free abelian group G, G^* is isomorphic to a direct sum of copies of the additive group $\mathbb{Q}$ [5, Theorem 23.1].

Lemma 4 *Let D be an integral domain, and let G be a torsion-free abelian group with divisible hull G^*. If $D[G]$ is an antimatter domain, then so is $D[G^*]$.*

Proof Suppose that $D[G]$ is an antimatter domain. For a positive integer m, let $G_m = \{g \in G^* : (m!)g \in G\}$. Clearly, each G_m is a subgroup of G^* that is isomorphic to G, $G_m \subseteq G_{m+1}$, and $G^* = \cup_{m=1}^{\infty} G_m$. Hence $D[G_m] \simeq D[G]$ is an antimatter domain for every positive integer m. We claim that $D[G^*] = \cup_{m=1}^{\infty} D[G_m]$ is an antimatter domain as union of an ascending chain of antimatter domains. This follows from the fact that the sets $U(D[G_m])$ of units of the domains $D[G_m]$ satisfy $U(D[G_{m+1}]) \cap D[G_m] = U(D[G_m])$ for every positive integer m. □

Remark 1 In Example 1, we showed that the group algebra $K[G]$ is not an antimatter domain, where $K = \mathbb{F}_2$ and $G = \langle \frac{1}{3^k} \mid k \in \mathbb{N} \rangle$. However $K[G^*] \simeq K[\mathbb{Q}]$ is an antimatter domain since $K = \mathbb{F}_2$ is a perfect field of positive characteristic and $(\mathbb{Q}, +)$ is divisible (cf. Proposition 5). Thus the converse of Lemma 4 is not true in general.

We next state our main result of this section, which characterizes the antimatter property of a group algebra $K[G]$ in characteristic 0, in terms of that of a group algebra with rational exponents studied in Sect. 1.

Theorem 3 *Let K be a field of characteristic 0, and let G be a torsion-free abelian group. The following conditions are equivalent.*

1. *$K[G]$ is an antimatter domain.*
2. *G is isomorphic to a subgroup G' of the additive group $\mathbb{Q}$ and $K[G']$ is an antimatter domain.*

Proof (1)⇒(2) Let G^* be the divisible hull of G. Suppose that $K[G]$ is an antimatter domain. By Lemma 4, it follows that $K[G^*]$ is also an antimatter domain. We write G^* as $\mathbb{Q} \oplus H$ for some subgroup H of G^*. Then we need only show that $H = \{0\}$. Note that H is a torsion-free divisible group. Thus, assume that $H = \mathbb{Q} \oplus H_1$ for some subgroup H_1 of G^*. Since $K[G^*] \simeq K[\mathbb{Q} \oplus \mathbb{Q}][H_1]$ and $K[G^*]$ is antimatter, then $K[\mathbb{Q} \oplus \mathbb{Q}]$ must be an antimatter domain, which is a contradiction by Corollary 4. Hence $H = \{0\}$, and so G^* is isomorphic to $\mathbb{Q}$. Consequently, G is isomorphic to a subgroup G' of the additive group of rational numbers $\mathbb{Q}$.

(2)⇒(1) This is obvious. □

We end this section with some results in the positive characteristic side. As a consequence of Corollary 4 (see also Theorem 3), if K is a field of characteristic 0 and G is a torsion-free divisible abelian group of *rank* ≥ 2, $K[G]$ is never antimatter. However, in the case of positive characteristic the group algebra $K[G]$ may be antimatter.

Proposition 5 *Let K be a field of characteristic $\pi > 0$, and let G be a torsion-free abelian group. If K is perfect and G is $\pi-$divisible, then $K[G]$ is an antimatter domain.*

Proof Let $f = \sum_i a_i X^{g_i} \in K[G]$ be a nonzero nonunit of $K[G]$. Since K is perfect, $a_i^{\frac{1}{\pi}} \in K$ for every i. On the other hand, since G is $\pi-$divisible, $\frac{g_i}{\pi} \in G$. Then $f = \sum_i a_i X^{g_i} = \left(\sum_i a^{\frac{1}{\pi}} X^{\frac{g_i}{\pi}}\right)^{\pi}$ is reducible in $K[G]$, and hence $K[G]$ is an antimatter domain. □

Note that the converse of Proposition 5 is not true in general as the following example shows.

Example 3 Let K be any algebraically closed field of characteristic $\pi > 0$. Let G be the subgroup of the additive group of rational numbers generated by the set $\{\frac{1}{p^n} \mid n \in \mathbb{Z}_+\}$, where p is a prime number such that $p \neq \pi$. By Gotti [11, Corollary 5.4] and Theorem 2, $K[G]$ is an antimatter domain. However, it is clear that G is not $\pi-$divisible.

Acknowledgments We would like to thank the referee for several helpful comments and suggestions.

References

1. Anderson, D.D., Coykendall, J., Hill, L., Zafrullah, M: Monoid domain constructions of antimatter domains. Comm. Alg. **35**, 3236–3241(2007)
2. Benelmekki, M., El Baghdadi, S.: Factorization of polynomials with rational exponents over a field. Comm. Alg. **50(10)**, 4345–4355(2022).
3. Coykendall, J., Dobbs, D. E., Mullins, B.: On integral domains with no atoms. Comm. Alg. **27(12)**, 5813–5831(1999)
4. Coykendall, J., Gotti, F.: On the atomicity of monoid algebras. Journal of Algebra **539**, 138–151(2019)
5. Fuchs, L.: Infinite abelian groups, vol 1, Academic press, New York (1970)
6. Fuchs, L.: Infinite abelian groups, vol 2, Academic press, New York (1973)
7. Gilmer, R.: Multiplicative Ideal Theory, Marcel Dekker, New York (1972)
8. Gilmer, R.: Commutative Semigroup Rings, The Univ. of Chicago Press, Chicago and London (1984)
9. Gilmer, R. Parker, T.: Divisibility properties in semigroup rings. Michigan Math. J.**21**, 65–86(1974)
10. Gotti, F., Gotti, M.: On the molecules of numerical semigroups, Puiseux monoids, and Puiseux algebras. In: Barucci V., Chapman S., D'Anna M., Fröberg R. (eds) Numerical Semigroups. Springer INdAM Series, vol 40, pp. 141–161. Springer, Cham (2020)
11. Gotti, F.: On semigroup algebras with rational exponents. Comm. Alg. **50(1)**, 3–18(2022).
12. Schinzel, A.: Polynomials with special regard to reducibility, Cambridge University Press, Cambridge, UK (2000)

Yosida, Martínez, and $A + B$ Rings

Papiya Bhattacharjee, Lee Klingler, and Warren Wm. McGovern

1 Introduction

Throughout, R denotes a commutative ring with identity, usually assumed to be reduced (i.e., semiprime) unless otherwise noted. We are interested in rings R of type $A + B$ (see the next section for details of the construction). For a ring R, we denote by $\text{Rad}(R)$ the collection of radical ideals of R. Recall that an ideal I is radical if $x \in I$ whenever $x^n \in I$ for some $n \in \mathbb{N}$. Under the Axiom of Choice, which we shall assume, this condition is equivalent to saying that I is an intersection of prime ideals. When ordered by inclusion, $\text{Rad}(R)$ is a coherent frame (see [5] for more information on radical ideals and coherent frames). We let $\text{Spec}(R)$ denote the space of prime ideals of the ring R. We also use $\text{Max}(R)$ and $\text{Min}(R)$ to denote the subspaces of $\text{Spec}(R)$ consisting of the maximal and minimal prime ideals of R, respectively.

Our aim here is to study the frame of radical ideals of rings of type $A + B$. In particular, we are interested in characterizing when such a ring satisfies one of the following three conditions:

1. Every prime ideal of R is an intersection of maximal ideals.
2. Every prime ideal of R is an mz-ideal.
3. Every prime ideal of R is a strong d-ideal.

The $A + B$ construction, Popularized by Huckaba and Lucas, is quite useful for finding interesting examples of rings with zero-divisors (see, e.g., [11] and [13]).

P. Bhattacharjee · L. Klingler (✉)
Department of Mathematical Sciences, Florida Atlantic University, Boca Raton, FL, USA
e-mail: pbhattacharjee@fau.edu; klingler@fau.edu

W. W. McGovern
H. L. Wilkes Honors College, Florida Atlantic University, Jupiter, FL, USA
e-mail: warren.mcgovern@fau.edu

J.-L. Chabert et al. (eds.), *Algebraic, Number Theoretic, and Topological Aspects of Ring Theory*, https://doi.org/10.1007/978-3-031-28847-0_7

It shall be useful to recall the hull-kernel (or Zariski) topology on the prime spectrum $\mathrm{Spec}(R)$ and its subspaces. For any subset $S \subseteq R$, we let $V(S) = \{P \in \mathrm{Spec}(R) : S \subseteq P\}$. Set $U(S) = \mathrm{Spec}(R) \setminus V(S)$. When $S = \{s_1, \dots, s_n\}$ we instead write $V(s_1, \dots, s_n)$. The subsets of the form $V(S)$ form the closed sets (and hence $U(S)$ form the open sets) of the hull-kernel topology. We set $V_M(S) = V(S) \cap \mathrm{Max}(R)$ and $V_{\mathrm{Min}(R)}(S) = V(S) \cap \mathrm{Min}(R)$.

A ring satisfying (1) is called a **Hilbert ring**, and this class of rings is well-studied. A reduced Hilbert ring is necessarily Jacobson semisimple, that is, the Jacobson radical of R equals $\{0\}$. Examples of Hilbert rings include one-dimensional Jacobson semisimple integral domains. Obviously, a von Neumann regular ring is a Hilbert ring. If R is a Hilbert ring, then so is $R[X]$, and conversely. In general, even if not reduced, the ring R is a Hilbert ring if and only if $R/n(R)$ is a Hilbert ring, where $n(R)$ denotes the nilradical of R. A more esoteric example of a Hilbert ring is the ring $C(X, \mathbb{Z})$ consisting of all integer-valued continuous functions for some compact zero-dimensional Hausdorff space X (see [2]).

To understand (2), we first need to recall some notation. An ideal I is called a z-**ideal** if whenever $f \in I$ and $g \in R$, then $V_M(f) \subseteq V_M(g)$ implies that $g \in I$. Replacing $f \in I$ with a finite subset $f_1, \dots, f_n \in I$ produces the definition of an **mz-ideal**. We recall also that an ideal I is said to be a **strong** z-**ideal** (or **sz-ideal**) if it is an intersection of maximal ideals. It is apparent that every sz-ideal is an mz-ideal, and that every mz-ideal is a z-ideal. A z-ideal in a reduced ring is necessarily a radical ideal.

From a frame-theoretic point of view, the work of Martinez and Zenk [15], in the context of the frame $\mathrm{Rad}(R)$, is aimed at studying the mz-ideals of a ring. The paper [12] fleshes out the differences between z-ideal and mz-ideal. We do wish to highlight that the zero ideal of R is a z-ideal if and only if the ring R is Jacobson semisimple. What we can say is that the ring satisfies condition (2) if and only if $\mathrm{Rad}(R)$ is a Yosida frame, and hence we make the following definition.

Definition 1 We call a ring satisfying (2) a **Yosida ring**; that is, the ring R is a Yosida ring if every prime ideal of R is an mz-ideal.

A reduced Yosida ring is necessarily Jacobson semisimple, and a Hilbert ring is necessarily a Yosida ring.

Remark 1 According to [15, Definition 2.1], a coherent frame L is a Yosida frame if every compact element is the meet of maximal elements. For a coherent frame L one can define a nucleus z on L by first defining it for compact elements. If $c \in L$ is compact, then $z(c)$ is the meet of the maximal elements of L above c. Then for an arbitrary $x \in L$, $z(x)$ is the join of $z(c)$ ranging over all compact $c \leq x$. Then $zL = \{x \in L : z(x) = x\}$. In [15, Proposition 2.5], it is shown that a coherent frame is Yosida if and only if $L = zL$. Since $\mathrm{Rad}(R)$ is coherent, this is tantamount to saying that every radical ideal of R is an mz-ideal, which is equivalent to condition (2), as mentioned above.

It is straightforward to check that an intersection of mz-ideals (respectively sz-ideals, z-ideals) is again an mz-ideal (respectively sz-ideal, z-ideal).

Next, to understand condition (3), we recall that the notation $\mathrm{Ann}_R(S)$ is used to symbolize the annihilator of the set S. We are interested in the double annihilator, and since the notation is somewhat cumbersome, we shall use the notation $S^{\perp}$ and $S^{\perp\perp}$ for the annihilator and double annihilator of S. An ideal I is called a d-**ideal** if for all $f \in I$, $f^{\perp\perp} \subseteq I$. We say that the ideal I is a **strong** d-**ideal** (or sd-**ideal** for short) if for every finite subset $F \subseteq I$, $F^{\perp\perp} \subseteq I$. An sd-ideal is obviously a d-ideal. Each minimal prime ideal is an sd-ideal and so is the intersection of any collection of sd-ideals. We shall use $d\mathrm{Rad}(R)$ to denote the collection of sd-ideals of R. From a frame-theoretic point of view, this is the frame of d-elements of $\mathrm{Rad}(R)$.

Theorem 1 (Theorem 2.3 [4]) *The ring R is Jacobson semisimple if and only if every sd-ideal is an mz-ideal.*

Corollary 1 *If the ring R is Jacobson semisimple and satisfies condition (3), then R is a Yosida ring.*

Remark 2 In [4] Definition 2.1, the authors define the ideal I to be a ζ-**ideal** if whenever $F \subseteq I$ is a finite subset and $U(F^{\perp}) \subseteq V(a)$, then $a \in I$. An ideal is an sd-ideal if and only if it is a ζ-ideal; recall that we are assuming our rings are reduced. In [1] and the references mentioned there, the authors study what they call strong z°-ideals. The ideal I is a **strong** z°-**ideal** if for all finite $F \subseteq I$ and for all $a \in R$, if $V_{\mathrm{Min}(R)}(F) \subseteq V_{\mathrm{Min}(R)}(a)$, then $a \in I$. Interestingly, in a reduced ring, all three conditions, ζ-ideal, sd-ideal, and strong z°-ideals, are equivalent.

We are also interested in the class of rings satisfying condition (3), and we make the following definition.

Definition 2 We call a ring satisfying (3) a **Martínez ring**; that is, the ring R is a Martínez ring if every prime ideal of R is an sd-ideal.

Clearly, every von Neumann regular ring is a Martínez ring. As we already pointed out in Corollary 1, if we confine ourselves to the class of Jacobson semisimple rings, then we can say that every Martínez ring is a Yosida ring.

Remark 3 The following diagram illustrates the relationship for Jacobson semisimple rings.

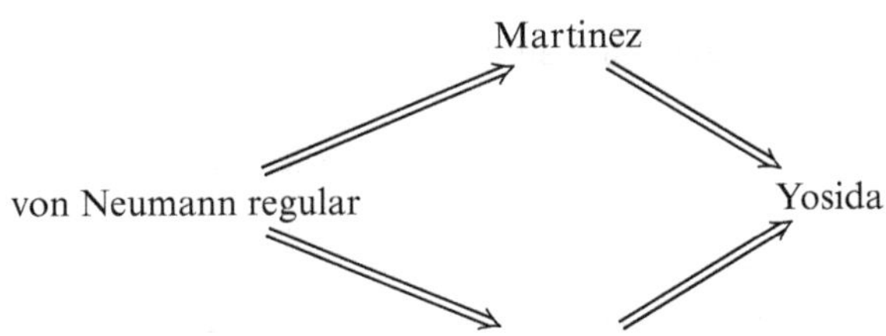

Observe that any Hilbert domain that is not a field is an example of a Hilbert ring that is not a Martínez ring, e.g., $R = \mathbb{Z}$. In section 3, we construct an example of a Yosida (and hence Jacobson semisimple) ring that is not Hilbert. We then use this example to construct a Jacobson semisimple Martínez ring that is not Hilbert.

As we already mentioned, the reduced ring R is a Martínez ring if and only if $d\mathrm{Rad}(R) = \mathrm{Rad}(R)$. This turns out to be a very strong condition. For example, in a Martínez ring every maximal ideal is a maximal sd-ideal, and since no sd-ideal can contain a regular element it follows that a Martínez ring is classical, that is, equals its classical ring of quotients. The converse of this statement is not true.

Example

Let $C(X)$ denote the ring of real-valued continuous functions. It is well-known that if $C(X)$ is not von Neumann regular, then there are prime ideals which are not z-ideals ([9, 14.13]). It follows that $C(X)$ is a Yosida ring if and only if $C(X)$ is von Neumann regular if and only if $C(X)$ is a Martínez ring. However, it is well-known that there are spaces X for which $C(X)$ is a classical ring yet not von Neumann regular. (In topological parlance, there are almost P-spaces which are not P-spaces.)

The interested reader is pointed to [8] where the author delved into the $A + B$ construction and some of its variants, and examines Example 4.1 of [1], which uses the $A + B$ construction to exhibit an example of a ring with a d-ideal which is not an sd-ideal. Our hope was to find a similar example of a ring with a z-ideal that is not an mz-ideal. Our attempts have been unsuccessful.

2 $A + B$ Rings

Throughout this section we assume that A is a reduced ring. We shall fix a collection of prime ideals, enumerated by the index set $\mathcal{A}$: $\mathcal{P} = \{P_\alpha\}_{\alpha \in \mathcal{A}} \subseteq \mathrm{Spec}(A)$. We further assume that $\cap\mathcal{P} = \{0\}$. We let $I = \mathcal{A} \times \mathbb{N}$, and for $i = (\alpha, n)$, let $P_i = P_\alpha$ and $A_i = A/P_i$. Set

$$B = \sum_{i \in I} A_i \subseteq \prod_{i \in I} A_i.$$

We let $e_i \in B$ denote the element defined by

$$e_i(i') = \begin{cases} 1 + P_{i'}, & \text{if } i = i' \\ 0 + P_{i'}, & \text{otherwise.} \end{cases}$$

Obviously, B is generated as a ring by the collection of e_i.

We identify A with its image inside of $\prod_{i \in I} A_i$ via $a \mapsto (a_i)$, where for a given $i = (\alpha, n)$, $a_i = a + P_\alpha$. We shall also use the notation $a(i)$ to mean the i-th coordinate of a, i.e., $a(i) = a_i$. Then $A + B$ is the ring consisting of elements of the form $a + b$ inside of $\prod_{i \in I} A_i$, and we note that each element has a unique such representation, because $A \cap B = \{0\}$.

Characterizing the prime ideals of $A + B$ is straightforward. A prime of $A + B$ either contains the ideal B or not. The primes containing B have the form $P + B$ for some $P \in \mathrm{Spec}(A)$. The primes not containing B are of a different type. First of all, for $i = (\alpha, n) \in \mathcal{I}$, the set

$$M_i = \{r \in R : r(i) = 0\}$$

as the kernel of the map from $A + B$ onto A_i, is a minimal prime ideal. Moreover, the minimal prime ideals of $A + B$ are precisely of the form M_i for some $i \in \mathcal{I}$ or of the form $P + B$ for $P \in \mathrm{Min}(A)$. One easily checks that the minimal primes M_i are pairwise comaximal, and each is comaximal with the ideal B. Thus, since $R/M_i \cong A/P_i$, it follows then that the prime ideals of $A+B$ not containing B have the form

$$P + M_i = \{r \in A + B : r(i) \in P/P_i\}$$

where $i \in \mathcal{I}$ and $P \in \mathrm{Spec}(A)$ is such that $P_i \subseteq P$.

Since a homomorphic image of a Hilbert ring is a Hilbert ring, characterizing Hilbert rings of the form $A + B$ turns out to be quite easy.

Theorem 2 *The ring $A + B$ is a Hilbert ring if and only if A is a Hilbert ring.*

Proof If A is a Hilbert ring, then for any $Q \in Spec(A)$, the prime ideal $Q + B$ is the intersection of ideals of the form $P + B$ for $P \in V_M(Q)$, and for any $i \in \mathcal{I}$ and $Q \in \mathrm{Spec}(A)$ such that $P_i \subseteq Q$, the ideal $Q + M_i$ is the intersection of ideals of the form $P + M_i$ for $P \in V_M(Q)$.

The converse is clear, since $A \cong A + B/B$. □

Remark 4 Notice that Theorem 2 makes no mention of the set of primes $\mathcal{P}$, essentially because the class of Hilbert rings is closed under homomorphic images.

Lemma 1 *If A is a reduced ring and $P \in \mathrm{Spec}(A)$, then the prime ideal $P + B$ of $A + B$ is an mz-ideal (respectively sz-ideal, z-ideal) if and only if P is an mz-ideal (respectively sz-ideal, z-ideal) of A.*

Proof We provide a proof for the case of mz-ideals and leave the other proofs to the interested reader.

We begin by proving that if $P + B$ is an mz-ideal of $A + B$, then P is an mz-ideal of A. To that end, let $x_1, \ldots, x_n \in P$ and $a \in A$ and suppose that $V_M(x_1, \ldots, x_n) \subseteq V_M(a)$; this takes place in A. One checks that the collection of maximal ideals of $A + B$ containing $x_1, \ldots, x_n$ is contained in the collection of maximal ideals of $A + B$ containing a. Thus, $a \in P + B$, whence $a \in P$.

For the reverse direction suppose that P is an mz-ideal of A. Let $x_1 + b_1, \ldots, x_n + b_n \in P + B$ and $V_M(x_1 + b_1, \ldots, x_n + b_n) \subseteq V_M(a + b)$, where this takes place in $A + B$. Obviously, $x_1, \ldots, x_n \in P$. Let M be a maximal ideal of A that contains $x_1, \ldots, x_n$. But then $M + B \in V_M(x_1 + b_1, \ldots, x_n + b_n)$, whence $a \in M$, so that $a \in P$, and hence $a + b \in P + B$. □

It follows that if $A+B$ is a Yosida ring, then so is A. However, as we shall see in the next section, the class of Yosida rings is not closed under homomorphic images. In fact, if every homomorphic image of A (including A itself) is a Yosida ring, then A is a Hilbert ring, which follows from the fact that a Yosida domain is Jacobson semisimple.

Proposition 1 *For index $i \in I$, the following statements are equivalent.*

1. *M_i is an sz-ideal of $A+B$.*
2. *M_i is an mz-ideal of $A+B$.*
3. *M_i is a z-ideal of $A+B$.*
4. *P_i is an sz-ideal of A.*

Proof As noted before, (1) $\Rightarrow$ (2) $\Rightarrow$ (3), and (1) $\Leftrightarrow$ (4) follows from $A+B/M_i \cong A/P_i$.

To complete the proof, suppose that M_i is a z-ideal; we need to show that P_i is the intersection of the maximal ideals of A which contain P_i. To this end, let a be an element of the intersection of the maximal ideals of A which contain P_i, and consider the element ae_i of $A+B$. Then by construction $ae_i \in P+M_i$ for every maximal ideal $P+M_i$ of $A+B$ containing M_i, and since $ae_i(i')=0$ for every index $i' \neq i$, it follows that $ae_i \in P+M_{i'}$ for every maximal ideal $P+M_{i'}$ of $A+B$ containing $M_{i'}$. Moreover, $ae_i \in B$, so $ae_i \in P+B$ for every maximal ideal $P+B$ of $A+B$ containing B, and hence $V_M(ae_i)=\mathrm{Max}(A+B)=V_M(0)$. Since M_i is a z-ideal of $A+B$, we conclude that $ae_i \in M_i$, which forces $a \in P_i$. □

Theorem 3 *The ring $A+B$ is a Yosida ring if and only if A is a Yosida ring and for all indices i, A/P_i is a Yosida ring.*

Proof Suppose first that $A+B$ is a Yosida ring. By Lemma 1, A is also a Yosida ring.

For index i and prime ideal $Q \in \mathrm{Spec}(A)$ such that $P_i \subseteq Q$, let $S=\{a_1+P_i,\dots,a_n+P_i\} \subseteq Q/P_i$, and suppose that $V_M(S) \subseteq V_M(a+P_i)$ in $\mathrm{Max}(A/P_i)$ for some $a \in A$. Consider the element $ae_i \in A+B$, and $V_M(ae_i)$ in $\mathrm{Max}(A+B)$. Since $ae_i \in B$, $V_M(ae_i)$ contains every maximal ideal of the form $P+B$ for $P \in \mathrm{Max}(A)$, and since $(ae_i)(i')=0$ for all indices $i' \neq i$, $V_M(ae_i)$ contains every maximal ideal of the form $P+M_{i'}$ for $P \in \mathrm{Max}(A)$ such that $P_{i'} \subseteq P$ and $i' \neq i$. If $P \in \mathrm{Max}(A)$ such that $P_i \subseteq P$ and $a_1,\dots,a_n \in P+M_i$, then $a_1(i),\dots,a_n(i) \in P/P_i$ implies $P/P_i \in V_M(S) \subseteq V_M(a+P_i)$ in $\mathrm{Max}(A/P_i)$, and therefore $(ae_i)(i)=a(i) \in P/P_i$, that is, $ae_i \in P+M_i$. It follows that $V_M(a_1,\dots,a_n) \subseteq V_M(ae_i)$ in $\mathrm{Max}(A+B)$, which in turn implies that $ae_i \in Q+M_i$, since $a_1,\dots,a_n \in Q$ and $A+B$ is a Yosida ring. Thus $a(i)=(ae_i)(i) \in Q/P_i$ shows that $a+P_i \in Q/P_i$, as required.

Conversely, suppose that A is a Yosida ring and for all indices i, A/P_i is a Yosida ring. By Lemma 1, for each $Q \in \mathrm{Spec}(A)$, the prime ideal $Q+B$ is an mz-ideal of $A+B$.

For index i and prime ideal $Q \in \mathrm{Spec}(A)$ such that $P_i \subseteq Q$, let $S=\{r_1,\dots,r_n\} \subseteq Q+M_i$ and $r \in A+B$ such that $V_M(S) \subseteq V_M(r(i))$ in

$\text{Max}(A + B)$. For all $P \in \text{Max}(A)$ such that $P_i \subseteq P$, if $r_1(i), \ldots, r_n(i) \in P/P_i$, then $r_1, \ldots, r_n \in P + M_i$, that is, $P + M_i \in V_M(S) \subseteq V_M(r)$, so that $r(i) \in P/P_i$. Therefore, $V_M(r_1(i), \ldots, r_n(i)) \subseteq V_M(r)$ in $\text{Max}(A/P_i)$, so that $r(i) \in Q/P_i$, since A/P_i is a Yosida ring, and hence $r \in Q + M_i$, as required. □

As we shall see shortly, if $R = A + B$ is a Martínez ring, then $\mathcal{P} \subseteq \text{Max}(A)$, from which it follows then each M_i is a maximal ideal, and therefore R is Jacobson semisimple. Consequently, the $A + B$ construction will not produce an example of a Martínez ring that is not Yosida. In the last section, we discuss a different construction and supply such an example.

A useful characterization of when $A + B$ is classical is the following.

Lemma 2 ([8]) *The ring $A + B$ is a classical ring if and only if $\mathcal{P} \subseteq \text{Max}(A)$ and $\cup\mathcal{P} = \cup\text{Max}(A)$.*

Next, we provide a useful characterization of the double annihilator of a subset of $A + B$.

Lemma 3 ([8]) *Let $R = A + B$ and $S \subseteq R$. Then*

$$S^{\perp\perp} = \{r \in A + B : r(i) = 0 \textit{ whenever } s(i) = 0 \textit{ for all } s \in S\}.$$

Next, we would like to characterize the prime sd-ideals of $R = A + B$. The following definition shall be useful.

Definition 3 Let A be a reduced ring and $\mathcal{P} \subseteq \text{Spec}(A)$ satisfying $\cap\mathcal{P} = \{0\}$. We call an ideal I of A a **$\mathcal{P}$-ideal** if whenever $F \subseteq I$ is a finite subset and $V_{\mathcal{P}}(F) \subseteq V_{\mathcal{P}}(a)$, then $a \in I$, where $V_{\mathcal{P}}(F) = V(F) \cap \mathcal{P} = \{P_\alpha \in \mathcal{P} : F \subseteq P\}$.

For example, when $\mathcal{P} = \text{Max}(A)$ then a $\mathcal{P}$-ideal is precisely an mz-ideal of A. On a different note, when R is reduced and $\mathcal{P} = \text{Min}(R)$, then a $\mathcal{P}$-ideal is an sd-ideal [1].

Lemma 4 *If A and B are as in the previous definition and I is an ideal of A, then $I + B$ is an sd-ideal of $A + B$ if and only if I is a $\mathcal{P}$-ideal of A.*

Proof Suppose first that I is a $\mathcal{P}$-ideal of A, and let $S = \{a_1 + b_1, \ldots, a_n + b_n\} \subseteq I + B$, with each $a_j \in I$ and $b_j \in B$, and let $r = a + b \in S^{\perp\perp}$, with $a \in A$ and $b \in B$. If $P_\alpha \in V_{\mathcal{P}}(a_1, \ldots, a_n)$, then $a_1, \ldots, a_n \in P_\alpha$, implies that each $(a_j + b_j)(i) = 0$ for infinitely many $i = (\alpha, n)$, where $n \in \mathbb{N}$. Hence by Lemma 3, $r(i) = 0$ for infinitely many such n, so that $a \in P_\alpha$. Therefore, $V_{\mathcal{P}}(a_1, \ldots, a_n) \subseteq V_{\mathcal{P}}(a)$, and since I is a $\mathcal{P}$-ideal, it follows that $a \in I$. Thus, $r \in I + B$.

Conversely, suppose that $I + B$ is an sd-ideal of $A + B$ for some ideal Q of A, and let $a_1, \ldots, a_n \in I$ and $a \in A$ such that $V_{\mathcal{P}}(a_1, \ldots, a_n) \subseteq V_{\mathcal{P}}(a)$. For each index $i = (\alpha, n) \in \mathcal{I}$, if $a_j(i) = 0$ for all j, then $a_1, \ldots, a_n \in P_\alpha$ implies $a \in P_\alpha$, by assumption, so that $a(i) = 0$ also. Hence by Lemma 3, $a \in \{a_1, \ldots, a_n\}^{\perp\perp} \subseteq I + B$, since $I + B$ is assumed to be an sd-ideal of $A + B$. Thus, $a \in I$. □

This lemma allows us to characterize the $A + B$ rings which are Martínez rings.

Theorem 4 *Let $R = A + B$. The following statements are equivalent.*

1. *The ring R is a Martínez ring.*
2. $d\mathrm{Rad}(R) = \mathrm{Rad}(R)$.
3. $\mathcal{P} \subseteq \mathrm{Max}(A)$*, and every prime ideal of A is a $\mathcal{P}$-ideal.*

Proof Clearly, (1) and (2) are equivalent.

Suppose that R is a Martínez ring; then as remarked above, R is classical, and hence $\mathcal{P} \subseteq \mathrm{Max}(A)$. Moreover, if $Q \in \mathrm{Spec}(A)$, then $Q + B \in Spec(R)$ is an sd-ideal by assumption, so that Q is a $\mathcal{P}$-ideal of A by Lemma 4.

Conversely, suppose that $\mathcal{P} \subseteq \mathrm{Max}(A)$ and every prime ideal of A is a $\mathcal{P}$-ideal. Then for each index $i \in \mathcal{I}$, the ideal $M_i = A/P_i$ is both maximal and minimal, so an sd-ideal. The only other prime ideals of R have the form $Q + B$ for some $Q \in \mathrm{Spec}(A)$, and since Q is assumed to be a $\mathcal{P}$-ideal, $Q + B$ is an sd-ideal by Lemma 4. □

As noted above, if $\mathcal{P} = \mathrm{Max}(A)$, then a $\mathcal{P}$-ideal is precisely an mz-ideal of A, so we obtain the following immediate consequence.

Corollary 2 *If A is a Jacobson semisimple Yosida ring and $\mathcal{P} = \mathrm{Max}(A)$, then every prime ideal of $R = A + B$ is an sd-ideal, i.e., R is a Martínez ring.*

Using Corollary 1, we obtain the following.

Corollary 3 *If $R = A + B$ is a Martínez ring, then R is Jacobson semisimple and hence a Yosida ring.*

Example

Observe that if $A = \mathbb{Z}$ and $\mathcal{P} = \mathrm{Max}(A)$, then $A+B$ is a Hilbert ring and a Martínez ring. However, $A + B$ is not von Neumann regular.

3 Lattice-Ordered Groups

To produce our examples we shall use the Jaffard-Ohm-Kaplansky theorem. We cite three references for the interested reader: [3, 7, 10]. Another good source for our discussion is [16].

For a domain A, we let $U(A)$ denote its set of units and $q(A)$ its classical field of fractions. The set $U(A)$ is a subgroup of $q(A)^* = q(A) \setminus \{0\}$. The factor group $q(A)^*/U(A)$ is called the *group of divisibility of* A and is denoted by $G(A)$. The group $G(A)$ can be partially ordered by setting $q_1U(A) \leq q_2U(A)$ if and only if $\frac{q_2}{q_1} \in A$. This partial order makes $G(A)$ into a partially ordered group. The order becomes a lattice-order precisely when A is a GCD-domain. In particular, if A is a Bézout domain, then $G(A)$ is a lattice-ordered group, or ℓ-group. Observe that for nonzero $a, b \in A$, $aU(A) = bU(A)$ if and only if $aA = bA$, and that $aU(A) \leq bU(A)$ if and only if $a|b$. This last fact is the reason that $G(A)$ is called the group of divisibility.

The Jaffard-Ohm-Kaplansky Theorem (JOK) states that if G is a lattice-ordered group, then there is a Bézout domain A for which $G(A) \cong G$. The main point here is that one can translate certain ring-theoretic properties covering a Bézout domain into a property about ℓ-groups. Thus, if one can construct an ℓ-group with or without certain properties, then there are Bézout domains with or without said ring-theoretic properties.

In [15], the authors used the JOK method to characterize when the frame of radical ideals of a Bézout domain is a Yosida frame. Unfortunately, there is a slight error in [15, Theorem 6.5]. One needs to include that the Jacobson radical of the domain is 0. The same needs to be done in Corollary 6.6, but then the result is trivial as such a valuation domain is in fact a field. In particular, if (A, M) is a valuation domain that is not a field, then $\text{Rad}(A)$ is not a Yosida frame as the zero-ideal is not a meet of maximal elements.

Throughout this section we assume that A is a Bézout domain. Observe that A is a valuation domain if and only if $G(A)$ is a totally ordered group. Examples of totally ordered groups include the real numbers and each of its subgroups.

The following theorem is well-known and can be found in several articles, e.g., [10, 16], and [3].

Theorem 5 *Let A be a Bézout domain. There is a one-to-one order-reversing correspondence between nonzero prime ideals of A and prime subgroups of $G(A)$, where a nonzero prime ideal $P \in \text{Spec}(A)$ corresponds to the convex ℓ-subgroup generated by $\{aU(A) : a \in A \setminus P\}$.*

Thus, it is not surprising that we can characterize when a Bézout domain is a Hilbert domain via its group of divisibility. Before we do that, we present some other useful notation for ℓ-groups.

Definition 4 Let $(G, +, \leq)$ be an ℓ-group and $0 < g$.

1. The polar of g is the set $g^{\perp} = \{h \in G : g \wedge |h| = 0\}$.
2. The element g is a weak order unit if $g^{\perp} = \{0\}$.
3. For $g \in G$, we set $U_m(g) = \{Q \in \text{Min}(G) : g \notin Q\}$, $V_m(g) = \{Q \in \text{Min}(G) : g \in Q\}$, and $V_g = \cap V_m(g)$.

We remark that $g^{\perp\perp} = V_g$. We also note that an element of G is a weak order unit if and only if it does not belong to any minimal prime subgroup of G. Thus, we obtain the following useful consequence.

Proposition 2 *Let A be a Bézout domain and $G(A)$ its group of divisibility. Then weak order units of $G(A)$ correspond to elements that belong to every maximal ideal of A, and hence A is Jacobson semisimple if and only if $G(A)$ has no weak order units.*

The following theorem should now be apparent.

Theorem 6 *Suppose A is a Bézout domain and let $G = G(A)$ be its group of divisibility. A is a Hilbert domain if and only if G does not have any weak order units and every prime subgroup of G is a union of minimal prime subgroups.*

We next turn to characterizing when a Bézout domain is a Yosida ring. We need the following ℓ-group theorem.

Proposition 3 *Let G be an ℓ-group. Every $Q \in \mathrm{Spec}(G)$ has the property that $g \notin Q$ and $U_m(g) \subseteq U_m(x)$ implies $x \notin Q$ if and only if for every $g \in G$, $G(g) = g^{\perp\perp}$.*

Proof Suppose every prime subgroup of G has the desired property and let $g \in G$. Without loss of generality, we assume that $0 < g$. Suppose $x \in g^{\perp\perp}$. If $x \notin G(g)$, then there is some prime subgroup, say Q, maximal with respect to the property that $G(g) \subseteq Q$ and $x \notin Q$. Observe that $U_m(x) \subseteq U_m(g)$, since $x \in V_g$ by the above remark. The hypothesis forces $g \notin Q$, a contradiction.

Conversely, suppose every principal convex ℓ-subgroup is a polar and let $Q \in \mathrm{Spec}(G)$. Let $g \in G^+ \setminus Q$ and suppose $U_m(g) \subseteq U_m(x)$. Observe then that $V_m(x) \subseteq V_m(g)$, and so consequently, $V_g \subseteq V_x$ which means that $g \in V_x = x^{\perp\perp} = G(x)$. If $x \in Q$, then so is $g \in Q$, whence $x \notin Q$. □

Theorem 7 *Suppose A is a Bézout domain and let $G = G(A)$ be its group of divisibility. Every nonzero prime ideal of A is a z-ideal if and only if every principal convex ℓ-subgroup of $G(A)$ is a polar. Moreover, A is a Yosida ring if and only if A is Jacobson semisimple and every principal convex ℓ-subgroup of $G(A)$ is a polar.*

Proof Suppose every nonzero prime ideal of A is a z-ideal. Let Q be a prime subgroup of G, $g \in G^+ \setminus Q$, and $x \in G^+$ for which $U_m(g) \subseteq U_m(x)$. We aim to show that $x \notin Q$. We would then be able to apply Proposition 3 to conclude that every principal convex ℓ-subgroup is a polar. Choose any $a, b \in A$ such that $g = aU(A)$ and $x = bU(A)$. Let $P \in \mathrm{Spec}(A)$ corresponding to Q and observe that $a \in P$. Note that $V_M(a) \subseteq V_M(b)$. So since P is a z-ideal, it follows that $b \in P$, whence $x \notin Q$.

We leave the proof of the converse to the interested reader. □

Example

We provide an example of an ℓ-group G for which any Bézout domain A whose group of divisibility is isomorphic to G is a Yosida ring (i.e., every prime is an mz-ideal) but is not a Hilbert domain.

In fact, the example of an ℓ-group that we have in mind will be without weak order units, and also have the properties that every principal convex ℓ-subgroup is a polar and not every prime subgroup is a union of minimal prime subgroups. Then, equipped with such a G we invoke JOK to obtain a Bézout domain whose group of divisibility is G. Such an A will be a Yosida ring that is not a Hilbert domain.

Our example is a modification of the example used in [17]. Let K be an infinite compact zero-dimensional space and set $K_n = K$ for $n \in \mathbb{Z}^+$. (For example, let

K be the Cantor space.) Let X be the disjoint union of the spaces K_n. We denote $\mathbb{Q}$ with the discrete topology by $\mathbb{Q}_d$. Let $\mathscr{C} = C_0(X, \mathbb{Q}_d)$ be the set of continuous rational-valued functions on X with compact support (i.e., for each $f \in \mathscr{C}$, the closure of the set of points where f takes nonzero values, $\operatorname{cl}\operatorname{coz}(f)$, is compact). It is known that $\mathscr{C}$ is a hyper-archimedean ℓ-group and that the prime subgroups are classified as $M_x = \{f \in \mathscr{C} : f(x) = 0\}$ for some $x \in X$.

It is important to know that hyper-archimedean means that every ℓ-homomorphism is archimedean, or equivalently, that every prime subgroup is minimal. One more equivalent condition about hyper-archimedean ℓ-groups is that every principal convex ℓ-subgroup is a summand and is a polar. (See [6] for more information as well as proofs.)

Next, define the functions

$$f(x) = \begin{cases} 1, & \text{if } x \in K_{2n} \\ 0, & \text{otherwise.} \end{cases}$$

and

$$e(x) = \begin{cases} \frac{1}{4n}, & \text{if } x \in K_{4n} \\ 0, & \text{otherwise.} \end{cases}$$

Finally, let $\mathscr{A}$ be the ℓ-subgroup of $C(X, \mathbb{Q}_d)$ generated by $\mathscr{C}$, f, and e. Observe that $0 \le e \le f$. Furthermore, every element of $\mathscr{A}$ can be written in the form $g = c + nf + me$ for some $c \in \mathscr{C}$ and $n, m \in \mathbb{Q}$.

The argument in [17] shows that a (proper) prime subgroup of $\mathscr{A}$ is either of the form M_x for some $x \in X$ or is $\mathscr{C}$ or is the subgroup generated by $\mathscr{C}$ and e. Denote the last of these by P. It follows that the minimal prime subgroups all have the form M_x. The union of these minimal prime subgroups is $\mathscr{C}$. Thus, P is not a union of minimal prime subgroups.

Clearly, $\mathscr{A}$ has no weak order units, and it is shown in [17] that $\mathscr{A}$ is an example of an existentially closed ℓ-group. Furthermore, one of the criteria for an ℓ-group to be existentially closed is that every principal convex ℓ-subgroup is a polar (see [18]). Therefore, $\mathscr{A}$ is our desired example.

Example

Here is an example of a Martínez ring that is not a Hilbert ring. Let A be a Yosida ring that is not a Hilbert ring, and set $\mathcal{P} = \operatorname{Max}(A)$. Then the ring $R = A + B$ is a Martínez ring by Theorem 4, because $\mathcal{P} = \operatorname{Max}(A)$, so a $\mathcal{P}$-ideal is an mz-ideal. Since A is not a Hilbert ring, neither is R. Observe that R is also not von Neumann regular.

4 $A \times QB$

We finish with an example of a Martínez ring that is not a Yosida ring. We notice that, if $R = A + B$ is a Martínez ring, then by Theorem 4 it follows that $\mathcal{P} \subseteq \mathrm{Max}(A)$, so that each $M_i \in \mathrm{Max}(R)$. Moreover, $\cap M_i = \{0\}$, so a Martínez $A + B$-ring is Jacobson semisimple. Consequently, if $R = A + B$ is a Martínez ring, then by Corollary 1, it is automatically a Yosida ring. Therefore, to construct an example of a Martínez ring that is not a Yosida ring, we need a different construction.

We use a variation of the $A+B$-construction due to Lucas [14] (see also Chapter 2 of [8]), which we shall call an *$A \times QB$-ring*, defined as follows. In the construction of the ring $A + B$, we drop the assumption that $\cap\mathcal{P} = \{0\}$. Let K_i be the field of fractions of the integral domain $A_i = A/P_i$ for each $i \in \mathcal{I}$, and let $QB = \sum_{i \in \mathcal{I}} K_i$. Finally, we set $R = A \times QB$, with addition defined component-wise, and multiplication by $(r, a)(s, b) = (rs, rb + sa + ab)$. One easily checks that R is a commutative ring with identity $(1, 0)$. As before, for $b \in QB$, let $b(i)$ be the i-th coordinate of b, and for $a \in A$, write $a(i)$ for the image of a in K_i.

From [14, Theorem 8.4], the ideal $M_i = \{(a, b) \in R : a(i) = -b(i)\}$ is both a maximal ideal and minimal prime ideal of R, and the only other prime ideals of R are of the form $P \times B$, where P ranges over the prime ideals of A. (In the statement of the theorem, Lucas assumes that A is an integral domain, but in fact the proof holds for any commutative ring A with identity.) Using this fact, we construct a Martínez ring that is not a Yosida ring.

Example

Let $R = A \times QB$, where A is a semilocal principal ideal domain with exactly two maximal ideals N_1 and N_2, and $\mathcal{P} = \{N_1, N_2\}$. From the above remarks, we see that $\mathrm{Max}(R) = \{M_i : i \in \mathcal{I}\} \cup \{N_1 \times B, N_2 \times B\}$, so that the Jacobson radical of R is $N_1 N_2 \times \{0\}$, and therefore R is not Jacobson semisimple. On the other hand, the above remarks imply that $\mathrm{Spec}(R)$ contains exactly one non-maximal prime ideal, $\{0\} \times B$, from which it follows that R is reduced. Therefore, R cannot be a Yosida ring. (Note that $\{0\} \times B$ is a minimal prime ideal.)

Since every minimal prime ideal is an sd-ideal, to show that R is a Martínez ring, we need only check that $N_1 \times B$ and $N_2 \times B$ are sd-ideals. Using the fact that R is reduced, it suffices to show that $N_1 \times B$ and $N_2 \times B$ are $\mathrm{Min}(R)$-ideals (see Definition 3). That is, if F is a finite subset of $N_j \times B$, and $V_{\mathrm{Min}(R)}(F) \subseteq V_{\mathrm{Min}(R)}(r)$ for some $r \in R$, then r must be an element of $N_j \times B$.

Note that $(a, b) \in \{0\} \times B$ if and only if $a = 0$, while $(a, b) \in M_i$, for index $i \in \mathcal{I}$, if and only if $a(i) + b(i) = 0$. Suppose that $(a_1, b_1), \ldots, (a_m, b_m) \in N_j \times B$ and

$$V_{\mathrm{Min}(R)}(\{(a_1, b_1), \ldots, (a_m, b_m)\}) \subseteq V_{\mathrm{Min}(R)}((a, b))$$

for some $(a, b) \in R$. Then $a_1, \ldots, a_m \in N_j$, so there are infinitely many indices $i = (j, n)$ such that $b_1(i) = \ldots = b_m(i) = 0$, so infinitely many such indices i that

$$M_i \in V_{\text{Min}(R)}(\{(a_1, b_1), \ldots, (a_m, b_m)\})$$

and hence infinitely many such that $M_i \in V_{\text{Min}(R)}((a, b))$. Then $b(i) = 0$ for at least one such index i, which implies that $a \in N_j$, and therefore $(a, b) \in N_j \times B$. Thus, $N_j \times B$ is an sd-ideal.

Acknowledgments The authors would like to thank the referee for the very helpful suggestions in improving the clarity of this paper.

References

1. Aliabad, A.R., Mohamadian, R.: On sz°-ideals in polynomial rings. Comm. Alg. **39**, 701–717 (2011)
2. Alling, N. L.: Rings of continuous integer-valued functions and nonstandard arithmetic. Trans. Am. Math. Soc. **118**, 498–525 (1965)
3. Anderson, M., Feil, T.: Lattice ordered groups. In: An Introduction. Reidel, Dordrecht (1988)
4. Artico, G., Marconi, U., Morescu, R.: A subspace of Spec(A) and its connexions with the maximal ring of quotients. Rendiconti del Seminario Matematico della Università di Padova, tome **64**, 93–107 (1981)
5. Banaschewski, B.: Radical ideals and coherent frame. Comment. Math. Univ. Carolin. **37**(2), 349–370 (1996)
6. Conrad, P.: Epi-archimedean groups. Czechoslovak Math. J. **24**(99), 192–218 (1974)
7. Darnel, M.: Theory of lattice-ordered groups. In: Monographs and Textbooks in Pure and Applied Mathematics, vol. 187. Marcel Dekker Inc, New York (1995)
8. Epstein, A. N.: Annihilators and $A + B$ Rings. Thesis (Ph.D.)–Florida Atlantic University, Boca Raton (2020)
9. Gillman, L., Jerrison, M.: Rings of continuous functions. In: The University Series in Higher Mathematics. Princeton, D. Van Nostrand (1960)
10. Gilmer, R.: Multiplicative Ideal Theory. Marcel Dekker, New York (1972)
11. Huckaba, J.: Commutative Rings with Zero Divisors. Marcel Dekker Inc, New York (1988)
12. Ighedo, O., McGovern, W. Wm.: On the lattice of z-ideals of a commutative ring. Top. Appl. **273**, 106969 (2020)
13. Lucas, T.: Two annihilator conditions: Property (A) and (A.C.). Commun. Alg. **14**, 557–580 (1986)
14. Lucas, T.: Prüfer v-multiplication rings and the ring of finite fractions. Rocky Mountain J. Math. **35**(4), 1251–1325 (2005)
15. Martinez, J., Zenk, E.: Yosida frames. J. Pure Appl. Alg. **204**(3), 473–492 (2006)
16. McGovern, W. Wm.: Neat rings. J. Pure Applied Alg. **205**(2), 243–265 (2006)
17. Saracino, D., Wood, C.: An example in the model theory of Abelian lattice-ordered groups. Algebra Univers. **19**, 34–37 (1984)
18. Wynne, B.: Construction of existentially closed Abelian lattice-ordered groups using upper extensions. Algebra Univers. **79**(3), Paper No. 51 (2018)

Functional Identities and Maps Preserving Two-Sided Zero Products

Matej Brešar

2020 Mathematics Subject Classification 16R60, 16N60, 46H05, 46H15

1 Introduction

A *functional identity* (FI for short) is an identical relation involving arbitrary elements of (a subset of) a ring along with some functions that are considered as unknowns. This is not a formal definition, but actually there is no formal definition. The concept of an FI can be understood through examples. We will give a very brief account of the FI theory in Sect. 2, focusing only on the most applicable aspects of the theory (but without discussing concrete applications). This theory has been developed since the 1990's and is surveyed in the 2007 book [8]. There has been further development since its publication, but we will stick within the basic context.

The main motivation for the development of the FI theory have always been applications to various areas. We will present one such application in Sect. 3. The result we will obtain is not entirely new, that is, it is a variation of the result obtained in the recent paper [9]. However, the proof we will give here is essentially different. The setting in which we will work is that of Banach algebras. Using some functional analytic tools is therefore unavoidable, but nevertheless our proof will be almost entirely algebraic. More precisely, using some standard analytic results we will reduce the problem to the situation where the methods of FI's, presented in Sect. 2, will be applicable.

Let us now introduce the problem we will study in Sect. 3. Let A and B be associative, but not necessarily commutative algebras and let $T : A \to B$ be a linear map. We say that T *preserves two-sided zero products* if for all $x, y \in A$,

M. Brešar (✉)
Faculty of Mathematics and Physics, University of Ljubljana, Ljubljana, Slovenia

Faculty of Natural Sciences and Mathematics, University of Maribor, Maribor, Slovenia
e-mail: matej.bresar@fmf.uni-lj.si

J.-L. Chabert et al. (eds.), *Algebraic, Number Theoretic, and Topological Aspects of Ring Theory*, https://doi.org/10.1007/978-3-031-28847-0_8

$$xy = yx = 0 \implies T(x)T(y) = T(y)T(x) = 0.$$

Homomorphisms as well as antihomomorphisms are obvious examples of such maps. Their natural generalization are *Jordan homomorphisms*. These are linear maps $J : A \to B$ satisfying

$$J(xy + yx) = J(x)J(y) + J(y)J(x)$$

for all $x, y \in A$. Under mild assumptions, Jordan homomorphisms also preserve two-sided zero products, see [7, Lemma 7.20]. Further, we say that a linear map $W : B \to B$ is a *centralizer* if

$$W(xy) = W(x)y = xW(y)$$

for all $x, y \in B$ (that is, W is an element of the *centroid* of B). If B is unital, then clearly W is of the form $W(x) = cx$ with $c = W(1)$ belonging to the center of B. However, we will work with algebras that do not necessarily possess unities. We say that $T : A \to B$ is a *weighted Jordan homomorphism* if there exist an invertible centralizer W of B and a Jordan homomorphism $J : A \to B$ such that $T = WJ$. If J preserves two-sided zero products, then clearly so does T.

Assume now that A and B are Banach algebras. The question that has been considered in [3, 9] is whether a surjective continuous linear map $T : A \to B$ that preserves two-sided zero products is a weighted Jordan homomorphism. We will show in Theorem 3.4 that the answer is positive under certain assumptions that are very similar to those employed in [9]. Still, the result in [9] is slightly better (on the other hand, our theorem considerably improves the result from [3]). Anyhow, our main purpose is to present the applicability of FI's to the problem that has occurred in an apparently entirely different mathematical area.

Let us finally mention that our basic assumption on A will be that A is *zero product determined*. These algebras are the subject of the recent book [7]. We will introduce them in Sect. 3.

2 Functional Identities

By a ring we will always mean an associative ring which may not be unital and may not be commutative. If fact, the theory that we will expose is almost vacuous for commutative rings.

Throughout this section we assume that A is a subring of a unital ring Q with center C. The reason for the notation Q is that we are particularly interested in the case where Q is one of the rings of quotients of A. However, this is not assumed in the definitions which we are about to present.

We will consider functions $E_i, F_j : A^k \to Q$ (where A^k denotes the Cartesian product of k copies of A). These functions are entirely arbitrary, i.e., they are just functions in the set theoretic sense. We will assume that they satisfy certain identities, which we call *functional identities* (FI's), and our goal is to describe their form.

We start by giving several examples.

Example 2.1 An extremely simple example of an FI satisfied by two functions $E_1, F_2 : A \to Q$ is

$$E_1(x_2)x_1 + x_2F_2(x_1) = 0 \tag{2.1}$$

for all $x_1, x_2 \in A$. What is the form of E_1 and F_2? By our assumption, Q has a unity 1, but we are not assuming that A contains 1. However, let us assume temporarily that $1 \in A$. Then (2.1) implies that $E_1(x_2) = x_2p$ where $p = -F_2(1) \in Q$. Writing 1 for x_2 in (2.1) it follows that $F_2(x_1) = -px_1$. Thus, a simple possibility when (2.1) holds is that there exists a $p \in Q$ such that

$$E_1(x_2) = x_2p, \quad F_2(x_1) = -px_1 \tag{2.2}$$

for all $x_1, x_2 \in A$. If $1 \in A$, then this is also the only possibility. There may be other possibilities if 1 does not belong to A. For example, if A contains a nonzero element a such that $aA = \{0\}$, then we may take

$$E_1(x_2) = a \quad \text{and} \quad F_2(x_1) = 0$$

for all $x_1, x_2 \in A$. However, we are not interested in various special situations at this stage. We call (2.2) the *standard solution* of the FI (2.1). This is a formal solution of (2.1) which makes sense for any rings $A \subseteq Q$, while other possible solutions may exist only if A has some special properties.

Example 2.2 Another similarly simple FI is

$$E_1(x_2)x_1 + E_2(x_1)x_2 = 0 \tag{2.3}$$

for all $x_1, x_2 \in A$. How to define the standard solution of (2.3)? To answer this, we again assume that A is unital. Setting $x_1 = 1$ in (2.3) we obtain $E_1(x_2) = ax_2$ for all $x_2 \in A$, where $a = -E_2(1)$. Writing 1 for x_2 in (2.1) we thus obtain $E_2(x_1) = -ax_1$ and so (2.3) becomes

$$a(x_2x_1 - x_1x_2) = 0$$

for all $x_1, x_2 \in A$. If A is commutative, then this is fulfilled for every $a \in Q$. However, if A is not commutative and Q has no zero-divisors, then this is possible only if $a = 0$. Since standard solutions should not depend on the structure of the ring, we define the standard solution of (2.3) as

$$E_1 = E_2 = 0.$$

As is evident from the above discussion, the existence of a nonstandard solution implies that A and Q have special properties.

Example 2.3 Now consider $E_1, E_2, E_3 : A^2 \to Q$ satisfying

$$E_1(x_2, x_3)x_1 + E_2(x_1, x_3)x_2 + E_3(x_1, x_2)x_3 = 0 \tag{2.4}$$

for all $x_1, x_2, x_3 \in A$. Observe that this FI is an extended version of (2.3) involving an additional variable. We thus define the standard solution of (2.4) as

$$E_1 = E_2 = E_3 = 0.$$

It is interesting to add that, unlike in Example (2.2), the class of rings admitting nonstandard solutions includes $A = Q = M_2(F)$, the ring of 2×2 matrices over a field F. Indeed, the Cayley-Hamilton Theorem tells us that, for each $x \in A$, $x^2 - \mathrm{tr}(x)x$ lies in the center C of A (=the set of scalar matrices). The linear map

$$H(x) = x - \mathrm{tr}(x)1$$

thus satisfies $H(x)x \in C$ for all $x \in A$, and hence, by linearizing,

$$H(x_1)x_2 + H(x_2)x_1 \in C \tag{2.5}$$

for all $x_1, x_2 \in A$. That is,

$$[H(x_1)x_2 + H(x_2)x_1, x_3] = 0 \tag{2.6}$$

for all $x_1, x_2, x_3 \in A$, where, as usual,

$$[x, y] = xy - yx.$$

Observe that (2.6) gives (2.4) with

$$E_1(x_2, x_3) = -x_3H(x_2),$$
$$E_2(x_1, x_3) = -x_3H(x_1),$$
$$E_3(x_1, x_2) = H(x_1)x_2 + H(x_2)x_1.$$

These functions E_i are nonzero, so they present a nonstandard solution of (2.4). Note that this solution could not be discovered by assuming that $1 \in A$ and then making some simple substitutions as in Examples 2.1 and 2.2. The question of the existence of nonstandard solutions is usually more subtle than these two examples suggest.

Example 2.4 A slightly more complicated FI is

$$E_1(x_2, x_3)x_1 + E_2(x_1, x_3)x_2 + x_1F_1(x_2, x_3) + x_3F_3(x_1, x_2) = 0 \tag{2.7}$$

for all $x_1, x_2, x_3 \in A$.

Assume for a moment that the terms involving E_1 and F_3 are absent. Then we are in the same situation as in Example 2.1 (with x_3 playing only a formal role), so we know how to define the standard solution. The cases where either the terms involving E_2 and F_1 or the terms involving E_1 and F_1 are absent are similar. Now assume that the terms involving E_2 and F_3 are absent. If $1 \in A$, then it follows immediately that $E_1(x_2, x_3)$ is equal to $-F_1(x_2, x_3)$ and lies in C.

These four special cases help us to discover that a natural solution of (2.7), which we define to be standard, is given by

$$\begin{aligned}
E_1(x_2, x_3) &= x_3p(x_2) + \lambda(x_2, x_3),\\
E_2(x_1, x_3) &= x_1p'(x_3) + x_3p''(x_1),\\
F_1(x_2, x_3) &= -p'(x_3)x_2 - \lambda(x_2, x_3),\\
F_3(x_1, x_2) &= -p(x_2)x_1 - p''(x_1)x_2,
\end{aligned}$$

where p, p', p'' are arbitrary functions from A to Q and λ is an arbitrary function from A^2 to the center C.

These examples should help the reader to grasp the definitions that we are about to give. Let us first introduce the necessary notation. Let m be a positive integer. For any $x_1, \ldots, x_m \in A$ we write

$$\begin{aligned}
\overline{x}_m &= (x_1, \ldots, x_m) \in A^m,\\
\overline{x}_m^i &= (x_1, \ldots, x_{i-1}, x_{i+1}, \ldots, x_m) \in A^{m-1},\\
\overline{x}_m^{ij} = \overline{x}_m^{ji} &= (x_1, \ldots, x_{i-1}, x_{i+1}, \ldots, x_{j-1}, x_{j+1} \ldots, x_m) \in A^{m-2}.
\end{aligned}$$

Let I, J be subsets of $\{1, 2, \ldots, m\}$ and let let

$$E_i : A^{m-1} \to Q \quad \text{and} \quad F_j : A^{m-1} \to Q,$$

where $i \in I$ and $j \in J$, be arbitrary functions. (If $m = 1$, then E_i and F_j are elements in Q.) The following are the fundamental FI's upon which the general theory is based:

$$\sum_{i\in I} E_i(\overline{x}_m^i)x_i + \sum_{j\in J} x_jF_j(\overline{x}_m^j) = 0 \tag{2.8}$$

for all $\overline{x}_m \in A^m$, and

$$\sum_{i\in I} E_i(\overline{x}_m^i)x_i + \sum_{j\in J} x_j F_j(\overline{x}_m^j) \in C \tag{2.9}$$

for all $\overline{x}_m \in A^m$. Observe that the FI's (2.1), (2.3), (2.4), and (2.7) are special cases of (2.8), and (2.5) is a special case of (2.9).

Of course, (2.8) trivially implies (2.9), so these two FI's should be treated separately; that is, we are not assuming that the same functions E_i and F_j simultaneously satisfy both (2.8) and (2.9).

We define the *standard solution* of the FI's (2.8) and (2.9) as

$$\begin{aligned} E_i(\overline{x}_m^i) &= \sum_{\substack{j\in J,\\ j\neq i}} x_j p_{ij}(\overline{x}_m^{ij}) + \lambda_i(\overline{x}_m^i), \quad i \in I, \\ F_j(\overline{x}_m^j) &= -\sum_{\substack{i\in I,\\ i\neq j}} p_{ij}(\overline{x}_m^{ij})x_i - \lambda_j(\overline{x}_m^j), \quad j \in J, \\ \lambda_k &= 0 \quad \text{if} \quad k \notin I \cap J, \end{aligned} \tag{2.10}$$

where

$$\begin{aligned} p_{ij} &: A^{m-2} \to Q, \quad i \in I, \ j \in J, \ i \neq j, \\ \lambda_k &: A^{m-1} \to C, \quad k \in I \cup J, \end{aligned}$$

are arbitrary functions (if $m = 1$ this should be understood as that $p_{ij} = 0$ and λ_k is an element in C). A straightforward verification shows that (2.10) is indeed a solution of (2.8), and therefore also of (2.9).

The case when one of the sets I and J is empty is of special interest. Following the convention that the sum over $\emptyset$ is 0, (2.8) with $J = \emptyset$ reads as

$$\sum_{i\in I} E_i(\overline{x}_m^i)x_i = 0$$

for all $\overline{x}_m \in A^m$, and the standard solution (2.10) is simply $E_i = 0$ for each $i \in I$ (see Examples 2.2 and 2.3). Similarly, $E_i = 0$ is the standard solution of the slightly more general FI

$$\sum_{i\in I} E_i(\overline{x}_m^i)x_i \in C$$

for all $\overline{x}_m \in A^m$ (see (2.5) in Example 2.3).

The following is the central notion of the FI theory.

Definition 2.5 Let d be a positive integer. We say that A is a *d-free* subset of Q if the following hold for all $m \geq 1$ and all $I, J \subseteq \{1, 2, \ldots, m\}$:

(a) If $\max\{|I|, |J|\} \le d$, then (2.8) implies (2.10).
(b) If $\max\{|I|, |J|\} \le d - 1$, then (2.9) implies (2.10).

Thus, roughly speaking, A is a d-free subset of Q (or simply "A is d-free" as we say slightly inaccurately) if the FI's (2.8) and (2.9) have only standard solutions whenever the sets I and J are sufficiently small. It is noteworthy that the definition implies that these standard solutions are unique.

Note that the above definition makes sense if A is any nonempty subset (and not necessarily a subring) of Q. We have restricted ourselves to the case where A is a ring since this is the most basic case and the only one that will be needed in Sect. 3. There are other variations of the notion of d-freeness, but we wil not discuss them here.

It is not immediately clear that d-free subrings actually exist. It turns out that not only that they do exist, but that many (if not most) noncommutative rings are d-free subsets of certain slightly larger rings, provided that they are "far enough" from commutativity. In order to state one of the fundamental results of the theory, we recall some facts about prime rings.

First of all, recall that a ring A is said to be *prime* if the product of two nonzero ideals of A is always nonzero. Equivalently, for all $a, b \in$A, $aAb = \{0\}$ implies $a = 0$ or $b = 0$. A commutative ring is prime if and only if it is an integral domain, and prime rings play a similar role in noncommutative algebra as integral domains do in commutative algebra. Simple rings as well as primitive rings are all prime.

For any prime ring A, one can define the *maximal left ring of quotients* $Q_{ml}(A)$ of A. We refer the reader to [4, Section 2.1] or [8, Appendix A] for details. The center C of $Q_{ml}(A)$ is a field, called the *extended centroid* of A. We write $\deg(x) = n$ if $x \in A$ is algebraic of degree n over C, and $\deg(x) = \infty$ if x is not algebraic over C. Set

$$\deg(A) = \sup\{\deg(x) \mid x \in A\}.$$

It is well known that $\deg(A) \le n < \infty$ if and only if A satisfies the standard polynomial identity of degree $2n$, or equivalently, A can be embedded into the ring of $n \times n$ matrices over a field.

We can now state [8, Corollary 5.12].

Theorem 2.6 *A prime ring A is a d-free subset of $Q_{ml}(A)$ if and only if* $\deg(A) \ge d$.

If A is not a PI-ring (i.e., a ring satisfying a nontrivial polynomial identity), then A is d-free for every d. This is of course the most desirable situation from the point of view of the FI theory. It is interesting that the theory becomes more difficult in PI-rings, which are generally considered simpler and easier to handle than prime rings not satisfying polynomial identities.

The following example is illustrative.

Example 2.7 A prototype of a noncommutative ring is $M_n(F)$, the ring of $n \times n$ matrices over a field F. It is simple and hence prime, so we may ask what does

Theorem 2.6 say for this ring. It is easy to see $Q_{ml}(M_n(F)) = M_n(F)$, the extended centroid C of $M_n(F)$ is just F, and $\deg(A) = n$. Theorem 2.6 thus states that $M_n(F)$ is d-free (as a subset of itself) if and only if $d \le n$. For example, $M_2(F)$ is 2-free, but not 3-free. The latter has been actually observed in Example 2.3. That is, we saw that the Cayley-Hamilton Theorem yields the FI (2.5) which has a nonstandard solution. Similarly we see that the Cayley-Hamilton Theorem for $n \times n$ matrices implies that $M_n(F)$ is not $(n+1)$-free (and neither d-free for any $d > n$). The nontrivial part of the theorem is that $M_n(F)$ is n-free (and in fact d-free for any $d \le n$).

We proceed to a certain more general type of FI's. Until the rest of the section, we fix a set S and a *surjective* function

$$T : S \to A.$$

Similarly as above, let $m \ge 1$, let I, J be subsets of $\{1, 2, \ldots, m\}$, and let

$$E_i : S^{m-1} \to Q \quad \text{and} \quad F_j : S^{m-1} \to Q,$$

where $i \in I$ and $j \in J$, be arbitrary functions. We now consider the FI's

$$\sum_{i \in I} E_i(\overline{x}_m^i)T(x_i) + \sum_{j \in J} T(x_j)F_j(\overline{x}_m^j) = 0 \tag{2.11}$$

for all $\overline{x}_m \in S^m$, and

$$\sum_{i \in I} E_i(\overline{x}_m^i)T(x_i) + \sum_{j \in J} T(x_j)F_j(\overline{x}_m^j) \in C \tag{2.12}$$

for all $\overline{x}_m \in S^m$. If $S = A$ and $T = \mathrm{id}_A$, these are the FI's (2.8) and (2.9). The standard solutions of (2.11) and (2.12) are thus defined similarly as (2.10), that is,

$$\begin{aligned} E_i(\overline{x}_m^i) &= \sum_{\substack{j \in J,\\ j \ne i}} T(x_j)p_{ij}(\overline{x}_m^{ij}) + \lambda_i(\overline{x}_m^i), \quad i \in I,\\ F_j(\overline{x}_m^j) &= -\sum_{\substack{i \in I,\\ i \ne j}} p_{ij}(\overline{x}_m^{ij})T(x_i) - \lambda_j(\overline{x}_m^j), \quad j \in J, \\ \lambda_k &= 0 \quad \text{if} \quad k \notin I \cap J, \end{aligned} \tag{2.13}$$

where

$$\begin{aligned} p_{ij} &: S^{m-2} \to Q, \quad i \in I,\ j \in J,\ i \ne j,\\ \lambda_k &: S^{m-1} \to C, \quad k \in I \cup J, \end{aligned}$$

are arbitrary functions.

We will now state a special case of [8, Theorem 4.3], which shows that treating (2.11) and (2.12) is no harder than treating (2.8) and (2.9). That is, if $A = T(S)$ is d-free, than, under the natural assumptions on $|I|$ and $|J|$, (2.11) and (2.12) have only standard solutions.

Theorem 2.8 *If A is a d-free subset of Q, then the following hold for all $m \geq 1$ and all $I, J \subseteq \{1, 2, \ldots, m\}$:*

(a) *If $\max\{|I|, |J|\} \leq d$, then (2.11) implies (2.13).*
(b) *If $\max\{|I|, |J|\} \leq d - 1$, then (2.12) implies (2.13).*

The last notion we have to introduce is that of a *quasi-polynomial* relative to our fixed function $T : S \to A$. A quasi-polynomial of degree 1 is a function $P : S \to Q$ of the form

$$P(x) = \alpha T(x) + \mu(x),$$

where $\alpha \in C$, $\mu : S \to C$, and at least one of α and μ is nonzero. A quasi-polynomial of degree 2 is a function $P : S^2 \to Q$ of the form

$$\begin{aligned} P(x_1, x_2) =& \alpha_1 T(x_1)T(x_2) + \alpha_2 T(x_2)T(x_1) \\ &+ \mu_1(x_1)T(x_2) + \mu_2(x_2)T(x_1) + \nu(x_1, x_2), \end{aligned}$$

where $\alpha_i \in C$, $\mu_i : S \to C$, $\nu : S^2 \to C$, and at least one of α_i, μ_i, ν is nonzero. In general, a quasi-polynomial of degree $d \geq 1$ is a function $P : S^d \to Q$ that can be written as a sum of terms of the form

$$\mu(x_{i_1}, \ldots, x_{i_k})T(x_{j_1}) \ldots T(x_{j_l})$$

where $i_1 < \cdots < i_k$, $\{i_1, \ldots, i_k\} \cup \{j_1, \ldots, j_l\}$ is a partition of $\{1, \ldots, d\}$, and at least one of the functions $\mu : S^k \to C$ is nonzero. The term $\mu(x_1, \ldots, x_d)$, i.e., the term corresponding to $\{j_1, \ldots, j_l\} = \emptyset$, is called the *central coefficient* of this quasi-polynomial.

The following lemma is a slightly simplified version of [8, Lemma 4.4]. It follows easily from Theorem 2.8 and is one of the most frequently used results in the FI theory.

Lemma 2.9 *Let $P : S^{d-1} \to Q$ be a quasi-polynomial of degree $d - 1$. Suppose one of the following two conditions holds:*

(a) *A is a d-free subset of Q, or*
(b) *A is a $(d - 1)$-free subset of Q and the central coefficient of P is 0.*

Then $P(x_1, \ldots, x_{d-1}) \neq 0$ for some $x_1, \ldots, x_{d-1} \in S$.

The desired conclusion when considering some FI's is that the involved functions are quasi-polynomials. The following is a very special but also very useful case of [8, Theorem 4.13].

Theorem 2.10 *Let $E_i : S^{d-1} \to Q$, $i = 1, \dots, d$, be arbitrary functions. Suppose there exist $\gamma_i \in C$, $i = 1, \dots, d$, such that*

$$\sum_{i=1}^{d} E_i(\overline{x}_d^i)T(x_i) + \gamma_i T(x_i)E_i(\overline{x}_d^i) = 0$$

for all $\overline{x}_d \in S^d$. If A is a d-free subset of Q, then all E_i are either 0 or quasi-polynomials of degree $d-1$.

To illustrate how Theorem 2.10 can be applied, we consider in the next example an extremely important FI (2.14) that naturally occurs in a number of applications of the FI theory and has in fact motivated its early development (most often one encounters the simplest case where $S = A$ and $T = \mathrm{id}_A$).

Example 2.11 Assume that S is now an additive group. Let $F : S^2 \to Q$ be a biadditive map satisfying

$$[F(x, x), T(x)] = 0 \tag{2.14}$$

for all $x \in S$. Using the standard linearization process, based on replacing one variable by the sum of two variables, we derive from (2.14) that the symmetric biadditive map $E : S^2 \to Q$ given by

$$E(x_1, x_2) = F(x_1, x_2) + F(x_2, x_1)$$

satisfies

$$\begin{aligned} &E(x_2, x_3)T(x_1) + E(x_1, x_3)T(x_2) + E(x_1, x_2)T(x_3) \\ -&T(x_1)E(x_2, x_3) - T(x_2)E(x_1, x_3) - T(x_3)E(x_1, x_2) = 0 \end{aligned} \tag{2.15}$$

for all $x_1, x_2, x_3 \in S$. Assuming that A is a 3-free subset of Q it follows from Theorem 2.10 that there exist $\alpha_1, \alpha_2 \in C$, $\mu_1, \mu_2 : S \to C$, and $\nu : S^2 \to C$ such that

$$\begin{aligned} E(x_1, x_2) =& \alpha_1 T(x_1)T(x_2) + \alpha_2 T(x_2)T(x_1) \\ &+ \mu_1(x_1)T(x_2) + \mu_2(x_2)T(x_1) + \nu(x_1, x_2), \end{aligned} \tag{2.16}$$

for all $x_1, x_2 \in S$, and hence

$$2F(x, x) = \alpha T(x)^2 + \mu(x)T(x) + \nu(x, x) \tag{2.17}$$

for all $x \in S$, where $\alpha = \alpha_1 + \alpha_2$ and $\mu(x) = \mu_1(x) + \mu_2(x)$. Assuming that $\frac{1}{2} \in C$ (i.e., $1 + 1$ is invertible in C), (2.17) gives the most complete conclusion that can be drawn from (2.14).

Let us add that using Lemma 2.9 one can easily derive from (2.16) that μ_1 and μ_2 (and hence also μ) are additive and ν is biadditive if T is additive.

With this example we conclude our brief and somewhat fragmentary survey of the FI theory. We have focused only on those results that will be needed in the next section, but nevertheless we hope this was enough to give the reader a taste of the theory.

3 Maps Preserving Two-Sided Zero Products

We now change the setting: in this section we consider Banach algebras (we recommend [11] as a general reference book). Throughout, we assume that A is a complex Banach algebra satisfying the following conditions which will be explained below:

(a) A is zero product determined (zpd),
(b) A is weakly amenable,
(c) A has a bounded approximate identity.

We now give the definitions.

(a) We say that a Banach algebra A is *zero product determined* (*zpd* for short) if every continuous bilinear functional $\varphi : A \times A \to \mathbb{C}$ satisfying $\varphi(x, y) = 0$ whenever $xy = 0$ is of the form $\varphi(x, y) = \tau(xy)$ for some τ in A', the dual space of A (i.e., the space of all continuous linear functionals on A). The theory of these algebras is surveyed in the recent book [7].

(b) If A is any Banach algebra, the dual space A' becomes a Banach A-bimodule by defining

$$(x \cdot f)(y) = f(yx), \quad (f \cdot x)(y) = f(xy)$$

for all $x, y \in A$, $f \in A'$. We can therefore consider derivations from A to A'; recall that these are linear maps $\delta : A \to A'$ satisfying $\delta(xy) = \delta(x) \cdot y + x \cdot \delta(y)$ for all $x, y \in A$. We call δ an inner derivation if there exists a $\tau \in A'$ such that $\delta(x) = \tau \cdot x - x \cdot \tau$ for every $x \in A$. If every continuous derivation from A to A' is inner, then A is said to be *weakly amenable.*

(c) A *bounded approximate identity* in A is a net $(e_\lambda)_{\lambda \in \Lambda}$ such that

$$\lim_{\lambda \in \Lambda} e_\lambda x = \lim_{\lambda \in \Lambda} x e_\lambda = x$$

for every $x \in A$ and $\|e_\lambda\| \leq M$ for some $M > 0$ and all $\lambda \in \Lambda$.

The class of Banach algebras satisfying conditions (a), (b), (c) is fairly large. In particular, it contains C^*-algebras and group algebras $L^1(G)$ where G is any locally compact group.

We also assume throughout this section that B is another Banach algebra and $T : A \to B$ is a *surjective continuous linear map which preserves two-sided zero products*. Recall from the introduction that this means that $T(x)T(y) = T(y)T(x) = 0$ whenever $x, y \in A$ are such that $xy = yx = 0$. We will show that, under some additional assumptions on B, T is a weighted Jordan homomorphism. Our approach is based on [7, Theorem 6.6] which states that if a continuous bilinear functional $\varphi : A \times A \to \mathbb{C}$ satisfies $\varphi(x, y) = 0$ whenever $xy = yx = 0$, then there exist $\tau_1, \tau_2 \in A'$ such that $\varphi(x, y) = \tau_1(xy) + \tau_2(yx)$ for all $x, y \in A$. We remark that all the assumptions (a), (b), (c) are needed for proving this result (which was originally established in [2]). A simple consequence is the following basic lemma.

Lemma 3.1 *For each $\omega \in B'$ there exist $\tau_1, \tau_2 \in A'$ such that*

$$\omega\big(T(x)T(y)\big) = \tau_1(xy) + \tau_2(yx) \tag{3.1}$$

for all $x, y \in A$.

Proof Observe that $\varphi(x, y) = \omega\big(T(x)T(y)\big)$ is a continuous bilinear functional on A with the property that $xy = yx = 0$ implies $\varphi(x, y) = 0$. The lemma therefore follows immediately from the result just stated. □

From now on we will not use the assumptions (a) and (b) anymore.

Let us first handle the case where B is finite-dimensional and unital. The author would like to thank Armando Villena for his help in proving the following lemma.

Lemma 3.2 *If B is finite-dimensional and unital, then there exists an invertible element c from the center of B such that $T(x)^2 = cT(x^2)$ for every $x \in A$ (and hence T is a weighted Jordan homomorphism).*

Proof Let $(e_\lambda)_{\lambda\in\Lambda}$ be as above. Then $\big(T(e_\lambda)\big)_{\lambda\in\Lambda}$ is a bounded net in the finite-dimensional Banach space B, and so it has a convergent subnet. Hence, by passing to a subnet we can assume that $(e_\lambda)_{\lambda\in\Lambda}$ is a bounded approximate identity for A and that there exists a $c \in B$ such that

$$\big(T(e_\lambda)\big)_{\lambda\in\Lambda} \to c.$$

Take an $\omega \in B'$. According to Lemma 3.1, there exist $\tau_1, \tau_2 \in A'$ such that (3.1) holds. Writing e_λ for y in (3.1) we obtain

$$\omega\big(T(x)T(e_\lambda)\big) = \tau_1(xe_\lambda) + \tau_2(e_\lambda x). \tag{3.2}$$

Similarly, writing e_λ for x and x for y in (3.1), we obtain

$$\omega\big(T(e_\lambda)T(x)\big) = \tau_1(e_\lambda x) + \tau_2(xe_\lambda). \tag{3.3}$$

Taking limits in (3.2) and (3.3) we arrive at

$$\omega\big(T(x)c\big) = \omega\big(cT(x)\big) = (\tau_1 + \tau_2)(x). \tag{3.4}$$

We thus get

$$\omega\big(T(x)c\big) = \omega\big(cT(x)\big)$$

for every $\omega \in B'$ and every $x \in A$, and hence

$$T(x)c = cT(x)$$

for every $x \in A$. Thus, c lies in the center of B. From (3.1) and (3.4) we also see that

$$\omega\big(T(x)^2\big) = \tau_1(x^2) + \tau_2(x^2) = \omega\big(cT(x^2)\big)$$

and therefore

$$T(x)^2 = cT(x^2)$$

for every $x \in A$. Now, let $b \in A$ be such that $T(b) = 1$. Writing b for x we see that c is invertible (and $c^{-1} = T(b^2)$).

Observe that $T(x)^2 = cT(x^2)$ implies that $J(x) = c^{-1}T(x)$ is a Jordan homomorphism, and hence $T(x) = cJ(x)$ is a weighted Jordan homomorphism (i.e., $T = WJ$ with $W(x) = cx$). □

In the proof of the next lemma we will use the machinery of FI's presented in Sect. 2. The lemma considers the case where B is a primitive Banach algebra. Such an algebra B has the following properties:

(1) B is prime,
(2) the extended centroid of B is $\mathbb{C}$,
(3) if B is infinite-dimensional, then it is not a PI-ring,
(4) if B is finite-dimensional, then it is isomorphic to $M_n(\mathbb{C})$.

The property (1) is standard, so is (2) (it can be, for example, deduced from [4, Corollary 4.1.2]), (3) follows from Kaplansky's theorem on primitive PI-rings (see [6, Theorem 7.54]), and (4) is a version of the classical Wedderburn's structure theorem.

We remark that the idea to use FI's for solving a similar problem on maps preserving zero Jordan products appeared in [10].

Lemma 3.3 *If B is primitive, then there exists a nonzero $\alpha \in \mathbb{C}$ such that either $T(xy) = \alpha T(x)T(y)$ for all $x, y \in A$ or $T(xy) = \alpha T(y)T(x)$ for all $x, y \in A$ (i.e., T is a scalar multiple of a homomorphism or an antihomomorphism).*

Proof If B is finite-dimensional, then the lemma follows from (4), Lemma 3.2, and the well known fact that surjective Jordan homomorphisms onto prime (complex) algebras are necessarily homomorhisms or antihomomorphisms [12]. We may therefore assume that B is infinite-dimensional and hence, by (1), (3), and Theorem 2.6, B is a d-free subset of $Q = Q_{ml}(B)$ for every $d \geq 1$.

Take an $\omega \in B'$ and let τ_1 and τ_2 be linear functionals from Lemma 3.1. Observing that

$$\tau_1\big(x(yxy)\big) + \tau_2\big((yxy)x\big) = \tau_1\big((xyx)y\big) + \tau_2\big(y(xyx)\big),$$

we see from (3.1) that

$$\omega\big(T(x)T(yxy)\big) = \omega\big(T(xyx)T(y)\big)$$

for all $x, y \in A$. Since ω is an arbitrary element of B' it follows that

$$T(x)T(yxy) = T(xyx)T(y) \tag{3.5}$$

for all $x, y \in A$. This is the first FI upon which our proof is based. To derive the second one, observe that

$$\omega\big([T(x), T(y)]\big) = (\tau_1 - \tau_2)([x, y])$$

for all $x, y \in A$. Since $[xy, z] + [zx, y] + [yz, x] = 0$ it follows that

$$\omega\big([T(xy), T(z)] + [T(zx), T(y)] + [T(yz), x]\big) = 0$$

for all $x, y, z \in A$ and all $\omega \in B'$, which implies that

$$[T(xy), T(z)] + [T(zx), T(y)] + [T(yz), T(x)] = 0 \tag{3.6}$$

for all $x, y, z \in A$. This is our second crucial FI.

We remark that (3.5) is similar to an FI studied in [10], and (3.6) is a standard FI that was already encountered in Example 2.11 (see (2.15)). Some of the arguments that follow are therefore repetitions of the known ones.

We start by considering (3.5). Incidentally, we could handle this FI by using a result on quasi-polynomials which is more general than Theorem 2.10. However, since this is a semi-expository article, we will use another approach which takes a bit more space but nicely illustrates how the d-freeness is used.

Linearizing (3.5) we obtain

$$\begin{aligned} T(x)T(yzw+wzy) + T(z)T(yxw + wxy)& \\ =T(xyz + zyx)T(w) + T(xwz + zwx)T(y)& \end{aligned} \tag{3.7}$$

for all $x, y, z, w \in A$. Since B is a 4-free subset of Q it follows from Theorem 2.8 that there exist functions $p_1, p_2, p_3, p_4 : A \times A \to Q$ such that

$$T(yzw + wzy) = p_1(z, y)T(w) + p_2(z, w)T(y) \tag{3.8}$$

$$T(yxw + wxy) = p_3(x, y)T(w) + p_4(x, w)T(y) \tag{3.9}$$

$$T(xyz + zyx) = T(x)p_1(z, y) + T(z)p_3(x, y) \tag{3.10}$$

$$T(xwz + zwx) = T(x)p_2(z, w) + T(z)p_4(x, w) \tag{3.11}$$

for all $x, y, z, w \in A$.

Writing x for z in (3.8) and comparing the obtained identity with (3.9) we get

$$(p_1(x, y) - p_3(x, y))T(w) + (p_2(x, w) - p_4(x, w))T(y) = 0$$

and hence, since B is 2-free,

$$p_1(x, y) = p_3(x, y) \quad \text{and} \quad p_2(x, w) = p_4(x, w)$$

for all $x, y, w \in A$. Similarly, by comparing (3.10) and (3.11) we obtain

$$p_1(z, y) = p_2(z, y) \quad \text{and} \quad p_3(x, y) = p_4(x, y)$$

for all $x, y, z \in A$. Hence,

$$p := p_1 = p_2 = p_3 = p_4.$$

By changing the notation, we can write (3.8) as

$$T(xyz + zyx) = p(y, x)T(z) + p(y, z)T(x), \tag{3.12}$$

so comparing it with (3.10) we arrive at

$$p(y, x)T(z) + p(y, z)T(x) = T(x)p(z, y) + T(z)p(x, y) \tag{3.13}$$

for all $x, y, z \in A$. Since B is 2-free and since (2) holds, there exist functions $q_1, q_2 : A \to Q$ and $\lambda_1, \lambda_2 : A^2 \to \mathbb{C}$ (here we identify $\mathbb{C}$ with $\mathbb{C}1$) such that

$$p(y, x) = T(x)q_1(y) + \lambda_1(x, y) \tag{3.14}$$

$$p(y, z) = T(z)q_2(y) + \lambda_2(z, y) \tag{3.15}$$

$$p(z, y) = q_1(y)T(z) + \lambda_2(z, y) \tag{3.16}$$

$$p(x, y) = q_2(y)T(x) + \lambda_1(x, y). \tag{3.17}$$

for all $x, y, z \in A$. Compare (3.14) and (3.15) (where we write x for z). Since B is 2-free it follows that

$$q := q_1 = q_2 \quad \text{and} \quad \lambda := \lambda_1 = \lambda_2.$$

Further, comparing (3.14) and (3.16) (where we write x for y and y for z) we obtain

$$T(x)q(y) - q(x)T(y) = \lambda(y, x) - \lambda(x, y)$$

for all $x, y \in A$. As B is 3-free it follows that

$$\lambda(y, x) = \lambda(x, y)$$

and there exists a $\beta \in Q$ such that

$$q(y) = \beta T(y) \quad \text{and} \quad q(x) = T(x)\beta$$

for all $x, y \in A$. Therefore, $\beta T(x) = T(x)\beta$ for every $x \in A$. It is a well known property of prime rings that elements in Q that commute with every element in B lie in the extended centroid. Thus, $\beta \in \mathbb{C}$. We thus have

$$p(y, x) = \beta T(x)T(y) + \lambda(x, y)$$

and hence, by (3.12),

$$T(xyz+zyx) = \beta\big(T(x)T(y)T(z) + T(z)T(y)T(x)\big) + \lambda(x, y)T(z) + \lambda(y, z)T(x)$$

for all $x, y, z \in A$. Setting $z = x$ we get

$$T(xyx) = \beta T(x)T(y)T(x) + \lambda(x, y)T(x) \tag{3.18}$$

for all $x, y \in A$.

We now turn our attention to (3.6). Since B is 3-free, we may (as in Example 2.11) use Theorem 2.10 to conclude that there exist $\alpha_1, \alpha_2 \in \mathbb{C}$ and functions $\mu_1, \mu_2 : A \to \mathbb{C}$, $\nu : A^2 \to \mathbb{C}$ such that

$$T(xy) = \alpha_1 T(x)T(y) + \alpha_2 T(y)T(x) + \mu_1(x)T(y) + \mu_2(y)T(x) + \nu(x, y)$$

for all $x, y \in A$. Moreover, applying Lemma 2.9 one easily shows that μ_1, μ_2 are linear and ν is bilinear. From (3.6) we now obtain

$$(\mu_1 - \mu_2)(x)[T(y), T(z)] + (\mu_1 - \mu_2)(y)[T(z), T(x)] + (\mu_1 - \mu_2)(z)[T(x), T(y)] = 0$$

for all $x, y, z \in A$, and hence $\mu := \mu_1 = \mu_2$ by Lemma 2.9. Therefore,

$$T(xy) = \alpha_1 T(x)T(y) + \alpha_2 T(y)T(x) + \mu(x)T(y) + \mu(y)T(x) + \nu(x, y) \tag{3.19}$$

for all $x, y \in A$.

We will now compute $T(xyz)$ in two different ways. Firstly, (3.19) gives

$$\begin{aligned}
&T((xy)z)\\
=&\alpha_1 T(xy)T(z) + \alpha_2 T(z)T(xy) + \mu(xy)T(z)\\
&+ \mu(z)T(xy) + \nu(xy, z)\\
=&\alpha_1^2 T(x)T(y)T(z) + \alpha_1\alpha_2 T(y)T(x)T(z) + \alpha_1\mu(x)T(y)T(z)\\
&+ \alpha_1\mu(y)T(x)T(z) + \alpha_1\nu(x, y)T(z) + \alpha_1\alpha_2 T(z)T(x)T(y)\\
&+ \alpha_2^2 T(z)T(y)T(x) + \alpha_2\mu(x)T(z)T(y) + \alpha_2\mu(y)T(z)T(x)\\
&+ \alpha_2\nu(x, y)T(z) + \mu(xy)T(z) + \alpha_1\mu(z)T(x)T(y)\\
&+ \alpha_2\mu(z)T(y)T(x) + \mu(x)\mu(z)T(y) + \mu(y)\mu(z)T(x)\\
&+ \mu(z)\nu(x, y) + \nu(xy, z).
\end{aligned} \tag{3.20}$$

Secondly, (3.19) gives

$$\begin{aligned}
&T(x(yz))\\
=&\alpha_1 T(x)T(yz) + \alpha_2 T(yz)T(x) + \mu(x)T(yz)\\
&+ \mu(yz)T(x) + \nu(x, yz)\\
=&\alpha_1^2 T(x)T(y)T(z) + \alpha_1\alpha_2 T(x)T(z)T(y) + \alpha_1\mu(y)T(x)T(z)\\
&+ \alpha_1\mu(z)T(x)T(y) + \alpha_1\nu(y, z)T(x) + \alpha_1\alpha_2 T(y)T(z)T(x)\\
&+ \alpha_2^2 T(z)T(y)T(x) + \alpha_2\mu(y)T(z)T(x) + \alpha_2\mu(z)T(y)T(x)\\
&+ \alpha_2\nu(y, z)T(x) + \alpha_1\mu(x)T(y)T(z) + \alpha_2\mu(x)T(z)T(y)\\
&+ \mu(x)\mu(y)T(z) + \mu(x)\mu(z)T(y) + \mu(x)\nu(y, z)\\
&+ \mu(yz)T(x) + \nu(x, yz).
\end{aligned} \tag{3.21}$$

Comparing (3.20) and (3.21) we obtain

$$\begin{aligned}
&\alpha_1\alpha_2[T(y), [T(x), T(z)]]\\
&+\big((\alpha_1 + \alpha_2)\nu(x, y) + \mu(xy) - \mu(x)\mu(y)\big)T(z)\\
&-\big((\alpha_1 + \alpha_2)\nu(y, z) + \mu(yz) - \mu(y)\mu(z)\big)T(x) \in \mathbb{C}
\end{aligned} \tag{3.22}$$

for all $x, y, z \in A$. We may now again use Lemma 2.9 to conclude that

$$(\alpha_1 + \alpha_2)\nu(x, y) + \mu(xy) - \mu(x)\mu(y) = 0 \tag{3.23}$$

for all $x, y \in A$ and

$$\alpha_1\alpha_2 = 0.$$

Thus, either $\alpha_1 = 0$ or $\alpha_2 = 0$.

Assume first that $\alpha_2 = 0$. Then (3.19) reduces to

$$T(xy) = \alpha_1 T(x)T(y) + \mu(x)T(y) + \mu(y)T(x) + \nu(x, y) \tag{3.24}$$

for all $x, y \in A$, and from (3.20) we see that

$$\begin{aligned} T(xyx) =& \alpha_1^2 T(x)T(y)T(x) + \alpha_1\mu(x)T(y)T(x) + \alpha_1\mu(y)T(x)^2 \\ &+ \big(\alpha_1\nu(x, y) + \mu(xy) + \mu(y)\mu(x)\big)T(x) + \alpha_1\mu(x)T(x)T(y) \\ &+ \mu(x)^2 T(y) + \mu(x)\nu(x, y) + \nu(xy, x). \end{aligned}$$

Comparing this identity with (3.18) we obtain

$$\begin{aligned} &(\alpha_1^2 - \beta)T(x)T(y)T(x) + \alpha_1\mu(x)T(y)T(x) + \alpha_1\mu(y)T(x)^2 \\ &+ \big(\alpha_1\nu(x, y) + \mu(xy) + \mu(y)\mu(x) - \lambda(x, y)\big)T(x) + \alpha_1\mu(x)T(x)T(y) \\ &+ \mu(x)^2 T(y) + \mu(x)\nu(x, y) + \nu(xy, x) = 0 \end{aligned}$$

for all $x, y \in A$. Linearizing and invoking Lemma 2.9 it follows, in particular, that $\mu(x)^2$, the coefficient at $T(y)$, is 0 for every $x \in A$. This means that $\mu = 0$. Assuming that $\alpha_1 = 0$ it follows from (3.24) that $T(xy)$ is always a scalar multiple of 1 which is impossible as every element in A can be written as xy since A is has an approximate identity (this is the content of Cohen's Factorization Theorem). Therefore, $\alpha_1 \neq 0$ and so, since $\alpha_2 = 0$, (3.23) implies that $\nu = 0$. Setting $\alpha = \alpha_1$ we thus have $T(xy) = \alpha T(x)T(y)$ for all $x, y \in A$.

Similarly we see that $\alpha_1 = 0$ implies $T(xy) = \alpha T(y)T(x)$ for all $x, y \in A$, where $\alpha = \alpha_2 \neq 0$. □

Recall that a Banach algebra B is said to be *semisimple* (or semiprimitive) if it has trivial Jacobson radical. Equivalently, the intersection of primitive ideals of B, i.e., ideals P such that B/P is a primitive Banach algebra, is trivial.

A semisimple algebra B is in particular *semiprime*, meaning that it has no nonzero nilpotent ideals. We can therefore define the extended centroid C of B, which, however, is a field only if B is prime. One of the important properties of C is that if I is an ideal of B such that $bI \neq \{0\}$ for every nonzero $b \in B$, then also $\lambda I \neq \{0\}$ for every nonzero $\lambda \in C$ (in particular, $\lambda B = \{0\}$ with $\lambda \in C$ implies $\lambda = 0$). In the proof that follows we will use another purely algebraic result [5, Theorem 3.1] which states that if f and g are functions from a set S to a semiprime ring B such that

$$f(s)yg(t) = g(s)yf(t) \tag{3.25}$$

for all $s, t \in S$ and $y \in B$, then there exist idempotents $\varepsilon_1, \varepsilon_2, \varepsilon_3 \in C$ and an invertible element $\gamma \in C$ such that $\varepsilon_i\varepsilon_j = 0$, $i \neq j$, $\varepsilon_1 + \varepsilon_2 + \varepsilon_3 = 1$, $\varepsilon_1 f(s) = \gamma\varepsilon_1 g(s)$, $\varepsilon_2 f(s) = 0$, and $\varepsilon_3 g(s) = 0$ for all $s \in S$.

We are now ready to establish our main theorem.

Theorem 3.4 *Let A and B be Banach algebras. Suppose that A is weakly amenable, zpd, and has a bounded approximate identity, and suppose that B is semisimple and is equal to its ideal generated by $\{y^2 \mid y \in B\}$. Then a surjective continuous linear map $T : A \to B$ preserves two-sided zero products if and only if T is a weighted Jordan homomorphism.*

Proof From [7, Lemma 7.20] it follows that a Jordan homomorphism onto a semiprime algebra preserves two-sided zero products. The same is then true for a weighted Jordan homomorphism. This proves the "if" part.

We start the proof of the "only if" part by taking a primitive ideal P of B. Define $T_P : A \to B/P$ by

$$T_P(x) = T(x) + P.$$

Obviously, T_P is a surjective continuous linear map which preserves two-sided zero products. Lemma 3.3 tells us that there is an $\alpha_P \in \mathbb{C}$ such that $T_P(x^2) = \alpha_P T_P(x)^2$ for every $x \in A$. In particular,

$$T_P(x^2)(y + P)T_P(z)^2 = T_P(x)^2(y + P)T_P(z^2)$$

for all $x, z \in A$ and $y \in B$. That is,

$$T(x^2)yT(z)^2 + P = T(x)^2yT(z^2) + P.$$

Since B is semisimple and P is an arbitrary primitive ideal it follows that

$$T(x^2)yT(z)^2 = T(x)^2yT(z^2)$$

for all $x, z \in A$, $y \in B$.

The functions $x \mapsto T(x^2)$ and $x \mapsto T(x)^2$ thus satisfy the condition (3.25) of the aforementioned result. Therefore, there exist orthogonal idempotents $\varepsilon_1, \varepsilon_2, \varepsilon_3 \in C$ whose sum is 1 and an invertible element $\gamma \in C$ such that for all $x \in A$,

$$\varepsilon_1 T(x)^2 = \gamma\varepsilon_1 T(x^2), \quad \varepsilon_2 T(x)^2 = 0, \quad \varepsilon_3 T(x^2) = 0.$$

The second identity can be read as $\varepsilon_2 y^2 = 0$ for all $y \in B$. Together with our assumption on B this gives $\varepsilon_2 B = \{0\}$, and therefore, $\varepsilon_2 = 0$.

We claim that the existence of an approximate identity implies that A is equal to $\operatorname{span}\{x^2 \mid x \in A\}$ (the linear span of all x^2, $x \in A$). Indeed, by [1, Theorem II.16], every $w \in A$ can be written as $w = yzy$ for some $y, z \in B$. Writing $u \circ v$ for $uv+vu$ we thus have

$$w = \tfrac{1}{2} y \circ (y \circ z) - \tfrac{1}{2} y^2 \circ z.$$

This proves that A is equal to $\mathrm{span}\{x \circ y \mid x, y \in A\}$, which in turn is obviously equal to $\mathrm{span}\{x^2 \mid x \in A\}$. Our claim is thus proved. As T is surjective, it follows that $\varepsilon_3 = 0$.

Accordingly, $\varepsilon_1 = 1$ and

$$T(x)^2 = \gamma T(x^2) \tag{3.26}$$

for every $x \in A$. This in particular shows that $\gamma B \subseteq B$, so γ actually lies in the centroid of B, i.e., $W(x) = \gamma x$ is a centralizer. Similarly, $\gamma^{-1} T(x)^2 = T(x^2)$ shows that $\gamma^{-1} B \subseteq B$ and so γ^{-1} also lies in the centroid; thus, γ is an invertible centralizer.

Observe that (3.26) implies that $J(x) = \gamma^{-1} T(x)$ is a Jordan homomorphism. Accordingly,

$$T(x) = \gamma J(x) = (WJ)(x)$$

is a weighted Jordan homomorphism. □

Concluding Remarks

1. If either A has no zero-divisors or if B has trivial multiplication ($xy = 0$ for all $x, y \in B$), then every linear map $T : A \to B$ preserves two-sided zero products. By taking direct products one obtains slightly more sophisticated examples. Some restrictions on A and B are therefore necessary.
2. The conclusion of Theorem 3.4 is that $T = WJ$ where W is an invertible centralizer and J is a Jordan homomorphism. Using the Closed Graph Theorem it is easy to see that W and W^{-1} are automatically continuous. Since T is continuous, so is J.
3. As mentioned in the introduction, a version of Theorem 3.4 was proved in [9]. More precisely, the main result in that paper differs from Theorem 3.4 only in the assumptions concerning B: instead of being semisimple and equal to its ideal generated by $\{x^2 \mid x \in B\}$, B is assumed to have a bounded approximate identity. However, from the proof, which is different and much more analytic, it is evident that Theorem 3.4 could be also established in [9]. Anyway, Theorem 3.4 covers the most important cases where A and B are either

   - C^*-algebras or
   - group algebras $L^1(G)$ for any locally compact group G.

   In [3], the result on $L^1(G)$ was proved only for some special groups G. It is interesting that the complete result covering all locally compact groups can be obtained by an essentially algebraic method. What seems particularly striking in our proof is that it involves the maximal left ring of quotients $Q_{ml}(B)$ and the extended centroid C of the Banach algebra B, which are apparently unnatural objects in the functional analytic context. However, they have proved useful.

Acknowledgments The author was supported by the Slovenian Research Agency (ARRS) Grant P1-0288.

References

1. M. Akkar, M. Laayouni, Théorèmes de factorisation dans les algèbres completes de Jordan, *Collect. Math.* **46** (1995), 239–254.
2. J. Alaminos, M. Brešar, J. Extremera, A. R. Villena, Zero Lie product determined Banach algebras, II, *J. Math. Anal. Appl.* **474** (2019), 1498–1511.
3. J. Alaminos, J. Extremera, A. R. Villena, Orthogonality preserving linear maps on group algebras, *Math. Proc. Cambridge Philos. Soc.* **158** (2015), 493–504.
4. K. I. Beidar, W. S. Martindale 3rd, A. V. Mikhalev, *Rings with Generalized Identities*, Marcel Dekker, Inc., 1996.
5. M. Brešar, On certain pairs of functions of semiprime rings, *Proc. Amer. Math. Soc.* **120** (1994), 709–713.
6. M. Brešar, *Introduction to Noncommutative Algebra*, Springer, 2014.
7. M. Brešar, *Zero Product Determined Algebras*, Birkhäuser/Springer, 2021.
8. M. Brešar, M. A. Chebotar, W. S. Martindale 3rd, *Functional Identities*, Birkhäuser/Springer, 2007.
9. M. Brešar, M. L. C. Godoy, A. R. Villena, Maps preserving two-sided zero products on Banach algebras, *J. Math. Anal. Appl.* **515** (2022), no. 1, Paper No. 126372, 16 pp.
10. M. A. Chebotar, W.-F. Ke, P.-H. Lee, R. Zhang, On maps preserving zero Jordan products, *Monatsh. Math.* **149** (2006), 91–101.
11. H. G. Dales, *Banach Algebras and Automatic Continuity,* Oxford University Press, 2000.
12. I. N. Herstein, Jordan homomorphisms, *Trans. Amer. Math. Soc.* **81** (1956), 331–341.

Bounded Factorization and the Ascending Chain Condition on Principal Ideals in Generalized Power Series Rings

H. E. Bruch, J. R. Juett, and Christopher Park Mooney

Mathematics Subject Classification (2020) 13F15, 13A05, 13F25, 20M25

1 Introduction

Throughout, all rings will be commutative with $1 \neq 0$ and all monoids and groups will be commutative.

In a landmark paper for the modern theory of factorization in integral domains, Anderson et al. [2] studied several properties weaker than factoriality, considering (among other things) whether each property is preserved by polynomial/power series extensions. For example, if an integral domain D satisfies the ascending chain condition on principal ideals (ACCP) or is a *bounded factorization ring* (BFR) (i.e., every nonzero nonunit has a finite upper bound on the lengths of its factorizations into nonunits), then the same holds for $D[X]$ and $D[[X]]$ [2, p. 5 and Proposition 2.5]. (We refer the interested reader to the work of authors such as Chapman, Geroldinger, F. Gotti, M. Gotti, and Halter-Koch, e.g., [21, 28, 31], for generalizations of concepts from [2] to the broader context of cancellative monoids.) Monoid rings and generalized power series rings are respectively natural generalizations of polynomial rings and power series rings. Anderson and Juett [5], Gilmer and Parker [30], and Kim [35] obtained many results about the ACCP and the bounded factorization property (BF-property) in these expanded contexts. We refer

H. E. Bruch
Goodwill Excel Center, Lockhart, TX, USA
e-mail: heather.bruch@excelcenterhighschool.org

J. R. Juett
University of Dubuque, Dubuque, IA, USA
e-mail: jjuett@dbq.edu

C. P. Mooney (✉)
University of Wisconsin-Stout, Menomonie, WI, USA
e-mail: mooneych@uwstout.edu

J.-L. Chabert et al. (eds.), *Algebraic, Number Theoretic, and Topological Aspects of Ring Theory*, https://doi.org/10.1007/978-3-031-28847-0_9

the reader to a recent survey paper by Anderson and Gotti [10] for a reference on the current state of the art in the theory of bounded factorization in integral domains.

The study of factorization is considerably more complex in rings in general than in integral domains. For example, in the presence of proper zero divisors, the ACCP and BF-property do not generally ascend from R to $R[X]$ or $R[[X]]$ [4, Section 3; 25, Example; 34, Example] and the question of exactly when these properties do ascend remains open. Results about the ACCP and BF-property in polynomial/power series rings (which may have proper zero divisors) can be found in the work of Anderson et al. [4,8, p. 447], Frohn [25–27], and Heinzer and Lantz [33, 34]. Edmonds and Juett [23] have recently explored the ACCP and BF-property within the realm of monoid rings (with zero divisors). In the present paper, we will examine these properties within the context of Ribenboim's generalized power series rings (with zero divisors), of which all the aforementioned kinds of extensions are special cases. We will generalize much of the known theory of the above factorization-theoretic topics, but at the same time many of our results are new even when specialized to (Laurent) power series rings or the "large (Laurent) polynomial rings" of Halter-Koch [32]. We have taken care to always make an explicit note whenever a special case of one of our results has already appeared in the literature.

For the reader's convenience and ease of reference, in Sect. 2 we carefully review several definitions, conventions, and basic facts that are necessary for understanding our results in the later sections. We pay particular attention to the relationships among generalized power series rings and related algebraic constructions such as (Laurent) power series rings and large (Laurent) polynomial rings. A lot of subtle but necessary behind-the-scenes work is done in establishing these connections, which we will be implicitly using in proofs regarding special cases of our generalized power series results. A detailed development of the facts we cite as general background can be found in the work of Ribenboim et al. (e.g., [24, 36–41]) and Aylesworth and Juett [12, Section 2]. The latter paper is similar to our present one in that it studies factorization in generalized power series rings, but there is little overlap between the two besides general background, because the two papers study different factorization properties.

It is easy to see that a BFR is présimplifiable [8, p. 456]. Additionally, while présimplifiability is certainly not a necessary precondition for the ACCP, the former property does sometimes simplify the study of the latter property. Thus we will begin our foray into factorization theory with présimplifiability. In Sect. 3 we will investigate this property, together with the related properties of weakly présimplifiable and domainlike, in the context of generalized power series rings. We provide our general results about these properties in Theorem 3 and deduce several special cases of interest in Corollary 4. We give an example showing that, at least for the présimplifiable and domainlike properties, none of the hypotheses in our results are superfluous.

Section 4 will build upon the groundwork laid in Sect. 3 to examine generalized power series rings that are BFRs or satisfy the ACCP. We extend two theorems of Juett et al. [5, 23] about monoid domains and generalized power series domains to analogous results about generalized power series rings with (in some sense) "not too many zero divisors." Many of our results in this section are new even in the (Laurent) power series and large (Laurent) polynomial ring special cases. Additionally, we will provide sharper versions of two past results about factorization in generalized power series domains. We will close the paper with a few examples illustrating our results and exploring the necessity of our hypotheses.

2 Background

This section develops the background necessary for understanding our results in Sects. 3 and 4. We remind the reader that all rings are commutative with $1 \neq 0$ and all monoids and groups are commutative.

Let R be a ring. We respectively use $\mathrm{Nil}(R)$, $\mathcal{J}(R)$, $\mathcal{Z}(R)$, and $U(R)$ to denote the nilradical, Jacobson radical, set of zero divisors, and group of units of R. The *regular* elements of R are those in $\mathrm{Reg}(R) := R \setminus \mathcal{Z}(R)$. A ring is *quasilocal* if it has only one maximal ideal, *semi-quasilocal* if it has only finitely many maximal ideals, and *(semi-)local* if it is Noetherian and (semi-)quasilocal. For ideals I and J of R, we define $(I :_R J) := \{x \in R \mid xJ \subseteq I\}$.

Let $(S, +)$ be a monoid. We use $U(S)$ to denote the group of units of S. We say S is *reduced* if $U(S)$ is trivial, *torsion-free* if $ns = nt$ with $s, t \in S$ and n a positive integer implies $s = t$, and *cancellative* if $s + t = s + u$ with $s, t, u \in S$ implies $t = u$.

We call a monoid $(S, +)$ *ordered* if it is equipped with an order $\leq$ that is *compatible* in the sense that $s \leq t$ with $s, t \in S$ implies $s + u \leq t + u$ for all $u \in S$. (Following Ribenboim's convention [39, p. 72], we do not assume all orders are total.) If S is ordered, we abbreviate $S^+ := \{s \in S \mid s > 0\}$, $S_0^+ := \{s \in S \mid s \geq 0\}$, $S^- := \{s \in S \mid s < 0\}$, and $S_0^- := \{s \in S \mid s \leq 0\}$. When we write "$\mathbb{Z}^+$," etc., we are always referring to that notation as defined with respect to the standard order. We say a compatible order $\leq$ on S is *cancellative* if $s + t \leq s + u$ with $s, t, u \in S$ implies $t \leq u$ (cf. [40, p. 568]), *negative* if $S = S_0^-$, *potentially negative* if it can be refined to a compatible negative order on S [12, Section 2], *positive* if $S = S_0^+$ (cf. [39, p. 76]), *subpositive* if $\mathbb{Z}^+ s \cap S^+ \neq \emptyset$ for all $0 \neq s \in S$ [12, Subsection 2.2], *potentially positive* if it can be refined to a positive order on S, *total* if all elements of S are comparable, and *subtotal* if for all $s, t \in S$ we have $ks \leq kt$ or $ks \geq kt$ for some $k \in \mathbb{Z}^+$ [39, p. 76]. We use phrases of the form "$\mathcal{P}$ ordered" to indicate that a monoid is equipped with a compatible order with property $\mathcal{P}$. In the context of ordered cancellative monoids, (i) total $\Rightarrow$ cancellative + subtotal, (ii) positive $\Rightarrow$ subpositive $\Rightarrow$ potentially positive, and (iii) negative $\Rightarrow$ potentially negative, but none of the implications reverse [12, Section 2.2]. Note that a (potentially) positively or (potentially) negatively ordered cancellative monoid is reduced [12, Section 2.2].

We now begin our review of definitions directly related to generalized power series. We refer the reader to the extensive work done by Ribenboim et al. (e.g.,

[24, 36–41]) and Aylesworth and Juett [12, Section 2] for further information. Let R be a ring and $(S, +, \leq)$ be a cancellative ordered monoid. The *generalized power series ring* $R[[S^{\leq}]]$ consists of all formal sums $f = \sum_{s \in S} a_s X^s$ $(a_s \in R)$ such that the *support* of f, namely $\operatorname{supp}(f) := \{s \in S \mid a_s \neq 0\}$, is *Artinian* (i.e., contains no infinite strictly decreasing sequence) and *narrow* (i.e., contains no infinite antichain), with operations defined exactly analogously to the usual addition and multiplication of power series [36, p. 273]. (The fact that S is cancellative, coupled with the Artinian and narrow conditions, ensures that the operations are well defined. That is, for $f, g \in R[[S^{\leq}]]$, the computation of the coefficients of fg does not involve any infinite sums [39, 1.16] and the supports of $f + g$ and fg are indeed Artinian and narrow [24, pp. 366-367].) For example, if $\leq$ is the standard order on $\mathbb{Z}$, then $R[[(\mathbb{Z}_0^+)^{\leq}]] = R[[X]]$ (the usual power series ring over R), $R[[\mathbb{Z}^{\geq}]] \cong R[[\mathbb{Z}^{\leq}]] = R[[X]][X^{-1}]$ (the Laurent power series ring over R), $R[[(\mathbb{Z}_0^+)^{\geq}]] = R[[(\mathbb{Z}_0^+)^{=}]] = R[X]$ (the polynomial ring over R), $R[[\mathbb{Z}^{=}]] = R[X, X^{-1}]$ (the Laurent polynomial ring over R), and $R[[S^{=}]] = R[S]$ (the *monoid ring* with coefficients in R and exponents in S). We will discuss the much more intricate matter of defining (Laurent) power series rings over an arbitrary set of variables below. The *degree* and *trailing degree* of $f \in R[[S^{\leq}]] \setminus (0)$ are respectively the maximum and minimum element of $\operatorname{supp}(f)$, provided that they exist. We analogously define notions like *leading coefficient*, *trailing coefficient*, *trailing term*, and *constant term* in the obvious way. If S is reduced, then $R[[S^{\leq}]] \to R : f \mapsto f(0)$ is a ring epimorphism, where $f(0)$ denotes the constant term of f. For $f \in R[[S^{\leq}]]$, the *content* of f, denoted $c(f)$, is the ideal of R generated by the coefficients of f. For ideals I of R and J of S, we define $I[[J^{\leq}]] := \{f \in R[[S^{\leq}]] \mid c(f) \subseteq I \text{ and } \operatorname{supp}(f) \subseteq J\}$, which is an ideal of $R[[S^{\leq}]]$. When there is no danger of confusion, we will drop the superscript "$\leq$" from the preceding notation.

The remainder of this section will discuss the relationships among generalized power series rings and the various kinds of (Laurent) power series rings and large (Laurent) polynomial rings. The literature contains several distinct ways to define a power series ring over a ring R and a possibly infinite family $\{X_\lambda\}_{\lambda \in \Lambda}$ of analytically independent indeterminates. The first and perhaps most natural definition is due to Cashwell and Everett, whose power series ring $R[[\{X_\lambda\}_{\lambda \in \Lambda}]]_3$ consists of all formal sums $\sum_s a_s \prod_{\lambda \in \Lambda} X_\lambda^{s(\lambda)}$ $(a_s \in R)$, where s ranges over $\bigoplus_{\lambda \in \Lambda} \mathbb{Z}_0^+$, with the operations defined in the obvious way [20, pp. 46-47]. The *total degree* of a monomial is the sum of the exponents of the indeterminates appearing in the monomial; the *total degree* of a series is the supremum of the total degrees of its terms. For a power series f and an $n \in \mathbb{Z}_0^+$, the *homogeneous part* of f of *total degree* n is the sum of all the terms of f that have total degree n. Gilmer studied the ring $R[[\{X_\lambda\}_{\lambda \in \Lambda}]]_2$, which consists of the series whose homogeneous parts contain only finitely many indeterminates [29, p. 6]. For each infinite cardinal α, Aylesworth and Juett defined $R[[\{X_\lambda\}_{\lambda \in \Lambda}]]_{<\alpha}$ to be the subring of $R[[\{X_\lambda\}_{\lambda \in \Lambda}]]_3$ consisting of the series whose defining set of indeterminates has cardinality strictly less than α [12, Subsection 2.3]. Special cases of this definition include Gilmer's ring $R[[\{X_\lambda\}_{\lambda \in \Lambda}]]_1 := R[[\{X_\lambda\}_{\lambda \in \Lambda}]]_{<\aleph_0}$ [29, p. 6], Benhissi and Eljeri's ring

$R[[\{X_\lambda\}_{\lambda\in\Lambda}]]_{\leq\aleph_0} := R[[\{X_\lambda\}_{\lambda\in\Lambda}]]_{<\aleph_1}$ [14, Proposition 3.1], and Chang's ring $R[[\{X_\lambda\}_{\lambda\in\Lambda}]]_{\leq\alpha} := R[[\{X_\lambda\}_{\lambda\in\Lambda}]]_{<\mathfrak{s}(\alpha)}$, where $\mathfrak{s}(\alpha)$ is the cardinal successor of α. The choice of i in $R[[\{X_\lambda\}_{\lambda\in\Lambda}]]_i$ makes no difference when Λ is finite, so in that case we often drop the subscript i. Furthermore, we will drop the braces when it is more convenient to list the indeterminates than represent them as an indexed set.

Unfortunately, many further complications are introduced when defining Laurent power series rings. In this paper, we will confine our attention to Laurent power series rings of the forms $R\langle\langle\{X_\lambda\}_{\lambda\in\Lambda}\rangle\rangle$ and $R[[\{X_\lambda\}_{\lambda\in\Lambda}]]_i[\{X_\lambda^{-1}\}_{\lambda\in\Lambda}]$ for $i \in \{1, 2, 3, <\alpha\}$, where R is a ring, $\{X_\lambda\}_{\lambda\in\Lambda}$ is a family of analytically independent indeterminates, and α is an infinite cardinal number. Here Anderson and Juett defined $R\langle\langle\{X_\lambda\}_{\lambda\in\Lambda}\rangle\rangle$ (for Λ well ordered) to be the ring consisting of all formal sums $\sum_s a_s \prod_{\lambda\in\Lambda} X_\lambda^{s(\lambda)}$ ($a_s \in R$), where s ranges over a subset of $\bigoplus_{\lambda\in\Lambda}\mathbb{Z}$ that is well ordered with respect to the colexicographic order [5, p. 338]—see Proposition 1 below for the definition of this order. For Λ finite, this definition coincides with Xin's "iterated power series rings" [42, p. 3], i.e., $R\langle\langle X_1, \ldots, X_n\rangle\rangle = R[[X_1]][X_1^{-1}]\cdots[[X_n]][X_n^{-1}]$ [38, 4.5(i)]. We note that well orders of Λ with the same order type lead to isomorphic definitions of $R\langle\langle\{X_\lambda\}_{\lambda\in\Lambda}\rangle\rangle$, but different order types generally lead to non-isomorphic definitions. Other kinds of Laurent power series rings occur in the literature (e.g., the ring $R(((\{X_\lambda\}_{\lambda\in\Lambda})))$ introduced by Cheng et al. [22, p. 1770]), but we will not be considering them. If α is uncountable, then $R[[\{X_\lambda\}_{\lambda\in\Lambda}]]_1 \subseteq R[[\{X_\lambda\}_{\lambda\in\Lambda}]]_2 \subseteq R[[\{X_\lambda\}_{\lambda\in\Lambda}]]_{<\alpha} \subseteq R[[\{X_\lambda\}_{\lambda\in\Lambda}]]_3$, where the first two inclusions are proper if and only if $|\Lambda| \geq \aleph_0$ and the last inclusion is proper if and only if $|\Lambda| \geq \alpha$ [12, Subsection 2.3]. The corresponding statement about inclusions among Laurent power series rings holds [12, Subsection 2.3]. Additionally, we have $R[[\{X_\lambda\}_{\lambda\in\Lambda}]]_3[\{X_\lambda^{-1}\}_{\lambda\in\Lambda}] \subseteq R\langle\langle\{X_\lambda\}_{\lambda\in\Lambda}\rangle\rangle$ and this inclusion is proper if and only if $|\Lambda| \geq 2$ [12, Subsection 2.3].

Let R be a ring, $\{X_\lambda\}_{\lambda\in\Lambda}$ be a family of analytically independent indeterminates, and α be an infinite cardinal number. Halter-Koch defined two kinds of "large polynomial rings" [32, p. 124], namely the subring $R[\{X_\lambda\}_{\lambda\in\Lambda}]_L$ of $R[[\{X_\lambda\}_{\lambda\in\Lambda}]]_3$ consisting of the power series with only finitely many homogeneous parts and $R[\{X_\lambda\}_{\lambda\in\Lambda}]_{L_{<\alpha}} := R[\{X_\lambda\}_{\lambda\in\Lambda}]_L \cap R[[\{X_\lambda\}_{\lambda\in\Lambda}]]_{<\alpha}$—here we use Aylesworth and Juett's notation [12, Subsection 2.3]. These "large polynomials" in many ways resemble polynomials more than power series, e.g., [12, Proposition 2.4.5(3); 32, Proposition 2]. Note that the class of (Laurent) polynomial rings is a subclass of the large (Laurent) polynomial rings, i.e., $R[\{X_\lambda\}_{\lambda\in\Lambda}] = R[\{X_\lambda\}_{\lambda\in\Lambda}]_{L_{<\aleph_0}}$ [32, p. 124] and $R[\{X_\lambda, X_\lambda^{-1}\}_{\lambda\in\Lambda}] = R[\{X_\lambda\}_{\lambda\in\Lambda}]_{L_{<\aleph_0}}[\{X_\lambda^{-1}\}_{\lambda\in\Lambda}]$.

The following proposition from Aylesworth and Juett's paper [12] shows how various kinds of (Laurent) power series rings and large (Laurent) polynomial rings can be realized as generalized power series rings in some "nice" way. We will be implicitly using these facts throughout the rest of the paper to translate statements about (Laurent) power series or large (Laurent) polynomials into generalized power series terminology. We note that this proposition from Aylesworth and Juett's paper [12] corrects a couple of erroneous past statements from other authors about which order corresponds to which power series ring—see [12, Subsection 2.3] for details.

Proposition 1 ([12, Proposition 2.3.1]) Let R be a ring, $\{X_\lambda\}_{\lambda\in\Lambda}$ be a nonempty family of analytically independent indeterminates, $F := \bigoplus_{\lambda\in\Lambda} \mathbb{Z}_0^+$, and $G := \bigoplus_{\lambda\in\Lambda} \mathbb{Z}$. Well order Λ. For $s \in G$, abbreviate $\sigma(s) := \sum_{\lambda\in\Lambda} s(\lambda)$. For distinct $s, t \in G$, define

- $\phi(s, t)$ to be the first coordinate where s and t differ,
- $\psi(s, t)$ to be the last coordinate where s and t differ,
- $s <_2 t$ (the "graded lexicographic order") to hold if $\sigma(s) < \sigma(t)$ or $\sigma(s) = \sigma(t)$ and $s(\phi(s, t)) < t(\phi(s, t))$,
- $s <_3 t$ (the "colexicographic order") to hold if $s(\psi(s, t)) < t(\psi(s, t))$, and
- $s <_L t$ (the "reversely graded colexicographic order") to hold if $\sigma(s) > \sigma(t)$ or $\sigma(s) = \sigma(t)$ and $s <_3 t$.

1. $\leq_2$, $\leq_3$, and $\leq_L$ are compatible total orders on G, $\leq_2$ and $\leq_3$ are positive on F, $\leq_L$ is negative on F, and F is well ordered by $\leq_3$.
2. The map $\sum_s a_s X^s \mapsto \sum_s a_s \prod_{\lambda\in\Lambda} X_\lambda^{s(\lambda)}$ ($a_s \in R$) is a ring isomorphism from $R[[F^{\leq_i}]]$ onto $R[[\{X_\lambda\}_{\lambda\in\Lambda}]]_i$ for $i \in \{2, 3\}$, from $R[[G^{\leq_3}]]$ onto $R\langle\langle\{X_\lambda\}_{\lambda\in\Lambda}\rangle\rangle$, from $R[[F^{\geq_3}]] = R[F]$ onto $R[\{X_\lambda\}_{\lambda\in\Lambda}]$, from $R[G]$ onto $R[\{X_\lambda, X_\lambda^{-1}\}_{\lambda\in\Lambda}]$, and from $R[[F^{\leq_L}]]$ onto $R[\{X_\lambda\}_{\lambda\in\Lambda}]_L$.

3 Présimplifiability and Related Notions in Generalized Power Series Rings

In this section, we study generalized power series rings that possess properties related to présimplifiability.

Following Bouvier [16–19], we call a ring R *présimplifiable* if $x = \lambda x$ with $x, \lambda \in R$ implies $x = 0$ or $\lambda \in U(R)$. It is not hard to see that R is présimplifiable if and only if $\mathcal{Z}(R) \subseteq \mathcal{J}(R)$ [9, p. 203], so R is présimplifiable if it is *domainlike* (i.e., $\mathcal{Z}(R) = \mathrm{Nil}(R)$ [15, Definition 10]). More recently, Anderson and Chun defined R to be *weakly présimplifiable* if $x = \lambda x$ with $x, \lambda \in R$ implies $x = 0$ or $\lambda \in \mathrm{Reg}(R)$ [3, Definition 5]. Examples of (weakly) présimplifiable rings include (polynomial rings over) quasilocal rings; examples of domainlike rings include domains, (polynomial rings over) zero-dimensional quasilocal rings, and power series rings over Artinian local rings [11, p. 153]. None of the following implications reverse: domain $\Rightarrow$ domainlike $\Rightarrow$ présimplifiable $\Rightarrow$ weakly présimplifiable. For instance, the ring $\mathbb{Z}/4\mathbb{Z}$ is domainlike but not a domain, $\mathbb{Z}_{(2)}[\mathbb{Z}/2\mathbb{Z}]$ is présimplifiable but not domainlike [1, Theorem 3.8], and $\mathbb{Q}[X, Y, Z]/(X - XYZ)$ is weakly présimplifiable but not présimplifiable [3, Example 11(3)].

The following facts from Aylesworth and Juett's paper [12] will be useful in many of our proofs.

Lemma 2 ([12, Proposition 2.4.5]) Let R be a ring, $(S, +, \leq)$ be a torsion-free cancellative ordered monoid, and $\mu \in R[[S]]$.

1. If $\text{Nil}(R)$ is nilpotent and $\leq$ is subtotal, then $\mu \in U(R[[S]])$ if and only if the $\leq'$-trailing term of $\overline{\mu}$ is a unit, where $\overline{R} := R/\text{Nil}(R)$ and $\leq'$ is the unique refinement of $\leq$ to a compatible total order on S [12, p. 6].
2. If $\leq$ is subpositive, then $U(R[[S]]) = U(R) + R[[S \setminus \{0\}]]$.
3. If $\text{Nil}(R)$ is nilpotent and $\leq$ is potentially negative, then $U(R[[S]]) = U(R) + \text{Nil}(R)[[S \setminus \{0\}]]$.

Let R be a ring and $(S, +)$ be a monoid. Edmonds and Juett have completely characterized when $R[S]$ is présimplifiable or domainlike [23, Theorem 2.3]. Under the hypothesis that S is torsion-free cancellative, they also characterized when $R[S]$ is weakly présimplifiable [23, Theorem 2.2(3)]. The situation with generalized power series rings is considerably more complex, but the following theorem gives a partial answer that will be good enough for our purposes. To state this result and its corollary in the strongest possible form, we recall that a *strongly primary decomposition* of an ideal I is a primary decomposition $I = \bigcap_{i=1}^n Q_i$ where each Q_i contains a power of $\sqrt{Q_i}$. By the Lasker-Noether theorem, every ideal in a Noetherian ring has a strongly primary decomposition.

Theorem 3 Let R be a ring and $(S, +, \leq)$ be a nontrivial torsion-free cancellative ordered monoid.

1. If $\leq$ is potentially positive or (0) has a strongly primary decomposition in R, then $R[[S]]$ is weakly présimplifiable if and only if R is weakly présimplifiable.
2. If $\leq$ is subpositive, then $R[[S]]$ is présimplifiable if and only if R is présimplifiable.
3. If $\text{Nil}(R)$ is nilpotent and $\leq$ is subtotal or cancellative, then $R[[S]]$ is présimplifiable if and only if either (i) R is domainlike or (ii) R is présimplifiable and $\leq$ is subpositive.
4. If $\text{Nil}(R)$ is nilpotent and $\leq$ is potentially negative, then $R[[S]]$ is présimplifiable (or equivalently domainlike) if and only if R is domainlike.
5. If $\text{Nil}(R)$ is nilpotent, then $R[[S]]$ is domainlike if and only if R is domainlike.

Proof Because S is torsion-free cancellative, there is a compatible total order $\leq'$ on S that refines $\leq$ [24, Proposition 2] and $R[[S^{\leq}]] \subseteq R[[S^{\leq'}]]$ [24, p. 368].

($\Rightarrow$): If $R[[S]]$ is domainlike or weakly présimplifiable, then so is its subring R. If $\text{Nil}(R)$ is nilpotent, $\leq$ is potentially negative, and $R[[S]]$ is présimplifiable, then $R[[S]]$ is in fact domainlike since $\text{Nil}(R[[S]]) = \mathcal{J}(R[[S]])$ [12, Proposition 2.4.5(3)]. If $R[[S]]$ is présimplifiable, then so is R since $U(R) = R \cap U(R[[S]])$. All that remains is the "only if" part of (3). Assume $R[[S]]$ is présimplifiable and $\leq$ is subtotal or cancellative. We may assume R is présimplifiable but not domainlike, so $ab = 0$ for some $a \in R \setminus \text{Nil}(R)$ and $b \in R \setminus (0)$. To show that $\leq$ is subpositive, pick $0 \neq s \in S$. Then $0 \neq b = b(1 - aX^s)$, so $1 - aX^s \in U(R[[S^{\leq}]]) \subseteq U(R[[S^{\leq'}]])$ since $R[[S^{\leq}]]$ is présimplifiable.

Suppose that $s <' 0$. Because $a \in R \setminus \mathrm{Nil}(R)$, there is a prime ideal P of R with $a \notin P$. Let $\overline{R} := R/P$. Then $\overline{a}X^s$ is the $\leq'$-trailing term of $\overline{1} - \overline{a}X^s \in U(\overline{R}[[S^{\leq'}]])$, so $\overline{a} \in U(\overline{R})$ by Lemma 2(1). Therefore $R = P + aR \subseteq P + \mathcal{Z}(R) \subseteq P + \mathcal{J}(R)$, where the last inclusion holds since R is présimplifiable. Thus $P = R$, a contradiction.

Therefore $s >' 0$, so the inverse of $1 - aX^s$ in $R[[S^{\leq'}]]$ (and consequently in $R[[S^{\leq}]]$) is $\sum_{n=0}^{\infty} a^n X^{ns}$. Because $a \notin \mathrm{Nil}(R)$, we have $\mathrm{supp}(\sum_{n=0}^{\infty} a^n X^{ns}) = \{ns\}_{n=0}^{\infty}$, so ms and ns are $\leq$-comparable for some $m < n$ in $\mathbb{Z}_0^+$. But $ms <' ns$ since $s >' 0$, so $ms < ns = ms + (n-m)s$. Thus $(n-m)s > 0$ if $\leq$ is cancellative. So let us assume $\leq$ is subtotal. Then ks and 0 are $\leq$-comparable for some $k \in \mathbb{Z}^+$. If $ks \leq 0$, then $(n-m)ks \leq 0$, so $kns \leq kms$, contradicting the fact that $ns > ms$. Therefore $ks > 0$, as desired.

($\Leftarrow$): If R is domainlike and $\mathrm{Nil}(R)$ is nilpotent, then $R[[S^{\leq'}]]$ (and therefore $R[[S^{\leq}]]$) is domainlike [39, 3.4(ii)]. The "if" parts of (4) and (5) are now established.

For the "if" parts of (2) and (3), assume R is présimplifiable and $\leq$ is subpositive. Pick $f, \lambda \in R[[S^{\leq}]]$ with $0 \neq f = \lambda f$. Because $\leq'$ is positive, the $\leq'$-trailing coefficient of $f = \lambda f$ is $a = \lambda(0)a$, where a is the $\leq'$-trailing coefficient of f. So $\lambda(0) \in U(R)$ since R is présimplifiable. Therefore $\lambda \in U(R[[S^{\leq}]])$ by Lemma 2(2).

All that remains is the "if" part of (1). Assume R is weakly présimplifiable. We need to show that $f - 1 \in \mathrm{Reg}(R[[S]])$ for each $f \in \mathcal{Z}(R[[S]])$ [3, Theorem 6]. Pick $g \in R[[S]]$ with $fg = 0 \neq g$.

First consider the case where $\leq$ is potentially positive. Then we can choose $\leq'$ to be positive [24, Proposition 2]. Because $fg = 0$, we have $f(0)b = 0$, where b is the $\leq'$-trailing coefficient of g. Thus $f(0) \in \mathcal{Z}(R) \subseteq 1 - \mathrm{Reg}(R)$ [3, Theorem 6] since R is weakly présimplifiable. So $f - 1$ is regular since its $\leq'$-trailing coefficient $f(0) - 1$ is regular, as desired.

Finally, assume there is an irredundant strongly primary decomposition $(0) = \bigcap_{i=1}^n Q_i$ in R. Then $(0) = \bigcap_{i=1}^n Q_i[[S]]$ is an irredundant strongly primary decomposition in $R[[S]]$ with each $\sqrt{Q_i[[S]]} = \sqrt{Q_i}[[S]]$ [39, 1.21]. We have $g \notin Q_j[[S]]$ for some j since $g \neq 0$, so $f \in \sqrt{Q_j}[[S]]$. Note that $Q_j \subseteq \mathcal{Z}(R)$ (and therefore $\sqrt{Q_j} \subseteq \mathcal{Z}(R)$) since $Q_j \left(\bigcap_{i \neq j} Q_i \right) = (0) \neq \bigcap_{i \neq j} Q_i$. This shows that every element of $R[[S]]$ with a regular coefficient is regular. Thus the constant term of $f - 1$ is in $-1 + \mathcal{Z}(R) \subseteq \mathrm{Reg}(R)$, where the inclusion holds [3, Theorem 6] since R is weakly présimplifiable, so $f - 1 \in \mathrm{Reg}(R[[S]])$, as desired. □

Ribenboim proved special cases of parts (2) and (5) of Theorem 3 [39, 3.4 and 3.9]. We collect other special cases of interest in the following corollary. For completeness and purposes of comparison, we also collect the known results about weakly présimplifiable, présimplifiable, or domainlike (Laurent) polynomial rings.

Corollary 4 Let R be a ring, $\{X_\lambda\}_{\lambda \in \Lambda}$ be a nonempty family of analytically independent indeterminates, α be an infinite cardinal number, $i \in \{1, 2, 3, <\alpha\}$, $j \in \{L, L_{<\alpha}\}$, $A := R[[\{X_\lambda\}_{\lambda \in \Lambda}]]_i$, $A' := A[\{X_\lambda^{-1}\}_{\lambda \in \Lambda}]$, $B_0 := R[\{X_\lambda\}_{\lambda \in \Lambda}]$,

$B_0' := R[\{X_\lambda, X_\lambda^{-1}\}_{\lambda\in\Lambda}]$, $B := R[\{X_\lambda\}_{\lambda\in\Lambda}]_j$, $B' := B[\{X_\lambda^{-1}\}_{\lambda\in\Lambda}]$, and $C := R\langle\langle\{X_\lambda\}_{\lambda\in\Lambda}\rangle\rangle$.

1. If $D \in \{A, B_0, B_0', B\}$, then D is weakly présimplifiable if and only if R is weakly présimplifiable.
2. If $D \in \{A', B', C\}$ and (0) has a strongly primary decomposition in R, then D is weakly présimplifiable if and only if R is weakly présimplifiable.
3. A is présimplifiable if and only if R is présimplifiable.
4. If $D \in \{B_0, B_0'\}$, then D is présimplifiable (or equivalently domainlike) if and only if R is domainlike [1, Theorem 3.11(b)].
5. If $D \in \{A', B, B', C\}$ and $\mathrm{Nil}(R)$ is nilpotent, then D is présimplifiable (or equivalently domainlike) if and only if R is domainlike.
6. If $\mathrm{Nil}(R)$ is nilpotent, then A is domainlike if and only if R is domainlike.

Proof All statements about B_0 and B_0' are special cases of results of Anderson et al. [3, Theorem 18; 1, Theorem 3.11(b)] and/or Edmonds and Juett [23, Theorems 2.2(3) and 2.3].

($\Rightarrow$): Let $D \in \{A, A', B, B', C\}$. If D is weakly présimplifiable or domainlike, then so is its subring R. If D is présimplifiable, then so is R since $R \cap U(D) = U(R)$. All that remains is the "only if" part of (5). Assume $\mathrm{Nil}(R)$ is nilpotent. Pick $\gamma \in \Lambda$. If B or C is présimplifiable, then R is domainlike by Theorem 3. If B' is présimplifiable, then so is B_0' since $B_0' \cap U(B') = U(B_0')$, so R is domainlike [1, Theorem 3.11(b)]. (To demonstrate the equality, it suffices to show that each $f \in B_0 \cap U(B')$ is invertible in B_0'. There is a $g \in B$ with $h := fg$ a monic monomial. Let $X_{\lambda_1}, \dots, X_{\lambda_n}$ be the indeterminates appearing in f or h. Then $h = f\pi(g)$, where π is the canonical ring epimorphism of B onto $R[X_{\lambda_1}, \dots, X_{\lambda_n}]$, so $f \in U(B_0')$, as desired.) Thus we may assume A' is présimplifiable. We claim that $R\langle\langle X_\gamma\rangle\rangle \cap U(A') = U(R\langle\langle X_\gamma\rangle\rangle)$. It suffices to show that each $f \in R[[X_\gamma]] \cap U(A')$ is invertible in $R\langle\langle X_\gamma\rangle\rangle$. With a trivial adjustment to the above parenthetical argument, we see that $f \in U(R[[X_\gamma, X_{\lambda_1}, \dots, X_{\lambda_n}]][X_\gamma^{-1}, X_{\lambda_1}^{-1}, \dots, X_{\lambda_n}^{-1}])$ for some $\lambda_1, \dots, \lambda_n \in \Lambda$ ($n \geq 0$). Thus $f \in R\langle\langle X_\gamma\rangle\rangle \cap U(R\langle\langle X_\gamma\rangle\rangle\langle\langle X_{\lambda_1}, \dots, X_{\lambda_n}\rangle\rangle) = U(R\langle\langle X_\gamma\rangle\rangle)$, as desired. Therefore $R\langle\langle X_\gamma\rangle\rangle$ is présimplifiable, so R is domainlike by Theorem 3(5).

($\Leftarrow$): Assume R is weakly présimplifiable. Then $A_3 := R[[\{X_\lambda\}_{\lambda\in\Lambda}]]_3$ is weakly présimplifiable by Theorem 3(1), so its subrings A and B are weakly présimplifiable. If (0) has a strongly primary decomposition in R, then C and its subrings A' and B' are weakly présimplifiable by Theorem 3(1). If R is présimplifiable, then A_3 is présimplifiable by Theorem 3(2) and $A \cap U(A_3) = U(A)$ by Lemma 2(2), so A is présimplifiable. If $\mathrm{Nil}(R)$ is nilpotent and R is domainlike, then C and its subrings A, A', B, and B' are domainlike by Theorem 3(5), as desired. □

We note that some special cases of parts of Corollary 4 are already known. The statements about B_0 and B_0' have already been proven by Anderson et al. [3, Theorem 18; 1, Theorem 3.11(b); 8, pp. 471–472], Bouvier [17], and/or Edmonds and Juett [23, Theorems 2.2(3) and 2.3]. Ribenboim proved the "if" direction of the

$D = C$ part of (5) [39, 3.4]. The remaining assertions of Corollary 4 appear to be new, although slightly weaker versions of a few of them have already appeared in the literature. The results regarding the (weak) présimplifiability of power series, in the special case with only finitely many indeterminates, can be found in the work of Anderson et al. [3, Theorem 20(1); 8, p. 471] and/or Ribenboim [39, 3.10]. Part (5), in the special case where R is Noetherian and Λ is finite, was proven by Axtell et al. [11, Proposition 9].

We close the section with an example showing that the nilpotency hypotheses cannot be removed from Theorem 3 parts (3)–(5) or Corollary 4 parts (5) and (6). We do not know if the strongly primary decomposition hypotheses in Theorem 3(1) and Corollary 4(2) are necessary.

Example 5 (A Zero-Dimensional Quasilocal Ring R for Which Neither $R[[X]]$ nor $R\langle\langle X\rangle\rangle$ Is Domainlike and Neither $R[X_1, X_2, \ldots]_L$ nor $R[X_1, X_2, \ldots]_L[X_1^{-1}, X_2^{-1}, \ldots]$ Is Présimplifiable.)) Let $R := \mathbb{Z}_2[Y, Z, Z^{1/2}, Z^{1/3}, \ldots]/(Z, Y^2, YZ^{1/2}, YZ^{1/3}, \ldots)$, where $\mathbb{Z}_2$ is the field of cardinality 2. Then R is domainlike since it has a unique prime ideal. We know from Corollary 4 that $R[[X]]$ and $R\langle\langle X\rangle\rangle$ are at least présimplifiable. However, they are not domainlike. To see this, let $f := \sum_{n=1}^{\infty} \overline{Z^{1/2^n}} X^n \in R[[X]]$. Then $\overline{Y} f = \overline{0} \neq \overline{Y}$, yet $f^{(2^m)} = \sum_{n=m+1}^{\infty} \overline{Z^{1/2^{n-m}}} X^{(n2^m)} \neq \overline{0}$ for each $m \in \mathbb{Z}^+$, as desired. We also know from Corollary 4 that every (Laurent) polynomial ring over R is présimplifiable. But neither $B := R[X_1, X_2, \ldots]_L$ nor $B' := B[X_1^{-1}, X_2^{-1}, \ldots]$ is présimplifiable. To see this, let $g := \sum_{n=1}^{\infty} \overline{Z^{1/2^n}} X_n \in B$. Then $\overline{Y} g = \overline{0} \neq \overline{Y}$ and $g^{(2^m)} = \sum_{n=m+1}^{\infty} \overline{Z^{1/2^{n-m}}} X_n^{(2^m)} \neq \overline{0}$ for each $m \in \mathbb{Z}^+$, so $g \in \mathcal{Z}(B) \setminus \mathrm{Nil}(B)$. Suppose that $g \in \mathcal{J}(B)$ or $g \in J(B')$. Then there is an $h \in B$ with $w := (\overline{1} + g)h$ a non-constant monic monomial. We will contradict the fact that h has bounded total degree by showing that $h_n = 0$ for all $0 \leq n < m$ and $h_n = wg^{n-m} \neq 0$ for all $n \geq m$, where m is the total degree of w and h_n is the homogeneous part of h of total degree n. The homogeneous part of $w = (\overline{1} + g)h$ of total degree 0 is $0 = \overline{1} \cdot h_0 = h_0$. So let us assume $n > 0$. If $n < m$, then $h_{n-1} = 0$ by induction on n, so the homogeneous part of $w = (\overline{1} + g)h$ of total degree n is $0 = \overline{1} \cdot h_n + gh_{n-1} = h_n$. So let us assume $n \geq m$. If $n = m$, then $h_{n-1} = 0$ by induction on n, so the homogeneous part of $w = (\overline{1} + g)h$ of total degree n is $w = \overline{1} \cdot h_n + gh_{n-1} = h_n$. If $n > m$, then $h_{n-1} = wg^{n-1-m}$ by induction on n, so the homogeneous part of $w = (\overline{1} + g)h$ of total degree n is $\overline{0} = \overline{1} \cdot h_n + gh_{n-1} = h_n + wg^{n-m}$, so $h_n = wg^{n-m}$, as desired.

4 Bounded Factorization and the ACCP in Generalized Power Series Rings

In this final section we will consider the question of when a generalized power series ring is a BFR or satisfies the ACCP.

One of the main tools we will use is the theory of recursive factorization length developed by Juett et al. [5–7, 23]. Let R be a ring and $a \in R$. If every sequence of ascending principal ideals of R starting with (a) stabilizes, then the *(recursive) factorization length* of a, denoted $L_R(a)$, is the least ordinal number strictly greater than $L_R(b)$ for each proper divisor b of a; otherwise $L_R(a)$ is left undefined [23, p. 1845]. If R satisfies the ACCP, then $L(R)$ is defined to be the least ordinal number strictly greater than $L_R(a)$ for all $0 \neq a \in R$ [6, pp. 1590–1591]. We sometimes drop the subscript R in the preceding notation if there is no danger of confusion. The ACCP, the BF-property, recursive factorization length, etc., are defined analogously for monoids. Examples of BFRs include présimplifiable Noetherian rings [8, Theorem 3.9] and factorial (or more generally Krull [2, Proposition 2.2]) domains; examples of bounded factorization monoids (BFMs) include finitely generated cancellative monoids and factorial (or more generally Krull) cancellative monoids [28, Proposition 2.7.8(4)]. Clearly the BF-property implies the ACCP, but the converse is false even in the integral domain and reduced cancellative monoid cases [2, Example 2.1].

We will need the following facts from the work of Juett et al. [5–7, 23].

Lemma 6 ([6, Proposition 4.1; 23, Proposition 3.2]) Let R be a ring and $a \in R$.

1. $L(a)$ exists if and only if there is an ordinal-valued function ϕ on $\{b \in R \mid (a) \subseteq (b)\}$ such that $\phi(c) < \phi(b)$ whenever $(a) \subseteq (b) \subsetneq (c)$, in which case $L(a) \leq \phi(a)$. Therefore R satisfies the ACCP if and only if there is an ordinal-valued function ϕ on R such that $\phi(b) < \phi(a)$ whenever $(a) \subsetneq (b)$.
2. If R is présimplifiable and $0 \neq a \in R$, then $L(a) = \overline{L}(a) := \sup\{n \in \mathbb{Z}_0^+ \mid (a) = (a_1 \cdots a_n)$ with each $a_i \in R \setminus U(R)\}$, where we interpret $L(a) = \infty$ to hold if $L(a)$ is an infinite ordinal or does not exist. Therefore R is a BFR if and only if it is présimplifiable and there is a function $\phi : R \to \mathbb{Z}_0^+$ such that $\phi(b) < \phi(a)$ whenever $(a) \subsetneq (b)$.
3. If R is présimplifiable and $a, b \in R \setminus (0)$, then $L(ab) \geq L(a) \oplus L(b)$, where $\oplus$ is the Hessenberg sum (see [5, p. 329] for the definition).

Analogous statements hold for recursive factorization length in monoids.

Although we are mainly interested in generalized power series rings with proper zero divisors in this paper, we will still need to understand the integral domain case, as our proofs will often reduce to it. To this end, we present the following sharpened version of two results from Anderson and Juett's paper [5]. They proved the following proposition under the stronger hypothesis that $\leq$ is total or positive [5, Theorem 17]. They also proved a related result for monoid rings $R[S]$ with S reduced torsion-free cancellative [5, Theorem 13]. This is also a special case of our next result. Indeed, if $(S, +)$ is a reduced torsion-free cancellative monoid, then $=$ is potentially negative on S [12, Proposition 2.2.1(3)] and $R[S] = R[[S^{=}]]$.

Proposition 7 Let D be a domain and $(S, +, \leq)$ be a torsion-free cancellative ordered monoid. Further assume that $\leq$ is subpositive, subtotal, or potentially negative. Then $D[[S]]$ satisfies the ACCP or has the BF-property if and only if the same holds for D and S, in which case $L(D[[S]]) = \max\{L(D), L(S)\}$.

Proof Because $D[[S]]$ is a domain [39, 1.20], it suffices to prove the ACCP part—see Lemma 6(2). Since S is torsion-free cancellative, there is a compatible total order $\leq'$ on S that refines $\leq$ [24, Proposition 2] and $D[[S^{\leq}]] \subseteq D[[S^{\leq'}]]$ [24, p. 368]. If $\leq$ is potentially negative, then we can choose $\leq'$ to be negative. In all cases we have $D[[S^{\leq}]] \cap U(D[[S^{\leq'}]]) = U(D[[S^{\leq}]])$ by Lemma 2. Therefore $D[[S^{\leq}]]$ satisfies the ACCP if $D[[S^{\leq'}]]$ satisfies the ACCP, in which case $L(D[[S^{\leq}]]) \leq L(D[[S^{\leq'}]])$ [5, p. 330]. On the other hand, we have $D \cap U(D[[S^{\leq}]]) = U(D)$ and $\psi[S] \cap U(D[[S^{\leq}]]) = U(\psi[S])$, where $\psi : S \to D[[S^{\leq}]]$ is the monoid monomorphism $s \mapsto X^s$. Thus D and S satisfy the ACCP if $D[[S^{\leq}]]$ satisfies the ACCP, in which case $\max\{L(D), L(S)\} \leq L(D[[S^{\leq}]])$ [5, p. 330]. With the above observations in mind, we may pass from $\leq$ to $\leq'$ and assume $\leq$ is total. This case is covered by [5, Theorem 17]. □

Now we open our consideration to rings with proper zero divisors. The following weaker version of our main result will prove to be a useful stepping stone.

Lemma 8 Let R be a quasilocal ring with $\mathcal{J}(R)$ nilpotent and $(S, +, \leq)$ be a torsion-free cancellative ordered monoid. Further assume $\leq$ is subtotal, subpositive, or potentially negative. Then $R[[S]]$ satisfies the ACCP or has the BF-property if and only if the same holds for S.

Proof The "only if" part is clear. For the converse, let $\mathcal{P} \in \{\text{ACCP, BF-property}\}$ and assume S has the $\mathcal{P}$. We proceed by induction on the nilpotency N of $M := \mathcal{J}(R)$. The $N = 1$ case is covered by Proposition 7, so let us assume $N > 1$. Because $R[[S]]$ is domainlike (see Theorem 3(5)), we just need to show that $L_{R[[S]]}(f)$ exists (and additionally that $L_{R[[S]]} < \omega$ if $\mathcal{P} =$ BF-property) for each nonzero nonunit $f \in R[[S]]$.

First consider the case where $f \notin M^{N-1}[[S]]$. Let $\overline{R} := R/M^{N-1}$. By induction on N, the ring $\overline{R}[[S]]$ has the $\mathcal{P}$. Note that $R[[S]]$ and $\overline{R}[[S]]$ are domainlike and nonunits of $R[[S]]$ have nonunit images in $\overline{R}[[S]]$ (since $M^{N-1}[[S]] \subseteq \text{Nil}(R[[S]])$). Thus, if $fR[[S]] \subseteq gR[[S]] \subsetneq hR[[S]]$, then $\overline{f}\overline{R}[[S]] \subseteq \overline{g}\overline{R}[[S]] \subsetneq \overline{h}\overline{R}[[S]]$ and therefore $L_{\overline{R}[[S]]}(\overline{g}) > L_{\overline{R}[[S]]}(\overline{h})$. By Lemma 6(1), it follows that $L_{R[[S]]}(f)$ exists and is bounded above by $L_{\overline{R}[[S]]}(\overline{f})$, as desired.

Now assume $f \in M^{N-1}[[S]]$. Let $\overline{R} := R/M$ and a be a nonzero coefficient of f. Then $\overline{R}[[S]]$ is a domain [39, 1.20] and it has the $\mathcal{P}$ by induction on N. Because $((0) :_R M)$ is a semisimple ideal of R containing a, we have $((0) :_R M) = aR \oplus I$ for some ideal I of R.

Suppose that there is an infinite properly ascending sequence $fR[[S]] \subsetneq f_1R[[S]] \subsetneq f_2R[[S]] \subsetneq f_3R[[S]] \subsetneq \cdots$ ($f_i \in R[[S]]$). Then each $f_i \in M^{N-1}[[S]] \subseteq ((0) :_R M)[[S]] = aR[[S]] \oplus I[[S]]$ by the previous case. Thus for each $n \in \mathbb{Z}^+$ there is an $h_n \in R[[S]]$ with $f_n - ah_n \in I[[S]]$. Note that each $h_n \notin M[[S]]$, for otherwise $a \in c(f) \subseteq c(f_n) = c(f_n - ah_n) \subseteq I$. Write $f_n = g_n f_{n+1}$ with $g_n \in R[[S]] \setminus U(R[[S]])$. Then each $a(h_n - g_n h_{n+1}) \equiv$

$f_n - g_n f_{n+1} \equiv 0$ modulo $I[[S]]$, so $\overline{h_n - g_n h_{n+1}} \in \overline{(I[[S]] :_{R[[S]]} aR[[S]])} = \overline{(I :_R aR)[[S]]} \subseteq \overline{M[[S]]}$. Therefore $\overline{(0)} \subsetneq \overline{h_1 R[[S]]} \subsetneq \overline{h_2 R[[S]]} \subsetneq \cdots$, where the inclusions are proper since $\overline{R}[[S]]$ is a domain and each $\overline{g_n} \notin U(\overline{R}[[S]])$ (because $M[[S]] \subseteq \mathrm{Nil}(R[[S]])$). But this contradicts the fact that $\overline{R}[[S]]$ satisfies the ACCP.

We are done unless $\mathcal{P}$ = BF-property. As above, there is a $g \in R[[S]] \setminus M[[S]]$ with $f - ag \in I[[S]]$. Every factorization of f into nonunits of $R[[S]]$ has the form $f = f_1 \cdots f_m$ with $f_1, \ldots, f_k \in M[[S]]$ and $f_{k+1}, \ldots, f_m \in R[[S]] \setminus M[[S]]$ $(0 \leq k \leq m)$. Then $k \leq N$ since $f \neq 0$ and $f_{k+1}, \ldots, f_m \in \mathrm{Reg}(R[[S]])$ since $R[[S]]$ is domainlike and $\mathrm{Nil}(R[[S]]) = M[[S]]$ [37, 3.2]. Because $f_1 \cdots f_m M[[S]] = fM[[S]] = (0)$, it follows that $f_1 \cdots f_k \in ((0) :_{R[[S]]} M[[S]]) = ((0) :_R M)[[S]] = aR[[S]] \oplus I[[S]]$ and therefore $f_1 \cdots f_k - ah \in I[[S]]$ for some $h \in R[[S]]$. Thus $ag \equiv f \equiv f_1 \cdots f_m \equiv ahf_{k+1} \cdots f_m$ modulo $I[[S]]$, so $\overline{g - hf_{k+1} \cdots f_m} \in \overline{(I[[S]] :_{R[[S]]} aR[[S]])} = \overline{(I :_R aR)[[S]]} \subseteq \overline{M[[S]]}$. So $\overline{g}\overline{R}[[S]] \subseteq \overline{f_{k+1}} \cdots \overline{f_m}\overline{R}[[S]]$, where each $\overline{f_i} \notin U(\overline{R}[[S]])$ since $M[[S]] = \mathrm{Nil}(R[[S]])$, and therefore $\overline{L}_{\overline{R}[[S]]}(\overline{g}) \geq m - k$. Thus $m = k + (m - k) \leq N + \overline{L}_{\overline{R}[[S]]}(\overline{g}) < \infty$, as desired. □

We are now ready to present the main result of our paper. Edmonds and Juett have proven two simlarly-flavored monoid ring results [23, Theorems 4.4 and 4.7].

Theorem 9 Let R be a ring with $\mathrm{Nil}(R)$ nilpotent and $(S, +, \leq)$ be a torsion-free cancellative monoid.

1. If R is a finite direct product of domainlike rings and $\leq$ is subtotal, subpositive, or potentially negative, then $R[[S]]$ satisfies the ACCP if and only if R and S satisfy the ACCP.
2. If $\leq$ is subpositive and R is domainlike, then $R[[S]]$ is a BFR if and only if R is a BFR and S is a BFM.
3. If $\leq$ is subtotal or potentially negative, then $R[[S]]$ is a BFR if and only if R is a domainlike BFR and S is a BFM.

Proof Because S is torsion-free cancellative, there is a compatible total order $\leq'$ on S that refines $\leq$ [24, Proposition 2] and $R[[S^{\leq}]] \subseteq R[[S^{\leq'}]]$ [24, p. 368]. We may arrange for $\leq'$ to be negative if $\leq$ is potentially negative. Lemma 2 implies that $R[[S^{\leq}]] \cap U(R[[S^{\leq'}]]) = U(R[[S^{\leq}]])$, so we may pass from $\leq$ to $\leq'$ and assume $\leq$ is total. Let K be the total quotient ring of R. For $0 \neq g \in R[[S]]$, define $\psi(g) := \max\{n \in \mathbb{Z}_0^+ \mid g\mathrm{Nil}(R)^n[[S]] \neq (0)\}$. The "only if" parts of (1) and (2) are clear, while the "only if" part of (3) follows from Theorem 3.

For the converses, we begin by assuming R and S satisfy the ACCP and $R = \prod_{i=1}^n R_i$ with each R_i domainlike. Then $R[[S]] \cong \prod_{i=1}^n R_i[[S]]$ [12, Proposition 2.4.1(1)] satisfies the ACCP if and only if each $R_i[[S]]$ satisfies the ACCP [8, Theorem 3.4]. Thus we may pass from R to an R_i and assume R is domainlike. Then K is quasilocal with $\mathcal{J}(K)$ nilpotent since $P := \mathrm{Nil}(R) = \mathcal{Z}(R)$ is nilpotent. So $R[[S]]$ and $K[[S]]$ are domainlike by Theorem 3(5) and $K[[S]]$ satisfies the ACCP by Lemma 8.

Suppose that there is a sequence $\{f_n\}_{n=1}^{\infty} \subseteq R[[S]] \setminus (0)$ with each $f_n = u_n f_{n+1}$ for some $u_n \in R[[S]] \setminus U(R[[S]])$. Because $0 \leq \psi(f_1) \leq \psi(f_2) \leq \cdots \leq N-1$, where N is the nilpotency of P, we may pass to a suitable subsequence of $\{f_n\}_{n=1}^{\infty}$ and assume $\psi(f_1) = \psi(f_2) = \cdots$. Since $K[[S]]$ is domainlike and satisfies the ACCP, we may again pass to a subsequence of $\{f_n\}_{n=1}^{\infty}$ and assume each $u_n \in U(K[[S]])$. Pick $z \in P^{\psi(f_1)}$ with $zf_1 \neq 0$. Each $zf_{n+1} \in ((0) :_{R[[S]]} P[[S]])$, so $zf_n = u_n z f_{n+1} = v_n z f_{n+1}$, where v_n is the sum of the terms of u_n that have non-nilpotent (or equivalently regular) coefficients. For each $n \in \mathbb{Z}^+$, let $a_n X^{s_n}$ and $b_n X^{t_n}$ ($a_n, b_n \in R$ and $s_n, t_n \in S$) respectively be the trailing terms of zf_n and v_n. Since each $b_n \in \mathrm{Reg}(R)$, it follows that $a_n = b_n a_{n+1}$ and $s_n = t_n + s_{n+1}$. Also note that each $L_R(b_n) \oplus L_S(t_n) > 0$, for otherwise $v_n \in U(R[[S]])$ (see Lemma 2), which would imply $u_n \in U(R[[S]]) + P[[S]] = U(R[[S]])$ (since P is nilpotent). Therefore each $L_R(a_n) \oplus L_R(s_n) \geq L_R(b_n) \oplus L_R(a_{n+1}) \oplus L_S(t_n) \oplus L_S(s_{n+1}) > L_R(a_{n+1}) \oplus L_S(s_{n+1})$ by Lemma 6(4), leading to an absurd infinite strictly decreasing sequence of ordinals.

It only remains to show that $R[[S]]$ is a BFR if R is a domainlike BFR and S is a BFM. Then $K[[S]]$ is a BFR by Lemma 8. Let f be a nonzero nonunit of $R[[S]]$. Each factorization of f into nonunits of $R[[S]]$ has the form $f = f_1 \cdots f_m$ with $f_1, \ldots, f_k \in (R[[S]] \cap U(K[[S]])) \setminus U(R[[S]])$ and $f_{k+1}, \ldots, f_m \in R[[S]] \setminus U(K[[S]])$ $(0 \leq k \leq m)$. Then $\overline{L}_{K[[S]]}(f) \geq m-k$. Pick $z \in P^{\psi(f)}$ with $zf \neq 0$. Then $zf_1 \cdots f_m = zf \in ((0) :_{R[[S]]} P[[S]])$, so $zf_{k+1} \cdots f_m \in ((0) :_{R[[S]]} P[[S]])$ since $f_1 \cdots f_k \in \mathrm{Reg}(R[[S]])$. Thus $zf = v_1 \cdots v_k z f_{k+1} \cdots f_m$, where each v_i is the sum of the terms of f_i that have non-nilpotent (or equivalently regular) coefficients. Let aX^s, $b_i X^{t_i}$, and cX^w respectively be the trailing terms of zf, v_i, and $zf_{k+1} \cdots f_m$. Then $a = b_1 \cdots b_k c$ and $s = t_1 + \cdots + t_k + w$ since the b_i's are regular. Also note that each $\overline{L}_R(b_i) + \overline{L}_S(t_i) > 0$, for otherwise $v_i \in U(R[[S]])$ (see Lemma 2), which would imply $f_i \in U(R[[S]]) + P[[S]] = U(R[[S]])$ (since P is nilpotent). Therefore $m = (m-k) + k \leq \overline{L}_{K[[S]]}(f) + \sum_{i=1}^{k} (\overline{L}_R(b_i) + \overline{L}_S(t_i)) \leq \overline{L}_{K[[S]]}(f) + \overline{L}_R(a) + \overline{L}_S(s) < \infty$, as desired. □

We now collect several results about the ACCP and BF-property in rings of (Laurent) polynomials, large polynomials, and (Laurent) power series. We have focused here on the kinds of extensions for which we were able to determine nice results, omitting large Laurent polynomial rings and only considering Laurent power series rings of the form $R\langle\langle \{X_\lambda\}_{\lambda \in \Lambda} \rangle\rangle$. There are several known results about the ACCP in the polynomial/power series special case that do not appear to have analogs in the general setting we are exploring. There one can achieve more if one assumes some stronger/different hypotheses on the coefficient ring than the kinds of conditions we consider here. We refer the reader to the work of Frohn [25–27] and Heinzer and Lantz [33, 34] for further information on that topic. In order to state some of the parts of Corollary 10 in the strongest and most useful way, we need to review the notion of ordinal powers of ideals. Let I be an ideal of a ring R. Define $I^0 := R$ and for each positive ordinal β recursively define $I^\beta := \bigcap_{\alpha < \beta} II^\alpha$. Then I^n has the usual meaning for $0 \leq n < \omega$. In addition, we have $I^\omega = \bigcap_{n=1}^{\infty} I^n$,

$I^{\omega+1} = I\left(\bigcap_{n=1}^{\infty} I^n\right)$, etc. We refer the reader to [13, Section 6] and [7, Lemma 2.9] for further information on properties of ordinal powers of ideals.

Corollary 10 Let R be a ring, $\{X_\lambda\}_{\lambda\in\Lambda}$ be a nonempty family of analytically independent indeterminates, α be an infinite cardinal number, $i \in \{1, 2, 3, <\alpha\}$, $j \in \{L, L_{<\alpha}\}$, $A := R[[\{X_\lambda\}_{\lambda\in\Lambda}]]_i$, $B_0 := R[\{X_\lambda\}_{\lambda\in\Lambda}]$, $B_0' := R[\{X_\lambda, X_\lambda^{-1}\}_{\lambda\in\Lambda}]$, $B := R[\{X_\lambda\}_{\lambda\in\Lambda}]_j$, and $C := R\langle\langle\{X_\lambda\}_{\lambda\in\Lambda}\rangle\rangle$.

1. If R is a finite direct product of présimplifiable rings, then A satisfies the ACCP if and only if R satisfies the ACCP.
2. If R is a finite direct product of domainlike rings and $\text{Nil}(R)^\mu = (0)$ for some ordinal μ, then B_0 satisfies the ACCP if and only if R satisfies the ACCP.
3. If $D \in \{B_0', B, C\}$, R is a finite direct product of domainike rings, and $\text{Nil}(R)$ is nilpotent, then D satisfies the ACCP if and only if R satisfies the ACCP.
4. If $\text{Nil}(R)^\omega = (0)$, then B_0 is a BFR if and only if R is a domainlike BFR [23, Corollary 4.12(2)].
5. If $\text{Nil}(R)$ is nilpotent and R is domainlike, then A is a BFR if and only if R is a BFR.
6. If $D \in \{B_0', B, C\}$ and $\text{Nil}(R)$ is nilpotent, then D is a BFR if and only if R is a domainlike BFR.

Proof The "only if" parts are each either clear or follow from Corollary 4. For the converses, we first argue as in the second paragraph of Theorem 9 to reduce (if necessary) to the case where R is indecomposable. (Note that, if $R = \prod_{i=1}^n R_i$, then an easy inductive proof shows that $\text{Nil}(R)^\beta = \prod_{i=1}^n \text{Nil}(R_i)^\beta$ for each ordinal β.) The statements about B_0 and B_0' now follow from [23, Corollary 4.12]. If $(D, D') \in \{(A, R[[\{X_\lambda\}_{\lambda\in\Lambda}]]_3), (B, R[\{X_\lambda\}_{\lambda\in\Lambda}]_L)\}$ and D' is a BFR, then so is D since $U(D) = D \cap U(D')$ (see Lemma 2). If R is domainlike, $\text{Nil}(R)$ is nilpotent, and $R[\{X_\lambda\}_{\lambda\in\Lambda}]_L$ satisfies the ACCP, then B satisfies the ACCP since $R[\{X_\lambda\}_{\lambda\in\Lambda}]_L$ is domainlike and $U(B) = B \cap U(R[\{X_\lambda\}_{\lambda\in\Lambda}]_L)$ (see Corollary 4(5) and Lemma 2(3)). With the above comments in mind, Theorem 9 finishes all the remaining "ifs" except for (1). Assume R is présimplifiable and satisfies the ACCP. Suppose that there is a sequence $\{f_n\}_{n=1}^\infty \subseteq A \setminus (0)$ with each $f_n = g_n f_{n+1}$ for some $g_n \in A \setminus U(A)$. Then each $g_n(0) \notin U(R)$ by Lemma 2(2). Well order Λ. Let $X_{\lambda_1}, \ldots, X_{\lambda_k}$ be the variables appearing in the trailing term of f_1. Let π be the canonical ring epimorphism from A onto $A_0 := R[[X_{\lambda_1}, \ldots, X_{\lambda_n}]]$. Then each $\pi(f_n) = \pi(g_n)\pi(f_{n+1})$, where $\pi(g_n) \notin U(A_0)$ since $\pi(g_n)(0) = g_n(0) \notin U(R)$, so $\pi(f_n)A_0 \subsetneq \pi(f_{n+1})A_0$ since A_0 is présimplifiable (see Corollary 4(3)). Thus we may pass from A to A_0 and assume Λ is finite. Induction then reduces our proof to the case where $A = R[[X]]$. Let X^{m_n} ($m_n \geq 0$) be the largest power of X dividing f_n. Then $m_1 \geq m_2 \geq \cdots \geq 0$, so we can pass to a suitable subsequence of $\{f_n\}_{n=1}^\infty$ and assume $m_1 = m_2 = \cdots$. Then we can pass from $\{f_n\}_{n=1}^\infty$ to $\{X^{-m_1} f_n\}_{n=1}^\infty$ and assume each $f_n(0) \neq 0$. Then each $f_n(0) = g_n(0) f_{n+1}(0)$, where $g_n(0) \notin U(R)$, so $f_n(0)R \subsetneq f_{n+1}(0)R$ since R is présimplifiable, contradicting the fact that R satisfies the ACCP. □

We review which (special cases of) parts of Corollary 10 have already appeared in the literature. The key présimplifiable special case of (1) was an observation of Frohn [25, p. 2961], although he did not give a proof or an elaboration on which value(s) of i he had in mind. Edmonds and Juett (at least implicitly) proved all the parts about B_0 and B_0' [23, Corollary 4.12]. The various remaining assertions of Corollary 10 and its generalizations of the aforementioned results all appear to be new.

We close the paper with some examples exploring the degree to which the hypotheses in Corollary 10 are necessary or might be improved.

Example 11 (An Example Where R Satisfies the ACCP but $R[[X]]$ and $R\langle\langle X\rangle\rangle$ Do Not) Let $D := \mathbb{Q}[\{Y_n, Z_n\}_{n=0}^{\infty}]$, $I := \sum_{m\leq n} Y_m(Z_n - 1)D$, $P_i := \sum_{n=0}^{i-1} Y_i D + \sum_{n=i}^{\infty}(Z_n - 1)D$, $P := \sum_{n=0}^{\infty} Y_n D$, $M := P + \sum_{n=0}^{\infty} Z_n D$, $Q := P + \sum_{n=0}^{\infty}(Z_n - 1)D$, $\overline{D} := D/I$, $\overline{S} := \overline{D} \setminus (\overline{M} \cup \overline{Q})$, and $R := \overline{D}_{\overline{S}}$. Frohn showed that R satisfies the ACCP but $R[[X]]$ does not [25, Example]. We will build upon his proof to show that $R\langle\langle X\rangle\rangle$ also fails the ACCP. Frohn showed that $f_1 R[[X]] \subsetneq f_2 R[[X]] \subsetneq \cdots$, where $f_n := \sum_{i=0}^{\infty}\left(\prod_{j=n}^{i-1}\overline{Z_j}\right)\overline{Y_i}X^i$ [25, Example(2)]. Suppose that some $f_n R\langle\langle X\rangle\rangle = f_{n+1} R\langle\langle X\rangle\rangle$. Then $X^m f_{n+1} = g f_n$ for some $m \in \mathbb{Z}_0^+$ and $g \in R[[X]]$. In $(R/\overline{P_i}_{\overline{S}})[[X]]$, the images of f_n and f_{n+1} both have trailing degree i, so the image of g in $(R/\overline{P_i}_{\overline{S}})[[X]]$ must have trailing degree m. Because $\bigcap_{i=1}^{\infty} P_i = I$ [25, Example(1)], it follows that $\mathrm{Nil}(R) = (\overline{0})$ and g has trailing degree m in $R[[X]]$. Therefore $X^{-m} g \in R[[X]]$ and $f_{n+1} = (X^{-m} g) f_n$, a contradiction. This example shows that the hypotheses on R cannot be dropped for the portions of Corollary 10 discussing the ACCP in the extensions A or C.

Example 12 (A Quasilocal BFR R with $\mathcal{J}(R)^{\omega} = (0)$ for Which Neither $R[X]$ nor $R[X, X^{-1}]$ Satisfies the ACCP) Let $D := \mathbb{Q}[Y_1, Y_2, \ldots]$, $I := \sum_{n=1}^{\infty} Y_{n+1}(Y_n - Y_{n+1})D$, $M := \sum_{n=1}^{\infty} Y_i D$, $\overline{D} := D/I$, and $R := \overline{D}_{\overline{M}}$. Heinzer and Lantz showed that R satisfies the ACCP while $R[X]$ does not [34, Example]; a careful reading of their proof and their comments in [34, Remark] shows that R is in fact a quasilocal BFR with $\mathcal{J}(R)^{\omega} = (0)$. We will adapt their proof to show that $R[X, X^{-1}]$ also fails the ACCP. Setting $f_n := \overline{Y_n}X + \overline{1}$, we have $f_1 R[X] \subsetneq f_2 R[X] \subsetneq \cdots$ [34, p. 976], so it will suffice to show that each $f_n R[X, X^{-1}] \subsetneq f_{n+1} R[X, X^{-1}]$. Suppose to the contrary that $X^m f_{n+1} = g f_n$ for some $m, n \in \mathbb{Z}^+$ and $g \in R[X]$. Let a_k denote the coefficient of X^k in g. Because f_n and f_{n+1} each have a trailing coefficient of $\overline{1}$, the equation $X^m f_{n+1} = g f_n$ implies $a_0 = \cdots = a_{m-1} = \overline{0}$. Therefore the coefficient of X^m in $X^m f_{n+1} = g f_n$ is $\overline{1} = a_m + a_{m-1}\overline{Y_n} = a_m$. We claim that $a_k = (-1)^{k-m-1}\overline{Y_n}^{k-m-1}(\overline{Y_{n+1}} - \overline{Y_n})$ for each $k > m$. The coefficient of X^{m+1} in $X^m f_{n+1} = g f_n$ is $\overline{Y_{n+1}} = a_{m+1} + a_m\overline{Y_n} = a_{m+1} + \overline{Y_n}$, which implies $a_{m+1} = \overline{Y_{n+1}} - \overline{Y_n}$. So let us assume $k > m + 1$. Then the coefficient of X^k in $X^m f_{n+1} = g f_n$ is $\overline{0} = a_k + a_{k-1}\overline{Y_n}$, so $a_k = -a_{k-1}\overline{Y_n} = -(-1)^{(k-1)-m-1}\overline{Y_n}^{(k-1)-m-1}(\overline{Y_{n+1}} - \overline{Y_n})\overline{Y_n} = (-1)^{k-m-1}\overline{Y_n}^{k-m-1}(\overline{Y_{n+1}} - \overline{Y_n})$ by induction on k, as desired. But then $a_k \neq \overline{0}$ for all $k \geq m$ [34, p. 975], contradicting the fact that $g \in R[X]$. This example shows that, at least for the assertions about B_0, B_0', and B in Corollary 10 parts (2) and (3), the "domainlike" in the hypotheses cannot be weakened to "présimplifiable." We do not know to what extent the hypotheses about $\mathrm{Nil}(R)$ in those assertions are necessary.

Example 13 (A Zero-Dimensional Quasilocal BFR R with $\mathcal{J}(R)^{\omega+1} = (0)$ for Which $R[X]$, $R[X, X^{-1}]$, $R[[X]]$, and $R\langle\langle X\rangle\rangle$ Are Not BFRs) Let $D := \mathbb{Z}_2[Y_1, Y_2, \ldots]$, where $\mathbb{Z}_2$ is the field of cardinality 2. Partition $\{Y_k\}_{k=1}^{\infty} = \bigsqcup_{n=0}^{\infty} B_n$, where $B_0 := \{Y_1\}$ and $B_n := \{Y_i\}_{i=n^2+3n-2}^{n^2+5n+1}$ for $n \geq 1$. For each $n \geq 1$ and $0 \leq k \leq n+2$, let $g_{n,k} \in D$ be the coefficient of X^k in $\prod_{k=0}^{n+1}(Y_{n^2+3n-2+2k} + Y_{n^2+3n-1+2k}X)$. Let J be the ideal of D generated by elements of the forms (i) Y_iY_j with Y_i and Y_j in different B_n's, (ii) Y_1^2, (iii) $\prod_{i=1}^{n+3} Y_{k_i}$ with $n \geq 1$ and each $Y_{k_i} \in B_n$, (iv) $g_{n,k}$ with $k \notin \{1, 2\}$, and (v) $Y_1 - g_{n,k}$ with $k \in \{1, 2\}$. (See [4, Section 3] for a more detailed discussion of the definitions.) Anderson and Ganatra showed that $R := D/J$ is a zero-dimensional quasilocal BFR for which neither $R[X]$ nor $R[[X]]$ is a BFR [4, Section 3]; Juett and Edmonds later observed that $\mathcal{J}(R)^{\omega+1} = (\overline{0}) \neq \mathcal{J}(R)^{\omega}$ [23, p. 1855]. We will show that $R[X, X^{-1}]$ and $R\langle\langle X\rangle\rangle$ also fail to be BFRs. Conditions (iv) and (v) imply that $\overline{0} \neq \overline{Y}_1X + \overline{Y}_1X^2 = \prod_{k=0}^{n+1}(\overline{Y_{n^2+3n-2+2k}} + \overline{Y_{n^2+3n-1+2k}}X)$ for each $n \in \mathbb{Z}^+$ [4, p. 3897]. Each of the factors in the preceding equation is in $\mathcal{J}(R)\langle\langle X\rangle\rangle$ and therefore not invertible in $R\langle\langle X\rangle\rangle$. Thus neither $R[X]$, nor $R[[X]]$, nor $R[X, X^{-1}]$, nor $R\langle\langle X\rangle\rangle$ is a BFR, but they do satisfy the ACCP. Indeed, the rings $R[[X]]$ and $R[X]$ satisfy the ACCP by Corollary 10, while $R[X, X^{-1}]$ satisfies the ACCP by [23, Lemma 4.5(1)] since R is a zero-dimensional quasilocal ring with $\mathcal{J}(R)^{\omega+1} = (\overline{0})$. Our first step in showing that $R\langle\langle X\rangle\rangle$ satisfies the ACCP is to demonstrate that it is quasilocal with maximal ideal $\mathcal{J}(R)\langle\langle X\rangle\rangle$. That ideal is indeed maximal by [12, Proposition 2.4.2(2)]. So we need to show that $\overline{1} - gf \in U(R\langle\langle X\rangle\rangle)$ for each $f \in \mathcal{J}(R)\langle\langle X\rangle\rangle$. Because $\mathrm{supp}(\overline{1} - gf)$ is well ordered and $(\overline{1} - gf)(\overline{0}) \in \overline{1} - \mathcal{J}(R) \subseteq U(R)$, there is a smallest $m \in \mathbb{Z}$ such that X^m has a unit coefficient in $\overline{1} - gf$. Then the sum of the terms of $\overline{1} - gf$ of degree at least m is a unit in $R\langle\langle X\rangle\rangle$ [39, 2.6(ii)] and the finitely many terms of $\overline{1} - gf$ of degree less than m are nilpotent, so $\overline{1} - gf \in U(R\langle\langle X\rangle\rangle)$, as desired. An easy inductive proof now shows that $\mathcal{J}(R\langle\langle X\rangle\rangle)^{\beta} \subseteq \mathcal{J}(R)^{\beta}\langle\langle X\rangle\rangle$ for each ordinal β. Because $R\langle\langle X\rangle\rangle$ is a quasilocal ring with $\mathcal{J}(R\langle\langle X\rangle\rangle)^{\omega+1} = (\overline{0})$, it satisfies the ACCP [7, Theorem 3.22]. This example shows that the statement of Corollary 10(4) is in some sense the best possible and places limits on how much the results in Corollary 10 parts (5) and (6) could potentially be improved.

Acknowledgments On the occasion of Professor Dan Anderson's retirement, we would like to acknowledge the mentorship that he has provided us and many others throughout his career. We would also like to acknowledge the anonymous referee for carefully reading our manuscript and offering several corrections/clarifications to our writing.

References

1. D.D. Anderson and O.A. Al-Mallah. Commutative group rings that are présimplifiable or domainlike. *J. Algebra Appl.*, 16:1750019, 2017.
2. D.D. Anderson, D.F. Anderson, and M. Zafrullah. Factorization in integral domains. *J. Pure Appl. Algebra*, 69:1–19, 1990.
3. D.D. Anderson and S. Chun. Associate elements in commutative rings. *Rocky Mountain J. Math.*, 44:717–731, 2014.
4. D.D. Anderson and A. Ganatra. Bounded factorization rings. *Comm. Algebra*, 35:3892–3903, 2007.
5. D.D. Anderson and J.R. Juett. Long length functions. *J. Algebra*, 426:327–343, 2015.
6. D.D. Anderson and J.R. Juett. Length functions in commutative rings with zero divisors. *Comm. Algebra*, 45:1584–1600, 2017.
7. D.D. Anderson, J.R. Juett, and C.P. Mooney. Factorization of ideals. *Comm. Algebra*, 47:1742–1772, 2019.
8. D.D. Anderson and S. Valdes-Leon. Factorization in commutative rings with zero divisors. *Rocky Mountain J. Math.*, 26:439–480, 1996.
9. D.D. Anderson and S. Valdes-Leon. Factorization in commutative rings with zero divisors, II. In D.D. Anderson, editor, *Factorization in integral domains*, pages 197–219. Marcel Dekker, 1997.
10. D.F. Anderson and F. Gotti. Bounded and finite factorization domains. In A. Badawi and J. Coykendall, editors, *Rings, monoids, and module theory*. Springer-Verlag, Singapore, 2022.
11. M. Axtell, S.J. Forman, and J. Stickles. Properties of domainlike rings. *Tamkang J. Math.*, 40:151–164, 2009.
12. N.P. Aylesworth and J.R. Juett. Generalized power series with a limited number of factorizations. *J. Commut. Algebra*, 14:471–507, 2022.
13. R. Baer. Radical ideals. *Amer. J. Math.*, 65:537–568, 1943.
14. A. Benhissi and M. Eljeri. MC-extensions: examples, zero-divisors graph, and colorability. *Comm. Algebra*, 40:104–124, 2012.
15. G.M. Benkart, A.M. Gaglione, W.D. Joyner, M.E. Kidwell, M.D. Meyerson, D. Spellman, and W.P. Wardlaw. Principal ideals and associate rings. *JP J. Algebra Number Theory Appl.*, 2:181–193, 2002.
16. A. Bouvier. Anneaux présimplifiables et anneaux atomiques. *C.R. Acad. Sci. Paris*, 272:992–994, 1971.
17. A. Bouvier. Anneaux présimplifiables. *C.R. Acad. Sci. Paris*, 274:A1605–A1607, 1972.
18. A. Bouvier. Résultats nouveaux sur les anneaux présimplifiables. *C.R. Acad. Sci. Paris*, 275:A955–A957, 1972.
19. A. Bouvier. Anneaux présimplifiables. *Rev. Roumaine Math. Pures Appl.*, 19:713–724, 1974.
20. E.D. Cashwell and C.J. Everett. Formal power series. *Pacific J. Math.*, 13:45–64, 1963.
21. S.T. Chapman, F. Gotti, and M. Gotti. Factorization invariants of Puiseux monoids generated by geometric sequences. *Comm. Algebra*, 48:380–396, 2020.
22. C.C. Cheng, J.H. McKay, J. Towber, S.S. Wang, and D.L. Wright. Reversion of power series and the extended Raney coefficients. *Trans. Amer. Math. Soc.*, 349:1769–1782, 1997.
23. R.A.C. Edmonds and J.R. Juett. Associates, irreducibility, and bounded factorization in monoid rings with zero divisors. *Comm. Algebra*, 49:1836–1860, 2021.
24. G.A. Elliott and P. Ribenboim. Fields of generalized power series. *Arch. Math.*, 54:365–371, 1990.
25. D. Frohn. A counterexample concerning ACCP in power series rings. *Comm. Algebra*, 30:2961–2966, 2002.
26. D. Frohn. Modules with n-acc and the acc on certain types of annihilators. *J. Algebra*, 256:467–483, 2002.
27. D. Frohn. ACCP rises to the polynomial ring if the ring has only finitely many associated primes. *Comm. Algebra*, 32:1213–1218, 2004.

28. A. Geroldinger and F. Halter-Koch. *Non-Unique Factorizations: Algebraic, Combinatorial and Analytic Theory*, volume 278 of *Pure and Applied Mathematics (Boca Raton)*. Chapman & Hall/CRC, Boca Raton, FL, 2006.
29. R. Gilmer. *Multiplicative Ideal Theory*, Queen's Papers in Pure and Appl. Math., vol. 90. *Queen's University, Kingston, Ontario*, 1992.
30. R. Gilmer and T. Parker. Divisibility properties in semigroup rings. *Michigan Math. J.*, 21:65–86, 1974.
31. F. Gotti. Systems of sets of lengths of Puiseux monoids. *J. Pure Appl. Algebra*, 223:1856–1868, 2019.
32. F. Halter-Koch. On the algebraic and arithmetical structure of generalized polynomial algebras. *Rend. Semin. Mat. Univ. Padova*, 90:121–140, 1993.
33. W.J. Heinzer and D.C. Lantz. Commutative rings with ACC on n-generated ideals. *J. Algebra*, 80:261–278, 1983.
34. W.J. Heinzer and D.C. Lantz. ACCP in polynomial rings: a counterexample. *Proc. Amer. Math. Soc.*, 121:975–977, 1994.
35. H. Kim. Factorization in monoid domains. *Comm. Algebra*, 29:1853–1869, 2001.
36. P. Ribenboim. Generalized power series rings. In *Lattices, semigroups, and universal algebra (Lisbon, 1988)*, pages 271–277. Plenum, New York, 1990.
37. P. Ribenboim. Rings of generalized power series: nilpotent elements. *Abh. Math. Sem. Univ. Hamburg*, 61:15–33, 1991.
38. P. Ribenboim. Noetherian rings of generalized power series. *J. Pure Appl. Algebra*, 79:293–312, 1992.
39. P. Ribenboim. Rings of generalized power series II: Units and zero-divisors. *J. Algebra*, 168:71–89, 1994.
40. P. Ribenboim. Special properties of rings of generalized power series. *J. Algebra*, 173:566–586, 1995.
41. P. Ribenboim. Semisimple rings and von Neumann regular rings of generalized power series. *J. Algebra*, 198:327–338, 1997.
42. G. Xin. A fast algorithm for MacMahon's partition analysis. *Electron. J. Combin.*, 11, 2004.

Probabilities and Fixed Divisors of Integer Polynomials

Jean-Luc Chabert

1 Turk's Formula

Turk [6] stated the following nice formula which gives the probability that the fixed divisor of a polynomial with integer coefficients is equal to one:

$$\text{Prob}\,(\,d(f) = 1 \mid f \in \mathbb{Z}[X]\,) = \prod_{p\in\mathbb{P}} (1 - p^{-p}) \qquad \text{[Turk, 1986]} \tag{1}$$

Fixed Divisor Recall that the *fixed divisor* of a polynomial $f \in \mathbb{Z}[X]$, denoted by $d(f)$, is defined as the greatest common divisor of all the values of f on $\mathbb{Z}$.

$$d(f) = \gcd\{\, f(n) \mid n \in \mathbb{N}\,\} = \max\{\, d \in \mathbb{N} \mid f(\mathbb{Z}) \subseteq d\mathbb{Z}\,\}\,.$$

More generally, if f is an integer-valued polynomial, that is, a polynomial f such that $f(\mathbb{Z}) \subseteq \mathbb{Z}$, we can still consider the fixed divisor $d(f)$ of f as the gcd of the values of f on $\mathbb{Z}$. It is known that every such polynomial may be written as

$$f(X) = \sum_{k=0}^{d} b_k \binom{X}{k} \quad \text{where } b_k \in \mathbb{Z}\,.$$

It is then easy to see that, if $\deg f = d$, then

J.-L. Chabert (✉)
LAMFA, Université de Picardie Jules Verne, Amiens, France
e-mail: jean-luc.chabert@u-picardie.fr

J.-L. Chabert et al. (eds.), *Algebraic, Number Theoretic, and Topological Aspects of Ring Theory*, https://doi.org/10.1007/978-3-031-28847-0_10

$$d(f) = \gcd\{f(0), f(1), \ldots, f(d)\} = \gcd\{b_0, b_1, \cdots, b_d\}.$$

Probability and Asymptotic Density In fact, the probability in Formula (1) is not really a probability since there is no probability neither on $\mathbb{Z}$ nor on $\mathbb{Z}[X]$ which corresponds to the intuition of an arithmetician (see for instance [5, § III.1.1]). This is an asymptotic density obtained by means of the height.

Recall that the *height* $H(f)$ of a polynomial $f(X) = \sum_{k=0}^{d} c_k X^k \in \mathbb{Z}[X]$ is defined by:

$$H(f) = \max\{\, |c_k| \mid 0 \leq k \leq d\,\}\,.$$

Let T be a subset of $\mathbb{Z}[X]$. The probability that a polynomial $f \in \mathbb{Z}[X]$ of degree $\leq d$ and height $\leq H$ belongs to T is clearly the rational number:

$$\frac{\mathrm{Card}\{\, f \in T \mid \deg f \leq d,\, H(f) \leq H\,\}}{\mathrm{Card}\{f \in \mathbb{Z}[X] \mid \deg f \leq d,\, H(f) \leq H\}}\,.$$

Then, for a fixed integer d, the density of T in the set of integer polynomials with degree less or equal to d is defined as the limit of the previous rational numbers when the height tends to infinity, of course if such a limit exists:

$$\text{“Prob”}\,(f \in T \mid \deg f \leq d\,) = \lim_{H\to+\infty} \frac{\mathrm{Card}\{\, f \in T \mid \deg f \leq d,\, H(f) \leq H\,\}}{\mathrm{Card}\{f \in \mathbb{Z}[X] \mid \deg f \leq d,\, H(f) \leq H\}}\,.$$

Finally, the *asymptotic density* of T in $\mathbb{Z}[X]$ is defined as the limit of the previous density when the degree d tends to infinity, once more if such a limit exists:

$$\text{“Prob”}\,(\, f \in T \mid f \in \mathbb{Z}[X]\,) = \lim_{d\to+\infty} \text{“Prob”}\,(\, f \in T \mid \deg f \leq d\,)\,.$$

Notation We will systematically write Prob to designate a classical probability while “Prob” will designate an asymptotic density obtained by means of a limit.

Of course, Turk was aware of this fact, but at some step of his proof he argues as if he considered ‘real probabilities’. Let us be more precise. He proved first by means of the height that whatever the positive integers d and δ:

$$\text{“Prob”}(f(\mathbb{Z}) \subseteq \delta\mathbb{Z} \mid f \in \mathbb{Z}[X],\, \deg f \leq d\,) = \prod_{k=0}^{d} \frac{\gcd(k!, \delta)}{\delta}\,. \tag{2}$$

Then, from the formula

$$\mathrm{Prob}\,(\, f(\mathbb{Z}) \subseteq \delta\mathbb{Z} \mid \deg f \leq d\,) = \sum_{t=1}^{\infty} \mathrm{Prob}\,(\, d(f) = t \times \delta \mid \deg f \leq d\,) \tag{3}$$

and, using Möbius inversion formula, he obtained that:

$$\text{Prob}\,(d(f) = \delta \mid \deg f \leq d) = \sum_{t=1}^{+\infty} \mu(n)\ \text{Prob}\,(\,f(\mathbb{Z}) \subseteq t \times \delta\mathbb{Z} \mid \deg f \leq d\,) \tag{4}$$

where μ denotes the Möbius function $\mu(t) = \begin{cases} 1 & \text{if } t = 1 \\ (-1)^r & \text{if } t = p_1 \cdots p_r . \\ 0 & \text{else} \end{cases}$

But, as in Formula (3), we do not consider real probabilities, we don't have in general the countable additivity. Thus, we have to be careful when considering limits as countable sums of limits.

Second Comment on Turk's Proof In the right hand side of Formula (2), we can recognize the inverse of Kempner's formula which gives the number of polynomial functions $\varphi : \mathbb{Z}/\delta\mathbb{Z} \to \mathbb{Z}/\delta\mathbb{Z}$ of degree $\leq d$, namely [3]:

$$\prod_{k=0}^{d} \frac{\delta}{\gcd(k!, \delta)} \qquad \text{[Kempner, 1921]} \tag{5}$$

where the functions φ are induced by polynomials $f \in \mathbb{Z}[X]$ of degree $\leq d$.

There is a generalization of Kempner's formula obtained by Manjul Bhargava which extends this formula, first by replacing the ring $\mathbb{Z}$ by the ring of integers O_K of a number field K and the positive integer δ by a proper ideal $\mathfrak{I}$ of O_K, and then, by considering polynomial functions not only from $O_K/\mathfrak{I}$ to itself but also polynomial functions from any subset S of O_K to $O_K/\mathfrak{I}$. This suggests us to extend Turk's formula by means of Kempner-Bhargava's formula, and we obtain:

$$\text{“Prob”}\ (\,d(S, f) = O_K \mid f \in O_K[X]\,) = \prod_{\mathfrak{p} \in \text{Max}(O_K)} \left(1 - \frac{1}{N(\mathfrak{p})^{\nu_{\mathfrak{p}}(S)}}\right) \tag{6}$$

where $N(\mathfrak{p})$ denotes the norm of $\mathfrak{p}$ that is, Card$(O_K/\mathfrak{p})$, while $\nu_{\mathfrak{p}}(S)$ denotes the number of classes of S modulo $\mathfrak{p}$.

Recall that the *fixed divisor* of a polynomial $f \in O_K[X]$ on a subset S of O_K is defined as the ideal of O_K generated by the values of f on S :

$$d(S, f) = (f(S))\,.$$

So, we first recall Kempner-Bhargava's formula and then, step by step, we compute the aymptotic density of the polynomial functions whose fixed divisor on S is the whole ring O_K.

2 Kempner-Bhargava's Formula

Notation Let K be a number field of degree n and denote by O_K its ring of integers, by $\mathfrak{I}$ a nonzero ideal of O_K, and by S an infinite subset of O_K.

The first natural generalization of Kempner's formula (5) is obtained by replacing

- the positive integer δ by a proper ideal $\mathfrak{I}$ of O_K,
- the factorial number $k!$ by the factorial ideal $k!_{O_K}$ introduced by Zantema

$$k!_{O_K} = \prod_{\mathfrak{p}} \mathfrak{p}^{\sum_{t\geq 1} \left\lfloor \frac{k}{N(\mathfrak{p})^t} \right\rfloor} \qquad \text{[Zantema, 1982]}$$

analogously to Legendre's formula: $k! = \prod_{p\in\mathbb{P}} p^{\sum_{t\geq 1} \left\lfloor \frac{k}{p^t} \right\rfloor}$ [Legendre, 1808],
- the gcd of the numbers δ and $k!$ by the gcd of the ideals $\mathfrak{I}$ and $k!_{O_K}$, that is, the ideal generated by these ideals,
- the integers in the previous rational fractions by the norm of the corresponding ideals, that is the cardinality of the corresponding residue rings.

Finally, the number of functions $\varphi : O_K/\mathfrak{I} \to O_K/\mathfrak{I}$ induced by polynomials $f \in O_K[X]$ of degree $\leq d$ should be

$$\prod_{k=0}^{d} \frac{N(\mathfrak{I})}{N(\mathfrak{I}, k!_{O_K})} \qquad \text{[Bhargava, 1997]}$$

where $N(\mathfrak{I}, k!_{O_K})$ denotes the norm of the ideal generated by $\mathfrak{I}$ and $k!_{O_K}$.

In fact, Manjul Bhargava gave a powerful generalization of the previous factorial ideals $k!_{O_K}$. For every infinite subset S of O_K, he defined factorial ideals $k!_S$ associated to S which have many properties of the classical factorials. In general, there is no more an explicit formula for a subset as for the whole ring O_K. Bhargava's definition is of combinatorial nature but we can introduce these factorial ideals more easily by means of integer-valued polynomials. Let us consider the O_K-algebra formed by the integer-valued polynomials on S:

$$\mathrm{Int}(S, O_K) = \{f \in K[X] \mid f(S) \subseteq O_K\}$$

and denote by $\mathfrak{I}_k(S, O_K)$ the fractional ideal formed by the leading coefficients of the polynomials of $\mathrm{Int}(S, O_K)$ of degree $\leq k$. Then, the factorial ideal $k!_S$ may be defined as the inverse of $\mathfrak{I}_k(S, O_K)$:

$$k!_S = \mathfrak{I}_k(S, O_K)^{-1} = \{t \in K \mid t \times \mathfrak{I}_k(S, O_K) \subseteq O_K\}.$$

With these factorial ideals, Bhargava obtained the following strong generalization of Kempner's formula.

Theorem 2.1 ([1, Theorem 5]) *The number of functions $\varphi : S \to O_K/\mathfrak{I}$ induced by polynomials $f \in O_K[X]$ of degree $\leq d$ is equal to*

$$\prod_{k=0}^{d} \frac{N(\mathfrak{I})}{N(\mathfrak{I}, k!_S)} \qquad \text{(Bhargava, 1997)}. \tag{7}$$

Note that, if $\nu_{\mathfrak{I}}(S)$ denotes the number of classes of S modulo $\mathfrak{I}$, then $k!_S \subseteq \mathfrak{I}$ for $k \geq \nu_{\mathfrak{I}}(S)$. Consequently, for $d \geq \nu_{\mathfrak{I}}(S) - 1$, the formula does not depend on d and is equal to

$$\prod_{k=0}^{\nu_{\mathfrak{I}}(S)-1} \frac{N(\mathfrak{I})}{N(\mathfrak{I}, k!_S)}.$$

As it is very difficult to recognize this extension of Kempner's formula in the statement of Theorem 5 of [1], it seems simpler to us to sketch a direct proof.

Proof For every maximal ideal $\mathfrak{p}$ dividing the ideal $d!_S \times \mathfrak{I}$, consider a $\mathfrak{p}$-ordering of S of length d, that is, a sequence $a_{\mathfrak{p},0}, a_{\mathfrak{p},1}, \ldots, a_{\mathfrak{p},d}$ of elements of S such that,

$$v_{\mathfrak{p}}\left(\prod_{l=0}^{k-1}(a_{\mathfrak{p},k} - a_{\mathfrak{p},l})\right) = \min_{x \in S} v_{\mathfrak{p}}\left(\prod_{l=0}^{k-1}(x - a_{\mathfrak{p},l})\right) \quad (1 \leq k \leq d)$$

where $v_{\mathfrak{p}}$ denotes the valuation of K associated to the ideal $\mathfrak{p}$.

By the Chinese remainder theorem, there exists a sequence $b_0, b_1, \ldots, b_d$ of elements of O_K such that

$$\forall k \in \{0, 1, \ldots, d\} \;\; \forall \mathfrak{p} \mid d!_S \times \mathfrak{I} \quad v_{\mathfrak{p}}(b_k - a_{\mathfrak{p},k}) > v_{\mathfrak{p}}(d!_S \times \mathfrak{I}).$$

Now, consider the polynomials

$$g_k(X) = (X - b_0)(X - b_1) \cdots (X - b_{k-1}) \quad (0 \leq k < d).$$

By definition of $\mathfrak{p}$-orderings (cf. [2, § 2]),

$$\forall \mathfrak{p} \mid d!_S \times \mathfrak{I} \quad \forall x \in S \quad v_{\mathfrak{p}}(g_k(x)) \geq v_{\mathfrak{p}}(k!_S) \text{ and } v_{\mathfrak{p}}(g_k(a_{\mathfrak{p},k}) = v_{\mathfrak{p}}(k!_S).$$

Every polynomial $g(X) \in O_K[X]$ of degree $\leq d$ may be written

$$g(X) = \sum_{k=0}^{d} c_k\, g_k(X) \text{ with } c_k \in O_K. \tag{8}$$

Clearly, the following conditions (9) on the coefficients c_k in (8)

$$\forall k \leq d \ \ \forall \mathfrak{p} \mid k!_S \times \mathfrak{J} \quad v_{\mathfrak{p}}(c_k) \geq v_{\mathfrak{p}}(\mathfrak{J}) - v_{\mathfrak{p}}(k!_S) \tag{9}$$

are sufficient to imply that $g(S) \subseteq \mathfrak{J}$. Conversely, one may prove by induction on the degree of g that these conditions are necessary to have $g(S) \subseteq \mathfrak{J}$.

Note that conditions (9) may also be written

$$v_{\mathfrak{p}}(c_k) \geq \max(0, v_{\mathfrak{p}}(\mathfrak{J}) - v_{\mathfrak{p}}(k!_S)) = v_{\mathfrak{p}}(\mathfrak{J}) - \min(v_{\mathfrak{p}}(\mathfrak{J}), v_{\mathfrak{p}}(k!_S)),$$

that is,

$$c_k \in \mathfrak{J} \times (\mathfrak{J}, k!_S)^{-1}.$$

Thus, the functions from S to $O_K/\mathfrak{J}$ induced by two polynomials of $O_K[X]$ whose coefficients are congruent modulo $\mathfrak{J} \times (\mathfrak{J}, k!_S)^{-1}$ are equal. We may impose to the c_k's to belong to a fixed system of representatives of O_K modulo $\mathfrak{J} \times (\mathfrak{J}, k!_S)^{-1}$ and then the representation of polynomial functions is unique. To count the cardinality of the set of polynomial functions it suffices to count, for each $k \leq d$, the cardinality of the values taken by c_k modulo $\mathfrak{J} \times (\mathfrak{J}, k!_S)^{-1}$, that is, $N(\mathfrak{J} \times (\mathfrak{J}, k!_S)^{-1}) = \frac{N(\mathfrak{J})}{N((\mathfrak{J}, k!_S))}$. □

3 "Prob" $(f(S) \subseteq \mathfrak{J} \mid f \in O_K[X], \deg f \leq d)$

We are going to compute asymptotic densities on the polynomial ring $O_K[X]$. As said previously, we should use the notion of height. In fact, we will use a notion which extends the height in the case of $\mathbb{Z}$ and which is more manageable than the classical notion of height in number fields when computing our asymptotic densities. We call it the broadness. Using the fact that O_K is a free $\mathbb{Z}$-module, we define it by means of a $\mathbb{Z}$-basis of O_K.

Definition 3.1 Let us fix some basis $(e_i)_{i=1}^n$ of the $\mathbb{Z}$-module O_K. With respect to this basis, the *broadness* of

- an element $x = \sum_{i=1}^n x_i e_i$ of O_K is the integer $B(x) = \max_{1 \leq i \leq n} |x_i|$,
- a polynomial $f(X) = \sum_{k=0}^d c_k X^k \in O_K[X]$ is the integer $B(f) = \max_{0 \leq k \leq d} B(c_k)$.

If we change the basis, we have another broadness B'. But these two measures of the size of an element x of O_K are equivalent since there exist constants α and β such that

$$\forall x \in O_K \ \ \alpha\, B'(x) \leq B(x) \leq \beta\, B'(x).$$

Thus, in the following, we will forget the choice of the basis and we consider a fixed basis.

For elements, there are links with the norm: there exists a constant γ_K only depending on K such that

$$\forall x \in O_K \quad |N(x)| \leq \gamma_K \times B(x)^n \,. \tag{10}$$

Indeed, $N(\sum_{i=1}^n x_i e_i)$ is a homogeneous polynomial of degree n in the x_i's.

For polynomials, there are links with the norm of the content:

$$N(\mathrm{cont}(f)) \leq \gamma_K \times B(f)^n \,. \tag{11}$$

Recall that the *content* of a polynomial f is the ideal generated by its coefficients. Indeed, if $f(X) = \sum_{k=0}^d c_k X^d$ then, for every k, one has $(c_0, c_1, \ldots, c_d) \subseteq (\mathrm{cont}(f))$, and hence, $N(\mathrm{cont}(f)) \leq |N(c_k)|$.

Proposition 3.2 *For every nonzero ideal $\mathfrak{I}$ of O_K and positive integers d and B,*

$$\mathrm{Prob}(f(S) \subseteq \mathfrak{I} \mid f \in O_K[X], \deg f \leq d, B(f) \leq B\,) = \prod_{k=0}^{d} \frac{N(\mathfrak{I}, k!_S)}{N(\mathfrak{I})} + \epsilon(\mathfrak{I}, d, B).$$

where

$$|\epsilon(\mathfrak{I}, d, B)| \leq \frac{\delta_d}{N(\mathfrak{I})^d} \times \frac{1}{B} \quad (\delta_d \textit{ depends only on } d,\ K \textit{ and } S\,).$$

Notation (for our Proofs)

$$P(\mathfrak{I}, d, B) = \mathrm{Prob}\,(\,f(S) \subseteq \mathfrak{I} \mid f \in O_K[X], \deg f \leq d, B(f) \leq B\,)\,.$$

Proof For the time of proof, as K and d are fixed, we will forget to say, in the notation, that the polynomials we consider have coefficients in O_K and degrees $\leq d$. Moreover, we let $N = N(\mathfrak{I})$.

In order to compute $P(\mathfrak{I}, d, B)$, we let

$$\eta(\mathfrak{I}, B) = \#\{f \in O_K[X] \mid \deg f \leq d, B(f) \leq B, f(S) \subseteq \mathfrak{I}\}.$$

Clearly,

$$P(\mathfrak{I}, d, B) = \frac{\eta(\mathfrak{I}, B)}{\eta(O_K, B)} = \frac{\eta(\mathfrak{I}, B)}{(2B+1)^{n(d+1)}}\,.$$

On the other hand, we compute now the number of polynomial functions from S to $O_K/\mathfrak{I}$ of degree $\leq d$. To add elements of the ideal NO_K to the coefficients of a polynomial f does not change the values of f modulo $\mathfrak{I}$ since N belongs to

$\mathfrak{I}$. Indeed, for each prime $\mathfrak{p}$ of O_K lying over a prime number p, p belongs to $\mathfrak{p}$, and hence, $N(\mathfrak{p}) = p^f \in \mathfrak{p}$. Consequently, if $B \geq N$, the polynomials f such that $B(f) \leq B$ induce all the polynomial functions from S to $O_K/\mathfrak{I}$.

Thus, when B is a multiple of N, say λN where $\lambda \in \mathbb{N}^*$, every polynomial function from S to $O_K/\mathfrak{I}$ of degree $\leq d$ is induced by a polynomial $\overline{f}$ with coefficients in $O_K/(\lambda N)$, that is, a polynomial $\overline{f}$ belonging to the O_K-module $M_\lambda = (O_K/(\lambda N))[X]_d$ formed by the polynomials of degree $\leq d$. Two such polynomials $\overline{f}$ and $\overline{g}$ induce the same polynomial function if and only if their difference induces the null function, that is, belongs to the sub-O_K-module N_λ formed by the polynomials with values in $\overline{\mathfrak{I}} = \mathfrak{I}/(\lambda N)$.

Consequently, the cardinality of the set of polynomial functions from S to $O_K/\mathfrak{I}$ of degree $\leq d$ is equal to $\frac{\#M_\lambda}{\#N_\lambda}$. Let us introduce another notation:

$$\eta^*(\mathfrak{I}, \lambda N) = \#N_\lambda = \#\{\overline{f} \in (O_K/(\lambda N)[X]_d \mid \overline{f}(S) \subseteq \overline{\mathfrak{I}}\}.$$

In particular,

$$\eta^*(O_K, \lambda N) = \#M_\lambda = (\lambda N)^{n(d+1)}.$$

Thanks to Kempner-Bhargava's formula (7), we then have:

$$\forall \lambda \in \mathbb{N}^* \qquad \frac{\eta^*(O_K, \lambda N)}{\eta^*(\mathfrak{I}, \lambda N)} = \prod_{k=0}^{d} \frac{N(\mathfrak{I})}{N(\mathfrak{I}, k!_S)}.$$

But, there are obvious links between $\eta(\mathfrak{I}, B)$ and $\eta^*(\mathfrak{I}, \lambda N)$. Indeed,

$$\lambda N \leq B < (\lambda + 1)N \Rightarrow \eta^*(\mathfrak{I}, 2\lambda N) \leq \eta(\mathfrak{I}, B) \leq \eta^*(\mathfrak{I}, 2(\lambda + 1)N).$$

Consequently, assuming that $\lambda N \leq B < (\lambda + 1)N$, we have

$$\frac{\eta^*(\mathfrak{I}, 2\lambda N)}{\eta^*(O_K, 2\lambda N)} \cdot \frac{\eta^*(O_K, 2\lambda N)}{\eta(O_K, B)} \leq \frac{\eta(\mathfrak{I}, B)}{\eta(O_K, B)} \leq \frac{\eta^*(\mathfrak{I}, 2(\lambda+1)N)}{\eta^*(O_K, 2(\lambda+1)N)} \cdot \frac{\eta^*(O_K, 2(\lambda+1)N)}{\eta(O_K, B)},$$

that is,

$$\prod_{k=0}^{d} \frac{N(\mathfrak{I}, k!_S)}{N(\mathfrak{I})} \times \left(\frac{2\lambda N}{2B+1}\right)^{n(d+1)} \leq P(\mathfrak{I}, d, B) \leq \prod_{k=0}^{d} \frac{N(\mathfrak{I}, k!_S)}{N(\mathfrak{I})} \times \left(\frac{2(\lambda+1)N}{2B+1}\right)^{n(d+1)}.$$

Letting

$$P(\mathfrak{I}, d, B) = \prod_{k=0}^{d} \frac{N(\mathfrak{I}, k!_S)}{N(\mathfrak{I})} + \epsilon(\mathfrak{I}, d, B), \tag{12}$$

we get

$$-1 + \left(\frac{2\lambda N}{2B+1}\right)^{n(d+1)} \leq \frac{\epsilon(\mathfrak{I}, d, B)}{\prod_{k=0}^{d} \frac{N(\mathfrak{I}, k!_S)}{N(\mathfrak{I})}} \leq -1 + \left(\frac{2(\lambda+1)N}{2B+1}\right)^{n(d+1)}$$

As the extreme sides of these inequalities are bounded by:

$$n(d+1) \times 2N \times \frac{(2(B+N))^{n(d+1)-1}}{(2B+1)^{n(d+1)}} \leq 2^{n(d+1)} n(d+1) \frac{N}{B},$$

we have $|\varepsilon(\mathfrak{I}, d, B)| \leq \frac{\delta_d}{N^d} \times \frac{1}{B}$ where $\delta_d = 2^{n(d+1)} n(d+1) \times \prod_{k=0}^{d} N(k!_S)$. □

Corollary 3.3 *For every nonzero ideal $\mathfrak{I}$ of O_K and every positive integer d,*

$$\text{“Prob”}(f(S) \subseteq \mathfrak{I} \mid f \in O_K[X],\ \deg f \leq d) = \prod_{k=0}^{d} \frac{N(\mathfrak{I}, k!_S)}{N(\mathfrak{I})}.$$

4 “Prob”$(d(S, f) = \mathfrak{I} \mid f \in O_K[X], \deg f \leq d)$

Notation

$$Q(\mathfrak{I}, d, B) = \text{Prob}(\, d(S, f) = \mathfrak{I} \mid f \in O_K[X],\ \deg f \leq d,\, B(f) \leq B\,).$$

Now we are going to compute $Q(\mathfrak{I}, d, B)$ by means of $P(\mathfrak{I}, d, B)$.

Lemma 4.1 *For every nonzero ideal $\mathfrak{I}$ and every positive integers d and B,*

$$Q(\mathfrak{I}, d, B) = \sum_{J \text{ finite} \subset \text{Max}(O_K)} (-1)^{\text{Card}(J)}\, P(\mathfrak{I} \prod_{\mathfrak{p} \in J} \mathfrak{p}, d, B) \tag{13}$$

Moreover, this sum has only finitely many nonzero terms.

Proof Clearly,

$$d(f, S) = \mathfrak{I} \Leftrightarrow f(S) \subseteq \mathfrak{I} \text{ and, for all } \mathfrak{p} \in \text{Max}(O_K),\ f(S) \nsubseteq \mathfrak{I}\mathfrak{p}\,.$$

Consequently,

$$Q(\mathfrak{I}, d, B) = P(\mathfrak{I}, d, B,) - \sum_{\mathfrak{p}} P(\mathfrak{I}\mathfrak{p}, d, B) + \sum_{\mathfrak{p}\neq\mathfrak{q}} P(\mathfrak{I}\mathfrak{p}\mathfrak{q}, d, B) - \cdots$$

that is,

$$Q(\mathfrak{I}, d, B) = \sum_{J\subset \mathrm{Max}(O_K)} (-1)^{\mathrm{Card}(J)}\, P(\mathfrak{I}\prod_{\mathfrak{p}\in J}\mathfrak{p}, d, B)$$

where the subsets J are finite.

Let us prove that this sum on the subsets J is itself a finite sum. We use Bhargava's following result [2, Theorem 2] concerning the properties of its generalized factorials:

$$\forall f \in O_K[X]\ \ [\deg(f) \leq d \ \Rightarrow\ d(S, f) \mid d!_S \times \mathrm{cont}(f)\,]\,. \tag{14}$$

If follows from (14) that

$$f(S) \subseteq \mathfrak{I}\prod_{\mathfrak{p}\in J}\mathfrak{p} \ \Rightarrow\ \mathfrak{I}\prod_{\mathfrak{p}\in J}\mathfrak{p} \mid d(S, f) \ \Rightarrow\ \mathfrak{I}\prod_{\mathfrak{p}\in J}\mathfrak{p} \mid d!_S \times \mathrm{cont}(f)\,.$$

Thus, by (10):

$$N(\mathfrak{I}\prod_{\mathfrak{p}\in J}\mathfrak{p}) \leq N(d!_S \times \mathrm{cont}(f)) \leq N(d!_S) \times \gamma_K \times B(f)^n\,.$$

Finally, if $P(\mathfrak{I}\prod_{\mathfrak{p}\in J}\mathfrak{p}, d, B) \neq 0$, then there exists at least a polynomial $f \in O_K[X]$ with $\deg f \leq d$ and $B(f) \leq B$ such that $f(S) \subseteq \mathfrak{I}\prod_{\mathfrak{p}\in J}\mathfrak{p}$ which shows that

$$P(\mathfrak{I}\prod_{\mathfrak{p}\in J}\mathfrak{p}, d, B) \neq 0 \ \Rightarrow\ N\left(\prod_{\mathfrak{p}\in J}\mathfrak{p}\right) \leq \gamma_K \times \frac{N(d!_S)}{N(\mathfrak{I})} \times B^n\,.$$

Consider the subset

$$\mathcal{J}_B = \left\{ J \subset \mathrm{Max}(O_K) \mid N\left(\prod_{\mathfrak{p}\in J}\mathfrak{p}\right) \leq \gamma_K \times \frac{N(d!_S)}{N(\mathfrak{I})} \times B^n \right\}. \tag{15}$$

The set $\mathcal{J}_B$ is finite since there are at most finitely many prime ideals $\mathfrak{p}$ of O_K such that $N(\mathfrak{p}) \leq \gamma_K \times \frac{N(d!_S)}{N(\mathfrak{I})} \times B^n$ and we have

$$Q(\mathfrak{I}, d, B) = \sum_{J\in\mathcal{J}_B} (-1)^{\mathrm{Card}(J)}\, P(\mathfrak{I}\prod_{\mathfrak{p}\in J}\mathfrak{p}, d, B)\,. \tag{16}$$

□

We then deduce from (13) a second asymptotic density.

Proposition 4.2 *For every d, "Prob"*$(d(S, f) = \mathfrak{I} \mid f \in O_K[X], \deg f \leq d)$

$$= \sum_{J \text{finite}} (-1)^{\mathrm{Card}(J)} \prod_{k=0}^{d} \frac{N((\mathfrak{I} \prod_{\mathfrak{p} \in J} \mathfrak{p}, k!_S))}{N(\mathfrak{I} \prod_{\mathfrak{p} \in J} \mathfrak{p})}. \quad (17)$$

Proof First, note that the family on the right hand side is summable. Indeed, the general term satisfies the inequality:

$$\prod_{k=0}^{d} \frac{N((\mathfrak{I} \prod_{\mathfrak{p} \in J} \mathfrak{p}, k!_S))}{N(\mathfrak{I} \prod_{\mathfrak{p} \in J} \mathfrak{p})} \leq \frac{1}{N(\mathfrak{I})^{d+1}} \times \prod_{k=0}^{d} N(k!_S) \times \prod_{\mathfrak{p} \in J} \frac{1}{N(\mathfrak{p})^{d+1}}.$$

As

$$\sum_J \prod_{\mathfrak{p} \in J} \frac{1}{N(\mathfrak{p})^{d+1}} \leq \sum_{\mathfrak{a} \in I_K} \frac{1}{N(\mathfrak{a})^{d+1}} = \zeta_K(d+1)$$

where I_K denotes the set of entire ideals of O_K and ζ_K denotes the Dedekind zeta function of the number field K, the sum

$$\sum_J (-1)^{\mathrm{Card}(J)} \prod_{k=0}^{d} \frac{N((\mathfrak{I} \prod_{\mathfrak{p} \in J} \mathfrak{p}, k!_S))}{N(\mathfrak{I} \prod_{\mathfrak{p} \in J} \mathfrak{p})}$$

is well defined. By Equalities (12) and (16), we have

$$Q(\mathfrak{I}, d, B) = \sum_{J \in \mathcal{J}_B} (-1)^{\mathrm{Card}(J)} \left[\prod_{k=0}^{d} \frac{N((\mathfrak{I} \prod_{\mathfrak{p} \in J} \mathfrak{p}, k!_S))}{N(\mathfrak{I} \prod_{\mathfrak{p} \in J} \mathfrak{p})}) + \epsilon(\mathfrak{I} \prod_{\mathfrak{p} \in J} \mathfrak{p}, d, B) \right]$$

and

$$\left| Q(\mathfrak{I}, d, B) - \sum_{J \in \mathcal{J}_B} (-1)^{\mathrm{Card}(J)} \prod_{k=0}^{d} \frac{N((\mathfrak{I} \prod_{\mathfrak{p} \in J} \mathfrak{p}, k!_S))}{N(\mathfrak{I} \prod_{\mathfrak{p} \in J} \mathfrak{p})} \right| \leq$$

$$\left| \sum_{J \in \mathcal{J}_B} \epsilon(\mathfrak{I} \prod_{\mathfrak{p} \in J} \mathfrak{p}, d, B) \right| \leq \frac{\delta_d}{N(\mathfrak{I})^d} \times \frac{1}{B} \times \sum_J \frac{1}{\prod_{\mathfrak{p} \in \mathcal{J}_B} N(\mathfrak{p})^d} \leq \frac{\delta_d}{N(\mathfrak{I})^d} \times \frac{1}{B} \times \zeta_K(d).$$

This last inequality, that is true for $d > 1$, shows that, when $B \to +\infty$, $Q(\mathfrak{I}, d, B)$ tends to the limit that we gave since $\zeta_K(d)$ is finite.

For $d = 1$, we have to take care because $\zeta_K(1)$ is infinite. Using (15), we compute a more precise bound. First, we have:

$$\left|\sum_{J\in\mathcal{J}_B}\epsilon(\mathfrak{I}\prod_{\mathfrak{p}\in J}\mathfrak{p},d,B)\right| \le \frac{\delta_1}{N(\mathfrak{I})}\times\frac{1}{B}\times\sum_{J\in\mathcal{J}_B}\frac{1}{\prod_{\mathfrak{p}\in J}N(\mathfrak{p})}$$

and

$$\sum_{J\in\mathcal{J}_B}\frac{1}{\prod_{\mathfrak{p}\in J}N(\mathfrak{p})} \le \sum_{\{\mathfrak{a}\in I_K \mid N(\mathfrak{a})\le\frac{\gamma_K}{N(\mathfrak{I})}B^n\}}\frac{1}{N(\mathfrak{a})}.$$

By [4, VI § 3 Theorem 3], there exits β_K depending only on K such that

$$\forall n\in\mathbb{N}\quad a_n = \#\{\mathfrak{a}\in I_K \mid N(\mathfrak{a})\le n\}\le\beta_K\, n.$$

Consequently, for every M,

$$\sum_{\{\mathfrak{a}\mid N(\mathfrak{a})\le M\}}\frac{1}{N(\mathfrak{a})} = \sum_{1\le n\le M}\frac{a_n-a_{n-1}}{n} = \frac{a_M}{M}+\sum_{1\le n\le M-1}\frac{a_n}{n(n+1)}\le\beta_K(1+\ln M).$$

Finally, there exists a constant μ which depends only on K, S, and $\mathfrak{I}$ such that

$$\left|\sum_{J\in\mathcal{J}_B}\epsilon(\mathfrak{I}\prod_{\mathfrak{p}\in J}\mathfrak{p},d,B)\right| \le \frac{\delta_1}{N(\mathfrak{I})}\times\frac{1}{B}\times\beta_K\times\left(1+\ln\left(\frac{\gamma_K}{N(\mathfrak{I})}B^n\right)\right)\le\mu\times\frac{\ln B}{B}$$

and we may conclude in the same way. □

By means of the extension of Möbius function to the ideals of number fields, Formula (17) may also be written in the following way:

Proposition 4.3 *For every d, one has*

$$\text{“Prob”}(d(S,f)=\mathfrak{I}\mid f\in O_K[X],\ \deg f\le d) = \sum_{\mathfrak{a}}\mu(\mathfrak{a})\prod_{k=0}^{d}\frac{N(\mathfrak{I}\mathfrak{a},k!_S)}{N(\mathfrak{I}\mathfrak{a})} \tag{18}$$

where

$$\mu(\mathfrak{a}) = \begin{cases} 1 & \text{if } \mathfrak{a}=O_K \\ (-1)^r & \text{if } \mathfrak{a}=\mathfrak{p}_1\cdots\mathfrak{p}_r \text{ where the } \mathfrak{p}_i\in\mathrm{Max}(O_K) \text{ are distinct} \\ 0 & \text{if not.}\end{cases}$$

5 "Prob" ($d(S, f) = O_K \mid f \in O_K[X], \deg f \leq d$)

We were interested in the case where the fixed divisor is one, that is, in fact, the whole ring O_K. In this case, the previous formula can be more easily written.

Proposition 5.1 *For every d, one has*

$$\text{“Prob”}\ (d(S,f)=O_K \mid f\in O_K[X],\ \deg f\le d)=\prod_{\mathfrak{p}}\left(1-\frac{1}{N(\mathfrak{p})^{\min(d+1,\,\nu_{\mathfrak{p}}(S))}}\right)$$

$$=\zeta_K^{-1}(d+1)\times\prod_{\mathfrak{p}|d!_S}\frac{1-\frac{1}{N(\mathfrak{p})^{\nu_{\mathfrak{p}}(S)}}}{1-\frac{1}{N(\mathfrak{p})^{d+1}}}.$$

where $\nu_{\mathfrak{p}}(S)$ denotes the number of classes of S modulo $\mathfrak{p}$.

Note that the second factor is a finite product that can be quite easily computed.

Proof Following Turk's proof for the case of $\mathbb{Z}$, we first note that the arithmetic application:

$$\mathfrak{I}\mapsto\prod_{k=0}^{d}\frac{N(\mathfrak{I},k!_S)}{N(\mathfrak{I})}$$

is multiplicative with respect to $\mathfrak{I}$, that is,

$$(\mathfrak{I}_1,\mathfrak{I}_2)=O_K\ \Rightarrow\ \prod_{k=0}^{d}\frac{N(\mathfrak{I}_1\mathfrak{I}_2,k!_S)}{N(\mathfrak{I}_1\mathfrak{I}_2)}=\prod_{k=0}^{d}\frac{N(\mathfrak{I}_1,k!_S)}{N(\mathfrak{I}_1)}\times\prod_{k=0}^{d}\frac{N(\mathfrak{I}_2,k!_S)}{N(\mathfrak{I}_2)}.$$

Applying this result to the asymptotic formula (17), we obtain:

$$\text{“Prob”}(d(S,f)=O_K \mid f\in O_K[X],\ \deg f\le d)$$

$$=\sum_J(-1)^{\mathrm{Card}(J)}\prod_{k=0}^{d}\frac{N(\prod_{\mathfrak{p}\in J}\mathfrak{p},k!_S)}{N(\prod_{\mathfrak{p}\in J}\mathfrak{p})}=\prod_{\mathfrak{p}}\left(1-\prod_{k=0}^{d}\frac{N(k!_S,\mathfrak{p})}{N(\mathfrak{p})}\right)$$

provided that the infinite product is absolutely convergent.

Noticing that

$$(k!_S,\mathfrak{p})=\begin{cases}\mathfrak{p} & if\ \ \mathfrak{p}|k!_S\\ O_K & if\ \ \mathfrak{p}\nmid k!_S\end{cases}$$

and that, by definition of the factorial ideal $k!_S$ (cf [1, § 2]),

$$\mathfrak{p} \mid k!_S \;\Leftrightarrow\; k \geq \nu_{\mathfrak{p}}(S)\,,$$

we have

$$\prod_{k=0}^{d} \frac{N(k!_S,\mathfrak{p})}{N(\mathfrak{p})} = \prod_{0\leq k\leq d,\, k<\nu_{\mathfrak{p}}(S)} \frac{1}{N(\mathfrak{p})} = \frac{1}{N(\mathfrak{p})^{\min(d+1,\,\nu_{\mathfrak{p}}(S))}}\,.$$

As there is only a finite number of $\mathfrak{p}$ dividing $d!_S$, $\prod_{k=0}^{d} \frac{N(k!_S,\mathfrak{p})}{N(\mathfrak{p})} = \frac{1}{N(\mathfrak{p})^{d+1}}$ for almost all $\mathfrak{p}$. Consequently, the infinite product is absolutely convergent and is equal to:

$$\prod_{\mathfrak{p}} \left(1 - \frac{1}{N(\mathfrak{p})^{\min(d+1,\,\nu_{\mathfrak{p}}(S))}}\right) = \prod_{\mathfrak{p}} \left(1 - \frac{1}{N(\mathfrak{p})^{d+1}}\right) \times \prod_{\mathfrak{p}\mid d!_S} \frac{1 - \frac{1}{N(\mathfrak{p})^{\nu_{\mathfrak{p}}(S)}}}{1 - \frac{1}{N(\mathfrak{p})^{d+1}}} =$$

$$\zeta_K^{-1}(d+1) \times \prod_{\mathfrak{p}\mid d!_S} \frac{1-\frac{1}{N(\mathfrak{p})^{\nu_{\mathfrak{p}}(S)}}}{1-\frac{1}{N(\mathfrak{p})^{d+1}}}\,. \qquad \square$$

Corollary 5.2

$$\text{“Prob”}(\,(a,b) = O_K \mid a,b \in O_K\,) = \zeta_K^{-1}(2)\,.$$

Proof To say that a et $b \in O_K$ generate the ideal O_K is equivalent to say that the fixed divisor on O_K of the polynomial $aX + b$ of degree ≤ 1 is O_K. By Proposition 5.1, this “probability” is equal to

$$\prod_{\mathfrak{p}} \left(1 - \frac{1}{N(\mathfrak{p})^{\min(2,\,\nu_{\mathfrak{p}}(O_K))}}\right) = \prod_{\mathfrak{p}} \left(1 - \frac{1}{N(\mathfrak{p})^2}\right) = \zeta_K^{-1}(2)\,.$$

$\square$

When d tends to infinity, we obtain:

Theorem 5.3 *For every infinite subset S of O_K,*

$$\text{“Prob”}\ (d(S,f) = O_K] \mid f \in O_K[X]) = \prod_{\mathfrak{p}\in \mathrm{Max}(O_K)} \left(1 - \frac{1}{N(\mathfrak{p})^{\nu_{\mathfrak{p}}(S)}}\right)$$

where $\nu_{\mathfrak{p}}(S)$ denotes the number of classes of S modulo $\mathfrak{p}$. In particular,

$$\text{“Prob”}\ (d(O_K,f) = O_K \mid f \in O_K[X]) = \prod_{\mathfrak{p}\in \mathrm{Max}(O_K)} \left(1 - \frac{1}{N(\mathfrak{p})^{N(\mathfrak{p})}}\right)\,.$$

Proof Let $s \neq s' \in S$. If $\nu_{\mathfrak{p}}(S) = 1$, then $s - s' \in \mathfrak{p}$. Thus, $\nu_{\mathfrak{p}}(S) \geq 2$ for all but finitely many $\mathfrak{p}$ and the infinite product $\prod_{\mathfrak{p}} \left(1 - \frac{1}{N(\mathfrak{p})^{\nu_{\mathfrak{p}}(S)}}\right)$ is absolutely convergent. Consequently,

$$\lim_{d \to +\infty} \prod_{\mathfrak{p} | d!_S} \left(1 - \frac{1}{N(\mathfrak{p})^{\nu_{\mathfrak{p}}(S)}}\right) = \prod_{\mathfrak{p}} \left(1 - \frac{1}{N(\mathfrak{p})^{\nu_{\mathfrak{p}}(S)}}\right).$$

On the other hand, $\lim_{d \to +\infty} \prod_{\mathfrak{p} \nmid d!_S} \left(1 - \frac{1}{N(\mathfrak{p})^{d+1}}\right) = 1$ since $\prod_{\mathfrak{p}} \left(1 - \frac{1}{N(\mathfrak{p})^{d+1}}\right)$ converges. Finally,

$$\prod_{\mathfrak{p}} \left(1 - \frac{1}{N(\mathfrak{p})^{\min(d+1,\, \nu_{\mathfrak{p}}(S))}}\right) = \prod_{\mathfrak{p} | d!_S} \left(1 - \frac{1}{N(\mathfrak{p})^{\nu_{\mathfrak{p}}(S)}}\right) \times \prod_{\mathfrak{p} \nmid d!_S} \left(1 - \frac{1}{N(\mathfrak{p})^{d+1}}\right)$$

tends to $\prod_{\mathfrak{p}} \left(1 - \frac{1}{N(\mathfrak{p})^{\nu_{\mathfrak{p}}(S)}}\right)$. □

Example 5.4 Let $\mathbb{P}$ be the subset of $\mathbb{Z}$ formed by the prime numbers, then $\nu_p(\mathbb{P}) = p$ for every prime p, and hence,

$$\text{“Prob”}\ (d(\mathbb{P}, f) = 1 \mid f \in \mathbb{Z}[X]) = \prod_{\mathfrak{p} \in \mathbb{P}} \left(1 - \frac{1}{p^p}\right).$$

References

1. M. Bhargava, P-orderings and polynomial functions on arbitrary subsets of Dedekind rings, *J. reine angew. Math.*, **490** (1997), 101-127.
2. M. Bhargava, Generalized Factorials and Fixed Divisors over Subsets of a Dedekind Domain, *J. Number Theory* **72** (1998), 67–75.
3. A.J. Kempner, Polynomials and their residue systems, *Amer. Math. Soc. Trans.*, **22** (1921), 240–288.
4. S. Lang, *Algebraic Number Theory*, Addison-Wesley, 1970.
5. G. Tenenbaum, Introduction to analytic and probabilistic number theory, Cambridge University Press, 1995.
6. J. Turk, The fixed divisor of a polynomial, *Amer. Math. Monthly* **93** (1986), 282–286.

Modules over Trusses vs. Modules over Rings: Internal Direct Sums

Devi Fitri Ferdania, Irawati, and Hanni Garminia

1 Introduction

In algebra, it is well-studied that many algebraic structures have many applications in real life, such as groups, rings, vector spaces, and modules over rings. Among each of the structures mentioned above, the existence of the neutral element (identity element) is essential. But, the fact that sometimes it is not easy to have or to find the neutral element of an algebraic structure occasionally restrains the construction of a structure. R. Baer [2] and H. Prufer [8] introduced a group-like structure without specifying the neutral element by changing the associative binary operation to the associative ternary operation that satisfies the Mal'cev identities. That structure is called a heap or herd. Moreover, they explained that a choice of any element in a heap is needed to convert its ternary operation into a binary operation and make the underlying set into a group structure. Later, the chosen element will play a role as the neutral element of the group retracts from its heap. By following the progression from a group to a ring, one can augment a heap by equipping it with an associative binary operation that distributes over the ternary heap operation. This structure is called a truss and was introduced by Brzezinski [3].

From an algebraist point of view, the introduction of trusses seems to be a natural progression when there is a concept of a group-like structure. Aforementioned, a choice of any element in a truss converts the underlying Abelian heap addition into a binary Abelian group addition. Interestingly, only some trusses can be viewed as rings that have no specified neutral element of their Abelian group addition (later, this type of truss will be called a ring-type truss and the chosen element will be its zero element), but also some of them can be viewed as a brace structure (later, this

D. F. Ferdania (✉) · Irawati · H. Garminia
Institut Teknologi Bandung, Bandung, Indonesia
e-mail: fitriferdania@students.itb.ac.id

J.-L. Chabert et al. (eds.), *Algebraic, Number Theoretic, and Topological Aspects of Ring Theory*, https://doi.org/10.1007/978-3-031-28847-0_11

type of truss will be called a unital truss or braceable truss) [4]. A set B together with two group operations: a commutative $+$ operation and $*$ operation is called a left brace if it satisfies the brace distributive law. Specifically, for all $x, y, z \in B$

$$x * (y + z) = x * y - x + y * z.$$

Braces, as introduced in [9], play an important role in the existence of the non-degenerate solution of the Yang–Baxter equation. As one of the basic equations in mathematical physics, the Yang–Baxter equation is the foundation of the theory of quantum groups [7]. By having a truss structure that contains an identity element under its binary operation, one can convert it to have a brace structure. Consequently, the study of trusses turns out to be necessary and beneficial.

As a natural continuation of truss studies, the definition of a module over a truss was introduced in [4]. This categorical construction on modules over trusses and its contrast to modules over rings are explored in [6]. Precisely, the product and coproduct of modules over trusses are constructed in [5]. It is shown that the direct sum of two non-empty Abelian heaps is isomorphic to the heap associated with the group's direct sum retract of both heaps and $\mathbb{Z}$. This condition affects the internal direct sum of a module over a truss. As a consequence, the concept of internal direct sum of modules over trusses will have some differences from the one that arises in modules over rings. On modules over rings, internal direct sums are fundamental in the concept of decomposition of a module from its submodules. Moreover, the decomposition of a ring was derived from the internal direct sum of its regular modules. Hence, a rigorous definition of internal direct sum of modules over trusses is needed not only to know the difference between modules over rings but can be seen as the first step to building a concept of the decomposition of a truss.

This chapter aims to construct a rigorous definition of the internal direct sum of modules over trusses, how it behaves, and contrast it to the internal direct sum of modules over rings. In general, the objective is to continue the study of the categorical properties of modules over trusses.

The organization of this chapter is as follows. The first section is the introduction. Next, in the second section, we present some necessary definitions, notation, lemmas, and theorems that are used to construct the internal direct sums concept. The main results are in the third section and are divided into 2 parts. In its first subsection, the definition of internal direct sums of modules over trusses is presented together with an example. Some characteristics of the internal direct sum of modules over trusses are brought in the next subsection and are contrasted with what we usually have in modules over rings.

2 Preliminaries

It is assumed that the readers are familiar enough with the basic notions of various algebraic structures, such as groups, rings, and modules over rings. Below, some important preliminary definitions and terms are explained for convenience.

2.1 Heaps

Referring to [7, 8], and [3], a heap is a pair $(H, [-,-,-])$ that consists of a non-empty set H equipped with a ternary operation $[-,-,-] : H \times H \times H \to H$ that satisfies the following two conditions, for all $v, w, x, y, z \in H$

$$[v, w, [x, y, z]] = [[v, w, x], y, z] \quad \text{and} \quad [x, x, y] = y = [y, x, x]. \tag{1}$$

The above equations are known as the associativity property and the Mal'cev identities, respectively. Regarding the commutativity, a heap $(H, [-,-,-])$ is said to be Abelian if it is commutative, i.e., if for all $x, y, z \in H$,

$$[x, y, z] = [z, y, x].$$

A heap morphism between two heaps $(H_1, [-,-,-]_1)$ and $(H_2, [-,-,-]_2)$ is a function $\phi : H_1 \to H_2$ preserving the ternary operation, or simply $\phi([x, y, z]_1) = [\phi(x), \phi(y), \phi(z)]_2$ for all $x, y, z \in H_1$.

A subset S of a heap $(H, [-,-,-])$ is called a sub-heap if for $x, y, z \in S$, $[x, y, z] \in S$, i.e., it is closed under its ternary operation. In addition, a sub-heap S is said to be a normal sub-heap if there exists $e \in S$ such that for all $x \in H$ and $s \in S$ there exists $t \in S$ such that $[x, e, s] = [t, e, x]$. Similar to groups, every sub-heap of an Abelian heap is normal. The definition of quotient heaps is constructed by defining a relation $\sim_S$ on a sub-heap S of a heap H as $x \sim_S y$ if and only if there exists $s \in S$ such that $[x, y, s] \in S$. This idea is introduced in [3], and it is proved there that this relation is an equivalence relation.

Interestingly, a heap can be seen as a group without specifying its neutral element. This corresponds to the following lemma:

Lemma 1 *[7], [8]*

1. *Let $(G, \diamond, 1_\diamond)$ be a group. If a ternary operation is defined by*

$$[-,-,-]_\diamond : G \times G \times G \to G \quad , \quad [x, y, z] = x \diamond y^{-1} \diamond z, \tag{2}$$

then $(G, [-,-,-]_\diamond)$, often notated as $H(G)$, forms a heap. Moreover, every homomorphism of groups is a homomorphism of corresponding heaps.

2. *Let $(H, [-,-,-])$ be a heap and $e \in H$. If a binary operation is defined by*

$$-\diamond_e- : H \times H \to H \quad x, \quad x \diamond_e y = [x, e, y], \tag{3}$$

then $(H, \diamond_e, e)$, often notated as $G(H, e)$, forms a group. Moreover, if φ is a morphism of heaps from $(H_1, [-,-,-])$ to $(H_2, [-,-,-])$, then for all $e_1 \in H_1$ and $e_2 \in H_2$, the functions below

$$\hat{\varphi} : H_1 \to H_2 \qquad \text{and} \qquad \hat{\varphi}^{\circ} : H_1 \to H_2$$
$$x \mapsto [\varphi(x), \varphi(e_1), e_2] \qquad\qquad x \mapsto [e_2, \varphi(e_1), \varphi(x)]$$

are homomorphism of groups from $G(H_1, e_1)$ to $G(H_2, e_2)$.

Lemma 1 leads to the concept of converting a group into a heap by specifying the ternary operation and, conversely, converting a heap into a group by specifying the binary operation and by choosing a specific element to be its neutral element. Thus, a heap can be seen as a group that has not specified its neutral element. Later, Lemma 1 will often be used in the main results.

2.2 Trusses

Following [3] and [4], a truss $(T, [-, -, -])$ consists of a non-empty set T, a ternary operation $[-, -, -]$ making T into an Abelian heap, and an associative binary operation $\cdot$ (denoted by juxtaposition) that distributes over $[-, -, -]$, that is, for all $w, x, y, z \in T$ satisfy

$$w[x, y, z] = [wx, wy, wz] \quad \text{and} \quad [x, y, z]w = [xw, yw, zw]. \tag{4}$$

Furthermore, a commutative truss is obtained if its binary operation is commutative. Let $(T_1, [-, -, -], .)$ and $(T_2, [-, -, -], .)$ be trusses and $\varphi : T_1 \to T_2$ be a function. The function φ is called a truss homomorphism if φ is both a morphism of heaps with respect to the ternary operation and a morphism of semigroup with respect to the binary operation. Also, a sub-truss of a truss $(T, [-, -, -], .)$ is defined as a non-empty subset that is closed under both operations.

Recall from [3], a truss $(T, [-, -, -], .)$ is called a unital truss, if $(T, .)$ forms a monoid. Additionally, an element $0 \in T$ is said to be an absorber if it satisfies $x0 = 0 = 0x$ for all $x \in T$. It is easy to prove that if a truss has an absorber, then it is a unique element.

It has already been mentioned in the introduction section that trusses interpolate between two other algebraic structures, rings and braces. Ergo, it is interesting to observe the types of trusses that have a close relationship to either rings or braces.

Lemma 2 *[4, Lemma 3.9] Let $(T, [-, -, -], .)$ be a truss, and $+_e$ denotes a binary operation obtained from the ternary operation such that $x +_e y = [x, e, y]$.*

1. *If T is a unital truss with 1 as its identity element, then for all $x, y, z \in T$ the binary operations $+_1$ and $\cdot$ satisfy*

$$x(y +_1 z) = (xy) -_1 x +_1 (xz) \text{ and } (y +_1 z)x = (yx) -_1 x +_1 (zx), \tag{5}$$

which are the left and right brace distributive laws.

2. *If T contains an absorber* 0*, then for all $x, y, z \in T$ the binary operations $+_0$ and $\cdot$ satisfy*

$$x(y +_0 z) = (xy) +_0 (xz) \text{ and } (y +_0 z)x = (yx) +_0 (zx). \tag{6}$$

Thus, $(T, +_0, .)$ forms a ring structure.

By the above lemma, a unital truss can be seen as braceable or a brace-type structure, whereas a truss that contains an absorber is a ring-type structure.

Let $(H, [-, -, -])$ be an Abelian heap. According to [3, Proposition 3.38], the set of all endomorphism of $(H, [-, -, -])$, denoted as $E(H)$, is a truss with the following operation, for all $\alpha, \beta, \gamma \in E(H)$

$$[\alpha, \beta, \gamma] : H \to H \quad \text{and} \quad x \mapsto [\alpha(x), \beta(x), \gamma(x)]$$

and the composition function. By this fact, the action from a truss to an Abelian heap is feasible and leads to the concept of modules over trusses.

2.3 Modules over Trusses

From the previous subsections, it can be seen that rings and trusses have many similarities. Interestingly, the behaviors of modules over them are not alike. To begin with, let $(T, [-, -, -], .)$ be a truss. From [3, Definition 4.1], an Abelian heap $(M, [-, -, -])$ together with a morphism of trusses $\varphi_M : T \to E(M)$ resulting in a triple $(M, [-, -, -], \varphi_M)$ is called a left T-module. In another way, if $t \triangleright m$ denotes $\varphi(t, m)$, then for all $t_1, t_2, t_3 \in T$ and $m_1, m_2, m_3 \in M$:

1. $(t_1 t_2) \triangleright m_1 = t_1 \triangleright (t_2 \triangleright m_1)$.
2. $t_1 \triangleright [m_1, m_2, m_3] = [t_1 \triangleright m_1, t_1 \triangleright m_2, t_3 \triangleright m_3]$.
3. $[t_1, t_2, t_3] \triangleright m_1 = [t_1 \triangleright m_1, t_2 \triangleright m_1, t_3 \triangleright m_1]$.

In addition, the morphism φ_M satisfying properties 1, 2, 3 is called the action of truss T on M. Thus, for convenience, a left T-module M can be written as a module (M, φ_M) over T or simply a module M over T if the morphism is clear.

The definition of a right T-module follows symmetrically by changing the morphism to be $\varphi_M^\circ : T^{op} \to E(M)$, where T^{op} is the opposite truss of T in which its binary operation is opposite to the binary operation of T.

Now, let M be a left T-module. An element $e \in M$ is said to be an absorber if it satisfies

$$t.e = e$$

for all $t \in T$. Further, denote by $Abs(M)$ the set of all absorbers contained in M.

Fig. 1 Homomorphism of modules diagram

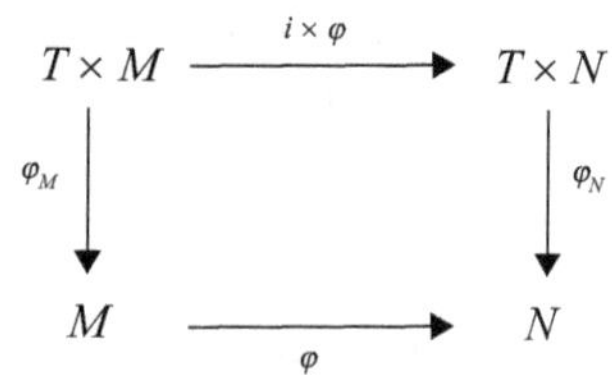

Let (M, φ_M) and (N, φ_N) be left T-modules. A module homomorphism from M to N is a homomorphism of heaps $\varphi : M \to N$ that makes the above diagram commutative (Fig. 1).

The set of all module homomorphisms from M to N will be denoted as $Hom_T(M, N)$.

Now, to define the coproduct of modules over trusses, a clear construction of the coproduct of Abelian heaps is needed, and this is explained in [4]. First, let X be a non-empty set and define $W(X)$ as a set of reduced words in X. A reduced word from set X is an odd-length word whose letters are from the elements of X and which contains no consecutive identical letters.

$$W(X) := \{x_1 x_2 \ldots x_{2n+1} | x_i \neq x_{i+1}\}, \tag{7}$$

where $x_i \in X$, for all $1 \leq i \leq 2n + 1$ and $n \in \mathbb{N}$. Every word should contain an odd number of elements since we will equip this set with a ternary operation. Moreover, it cannot have consecutive letters with the same elements since it should satisfy the Mal'cev identities property. Also, define the opposite of a reduced word $w = x_1 x_2 \ldots x_{2n+1}$ as $w^\circ = x_{2n+1} \ldots x_2 x_1$.

For the ternary operation, define $[-, -, -]$ on $W(X)$ as a grafting and pruning operation. This means, given words $u, v, w \in W(X)$, then $[u, v, w]$ is obtained by concatenating u, v°, w and systematically removing all pairs of the consecutive similar letter. Furthermore, one can easily check that this operation satisfies associativity and the Mal'cev identities properties. Thus $(W(X), [-, -, -])$, denoted as $\mathcal{H}(X)$, forms a heap, and it is proven in [4, Lemma 3.1] that this heap is a free heap.

The next construction involves making the free heap $\mathcal{H}(X)$ to be commutative to form an Abelian free heap. A free Abelian heap of a set X, denoted as $\mathcal{A}(X)$, is constructed by defining the set of all symmetric words of odd length in which no two consecutive letters are the same. Consequently, a symmetric word can be defined as follows:

$$\begin{aligned} w = &: x_1 y_1 x_2 y_2 \ldots y_n x_{n+1} : \\ &= \{x_{\tau(1)} y_{\sigma(1)} x_{\tau(2)} y_{\sigma(2)} \ldots y_{\sigma(n)} x_{\tau(n+1)} | \tau \in S_{n+1} \text{ and } \sigma \in S_n\}, \end{aligned} \tag{8}$$

where $x_i, y_i \in X$. This construction makes $: w^\circ :=: w :$, and this will lead to the commutative property. Furthermore, a symmetric word is said to be a symmetric reduced word if the set contains only reduced words. The set of all symmetric reduced words of a set X is denoted by $\overline{W}(X)$. Thus, together with a ternary

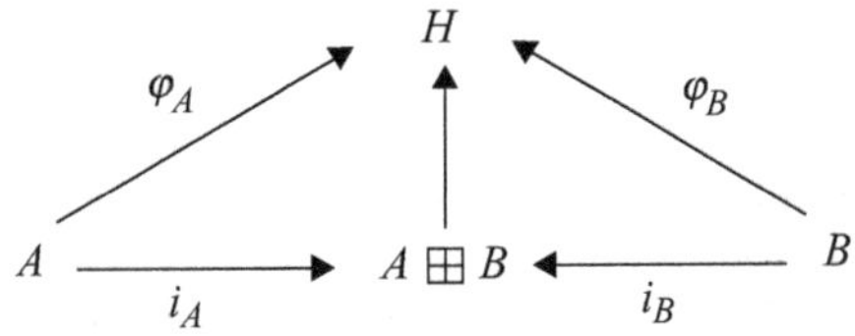

Fig. 2 Universal property of coproduct diagram

operation obtained by concatenating the representatives of symmetric reduced words and symmetric pruning, it results in a free Abelian heap $\mathcal{A}(X)$.

The direct sum or coproduct of Abelian heaps can be constructed as follows. Let $(A, [-,-,-]_A)$ and $(B, [-,-,-]_B)$ be Abelian heaps. Apply the ternary operations of A and B to the free Abelian heap $\mathcal{A}(A \sqcup B)$ of disjoint union of sets A and B whenever possible to reduce the words. Specifically, fix any element $e \in \mathcal{A}(A \sqcup B)$ and form a sub-heap C_e of the $\mathcal{A}(A \sqcup B)$ generated by

$$[[a, a', a''], [a, a', a'']_A, e] \qquad [[b, b', b''], [b, b', b'']_B, e],$$

where $a, a', a'' \in A$, $b, b', b'' \in B$, and $[-,-,-]$ is the ternary operation in $\mathcal{A}(A \sqcup B)$. One can prove that C_e forms a normal sub-heap; thus we can form a quotient heap $\mathcal{A}(A \sqcup B)/C_e$. This quotient heap is the coproduct of A and B since it satisfies the universal property of coproduct as stated below.

Proposition 1 *[5, Proposition 3.7] Let A, B and H be Abelian heaps. Suppose that $i_A : A \to A \boxplus B$ and $i_B : B \to A \boxplus B$ are the inclusion maps. Then $A \boxplus B$ is a coproduct in the category of Abelian heaps (Fig. 2).*

Proof Choose $\varphi : A \boxplus B \to H$ to be $\varphi_A \boxplus \varphi_B$ in the sense $\varphi_A \boxplus \varphi_B(a) = \varphi_A(a)$ whenever $a \in A$ and $\varphi_A \boxplus \varphi_B(b) = \varphi_B(b)$ whenever $b \in B$ and then extend this definition letter-by-letter for the remaining elements of $A \boxplus B$. □

Explicitly, the elements of $A \boxplus B$ as a quotient heap can be specified as follows.

Proposition 2 (5, Proposition 3.6) *Let A and B be Abelian heaps:*

1. *The direct sum of A and B contains only three types of elements (symmetric reduced words) that are:*
   a. *The single-letter words from the element of A or the element of B, i.e., $a \in A$ and $b \in B$.*
   b. *The triple-letter words in the form of either $\overline{:abb':}$ or $\overline{:aa'b}$, where $a, a' \in A$ and $b \in B$ with $a \neq a'$ and $b \neq b'$.*
   c. *The alternating words $\overline{:a_1b_1a_2 \dots a_nb_na_{n+1}}$ and $\overline{b_1a_1b_2 \dots b_na_nb_{n+1}}$, where $a_i \in A$ and $b_i \in B$.*
2. *By fixing any $e_A \in A$ and $e_B \in B$, the triple-letter words and the alternating words in the first statement can be written as:*
   a. *The triple words: $\overline{:abe_B:}$ and $\overline{:bae_A:}$.*

b. *The alternating words:* $\overline{abe_Ae_B \ldots e_Ae_Be_A :}$ *and* $\overline{: bae_Be_A \ldots e_Be_Ae_B :}$.

Interestingly, despite the seemingly similar construction between heaps and groups, the coproduct of heaps is not isomorphic to the heap associated with the coproduct of groups retract from those heaps, i.e., $A \boxplus B \ncong H(G(A, e_A) \oplus G(B, e_B))$. This follows the proposition below.

Proposition 3 *[5, Proposition 3.9] Let A and B be Abelian heaps, for any $e_A \in A$ and $e_B \in B$, and then*

$$A \boxplus B \cong H(G(A, e_A) \oplus G(B, e_B) \oplus \mathbb{Z}). \tag{9}$$

This fact makes a significant difference between the coproduct of modules over trusses and over rings. In addition, the existence of $\mathbb{Z}$ in Proposition 3 analogous to the types of elements in $A \boxplus B$ as shown in Proposition 2, especially for the existence of alternating elements. The alternating part of $A \boxplus B$ is often called the tail part.

By specifying the action from a truss, one can define the coproduct of modules over a truss. Let M and N be left modules over a truss T whose action is notated by a juxtaposition. The coproduct of T-modules M and N is defined by fixing any $e_m \in M$ and $e_n \in N$, and for all $t \in T$, and defining the action on the coproduct heap $M \boxplus N$ letter-by-letter as follows:

$$t.\overline{: mne_Me_Ne_M \ldots e_Me_Ne_M :} := \overline{\begin{array}{l} : (t.m)(t.n)(t.e_M)(t.e_N)(t.e_M)\ldots \\ (t.e_M)(t.e_N)(t.e_M) : \end{array}}.$$

Generally, if $(M_\alpha)_{\alpha \in A}$ is an indexed set of left modules over a truss T, then for each $t \in T$

$$\varphi_\alpha^t : M_\alpha \to \boxplus_{\alpha \in A} M_\alpha$$
$$m \mapsto t.m,$$

which is a homomorphism of heaps. Thus, the family $(\varphi_\alpha^t)_{\alpha \in A}$ can be extended to $\boxplus_{\alpha \in A} \varphi_\alpha^t : \boxplus_{\alpha \in A} M_\alpha \to \boxplus_{\alpha \in A} M_\alpha$. So that, there is an action from T that is

$$T \times \boxplus_{\alpha \in A} M_\alpha \mapsto \boxplus_{\alpha \in A} M_\alpha$$
$$(t, m) \mapsto \boxplus_{\alpha \in A} \varphi_\alpha^t(t, m),$$

and it makes $\boxplus_{\alpha \in A} M_\alpha$ a left T-module.

3 Main Results

Throughout this section, the symbol $\boxplus$ denotes the coproduct of modules over trusses. Meanwhile, the symbol $\oplus$ is related to the coproduct of modules over rings.

3.1 Internal Direct Sums of Modules over Trusses

Before the internal direct sums of modules over trusses are constructed, consider the definition of internal direct sums of modules over rings [1]. Let N be a left module over a ring R and N_1, N_2 be its R-submodules. Suppose that $\iota_1 : N_1 \to N$ and $\iota_2 : N_2 \to N$ are the inclusion maps. Then, the R-module N is said to be the internal direct sum of its submodules N_1 and N_2, denoted by $N = N_1 \oplus N_2$, if $\iota = \iota_1 \oplus \iota_2$ is an isomorphism. A diagram that illustrates this condition is given in Fig. 3.

Furthermore, if a module over a ring is the internal direct sum, then every element of the module can be written uniquely as the sum of elements from its submodules. This phenomenon also appears equivalently in the other algebraic structures. In this sense, the internal direct sum for a module over a truss is defined as follows.

Definition 1 Let M be a T-module and M_1, M_2 be its T-submodules. Suppose that $i_1 : M_1 \to M$ and $i_2 : M_2 \to M$ are the inclusion maps. Then, the T-module M is said to be the internal direct sum of its submodules M_1 and M_2 denoted by $M = M_1 \boxplus M_2$ if $i = i_1 \boxplus i_2$ is an isomorphism.

For convenience, see the diagram Fig. 4. More generally, suppose that $(M_\alpha)_{\alpha \in A}$ is an indexed set of submodules of a left T-module M, where $\alpha \in A$ a set of index. Let $i_\alpha : M_\alpha \to M$ be inclusion maps. Generalizing the definition stated above, M is the internal direct sum of its submodules $(M_\alpha)_{\alpha \in A}$ if the coproduct morphism

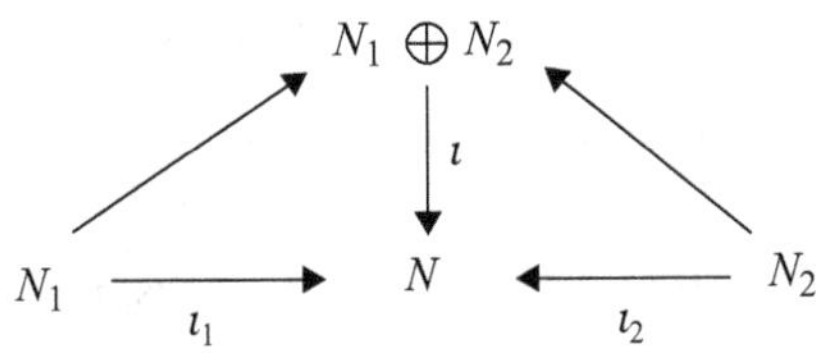

Fig. 3 Internal direct sums of modules over rings diagram

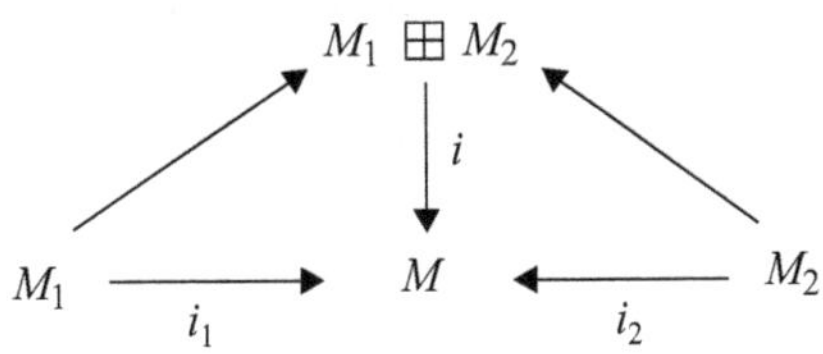

Fig. 4 Internal direct sums of a module over a truss

$$i = \boxplus_{A} i_\alpha : \boxplus_{A} M_\alpha \to M \tag{10}$$

is an isomorphism.

As previously mentioned, the existence of the tail part on the coproduct of heaps forces the coproduct of heaps to not be isomorphic to the heap associated with the coproduct of groups retracts from both heaps. The tail part is the main reason that makes the *behavior* of the internal direct sum of a module over a truss behave differently from what is familiar with a module over a ring.

To begin with, consider an interesting example below. Consider a module whose heap is associated with a group of integers over a unital truss. To be specific, consider a unital truss $(T, [-,-,-]_T, *)$ where its identity is denoted by 1 (it is still valid if $T = \{1\}$ is chosen). Let $H(\mathbb{Z})$ be an Abelian heap associated with the group of integers with the following ternary operation:

$$[x, y, z] = x - y + z \tag{11}$$

for all $x, y, z \in H(\mathbb{Z})$. Define the action of T to $H(\mathbb{Z})$ as

$$t \triangleright x = x \tag{12}$$

for all $t \in T$ and $x \in H(\mathbb{Z})$. Observe that for all $t, t', t'' \in T$ and $x, y, z \in H(\mathbb{Z})$:

- $(t * t') \triangleright x = x = t \triangleright x = t \triangleright (t' \triangleright x)$.
- $t \triangleright [x, y, z] = [x, y, z] = [t \triangleright x, t \triangleright y, t \triangleright z]$.
- $[t, t', t'']_T \triangleright x = x = [x, x, x] = [t \triangleright x, t' \triangleright x, t'' \triangleright x]$.
- $1 \triangleright x = x$.

Thus, $H(\mathbb{Z})$ is a T-module.

Now, consider two subsets of $H(\mathbb{Z})$ that are the singleton $\{0\}$ and $\{1\}$. Since every singleton forms a sub-heap then so for $\{0\}$ and $\{1\}$. It is clear that the action from T is closed on both sub-heaps. Thus, $\{0\}$ and $\{1\}$ are the T-submodules of $H(\mathbb{Z})$. Therefore, according to Proposition 2, the elements of $\{0\} \boxplus \{1\}$ are:

- Single elements: 0 and 1.
- Alternating elements: Let $n \in \mathbb{N}$:
  - Element 1 appears n times, i.e.,

$$\overline{:01010\ldots010:} = [[[[0, 1, 0], 1, 0], \ldots], 1, 0] = -n.$$

  - Element 0 appears n times, i.e.,

$$\overline{:10101\ldots101:} = [[[[1, 0, 1], 0, 1], \ldots], 1, 0] = n + 1.$$

The coproduct $\{0\} \boxplus \{1\}$ contains no triple elements type since by the Mal'cev identities $[0, 1, 1] = 0$ and $[1, 0, 0] = 1$ that are the single elements. Thus, there is an isomorphism from $\{0\} \boxplus \{1\}$ to $H(\mathbb{Z})$ as a T-module. Ergo, $H(\mathbb{Z})$ is the internal direct sum of its submodules $\{0\}$ and $\{1\}$.

The above example shows that a module of integer over a unital truss with the given action can be decomposed by its two singleton submodules. This phenomenon is rarely seen on modules over rings since the direct sum of two singleton submodules only produces finite elements. Meanwhile, on modules over trusses, it is possible to produce an infinity of elements from the coproduct. Moreover, the existence of the tail part in the direct sum of two singleton submodules over a truss makes it possible to reach all elements of its module. This example works well since all elements of T-module $H(\mathbb{Z})$ form absorbers. Other than that, this example cannot be converted into a module over a ring, since T is a unital truss. Consequently, it will not form a ring if the Abelian heap structure is converted into the Abelian group.

3.2 *Modules over Trusses vs. Modules over Rings*

It is well-known that in modules over rings, a left R-module N is the internal direct sum of its submodules N_1, N_2 if and only if N_1, N_2 are independent and span N. To be specific, N_1 and N_2 are said to be independent if $N_1 \cap N_2 = \{0\}$ where 0 is the neutral element of the Abelian group N, whereas it is said to be span N if $N_1 + N_2 = N$. This bi-implication works well since the independency of the submodules implies injectivity, while span submodules imply the surjectivity of the coproduct homomorphism.

Therefore, it is interesting to observe this characteristic on modules over trusses. Remember that the neutral element of an Abelian heap structure of a module is not specified, and the operation is ternary. In consequence, the definition of spanning submodule is slightly different compared to what we have on modules over trusses. So are independent submodules.

Note that the initial object in the category of modules over trusses is an empty set [3]. Hence, the submodules of a module over a truss are said to be independent modules if the intersection is empty. Moreover, define the addition of two submodules as the image of the coproduct of their inclusion maps. For example in the case of two submodules, if M_1, M_2 are T-submodules of M, then $M_1 + M_2 = Im(i_1 \boxplus i_2)$, where i_1, i_2 are the inclusion maps defined in Definition 1.

Definition 2 Let M be a left T-module and M_1, M_2 be its T-submodules. Submodules M_1 and M_2 are said to be independent if they satisfy $M_1 \cap M_2 = \emptyset$, whereas M_1 and M_2 are said to be span M if they satisfy $M_1 + M_2 = M$.

In general, a set of indexed submodules $(M_\alpha)_{\alpha \in A}$ of a left T-module M is said to be independent if for all $\alpha \in A$

$$M_\alpha \cap (\sum_{\beta \neq \alpha} M_\beta) = \emptyset.$$

Note that it is possible for $(M_\alpha)_{\alpha \in A}$ to be independent in pairs without being independent, whereas it is said to be span M if

$$\begin{aligned} M &= \sum_{\alpha \in A} M_\alpha \\ &= Im(\boxplus_{\alpha \in A} i_\alpha) = Im(i). \end{aligned}$$

Now, first observe the relation between internal direct sums with the above definition of spanning and independent submodules. For convenience, observe for the case of two submodules below.

Theorem 1 *Let M be a left T-module and M_1, M_2 be its T-submodules. If $M = M_1 \boxplus M_2$, then M_1 and M_2 are independent and span M.*

Proof It is clear that $M = M_1 + M_2$ since the coproduct morphism is onto. Then M_1 and M_2 span M. Suppose that $M_1 \cap M_2 \neq \emptyset$ and let $x \in M_1 \cap M_2$, $m_1 \in M_1$, and $m_2 \in M_2$, where $m_1 \neq x \neq m_2$. Consider the alternating element on $M_1 \boxplus M_2$. By Proposition 2, fix any $e_1 \in M_1$ and $e_2 \in M_2$, and then the tail part of any alternating words in $M_1 \boxplus M_2$ can be written as the alternating of e_1 and e_2. Let $e_1 = x$; then the construction of the coproduct of modules over a truss forces to reduce every 3-letter words that come from the same direct summand. Thus, x as an element of M_2 makes the alternating element $m = \overline{: m_1 m_2 x e_2 x :}$ to be

$$m = \overline{: m_1 m_2 x e_2 x :} = [m_1, m_2, [x, e_2, x]] = [m_1, m_2, m_2']$$

with $m_2' = [x, e_2, x] \in M_2$, whereas it can also be written as

$$m = \overline{: m_1 m_2 x e_2 x :} = [m_1, [m_2, x, e_2], x] = [m_1, m_2'', x]$$

with $m_2'' = [m_2, x, e_2] \in M_2$. As the consequence,

$$i([m_1, m_2, m_2']) = m = i([m_1, m_2'', x]),$$

where i be the coproduct morphism. Hence, there will be an element $m \in M$ that has more than one pre-image in $M_1 \boxplus M_2$. It is a contradiction since the coproduct morphism is injective. Then, $M_1 \cap M_2$ should be an empty set that means M_1 and M_2 are independent. □

An interesting question is, does the converse hold true? Observe the heap of integers $H(\mathbb{Z}, [-, -, -])$ and a unital truss $T = (\{1\}, [-, -, -], *)$ where $[-, -, -]$ is obtained by converting the binary addition operation into ternary operation (see Eq. 11) and $*$ is the ordinary binary multiplication operation. The action is similar to Eq. 12. Then, $H(\mathbb{Z})$ is a left T-module. Let $O(\mathbb{Z}, [-,-,-])$ be a heap of odd integers and $E(\mathbb{Z}, [-,-,-])$ be a heap of even integers and 0. It is clear that $O(\mathbb{Z})$ and $E(\mathbb{Z})$ are sub-heaps of $H(\mathbb{Z}, [-,-,-])$ and also they are closed under the T-action. Thus, $O(\mathbb{Z})$ and $E(\mathbb{Z})$ are T-submodules of $H(\mathbb{Z})$.

Now, one can observe that $O(\mathbb{Z}) \cap E(\mathbb{Z}) = \emptyset$ and $O(\mathbb{Z}) + E(\mathbb{Z}) = H(\mathbb{Z})$ since every element in $H(\mathbb{Z})$ can be reached by the single elements of $O(\mathbb{Z}) \boxplus E(\mathbb{Z})$.

However, we have $[1, 2, 6] = 1 - 2 + 6 = 5$ and $5 \in O(\mathbb{Z})$. This means 5 can be written as a single element in $O(\mathbb{Z})$ itself and also as a triple element $\overline{:126:}$. Since not unique, then $H(\mathbb{Z}) \neq O(\mathbb{Z}) \boxplus E(\mathbb{Z})$. This indicates that the coproduct morphism i is not injective. Summing up, the converse of Theorem 1 is not true.

Generalizing Theorem 1 for a family of T-submodules, we have the following.

Theorem 2 *Let $(M_\alpha)_{\alpha \in A}$ be an indexed set of submodules of a left module M over a truss T. If $M = \boxplus_{\alpha \in A} M_\alpha$, then:*

1. *$(M_\alpha)_{\alpha \in A}$ span M.*
2. *$(M_\alpha)_{\alpha \in A}$ is independent.*
3. *$(M_\alpha)_{\alpha \in A}$ is independent for every finite subset $F \subseteq A$.*
4. *For every pair of indexed set $B, C \subseteq A$, if $B \cap C = \emptyset$, then*

$$(\sum_{\beta \in B} M_\beta) \cap (\sum_{\gamma \in C} M_\gamma) = \emptyset.$$

Proof (1) It is clear that $(M_\alpha)_{\alpha \in A}$ span M since the coproduct morphism is onto. (2,3,4) If their intersection is not empty, then the coproduct morphism will not be injective. □

Another characteristic of the internal direct sum found in modules over rings is the unique expression of every element in the module. Specifically, let N be a left module over a ring R and N_1, N_2 be its R-submodules. If $N = N_1 \oplus N_2$, then every element $n \in N$ has a unique representation as

$$n = n_1 + n_2,$$

where $n_1 \in N_1$ and $n_2 \in N_2$.

Whereas, in modules over trusses, the expression is unique as one of the element types mentioned in Proposition 2. In this section, a unique representation without depending on the types of the element will be searched for. To this aim, let M be a left module over a truss T and M_1, M_2 be its submodules such that $M = M_1 \boxplus M_2$. Fix $e_1 \in M_1$ and $e_2 \in M_2$. Define

$$\begin{aligned} ne_i &:= \overline{: e_i e_j e_i e_j e_i \ldots e_i e_j :} \\ &= [[\ldots[[e_i, e_j, e_i], e_j, e_i], \ldots], e_i, e_j], \end{aligned}$$

where e_i appears n times, $i, j \in \{1, 2\}$, and $i \neq j$.

Based on Proposition 2, the element of $M_1 \boxplus M_2$ is one of the following types:

1. The single-letter words:
   It is either $m_1 \in M_1$ or $m_2 \in M_2$, and both elements can be written as
   $$m_1 = [m_1, e_2, e_2] = [m_1, e_2, 1e_2]$$
   and
   $$m_2 = [m_2, e_1, e_1] = [m_2, e_1, 1e_1].$$
2. The triple-letter words:
   It is in the form of either $\overline{: m_1e_1e_2 :}$ or $\overline{: m_2e_2e_1 :}$, and both elements can be written as
   $$\begin{aligned}\overline{: m_1e_1e_2 :} &= [[m_1, e_2, e_2], e_1, e_2]\\ &= [m_1, e_2, [e_2, e_1, e_2]]\\ &= [m_1, e_2, 2e_2]\end{aligned}$$
   and
   $$\begin{aligned}\overline{: m_2e_2e_1 :} &= [[m_2, e_1, e_1], e_2, e_1]\\ &= [m_2, e_1, [e_1, e_2, e_1]]\\ &= [m_2, e_1, 2e_1].\end{aligned}$$
3. The alternating element:
   It is in the form of either $\overline{: m_1m_2e_1e_2e_1...e_1e_2e_1 :}$, where e_1 appears n_1 times, or $\overline{: m_2m_1e_2e_1e_2...e_2e_1e_2 :}$, where e_2 appears n_2 times. Both elements can be written as
   $$\begin{aligned}\overline{: m_1m_2e_1e_2e_1...e_1e_2e_1 :} =&[[\dots[[m_1, m_2, \underbrace{e_1], e_2, e_1], ...], e_2, e_1]}_{e_1 \text{ appears } n_1\text{-times}}\\ =&[m_1, m_2, [[...[[e_1, e_2, e_1], e_2, e_1], ...]e_2, e_1]\\ =&[m_1, m_2, n_1e_1]\end{aligned}$$
   and
   $$\begin{aligned}\overline{: m_2m_1e_2e_1e_2...e_2e_1e_2 :} =&[[\dots[[m_2, m_1, \underbrace{e_2], e_1, e_2], \dots], e_1, e_2]}_{e_2 \text{ appears } n_2\text{-times}}\\ =&[m_2, m_1, [[...[[e_2, e_1, e_2], e_1, e_2], ...]e_1, e_2]\\ =&[m_2, m_1, n_2e_2].\end{aligned}$$

It can be seen that there is a pattern for every types of element. Thus, the unique expression has been found and stated as follows.

Theorem 3 *Let $M = M_1 \boxplus M_2$. Fix any $e_1 \in M_1$ and $e_2 \in M_2$; then every element $m \in M$ has a unique representation as*

$$m = [m_{i_1}, m_{i_2}, ne_i], \tag{13}$$

where $i_1, i_2 \in \{1, 2\}$, $m_{i_k} \in M_{i_k}$, and $i_1 \neq i_2$. □

The last observation of the internal direct sums of a module over a ring is that it does not preserve the internal direct sum if the structure is being converted into a module over its truss. To be more precise, let $(R, +, *)$ be a ring; then by using Lemma 2, $T(R) = (R, [-, -, -]_+, *)$ forms a truss. Also, let N be a left R-module. If the Abelian group structure on N is changed to be an Abelian heap, then N forms a left $T(R)$-module (see [4]), call it $T(N)$. Observe the following lemma.

Lemma 3 *Let R be a ring and N be a left R-module. If $N = N_1 \oplus N_2$, for some N_1 and N_2 the R-submodules of N, then $T(N) \neq T(N_1) \boxplus T(N_2)$ as a left $T(R)$-module.*

Proof If $N = N_1 \oplus N_2$, then N_1, N_2 are independent submodules. Hence, it implies that $N_1 \cap N_2 = \{0_N\}$, where 0_N is the identity element under addition of N. Thus, $T(N_1) \cap T(N_2) \neq \emptyset$. By Theorem 1, $T(N) \neq T(N_1) \boxplus T(N_2)$ as a T-module. □

Acknowledgments We would like to thank Bernard Rybolowicz from Swansea University for warmingly answering all author's questions and for the discussion. Also, we would like to thank P3MI ITB, as this research was fully funded by P3MI ITB (research and community service program of ITB).

References

1. Anderson, F. and Fuller, K.: Ring and Categories of Modules. Springer Verlag Second Edition, New York (1991)
2. Baer, R.: Zur Einfuhrung des Scharbegriffs. Journal fur die Reine und Angewandte Mathematik **160** (1929) 199–207
3. Brzezinski, T.: Trusses: Between braces and rings. Transactions of the American Mathematical Society **372**, 4149–4176 (2019)
4. Brzezinski, T.: Trusses: Paragons, ideals, and modules. Journal of Pure and Applied Algebra. **224** (2020)
5. Brzezinski, T. and Rybolowicz, B.: Modules over trusses vs. modules over rings: Direct sums and free modules. Algebras and Representation Theory **25(1)**, 1–23 (2022)
6. Brzezinski, T., Rybolowicz, B., and Saracco, P.: On functors between categories of modules over trusses. Journal of Pure and Applied Algebra. **226(11)** (2022)
7. Cedo, F., Jespers, E., and Okninski, J.: Braces and the Yang-Baxter Equation. Communications in Mathematical Physics. **327**, 101–116 (2014)
8. Prufer, H.: Theorie der Abelschen Gruppen. I. Grundeigenschaften, Mathematische Zeitschrift **20** (1924) 165–187
9. Rump, W. : Braces, radical rings, and the quantum Yang Baxter equation. Journal of Algebra **307** (2007), 153–170

A Survey on Essential-Like Properties of Prüfer v-Multiplication Domains

Carmelo Antonio Finocchiaro and Francesca Tartarone

2010 Mathematics Subject Classification Primary: 13A15, 13A18, 13F05, 13F20, 54A20

Introduction

A globalization of the concept of a valuation domain is obtained through the definition of a Prüfer domain (H. Prüfer, 1932), i.e., a domain whose localizations at prime ideals are valuation domains (recent developments on this topic are, for instance, [9, 16]).

Prüfer v-multiplication domains further generalize the concept of a Prüfer domain. To define this class of rings, we first recall some notation and terminology.

First, we recall some notation and terminology. Let D be an integral domain with quotient field K. Given a fractional ideal I of D, let $I^{-1} := \{x \in K \mid xI \subseteq D\}$.

We dedicate this paper to Paul-Jean Cahen, whose research on integer-valued polynomial rings greatly influenced our work.

The authors were partially supported by GNSAGA of Istituto Nazionale di Alta Matematica. The first author was partially supported by PIA.CE.RI. 2020–2022 University of Catania—Linea 2—progetto "Proprietà locali e globali di anelli e di varietà algebriche" and by the research project PRIN "Squarefree Gröbner degenerations, special varieties and related topics," by the research project "MTTAI - Metodi topologici in teoria degli anelli e loro ideali" of the University of Catania and by Fondazione Cariverona (Research project "Reducing complexity in algebra, logic, combinatorics - REDCOM" within the framework of the programme Ricerca Scientifica di Eccellenza 2018).

C. A. Finocchiaro
Dipartimento di Matematica e Informatica, Università di Catania, Catania, Italy
e-mail: cafinocchiaro@unict.it

F. Tartarone (✉)
Dipartimento di Matematica e Fisica, Università Degli Studi Roma Tre, Roma, Italy
e-mail: francesca.tartarone@uniroma3.it

J.-L. Chabert et al. (eds.), *Algebraic, Number Theoretic, and Topological Aspects of Ring Theory*, https://doi.org/10.1007/978-3-031-28847-0_12

The divisorial closure of I is the fractional ideal $I_v := (I^{-1})^{-1}$, and the t-closure of I is the fractional ideal $I_t := \bigcup J_v$, where J ranges among the finitely generated subideals of I. An ideal I is said to be *divisorial* (resp., *a t-ideal*) if either $I = (0)$ or $I = I_v$ (resp., either $I = 0$ or $I = I_t$). We say that an ideal I of D is t-invertible if $(IJ)_t = D$, for some fractional ideal J of D. A t-ideal I is said to be t-finite if there exists some finitely generated ideal J of D such that $I = J_t$.

An integral ideal I of D is called t-maximal if it is maximal among the proper t-ideals of D. It is well-known that t-maximal ideals are prime and for any domain D we have that $D = \bigcap_{M \in t-\mathrm{Max}(D)} D_M$, where $t - \mathrm{Max}(D)$ denotes the set of all t-maximal ideals of D. We will denote by $t - \mathrm{Spec}(D)$ the set of all prime ideals of D that are also t-ideals, and we will call them t-prime ideals.

The notion of *Prüfer v-multiplication domain* (briefly, PvMD) was introduced to enlarge the class of Prüfer domains (for instance, integrally closed Noetherian domains of dimension greater than 1 are PvMD but not Prüfer). More precisely, an integral domain is a PvMD if and only if it is t-locally a valuation domain, i.e., each localization at a t-prime ideals is a valuation domain. Equivalently, a domain D is a PvMD if and only if each t-finite t-ideal is t-invertible while a domain D is Prüfer if and only if each nonzero finitely generated ideal is invertible.

Interesting examples of PvMD's, besides Prüfer domains, are, for instance, Krull domains and polynomial rings over PvMDs.

We also recall that PvMDs are central to the theory of divisibility of integral domains; indeed a domain is GCD if and only if it is a PvMD with trivial class group (see [2]).

There is a wide and very interesting literature on PvMD; for instance, [11, 12, 14, 17]. In this short note we focus on some topological aspects of their t-prime spectrum and on the contribution that P-J. Cahen, among others, gave to the characterization of when the integer-valued polynomial ring on a domain is a PvMD (see, in particular, Theorem 2). The second section is entirely dedicated to the connections between integer-valued polynomial rings and PvMD's, with regard to the more general essential property.

1 Some General Properties and the t-Finite Character

A domain D is *essential* if it can be represented as an intersection of valuations overrings which are essential for D (i.e., they are localizations at some prime ideal of D). A prime ideal $\mathfrak{p}$ of D is called *essential* if $D_{\mathfrak{p}}$ is a valuation domain. Let $\mathcal{E}(D)$ denote the set of all the essential prime ideals of D.

From the definition it follows that PvMD's are essential domains. This immediately follows from the equality

$$D = \bigcap_{M \in t-\mathrm{Max}(D)} D_M$$

and the fact that D_M (where $M \in t-\mathrm{Max}(D)$) is a valuation domain whenever D is PvMD.

But the essential property for a domain D, in general, is too weak to force D to be a PvMD. Examples of this fact have been given in [14] and [5, Theorem 5.1]. In this last case, we will see in Sect. 2 that the authors use integer-valued polynomial rings.

One of the first questions raised around the PvMD notion is related to the characterization of PvMD's with the t-finite character. We recall that a domain D has the t-finite character if each nonzero element of D belongs at most to finitely many t-maximal ideals; this is equivalent to saying that the intersection $\bigcap_{M\in t-\mathrm{Max}(D)} D_M$ is *locally finite* (i.e., each nonzero element of D is invertible in all but finitely many rings D_M's). An important class of domains with the t-finite character is the class of Noetherian domains (see [15, Theorem 1.3]). M. Griffin in [11] showed that essential domains with the t-finite character are PvMDs. These domains are also known as Krull-type domains, i.e., domains of the type $D = \bigcap_{P\in\mathcal{P}} D_P$ where D_P is a valuation domain and $\mathcal{P}$ is a set of prime ideals of D such that and the intersection $\bigcap_{P\in\mathcal{P}} D_P$ is locally finite. More precisely Griffin's result ([11, Proposition 4, Theorems 5 and 7]) states the following equivalence:

Theorem 1 *Let D be an integral domain. The following conditions are equivalent.*

(i) *D is a Krull-type domain.*
(ii) *D is a PvMD with t-finite character.*

The following question then arises from what we have seen: are there other interesting extra conditions that allow an essential domain to be a PvMD? This is the focus of the paper [8] in which a topological approach to this question is carried on by using the tool of ultrafilter limits of prime ideals and the constructible topology. We briefly recall some notions and basic results.

For a ring R we denote, as usual, by

$$V(I) := \{\mathfrak{p} \in \mathrm{Spec}(R) \mid I \subseteq \mathfrak{p}\}$$

the typical closed set of the Zariski topology (here I runs in the set of all ideals of R). If $f : R \to S$ is a ring homomorphism, let $f^\star : \mathrm{Spec}(S) \to \mathrm{Spec}(R)$ denote the canonical mapping defined by setting $f^\star(\mathfrak{q}) := f^{-1}(\mathfrak{q})$, for every prime ideal $\mathfrak{q}$ of S.

The *constructible topology* is the coarsest topology on $\mathrm{Spec}(R)$ for which all quasi-compact open (in the Zariski topology) subsets of $\mathrm{Spec}(R)$ are clopen sets. The closed subsets of $\mathrm{Spec}(R)$, with respect to the constructible topology, are called *proconstructible*. By [1, Exercise 28 of Chapter 3] a subset X of $\mathrm{Spec}(R)$ is proconstructible if and only if $X = f^\star(\mathrm{Spec}(S))$, for some ring homomorphism $f : R \to S$. However, the previous characterization is not particularly useful in some applications. So in what follows we will recall an alternative way to describe proconstructible subsets of the spectrum of a ring, based on the notion of ultrafilter limit point.

Proposition 1 *Let R be a ring and let $Y \subseteq \mathrm{Spec}(R)$.*

1. *If $\mathscr{U}$ is an ultrafilter on Y, then the subset*

$$Y_{\mathscr{U}} := \{r \in R \mid V(r) \cap Y \in \mathscr{U}\}$$

of R is prime ideal of R, called the ultrafilter limit point of Y with respect to $\mathscr{U}$ *(see* [5, Lemma 2.4]*).*

2. *Y is proconstructible if and only if $Y_{\mathscr{U}} \in Y$, for every ultrafilter $\mathscr{U}$ on Y (see* [10, Theorem 8] *or* [7, Corollary 2.17]*).*
3. *Let $\mathrm{Cl}^{\mathrm{c}}(Y)$ denote the closure of Y, with respect to the constructible topology. Then*

$$\mathrm{Cl}^{\mathrm{c}}(Y) = \{Y_{\mathscr{U}} \mid \mathscr{U} \text{ ultrafilter on } Y\}$$

(see [7, Remark 2.7 and Proposition 2.13]*).*

4. *If R is an integral domain, then $t - \mathrm{Spec}(R)$ is proconstructible (see* [5, Proposition 2.5]*).*

In the case D is essential, we also have that the t-maximal ideals are ultrafilter limits of the prime ideals which are centers of the valuation overrings. In other words:

Proposition 2 ([5, Proposition 2.8]) *Let D be an essential domain and let $Y \subseteq \mathcal{E}(D)$ be such that $D = \bigcap_{\mathfrak{p} \in Y} D_{\mathfrak{p}}$. Then*

$$t - \mathrm{Max}(D) \subseteq \mathrm{Cl}^{\mathrm{c}}(Y).$$

The above proposition gives the intuition for the fact that an extra condition on essential domains which can make them to be PvMDs should involve ultrafilter limits points of families of t-ideals.

The following theorem gives a possible answer to this question.

Theorem 2 *Let D be an integral domain. Then, the following conditions are equivalent.*

(i) *D is a PvMD;*
(ii) *D is an essential domain and there is an essential representation $\mathcal{V} := \{D_{\mathfrak{p}} \mid \mathfrak{p} \in Y\}$ of D for some $Y \subseteq \mathrm{Spec}(D)$ such that $\mathrm{Cl}^{\mathrm{c}}(Y) \subseteq \mathcal{E}(D)$.*

As an example of application of the previous theorem, we show that any localization of a PvMD is still a PvMD.

Corollary 1 ([14, Proposition 1.8]) *Any ring of fractions of a PvMD is a PvMD.*

Proof First, recall that prime ideal of any domain D is *an associated prime of* D if it is minimal over an ideal of the type $(aD :_D bD) := \{x \in D \mid xb \in aD\}$, for some $a, b \in D$ satisfying $b \in D \setminus aD$. If $\mathrm{Ass}(D)$ denotes the set of associated prime ideals of D, it is clear that $\mathrm{Ass}(D) \subseteq t-\mathrm{Spec}(D)$, since every associated prime is minimal over an integral divisorial ideal of D (each ideal of the type $(aD :_D bD)$ is, by definition, divisorial).

In view of [3, Proposition 4], we have the equality

$$D_S = \bigcap\{D_{\mathfrak{p}} \mid \mathfrak{p} \in \mathrm{Ass}(D) \text{ and } \mathfrak{p} \cap S = \emptyset\},$$

and a fortiori $D_S = \bigcap\{D_{\mathfrak{p}} \mid \mathfrak{p} \in t-\mathrm{Spec}(D) \text{ and } \mathfrak{p} \cap S = \emptyset\}$, since $\mathrm{Ass}(D) \subseteq t-\mathrm{Spec}(D)$. Notice that the collection

$$\{D_{\mathfrak{p}} \mid \mathfrak{p} \in t-\mathrm{Spec}(D) \text{ and } \mathfrak{p} \cap S = \emptyset\}$$

is an essential representation of D_S: indeed for every $\mathfrak{p} \in t-\mathrm{Spec}(D)$, $D_{\mathfrak{p}}$ is a valuation domain, since D is a PvMD, and moreover $\mathfrak{p} \cap S = \emptyset$ implies $(D_S)_{\mathfrak{p}D_S} = D_{\mathfrak{p}}$. If $i : D \to D_S$ is the inclusion, then the mapping $i^\star : \mathrm{Spec}(D_S) \to \mathrm{Spec}(D)$ is continuous and closed with respect to the constructible topology, by [1, Exercise 29 of Chapter 3]. Thus

$$\{\mathfrak{p} \in t-\mathrm{Spec}(D) \mid \mathfrak{p} \cap S = \emptyset\} = i^\star(\mathrm{Spec}(D_S)) \cap t-\mathrm{Spec}(D)$$

is proconstructible, being an intersection of proconstructible subsets of $\mathrm{Spec}(D)$. By Theorem 2 the conclusion is now clear. □

2 Integer-Valued Polynomials

Given a domain D with quotient field K, the integer-valued polynomial ring over D is defined as $\mathrm{Int}(D) = \{f(X) \in K[X] \mid f(D) \subseteq D\}$. A prime ideal $\mathfrak{p}$ is called to be an *int prime* if $\mathrm{Int}(D) \nsubseteq D_{\mathfrak{p}}[X]$, and it is called a *polynomial prime* if $\mathrm{Int}(D) \subseteq D_{\mathfrak{p}}[X]$. In this case it is easily seen that $\mathrm{Int}(D)_{\mathfrak{p}} = D_{\mathfrak{p}}[X]$, where $\mathrm{Int}(D)_{\mathfrak{p}} := \mathrm{Int}(D)_{D\setminus\mathfrak{p}}$. It is well-known that if a prime ideal $\mathfrak{p}$ has infinite residue field, then it is a polynomial prime ([4, Proposition I.3.4]).

If $\mathrm{Int}(D)$ is a PvMD, then D is a PvMD too ([5, Proposition 1.10]).

In general, as we will see further, the fact that D is a PvMD is not sufficient to get that $\mathrm{Int}(D)$ is a PvMD but, however, $\mathrm{Int}(D)$ is essential.

Proposition 3 *Let D be a PvMD. Then* $\mathrm{Int}(D)$ *is essential.*

Proof We first recall the following well-known equality

$$\text{Int}(D) = \bigcap_{\mathfrak{p} \in t\text{-Max}(D)} \text{Int}(D)_{\mathfrak{p}}$$

(see [11, Proposition 4] and [6, Corollaire 1]).

For all the upper to zero of Int(D) (i.e., the prime ideals $\mathfrak{H}$ of Int(D) such that $\mathfrak{H} \cap D = 0$), the localizations of Int(D) at these primes are localizations of $K[X]$, whence they are valuation domains, and $K[X]$ is exactly the intersection of all localizations of Int(D) at the upper to zero primes.

If $\mathfrak{q}$ is a t-polynomial prime (so, $\text{Int}(D)_{\mathfrak{q}} = D_{\mathfrak{q}}[X]$)), then $\mathfrak{q}D_{\mathfrak{q}}[X]$ is a t-prime and its contraction to Int(D) is still a t-prime. Moreover, $\text{Int}(D)_{\mathfrak{q}D_{\mathfrak{q}}[X] \cap \text{Int}(D)} = D_{\mathfrak{q}}[X]_{\mathfrak{q}D_{\mathfrak{q}}[X]} = D_{\mathfrak{q}}(X)$, the *Nagata ring* of $D_{\mathfrak{q}}$, which is a valuation domain, and $\text{Int}(D)_{\mathfrak{q}} = D_{\mathfrak{q}}[X] = D_{\mathfrak{q}}(X) \cap K[X]$

If $\mathfrak{m}$ is an int prime, for each prime ideal $\mathfrak{M}$ of Int(D) above $\mathfrak{m}$, we have that $\text{Int}(D)_{\mathfrak{M}}$ is a valuation domain ([5, Lemma 31]).

Thus, Int(D) is the intersection of its localizations at the following primes:

(i) uppers to zero;
(ii) $\mathfrak{q}D_{\mathfrak{q}}[X] \cap \text{Int}(D)$, where $\mathfrak{q}$ is a t-maximal polynomial prime of D;
(iii) primes above int primes $\mathfrak{p}$.

Since these localizations are valuation domains, Int(D) is essential.

A first result which describes domains D such that Int(D) is a PvMD deals with domains D that are Krull-type.

Theorem 3 ([19, Theorem 3.2]) *Let D be a Krull-type domain. Then* Int(D) *is a PvMD if and only if $D_{\mathfrak{p}}$ is one dimensional for each int prime ideal $\mathfrak{p}$ of D.*

The condition for int primes to be height-one is also necessary for the general case, when D is PvMD not necessarily Krull-type ([5, Proposition 1.7]).

A complete characterization of domains D such that Int(D) is PvMD has been given by Cahen, Loper, and Tartarone as follows:

Theorem 4 ([5, Theorem 3.4]) *Let D be a domain. Then* Int(D) *is a PvMD if and only if the following conditions hold:*

(a) *D is a PvMD;*
(b) *each int prime ideal of D is height-one;*
(c) *each nonzero polynomial t-prime ideal of D contains a finitely generated ideal which is not contained in any int prime ideal.*

Remark 1 Conditions (a) and (b) in the above theorem are analogous to the conditions given in Theorem 3 and, if D is Krull-type, condition (c) is automatically satisfied. Indeed, if $\mathfrak{q}$ is a polynomial prime of D, take any $x \in \mathfrak{q}$. Since the int primes are t-maximal ideals and D has the t-finite character, there are only finitely many int prime ideals of D containing x, say $\mathfrak{p}_1, \cdots, \mathfrak{p}_s$. From (b), $\mathfrak{q}$ cannot be contained in any $\mathfrak{p}_i$, $i = 1, \cdots, s$. For each $i = 1, \cdots, s$, choose $y_i \in \mathfrak{q} \backslash \mathfrak{p}_i$. Thus $(x, x_1, \cdots, x_s) \subseteq \mathfrak{q}$ and it is not contained in any int prime.

Remark 2 Using the characterization given in Theorem 4, it is possible to get a domain $\mathrm{Int}(D)$ which is essential but not PvMD. This is what the authors do in [5, Example 5.1] by constructing a particular almost Dedekind domain D. By definition, an almost Dedekind domain always satisfies conditions (a) and (b) of Theorem 4, but this domain D does not verify condition (c); thus $\mathrm{Int}(D)$ is not PvMD. Neverthless, $\mathrm{Int}(D)$ is essential.

This last fact is claimed in [5], but it is proved in more detail in [18]. Before this, another essential ring which is not a PvMD was constructed by W. Heinzer and J. Ohm many years before (see [14]) by using a very technical construction involving polynomial rings.

The example given in [5, Example 5.1] also shows that conditions (a) and (b) of Theorem 4 do not imply condition (c).

Let Δ_0 be the set of int prime ideals of D and Δ_1 be the set of polynomial prime ideals of D. We put $D_0 := \bigcap_{\mathfrak{p}\in\Delta_0} D_{\mathfrak{p}}$ and $D_1 := \bigcap_{\mathfrak{p}\in\Delta_1} D_{\mathfrak{p}}$, whence we have that $D = D_0 \cap D_1$. By [5, Lemma 4.1], $\mathrm{Int}(D) = \mathrm{Int}(D_0) \cap D_1[X]$.

If D is a PvMD, then both D_0 and D_1 are PvMDs. Moreover $D_1[X]$ is a PvMD. From [5, Corollary 4.9], if $\mathrm{Int}(D)$ is a PvMD, then $\mathrm{Int}(D_0)$ is Prüfer. The domain constructed in [5, Example 5.1] shows that, assuming that D is a PvMD, $\mathrm{Int}(D_0)$ being Prüfer is not sufficient to get that $\mathrm{Int}(D)$ is a PvMD. However, this works if D is Krull-type.

Theorem 5 ([8, Theorem 3.1]) *Let D be a Krull-type domain. Then* $\mathrm{Int}(D)$ *is a PvMD if and only if* $\mathrm{Int}(D_0)$ *is Prüfer.*

It is possible to generalize Theorem 5 considering the *well-behavior under localization* (WBL) of $\mathrm{Int}(D)$, which means that $S^{-1}\mathrm{Int}(D) = \mathrm{Int}(S^{-1}D)$, if S is a multiplicative subset of D. In particular, when $S = D\setminus\mathfrak{p}$, we have that $\mathrm{Int}(D)_{\mathfrak{p}} = \mathrm{Int}(D_{\mathfrak{p}})$. In some cases, when this last equality holds, we are able to describe the primes of $\mathrm{Int}(D)$ above $\mathfrak{p}$. For instance, if $D_{\mathfrak{p}}$ is a valuation domain, we have a complete description of the prime spectrum of $\mathrm{Int}(D_{\mathfrak{p}})$ and a good control of its localizations. These kind of arguments fit very well to the case of essential domains.

It is well-known ([19, Proposition 2.3]) that if $D = \cap_{\mathfrak{p}\in\mathcal{P}} D_{\mathfrak{p}}$, where $\mathcal{P} \subseteq \mathrm{Spec}(D)$ and the intersection is locally finite, then $\mathrm{Int}(D)_{\mathfrak{p}} = \mathrm{Int}(D_{\mathfrak{p}})$, for each $\mathfrak{p} \in \mathcal{P}$. In particular, if D is Krull-type, then $\mathrm{Int}(D)_{\mathfrak{p}} = \mathrm{Int}(D_{\mathfrak{p}})$, for each t-maximal ideal $\mathfrak{p}$ of D.

By replacing the Krull-type hypothesis with the weaker WBL condition, it is possible to obtain an analogous of Theorem 3 as follows:

Theorem 6 ([8, Theorem 3.7]) *With the notation above, let D be an integral domain such that* $\mathrm{Int}(D)_{\mathfrak{p}} = \mathrm{Int}(D_{\mathfrak{p}})$*, for each t-maximal ideal $\mathfrak{p}$ of D. Then* $\mathrm{Int}(D)$ *is a PvMD if and only if D is a PvMD and* $\mathrm{Int}(D_0)$ *is a Prüfer domain.*

It is easily seen that a locally essential domain is essential, but the property of being essential for a domain is not, in general, a local property. As far as we know, examples of essential, but not locally essential domains are not trivial. For instance, the reader can look at [13].

An interesting characterization of locally essential domains has been given by J. Mott and M. Zafrullah in [17]. Here, the authors define a P-domain to be a domain such that its localizations at associated primes are valuation domains. Since $\mathrm{Ass}(D) \subseteq t-\mathrm{Spec}(D)$, P-domains generalize well the notion of PvMD. In particular, [17, Proposition 1.1] states the following equivalence:

Proposition 4 *For a domain D the following conditions are equivalent:*

(i) *D is a P-domain;*
(ii) *D is an essential domain such that each ring of fractions of D is essential.*

A characterization of integer-valued polynomials that are locally essential is given in [18, Theorem 2.9].

Theorem 7 *Let D be an integral domain. Then* $\mathrm{Int}(D)$ *is locally essential if and only if D is locally essential and each int prime ideal of D is height-one.*

In [18, Section 1], it is shown that the domain constructed in [5, Example 5.1] is locally essential too. This domain is an example of a P-domain that is not PvMD.

Acknowledgments The authors would like to thank the referee for his/her useful suggestions that helped to improve the presentation.

References

1. M. F. Atiyah and I. G. Macdonald, Introduction to Commutative Algebra, Addison-Wesley, Reading, 1969.
2. A. Bouvier, Le Groupe des Classes d'un anneau intégre 107 Congrès national des Sociétés Savantes vol. IV, Brest, (1982) 85–92
3. J. W. Brewer and W. J. Heinzer, Associated primes of principal ideals. *Duke Math. J.* **41** (1974), 1–7.
4. P.-J. Cahen and J.-L. Chabert, Integer-Valued Polynomials, Amer. Math. Soc. Surveys and Monographs, **48**, Providence, 1997.
5. P. J. Cahen, A. Loper and F. Tartarone, Integer-valued polynomials and Prüfer $v-$multiplication domains, *J. Algebra* **226** (2000), 765–787.
6. J.-L. Chabert, Anneaux de 'polynômes à valeurs entières' et anneaux de Fatou, *Bull. Sc. Math. France* **99** (1971), 273–283.
7. C. A. Finocchiaro, Spectral spaces and ultrafilters, *Comm. Algebra*, **42** (4) (2014), 1496-1508.
8. C. A. Finocchiaro and F. Tartarone, On a topological characterization of Prüfer v-multiplication domains among essential domains, *J. Commutative Algebra*, **8** (4) (2016), 513–536.
9. M. Fontana, E. Houston and T. Lucas, Toward a classification of prime ideals in Prüfer domains, *Forum Math.* **22** (2010), no. 4, 741–766.
10. M. Fontana and A. Loper, The patch topology and the ultrafilter topology on the prime spectrum of a commutative ring. *Comm. Algebra* **36** (2008), no. 8, 2917–2922.
11. M. Griffin, Some results on v-multiplication rings. *Canad. J. Math.* **19** (1967) 710–722.
12. M. Griffin, Rings of Krull type. *J. Reine Angew. Math.* 229 (1968) 1–27.

13. W. Heinzer, An essential integral domain with a nonessential localization, *Canad. J. Math.* **33** (1981), 400–403.
14. W. Heinzer and J. Ohm, An essential ring which is not a v-multiplication ring. *Canad. J. Math.* **25** (1973), 856–861.
15. E. Houston and M. Zafrullah, Integral domains in which each t-ideal is divisorial. *Michigan Math. J.* **35** (1988), no. 2, 291–300.
16. B. Olberding, Globalizing local properties of Prüfer domains, *J. Algebra* **205** (1998), pp. 480–504.
17. J. Mott and M. Zafrullah, On Prüfer v-multiplication domains, *Manuscripta Math.* **35** (1981), 1–26.
18. A. Tamoussit and F. Tartarone, Essential properties for rings of integer-valued polynomials, submitted for publication.
19. F. Tartarone, Integer-valued polynomials over Krull-type domains and Prüfer v-multiplication domains, *Proc. Amer. Math. Soc.* **128** (2000), no. 6, 1617–1625.

On the Subatomicity of Polynomial Semidomains

Felix Gotti and Harold Polo

2010 Mathematics Subject Classification Primary: 16Y60, 11C08; Secondary: 20M13, 13F05

1 Introduction

A semidomain is an additive submonoid of an integral domain that is closed under multiplication and contains a multiplicative identity element. Let S be a semidomain, and set $S^* := S \setminus \{0\}$; that is, S^* is the multiplicative monoid of S. We say that S is atomic provided that every non-invertible element of S^* can be written as a finite product of atoms (i.e., irreducible elements). Factorizations in atomic domains have been systematically studied for more than three decades, considerably motivated by the landmark paper [2] by D. D. Anderson, D. F. Anderson, and M. Zafrullah. However, factorizations in the more general context of atomic semidomains have been investigated just recently by N. R. Baeth, S. T. Chapman, and the first author [3]. In the present paper, we investigate atomic properties that are weaker than being atomic in the setting of semidomains. We put special emphasis on the ascent of such properties from the semidomain S to the polynomial semidomain $S[x]$ and the Laurent polynomial semidomain $S[x^{\pm 1}]$.

Special cases of polynomial semidomains and Laurent polynomial semidomains have been the focus of a great deal of attention lately in the factorization theory community. For instance, methods to factorize polynomials in $\mathbb{N}_0[X]$ were investigated by H. Brunotte in [7], and, more recently, F. Campanini and A. Facchini in [8] carried out a more systematic investigation of factorizations in the semidomain $\mathbb{N}_0[X]$. More generally, semigroup semirings were studied by

F. Gotti (✉)
Department of Mathematics, MIT, Cambridge, MA, USA
e-mail: fgotti@mit.edu

H. Polo
Department of Mathematics, University of Florida, Gainesville, FL, USA

J.-L. Chabert et al. (eds.), *Algebraic, Number Theoretic, and Topological Aspects of Ring Theory*, https://doi.org/10.1007/978-3-031-28847-0_13

V. Ponomarenko in [26] from the factorization perspective. The arithmetic of polynomial semidomains with coefficients in $\mathbb{R}_{\geq 0}$ has also been considered; for instance, P. Cesarz, S. T. Chapman, S. McAdam, and G. J. Schaeffer in [9] studied the elasticity of $\mathbb{R}_{\geq 0}[X]$, where $\mathbb{R}_{\geq 0}$ is the set of nonnegative real numbers.

Positive semirings, that is, subsemirings of $\mathbb{R}_{\geq 0}$, have been actively studied in the last few years. Factorizations in positive semirings consisting of rational numbers were considered in [10] by S. T. Chapman, M. Gotti, and the first author, and then in [1] by S. Albizu-Campos, J. Bringas, and H. Polo. The same semidomains were studied in [4] by Baeth and the first author in connection with factorizations of matrices. This, in turn, motivated the paper [3] by Baeth, Chapman, and the first author, where several examples of positive semirings were constructed. Positive semirings can also be produced as valuations of polynomial and Laurent polynomial semidomains, and such valuations have also been investigated recently: the arithmetic of factorizations of $\mathbb{N}_0[\alpha]$, where α is a positive algebraic number, was studied recently for rational valuations in [10] by Chapman, Gotti, and the first author and for algebraic valuations in [14] by J. Correa-Morris and the first author, and in [24] by N. Jiang, B. Li, and S. Zhu. On the other hand, the atomic structure of the algebraic valuations of the Laurent polynomial semidomain $\mathbb{N}_0[X^{\pm 1}]$ has been recently studied in [28] by Zhu.

Following the terminology introduced by P. Clark in [11], we say that the semidomain S is a Furstenberg semidomain if every nonunit element in S^* is divisible by an atom. It is clear that each atomic semidomain is a Furstenberg semidomain. Furstenberg domains have been studied by N. Lebowitz-Lockard in [25] in connection with the properties of almost atomicity and quasi-atomicity, which we define in the next two paragraphs. In addition, Furstenberg domains have been recently considered in [22] by the first author and Zafrullah in connection with idf-domains (i.e., integral domains whose elements have only finitely many irreducible divisors up to associates). Finally, Furstenberg domains have been considered in [21, Section 5] by B. Li and the first author in the context of integer-valued polynomials. In Sect. 3, we prove that the property of being Furstenberg ascends from the semidomain S to both $S[x]$ and $S[x^{\pm 1}]$. We also construct an example of a Furstenberg semidomain that is neither an integral domain nor an atomic semidomain.

The semidomain S is said to be almost atomic provided that, for every nonunit $b \in S^*$, there exist atoms $a_1, \dots, a_k$ of S^* such that $a_1 \cdots a_k b$ factors into atoms in S^*. Observe that each atomic semidomain is almost atomic. The notion of almost atomicity was introduced in [6] by J. G. Boynton and J. Coykendall, and it was later studied in parallel to various other subatomic properties in [25] by Lebowitz-Lockard. In Sect. 4, we study almost atomicity in the context of semidomains. Unlike the Furstenberg property, we do not know whether the property of being almost atomic ascends in general from the semidomain S to either $S[x]$ or $S[x^{\pm 1}]$ (see Question 4.3). We provide in Sect. 4 an example of an almost atomic semidomain that is not atomic as well as an example of an antimatter semidomain whose polynomial extension is almost atomic.

As the notion of almost atomicity, that of quasi-atomicity was introduced in [6] and further studied in [25] in the context of integral domains. Motivated by this, we say that the semidomain S is quasi-atomic provided that, for every nonunit $b \in S^*$, there exists an element a of S^* such that ab factors into atoms in S^*. It follows directly from definitions that each almost atomic semidomain is quasi-atomic. In Sect. 5, we provide a simple ideal-theoretical characterization of quasi-atomic semidomains. In addition, as for the property of being almost atomic, we prove that the property of being quasi-atomic ascends from the semidomain S to both $S[x]$ and $S[x^{\pm 1}]$ under the same divisibility conditions referred to in the previous paragraph.

2 Background

In this section, we introduce the notation and terminology necessary to follow our exposition. Reference material on factorization theory and semiring theory can be found in the monographs [16] by A. Geroldinger and F. Halter-Koch and [18] by J. Golan, respectively. Throughout this paper, we let $\mathbb{P}$, $\mathbb{Z}$, $\mathbb{Q}$, and $\mathbb{R}$ denote the set of primes, integers, rational numbers, and real numbers, respectively. Additionally, we let $\mathbb{N}$ denote the set of positive integers, and we set $\mathbb{N}_0 := \{0\} \cup \mathbb{N}$. Given $r \in \mathbb{R}$ and $S \subseteq \mathbb{R}$, we set $S_{<r} := \{s \in S \mid s < r\}$, and we define $S_{>r}$ and $S_{\geq r}$ in a similar way. For $m, n \in \mathbb{Z}$, we denote by $[\![m, n]\!]$ the discrete interval from m to n, that is, $[\![m, n]\!] := \{k \in \mathbb{Z} \mid m \leq k \leq n\}$.

2.1 Monoids

A *monoid*[1] is defined here to be a semigroup with identity that is cancellative and commutative. Since our interest lies in the multiplicative structure of certain semirings, we will use multiplicative notation for monoids unless we specify otherwise. For the rest of this section, let M be a monoid with identity 1. We set $M^\bullet := M \setminus \{1\}$, and we let $\mathscr{U}(M)$ denote the group of units (i.e., invertible elements) of M. In addition, we let M_{red} denote the quotient $M/\mathscr{U}(M)$, which is also a monoid. We say that M is *reduced* provided that $\mathscr{U}(M)$ is the trivial group, in which case we identify M_{red} with M. The *Grothendieck group* of M, denoted here by $\mathscr{G}(M)$, is the abelian group (unique up to isomorphism) satisfying the property that any abelian group containing a homomorphic image of M also contains a homomorphic image of $\mathscr{G}(M)$. For a subset S of M, we let $\langle S \rangle$ denote the smallest submonoid of M containing S, and if $M = \langle S \rangle$, then we say that S is a *generating set* of M.

[1] The standard definition of a monoid does not assume the cancellative and the commutative conditions.

Puiseux monoids and positive monoids are used several times throughout this paper to construct needed examples. Following [20], a *Puiseux monoid* is simply an additive submonoid of $\mathbb{Q}_{\geq 0}$. The class of Puiseux monoids is, therefore, a natural generalization of that consisting of numerical monoids (i.e., additive submonoids of $\mathbb{N}_0$ up to isomorphism). Puiseux monoids account up to isomorphism for all rank-1 torsion-free monoids that are not groups (see [15, Theorem 3.12.1]). Following [19], a *positive monoid* is an additive submonoid of $\mathbb{R}_{\geq 0}$. It follows from the definitions that every Puiseux monoid is a positive monoid.

For $b, c \in M$, it is said that b *divides* c *in* M if there exists $b' \in M$ such that $c = bb'$, in which case we write $b \mid_M c$, dropping the subscript precisely when $M = (\mathbb{N}, \times)$. We say that $b, c \in M$ are *associates* if $b \mid_M c$ and $c \mid_M b$. A submonoid N of M is *divisor-closed* if for each $b \in N$ and $d \in M$ the relation $d \mid_M b$ implies that $d \in N$. Let S be a nonempty subset of M. An element $d \in M$ is called a *common divisor* of S provided that $d \mid_M s$ for all $s \in S$. A common divisor d of S is called a *greatest common divisor* of S if d is divisible by all common divisors of S. Also, a common divisor of S is called a *maximal common divisor* if every greatest common divisor of the set $S/d := \{s/d \mid s \in S\}$ belongs to $\mathscr{U}(M)$. We let $\gcd_M(S)$ (resp., $\text{mcd}_M(S)$) denote the set consisting of all greatest common divisors (resp., maximal common divisors) of S. The monoid M is called a *GCD-monoid* (resp., an *MCD-monoid*) if each finite nonempty subset of M has a greatest common divisor (resp., a maximal common divisor). It is clear that the inclusion $\gcd_M(S) \subseteq \text{mcd}_M(S)$ holds, whence each GCD-monoid is an MCD-monoid. The converse does not hold in general, as the following easy example illustrates.

Example 2.1 Let M be the numerical monoid $\mathbb{N}_0 \setminus \{1\}$; that is, $M = \langle 2, 3 \rangle$. The set of common divisors of $\{5, 6\}$ in M is $\{0, 2, 3\}$. As $2 \nmid_M 3$ and $3 \nmid_M 2$, it follows that $\gcd_M(5, 6)$ is empty. However, the only common divisor of both sets $\{5-2, 6-2\}$ and $\{5-3, 6-3\}$ is 0 and, therefore, $\text{mcd}_M(5, 6) = \{2, 3\}$. Indeed, one can readily argue that every numerical monoid is an MCD-monoid.

An element $a \in M \setminus \mathscr{U}(M)$ is called an *atom* if for all $b, c \in M$ the equality $a = bc$ implies that either $b \in \mathscr{U}(M)$ or $c \in \mathscr{U}(M)$. We let $\mathscr{A}(M)$ denote the set consisting of all atoms of M. Following Cohn [12] we say that M is *atomic* if each element in $M \setminus \mathscr{U}(M)$ can be written as a (finite) product of atoms, while following Coykendall, Dobbs, and Mullin [13], we say that M is *antimatter* if $\mathscr{A}(M)$ is empty. One can readily check that M is atomic (resp., antimatter) if and only if M_{red} is atomic (resp., antimatter). Assume for the rest of this paragraph that M is atomic. We let $\mathsf{Z}(M)$ denote the free (commutative) monoid on $\mathscr{A}(M_{\text{red}})$. The elements of $\mathsf{Z}(M)$ are called *factorizations*, and if $z = a_1 \cdots a_\ell \in \mathsf{Z}(M)$ for some $a_1, \dots, a_\ell \in \mathscr{A}(M_{\text{red}})$, then ℓ is called the *length* of z, which is denoted by $|z|$. Let $\pi : \mathsf{Z}(M) \to M_{\text{red}}$ be the unique monoid homomorphism satisfying that $\pi(a) = a$ for all $a \in \mathscr{A}(M_{\text{red}})$. For each $b \in M$, the sets

$$\mathsf{Z}_M(b) := \pi^{-1}(b\mathscr{U}(M)) \subseteq \mathsf{Z}(M) \qquad \text{and} \qquad \mathsf{L}_M(b) := \{|z| : z \in \mathsf{Z}_M(b)\} \subseteq \mathbb{N}_0 \tag{2.1}$$

are of crucial importance to study the atomicity of M. When there seems to be no risk of ambiguity, we drop the subscript M from the notations in (2.1). Following [2] and [23], we say that M is a *bounded factorization monoid* if $\mathsf{L}(b)$ is finite for all $b \in M$.

Following [11], we say that a monoid is *Furstenberg* provided that every nonunit has a divisor that is an atom. On the other hand, extending the terminology in [6], a monoid M is called *almost atomic* (resp., *quasi-atomic*) provided that, for every nonunit $c \in M$, there exists $a_1, \ldots, a_k \in \mathscr{A}(M)$ (resp., $b \in M$) such that $a_1 \cdots a_k c$ (resp., bc) can be written as a product of atoms in M.

2.2 Semirings

A *semiring* S is a nonempty set endowed with two binary operations denoted by "+" and "·" and called *addition* and *multiplication*, respectively, such that the following conditions hold:

(1) $(S, +)$ is a monoid with its identity element denoted by 0;
(2) $(S, \cdot)$ is a commutative semigroup with an identity element denoted by 1;
(3) $b \cdot (c + d) = b \cdot c + b \cdot d$ for all $b, c, d \in S$.

With notation as in the previous definition and for any $b, c \in S$, we write bc instead of $b \cdot c$ when there seems to be no risk of confusion. It follows from conditions (1) and (3) in the definition of a semiring S that $0 \cdot b = 0$ for all $b \in S$. A more general notion of a "semiring" S does not assume that the semigroup $(S, \cdot)$ is commutative. However, this more general type of algebraic objects is not of interest in the scope of this paper. A subset S' of a semiring S is a *subsemiring* of S if $(S', +)$ is a submonoid of $(S, +)$ that contains 1 and is closed under multiplication. Observe that every subsemiring of S is a semiring.

Definition 2.2 We say that a semiring S is a *semidomain* provided that S is a subsemiring of an integral domain.

Let S be a semidomain. We set $S^* := (S \setminus \{0\}, \cdot)$ and call it the *multiplicative monoid* of S. It is worth emphasizing that a semiring S may not be a semidomain even if S^* is a monoid; for instance, consider $\{(0, 0)\} \cup (\mathbb{N} \times \mathbb{N})$ under the usual component-wise addition and multiplication. Following standard notation from ring theory, we refer to the units of the multiplicative monoid S^* simply as *units* of S, and we denote the set of units of S by $S^\times$. We never consider in this paper the units of the monoid $(S, +)$, so the use of the term "unit" in the context of the semidomain S should not generate any ambiguity. In addition, we write $\mathscr{A}(S)$ instead of $\mathscr{A}(S^*)$ for the set of atoms of the multiplicative monoid S^*, while we let $\mathscr{A}_+(S)$ denote the set of atoms of the additive monoid $(S, +)$. Finally, for any $b, c \in S$ such that b divides c in S^*, we write $b \mid_S c$ instead of $b \mid_{S^*} c$.

Lemma 2.3 *For a semiring S, the following conditions are equivalent.*

(a) *The multiplication of* S *extends to* $\mathscr{G}(S)$ *turning* $\mathscr{G}(S)$ *into an integral domain.*
(b) S *is a semidomain.*

Proof (a) $\Rightarrow$ (b): This is clear.

(b) $\Rightarrow$ (a): Let S be a semidomain, and suppose that S is embedded into an integral domain R. We can identify the Grothendieck group $\mathscr{G}(S)$ of $(S, +)$ with the subgroup $\{r - s \mid r, s \in S\}$ of the underlying additive group of R. It is easy to see then that $\mathscr{G}(S)$ is closed under the multiplication it inherits from R, and it contains the multiplicative identity because $0, 1 \in S$. Hence $\mathscr{G}(S)$ is an integral domain having S as a subsemiring. □

We say that a semidomain S is *atomic* (resp., *Furstenberg*, *almost atomic*, *quasi-atomic*) if its multiplicative monoid S^* is atomic (resp., Furstenberg, almost atomic, quasi-atomic). Similarly, we say that S is a *bounded factorization semidomain* (resp., a *unique factorization semidomain*) if S^* is a bounded factorization monoid (resp., a unique factorization monoid). We let BFS and UFS stand for bounded factorization semidomain and unique factorization semidomain, respectively. A subset I of S is an *ideal*[2] of S provided that $(I, +)$ is a submonoid of $(S, +)$ and $IS \subseteq I$. We say that an ideal I is *prime* if $I \neq S$ and, for $b, c \in S$, the containment $bc \in I$ implies that either $b \in I$ or $c \in I$. Although a semidomain S can be embedded into an integral domain R, the semidomain S may not inherit any (sub)atomic property from R as, after all, the integral domain $\mathbb{Q}[x]$ is a UFD but it contains as a subring the integral domain $\mathbb{Z}+x\mathbb{Q}[x]$, which is not even quasi-atomic (see [25, Lemma 17]).

The set consisting of all polynomial expressions with coefficients in the semiring S is also a semiring, which we denote by $S[x]$ and call the *semiring of polynomials over* S. Additionally, if S is a semidomain embedded into an integral domain R, then it is clear that $S[x]$ is also a semidomain, and the elements of $S[x]$ are, in particular, polynomials in $R[x]$. Consequently, when S is a semidomain all the standard terminology for polynomials can be applied to elements of $S[x]$, including *constant polynomial*, *degree*, *order*, and *leading coefficient*. Observe that S^* is a divisor-closed submonoid of $S[x]^*$ and, therefore, $S[x]^\times = S^\times$ and $\mathscr{A}(S[x]) \cap S = \mathscr{A}(S)$. Following [27], we say that a nonzero polynomial in $S[x]$ is *indecomposable* if it is not a product of two nonconstant polynomials in $S[x]$.

Following the terminology in [3], we call a subsemiring of $\mathbb{R}$ consisting of nonnegative numbers a *positive semiring*. The fact that underlying additive monoids of positive semirings are reduced makes them more tractable. The reader can check the recent paper [3] for several examples of positive semirings. The class of semidomains clearly contains those of integral domains and positive semirings.

---

[2] Golan [18] defines an ideal in a more restrictive way: if I is an ideal of a semiring S, then by definition $I \neq S$. Consequently, any result we cite from [18] is interpreted here as a statement about the proper ideals of a semiring.

3 Furstenbergness

In this section, we analyze under which conditions the Furstenberg property ascends from a semidomain to its semidomain of (Laurent) polynomials.

The Furstenberg property is a relaxation of being atomic, and the reader can find interesting examples of non-atomic Furstenberg domains in [21, Section 5] and [25, Section 4]. We now construct an example of a Furstenberg positive semiring that is not atomic. In the construction, we use Lindemann-Weierstrass Theorem from transcendental number theory (see [5, Chapter 1]), which states that, for distinct algebraic numbers $\alpha_1, \dots, \alpha_n$, the set $\{e^{\alpha_1}, \dots, e^{\alpha_n}\}$ is linearly independent over the algebraic numbers.

Example 3.1 Consider the Puiseux monoid $P = \langle \frac{1}{p} \mid p \in \mathbb{P} \rangle$, and set $M := P \cup \mathbb{Q}_{\geq 1}$. It is clear that M is also a Puiseux monoid. It is well known and not difficult to argue that $\mathscr{A}(P) = \{\frac{1}{p} \mid p \in \mathbb{P}\}$. This implies that $\{\frac{1}{p} \mid p \in \mathbb{P}\} \subseteq \mathscr{A}(M)$. Clearly, $1 \notin \mathscr{A}(M)$. In addition, for any $q \in M_{>1}$, we can find $p \in \mathbb{P}$ large enough so that $\frac{1}{p} \mid_M q$. Putting the three previous observations together, we conclude that $\mathscr{A}(M) = \{\frac{1}{p} \mid p \in \mathbb{P}\}$. This implies that M is not atomic as, for instance, $5/4$ cannot be written as a sum of atoms in M. On the other hand, it follows from our previous observations that M is a Furstenberg monoid.

Now consider the additive monoid $E(M) := \langle e^m \mid m \in M \rangle$, which is free on the set $\{e^m \mid m \in M\}$ by Lindemann-Weierstrass Theorem. Observe that $E(M)$ is closed under multiplication and, consequently, it is a positive semiring. Since $M \subseteq \mathbb{R}_{\geq 0}$, it follows that $\min E(M)^* = 1$, which implies that the multiplicative monoid $E(M)^*$ is reduced.

We argue that $E(M)$ is a Furstenberg semidomain that is not atomic. Clearly, the multiplicative submonoid $e(M) := \{e^m \mid m \in M\}$ of $E(M)^*$ is isomorphic to $(M, +)$, which implies that $e(M)$ is not atomic. Since $e(M)$ is a divisor-closed submonoid of $E(M)^*$, the semidomain $E(M)$ cannot be atomic. To argue that $E(M)$ is Furstenberg, take a nonunit $x \in E(M)^*$ (i.e., $x \in E(M)^* \setminus \{1\}$) and write $x = c_1 e^{q_1} + \cdots + c_k e^{q_k}$, where $c_1, \dots, c_k \in \mathbb{N}$ and $q_1, \dots, q_k \in M$. We split the rest of the argument into the following two cases.

Case 1: There exists a positive common divisor d of the elements $q_1, \dots, q_k$ in M. In this case, we can factor x in $E(M)^*$ as $x = e^d(c_1 e^{q_1 - d} + \cdots + c_k e^{q_k - d})$. Because $d > 0$ and M is a Furstenberg monoid, there exists $a \in \mathscr{A}(M)$ such that $a \mid_M d$, in which case, it is clear that e^a is an atom of $E(M)$ satisfying that $e^a \mid_{E(M)} x$.

Case 2: The only common divisor of the elements $q_1, \dots, q_k$ in M is 0. In this case, we can choose elements $y_1, \dots, y_m \in E(M)^* \setminus \{1\}$ satisfying that $x = y_1 \cdots y_m$. For each $i \in [\![1, m]\!]$, let ℓ_i be the length of y_i in the (additive) free monoid $E(M)$. Since no element of the form e^d with $d \in M^\bullet$ divides any of the factors $y_1, \dots, y_m$ in $E(M)^*$, we see that $\ell_i \geq 2$ for every $i \in [\![1, m]\!]$. Hence from the equality $y_1 \cdots y_m = c_1 e^{q_1} \cdots c_k e^{q_k}$, we deduce that $2^m \leq \ell_1 \cdots \ell_m = c_1 + \cdots + c_k$, and so $m \leq \log_2(c_1 +$

$\cdots + c_k)$. Now, we can assume that m has been taken as large as it can possibly be to conclude that y_1 is an atom of $E(M)$ such that $y_1 \mid_{E(M)} x$.

Thus, $E(M)$ is a Furstenberg semidomain, which concludes our construction.

Next we prove that the Furstenberg property ascends from a semidomain to its semidomain of (Laurent) polynomials.

Theorem 3.2 *For a semidomain S, the following statements are equivalent.*

(a) *S is Furstenberg.*
(b) *$S[x]$ is Furstenberg.*
(c) *$S[x^{\pm 1}]$ is Furstenberg.*

Proof

(a) $\Rightarrow$ (b): Suppose that S is a Furstenberg semidomain. Take a nonzero nonunit $f \in S[x]$. Suppose first that $f \in S$. Then the fact that S^* is a Furstenberg monoid guarantees the existence of $a \in \mathscr{A}(S)$ with $a \mid_S f$. As S^* is a divisor-closed submonoid of $S[x]^*$, it follows that a is also an atom of $S[x]$, and so f is divisible by an atom in $S[x]$. Suppose now that $\deg f \geq 1$. Take the largest $m \in \mathbb{N}$ such that $f = rg_1 \cdots g_m$ for some $r \in S^*$ and $g_1, \ldots, g_m \in S[x]$ with $\deg g_i \geq 1$ for every $i \in [\![1, m]\!]$. If $g_1 \in \mathscr{A}(S[x])$ we are done. If g_1 is reducible, then the maximality of m guarantees that $g_1 = s(g_1/s)$ for some nonunit element $s \in S^*$ dividing g_1 in $S[x]^*$. Because s is a nonunit of S^* and S^* is a Furstenberg monoid, s must be divisible by an atom b in S^*. Since S^* is a divisor-closed submonoid of $S[x]^*$, we see that b is an atom of $S[x]$ that divides f in $S[x]^*$. Hence $S[x]$ is also a Furstenberg semidomain.

(b) $\Rightarrow$ (c): First, observe that every irreducible f in $S[x]$ with $\operatorname{ord} f = 0$ is an irreducible in $S[x^{\pm 1}]$. Now take a nonzero nonunit $g \in S[x^{\pm 1}]$, and write $g = x^d h$ for some $d \in \mathbb{Z}$ and $h \in S[x]$ with $\operatorname{ord} h = 0$. As g is not a unit in $S[x^{\pm 1}]$, we see that h is not a unit in $S[x]$, and so there is an $a \in \mathscr{A}(S[x])$ such that $a \mid_{S[x]} h$. Note that $\operatorname{ord} a = 0$ because the same holds for h. Thus, a is an irreducible in $S[x^{\pm 1}]$ dividing g. Therefore $S[x^{\pm 1}]$ is also a Furstenberg semidomain.

(c) $\Rightarrow$ (a): This follows from the fact that $\{sx^n \mid s \in S^* \text{ and } n \in \mathbb{Z}\}$ is a divisor-closed submonoid of $S[x^{\pm 1}]^*$ whose reduced monoid is isomorphic to that of S^*.

□

Observe that Theorem 3.2 can help us identify Furstenberg semidomains that are not atomic. For instance, M. Roitman [27] provided the first example of an atomic domain D such that $D[x]$ is not atomic. By virtue of Theorem 3.2, we can now assert that $D[x]$ is a non-atomic Furstenberg domain.

4 Almost Atomicity

In this section, we focus on the property of being almost atomic which, as being Furstenberg, is a property weaker than being atomic. As the following example illustrates, there are almost atomic semidomains (indeed positive semirings) that are not atomic.

Example 4.1 Let $B := \{(1/2)^{n+2} \mid n \in \mathbb{N}\}$, and let $(p_n)_{n\in\mathbb{N}}$ be the strictly increasing sequence consisting of all primes greater than 4. Set

$$A := \left\{ \frac{1}{p_n}, \frac{1}{2^{n+2}} + \frac{1}{2} - \frac{1}{p_n} \;\middle|\; n \in \mathbb{N} \right\},$$

and consider the monoid $M = \langle A \cup B \rangle$. It is not hard to see that $\mathscr{A}(M) = A$, which implies that M is not atomic. Note that every element $x \in M$ can be written as $x = c(1/2)^N + x'$, where $c \in \mathbb{N}_0$, $N \in \mathbb{N}_{\geq 3}$, and x' factors into atoms in M. Thus, $x + c = x' + 2c\left((1/2)^{N+1} + 1/2\right)$. Therefore M is almost atomic. Now pick an arbitrary element $x \in M$. Clearly, we can write

$$x = \frac{c}{2^n} + \frac{c_1}{p_{n_1}} + \cdots + \frac{c_k}{p_{n_k}} + d_1\left(\frac{1}{2^{m_1+2}} + \frac{1}{2} - \frac{1}{p_{m_1}}\right) + \cdots + d_l\left(\frac{1}{2^{m_l+2}} + \frac{1}{2} - \frac{1}{p_{m_l}}\right), \tag{4.1}$$

where either $c = 0$ or $\gcd(c, 2^n) = 1$, $0 < c_i < p_{n_i}$, $0 < d_j < p_{m_j}$, and $p_{n_i} \neq p_{m_j}$ for $i \in [\![1, k]\!]$ and $j \in [\![1, l]\!]$. We also assume that if $p_{n_i} = p_{n_{i'}}$ (resp., $p_{m_j} = p_{m_{j'}}$) for $i, i' \in [\![1, k]\!]$ (resp., for $j, j' \in [\![1, l]\!]$), then $i = i'$ (resp., $j = j'$). We claim that $x \in M$ has finitely many representations of the form (4.1). Consider another representation of x having this form:

$$x = \frac{\alpha}{2^\ell} + \frac{\alpha_1}{p'_{n_1}} + \cdots + \frac{\alpha_t}{p'_{n_t}} + \beta_1\left(\frac{1}{2^{m'_1+2}} + \frac{1}{2} - \frac{1}{p_{m'_1}}\right) + \cdots + \beta_r\left(\frac{1}{2^{m'_r+2}} + \frac{1}{2} - \frac{1}{p_{m'_r}}\right). \tag{4.2}$$

After cancelling similar terms in expressions (4.1) and (4.2), we may assume that either $\alpha = 0$ or $c = 0$ and $p_{n_i} \neq p'_{n_j}$ for any $i \in [\![1, k]\!]$ and any $j \in [\![1, t]\!]$. Then $p_{m'_j} = p_{n_1}$ for some $j \in [\![1, r]\!]$; otherwise, we would obtain a contradiction after clearing denominators. This, in turn, implies that $\beta_j = p_{n_1} - c_1$. Using an inductive argument, it is not hard to see that, if we fix the representation (4.1), then the representation (4.2) is completely determined by the similar terms we can cancel in both expressions, which proves our claim. Consequently, M is an MCD-monoid.

Indeed, if $(\alpha_n)_{n\in\mathbb{N}}$ is a nonconstant sequence of common divisors of $x_1, \dots, x_m \in M$ for some $m \in \mathbb{N}_{>1}$ such that $\alpha_n \mid_M \alpha_{n+1}$ for every $n \in \mathbb{N}$, then x_j has infinitely many representations of the form (4.1) for some $j \in [\![1, m]\!]$.

As in Example 3.1, consider the positive semiring $E(M)$. As we mentioned before, the multiplicative monoid $e(M)$ is isomorphic to $(M, +)$, which implies that $e(M)$ is not atomic. Since $e(M)$ is a divisor-closed submonoid of $E(M)^*$, the monoid $E(M)^*$ is not atomic either. Let x be a nonzero nonunit element of $E(M)$, and write $x = c_1e^{q_1} + \cdots + c_ke^{q_k}$, where $c_1, \dots, c_k \in \mathbb{N}$ and $q_1, \dots, q_k \in M$. We can assume, without loss of generality, that $c_1, \dots, c_k$ are relatively prime positive integers. Now take $d \in \operatorname{mcd}_M(q_1, \dots, q_k)$, and then write

$$x = e^d(c_1e^{q_1-d} + \cdots + c_ke^{q_k-d}) = e^d y_1 \cdots y_m$$

for some $y_1, \dots, y_m \in E(M) \setminus \{0, 1\}$. Since no element of the form $e^{d'}$ (with $d' \in M^\bullet$) divides any of the factors $y_1, \dots, y_m$ in $E(M)$, the inequality $m \le \log_2(c_1 + \cdots + c_k)$ holds. Now, we can assume that m has been taken as large as it can possibly be to conclude that $y_1, \dots, y_m \in \mathscr{A}(E(M))$. Since M is almost atomic, there exists $a_1, \dots, a_t \in \mathscr{A}(M)$ such that $a_1 + \cdots + a_t + d$ is the sum of atoms in M. As a result, $e^{a_1+\cdots+a_t}x$ factors into atoms in $E(M)$. Hence $E(M)$ is almost atomic.

For almost atomic semidomains, we have a result similar to Theorem 3.2.

Theorem 4.2 *For a semidomain S, each of the following statements implies the next.*

(a) *S is almost atomic and* mcd$(s_1, \dots, s_n) \neq \emptyset$ *for any coefficients $s_1, \dots, s_n$ of an indecomposable polynomial in $S[x]$.*
(b) *$S[x]$ is almost atomic.*
(c) *$S[x^{\pm 1}]$ is almost atomic.*

Moreover, conditions (b) *and* (c) *are equivalent.*

Proof

(a) $\Rightarrow$ (b): Let f be a nonzero nonunit element of $S[x]$ such that $\deg f = n$ for some $n \in \mathbb{N}_0$. If $n = 0$, then our result follows from the fact that S is almost atomic and $\langle \mathscr{A}(S) \rangle \subseteq \langle \mathscr{A}(S[x]) \rangle$. Consequently, we may assume that $n > 0$. Write $f = f_1 \cdots f_m$, where $f_i \in S[x]$ and $\deg f_i > 0$ for each $i \in [\![1, m]\!]$. Without loss of generality, assume that m is maximal. Fix an arbitrary $j \in [\![1, m]\!]$. Since m is maximal, the polynomial f_j is indecomposable in $S[x]$. Now write $f_j = s_1x^{n_1} + \cdots + s_kx^{n_k}$ with coefficients $s_1, \dots, s_k \in S^*$ and exponents $n_1, \dots, n_k \in \mathbb{N}_0$. Take $s \in \operatorname{mcd}(s_1, \dots, s_k)$, and note that $s^{-1}f_j \in \mathscr{A}(S[x])$. Since S is almost atomic, there exists β_j, which is a product of atoms in S, such that $\beta_j s$ factors into atoms in S. Consequently, after setting $\beta := \prod_{i=1}^m \beta_i$, we see that β factors into atoms in $S[x]$ and satisfies that βf also factors into atoms in $S[x]$. Hence $S[x]$ is almost atomic.

(b) $\Rightarrow$ (c): First, observe that $\mathscr{A}(S[x]) \setminus \{x\} \subseteq \mathscr{A}(S[x^{\pm 1}])$. In fact, assume towards a contradiction that $f = gh$, where $f \in \mathscr{A}(S[x]) \setminus \{x\}$ and $g, h \in$

$S[x^{\pm 1}] \setminus S[x^{\pm 1}]^\times$. Then $f = x^n g'(x) \cdot x^m h'(x)$ for some $m, n \in \mathbb{Z}$ and $g', h' \in S[x]$ such that $\operatorname{ord} g' = \operatorname{ord} h' = 0$. It is not hard to see that $n + m = 0$, which implies that either g' or h' is a unit of $S[x]$. This, in turn, implies that either g or h is a unit of $S[x^{\pm 1}]$, a contradiction. Now let f be a nonzero nonunit element of $S[x^{\pm 1}]$, and write $f = x^k g$ for some $k \in \mathbb{Z}$ and $g \in S[x]$ such that $\operatorname{ord} g = 0$. Since $S[x]$ is almost atomic, there exist $a_1, \dots, a_t \in \mathscr{A}(S[x])$ such that $a_1 \cdots a_t g$ factors into atoms in $S[x]$. Observe that $\langle \mathscr{A}(S[x]) \rangle \subseteq \langle \mathscr{A}(S[x^{\pm 1}]) \rangle$; this is because $\mathscr{A}(S[x]) \setminus \{x\} \subseteq \mathscr{A}(S[x^{\pm 1}])$ and $\{x\} \cup S[x]^\times \subseteq S[x^{\pm 1}]^\times$. Since $x^k \in S[x^{\pm 1}]^\times$ for every $k \in \mathbb{Z}$, our result follows.

(c) $\Rightarrow$ (b): Let f be a nonzero nonunit element of $S[x]$. Given that $x \in \mathscr{A}(S[x])$, we can assume, without loss of generality, that $\operatorname{ord} f = 0$. Because $S[x^{\pm 1}]$ is almost atomic, there exists g, which is a product of atoms in $S[x^{\pm 1}]$ such that gf is also a product of atoms in $S[x^{\pm 1}]$. As $x^k \in S[x^{\pm 1}]^\times$ for every $k \in \mathbb{Z}$, we may assume $\operatorname{ord} g = 0$ (consequently, $\operatorname{ord} gf = 0$ because S contains no zero-divisors). If $g \in S[x^{\pm 1}]^\times$, then $g \in S^\times$, which trivially implies that g factors into atoms in $S[x]$. Otherwise, we can write $g = g_1 \cdots g_n$, where $g_1, \dots, g_n \in \mathscr{A}(S[x^{\pm 1}])$. Again, without loss of generality, we can assume that $\operatorname{ord} g_i = 0$ for every $i \in [\![1, n]\!]$ which, in turn, implies that $g_i \in \mathscr{A}(S[x])$ for all $i \in [\![1, n]\!]$. Hence g factors into atoms in $S[x]$. By the same argument, gf factors into atoms in $S[x]$. Therefore $S[x]$ is almost atomic.

□

In general, we do not know whether the polynomial extension of an almost atomic semidomain is almost atomic, so we pose the following question.

Question 4.3 Is there an almost atomic semidomain S such that $S[x]$ is not almost atomic?

We conclude this section by providing an example of an antimatter semidomain S whose polynomial extension $S[x]$ is almost atomic.

Example 4.4 Consider the positive semiring $S = \{0\} \cup \mathbb{Q}_{\geq 1}$, which is antimatter (see [3, Example 3.10]). We shall prove that $S[x]$ is almost atomic. Take an arbitrary nonzero nonunit element $f \in S[x]$, and observe that we can write $f = cg$, where $c \in \mathbb{Q}_{\geq 1}$ and $g = c_n x^n + \cdots + c_1 x + c_0$ with $c_j = 1$ for some $j \in [\![0, n]\!]$. Then our problem reduces to show that every element of $\mathbb{Q}_{\geq 1}$ and every polynomial $g = c_n x^n + \cdots + c_1 x + c_0$ with $c_j = 1$ for some $j \in [\![0, n]\!]$ can be expressed as a quotient of a (finite) product of atoms of $S[x]$. Let us start with the latter case: write $g = f_1 \cdots f_m$ as a product of indecomposable polynomials $f_1, \dots, f_m \in S[x]$. Note that every $c \in \mathbb{Q}_{\geq 1}$ dividing all coefficients $c_0, c_1, \dots, c_n$ is necessarily a unit of S, which means that f_i is an atom of $S[x]$ for every $i \in [\![1, m]\!]$. To tackle the first case, observe that every $c \in \mathbb{Q}_{\geq 1}$ can be written as

$$c = \frac{(cx + 1)(x + c)}{x^2 + \left(c + \frac{1}{c}\right)x + 1}, \tag{4.3}$$

where each of the polynomials in (4.3) factors into atoms in $S[x]$ by the previous argument. Thus $S[x]$ is almost atomic.

5 Quasi-atomicity

In this section, we provide an ideal-theoretical characterization of quasi-atomic semidomains, and then we study when quasi-atomicity ascends from a semidomain to its semidomain of (Laurent) polynomials.

The fact that almost atomic semidomains are quasi-atomic follows immediately from the corresponding definitions. Next we construct a semidomain that is quasi-atomic but not almost atomic, and our construction is based on that given in [25, Example 7] for integral domains. First, we need to introduce the notion of a semifield. A *semifield* is a semiring in which every nonzero element has a multiplicative inverse.

Example 5.1 Now let S be a BFS that is not a semifield (for instance, $\mathbb{N}_0$). Let K be a field properly containing the field of fractions of $\mathscr{G}(S)$, and then consider the semidomain $R = S[x] + x^2K[x] = S + Sx + x^2K[x]$. Take an arbitrary $f = c_0 + c_1x + \cdots + c_nx^n \in R^*$ with $n \in \mathbb{N}_0$, and suppose that $\operatorname{ord} f = m$ for some $m \in \mathbb{N}_0$.

We shall prove that f factors into atoms in R if and only if $c_m \in S$. Assume that $c_m \notin S$, and write $f = g_1 \cdots g_\ell$ with $g_1, \dots, g_\ell \in R^*$. As $c_m \notin S$, we see that for some $j \in [\![1, \ell]\!]$ the coefficient corresponding to the term $x^{\operatorname{ord} g_j}$ in g_j is not an element of S. This implies that $\operatorname{ord} g_j \geq 2$. Thus, $g_j \notin S$ and every element of S^* divides g_j in R. Observe that $R^\times = S^\times$. Since S is not a semifield, some nonunit of S divides g_j in R; in other words, f cannot factor into atoms in R. To argue the reverse implication, assume that $c_m \in S$, and then write $f = g_1 \cdots g_\ell$, where $g_i \notin R^\times$ for any $i \in [\![1, \ell]\!]$. Since S is a BFS, we see that if $m = 0$ (resp., $m = 1$), then the inequality $\ell \leq n + \max \mathsf{L}(c_0)$ (resp., $\ell \leq n + \max \mathsf{L}(c_1)$) holds: indeed, for each $i \in [\![1, \ell]\!]$, either $\deg g_i \geq 1$ or g_i is a divisor of c_0 (resp., c_1) in S that is not a unit. Consequently, if $m \in \{0, 1\}$, then f factors into atoms in R. On the other hand, if $m \geq 2$, then $f = x^{m-1}c_m g$ with $g := x + (c_{m+1}/c_m)x^2 + \cdots + (c_n/c_m)x^{n-m+1}$, and the reverse implication follows from the fact that x is an atom of R and g is the product of finitely many atoms.

Observe now that if $c_m \notin S$, then $(x^2/c_m) \cdot f$ factors into atoms in R. This, along with the argument in the previous paragraph, ensures that R is quasi-atomic. On the other hand, if c_m is not in the field of fractions of $\mathscr{G}(S)$, then for any $a_1, \dots, a_k \in \mathscr{A}(R)$ the element $h := a_1 \cdots a_k f$ does not factor into atoms in R as the constant coefficient of $x^{-\operatorname{ord} h}h$ does not belong to S. Consequently, the semidomain R is not almost atomic.

We proceed to characterize quasi-atomic semidomains. To do so, we mimic the proof of [25, Theorem 8].

Theorem 5.2 *A semidomain S is quasi-atomic if and only if every nonzero prime ideal of S contains an irreducible element.*

Proof For the direct implication, suppose that S is quasi-atomic. Let P be a nonzero prime ideal of S. Take a nonzero $x \in P$ (clearly, $x \notin S^\times$). Since S is quasi-atomic, there exist $b \in S^*$ and $a_1, \ldots, a_n \in \mathscr{A}(S)$ such that $a_1 \cdots a_n = bx \in P$. Because P is a prime ideal, $a_i \in P$ for some $i \in [\![1, n]\!]$. Thus, each nonzero prime ideal of S contains an irreducible element. For the reverse implication, assume towards a contradiction that there exists $x \in S^*$ such that none of the elements in Sx factors into irreducibles. Let A be the subset of S consisting of all elements that can be factored into irreducibles. Let P be an ideal of S that is maximal among those ideals disjoint from A. By virtue of [18, Proposition 7.12], the ideal P is prime, and it is clear that P contains no irreducible elements, which concludes the proof. □

The following result is a version of Theorem 4.2 in the context of quasi-atomicity.

Theorem 5.3 *For a semidomain S, each of the following statements implies the next.*

(a) *S is quasi-atomic and* $\text{mcd}(s_1, \ldots, s_n) \neq \emptyset$ *for any coefficients $s_1, \ldots, s_n$ of an indecomposable polynomial in $S[x]$.*
(b) *$S[x]$ is quasi-atomic.*
(c) *$S[x^{\pm 1}]$ is quasi-atomic.*

Moreover, conditions (b) *and* (c) *are equivalent.*

Proof

(a) ⇒ (b): Let f be a nonzero nonunit of $S[x]$ such that $\deg f = n$ for some $n \in \mathbb{N}_0$. If $n = 0$, then our result follows from the fact that S is quasi-atomic and $\langle \mathscr{A}(S) \rangle \subseteq \langle \mathscr{A}(S[x]) \rangle$. Consequently, we may assume that $n \geq 1$. Write $f = f_1 \cdots f_m$, where $f_i \in S[x]$ and $\deg f_i \geq 1$ for every $i \in [\![1, m]\!]$. Without loss of generality, assume that m has been taken as large as it can possibly be. Fix an arbitrary $j \in [\![1, m]\!]$. It follows from the maximality of m that the polynomial f_j is indecomposable. Now write $f_j = s_1 x^{n_1} + \cdots + s_k x^{n_k}$ with coefficients $s_1, \ldots, s_k \in S^*$ and exponents $n_1, \ldots, n_k \in \mathbb{N}_0$. Take $s \in \text{mcd}(s_1, \ldots, s_k)$, and note that $s^{-1} f_j \in \mathscr{A}(S[x])$. Since S is quasi-atomic, there exists $b_j \in S^*$ such that $b_j s$ factors into atoms in S. Now set $b := b_1 \cdots b_m$. It is clear that $b \in S[x]^*$ satisfies that bf factors into atoms in $S[x]$. Therefore $S[x]$ is quasi-atomic.

(b) ⇒ (c): We have already established that $\mathscr{A}(S[x]) \setminus \{x\} \subseteq \mathscr{A}(S[x^{\pm 1}])$. Now let f be a nonzero nonunit element of $S[x^{\pm 1}]$, which can be written as $f = x^k g$ for some $k \in \mathbb{Z}$ and $g \in S[x]$ such that $\text{ord}\, g = 0$. Since $S[x]$ is quasi-atomic, there exists $h \in S[x]^*$ such that hg factors into atoms in $S[x]$. From $\mathscr{A}(S[x]) \setminus \{x\} \subseteq \mathscr{A}(S[x^{\pm 1}])$ and $\{x\} \cup S[x]^\times \subseteq S[x^{\pm 1}]^\times$, we infer that $\langle \mathscr{A}(S[x]) \rangle \subseteq \langle \mathscr{A}(S[x^{\pm 1}]) \rangle$. Therefore the desired result follows from the fact that $x^k \in S[x^{\pm 1}]^\times$ for every $k \in \mathbb{Z}$.

(c) ⇒ (b): Let f be a nonzero nonunit element of $S[x]$. Given that $x \in \mathscr{A}(S[x])$, we can assume, without loss of generality, that $\text{ord}\, f = 0$. Since $S[x^{\pm 1}]$ is quasi-atomic, there exists $g \in S[x^{\pm 1}]^*$ such that gf factors into atoms in $S[x^{\pm 1}]$. As $x^k \in S[x^{\pm 1}]^\times$ for every $k \in \mathbb{Z}$, we may assume $\operatorname{ord} g = 0$, which, in turn, implies that $\operatorname{ord} gf = 0$. If $gf \in S[x^{\pm 1}]^\times$, then $gf \in S^\times$, and so gf is trivially a product of atoms in $S[x]$. Otherwise, we can write $gf = g_1 \cdots g_n$ for some $g_1, \dots, g_n \in \mathscr{A}(S[x^{\pm 1}])$. Once again, we can assume, without loss of generality, that $\operatorname{ord} g_i = 0$ for every $i \in [\![1, n]\!]$ which, in turn, implies that $g_i \in \mathscr{A}(S[x])$ for all $i \in [\![1, n]\!]$. Hence gf factors into atoms in $S[x]$. We can, therefore, conclude that $S[x]$ is quasi-atomic.

□

As a corollary of Theorem 5.3, we obtain that, in a GCD-semidomain S, quasi-atomicity ascends from S to its semidomain of (Laurent) polynomials. The following result sheds some light upon this observation.

Proposition 5.4 *Let M be a monoid. Then M is a UFM if and only if M is a quasi-atomic GCD-monoid.*

Proof The direct implication clearly holds. As for the reverse implication, it is well known that an atomic GCD-monoid is a UFM (see, e.g., [23, Section 10.7]). Thus, it suffices to show that M is atomic. Let x be a nonunit element of M. Since M is quasi-atomic, there exists $b \in M^\bullet$ such that bx factors into atoms in M. Write $bx = p_1 \cdots p_n$ for some $p_1, \dots, p_n \in \mathscr{A}(M)$. It follows from [17, Theorem 6.7(2)] that $p_1, \dots, p_n$ are primes. Thus, for each $i \in [\![1, n]\!]$, either $p_i \mid_M b$ or $p_i \mid_M x$. Therefore x must be the product of some of the factors $p_1, \dots, p_n$. Hence M is atomic. □

Corollary 5.5 *Let S be a semidomain. Then S is a UFS if and only if it is quasi-atomic and GCD.*

To ensure that a GCD-monoid is a UFM, some sort of subatomic property needs to be assumed as the following example illustrates.

Example 5.6 Let $\alpha = \frac{\sqrt{5}-1}{2}$, and consider the additive monoid $\mathbb{N}_0[\alpha]$. Observe that α is an algebraic number with minimal polynomial $m_\alpha(X) = X^2 + X - 1$. Since $1 \notin \mathscr{A}_+(\mathbb{N}_0[\alpha])$, the monoid $\mathbb{N}_0[\alpha]$ is antimatter by [14, Theorem 4.1]. Next we show that $\mathbb{N}_0[\alpha]$ is a GCD-monoid. We start by proving that $\gcd(m\alpha^n, k\alpha^{n+1}) = \min(m\alpha^n, k\alpha^{n+1})$ for all $k, m \in \mathbb{N}$ and $n \in \mathbb{N}_0$. Let

$$S = \left\{(m, k) \in \mathbb{N} \times \mathbb{N} \mid \gcd(m\alpha^n, k\alpha^{n+1}) \neq \min(m\alpha^n, k\alpha^{n+1}) \text{ for some } n \in \mathbb{N}_0\right\}.$$

By way of contradiction, assume that S is nonempty. Let $(m', k') \in S$ such that $m' + k'$ is minimal. Clearly, there exists $n' \in \mathbb{N}_0$ such that $\gcd(m'\alpha^{n'}, k'\alpha^{n'+1}) \neq \min(m'\alpha^{n'}, k'\alpha^{n'+1})$. Observe that $k' > m' > 0$ since the equality $\alpha^{n'} = \alpha^{n'+1} + \alpha^{n'+2}$ holds. Thus,

$$\gcd((k' - m')\alpha^{n'+1}, m'\alpha^{n'+2}) = \min((k' - m')\alpha^{n'+1}, m'\alpha^{n'+2}),$$

which in turn implies that $m'\alpha^{n'+1} + \min((k' - m')\alpha^{n'+1}, m'\alpha^{n'+2})$ is a common divisor of $m'\alpha^{n'}$ and $k'\alpha^{n'+1}$ in $\mathbb{N}_0[\alpha]$. Now if the inequality $(k' - m')\alpha^{n'+1} < m'\alpha^{n'+2}$ holds, then $\gcd(m'\alpha^{n'}, k'\alpha^{n'+1}) = k'\alpha^{n'+1}$, which is a contradiction. We obtain a similar contradiction if $(k' - m')\alpha^{n'+1} \geq m'\alpha^{n'+2}$. Consequently, S is an empty set. Let x, y be nonzero elements of $\mathbb{N}_0[\alpha]$. Since $\alpha^n = \alpha^{n+1} + \alpha^{n+2}$ for all $n \in \mathbb{N}_0$, it is not hard to see that there exist $m \in \mathbb{N}$ and $c_1, c_2, c_3, c_4 \in \mathbb{N}_0$ such that $x = c_1\alpha^m + c_2\alpha^{m+1}$ and $y = c_3\alpha^m + c_4\alpha^{m+1}$. We may assume that $c_1 \geq c_3$. If $c_2 \geq c_4$ then it follows readily that $\gcd(x, y) = \min(x, y)$. On the other hand, if $c_2 < c_4$ then $c_3\alpha^m + c_2\alpha^{m+1} + \gcd((c_1 - c_3)\alpha^m, (c_4 - c_2)\alpha^{m+1})$ is a common divisor of x and y in $\mathbb{N}_0[\alpha]$. Since $\gcd((c_1 - c_3)\alpha^m, (c_4 - c_2)\alpha^{m+1}) = \min((c_1 - c_3)\alpha^m, (c_4 - c_2)\alpha^{m+1})$, a simple computation shows that $\gcd(x, y) = \min(x, y)$. By [17, Corollary 6.3], $\mathbb{N}_0[\alpha]$ is a GCD-monoid.

Acknowledgments The authors would like to thank an anonymous referee for several comments and suggestions that helped improve an early version of this paper. During the preparation of this paper, the first author was supported by the NSF award DMS-1903069 and the second author by the University of Florida Mathematics Department Fellowship.

References

1. S. Albizu-Campos, J. Bringas, and H. Polo: *On the atomic structure of exponential Puiseux monoids and semirings*, Comm. Algebra **49** (2021) 850–863.
2. D. D. Anderson, D. F. Anderson, and M. Zafrullah: *Factorizations in integral domains*, J. Pure Appl. Algebra **69** (1990) 1–19.
3. N. R. Baeth, S. T. Chapman, and F. Gotti: *Bi-atomic classes of positive semirings*, Semigroup Forum **103** (2021) 1–23.
4. N. R. Baeth and F. Gotti: *Factorizations in upper triangular matrices over information semialgebras*, J. Algebra **562** (2020) 466–496.
5. A. Baker: *Transcendental Number Theory* (second ed.), Cambridge Mathematical Library, Cambridge University Press, 1990.
6. J. G. Boynton and J. Coykendall: *On the graph divisibility of an integral domain*, Canad. Math. Bull. **58** (2015) 449–458.
7. H. Brunotte: *On some classes of polynomials with nonnegative coefficients and a given factor*, Period. Math. Hungar. **67** (2013) 15–32.
8. F. Campanini and A. Facchini: *Factorizations of polynomials with integral non-negative coefficients*, Semigroup Forum **99** (2019) 317–332.
9. P. Cesarz, S. T. Chapman, S. McAdam, and G. J. Schaeffer: *Elastic properties of some semirings defined by positive systems.* In: Commutative Algebra and Its Applications (Eds. M. Fontana, S. E. Kabbaj, B. Olberding, and I. Swanson), pp. 89–101, Proceedings of the Fifth International Fez Conference on Commutative Algebra and its Applications, Walter de Gruyter, Berlin, 2009.
10. S. T. Chapman, F. Gotti, and M. Gotti: *Factorization invariants of Puiseux monoids generated by geometric sequences*, Comm. Algebra **48** (2020) 380–396.
11. P. L. Clark: *The Euclidean Criterion for irreducibles*, Amer. Math. Monthly **124** (2017) 198–216.
12. P. M. Cohn: *Bezout rings and and their subrings*, Proc. Cambridge Philos. Soc. **64** (1968) 251–264.

13. J. Coykendall, D. E. Dobbs, and B. Mullins: *On integral domains with no atoms*, Comm. Algebra **27** (1999) 5813–5831.
14. J. Correa-Morris and F. Gotti: *On the additive structure of algebraic valuations of polynomial semirings*, J. Pure Appl. Algebra **226** (2022) 107104.
15. A. Geroldinger, F. Gotti, and S. Tringali: *On strongly primary monoids, with a focus on Puiseux monoids*, J. Algebra **567** (2021) 310–345.
16. A. Geroldinger and F. Halter-Koch: *Non-unique Factorizations: Algebraic, Combinatorial and Analytic Theory*, Pure and Applied Mathematics Vol. 278, Chapman & Hall/CRC, Boca Raton, 2006.
17. R. Gilmer: *Commutative Semigroup Rings*, The University of Chicago Press, 1984.
18. J. S. Golan: *Semirings and their Applications*, Kluwer Academic Publishers, 1999.
19. F. Gotti: *Increasing positive monoids of ordered fields are FF-monoids*, J. Algebra **518** (2019) 40–56.
20. F. Gotti and M. Gotti: *Atomicity and boundedness of monotone Puiseux monoids*, Semigroup Forum **96** (2018) 536–552.
21. F. Gotti and B. Li: *Divisibility in rings of integer-valued polynomials*, New York J. Math **28** (2022) 117–139.
22. F. Gotti and M. Zafrullah: *Integral domains and the IDF property*, J. Algebra **614** (2023) 564–591.
23. F. Halter-Koch: *Ideal Systems. An Introduction to Multiplicative Ideal Theory*, Marcel Dekker Inc., 1998.
24. N. Jiang, B. Li, and S. Zhu: *On the primality and elasticity of algebraic valuations of cyclic free semirings*. International Journal of Algebra and Computations (to appear). Preprint on arXiv: https://arxiv.org/pdf/2201.01245.pdf
25. N. Lebowitz-Lockard: *On domains with properties weaker than atomicity*, Comm. Algebra **47** (2019) 1862–1868.
26. V. Ponomarenko: *Arithmetic of semigroup semirings*, Ukr. Math. J. **67** (2015) 213–229.
27. M. Roitman: *Polynomial extensions of atomic domains*, J. Pure Appl. Algebra **87** (1993) 187–199.
28. S. Zhu: *Factorizations in evaluation monoids of Laurent semirings*, Comm. Algebra **50** (2022) 2719–2730.

Invertibility, Semistar Operations, and the Ring of Finite Fractions

Kaiser A. Grap and Jason R. Juett

Mathematics Subject Classification (2020) 13F05, 13A15, 13F15, 13A05

Throughout, all rings will be commutative, unital, and nontrivial.

In this article we will investigate a notion of "Dedekindness" of a ring R defined with respect to a different "quotient ring" than its total quotient ring $T(R)$. Specifically, we wish to consider "invertibility" of ideals in the *ring of finite fractions* of R, which is $Q_0(R) := \{h \in T(R[X]) \mid hJ \subseteq R$ for some semiregular ideal J of $R\}$ (cf. [60, p. 1257]). (Here we call an ideal *regular* if it has a principal faithful subideal and *semiregular* if it has a finitely generated faithful subideal [5, p. 105].) Thus $T(R) = Q_0(R)$ in rings with *Property A* (i.e., where every semiregular ideal is regular [35, p. 4]), but in general the inclusion $T(R) \subseteq Q_0(R)$ may be proper [54, p. 60]. Defining "invertibility" with respect to $Q_0(R)$ rather than $T(R)$ leads to the kind of "Dedekind" ring we consider in this article, which Elliott has called a "Q_0-Dedekind" ring [22, p. xiv]. (Precise definitions of terms in this introductory paragraph will appear below.) The main popularizer of the ring of finite fractions has been Lucas (e.g., [52–62]), who has demonstrated its applicability to a number of important problems. The following are just a few natural questions that do not directly refer to the ring of finite fractions, but whose solution at least indirectly involves this ring and/or "invertibility" with respect to it—see [5, Theorem 3.2(1); 9, Theorem 3.11; 52, Corollary 4; 53, Theorem 1.6(2); 56, Corollary 15; 57, Theorem 14; 59, Theorem 6; 60, Theorems 6.7 and 7.9] for details. When is $R[X]$ or $R(X)$ (completely) integrally closed? When does the content formula $c(fg) = c(f)c(g)$ hold for all regular $f, g \in R[X]$? When is the lattice consisting of the zero or semiregular ideals of R an r-lattice? When is the Nagata ring $R(X)$ a Krull or Prüfer (v-multiplication) ring? With this background in mind, it is unsurprising that Q_0-Dedekind rings have several natural characterizations not directly referring to rings of finite fractions, justifying their study as more than a mere mental exercise in

K. A. Grap · J. R. Juett (✉)
Department of Computer Studies and Mathematics, University of Dubuque, Dubuque, IA, USA
e-mail: KGrap@dbq.edu; jjuett@dbq.edu

J.-L. Chabert et al. (eds.), *Algebraic, Number Theoretic, and Topological Aspects of Ring Theory*, https://doi.org/10.1007/978-3-031-28847-0_14

reformulating "Dedekindness." For instance, in Theorems 6 and 9(2) below we will see that a ring R is Q_0-Dedekind if and only if every semiregular proper ideal is a product of prime ideals, if and only if every semiregular proper ideal is a unique up to order product of unfactorable ideals, if and only if R/I is a principal ideal ring (PIR) for every semiregular proper ideal I, if and only if $R(X)$ is Dedekind. Thus, although Q_0-Dedekind rings have received little explicit attention in the previous literature, we find them to be a significant class of rings worthy of investigation. In order to make our results as broadly applicable as possible, we have taken a general approach involving Q_0-semistar operations, which enabled us to simultaneously prove new results about Q_0-Dedekind rings and Lucas's Q_0-Krull rings [60, p. 1253]. Although our focus has been on the extension $R \subseteq Q_0(R)$, we have also considered general ring extensions and the traditional extension $R \subseteq T(R)$. Thus, we also have several new results about traditional notions defined with respect to $T(R)$, e.g., Dedekind rings, Krull rings, and Prüfer (v-multiplication) rings. In particular, in Lemmas 7 and 8 we have answered a few (previously) open questions posed by Elliott concerning t-linked overrings [22, pp. 281 and 284]. Along the way to our main results, we have developed/refined several tools (e.g., Lemmas 2 and 4) for working with (Q_0-)semistar operations, which we expect will prove useful to many researchers even if they have no immediate interest in "Dedekindness." All results we provide proofs for are new (to the best of our knowledge), but we have taken care to document which past results we have built on, either with an explicit explanation or (more often) a simple "cf." citation to compare our result with the strongest/most general previously known special case.

Now that we have explained our motivations for investigating these topics, let us establish some general conventions and terminology and make the definitions of "Q_0-Dedekind" and "Q_0-Krull" precise. Let R be a ring. We respectively use $\mathcal{I}(R)$, $\mathcal{I}_r(R)$, $\mathcal{I}_{sr}(R)$, $\mathrm{Spec}(R)$, and $\mathrm{Max}(R)$ to denote the sets of ideals, regular ideals, semiregular ideals, prime ideals, and maximal ideals of R. By a *multiplicative* subset of R, we mean a submonoid of $(R, \cdot)$ that does not contain 0. For an R-module C and $A, B \subseteq C$, we set $\mathrm{Mod}_R(C) := \{R\text{-submodules of } C\}$ and $(A :_R B) := \{x \in R \mid xB \subseteq A\}$. The *zero divisors* of R are the elements of $\mathcal{Z}(R) := \bigcup_{0 \neq a \in R}((0) :_R Ra)$ and the *regular* elements are those in $\mathrm{Reg}(R) := R \setminus \mathcal{Z}(R)$. We say R is *additively regular* if for each $x \in T(R)$ there is a $y \in R$ with $x + y \in \mathrm{Reg}(T(R))$ [29, p. 421]. See [31, Proposition 3.1] for a list of equivalent ways one could define additive regularity. Additively regular rings are *Marot* [35, Theorem 7.2], which means their regular ideals are generated by regular elements [35, p. 31]. Let I be an R-submodule of $Q_0(R)$. We abbreviate $I^{-1} := (R :_{T(R)} I)$ and $I^{-1_0} := (R :_{Q_0(R)} I)$. Thus $I^{-1} \subseteq I^{-1_0}$, with equality if $I \cap \mathrm{Reg}(R) \neq \emptyset$, but in general the inclusion may be proper [60, p. 1257]. Following Lucas and Elliott, we call I *invertible* if $II^{-1} = R$ and Q_0-*invertible* if $II^{-1_0} = R$ and we say R is *(Q_0-)Dedekind* if every (semi)regular ideal is (Q_0-)invertible [22, p. xiv; 54, p. 64]. Thus Dedekind and Q_0-Dedekind are equivalent for rings with Property A (e.g., integral domains, Noetherian rings, and polynomial rings [35, Corollary 2.9; 45, Theorem 82]), but in general Q_0-Dedekind is a strictly stronger property. Indeed, while a total quotient ring is vacuously Dedekind, a total quotient ring (or even a ring of finite fractions) need not be Q_0-Dedekind [54, Example 12]. For an interesting example of a Q_0-Dedekind ring without Property A, we refer the reader to [54, Example 14].

One traditionally defines "(Q_0-)Krull" rings in valuation-theoretic terms [7, p. 23; 46, p. 132; 60, p. 1253], but for our purposes it will be most useful to define a ring R to be *(Q_0-)Krull* if $(II^{-1})^{t_0} = R$ for every (semi)regular $I \in \mathcal{I}(R)$ (cf. [44, Theorem 13; 60, Theorem 4.2]), where t_0 is a certain "(Q_0-)semistar operation" that we will precisely define below. We will now summarize the necessary terminology and concepts to set up this approach. Excellent textbooks with more details about such topics include those written by Elliott [22], Halter-Koch [33], Kaiser and Knesbusch [41], and Kim and Wang [51]. One can find thorough coverage of the above "t-invertibility" and its applications in a series of papers by Zafrullah et al. [11, 12, 63, 76, 77]. Let T be a ring extending R. By a *(unital) T-semistar operation* on R, we mean a closure operation $\star : \mathrm{Mod}_R(T) \to \mathrm{Mod}_R(T)$, $\star : I \mapsto I^\star$, such that $R^\star = R$ and $(IJ)^\star = (I^\star J^\star)^\star$ for all $I, J \in \mathrm{Mod}_R(T)$ [22, Definition 3.6.16]. Note that the restriction of $\star$ to $\mathcal{I}(R)$ is a *semiprime operation* on R (i.e., an R-semistar operation [22, Definition 4.1.1(1)]), a special case of what Halter-Koch called a "weak ideal system" on $(R, \cdot)$ [33, Definition 2.1(a)]. We respectively shorten "$T(R)$-semistar operation" and "$Q_0(R)$-semistar operation" to *semistar operation* [22, Definition 2.4.2] and *Q_0-semistar operation.* Two T-semistar operations induced by $\star$ are $\star_t : I \mapsto I^{\star_t} := \bigcup\{J^\star \mid J \in \mathrm{Mod}_R(I)$ is finitely generated$\}$ and $\overline{\star} : I \mapsto I^{\overline{\star}} := \bigcup\{(I :_T J) \mid J \in \mathcal{I}(R)$ and $J^\star = R\}$ [41, Proposition 1.6.3; 22, Corollary 5.7.10]. We will introduce a third induced T-semistar operation, namely, $\star_w : I \mapsto I^{\star_w} := \bigcup\{(I :_T J) \mid J \in \mathcal{I}(R)$ is finitely generated, $((0) :_T J) = (0)$, and $J^\star = R\}$, in Lemma 1(1) below. Note that, if T is a subring of $T(R[X])$ (e.g., $T = Q_0(R)$ or $T = T(R)$), then $\star_w$ is given by $I^{\star_w} = \bigcup\{(I :_T J) \mid J \in \mathcal{I}_{sr}(R)$ is finitely generated and $J^\star = R\}$ (see Lemma 1(1) below). We also note that Elliott uses "$\star_w$" to mean "$\overline{\star_t}$" in his book [22, pp. 428–429]; we have defined the symbol as we have in order to be consistent with how Wang et al. defined the "w-operation" (see Lemma 1(3) below). Observe that $I^\star = I^{\star_t}$ for finitely generated $I \in \mathrm{Mod}_R(T)$. We will show in Lemma 1(2) below that, if T is a subring of $Q_0(R)$, then $J^{\overline{\star_t}} = J^{\star_w}$ for all semiregular $J \in \mathrm{Mod}_R(T)$. (Here we call an R-submodule of T *(semi)regular* if it contains a (semi)regular ideal of R; this is not to be confused with the concept of "T-regular" as defined in [22, Definition 2.1.21; 40, p. 84].) Our results will mostly concern the case where $\star$ is *finite type* (i.e., $\star = \star_t$), *stable* (i.e., $\star = \overline{\star}$), and/or *reduced* (i.e., $(0)^\star = (0)$) (cf. [22, Definition 5.4.8 and Corollary 5.7.10; 41, p. 157]), because these assumptions are key to $\star$-ideals possessing a theory analogous to that of classical ideal theory. The following are some useful identities that we will frequently be implicitly using: (i) $(I^\star :_T J) = (I^\star :_T J)^\star = (I^\star :_T J^\star)$ for all $I, J \in \mathrm{Mod}_R(T)$ [41, Proposition 4.4], (ii) $\left(\bigcap_{\lambda\in\Lambda} I_\lambda^\star\right)^\star = \bigcap_{\lambda\in\Lambda} I_\lambda^\star$ for $\emptyset \subsetneq \{I_\lambda\}_{\lambda\in\Lambda} \subseteq \mathrm{Mod}_R(T)$ [22, Proposition 5.2.6(2)], (iii) $\left(\bigcap_{i=1}^n I_i\right)^{\overline{\star}} = \bigcap_{i=1}^n I_i^{\overline{\star}}$ for all $I_1, \ldots, I_n \in \mathrm{Mod}_R(T)$, and (iv) $(I :_T J)^{\overline{\star}} = (I^{\overline{\star}} :_T J)$ for all $I, J \in \mathrm{Mod}_R(T)$ with J finitely generated—here (iii) and (iv) can be readily proven directly. The T-semistar operations on R are partially ordered by declaring $\star_1 \leq \star_2$ to hold if

$I^{\star_1} \subseteq I^{\star_2}$ for each $I \in \mathrm{Mod}_R(T)$ [41, p. 142]. With this partial ordering, the smallest and largest T-semistar operations on R are respectively $d : I \mapsto I^d := I$ and $v_{R,T} : I \mapsto I^{v_{R,T}} := (R :_T (R :_T I))$ [41, Subexample 3.5 and Proposition 3.6]. The $\star_t$-, $\overline{\star}$-, $\overline{\star_t}$-, and $\star_w$-operations are respectively the largest finite type, largest stable, largest finite type stable, and largest reduced finite type stable T-semistar operations on R bounded above by $\star$—see [22, pp. 428–429 and Theorem 6.6.9(1)] and Lemma 1(1) below. Therefore $t_{R,T} := (v_{R,T})_t$ and $w_{R,T} := (v_{R,T})_w = (t_{R,T})_w$ are respectively the largest finite type and largest reduced finite type stable T-semistar operations on R. We abbreviate $v := v_{R,T(R)}$, $t := t_{R,T(R)}$, $w := w_{R,T(R)}$, $v_0 := v_{R,Q_0(R)}$, $t_0 := t_{R,Q_0(R)}$, and $w_0 := w_{R,Q_0(R)}$. We note that "w" in fact usually means "w_0" in the literature (see Lemma 1(3) below), but we will keep the subscript "0" for consistency with the rest of our notational conventions. We will verify in Lemma 1(4) below that $I^v = I^{v_0}$ for each *regular fractional ideal* I of R (meaning that $I \in \mathrm{Mod}_R(T(R))$ is regular and I^{-1} is also regular [22, Definition 2.2.19]) and that $J^{t_0} = J^t$ and $J^{w_0} = J^w$ for all regular $J \in \mathrm{Mod}_R(T(R))$. We call $I \in \mathrm{Mod}_R(T)$ *$\star$-invertible* if $(I(R :_T I))^\star = R$ [41, p. 147]. If one wishes to emphasize or clarify what the extension is, one can write "T-$\star$-invertible." For instance, the above notions of "invertible" and "Q_0-invertible" are respectively the same as $T(R)$-d-invertible and $Q_0(R)$-d-invertible. Finally, if $\star$ is a (Q_0-)semistar operation on R, then we call R *(Q_0-)$\star$-Dedekind* if every (semi)regular ideal is $\star$-invertible and *(strongly) $\star$-Prüfer* if every finitely generated (semi)regular ideal is $\star$-invertible (cf. [22, Definition 2.5.5]). A *(Q_0-)Prüfer v-multiplication ring* ((Q_0-)PVMR) is a (strongly) t_0-Prüfer ring and of course a *(strongly) Prüfer ring* is a (strongly) d-Prüfer ring [5, p. 103; 22, Definitions 2.5.2(2) and 2.5.9(1); 54, Theorem 8; 60, p. 1254]. Here we have sacrificed a little bit of internal consistency in our naming schema so that it is consistent with the literature's established terminology for the cases $\star = d$ and $\star = t_0$. Additionally, we avoid calling strongly $\star$-Prüfer rings "Q_0-$\star$-Prüfer" to avoid possible confusion between strongly Prüfer rings and Lucas's strictly weaker notion of "Q_0-Prüfer" rings [54, p. 61]. We will return to this topic in more detail in Lemma 8 and the paragraph above it.

In our first lemma, we prove the aforementioned properties of the $\star_w$-operation.

Lemma 1 Let $R \subseteq T$ be an extension of rings and $\star$ be a T-semistar operation on R.

1. The map $\star_w : \mathrm{Mod}_R(T) \to \mathrm{Mod}_R(T)$, $I \mapsto I^{\star_w} := \bigcup\{(I :_T J) \mid J \in \mathcal{I}(R)$ is finitely generated, $((0) :_T J) = (0)$, and $J^\star = R\}$, is the largest reduced finite type stable T-semistar operation on R bounded above by $\star$. If T is a subring of $T(R[X])$, then $I^{\star_w} = \bigcup\{(I :_T J) \mid J \in \mathcal{I}_{sr}(R)$ is finitely generated and $J^\star = R\}$ for all $I \in \mathrm{Mod}_R(T)$.
2. If T is a subring of $Q_0(R)$, then $J^{\overline{\star_t}} = J^{\star_w}$ for all semiregular $J \in \mathrm{Mod}_R(T)$.
3. An ideal J of R is a *Glaz-Vasconcelos ideal* (meaning that J is finitely generated and the natural homomorphism $R \to \mathrm{Hom}_R(J, R)$ is an isomorphism [18, Definition 1.1]) if and only if J is a finitely generated semiregular ideal with $J^{-1_0} = R$ (or equivalently with $J^{v_0} = R$). This shows that our $w_0 := (v_0)_w$ is

the same Q_0-semistar operation as the "w-operation" defined by Wang et al. [18, Definition 3.1; 70, Lemma 3.3].
4. $I^v = I^{v_0}$ for each regular fractional ideal I of R and $J^{t_0} = J^t$ and $J^{w_0} = J^w$ for all regular $J \in \text{Mod}_R(T(R))$. In particular, every finite type Q_0-semistar operation on R restricts to a finite type semistar operation on R.

Proof

1. The last sentence of (1) follows immediately from the fact that $((0) :_{T(R[X])} J) = ((0) :_R J)T(R[X])$ for all $J \in \mathcal{I}(R)$. We have $(0)^{\star_w} = (0)$ by definition and $R^{\star_w} = (R :_T R) = R$ since $(R :_T J) = (R :_T J^\star)$ for each $J \in \text{Mod}_R(T)$. Let $A, B \in \text{Mod}_R(T)$. Note that $A^{\star_w}$, being a directed union of R-submodules of T, is indeed an R-submodule of T. If $A \subseteq B$, then $A = (A :_T R) \subseteq A^{\star_w} \subseteq B^{\star_w} \subseteq B^{\overline{\star_t}} \subseteq B^\star$. Therefore $A^{(\star_w)_w} \subseteq A^{\star_w} \subseteq (A^{\star_w})^{\star_w}$, where $A^{(\star_w)_w}$ is defined in the obvious way. To show that those two inclusions are equalities, pick $b \in (A^{\star_w})^{\star_w}$. Then $b \in (A^{\star_w} :_T J_0)$ for some finitely generated $J_0 \in \mathcal{I}(R)$ with $((0) :_T J_0) = (0)$ and $J_0^\star = R$. Write $J_0 = (c_1, \ldots, c_n)$. Then there are finitely generated $J_1, \ldots, J_n \in \mathcal{I}(R)$ with each $bc_i \in (A :_T J_i)$, $((0) :_T J_i) = (0)$, and $J_i^\star = R$. So $J := J_0 \cdots J_n \in \mathcal{I}(R)$ is finitely generated, $((0) :_T J) = (0)$, and $J^\star = R$. Thus $R \subseteq (J :_T J) \subseteq J^{\star_w} \subseteq R$ and $b \in (A :_T J) \subseteq A^{(\star_w)_w}$, as desired.

   We have now shown that $\star_w$ is a closure operation on $\text{Mod}_R(T)$, so $(AB)^{\star_w} \subseteq (A^{\star_w} B^{\star_w})^{\star_w}$. For the reverse inclusion, it suffices to show that $ab \in (AB)^{\star_w}$ for each $a \in A^{\star_w}$ and $b \in B^{\star_w}$. Then $a \in (A :_T J_1)$ and $b \in (B :_T J_2)$ for some finitely generated $J_1, J_2 \in \mathcal{I}(R)$ with each $((0) :_T J_i) = (0)$ and $J_i^\star = R$. Therefore $J_1 J_2 \in \mathcal{I}(R)$ is finitely generated, $((0) :_T J_1 J_2) = (0)$, $(J_1 J_2)^\star = R$, and $ab \in (AB :_T J_1 J_2) \subseteq (AB)^{\star_w}$, as desired. We have now demonstrated that $\star_w$ is a reduced T-semistar operation on R bounded above by $\star$. We have also shown that $\star_w = (\star_w)_w \leq \overline{(\star_w)_t} \leq \star_w$, so $\star_w$ is finite type stable.

   Finally, let $\star'$ be a reduced finite type stable T-semistar operation on R with $\star' \leq \star$. Then $A^{\star'} = \bigcup\{(A :_T J) \mid J \in \mathcal{I}(R) \text{ is finitely generated with } J^{\star'} = R\}$ since $\star'$ is finite type stable. If J is an ideal with $J^{\star'} = R$, then $(0) \subseteq ((0) :_T J) \subseteq (((0) :_T J)J)^{\star'} = (0)^{\star'} = (0)$. Therefore $A^{\star'} = A^{(\star')_w} \subseteq A^{\star_w}$, as desired.
2. Assume T is a subring of $Q_0(R)$. Pick a $J \in \text{Mod}_R(T)$ containing a finitely generated $B \in \mathcal{I}_{sr}(R)$. We have already noted that $J^{\star_w} \subseteq J^{\overline{\star_t}}$. For the reverse inclusion, let $x \in J^{\overline{\star_t}}$. Then $x \in (J :_T A)$ for some finitely generated $A \in \mathcal{I}(R)$ with $A^\star = R$. Because $T \subseteq Q_0(R)$, we also have $x \in (R :_T C)$ for some finitely generated $C \in \mathcal{I}_{sr}(R)$. Then $x \in (J :_T A + BC)$, where $A + BC \in \mathcal{I}_{sr}(R)$ is finitely generated and $(A + BC)^\star = R$, as desired.
3. In view of [50, Lemma 4.7; 51, Exercise 6.10(3)], we only need to verify the minor detail that $J^{-1_0} = R$ if and only if $J^{v_0} = R$. If $J^{-1_0} = R$, then $J^{v_0} = R^{-1_0} = R$. Conversely, if $J^{v_0} = R$, then $J^{-1_0} = (J^{v_0})^{-1_0} = R^{-1_0} = R$, as desired.
4. We have $I^{-1_0} = I^{-1}$ since I is regular, so $I^{v_0} = (I^{-1})^{-1_0} = I_v$ since I^{-1} is regular. Because t, t_0, w, and w_0 are finite type, we may assume J is finitely

generated and therefore a regular fractional ideal. Then $J^{t_0} = J^{v_0} = J^v = J^t$. The second sentence of (4) now follows from the fact that $T(R)^{t_0} = T(R)$. So $J^{w_0} \subseteq J^w$ by (1). For the reverse inclusion, let $x \in J^w$. Then $x \in (J :_{T(R)} A)$ for some finitely generated $A \in \mathcal{I}_{sr}(R)$ with $A^t = R$. Pick $r \in \mathrm{Reg}(R) \cap J$ and $s \in \mathrm{Reg}(R) \cap (x)^{-1}$. Then $A + (sr) \in \mathcal{I}_r(R)$ is finitely generated, $(A+(sr))^{t_0} = (A + (sr))^t = R$, and $x \in (J :_{Q_0(R)} A + (sr)) \subseteq J^{w_0}$, as desired.

□

Before we can prove our classifications of "Dedekindness," we will need to spend four more lemmas establishing some tools for working with (Q_0-)semistar operations. The first of these lemmas involves localization. For its statement (and the remainder of the paper), we will need the following definitions. We begin with the following variant of the "generalized transform" studied by Arnold and Brewer [14, p. 255]. Let $R \subseteq T$ be an extension of rings, $\mathcal{S}$ be a nonempty multiplicatively closed set of ideals of R, and $I \in \mathrm{Mod}_R(T)$. We will verify in Lemma 2(2) below that $R^T_{\mathcal{S}} := \bigcup_{B \in \mathcal{S}} (R :_T B)$ is a T-overring of R and $I^T_{\mathcal{S}} := \bigcup_{B \in \mathcal{S}} (I :_T B) \in \mathrm{Mod}_{R^T_{\mathcal{S}}}(T)$. (We recall that a *$T$-overring* of R is a ring between R and T, an *overring* of R is a $T(R)$-overring, and a *Q_0-overring* of R is a $Q_0(R)$-overring [14, p. 254; 41, p. 37; 54, p. 61].) For a multiplicative subset S of R, we set $I^T_{[S]} := R^T_{\{(s)\}_{s \in S}}$ and we abbreviate $I^T_{[P]} := I^T_{[R \setminus P]}$ for $P \in \mathrm{Spec}(R)$. We note that $R^{T(R)}_{[S]}$ and $R^{Q_0(R)}_{[S]}$ are respectively the *large quotient ring* and *Q_0-quotient ring* introduced by Griffin and Lucas [32, p. 56; 54, p. 62]. We call I *$\star$-closed* if $I = I^\star$; a $\star$-closed ideal of R is called a *$\star$-ideal* (cf. [22, Definitions 0.2.5 and 4.2.14(1)]). The set $\mathrm{Mod}^\star_R(R)$ of $\star$-closed R-submodules of T is a commutative monoid under the *$\star$-product* $I \star J := (IJ)^\star$ [41, p. 147]. We respectively use $\mathcal{I}^\star(R)$, $\mathcal{I}^\star_r(R)$, and $\mathcal{I}^\star_{sr}(R)$ to denote the sets of integral, regular, and semiregular $\star$-ideals of R. We call I *$\star$-finite* if $I^\star = J^\star$ for some finitely generated $J \in \mathrm{Mod}_R(T)$ (cf. [22, Definition 2.6.9(1)]). We define $\mathrm{Max}^\star(R)$ to be the set of $\star$-maximal $\star$-ideals of R, where a $\star$-ideal is *$\star$-maximal* if it is a maximal element of $\mathcal{I}^\star(R) \setminus \{R\}$ (cf. [22, Definition 2.4.14]). For a property $\mathcal{P}$ of R-modules, we say an R-module A *($\star$-)locally* satisfies $\mathcal{P}$ if A_M satisfies $\mathcal{P}$ as an R_M-module for every ($\star$-)maximal ($\star$-)ideal M. We note that $\mathrm{Max}^\star(R) \subseteq \mathrm{Spec}^\star(R) := \mathrm{Spec}(R) \cap \mathcal{I}^\star(R)$ and that, if $\star$ has finite type, every proper $\star$-ideal is contained in a $\star$-maximal $\star$-ideal [33, Theorem 6.4]. Observe that $\mathrm{Max}^\star(R) = \mathrm{Max}^{\bar{\star}}(R)$ since $\star$ and $\bar{\star}$ map the same ideals to R. Also note that, if $J \in \mathcal{I}(R)$, $P \in \mathrm{Spec}(R)$, and $P^\star \neq R = J^\star$, then $(P :_T J) = (P :_R J) = P$. It follows that $\mathrm{Spec}^{\bar{\star}}(R) = \{P \in \mathrm{Spec}(R) \mid P^\star \neq R\}$. We will constantly be implicitly using the facts developed in this paragraph.

Lemma 2 Let $R \subseteq T$ be an extension of rings, $\star$ be a finite type stable T-semistar operation on R, $\mathcal{S}$ be a nonempty multiplicatively closed set of ideals of R, and $I, J \in \mathrm{Mod}_R(T)$.

1. $I^\star \subseteq J^\star$ if and only if $I_M \subseteq J_M$ for each ($\star$-maximal) $M \in \mathrm{Spec}^\star(R)$ (cf. [51, Theorem 6.2.17]).

2. $R_{\mathcal{S}}^T := \bigcup_{B\in\mathcal{S}}(R :_T B)$ is a T-overring of R, $I \subseteq I_{\mathcal{S}}^T := \bigcup_{B\in\mathcal{S}}(I :_T B) \in \mathrm{Mod}_{R_{\mathcal{S}}^T}(T)$, and $(I_{\mathcal{S}}^T)^\star = (I^\star)_{\mathcal{S}}^T$.
3. $I^\star = \bigcap_{P\in\mathrm{Spec}^\star(R)} I_{[P]}^T = \bigcap_{M\in\mathrm{Max}^\star(R)} I_{[M]}^T$ (cf. [22, Theorem 3.11.6(3)]).

Proof

1. ($\Rightarrow$): Assume $I^\star \subseteq J^\star$. Let $P \in \mathrm{Spec}^\star(R)$ and $a \in I$. Then $Ba \subseteq J$ for some $B \in \mathcal{I}(R)$ with $B^\star = R$. Thus $B \not\subseteq P$, so $(Ra)_P = (Ba)_P \subseteq J_P$, as desired.
   ($\Leftarrow$): By contrapositive. Assume $I^\star \not\subseteq J^\star$. Then $(J :_R Ra)^\star \neq R$ for some $a \in I$. Therefore there is a $\star$-maximal $\star$-ideal $M \supseteq (J :_R Ra)$ and $R_M \supsetneq (J :_R Ra)_M = (J_M :_{R_M} (Ra)_M)$, showing that $(Ra)_M \not\subseteq J_M$, as desired.
2. Of course $I \subseteq (I :_T B) \subseteq I_{\mathcal{S}}^T$ for each $B \in \mathcal{S}$. To prove the remainder of the first two assertions in (2), it suffices to show that $xa + yb \in I_{\mathcal{S}}^T$ for all $x, y \in R_{\mathcal{S}}^T$ and $a, b \in I_{\mathcal{S}}^T$. There are $B_1, B_2, B_3, B_4 \in \mathcal{S}$ with $xB_1 + yB_2 \subseteq R$ and $aB_3 + bB_4 \subseteq I$, so $B_1B_2B_3B_4 \in \mathcal{S}$ and $(xa + yb)B_1B_2B_3B_4 \subseteq (xB_1)(aB_3) + (yB_2)(bB_4) \subseteq I$, as desired. Finally, we have $(I_{\mathcal{S}}^T)^\star = \bigcup_{B\in\mathcal{S}}(I :_T B)^\star = (I^\star)_{\mathcal{S}}^T$ since finite type stable T-semistar operations distribute into directed unions and residuations.
3. Let $A := \bigcap_{M\in\mathrm{Max}^\star(R)} I_{[M]}^T$. Then $A_M \subseteq (I_{[M]}^T)_M = \bigcup_{s\in R\setminus M}(I :_T Rs)_M = (I_M :_{T_M} R_M) = I_M$ for each $M \in \mathrm{Max}^\star(R)$, so $A \subseteq A^\star \subseteq I^\star$ by (1). To show that $I^\star \subseteq \bigcap_{P\in\mathrm{Spec}^\star(R)} I_{[P]}^T$, let $x \in I^\star$ and $P \in \mathrm{Spec}^\star(R)$. Then $x \in (I :_T J)$ for some $J \in \mathcal{I}(R)$ with $J^\star = R$. Pick $s \in J \setminus P$. Then $x \in (I :_T Rs) \subseteq I_{[P]}^T$.

□

Several of our results and/or their proofs will require knowledge of a generalization of a classic construction of Nagata [68, p. 18]. We develop this notion and its basic properties in the following lemma. Further details on (special cases) of this topic can be found in [24, 25, 28, 35, 42, 43, 51]. In a couple of proofs, including that of the following lemma, we will make use of McCoy's classic theorem about zero divisors in polynomial rings. Let R be a ring and $\{X_\lambda\}_{\lambda\in\Lambda}$ be a family of algebraically independent indeterminates. McCoy's theorem asserts that a polynomial $f \in R[\{X_\lambda\}_{\lambda\in\Lambda}]$ is regular if and only if $c(f)$ is semiregular. Here $c(f)$ denotes the *content* of f, i.e., the ideal of R generated by the coefficients of f [27, p. 68].

Lemma 3 Let $R \subseteq T$ be an extension of rings, $\{X_\lambda\}_{\lambda\in\Lambda}$ be a nonempty family of algebraically independent indeterminates, and $\star$ be a reduced finite type stable T-semistar operation on R. Define $\mathcal{N}(\star) := \{f \in R[\{X_\lambda\}_{\lambda\in\Lambda}] \mid c(f)^\star = R\}$.

1. $\mathcal{N}(\star) = R[\{X_\lambda\}_{\lambda\in\Lambda}] \setminus \bigcup_{M\in\mathrm{Max}^{\star_t}(R)} MR[\{X_\lambda\}_{\lambda\in\Lambda}]$ is a multiplicative subset of $R[\{X_\lambda\}_{\lambda\in\Lambda}]$ consisting of regular elements (cf. [28, Proposition 33.1(1); 43, pp. 152–153; 51, Proposition 6.6.16(1)]). Therefore $R(\{X_\lambda\}_{\lambda\in\Lambda}, \star) := R[\{X_\lambda\}_{\lambda\in\Lambda}]_{\mathcal{N}(\star)}$, which we call the $\star$-*Nagata ring* of R in the variables $\{X_\lambda\}_{\lambda\in\Lambda}$, is an overring of $R[\{X_\lambda\}_{\lambda\in\Lambda}]$ and consequently additively regular with Property A [35, Corollary 2.6 and Theorems 2.7 and 7.5]. For future reference, the *Nagata ring* of R in the variables $\{X_\lambda\}_{\lambda\in\Lambda}$ is $R(\{X_\lambda\}_{\lambda\in\Lambda}) := R(\{X_\lambda\}_{\lambda\in\Lambda}, d)$ [68, p. 18] and we abbreviate $R(X, \star) := R(\{X\}, \star)$ and $R(X) := R(\{X\})$.

2. $\mathrm{Max}(R(\{X_\lambda\}_{\lambda\in\Lambda}, \star)) = \{MR(\{X_\lambda\}_{\lambda\in\Lambda}, \star)\}_{M\in\mathrm{Max}^\star(R)}$ (cf. [28, Proposition 33.1(3); 43, Proposition 2.1(1); 24, p. 4787]).
3. $R(\{X_\lambda\}_{\lambda\in\Lambda}, \star)_{PR(\{X_\lambda\}_{\lambda\in\Lambda}, \star)} = R_P(\{X_\lambda\}_{\lambda\in\Lambda})$ for each $P \in \mathrm{Spec}^\star(R)$ (cf. [35, pp. 86–87; 51, Proposition 6.6.17]).
4. $IR(\{X_\lambda\}_{\lambda\in\Lambda}, \star) = I^\star R(\{X_\lambda\}_{\lambda\in\Lambda}, \star)$ and $T \cap IR(\{X_\lambda\}_{\lambda\in\Lambda}, \star) = I^\star$ for each $I \in \mathrm{Mod}_R(T)$ (cf. [17, Lemma 2.3(2); 51, Theorem 6.6.19(1)]).

Proof Let $A := R[\{X_\lambda\}_{\lambda\in\Lambda}]$ and $B := R(\{X_\lambda\}_{\lambda\in\Lambda}, \star)$.

1. The equality follows immediately from the fact that every proper $\star$-ideal of R is contained in a $\star$-maximal $\star$-ideal. Therefore $\mathcal{N}(\star)$, being the complement of a union of a nonempty set of prime ideals of A, is multiplicative. If $f \in \mathcal{N}(\star)$ and $a \in ((0) :_R c(f))$, then $(a) \subseteq (ac(f))^\star = (0)^\star = (0)$. Therefore $\mathcal{N}(\star) \subseteq \mathrm{Reg}(A)$ by McCoy's theorem.
2. It suffices to show that each $J \in \mathcal{I}(A)$ with $J \cap \mathcal{N}(\star) = \emptyset$ is contained in MA for some $M \in \mathrm{Max}^\star(R)$. Note that $c(J) := \sum_{f\in J} c(f)$ is an ideal of R [51, Definition 1.7.11(2)] and $J \subseteq c(J)A$. Thus it will suffice to show that $c(J)^\star \neq R$. Suppose not. Then $c(I)^\star = R$ for some finitely generated $I \in \mathrm{Mod}_A(J)$ since $\star$ has finite type. Because I is finitely generated, we have $c(I) = c(f)$ for some $f \in I$ [51, Proposition 1.7.15]. So $f \in J \cap \mathcal{N}(\star)$, a contradiction.
3. We have $B_{PB} = (A_{\mathcal{N}(\star)})_{PA_{\mathcal{N}(\star)}} = A_{PA} = (A_{R\setminus P})_{PA_{R\setminus P}} = R_P(\{X_\lambda\}_{\lambda\in\Lambda})$, where the second and third equalities hold [28, Exercise 5.11] since $\mathcal{N}(\star) \cup (R \setminus P) \subseteq A \setminus PA$.
4. Combining (3) and Lemma 2(1), we see that $IB_{MB} = I_M R_M(\{X_\lambda\}_{\lambda\in\Lambda}) = I^\star B_{MB}$ for each $M \in \mathrm{Max}^\star(R)$. Part (2) now implies $IB = I^\star B$ holds locally and therefore globally. Finally, we compute $T \cap IB = \{a \in T \mid af \in IA$ for some $f \in \mathcal{N}(\star)\} = I^\star$, where the last equality holds since $\star$ is stable.

□

We will be implicitly using the facts discussed in Lemma 3 throughout the sequel. Our next lemma will provide several useful facts related to "invertibility."

Lemma 4 Let R be a ring, $T \in \{T(R), Q_0(R)\}$, $\star$ be a finite type T-semistar operation on R, and $I \in \mathrm{Mod}_R^\star(T)$.

1. If I is semiregular and t_0-finite, then $(I^{-1_0})_S = I_S^{-1_0}$ and $(I^{v_0})_S^{v_0} = I_S^{v_0}$ for each multiplicative $S \subseteq R$ (cf. [22, Proposition 3.3.2; 75, Lemma 4]).
2. If I is $\star_w$-locally flat, then $I = I^{w_0}$ (cf. [51, Theorem 6.7.24; 71, 1.1]). If I is semiregular and $\star_w$-locally flat, then $I = I^{t_0}$ (cf. [48, Lemma 1; 69, Theorem 1.4]).
3. An R-submodule of T is $\star$-invertible if and only if it is $\bar{\star}$-invertible (cf. [22, Lemma 2.5.15]). So a semiregular R-submodule of T is $\star$-invertible if and only if it is $\star_w$-invertible. If $T = Q_0(R)$ (resp., $T = T(R)$), then I is $\star$-invertible if and only if it is semiregular (resp., regular), $\star$-locally cyclic, and $\star$-finite (or equivalently t_0-finite) (cf. [33, Theorem 12.3; 50, Theorem 4.15; 63, Corollary 1.6; 54, p. 64]).

Proof

1. Let $S \subseteq R$ be multiplicative. Assume I is semiregular and t_0-finite. So $I^{t_0} = J^{t_0}$ for some finitely generated semiregular $J \in \mathrm{Mod}_R(T)$. Write $J = \sum_{i=1}^{k} Ra_i$. Because t_0 has finite type, each $a_i \in (I_i)^{t_0}$ for some finitely generated semiregular $I_i \in \mathrm{Mod}_R(I)$. By passing from J to $\sum_{i=1}^{k} I_i$, we may assume $J \subseteq I$. Because J is finitely generated and semiregular, we have $(J^{-1_0})_S = (R :_{T(R[X])} J)_S = (R_S :_{T(R[X])_S} J_S) \subseteq (R_S :_{T(R_S[X])} J_S) = J_S^{-1_0}$. For the reverse inclusion, pick $h \in J_S^{-1_0}$. Write $J = \sum_{i=1}^{n} Rf_i g_i^{-1}$ with $f_i \in R[X]$ and $g_i \in \mathrm{Reg}(R[X])$. For each i we have $(f_i/1)(g_i/1)^{-1}h = b_i/s_i$ for some $b_i \in R$ and $s_i \in S$, where for $f \in R[X]$ we use $f/1$ to denote the image of f in $R_S[X]$. Choose $m_1, \ldots, m_n \in \mathbb{Z}_0^+$ so that $g := \sum_{i=1}^{n} f_i X^{m_i} \prod_{j \neq i}^{n} g_j$ has content $\sum_{i=1}^{n} c(f_i \prod_{j \neq i} g_j)$, which is faithful since $J \prod_{i=1}^{n} g_i$ is faithful. So $g \in \mathrm{Reg}(R[X])$ by McCoy's theorem. We have $(sg/1)h = f/1$, where $s := \prod_{i=1}^{n} s_i$ and $f := g_1 \cdots g_n \sum_{i=1}^{n} b_i X^{m_i} \prod_{j \neq i} s_j$. So $h = (fg^{-1})/s \in (R_S :_{T(R[X])_S} J_S) = (J^{-1_0})_S$. Therefore $(J^{-1_0})_S = J_S^{-1_0}$. Thus $I_S^{v_0} \supseteq J_S^{v_0} = (J^{-1_0})_S^{-1_0} \supseteq (J^{v_0})_S^{v_0} = (I^{v_0})_S^{v_0} \supseteq I_S^{v_0}$ and $(I^{-1_0})_S = (J^{-1_0})_S = J_S^{-1_0} = (J_S^{v_0})^{-1_0} = (I_S^{v_0})^{-1_0} = I_S^{-1_0}$, as desired.
2. Let $B := R(X, \star_w)$. Assume I is $\star_w$-locally flat. Then $(IB)_{MB} = I_M R_M(X)$ is a flat module over $B_{MB} = R_M(X)$ for each $M \in \mathrm{Max}^{\star_w}(R)$ [65, Theorem 7.7 and Exercise 7.1] since $R_M(X)$ is a faithfully flat extension of R_M [5, p. 97]. Thus IB is B-flat [65, Theorem 7.1]. We need to show $I \supseteq (I :_{Q_0(R)} J)$ for each finitely generated $J \in \mathcal{I}_{sr}(R)$ with $J^{v_0} = R$. Since B has Property A, there is an $r \in \mathrm{Reg}(B) \cap JB$. Then $r(I :_{Q_0(R)} J)B \subseteq (rIB :_{T(B)} JB) \cap IB = (rB :_B JB)IB = (rB :_B (JB)^v)IB = (rB :_B (J[X]^v B)^v)IB = (rB :_B (J^{v_0}B)^v)IB = (rB :_B B)IB = rIB$, where the first equality holds [15, Exercise I.2.22; 22, Lemma 2.2.8] since IB is B-flat and JB is a finitely generated ideal of B, the third equality holds by (1), and the fourth equality holds [60, Lemma 5.1(b)] since $J \in \mathcal{I}_{sr}(R)$. Canceling the regular element r, we obtain $(I :_{Q_0(R)} J)B \subseteq IB$. So $(I :_{Q_0(R)} J) \subseteq Q_0(R) \cap IB = I^{\star_w} = I$, as desired.

   Now further assume I is semiregular. We need to show $I \supseteq J^{v_0}$ for each finitely generated faithful $J \in \mathrm{Mod}_R(I)$. Since B is Marot with Property A, we have $JB = \sum_{i=1}^{n} f_i B$ with each $f_i \in \mathrm{Reg}(T(B))$. Pick $g \in \mathrm{Reg}(B)$ with each $gf_i^{-1} \in \mathrm{Reg}(B)$. Then $I(JB)^{-1} = IB\left(\bigcap_{i=1}^{n} f_i^{-1}B\right) = g^{-1}IB\left(\bigcap_{i=1}^{n} gf_i^{-1}B\right) = \bigcap_{i=1}^{n} f_i^{-1}IB \supseteq B$, where the last equality holds [22, Lemma 2.2.8] since $g^{-1}IB \cong IB$ is B-flat and each $gf_i^{-1}B$ is an ideal of B. Therefore $J^{v_0}B = J[X]^v B \subseteq (JB)^v \subseteq I(JB)^{-1}(JB)^v \subseteq IB$, where the equality holds [60, Lemma 5.1(b)] since $J \in \mathcal{I}_{sr}(R)$ and the first inclusion is by (1). So $J^{v_0} \subseteq Q_0(R) \cap IB = I^{\star_w} = I$, as desired.
3. First note that $\star$-invertible and $\bar{\star}$-invertible are equivalent since $\star$ and $\bar{\star}$ map the same ideals to R. We prove (3)'s final assertion in the case where $T =$

$Q_0(R)$. The $T = T(R)$ case's proof can be obtained by simply replacing each "semiregular" with "regular."

($\Rightarrow$): Assume I is $\star$-invertible. Since $\star$ has finite type, we have $(\sum_{i=1}^{n} Rx_i y_i)^\star = R$ with each $x_i \in I$ and $y_i \in (R :_T I)$. Then $I = (\sum_{i=1}^{n} x_i(y_i I))^\star \subseteq (\sum_{i=1}^{n} Rx_i)^\star \subseteq I$. There is a semiregular $B \in \mathcal{I}(R)$ with each $y_i B \subseteq R$. So $I \supseteq (\sum_{i=1}^{n} x_i(y_i B))^\star \supseteq B$. To show that I is $\star$-locally cyclic, let $M \in \text{Max}^\star(R)$. Because R_M is quasilocal, it suffices to prove $J_M = (JI^{-1_0})_M I_M$ for each $J \in \text{Mod}_R(I)$ [4, Theorem 2.1]. This follows from Lemma 2(1) since $J^{\bar\star} = ((JI^{-1_0})I)^{\bar\star}$.

($\Leftarrow$): Assume I is semiregular, $\star$-locally cyclic, and t_0-finite. By (1), we have $(II^{-1_0})_M = I_M I_M^{-1_0} = R_M$ for each $M \in \text{Max}^\star(R)$, so $(II^{-1_0})^{\bar\star} = R$ by Lemma 2(1).

□

The last of our initial lemmas will develop some lattice-theoretic facts. We refer the reader to [1, pp. 131–134] for definitions related to multiplicative lattices.

Lemma 5 Let $R \subseteq T$ be an extension of rings, $\star$ be a reduced finite type stable T-semistar operation on R, and $\mathcal{I} \in \{\mathcal{I}_{sr}^\star(R), \mathcal{I}_r^\star(R)\}$.

1. $\mathcal{I}^\star(R)$ and $\{(0)\} \cup \mathcal{I}$, when endowed with the $\star$-product and the inclusion partial order, are compactly generated modular multiplicative lattices where the identity is compact (cf. [9, pp. 409–410; 33, p. 82]). Specifically, residuation in $\mathcal{I}^\star(R)$ and $\{(0)\} \cup \mathcal{I}$ is the usual residuation of ideals, the join in $\mathcal{I}^\star(R)$ or $\{(0)\} \cup \mathcal{I}$ is given by $\bigvee_{\lambda\in\Lambda} I_\lambda = \left(\sum_{\lambda\in\Lambda} I_\lambda\right)^\star$, the meet in $\mathcal{I}^\star(R)$ is given by $\bigwedge_{\lambda\in\Lambda} I_\lambda = \bigcap_{\lambda\in\Lambda} I_\lambda$, and the meet in $\{(0)\} \cup \mathcal{I}$ is $\bigwedge_{\lambda\in\Lambda} I_\lambda = \bigcap_{\lambda\in\Lambda} I_\lambda$ if $\bigcap_{\lambda\in\Lambda} I_\lambda \in \mathcal{I}$ and otherwise $\bigwedge_{\lambda\in\Lambda} I_\lambda = (0)$ (cf. [9, pp. 409–410; 33, p. 82]).
2. An element of $\mathcal{I}^\star(R)$ or $\{(0)\} \cup \mathcal{I}$ is compact if and only if it is $\star$-finite (cf. [9, Theorems 3.1(a) and 3.10(a); 33, Proposition 3.1(iii)]).
3. An element of $\mathcal{I}^\star(R)$ or $\{(0)\} \cup \mathcal{I}$ is principal (in the lattice-theoretic sense) if and only if it is $\star$-finite and $\star$-locally principal (cf. [33, Theorem 8.2(iv); 66, Theorem 2]). Therefore, if $T = Q_0(R)$ (resp., $T = T(R)$), then an element of $\{(0)\} \cup \mathcal{I}_{sr}^\star(R)$ (resp., $\{(0)\} \cup \mathcal{I}_r^\star(R)$) is principal if and only if it is zero or $\star$-invertible (cf. [9, Theorems 3.1(b) and 3.10(b)]). So $\mathcal{I}^\star(R)$ is an r-lattice (cf. [1, p. 134]) and, if $T = Q_0(R)$ (resp., $T = T(R)$), then $\{(0)\} \cup \mathcal{I}_{sr}^\star(R)$ (resp., $\{(0)\} \cup \mathcal{I}_r^\star(R)$) is an r-lattice if and only if every semiregular (resp., regular) $\star$-ideal has the form $(\sum_{\lambda\in\Lambda} I_\lambda)^\star$ with each I_λ a $\star$-invertible $\star$-ideal (cf. [9, Theorems 3.1(a) and 3.10(a)]).

Proof Let $\mathcal{L} \in \{\mathcal{I}^\star(R), \{(0)\} \cup \mathcal{I}\}$. We have already noted that (i) the intersection of a nonempty family of $\star$-ideals is a $\star$-ideal and (ii) $(I :_R J) \in \mathcal{L}$ for all $I, J \in \mathcal{L}$. With these two facts in mind, it follows directly from the definitions that $\mathcal{L}$ is a multiplicative lattice with the given meets, joins, and residuations. Because every element of $\mathcal{L}$ is the join of a family of $\star$-finite elements of L, it follows that every compact element of $\mathcal{L}$ is $\star$-finite. Conversely, the fact that $\star$-finite elements of $\mathcal{L}$ are compact follows from the fact that $\star$ has finite type. The fact that $\mathcal{L}$ is modular can be reduced to the known $\star = d$ special case [9, p. 409] by using the fact that stable T-semistar operations distribute over finite intersections. All that is left is the first sentence of (3). We recall that the *principal* elements of $\mathcal{L}$ are those $I \in \mathcal{L}$ for which $((A \cap (B :_R I))I)^\star = (AI)^\star \cap B$ and $((A + BI)^\star :_R I) = ((A :_R I) + B)^\star$ for all $A, B \in \mathcal{L}$ [19, p. 1]. Statements (1) and (2) imply principal elements of $\mathcal{L}$ are $\star$-finite [1, Theorem 1.3]. Thus, we can use Lemma 2(1) to see that an $I \in \mathcal{I}^\star(R)$ is a principal element of $\mathcal{I}^\star(R)$ if and only if I_M is a principal element of $\mathcal{I}(R_M)$ (or equivalently I_M is principal [66, Theorem 2]) for each $M \in \operatorname{Max}^\star(R)$. It only remains to show that a nonzero principal element I of $\{(0)\} \cup \mathcal{I}$ is $\star$-locally principal. Let M be a $\star$-maximal $\star$-ideal containing I. Because I is $\star$-finite, Lemma 2(1) implies $I_M = (a_1, \ldots, a_n)_M$ for some $a_1, \ldots, a_n \in I$. Choose so that n is as small as possible and let $J := (a_3, \ldots, a_n)$. Suppose $n \geq 2$. Because $I_M^3 = B_M I_M$, where $B := ((a_1^2 + a_2^2, a_1a_2) + J^2 + I^3)^\star \in \mathcal{I}$, we have $B_M \subseteq I_M^2 \subseteq (B_M I_M :_{R_M} I_M) = (((0) + BI)^\star :_R I)_M = ((((0) :_R I) + B)^\star)_M = B_M$. Nakayama's lemma implies $I_M^2 = (a_1^2 + a_2^2, a_1a_2)_M + J_M^2$, so $sa_1^2 = \lambda(a_1^2 + a_2^2) + b$ for some $s \in R \setminus M$, $\lambda \in R$, and $b \in (a_1a_2) + J^2$. Suppose $\lambda \in M$. Then $(a_1^2)_M = ((s - \lambda)a_1^2)_M \subseteq (a_2^2, a_1a_2)_M + J_M^2$. Thus $I_M^2 = C_M I_M$, where $C := ((a_2) + J + I^2)^\star \in \mathcal{I}$, and we argue as above to conclude $I_M = (a_2)_M + J_M$, a contradiction. Therefore $\lambda \notin M$. So $(a_2^2)_M = (-\lambda a_2^2)_M \subseteq (a_1^2, a_1a_2)_M + J_M^2$, leading to a contradiction as above. □

We are almost ready to present our first major theorem. Let $R \subseteq T$ be an extension of rings and $\star$ be a T-semistar operation on R. One of the major themes we wish to pursue, in line with earlier investigations of Juett et al. into factorization of ideals [6, 38, 39], is how one might characterize "Dedekindness" in terms of "unique $\star$-factorization." We call a $\star$-ideal I *$\star$-unfactorable* if it is an irreducible element of $(\mathcal{I}^\star(R), \star)$ in the sense of Anderson and Valdes-Leon [10, Definition 2.4], i.e., (i) $I \neq R$ and (ii) $I = (AB)^\star$ with $A, B \in \mathcal{I}^\star(R)$ implies $I \in \{A, B\}$. Thus d-unfactorable d-ideals are the same as the *unfactorable* or *weakly nonfactorable* ideals defined by Juett et al. [6, p. 1744; 39, p. 2102]. When studying factorization in non-cancellative monoids, there are multiple reasonable but generally unequivalent ways one might define "uniqueness" of factorizations. For simplicity, we will confine our attention to the two notions we find most natural. We say two $\star$-products $(I_1 \cdots I_m)^\star$ and $(J_1 \cdots J_n)^\star$ ($I_i, J_j \in \mathcal{I}^\star(R)$) are *equal up to order* if $m = n$ and each $I_i = J_i$ after a suitable reordering and *equal up to factors* if $\{I_1, \ldots, I_m\} = \{J_1, \ldots, J_n\}$ (cf. [10, Definition 4.1; 39, p. 2103]). We will now give our first characterization of "Dedekindness" and then discuss what special cases of the theorem are known afterwards.

Theorem 6 Let R be a ring, $T \in \{T(R), Q_0(R)\}$, and $\star$ be a reduced finite type stable T-semistar operation on R.

1. The following are equivalent if $T = Q_0(R)$.

  a. R is Q_0-$\star$-Dedekind.
  b. For all $I \subseteq P$ in $\mathcal{I}_{sr}^\star(R)$ (with P prime) we have $P = (I + (y))^\star$ for some $y \in P$.

c. R is integrally closed in $Q_0(R)$ and every semiregular prime $\star$-ideal of R is $\star$-maximal, $\star$-finite, and contains a $\star$-invertible $\star$-ideal.
d. Every ($\star$-finite) $I \in \mathcal{I}^\star_{sr}(R) \setminus \{R\}$ is a $\star$-product of prime $\star$-ideals.
e. Every ($\star$-finite) $I \in \mathcal{I}^\star_{sr}(R) \setminus \{R\}$ is a unique up to order (or equivalently factors) $\star$-product of $\star$-unfactorable $\star$-ideals.
f. R is strongly $\star$-Prüfer and every semiregular prime $\star$-ideal contains a $\star$-unfactorable $\star$-invertible $\star$-ideal.
g. $P \subsetneq PP^{-1_0}$ for each semiregular prime $\star$-ideal P that is either $\star$-maximal or directly below a $\star$-maximal $\star$-ideal.
h. There is a Euclidean domain D with $\mathcal{I}(D) \cong \{(0)\} \cup \mathcal{I}^\star_{sr}(R)$ as multiplicative lattices.

2. The following are equivalent if $T = T(R)$.

a. R is $\star$-Dedekind.
b. For all (prime) $P \in \mathcal{I}^\star_r(R)$ and $x \in \mathrm{Reg}(R) \cap P$ we have $P = (x, y)^\star$ for some $y \in P$.
c. R is integrally closed in $T(R)$ and every regular prime $\star$-ideal of R is $\star$-maximal and $\star$-finite.
d. Every $I \in \mathcal{I}^\star_r(R) \setminus \{R\}$ is a $\star$-product of prime $\star$-ideals.
e. $(a, b)^\star$ is a $\star$-product of prime $\star$-ideals for all $a \in \mathrm{Reg}(R)$ and $b \in R$.
f. Every ($\star$-finite) $I \in \mathcal{I}^\star_r(R)\setminus\{R\}$ is a unique up to order (or equivalently factors) $\star$-product of $\star$-unfactorable $\star$-ideals.
g. R is $\star$-Prüfer and every regular prime $\star$-ideal contains a $\star$-unfactorable $\star$-invertible $\star$-ideal.
h. $P \subsetneq PP^{-1}$ for each regular prime $\star$-ideal P that is either $\star$-maximal or directly below a $\star$-maximal $\star$-ideal.
i. There is a Euclidean domain D with $\mathcal{I}(D) \cong \{(0)\} \cup \mathcal{I}^\star_r(R)$ as multiplicative lattices.

Proof

1. Assume $T = Q_0(R)$.
(g) $\Rightarrow$ (b): Assume (g) holds. Pick $I \subseteq J \subseteq M$ in $\mathcal{I}^\star_{sr}(R)$ with M $\star$-maximal. Then $M \subsetneq (MM^{-1_0})^\star = R$ by (g). Pick $P \in \mathrm{Mod}_R(M)$ with $P_M = \bigcap_{n=1}^\infty M_M^n$. Then $P_M = M_M P_M$ is the largest prime ideal of R_M properly contained in M_M [8, Corollary 2.3] since M_M is regular and principal by Lemma 4(3). Adjust notation if necessary so that P is the largest prime ($\star$-)ideal properly contained in M.

We claim that P is not semiregular. Suppose it is. Then $P \subsetneq PP^{-1_0}$ (or equivalently $P_M \subsetneq (PP^{-1_0})_M$) by (g). Thus there is a maximum $n \in \mathbb{Z}_0^+$ with $(PP^{-1_0})_M \subseteq M_M^n$. Because M_M^n is a regular principal ideal, we have $(PP^{-1_0})_M = ((PP^{-1_0})_M :_{R_M} M_M^n)M_M^n$, where $M_M \nsubseteq ((PP^{-1_0})_M :_{R_M} M_M^n)$ (or equivalently $(PP^{-1_0})_M = M_M^n$) by the maximality of n. So $M_M^{n+1} = (MPP^{-1_0})_M = (PP^{-1_0})_M = M_M^n$, which is absurd.

It follows that $M_M = \sqrt{I_M}$. Because M_M is a regular principal ideal, there is a maximum $n \in \mathbb{Z}^+$ with $I_M \subseteq M_M^n$. In fact $I_M = M_M^n$ as in the previous paragraph. By the arbitrariness of I and M, we have shown that every semiregular $\star$-ideal is $\star$-locally principal and every semiregular prime $\star$-ideal is $\star$-maximal and $\star$-invertible (and therefore $\star$-finite by Lemma 4(3)). It follows that I is contained in only finitely many $\star$-maximal $\star$-ideals $M_1, \ldots, M_k$ [33, Proposition 6.6]. Each $J_{M_i} = (y_i)_{M_i}$ for some $y_i \in J$. Pick $x_i \in \left(\bigcap_{j\neq i} M_j\right) \setminus M_i$. Let $y := \sum_{i=1}^k x_i y_i$. Then each $J_{M_i} = (x_i y_i)_{M_i} = (y)_{M_i} + (M_i)_{M_i} J_{M_i} = (y)_{M_i}$ by Nakayama's lemma. Thus $J_M = (I + (y))_M$ holds for all $M \in \operatorname{Max}^\star(R)$, so $I = (J + (y))^\star$ by Lemma 2(1), as desired.

(b) $\Rightarrow$ (a): Assume the formally weaker version of (b) holds. We first show that each prime $P \in \mathcal{I}_{sr}^\star(R)$ is $\star$-invertible. By Lemma 4(3), we need to show P_M is cyclic for each $M \in \operatorname{Max}^\star(R)$. Pick a finitely generated semiregular $I \in \operatorname{Mod}_R(P)$. Then $P = ((I^2)^\star + (y))^\star$ for some $y \in P$, so $P_M = I_M^2 + (y)_M = (y)_M$ by Lemma 2(1) and Nakayama's lemma, as desired. By "(g) $\Rightarrow$ (b)," it follows that the formally stronger version of (b) holds. A trivial adjustment to the above argument now shows that R is Q_0-$\star$-Dedekind.

(a) $\Rightarrow$ (f): Follows from the fact that every prime $\star$-ideal is $\star$-unfactorable.

(f) $\Rightarrow$ (h): Assume (f) holds. Then Lemma 5(3) implies $\{(0)\} \cup \mathcal{I}_{sr}^\star(R)$ is an r-lattice. Because R is strongly $\star$-Prüfer, this r-lattice satisfies the *weak union condition*, i.e., for all $A, B, C \in \mathcal{I}_{sr}^\star(R)$ with $A \nsubseteq B$ and $A \nsubseteq C$, there is a $\star$-invertible $\star$-ideal $I \subseteq A$ with $I \nsubseteq B$ and $I \nsubseteq C$ [37, p. 1]. We claim that $\{(0)\} \cup \mathcal{I}_{sr}^\star(R)$ is also *distributive*, i.e., that $(A \cap B + A \cap C)^\star = A \cap (B + C)^\star$ for all $A, B, C \in \mathcal{I}_{sr}^\star(R)$ [74, p. 629]. Because $\star$ has finite type, it is enough to verify this for A, B, and C $\star$-finite. For each $M \in \operatorname{Max}^\star(R)$, Lemmas 2(1) and 4(3) imply $B_M + C_M = ((B + C)^\star)_M$ is a regular principal ideal of R_M, so B_M and C_M are comparable and therefore $(A \cap B + A \cap C)_M = A_M \cap B_M + A_M \cap C_M = A_M \cap (B_M + C_M) = (A \cap (B + C))_M$. So $(A \cap B + A \cap C)^\star = A \cap (B + C)^\star$ by Lemma 2(1), as desired.

Because $\{(0)\} \cup \mathcal{I}_{sr}^\star(R)$ is a distributive r-lattice satisfying the weak union condition, it is isomorphic to $\mathcal{I}(D)$ for some Prüfer domain D [1, Theorem 3.4]. In fact, if we carefully unravel the proof of [1, Theorem 3.4], we see that D is produced via the Krull-Kaplansky-Jaffard-Ohm theorem [36, p. 78]. Thus, due to Heinzer's observation about the proof of the latter theorem [34, pp. 1369–1370], we may take D to be a Bézout domain with stable rank 1. The isomorphism implies D is a Bézout domain where every nonzero prime ideal contains a nonzero irreducible (or equivalently prime [33, Proposition 10.5(1)]) element. So D is a factorial Bézout domain [45, Theorem 5] (or equivalently a principal ideal domain [33, Theorem 10.7]) with stable rank 1 and therefore Euclidean [23, Theorem 5.3].

(h) $\Rightarrow$ (d)+(e): Clear.

(d) $\Rightarrow$ (b): Assume the formally weaker version of (d) holds. It suffices to prove that for each $I \in \mathcal{I}_{sr}^\star(R) \setminus \{R\}$ and finitely generated faithful $J \in \operatorname{Mod}_R(I)$, we have $I = (J + (y))^\star$ for some $y \in I$.

We first claim that R_M/J_M^2 is an Artinian local PIR for each $M \in \text{Max}^\star(R)$ containing J. Nakayama's lemma implies $J_M \neq J_M^2$, so R_M/J_M^2 is a quasilocal ring with proper zero divisors. Thus it will suffice to show that $((x)_M + J_M^2)/J_M^2$ is a product of prime ideals of R_M/J_M^2 for each $x \in M$ [28, Theorem 46.11]. Because $((x) + J^2)^\star$ is a $\star$-product of prime $\star$-ideals, this follows from Lemma 2(1).

The above paragraph shows $\mathcal{I}_{sr}^\star(R) \cap \text{Spec}(R) \subseteq \text{Max}^\star(R)$. So J is a $\star$-product of $\star$-maximal $\star$-ideals, and consequently there are only finitely many $M_1, \ldots, M_n \in \text{Max}^\star(R)$ containing J. Each $R_{M_i}/J_{M_i}^2$ is a PIR, so $I_{M_i} = J_{M_i}^2 + (y_i)_{M_i}$ for some $y_i \in I$. Pick $x_i \in \left(\bigcap_{j \neq i} M_j\right) \setminus M_i$. Let $y := \sum_{i=1}^n x_i y_i$. Then each $I_{M_i} = (x_i y_i)_{M_i} + J_{M_i}^2 = (y)_{M_i} + (M_i)_{M_i} I_{M_i} = (y)_{M_i}$ by Nakayama's lemma. Thus $I_M = (J + (y))_M$ holds for all $M \in \text{Max}^\star(R)$, so $I = (J + (y))^\star$ by Lemma 2(1).

(e) $\Rightarrow$ (f): (We adapt Juett et al.'s proof of [39, Theorem 3.6].) By Lemma 4(3), we need to show the formally weakest version of (e) implies I_M is principal for each $M \in \text{Max}^\star(R)$ and $\star$-finite $I \in \mathcal{I}_{sr}^\star(R) \setminus \{R\}$. Throughout, we note that $\star$-finite semiregular proper $\star$-ideals have no redundant $\star$-factorizations. (To see this, combine Lemma 2(1) with Nakayama's lemma.) Let $I = (I_1^{m_1} \cdots I_k^{m_k})^\star$ be a $\star$-product of $\star$-unfactorable $\star$-ideals with the I_i's distinct. Because $\star$ has finite type and I is $\star$-finite, we have $I = (J_1^{m_1} \cdots J_k^{m_k})^\star$ with each $J_i \in \text{Mod}_R(I_i)$ $\star$-closed and $\star$-finite. By uniqueness up to factors, each $J_i = (I_1^{l_{i,1}} \cdots I_k^{l_{i,k}})^\star$ for some $l_{i,1}, \ldots, l_{i,k} \in \mathbb{Z}_0^+$. Because $I = (I_i^{m_i l_{i,i}} \prod_{j \neq i} I_j^{m_j + m_i l_{i,j}})^\star$ has no redundant $\star$-factorizations, we have $l_{i,j} = 0$ for $j \neq i$ and consequently $l_{i,i} = 1$. Because $I_M = (I_1)_M^{m_1} \cdots (I_k)_M^{m_k}$ (see Lemma 2(1)), we may pass from I to an I_i and assume I is $\star$-unfactorable. Because I is $\star$-finite, there is a minimum $n \in \mathbb{Z}^+$ with $I = ((x_1, \ldots, x_n) + MI)^\star$ for some $x_1, \ldots, x_n \in I$. Then $I_M = (x_1, \ldots, x_n)_M + M_M I_M = (x_1, \ldots, x_n)_M$ by Lemma 2(1) and Nakayama's lemma. Suppose $n > 1$. Let $A := ((x_3, \ldots, x_n) + MI)^\star$. It is straightforward to check that $(I^3)^\star = (IB)^\star$ for some finitely generated $B \in \text{Mod}_R((x_1^2 + x_2^2, x_1 x_2) + (x_1, x_2)A + A^2)$. Since $(I^3)^\star$ has a unique up to factors $\star$-factorization into $\star$-unfactorable $\star$-ideals and no redundant $\star$-factorizations, it follows that $B^\star = (I^2)^\star$. Because $\star$ is stable, this implies $x_1^2(y_1, \ldots, y_s) \subseteq B$ for some $y_1, \ldots, y_s \in R$ with $(y_1, \ldots, y_s)^\star = R$. Thus $(*)$: $x_1^2 y_i = \lambda_i(x_1^2 + x_2^2) + z_i$ for some $\lambda_i \in R$ and $z_i \in (x_1 x_2) + (x_1, x_2)A + A^2$.

Suppose that each $\lambda_i \in M$. Then each $\lambda_i x_1^2 \in MI^2 \subseteq AI \subseteq ((x_1, x_2)A + A^2)^\star$, so $x_1^2 y_i \in ((x_1 x_2, x_2^2) + (x_1, x_2)A + A^2)^\star$ by $(*)$. It is then straightforward to check that $(I^2)^\star = (I^2(y_1, \ldots, y_s))^\star = (IC)^\star$ for some finitely generated $C \in \text{Mod}_R((x_2) + A)$. Arguing as above yields $C^\star = I$, so $I = ((x_2, \ldots, x_n) + MI)^\star$, a contradiction. Thus some $\lambda_j \notin M$, so $(a_1 + w_1 \lambda_j, \ldots, a_r + w_r \lambda_j)^\star = R$ for some $a_1, \ldots, a_r \in M$ and $w_1, \ldots, w_r \in R$. Then each $a_i x_2^2 \in MI^2 \subseteq ((x_1 x_2) + (x_1, x_2)A + A^2)^\star$. By $(*)$, each $x_2^2(a_i + w_i \lambda_j) = w_i((y_j - \lambda_j)x_1^2 - z_i) + a_i x_2^2 \in ((x_1 x_2, x_1^2) + (x_1, x_2)A + A^2)^\star$. From here we obtain a contradiction as above.

(a) ⇒ (c): In view of "(a) ⇔ (h)" and Lemma 4, it suffices to note that R is integrally closed in $Q_0(R)$ if R is Q_0-Krull [60, Theorem 4.2].

(c) ⇒ (g): By contradiction. Suppose (c) holds but $M = MM^{-1_0}$ for some semiregular $M \in \text{Max}^\star(R)$. Then $M_M = M_M(M^{-1_0})_M$, where M_M is finitely generated and faithful by Lemma 2(1), so $(M^{-1_0})_M$ is integral over R_M [45, Theorem 12]. But R_M is integrally closed in $Q_0(R)_M$ [28, Proposition 10.2] since R is integrally closed in $Q_0(R)$, so $R_M = (M^{-1_0})_M = M_M^{-1_0}$, where the last equality is by Lemma 4(1). Pick a $\star$-invertible $\star$-ideal $I \subseteq M$. Since the minimal primes of I are $\star$-ideals [33, Proposition 6.6], we have $\sqrt{I_M} = M_M$, so there is a minimum $k \in \mathbb{Z}^+$ with $M_M^k \subseteq I_M$ since M_M is finitely generated. Lemma 4(3) implies I_M is a regular principal ideal but M_M is not, so $k \geq 2$. Thus $M_M^{k-1} = M_M^{k-1} I_M^{-1} I_M \subseteq M_M^{-1_0} I_M = I_M$, a contradiction.

2. This is virtually identical to the proof of (1): just make the obvious substitutions like changing each "semiregular" to "regular," etc., change the "finitely generated" to "principal" in the second sentence of the proof of "(1d) ⇒ (1b)" when constructing the proof of "(2e) ⇒ (2b)," and in the proof of "(2a) ⇒ (2c)" use [46, Proposition 2.2] to justify the fact that R is integrally closed in $T(R)$.

□

We note that in the special case $\star = d$ of Theorem 6(2), which characterizes Dedekind rings, the equivalence of (2a)–(2g) is essentially already known [39, Theorem 3.6]. If we further specialize to characterizations of Dedekind domains, then (2h) and (2i) are also known [2, Theorem 8(2); 5, Theorem 5.4; 16, Theorem]. We note that (2i) is really only a novel characterization for Dedekind rings with zero divisors, because a domain D is Dedekind if and only if $D(X)$ is a Euclidean domain [5, Theorem 5.4], in which case $\mathcal{I}(D) \cong \mathcal{I}(D(X))$ as multiplicative lattices [2, Theorem 8(2)]. The special case $\star = w$ of Theorem 6(2) characterizes Krull rings. Here "(2a) ⇔ (2d) " is implicit in [44, Theorem 13], but the other characterizations of Krull rings with zero divisors appear to be new. If one specializes further to integral domains, then the characterization of Krullness in (2b) and (2h) are essentially given by [49, Theorem 4.4; 67, Proposition 1.2] and the characterization in the formally stronger version of (2f) is given by [47, Theorem 3.6]. The remaining characterizations of Krullness appear to be new, even in the domain case. The special cases $\star = d$ and $\star = w_0$ of Theorem 6(1) respectively characterize Q_0-Dedekind rings and Q_0-Krull rings. There all parts are new. Of course, all the characterizations in Theorem 6 are new in the general case. We note that including the formally weaker versions of (1e) and (2f) considerably increases the strength of the results.

It is well known that every overring of a Dedekind ring is Dedekind [22, Exercise 3.8.1]. In Corollary 10 we will prove the analogous result that every Q_0-overring of a Q_0-Dedekind ring is Q_0-Dedekind. But it is far from true that every (Q_0-)overring of a (Q_0-)Krull ring is (Q_0-)Krull. Indeed, a domain is Dedekind if and only if each of its overrings is Krull [30, Theorem 18]. On the positive side, we can show that (Q_0-)Krullness is inherited by a special kind of (Q_0-)overring called "t_0-linked." Let us spend a moment developing the necessary terminology to describe such overrings. Let $R \subseteq T$ be an extension of rings and $\star$ be a finite type T-

semistar operation on R. We call a T-overring V of R *$\star$-linked* if $V = V^{\bar{\star}}$ (cf. [72, p. 337]). In Lemma 7 we will verify that this definition is consistent with the traditional definition of a "t-linked" overring [22, Definition 3.3.1] and develop several properties of $\star$-linked T-overrings. (We emphasize that we are only defining "$\star$-linked" for $\star$ of finite type, which is all we need for our purposes. So, for example, our results about $\star$-linked extensions do not apply to the "v-linked" extensions studied in [21, pp. 4139–4140].) Some examples of t_0-linked Q_0-overrings include Q_0-quotient rings (or more generally Q_0-generalized transforms) and flat Q_0-overrings—see Lemmas 2(2) and 4(2). We highlight a convention we will be using throughout the sequel. If V is a $\star$-linked T-overring of R, then by Lemma 7(2) we can regard $\bar{\star}$ as a finite type stable T-semistar operation on V. One point of caution about this: If V is a $w_{R,T}$-linked T-overring of R, then $w_{R,T} \leq w_{V,T}$ as T-semistar operations on V, but the inequality might be strict [73, Example].

Lemma 7 Let $R \subseteq T$ be an extension of rings and $\star$ be a finite type stable T-semistar operation on R.

1. A Q_0-overring V of R is t_0-linked (or equivalently w_0-linked) if and only if $(IV)^{t_{V,Q_0(R)}} = V$ for each (finitely generated semiregular) $I \in \mathcal{I}(R)$ with $I^{t_{R,Q_0(R)}} = R$ (cf. [72, Lemma 3.3]). An overring V of R is t-linked (or equivalently w-linked) if and only if $(IV)^{t_{V,T(R)}} = V$ for each (finitely generated regular) $I \in \mathcal{I}(R)$ with $I^{t_{R,T(R)}} = R$.
2. If V is a $\star$-linked T-overring of R, then $\star$ restricts to a finite type stable T-semistar operation on V (cf. [73, p. 1]).
3. The following are equivalent for a $\star$-linked T-overring V of R (cf. [14, Theorem 1.3; 32, Proposition 10; 70, Propositions 2.1 and 3.8]).
   a. V is $\star$-locally flat as an R-module.
   b. For each $P \in \operatorname{Spec}^\star(R)$, either $V = (PV)^\star$ or $V \subseteq R^T_{[P]}$.
   c. $V^T_{[M]} = R^T_{[R\cap M]}$ for each ($\star$-maximal) $M \in \operatorname{Spec}^\star(V)$.
   d. $V = \bigcap_{M\in \operatorname{Max}^\star(V)} R^T_{[R\cap M]} = \bigcap_{P\in \operatorname{Spec}^\star(V)} R^T_{[R\cap P]}$.
   e. $V = ((R :_R J)V)^\star$ for each finitely generated (cyclic) $J \in \operatorname{Mod}_R(V)$.
   f. $(AV :_T BV)^\star = ((A :_T B)V)^\star$ and $\left(\left(\bigcap_{i=1}^n A_i\right) V\right)^\star = \bigcap_{i=1}^n (A_i V)^\star$ for all $A, B, A_1, \ldots, A_n \in \operatorname{Mod}_R(T)$ with B finitely generated.
   g. $V = R^T_{\mathcal{S}}$ for some nonempty multiplicatively closed set $\mathcal{S}$ of (finitely generated) ideals of R with $V = (IV)^\star$ for each $I \in \mathcal{S}$.

Proof

1. We will prove the first sentence of (1). To prove the second sentence, just make the obvious substitutions, replacing "semiregular" with "regular," etc.

   ($\Rightarrow$): Assume $V = V^{w_{R,Q_0(R)}}$. Pick $I \in \mathcal{I}(R)$ with $I^{t_{R,Q_0(R)}} = R$. Then $J^{t_{R,Q_0(R)}} = R$ for some finitely generated $J \in \operatorname{Mod}_R(I)$ since $t_{R,Q_0(R)}$ has finite type. It suffices to show that $(V :_{Q_0(R)} JV) \subseteq V$, for then $V \supseteq (IV)^{t_{V,Q_0(R)}} \supseteq (JV)^{v_{V,Q_0(R)}} \supseteq V$. Let $x \in (V :_{Q_0(R)} JV)$. Then there is a finitely generated $B \in \mathcal{I}_{sr}(R)$ with $xB \subseteq R$. So $B + J \in \mathcal{I}_{sr}(R)$ is finitely generated, $(B + J)^{t_{R,Q_0(R)}} = R$, and $x(B + J) \subseteq V$. Thus $x \in V^{w_{R,Q_0(R)}} = V$, as desired.

($\Leftarrow$): Assume $(IV)^{t_{V,\mathcal{Q}_0(R)}} = V$ for each finitely generated semiregular $I \in \mathcal{I}(R)$ with $I^{t_{R,\mathcal{Q}_0(R)}} = R$. Let $x \in \overline{V^{t_{R,\mathcal{Q}_0(R)}}}$. Then $xJ \subseteq V$ for some finitely generated $J \in \mathcal{I}(R)$ with $J^{t_{R,\mathcal{Q}_0(R)}} = R$ and $xB \subseteq R$ for some finitely generated $B \in \mathcal{I}_{sr}(R)$. So $x \in (V :_{Q_0(R)} (B + J)V) = V$, as desired.

2. Assume $V = V^\star$. For each $I \in \mathrm{Mod}_V(T)$, we have $I^\star V \subseteq (IV)^\star = I^\star$, so $I^\star \in \mathrm{Mod}_V(T)$. Thus $\star$ is a T-semistar operation on V. To prove that it is a finite type stable T-semistar operation on V, we need to show that for each $I \in \mathrm{Mod}_V(T)$ and $x \in I^\star$ we have $xJ \subseteq I$ for some finitely generated $J \in \mathrm{Mod}_V(T)$ with $J^\star = V$. Since $\star$ is a finite type stable T-semistar operation on R, we have $xB \subseteq I$ for some finitely generated $B \in \mathcal{I}(R)$ with $B^\star = R$. So $BV \in \mathcal{I}(V)$ is finitely generated, $xBV \subseteq I$, and $(BV)^\star = V^\star = V$, as desired.
3. (e) $\Rightarrow$ (b): By contrapositive. Assume there is a $P \in \mathrm{Spec}^\star(R)$ with $V \neq (PV)^\star$ and $V \nsubseteq R^T_{[P]}$. Let $x \in V \setminus R^T_{[P]}$. Then $((R :_R xR)V)^\star \subseteq (PV)^\star \subsetneq V$.

(b) $\Rightarrow$ (c): Assume (b) holds. Let $P \in \mathrm{Spec}^\star(V)$. Then $((R \cap P)V)^\star \subseteq P \subsetneq V$, so $V \subseteq R^T_{[R\cap P]} \subseteq V^T_{[P]}$ by (b). Now let $x \in V^T_{[P]}$. Then $sx \in V \subseteq R^T_{[R\cap P]}$ for some $s \in V \setminus P \subseteq R^T_{[R\cap P]} \setminus P$. Therefore $ts, usx \in R$ for some $t, u \in R \setminus P$. Thus $(ts)ux \in R$, where $(ts)u \in R \setminus P$, so $x \in R^T_{[R\cap P]}$, as desired.

(c) $\Rightarrow$ (d): First note that the second equality in (d) always holds. This follows from the fact that each $P \in \mathrm{Spec}^\star(V)$ is contained in some $M \in \mathrm{Max}^\star(V)$ and $R^T_{[R\cap M]} \subseteq R^T_{[R\cap P]}$. With this in mind, the result follows from Lemma 2(3).

(d) $\Rightarrow$ (g): Assume (d) holds. Let $\mathcal{S} := \{I \in \mathcal{I}(R) \mid I \text{ is } d\text{-finite and } V = (IV)^\star\}$. Then $\mathcal{S}$ is nonempty and multiplicatively closed and $R^T_{\mathcal{S}} \subseteq \bigcup_{I\in\mathcal{S}}((R :_T I)V)^\star \subseteq \bigcup_{I\in\mathcal{S}}((R :_T I)(IV))^\star \subseteq V$. For the reverse inclusion, let $x \in V$. By (d), for each $M \in \mathrm{Max}^\star(V)$ there is a $b_M \in R \setminus M$ with $b_M x \in R$. Let $I := \sum_{M\in\mathrm{Max}^\star(V)} b_M R$. Then $(IV)^\star = V$ since $IV \nsubseteq M$ for each $M \in \mathrm{Max}^\star(V)$. Since $\star$ has finite type, we have $1 \in (\sum_{i=1}^n y_i z_i R)^\star$ with each $y_i \in I$ and $z_i \in V$. So $J := \sum_{i=1}^n R y_i \in \mathcal{S}$ and $x \in (R :_T I) \subseteq (R :_T J) \subseteq R^T_{\mathcal{S}}$, as desired.

(g) $\Rightarrow$ (f): Assume the formally weaker version of (g) holds. Let A, B, A_i be as in (f). Then $\bigcap_{i=1}^n (A_i V)^\star = \bigcap_{i=1}^n \bigcup_{I\in\mathcal{S}} (A_i(R :_T I))^\star = \bigcup_{I\in\mathcal{S}} \bigcap_{i=1}^n (A_i(R :_T I))^\star \subseteq \bigcup_{I\in\mathcal{S}} \left(\left(\bigcap_{i=1}^n A_i(R :_T I)\right) IV\right)^\star \subseteq \bigcup_{I\in\mathcal{S}} \left(\left(\bigcap_{i=1}^n A_i\right) V\right)^\star \subseteq \bigcap_{i=1}^n (A_i V)^\star$, where the second equality holds because $\{(R :_T I)\}_{I\in\mathcal{S}}$ is directed and the first equality holds since finite type T-semistar operations distribute into directed unions. We have $((A :_T B)V)^\star \subseteq ((AV)^\star :_T BV) = (AV :_T BV)^\star$, where the equality holds since $BV \in \mathrm{Mod}_V(T)$ is finitely generated and $\star$ is a stable T-semistar on V. For the reverse inclusion, pick $x \in ((AV)^\star :_T BV)$. We have $(AV)^\star = \bigcup_{I\in\mathcal{S}} (A(R :_T I))^\star$ as above, so $xB \subseteq (A(R :_T J))^\star$ for some $J \in \mathcal{S}$ since B is finitely generated and $\{(A(R :_T I))^\star\}_{I\in\mathcal{S}}$ is directed. So $x \in (xV)^\star = (xJV)^\star \subseteq ((A^\star :_T B)V)^\star = ((A :_T B)V)^\star$, where the last equality holds since $\star$ is stable, as desired.

(f) $\Rightarrow$ (a): Assume (f) holds. By Bourbaki's flatness criterion [15, Exercise I.2.22], we need to show that $(I_M V_M :_{V_M} (xV)_M) = (I_M :_{R_M} (xR)_M)V_M$ for each $M \in \mathrm{Max}^\star(R)$, $x \in R$, and finitely generated $I \in \mathcal{I}(R)$. Using (f) and Lemma 2(1), we compute $(I_M V_M :_{V_M} (xV)_M) = ((IV :_V xV)^\star)_M = ((IV :_T$

$xV)^\star \cap V)_M = (((I :_T xR)V)^\star \cap V)_M = ((((I :_T xR) \cap R)V)^\star)_M = ((I :_R xR)V)^\star)_M = (I_M :_{R_M} (xR)_M)V_M$, as desired.

(a) $\Rightarrow$ (e): Assume (a) holds. Let $x_1, \ldots, x_n \in V$. For each $M \in \mathrm{Max}^\star(R)$, we have $((R :_R \sum_{i=1}^n Rx_i)V)_M = \left(\bigcap_{i=1}^n (R_M :_{R_M} (Rx_i)_M)\right) V_M = \bigcap_{i=1}^n (R_M :_{R_M} (Rx_i)_M)V_M = \bigcap_{i=1}^n (V_M :_{V_M} (Vx_i)_M) = V_M$ [15, Exercise I.2.22; 22, Lemma 2.2.8] since V_M is R_M-flat. Thus $((R :_R \sum_{i=1}^n Rx_i)V)^\star = V$ by Lemma 2(1). □

The fact that Prüfer rings are characterized by having every overring integrally closed [32, Theorem 13] leads us to consider the following definitions. Let $R \subseteq T$ be an extension of rings and $\star$ be a finite type T-semistar operation on R. We call R *quasi-$\star$-Prüfer* if every $\star$-linked T-overring of R is integrally closed in T. If one wishes to emphasize or clarify what T is, one can write "quasi-T-$\star$-Prüfer." In particular, one calls R a *quasi-Q_0-Prüfer v-multiplication ring* (quasi-Q_0-PVMR) if it is quasi-$Q_0(R)$-t_0-Prüfer [70, p. 4028] and *quasi-Q_0-Prüfer* (or simply "Q_0-Prüfer" in Lucas's terminology [54, p. 61]) if it is quasi-$Q_0(R)$-d-Prüfer. In the case where R is a domain, it has long been known that R is a PVMR if and only if it is quasi-t-Prüfer [20, Theorem 2.10], in which case every t-linked overring of R has the form $\bigcap_{P\in\Omega} R_{[P]}^{T(R)}$ for some $\Omega \subseteq \mathrm{Spec}^t(R)$ [43, Theorem 3.8]. Elliott has asked whether this remains true without the domain hypothesis, and, if so, whether this generalizes to a characterization of $\star$-Prüfer rings [22, p. 284]. By combining Lemmas 7 and 8, we obtain affirmative answers to both questions. But (i) strongly Prüfer $\Rightarrow$ quasi-Q_0-Prüfer $\Rightarrow$ Prüfer and (ii) Q_0-PVMR $\Rightarrow$ quasi-Q_0-PVMR $\Rightarrow$ PVMR, with none of the implications reversing [54, p. 61; 70, pp. 4037–4038].

Lemma 8 (cf. [20, Theorem 2.10; 32, Theorem 13; 54, Theorem 7; 70, Lemma 4.5]) Let $R \subseteq T$ be an extension of rings and $\star$ be a finite type stable T-semistar operation on R. The following are equivalent.

1. R is quasi-$\star$-Prüfer.
2. Every $\star$-linked T-overring of R is quasi-$\star$-Prüfer.
3. Every $\star$-linked T-overring of R is $\star$-locally flat as an R-module.
4. $(R_{[P]}^T, P_{[P]}^T)$ is a valuation pair of T (see [64, p. 735]) for each $P \in \mathrm{Spec}^\star(R)$.

If $T = T(R)$, then R is quasi-$\star$-Prüfer if and only if it is $\star$-Prüfer. If R is quasi-$\star$-Prüfer, then each finitely generated $I \in \mathcal{I}(R)$ with $(IT)^\star = T$ is $\star$-invertible.

Proof The "only if" for the penultimate assertion will follow immediately from the lemma's final assertion since $IT(R) = T(R)$ for all $I \in \mathcal{I}_r(R)$. For the "if" part, assume $T = T(R)$ and R is $\star$-Prüfer. In view of our forthcoming proof of "(3) $\Rightarrow$ (1)," it suffices to show that each regular $I \in \mathrm{Mod}_R(T(R))$ is $\star$-locally flat. Because a directed union of flat modules is flat, we may assume I is finitely generated. Then $rI \in \mathcal{I}_r(R)$ is finitely generated for some $r \in \mathrm{Reg}(R)$, so rI (or equivalently $I^\star$) is $\star$-invertible. Lemmas 2(1) and 4(3) now imply I is $\star$-locally flat.

(3) $\Rightarrow$ (1): Assume (3) holds. Let V be a $\star$-linked T-overring of R and V' be the integral closure of V in T. Suppose that there is an $x \in V' \setminus V$. Then $(V :_V Vx)$ is contained in some $P \in \mathrm{Spec}^\star(V)$. By the lying-over theorem for integral extensions,

there is a $Q \in \mathrm{Spec}(V')$ with $V \cap Q = P$. Then $(V')^\star \supseteq Q^\star \supseteq ((V :_V Vx)V')^\star \supseteq ((R :_R Rx)V')^\star = (V')^\star$, where the equality holds by Lemma 7(3). Therefore $V = (V')^\star \cap V = Q^\star \cap V = (Q \cap V)^\star = P^\star = P$ since $\star$ is stable, a contradiction.

(1) $\Rightarrow$ (4): (We adapt the proof of "(1) $\Rightarrow$ (2)" in [70, Theorem 4.2].) Assume R is quasi-$\star$-Prüfer. Let $P \in \mathrm{Spec}^\star(R)$. We need to show that for each $z \in T \setminus R^T_{[P]}$ there is a $y \in P^T_{[P]}$ with $yz \in R^T_{[P]} \setminus P^T_{[P]}$ [64, p. 735]. Because $(R^T_{[P]}[z^2])^\star$ is integrally closed in T and $\star$ is stable, we have $Jz \subseteq R^T_{[P]}[z^2]$ for some $J \in \mathcal{I}(R)$ with $J^\star = R$. Then $J \nsubseteq P$, so $(*)$: $bz = \sum_{k=0}^n a_k z^{2k}$ for some $b \in R \setminus P$ and $a_0, \dots, a_n \in R^T_{[P]}$ with $a_n \neq 0$, where $n > 0$ since $z \notin R^T_{[P]}$. Choose so that n is as small as possible. Because R is integrally closed in T, the same holds for $R^T_{[P]}$. (To see this, pick $t \in T$ with $t^n + a_{n-1}t^{n-1} + \cdots + a_1 t + a_0 = 0$ for some $a_1, \dots, a_{n-1} \in R^T_{[P]}$. Multiply both sides by s^n, where s is an element of $R \setminus P$ with each $sa_i \in R$. The result is $(st)^n + sa_{n-1}(st)^{n-1} + \cdots + s^{n-1}a_1(st) + s^n a_0 = 0$, where $sa_{n-1}, \dots, s^{n-1}a_1, s^n a_0 \in R$. So $st \in R$ and therefore $t \in R^T_{[P]}$, as desired.) Multiplying both sides of $(*)$ by a_n^{2n-1} and rearranging, we obtain $(a_n z)^{2n} - ba_n^{2n-2}(a_n z) + \sum_{k=0}^{n-1} a_k a_n^{2(n-k)-1}(a_n z)^{2k} = 0$, so $a_n z \in R^T_{[P]}$. Thus $sa_n, (sa_n)z \in R$ for some $s \in R \setminus P$, so $sa_n \in P$. If $n > 1$, then we can multiply $(*)$ by a^{n-1} and rearrange to obtain $(a_n z^2)^n + (a_0 a_n^{n-1} - ba_n^{n-2}(a_n z)) + \sum_{k=1}^{n-1} a_k a_n^{n-k-1}(a_n z^2)^k$, conclude that $a_n z^2 \in R^T_{[P]}$, and then rearrange $(*)$ to obtain $bz = (a_{n-1} + a_n z^2)z^{2(n-1)} + \sum_{k=0}^{n-2} a_k z^{2k}$, contradicting the minimality of n. Therefore $n = 1$. Pick $t \in R \setminus P$ with $ta_0 \in R$. For each $u \in R \setminus P$, we have $stu(b - a_1 z) \in R$ and $stu(b - a_1 z)z = stua_0 \in R$, so $stu(b - a_1 z) \in P$ and therefore $usa_1 z \notin P$. Thus $sa_1 \in P^T_{[P]}$ and $(sa_1)z \in R^T_{[P]} \setminus P^T_{[P]}$, as desired.

(4) $\Rightarrow$ (3): Assume (4) holds. Let V be a $\star$-linked T-overring of R. By Lemma 7(3), we need to prove $V^T_{[M]} = R^T_{[R \cap M]}$ for each $M \in \mathrm{Max}^\star(V)$. Then $P := R \cap M \in \mathrm{Spec}^\star(R)$ since $\star$ is stable. Because $(R^T_{[P]}, P^T_{[P]})$ is a valuation pair of T and $R^T_{[P]} \subseteq V^T_{[M]}$, it suffices to show that $M^T_{[M]} \cap R^T_{[P]} = P^T_{[P]}$ [54, Theorem 1]. For the nontrivial inclusion, let $x \in M^T_{[M]} \cap R^T_{[P]}$. Then there are $s \in V \setminus M$ and $t \in R \setminus M$ with $sx \in M$ and $tx \in R$. Because $s(tx) \in M$, we have $tx \in P$, as desired.

Now we will adapt the proof of [70, Lemma 4.5] to show that each finitely generated $I \in \mathcal{I}(R)$ with $(IT)^\star = T$ is $\star$-invertible. We need to prove that $I(R :_T I) \nsubseteq M$ for each $M \in \mathrm{Max}^\star(R)$. Let ν be a valuation on T associated with the pair $(R^T_{[M]}, M^T_{[M]})$—see [54, p. 61]. By Lemma 2(2), there is an $a \in IT \setminus M^T_{[M]}$. Write $I = \sum_{i=1}^n Rx_i$ and $a = \sum_{i=1}^n x_i y_i$ with each $x_i \in I$ and $y_i \in T$. If $x_i y_j \in R^T_{[M]}$ for all i, j, then there is an $s \in R \setminus M$ with each $sx_i y_j \in R$, but $x_k(sy_l) \in I(R :_T I) \setminus M$ for some k, l since $sa \notin M$. So let us assume some $x_k y_l \notin R^T_{[M]}$; pick such a pair of indices with $\nu(x_k y_l)$ as small as possible and choose $z \in M^T_{[M]}$ with $\nu(z) = -\nu(x_k y_l)$. Then $zx_i y_j \in R^T_{[M]}$ for all i, j, but $zx_k y_l \notin M^T_{[M]}$. Thus there is an $s \in R \setminus M$ with $szx_i y_j \in R$ for all i, j, so $x_k(szy_l) \in I(R :_T I) \setminus M$, as desired.

(1) $\Leftrightarrow$ (2): Follows immediately from the fact that, if V is a $\star$-linked T-overring of R, then any $\star$-linked T-overring of V is also a $\star$-linked T-overring of R. □

Before arriving at our promised result about "Dedekindness" and (Q_0-)overrings, we pause to characterize "Q_0-Dedekindness" via Q_0-semistar Nagata rings.

Theorem 9 Let $\star$ be a reduced finite type stable Q_0-semistar operation on a ring R and $\{X_\lambda\}_{\lambda\in\Lambda}$ be a nonempty set of algebraically independent indeterminates.

1. The following are equivalent (cf. [5, Theorems 2.2(5), 3.2(1), and 3.3(2)]).

a. R is strongly $\star$-Prüfer.
b. $R(\{X_\lambda\}_{\lambda\in\Lambda}, \star)$ is (strongly) Prüfer.
c. Every finitely generated (semi)regular ideal of $R(\{X_\lambda\}_{\lambda\in\Lambda}, \star)$ is principal.
d. The map $\theta : \{(0)\} \cup \mathcal{I}^\star_{sr}(R) \to \{(0)\} \cup \mathcal{I}_r(R(\{X_\lambda\}_{\lambda\in\Lambda}, \star))$, $\theta : I \mapsto IR(\{X_\lambda\}_{\lambda\in\Lambda}, \star)$, is a multiplicative lattice isomorphism.
e. Every semiregular R-submodule of $Q_0(R)$ (or equivalently every finitely generated faithful ideal of R) is $\star$-locally flat.
f. R is quasi-$\star$-Prüfer and $(IQ_0(R))^\star = Q_0(R)$ for every $I \in \mathcal{I}_{sr}(R)$.
g. Every $\star$-linked Q_0-overring of R is strongly $\star$-Prüfer.

2. The following are equivalent (cf. [5, Theorem 5.4(1); 13, Theorem 6; 26, Theorem 2.2; 42, Lemma 3.8]).

a. R is Q_0-$\star$-Dedekind.
b. $R(\{X_\lambda\}_{\lambda\in\Lambda}, \star)$ is (Q_0-)Dedekind.
c. Every (semi)regular ideal of $R(\{X_\lambda\}_{\lambda\in\Lambda}, \star)$ is principal.

Proof Let $A := R[\{X_\lambda\}_{\lambda\in\Lambda}]$. Throughout, we recall that $B := R(\{X_\lambda\}_{\lambda\in\Lambda}, \star)$ is Marot with Property A. Part (2) will follow from (1) via Lemma 5.

(1a) $\Rightarrow$ (1c)+(1d): Assume R is strongly $\star$-Prüfer. Note that θ does indeed map $\{(0)\} \cup \mathcal{I}^\star_{sr}(R)$ into $\{(0)\} \cup \mathcal{I}_r(B)$ by McCoy's theorem. The fact that θ is an order-preserving monoid monomorphism follows nearly immediately from Lemma 3. Thus, since $\{(0)\}\cup\mathcal{I}^\star_{sr}(R)$ and $\{(0)\}\cup\mathcal{I}_r(B)$ are complete lattices, the only thing we need to verify for (1d) is the surjectivity of θ. Let $J \in \mathcal{I}_r(B)$. Lemma 4(3) implies $c(f)^\star$ is semiregular and $\star$-locally principal for each $f \in \mathrm{Reg}(A)$, so $c(f)B = fB$ holds locally [2, Theorem 7] and therefore globally. Since B is Marot, we have $J = \sum_{f\in J_0} fB = \sum_{f\in J_0} c(f)B = \theta\left(\left(\sum_{f\in J_0} c(f)\right)^\star\right)$, where $J_0 := J \cap \mathrm{Reg}(A)$, as desired. For (1c), assume J is finitely generated. Then $J = IB$ for some finitely generated regular ideal I of A and there is a $g \in \mathrm{Reg}(A) \cap I$ with $c(f) \subseteq c(g)$ for all $f \in I$ [51, Proposition 6.6.3(1)]. Therefore $gB \subseteq J \subseteq \theta(c(g)^\star) = gB$, as desired.

(1c) $\Rightarrow$ (1b); (1g) $\Rightarrow$ (1a): Clear.

(1b) $\Rightarrow$ (1a); (1d) $\Rightarrow$ (1a): By Lemma 4(3), we need to show that (1b) and (1d) each imply $c(f)^\star$ is $\star$-locally principal (or equivalently $c(f)B$ is locally principal) for each $f \in \mathrm{Reg}(A)$. We are done if (1b) holds, so let us assume (1d) holds. Then $fB = IB$ for some $I \in \mathcal{I}^\star_{sr}(R)$. Because IB is principal, the ideal I is $\star$-locally principal [3, Theorem 1]. We will show that $I = c(f)^\star$. Of course $I = I^\star =$

$R \cap IB \subseteq R \cap c(f)B = c(f)^{\star}$. For the reverse inclusion, we have $fg \in IA$ for some $g \in R[X]$ with $c(g)^{\star} = R$. By the Dedekind-Mertens lemma, there is an $n \in \mathbb{Z}^{+}$ with $c(f)c(g)^{n+1} = c(fg)c(g)^{n}$. Therefore $c(f)^{\star} = c(fg)^{\star} \subseteq I^{\star} = I$, as desired.

(1a) $\Rightarrow$ (1e): Assume R is strongly $\star$-Prüfer. Because a directed union of flat modules is flat, it suffices by Lemmas 2(1) and 4(3) to show that every finitely generated faithful $I \in \mathrm{Mod}_R(Q_0(R))$ is $\star$-invertible. Because I is finitely generated, there is a finitely generated $J \in \mathcal{I}_{sr}(R)$ with $JI \in \mathcal{I}_{sr}(R)$. Then JI (and therefore I) is $\star$-invertible since R is strongly $\star$-Prüfer.

(1e) $\Rightarrow$ (1a): This follows from Lemmas 2(1) and 4(3) and the fact that finitely generated flat ideals of quasilocal rings are principal [65, Theorem 7.10].

(1a) $\Rightarrow$ (1f): Assume R is strongly $\star$-Prüfer. Then R is quasi-$\star$-Prüfer by Lemma 8 and "(1a) $\Rightarrow$ (1e)." Let $I \in \mathcal{I}_{sr}(R)$. Then there is a finitely generated semiregular $J \in \mathrm{Mod}_R(I)$ and $Q_0(R) \supseteq (IQ_0(R))^{\star} \supseteq (JQ_0(R))^{\star} = (JJ^{-1_0}Q_0(R))^{\star} = Q_0(R)$ since J is $\star$-invertible, as desired.

(1f) $\Rightarrow$ (1g): Let V be a $\star$-linked Q_0-overring of R. By Lemma 8, we need to show that (1f) implies $(IQ_0(R))^{\star} = Q_0(R)$ for each finitely generated $I \in \mathcal{I}_{sr}(V)$. Write $I = \sum_{i=1}^{n} Vx_i$ with each $x_i \in I$. Then there is a finitely generated $J_0 \in \mathcal{I}_{sr}(R)$ with each $x_i J_0 \subseteq R$. So $J := \sum_{i=1}^{n} x_i J_0 \in \mathcal{I}_{sr}(R)$ and $Q_0(R) \supseteq (IQ_0(R))^{\star} \supseteq (JQ_0(R))^{\star} = Q_0(R)$ by (1f), as desired. □

We close the paper with our result on "Dedekindness" and (Q_0-)overrings. It is well known that overrings of Dedekind rings are Dedekind [22, Exercise 3.8.1] and that t-linked overrings of Krull domains are Krull [20, Corollary 2.22], but the following corollary appears to be new except in those two special cases. In particular, it affirmatively answers a (previously) open question posed by Elliott [22, p. 281]: is every t-linked overring of a Krull ring Krull?

Corollary 10 If R is a ring, $\star$ is a finite type (Q_0-)semistar operation on R, and R is (Q_0-)$\star$-Dedekind, then every $\star$-linked (Q_0-)overring of R is (Q_0-)$\star$-Dedekind.

Proof We prove the "Q_0" version. The proof of the other version can be obtained by just making the obvious substitutions, changing "semiregular" to "regular," etc., and replacing references to Theorem 9(1) with references to Lemma 8. We may pass from $\star$ to $\star_w$ and assume $\star$ is reduced finite type stable. Assume R is Q_0-$\star$-Dedekind. Let V be a $\star$-linked Q_0-overring of R. Lemma 7(3), Lemma 8, and Theorem 9(1) together imply $V = R_{\mathcal{S}}^{Q_0(R)}$ for some nonempty multiplicatively closed set $\mathcal{S}$ of $\star$-invertible ideals of R with $(IV)^{\star} = V$ for each $I \in \mathcal{S}$. By Theorem 9(1), it suffices to show each semiregular $B \in \mathcal{I}^{\star}(V)$ is $\star$-finite. Let $A := R \cap B \in \mathcal{I}^{\star}(R)$. Because B is a semiregular ideal of V, it has a finitely generated semiregular subideal B_0. Then $JB_0 \subseteq R$ for some semiregular ideal J of R. So A has a semiregular subideal (namely JB_0) and consequently is $\star$-finite by Lemma 4(3). We have $A_{\mathcal{S}}^{Q_0(R)} \subseteq \bigcup_{I \in \mathcal{S}} ((A :_{Q_0(R)} I)(IV))^{\star} \subseteq B$. On the other hand, if $x \in B$, then $xI \subseteq R$ for some $I \in \mathcal{S}$, so $x \in (A :_{Q_0(R)} I) \subseteq A_{\mathcal{S}}^{Q_0(R)}$. Thus $B = A_{\mathcal{S}}^{Q_0(R)} = \bigcup_{I \in \mathcal{S}} (I^{-1_0}A)^{\star} = (AV)^{\star}$ since each $I \in \mathcal{S}$ is $\star$-invertible,

$\{I^{-1_0}\}_{I \in S}$ is directed, and finite type Q_0-semistar operations distribute over directed unions. Since A is $\star$-finite, the same holds for $B = (AV)^\star$, as desired. □

Acknowledgments We would like to thank the referee for taking the time to write an exceptionally careful and thorough report. This thoughtful feedback has immeasurably improved the paper from its original version.

References

1. D.D. Anderson. Abstract commutative ideal theory without chain condition. *Algebra Universalis*, 6:131–145, 1976.
2. D.D. Anderson. Multiplication ideals, multiplication rings, and the ring $R(X)$. *Canad. J. Math.*, 28:760–768, 1976.
3. D.D. Anderson. Some remarks on the ring $R(X)$. *Canad. J. Math.*, 28:760–768, 1976.
4. D.D. Anderson. Some remarks on multplication ideals, II. *Comm. Algebra*, 28:2577–2583, 2000.
5. D.D. Anderson, D.F. Anderson, and R. Markanda. The rings $R(X)$ and $R\langle X\rangle$. *J. Algebra*, 95:96–115, 1985.
6. D.D. Anderson, J.R. Juett, and C.P. Mooney. Factorization of ideals. *Comm. Algebra*, 47:1742–1772, 2019.
7. D.D. Anderson and R. Markanda. Unique factorization rings with zero divisors. *Houston J. Math.*, 11:15–30, 1985.
8. D.D. Anderson, J. Matijevic, and W. Nichols. The Krull Intersection Theorem II. *Pacific J. Math.*, 66:15–22, 1976.
9. D.D. Anderson and J. Pascual. Regular ideals in commutative rings, sublattices of regular ideals, and Prüfer rings. *J. Algebra*, 111:404–426, 1987.
10. D.D. Anderson and S. Valdes-Leon. Factorization in commutative rings with zero divisors. *Rocky Mountain J. Math.*, 26:439–473, 1996.
11. D.D. Anderson and M. Zafrullah. On t-invertibility. III. *Comm. Algebra*, 21:1189–1201, 1993.
12. D.D. Anderson and M. Zafrullah. On t-invertibility. IV. In D.D. Anderson, editor, *Factorization in Integral Domains*, volume 189 of *Lecture Notes in Pure and Appl. Math.*, pages 221–225. Marcel Dekker, Inc., New York, 1997.
13. J.T. Arnold. On the ideal theory of the Kronecker function ring and the domain $D(X)$. *Canad. J. Math.*, 21:558–563, 1969.
14. J.T. Arnold and J.W. Brewer. On flat overrings, ideal transforms and generalized transforms of a commutative ring. *J. Algebra*, 18:254–263, 1971.
15. N. Bourbaki. *Commutative algebra. Chapters 1–7.* Elements of Mathematics (Berlin). Springer-Verlag, Berlin, 1989.
16. H.S. Butts. Quasi-invertible prime ideals. *Proc. Amer. Math. Soc.*, 16:291–292, 1965.
17. G.W. Chang. $*$-Noetherian domains and the ring $D[X]_{N_*}$. *J. Algebra*, 297:216–233, 2006.
18. Y. Chen, F. Wang, H. Yin, and X. Zhu. w-modules over commutative rings. *J. Korean Math. Soc.*, 48:207–222, 2011.
19. R.P. Dilworth. Abstract commutative ideal theory. *Pacific J. Math.*, 12:481–498, 1962.
20. D.E. Dobbs, E.G. Houston, T.G. Lucas, and M. Zafrullah. t-linked overrings and Prüfer v-multiplication domains. *Comm. Algebra*, 17:2835–2852, 1989.
21. J. Elliott. Integer-valued polynomial rings, t-closure, and associated primes. *Comm. Algebra*, 39:4128–4147, 2011.
22. J. Elliott. *Rings, modules, and closure operations*. Springer Monographs in Mathematics. Springer, Cham, 2019.

23. D. Estes and J. Ohm. Stable range in commutative rings. *J. Algebra*, 7:343–362, 1967.
24. M. Fontana and K.A. Loper. Nagata rings, Kronecker function rings, and related semistar operations. *Comm. Algebra*, 31:4775–4805, 2003.
25. M. Fontana and K.A. Loper. An historical overview of Kronecker function rings, Nagata rings, and related star and semistar operations. In J.W. Brewer, S. Glaz, W. Heinzer, and B. Olberding, editors, *Multiplicative ideal theory in commutative algebra*, pages 169–187. Springer, New York, 2006.
26. R. Gilmer. An embedding theorem for HCF-rings. *Proc. Camb. Phil. Soc.*, 68:583–587, 1970.
27. R. Gilmer. *Commutative Semigroup Rings*. University of Chicago Press, Chicago, 1984.
28. R. Gilmer. *Multiplicative Ideal Theory*, volume 90 of *Queen's Papers in Pure and Applied Mathematics*. Queen's University, Kingston, Ontario, 1992.
29. R. Gilmer and J. Huckaba. Δ-rings. *J. Algebra*, 28:414–432, 1974.
30. R.W. Gilmer and J.L. Mott. On proper overrings of integral domains. *Monatsh. Math.*, 72:61–71, 1968.
31. S. Glaz. Controlling the zero divisors of a commutative ring. In *Commutative ring theory and applications (Fez, 2001)*, volume 231 of *Lecture Notes in Pure and Appl. Math.*, pages 191–212. Marcel Dekker, Inc., New York, 2003.
32. M. Griffin. Prüfer rings with zero divisors. *J. Reine Angew. Math.*, 239/240:55–67, 1969.
33. F. Halter-Koch. *Ideal Systems: An Introduction to Multiplicative Ideal Theory*. Marcel Dekker, Inc., New York, 1998.
34. W.J. Heinzer. J-Noetherian integral domains with 1 in the stable range. *Proc. Amer. Math. Soc.*, 19:1369–1372, 1968.
35. J.A. Huckaba. *Commutative rings with zero divisors*, volume 117 of *Pure and Applied Mathematics*. Marcel Dekker, Inc., New York, 1988.
36. P. Jaffard. *Les Systemes d'Ideaux*. Dunod, Paris, 1960.
37. E.W. Johnson and J.P. Lediaev. Representable distributive Noether lattices. *Pacific J. Math.*, 28:561–564, 1989.
38. J.R. Juett and C.P. Mooney. U-factorization of ideals. *Comm. Algebra*, 47:878–895, 2019.
39. J.R. Juett, C.P. Mooney, and L.W. Ndungu. Unique factorization of ideals in commutative rings with zero divisors. *Comm. Algebra*, 49:2101–2125, 2021.
40. M. Knebusch and D. Zhang. *Manis valuations and Prüfer extensions I: A new chapter in commutative algebra*, volume 1791 of *Lecture Notes in Mathematics*. Springer, Cham.
41. T. Kaiser and M. Knebusch. *Manis valuations and Prüfer extensions. II*, volume 2103 of *Lecture Notes in Mathematics*. Springer, Cham, 2014.
42. B.G. Kang. On the converse of a well-known fact about Krull domains. *J. Algebra*, 124:284–299, 1989.
43. B.G. Kang. Prüfer v-multiplication domains and the ring $R[X]_{N_v}$. *J. Algebra*, 123:151–170, 1989.
44. B.G. Kang. Characterizations of Krull rings with zero divisors. *J. Pure Appl. Algebra*, 146:283–290, 2000.
45. I. Kaplansky. *Commutative Rings*. Allyn and Bacon, Inc., Boston, 1970.
46. R.E. Kennedy. Krull rings. *Pacific J. Math.*, 89:131–136, 1980.
47. H. Kim, M.O. Kim, and Y.S. Park. Some characterizations of Krull monoids. *Algebra Colloq.*, 14:469–477, 2007.
48. H. Kim and J.W. Lim. Integral domains in which every nonzero w-flat ideal is w-invertible. *Mathematics*, 8:247–257, 2020.
49. H. Kim and Y. Park. Some characterizations of Krull domains. *J. Pure Appl. Algebra*, 208:339–344, 2007.
50. H. Kim and F. Wang. Two generalizations of projective modules and their applications. *J. Pure Appl. Algebra*, 219:2099–2123, 2015.
51. H. Kim and F. Wang. *Foundations of commutative rings and their modules*, volume 22 of *Algebra and Applications*. Springer, Singapore, 2016.
52. T.G. Lucas. Characterizing when $R[X]$ is integrally closed, II. *J. Pure Appl. Algebra*, 61:49–52, 1989.

53. T.G. Lucas. The complete integral closure of $R[X]$. *Trans. Amer. Math. Soc.*, 330:757–768, 1992.
54. T.G. Lucas. Strong Prüfer rings and the ring of finite fractions. *J. Pure Appl. Algebra*, 84:59–71, 1993.
55. T.G. Lucas. The ring of finite fractions. In *Commutative Ring Theory*, volume 153 of *Lecture Notes in Pure and Appl. Math.*, pages 181–191. Marcel Dekker, Inc., New York, 1994.
56. T.G. Lucas. Characterizing when $R(X)$ is completely integrally closed. In D.D. Anderson, editor, *Factorization in Integral Domains*, volume 189 of *Lecture Notes in Pure and Appl. Math.*, pages 401–415. Marcel Dekker, Inc., New York, 1997.
57. T.G. Lucas. The integral closure of $R(X)$ and $R\langle X\rangle$. *Comm. Algebra*, 25:847–872, 1997.
58. T.G. Lucas. Integrality properties in rings with zero divisors. In D.D. Anderson, editor, *Ideal theoretic methods in commutative algebra*, volume 220 of *Lecture Notes in Pure and Appl. Math.*, pages 265–280. Marcel Dekker, Inc., New York, 2001.
59. T.G. Lucas. Gaussian polynomials and invertibility. *Proc. Amer. Math. Soc.*, 133:1881–1886, 2005.
60. T.G. Lucas. Krull rings, Prüfer v-multiplication rings and the ring of finite fractions. *Rocky Mountain J. Math.*, 34:1251–1325, 2005.
61. T.G. Lucas. Integrality properties of polynomials and semigroup rings. In J.W. Brewer, S. Glaz, W. Heinzer, and B. Olberding, editors, *Multiplicative ideal theory in commutative algebra*, pages 293–309. Springer, New York, 2006.
62. T.G. Lucas. The Mori property in rings with zero divisors, II. *Rocky Mountain J. Math.*, 37:1195–1228, 2007.
63. S. Malik, J.L. Mott, and M. Zafrullah. On t-invertibility. *Comm. Algebra*, 16:149–170, 1988.
64. M.E. Manis. Extension of valuation theory. *Bull. Amer. Math. Soc.*, 73:735–736, 1967.
65. H. Matsumura. *Commutative Ring Theory*. Cambridge University Press, New York, 1989.
66. P.J. McCarthy. Principal elements of lattices of ideals. *Proc. Amer. Math. Soc.*, 30:43–45, 1971.
67. J.L. Mott and M. Zafrullah. On Krull domains. *Arch. Math.*, 56:559–568, 1991.
68. M. Nagata. *Local Rings*. Interscience tracts in pure and applied mathematics. Interscience Publishers, 1962.
69. G. Picozza and F. Tartarone. Flat ideals and stability in integral domains. *J. Algebra*, 324, 2010.
70. L. Qiao and F. Wang. w-linked Q_0-overrings and Q_0-prüfer v-multiplication rings. *Comm. Algebra*, 44:4026–4040, 2016.
71. J.D. Sally and W.V. Vasconcelos. Flat ideals I. *Comm. Algebra*, 3:531–543, 1975.
72. Y. Tian, F. Wang, and L. Xie. On w-linked overrings. *J. Math. Res. Exposition*, 31:337–346, 2011.
73. F. Wang. On induced operations and UMT-domains. *J. Sichuan Normal Univ. Nat. Sci. Ed.*, 27:1–9, 2004.
74. M. Ward. Residuation in structures over which a multiplication is defined. *Duke Math. J.*, 3:627–636, 1937.
75. M. Zafrullah. On finite conductor domains. *Manuscripta Math.*, 24:191–204, 1978.
76. M. Zafrullah. On t-invertibility. II. *Comm. Algebra*, 17:1955–1969, 1989.
77. M. Zafrullah. Putting t-invertibility to use. pages 429–457. Kluwer Acad. Publ., Dordrecht, 2000.

The Quadratic Tree of a Two-Dimensional Regular Local Ring

William Heinzer, K. Alan Loper, Bruce Olberding, and Matthew Toeniskoetter

Dedicated to the memory of Paul-Jean Cahen.

1 Introduction

The birational algebra of two-dimensional regular local rings plays a fundamental role in the study of algebraic surfaces and embedded resolution of singularities of plane curves. A point on a variety is nonsingular if and only if the local ring of the point is a regular local ring. Singularities of a curve on a nonsingular algebraic surface can be resolved by blowing up appropriate closed points where the curve has a singularity and then iterating this process for the curve and surface that result.

Since blowing up is a birational transformation, the focus moves from intrinsic properties of a two-dimensional regular local ring D to the regular local rings between D and its quotient field F. The process of successively blowing up closed points leads to a sequence of regular local rings between D and F. There are a number of classical theorems regarding this process of quadratic transformation and the regular local rings that result, and many of these are special for dimension two; see Sect. 2.

W. Heinzer
Purdue University, Department of Mathematics, West Lafayette, IN, USA
e-mail: heinzer@purdue.edu

K. A. Loper
Ohio State University at Newark, Newark, OH, USA
e-mail: lopera@math.ohio-state.edu

B. Olberding (✉)
New Mexico State University, Department of Mathematical Sciences, Las Cruces, NM, USA
e-mail: bruce@nmsu.edu

M. Toeniskoetter
Oakland University, Department of Mathematics and Statistics, Rochester, MI, USA
e-mail: toeniskoetter@oakland.edu

J.-L. Chabert et al. (eds.), *Algebraic, Number Theoretic, and Topological Aspects of Ring Theory*, https://doi.org/10.1007/978-3-031-28847-0_15

Less is known about the non-regular integrally closed domains between a two-dimensional regular local ring and its quotient field. The restriction to dimension two in the classical theory of surfaces and their blowups motivates us to ask whether a classification of *all* integrally closed domains between a two-dimensional regular local ring D and its quotient field is possible. This is a rich and varied class of both Noetherian and non-Noetherian rings. As the work of Loper and Tartarone [23] shows, even describing the integrally closed rings between the very specific two-dimensional regular ring $D = \mathbb{Z}[X]_{(x,p)}$ (with p a prime) and the PID overring[1] $\mathbb{Q}D$ is a difficult task that draws on deep results in valuation theory and multiplicative ideal theory. For other special cases, see [26] and [28].

In a series of recent papers, as well as current work in progress, we have attempted to establish a framework for thinking about the integrally closed rings between a two-dimensional regular local ring D and its quotient field F. A first step in this direction, undertaken in [12, 14, 15], is to describe the intersection of a set of two-dimensional regular local rings between D and F. The remarkable explicitness of the two-dimensional regular local rings between D and F, as described in Sect. 2, motivated our study of this class of rings.

Some examples of the types of rings obtained as intersections of two-dimensional regular local rings are given in later sections. We also consider integrally closed overrings of a two-dimensional regular local ring that do not arise this way. In future work we will show how one-dimensional non-Noetherian integrally closed overrings of a two-dimensional regular local ring can arise from intersections of valuation rings. Such examples include a recasting into our setting of Nagata's difficult example from [24, 25] of a one-dimensional completely integrally closed local domain that is not a valuation ring, a ring Krull had conjectured in [19, p. 666] could not exist. A version of this example was also given by Ribenboim [29], and our variation on these ideas in the form of what we call a *Nagata-Ribenboim constructed ring* is inspired by their construction but is approached quite differently by means of the quadratic and valuative trees. There do also exist one-dimensional integrally closed local overrings of a two-dimensional regular local ring that are not completely integrally closed. Such an example was given by Krull in [19, p. 670] and is a pseudo-valuation ring in the sense of Houston and Hedstrom [18]. We will show in future work how the rings arising from his construction, are ubiquitous in the context of overrings of a two-dimensional regular local ring.

In a different direction, there exist many interesting classes of integrally closed overrings of a two-dimensional regular local ring that are not intersections of two-dimensional regular overrings. These include the Prüfer overrings of D. We mention one such class at the end of Sect. 5, but another example is the localization of the ring of integer-valued polynomials at a maximal ideal of $\mathbb{Z}[X]$. This is a particularly interesting class of Prüfer domains, and much of the theory for these rings and their generalizations was developed by Paul-Jean Cahen and Jean-Luc Chabert (see, e.g., [2, 3] and their references), and it is Paul-Jean to whose memory this article is dedicated.

[1] An *overring of* D is understood to be a subring of F that contains D.

2 Local Quadratic Transforms

One of the main tools in working with two-dimensional regular local rings is that of a local quadratic transform. These transforms can be defined more generally for a Noetherian local domain, with no restriction on dimension. They can be viewed algebraically as local rings obtained from a Noetherian local domain through a certain process of extension and localization, or geometrically as local rings of points in the exceptional fiber of the blowup along the maximal ideal of the base domain. If D is a Noetherian local domain with maximal ideal $\mathfrak{m}$ minimally generated by elements $x_1, \ldots, x_n$, then for each i we form the ring $D_i = D[\frac{x_1}{x_i}, \ldots, \frac{x_n}{x_i}]$. A *local quadratic transform* of D is a localization of any of the rings D_i at a prime ideal of D_i that contains $\mathfrak{m}$. Thus a local quadratic transform is the local ring of any point in the fiber over $\mathfrak{m}$ of the blowup of $\operatorname{Spec} D$ at the closed point $\mathfrak{m}$. Different choices of generators for $\mathfrak{m}$ produce different rings D_i but yield the same localizations at prime ideals as the original choice $x_1, x_2, \ldots, x_n$, and so the set of local quadratic transforms obtained from D is independent of the choice of generators of $\mathfrak{m}$.

Our focus is on the case in which D is a two-dimensional regular local ring, and we assume for the rest of the article that D denotes such a ring. In this case, the maximal ideal $\mathfrak{m}$ of D is generated by two elements, say x_1 and x_2. A local quadratic transform of D is a localization of either the ring $D_1 = D[\frac{x_2}{x_1}]$ or the ring $D_2 = D[\frac{x_1}{x_2}]$ at a prime ideal P containing $\mathfrak{m}$. This localization has Krull dimension 1 if P is a nonmaximal prime ideal, and in this case, the localization is a one-dimensional regular local ring, i.e., a discrete rank-one valuation ring (DVR). Otherwise, the localization at P is a two-dimensional regular local ring. The only nonmaximal prime ideal of D_1 that contains $\mathfrak{m}$ is $P_1 = \mathfrak{m}D_1 = x_1D_1$. Similarly, the only nonmaximal prime ideal of D_2 that contains $\mathfrak{m}$ is $P_2 = \mathfrak{m}D_2 = x_2D_2$. It is straightforward to see that $(D_1)_{P_1} = (D_2)_{P_2}$. Thus the only one-dimensional local quadratic transform of D is the DVR $(D_1)_{P_1} = (D_2)_{P_2}$. This DVR has an important interpretation as the valuation ring of the *order valuation* of D, which is the valuation that sends each nonzero element x of D to the largest nonnegative integer k such that $x \in \mathfrak{m}^k$.

The height-two maximal ideals in the rings D_1 and D_2 may also be described explicitly. All of these necessarily contain the maximal ideal $\mathfrak{m}$ of D. For each i,

$$D_i/x_iD_i = D_i/\mathfrak{m}D_i \cong (D/\mathfrak{m}D)[t],$$

where t is an indeterminate over the field $D/\mathfrak{m}D$. Since $(D/\mathfrak{m}D)[t]$ is a PID, there is for each height-two maximal ideal M_1 of D_1 a polynomial $f \in D[t]$ such that the image of f in $(D/\mathfrak{m})[t]$ is irreducible and $M_1 = \left(f\left(\frac{x_2}{x_1}\right), x_1\right)D_1$. Consequently, $(D_1)_{M_1}$ is a two-dimensional regular local ring. We may describe the height-two maximal ideals M_2 of D_2 similarly and obtain two-dimensional regular local rings $(D_2)_{M_2}$. In almost all cases, a localization of D_1 at a height-two maximal ideal is also a localization of D_2 at a height-two maximal ideal. The only localizations for which this is not true are the localization of D_1 at the maximal ideal $(x_1, \frac{x_2}{x_1})D_1$ and the localization of D_2 at the maximal ideal $(x_2, \frac{x_1}{x_2})D_2$. In other words, every local quadratic transform of D but one is obtainable as a localization of D_1. In summary, we have

Theorem 2.1 *The unique local quadratic transform of D of dimension 1 is $(D_1)_{x_1 D_1} = (D_2)_{x_2 D_2}$, the order valuation ring of D. The residue field of this ring is purely transcendental of degree 1 over $D/\mathfrak{m}$. Every other local quadratic transform of D has dimension 2 with residue field algebraic over $D/\mathfrak{m}$, and is of the form*

(a) $(D_2)_{M_2}$, *where* $M_2 = (x_2, \frac{x_1}{x_2})D_2$, *or*
(b) $(D_1)_{M_1}$, *where* $M_1 = (x_1, f(\frac{x_2}{x_1}))D_1$ *for some* $f \in D[t]$ *such that the image of f in $(D/\mathfrak{m})[t]$ is irreducible.*

We denote by $Q_1(D)$ the set of local quadratic transforms of D that have dimension 2. (The reason for the subscript 1 here is that in the next section we will iterate and consider local quadratic transformations of local quadratic transformations.) The set $Q_1(D)$ consists of the rings described in (a) and (b) of Theorem 2.1. Following Lipman [22], we view the rings in $Q_1(D)$ as points and denote these points with Greek letters, e.g., $\alpha \in Q_1(D)$.

There remain many localizations of D_1 and D_2 that are not captured by local quadratic transformations. These are the localizations at prime ideals that do not contain $\mathfrak{m}$. For such a nonzero prime ideal P of D_i, the contraction $P \cap D$ to D is necessarily of height 1 since D is local and P does not contain the maximal ideal of D. Since D is integrally closed, $D_{P\cap D}$ is then a DVR. Such a DVR, arising as a localization of D, is called an *essential prime divisor* of D. Since $D_{P\cap D} \subseteq (D_i)_P$ and $D_{P\cap D}$ is a DVR, so is $(D_i)_P$. It follows that the nonzero prime ideals of D_i are (a) the height-one prime ideals that contract to height-one prime ideals of D, (b) the only height-one prime ideal that does not contract to a height-one prime ideal of D (namely, $x_i D_i$), and (c) the height-two maximal ideals described in Theorem 2.1(b).

There is a more elegant way to express some of these ideas. We may glue $\operatorname{Spec} D_1$ and $\operatorname{Spec} D_2$ by identifying prime ideals $P_1 \in \operatorname{Spec} D_1$ and $P_2 \in \operatorname{Spec} D_2$ whenever $(D_1)_{P_1} = (D_2)_{P_2}$. The resulting topological space $\operatorname{Spec} D_1 \cup \operatorname{Spec} D_2$, equipped with the structure sheaf whose ring of sections over a nonempty open set, is the intersection of the localizations at the prime ideals in this set, is a projective scheme over $\operatorname{Spec} D$. With $\phi : \operatorname{Spec} D_1 \cup \operatorname{Spec} D_2 \to \operatorname{Spec} D$ the contraction mapping, the local quadratic transforms of D are the local rings of the points in the fiber $\phi^{-1}(\mathfrak{m})$ over the closed point $\mathfrak{m}$ of $\operatorname{Spec} D$. The set $Q_1(D)$ consisting of the two-dimensional local quadratic transforms of D is the set of closed points of $\operatorname{Spec} D_1 \cup \operatorname{Spec} D_2$. The closed subscheme $\phi^{-1}(\mathfrak{m})$ of $\operatorname{Spec} D_1 \cup \operatorname{Spec} D_2$, the *exceptional fiber of ϕ*, is isomorphic to the projective line $\operatorname{Proj}((D/\mathfrak{m})[U, V])$, where U and V are indeterminantes. Thus the set $Q_1(D)$ is parameterized by the projective line over $D/\mathfrak{m}$. With this interpretation, we can recast the closed point $(x_2, \frac{x_1}{x_2})D_2 \in \operatorname{Spec} D_2$ from Theorem 2.1 as the point at infinity for the closed points in $\operatorname{Spec} D_1$ described in Theorem 2.1(b).

One of the themes in later sections is the description of rings obtained as intersections of two-dimensional regular local overrings of D. For those rings in $Q_1(D)$, these intersections are Noetherian:

Theorem 2.2 ([14, Theorem 8.3]) *The intersection of rings in each nonempty subset $\mathcal{U}$ of $Q_1(D)$ is a Noetherian normal domain and is irredundant in the sense that no ring in $\mathcal{U}$ can be omitted from the intersection. Moreover, if $\mathcal{U}$ is a nonempty proper subset of $Q_1(D)$, then each ring in $\mathcal{U}$ is a localization of the intersection.*

One way to see that such an intersection is Noetherian is to observe that since each ring α in $Q_1(D)$ is an integrally closed Noetherian domain, each α is an intersection of its essential prime divisors. The essential prime divisors of α are the essential prime divisors of D that contain α and the localization of α at the unique height-one prime ideal of α that contains the maximal ideal of D. The set of essential prime divisors of D has finite character, meaning that each nonzero element in D is a unit in all but finitely many of the rings in the collection. It follows that the intersection of any rings in $Q_1(D)$ is a Krull domain. By [10, Theorem 9], a Krull overring of a two-dimensional Noetherian domain is a Noetherian ring.

This argument and Theorem 2.1 should be contrasted with the situations discussed in the rest of the article, where describing the intersections of two-dimensional regular local overrings of D not necessarily in $Q_1(D)$ is more subtle and produces complicated rings that need not be Noetherian.

3 The Quadratic Tree

For the rest of the article, we continue to assume D is a two-dimensional regular local ring, and we denote by $\mathfrak{m}$ the maximal ideal of D. As we saw in the last section, a local quadratic transform α of D is again a regular local ring, either of dimension 1 or of dimension 2. If α has dimension 1, then α is a DVR, and taking a local quadratic transform of α results in simply α again. However, if α has dimension 2, then a local quadratic transform of α produces a DVR or a new two-dimensional regular local ring, and the process can be iterated. In so doing, we obtain a sequence of regular local rings:

$$\alpha = \alpha_1 \subseteq \alpha_2 \subseteq \alpha_3 \subseteq \cdots \subseteq \alpha_i \subseteq \cdots .$$

This sequence is stationary as soon as α_i has dimension 1. Otherwise, each ring α_i is a distinct two-dimensional regular local ring, and the sequence is infinite. Remarkably, by a theorem of Abhyankar, every two-dimensional regular local overring of D is obtainable in this way:

Theorem 3.1 ([1, Theorem 3]) *Every two-dimensional regular local overring dominating D is obtained from a unique sequence of iterated local quadratic transforms.*

In light of the theorem, we define the *quadratic tree* $Q(D)$ of D to be the set of two-dimensional regular local overrings of D. Since each ring in $Q(D)$ is obtained by a unique sequence of iterated local quadratic transforms, the set $Q(D)$ is a tree with respect to set inclusion. We say that a regular local overring $\alpha \in Q(D)$ *occurs at level n* if there is a sequence $D = \alpha_0 \subseteq \alpha_1 \subseteq \cdots \subseteq \alpha_n = \alpha$ of iterated local quadratic transforms. We denote by $Q_n(D)$ the set of regular local overrings of D that occur at level n in the tree. Note that

$$Q_2(D) = \bigcup_{\alpha \in Q_1(D)} Q_1(\alpha),$$

and for greater choices of n, $Q_n(D)$ can be expressed similarly using the lower levels of the tree. Moreover, if $\alpha \in Q_n(D)$ for some $n > 0$, then $Q_1(\alpha)$ and $Q_1(D)$ have the same cardinality, namely, that of the algebraic closure of the residue field of D. In this sense, when viewed as a partially ordered set, the quadratic tree $Q(D)$ has a simple structure consisting of copies of $Q_1(D)$, and as a partially ordered set, the structure of $Q(D)$ is completely determined by the cardinality of the residue field.

The paths up the tree $Q(D)$ are significant also. If $\{\alpha_i\}$ is an infinite chain in $Q(D)$, then the union of the α_i is a valuation ring that dominates D:

Theorem 3.2 ([1, Lemma 12]) *Every dominating valuation overring V of D is the union of a chain of iterated local quadratic transforms. If the residue field of V is algebraic over $D/\mathfrak{m}$, the chain is infinite. If the residue field of V is not algebraic over $D/\mathfrak{m}$, then the chain is finite, and V is a DVR that is a local quadratic transform of its predecessor in the chain.*

In the last case of the theorem, V is said to be a *prime divisor dominating D*,[2] and V is the order valuation of its predecessor in the chain.

Otherwise, if V is a valuation overring that is residually algebraic over D, then V is by Theorem 3.2 a union of infinitely many regular local rings in $Q(D)$, namely, the local rings in $Q(D)$ that V dominates. Each such valuation ring can be viewed as an endpoint of the tree, and we may extend the partially ordered set $Q(D)$ to include the valuation rings that dominate D. The *extended quadratic tree* $Q^*(D)$ consists of $Q(D)$ and the valuation rings that dominate D, including the prime divisors dominating D. While $Q^*(D)$ remains a partially ordered set under inclusion, it is no longer a tree. This is because each prime divisor dominating D contains infinitely many valuation rings of Krull dimension 2 that dominate D, and these valuation rings are pairwise incomparable. In addition to the prime divisors dominating D, the set of valuation overrings dominating D includes also the discrete valuation rings of Krull dimension 2 and the residually algebraic valuation rings of Krull dimension 1.

[2] Zariski refers to these DVRs as "prime divisors of the second kind," whereas the essential prime divisors of D are "prime divisors of the first kind." Abhyankar refers to prime divisors dominating D as "hidden prime divisors" since these are prime divisors that come out on a blowup.

In higher dimensions—indeed, even in dimension 3—many of the results of this section break down. If D is a regular local ring of dimension $d > 2$, then the quadratic tree of d-dimensional iterated quadratic transforms still makes sense, but there are regular local overrings that are not obtainable from an iterated sequence of quadratic transforms, nor is every valuation ring dominating D obtainable as a union of regular local rings in $Q(D)$. However, much can still be said about the directed unions of iterated quadratic transforms in higher dimensions; see, for example, [5, 7–9, 13, 16, 30].

Extending in a different direction, if we require D only to be a normal Noetherian local domain of dimension 2, and not regular, then it is possible to obtain the dominating valuation overrings of D as unions of iterated local quadratic transforms alternating with normalization and localization; see [20, p. 202].

4 The Topology of the Quadratic Tree

So far we have described the quadratic tree $Q(D)$ and the extension $Q^*(D)$ of $Q(D)$ as partially ordered sets, both of which admit a simple structure. In fact, as a poset, $Q(D)$ is entirely determined by the cardinality of the algebraic closure of the residue field of D. Now we enrich the structure of $Q(D)$ and $Q^*(D)$ by considering these as topological spaces. The *Zariski topology* on $Q^*(D)$ has as a basis of open sets the sets of the form

$$\mathcal{U}(x_1, \ldots, x_n) = \{R \in Q^*(D) : x_1, \ldots, x_n \in R\},$$

where $x_1, \ldots, x_n$ are in the quotient field of D.

Theorem 4.1 ([12, Corollary 4.9]) *$Q^*(D)$ is a spectral space with respect to the Zariski topology; i.e., $Q^*(D)$ is a quasicompact T_0 space such that each irreducible closed set has a unique generic point and the intersection of any two quasicompact open sets is quasicompact.*

Hochster showed in [17] that a spectral space is homeomorphic to the prime spectrum of a commutative ring. While $Q^*(D)$ is therefore homeomorphic to Spec R for some ring R, we do not know of a natural choice of R for $Q^*(D)$. However, there is a natural choice for the spectral subspace consisting of the valuation rings in $Q^*(D)$. (For a proof that this subspace is spectral, see, for example, [14, Lemma 3.2].) In this case, the Kronecker function ring defined via the valuations rings in $Q^*(D)$ has prime spectrum homeomorphic to the space of valuation rings in $Q^*(D)$. This follows, for example, from [27, Proposition 5.6].

The Zariski topology on $Q(D)$ is completely determined by the order-theoretic properties of the poset $Q^*(D)$. To formulate this more precisely, for a nonempty subset S of $Q(D)$, let S^∞ be the set of prime divisors dominating D that contain infinity many incomparable rings in S. For a collection T of rings in the extended quadratic tree $Q^*(D)$, denote by $\downarrow T$ the set of rings in the quadratic tree $Q(D)$ that are contained in rings in T.

Theorem 4.2 ([12, Theorem 5.2]) *If S is a nonempty subset of $Q(D)$, then the Zariski closure of S in $Q(D)$ is $\downarrow (S \cup S^\infty)$.*

The Zariski topology on $Q^*(D)$ is not Hausdorff but can be refined to a Hausdorff topology, the *patch topology*, whose basic open sets are of the form

$$\mathcal{U}(x_1, \dots, x_n; y) = \{R \in Q^*(D) : x_1, \dots, x_n \in R,\ y \notin R\},$$

where $x_1, \dots, x_n, y$ are in the quotient field of D.

The complement of such a basic open set is also open, so the sets $\mathcal{U}(x_1, \dots, x_n; y)$ form a clopen basis of $Q^*(D)$, making $Q^*(D)$ a *Stone space*, a zero-dimensional compact Hausdorff space.

Theorem 4.3 ([12, Corollary 4.11]) *In the patch topology, $Q(D)$ is an open discrete dense subspace of $Q^*(D)$, and thus each valuation ring in $Q^*(D)$ is a patch limit point of $Q(D)$.*

A topological space is *Noetherian* if the open subsets of the space satisfy the ascending chain condition. The next theorem shows that whether a subspace of $Q(D)$ is Noetherian is determined by the order-theoretic properties of the extended quadratic tree $Q^*(D)$. We will see in the next section that the intersection of rings in an open subset of a Noetherian subset of $Q^*(D)$ is a Noetherian domain.

Theorem 4.4 ([12, Theorem 5.6]) *The following are equivalent for a nonempty subset S of $Q^*(D)$.*

(1) *S is a Noetherian space in the subspace Zariski topology.*
(2) *There are valuation rings $V_1, \dots, V_n \in Q^*(D)$ such that each ring in S is contained in one of the V_i.*
(3) *The Zariski closure of S in $Q^*(D)$ is a Noetherian space.*

There is a special case of the theorem that is worth singling out because of the role proximity plays in the study of quadratic transforms. If $\alpha, \beta \in Q(D)$ such that $\alpha \subseteq \beta$ and β is contained in the order valuation ring of α, then β is *proximate* to α. The set of rings β in $Q(D)$ that are proximate to α is denoted $P(\alpha)$.

Corollary 4.5 ([12, Remark 3.4.3 and Corollary 5.11]) *Let $\alpha \in Q(D)$. Then $P(\alpha)$ is a Noetherian subspace of $Q(D)$ that consists of α, the rings in $Q_1(\alpha)$, and, for each $\beta \in Q_1(\alpha)$, the rings in the infinite ascending ray of points in $Q(\beta)$ that are contained in the order valuation ring of α.*

With notation as in Corollary 4.5, the rank-two valuation domains contained in the order valuation ring of α are precisely the valuation rings associated to these rays.

Using ideas from [4, 5], there is another way to view the space of valuation rings birationally dominating D with a tree structure. Take the quadratic tree

$Q(D)$, and in place of each two-dimensional regular local ring, instead consider the order valuation ring associated to it. This yields a $\mathbb{N}$-tree consisting of the prime divisors dominating D. The order of the $\mathbb{N}$-tree is inherited from the order on the quadratic tree. The remaining valuation rings birationally dominating D, which are the minimal valuation rings, are in one-to-one correspondence with the infinite paths up the tree. With $\overline{\mathbb{N}} = \mathbb{N} \cup \{\infty\}$, add these points to the tree to obtain a $\overline{\mathbb{N}}$-tree that is a subspace of the space of all valuation overrings of D. Its patch topology is precisely the topology induced by the $\overline{\mathbb{N}}$-tree structure, where the sets corresponding to subtrees $Q(\alpha)$ and their complements form a subasis of open sets [4, Proposition 5.28].

By considering proximity, one can recover the Zariski topology on the $\overline{\mathbb{N}}$-tree. Given a patch-closed set in the $\overline{\mathbb{N}}$-tree, its Zariski closure is obtained by adding the rank-two valuations associated to each prime divisor in the set. Intuitively, for a path up the tree, following a ray of proximity for a prime divisor V as in Corollary 4.5 is moving "closer" to V in the Zariski topology, whereas leaving that ray creates a separation from V.

By gluing together each prime divisor dominating D with the set of rank-two valuation rings contained in it using the quotient topology, one obtains the *valuative tree* as in [4, Theorem 5.24]. The valuative tree can be viewed as folding all rays of proximity in on themselves. Let $\alpha \in Q(D)$, $\beta \in Q_1(\alpha)$, and let δ be the unique point in $Q_1(\beta)$ which is proximate to α. Then δ has a regular system of parameters (x, y) where $\delta_{x\delta} = V_\alpha$ is the order valuation ring of α and $\delta_{y\delta} = V_\beta$ is the order valuation ring of β. Using continued fractions, one can view the tree of monomial local quadratic transforms of δ as the rational numbers on an open real line segment, say $(1, 2)$. The endpoints 1 and 2 are the limits of the rays of proximity of α and β, and each other rational point is similarly the limits of two rays of proximity. The paths up the monomial tree that don't follow a ray of proximity correspond to irrational monomial valuations and to the irrational numbers in $(1, 2)$. The following diagram illustrates this process, where γ_1 is the unique point in $Q_1(\delta)$ proximate to α and γ_2 is the unique point in $Q_1(\delta)$ proximate to β and where dashed lines indicate this proximity.

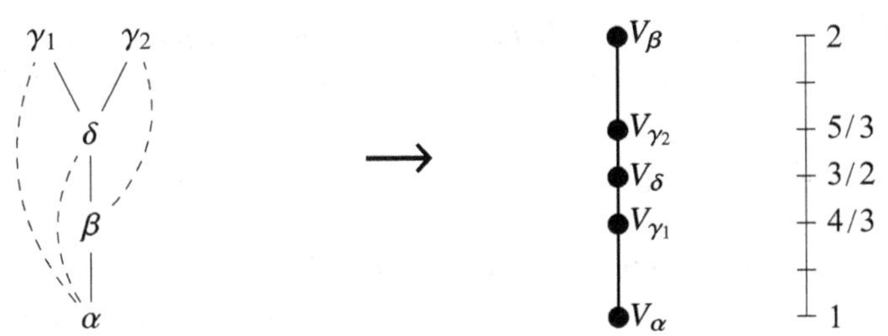

In this way, the discrete $\overline{\mathbb{N}}$-structure of the quadratic tree gives way to the continuous structure of the valuative tree. In future work we will draw out connections between the quadratic tree and the valuative tree to help describe integrally closed overrings between D and F.

5 Intersections of Rings in the Quadratic Tree

The articles [12, 14] examine the structure of overrings of D that arise as the intersection of a nonempty subset of the quadratic tree $Q(D)$. Since the regular local rings in $Q(D)$ are the intersections of DVRs, a ring having this form is necessarily completely integrally closed. It also has Krull dimension 2:

Proposition 5.1 *If R is the intersection of the two-dimensional regular local rings in a nonempty subset of $Q(D)$, then* $\dim R = 2$.

Proof Let T be a two-dimensional regular local ring in $Q(D)$ such that $D \subseteq R \subseteq T$. Since D is a Noetherian domain of Krull dimension 2, the ring R has dimension at most 2. To show that $\dim R = 2$, consider the inclusion map $R \hookrightarrow T$ and the spectral map $\operatorname{Spec} T \to \operatorname{Spec} R$. Since T is a two-dimensional regular local ring. T has infinitely many height-one prime ideals and a nonzero element of T is contained in only finitely many height-one primes. Thus the height-one primes of T intersect in zero.

Let $\mathfrak{m}_T$ denote the maximal ideal of T, and let $\mathfrak{p} = \mathfrak{m}_T \cap R$. Since T is an overring of R, the map $\operatorname{Spec} T \to \operatorname{Spec} R$ sends every nonzero prime ideal of T to a nonzero prime ideal of R. Since $\mathfrak{m}_T$ is the unique maximal ideal of T, every nonzero prime ideal of T maps to a nonzero prime ideal of R that is contained in $\mathfrak{p}$.

If $\mathfrak{p}$ has height one, then every height-one prime of T contains $\mathfrak{p}$. But as we observed above, the height-one prime ideals of T intersect in zero. This contradiction implies that $\dim R = 2$ for every ring R that is the intersection of rings in the quadratic tree $Q(D)$. □

It is not necessarily the case that every maximal ideal has height 2 in a ring that is an intersection of rings from $Q(D)$. The next theorem can be used to show such examples exist. Specifically, if V is a rank-one minimal valuation overring of D, then the following theorem asserts that V occurs as the localization at a maximal ideal M of a ring obtained by intersecting local rings in $Q(D)$. Since V has rank-one, this maximal ideal has height 1.

Theorem 5.2 *[[12, Theorem 6.8]] Let V be a minimal valuation overring of D. There is $X \subseteq Q(D)$ such that V is a localization of the intersection of the rings in X.*

While there exist interesting Prüfer overrings of D (see [15] for example), these rings are never obtained as an intersection of two-dimensional regular overrings, since every overring of a Prüfer ring is Prüfer and a Noetherian normal ring is Prüfer if and only if it has dimension 1. Beyond these constraints, the intersections of rings in $Q(D)$ can exhibit diverse behavior as Theorem 5.2 suggests. We collect more evidence for this in this section. However, at least in the case of finite subsets of $Q(D)$, the intersection has a transparent structure:

Theorem 5.3 ([14, Corollary 5.6]) *If $\alpha_1, \ldots, \alpha_n$ are incomparable rings in $Q(D)$, then the ring $R = \alpha_1 \cap \cdots \cap \alpha_n$ is a regular Noetherian domain of Krull dimension 2 having finitely many maximal ideals $\mathfrak{m}_1, \ldots, \mathfrak{m}_n$. For each i, $\alpha_i = R_{\mathfrak{m}_i}$.*

Not every normal Noetherian overring of D is an intersection of rings from $Q(D)$. This is a consequence of the following theorem.

Theorem 5.4 ([14, Theorem 7.4]) *A normal Noetherian overring R of D is the intersection of rings in $Q(D)$ if and only if every maximal ideal of R has height 2.*

Thus every Noetherian normal local overring of R of dimension 2 is obtained as an intersection of rings in $Q(D)$. On the other hand, no Noetherian local overring of dimension 1 can be obtained as such an intersection, since as discussed above no Prüfer overring can be obtained.

What is lacking from Theorem 5.4 is a criterion for determining when an intersection of regular local overrings of D is Noetherian. To fully address this, the language of projective models is needed, and so we postpone a definitive answer until the next section, where such models are discussed. For the moment, we mention in keeping with the spirit of the last section a topological criterion for when an intersection is Noetherian. The following theorem is a consequence of Theorem 5.4 and [12, Theorem 5.10].

Theorem 5.5 *Let $V_1, \ldots, V_n$ be valuation overrings of D, and let $\mathcal{U}$ be a subset of a Zariski open subset of $Q(D)$ such that each $\alpha \in \mathcal{U}$ is contained in V_i for some i. Then the intersection of the rings in $\mathcal{U}$ is a Noetherian normal domain for which each maximal ideal has height 2.*

By either Theorem 5.4 or 5.5, in order for an intersection of rings in $Q(D)$ to be non-Noetherian, the intersection must involve infinitely many rings in $Q(D)$ and, as we will see in the next section, must also involve a collection of rings that cannot be captured on a blowup of an ideal of D. Theorem 5.2 shows that not only can an intersection of rings in $Q(D)$ fail to be Noetherian, but it can fail in a strong way to be locally Noetherian also.

The next theorem can be compared to Theorem 5.3, in which we saw that an intersection R of finitely many incomparable rings $\alpha_1, \ldots, \alpha_n$ in $Q(D)$ is Noetherian and each α_i is obtained as a localization of R. Theorem 5.6 shows that it is possible to intersect rings in $Q(D)$ in such a way that each of the rings is obtainable as a localization yet the intersection is not a Noetherian ring.

Theorem 5.6 ([12, Theorem 6.15]) *If $D = k + \mathfrak{m}_D$ for an algebraically closed field k, then there exists a subset X of $Q(D)$ such that the intersection R of the rings in X is non-Noetherian yet every ring in X is a localization of R.*

The proof of Theorem 5.6 is technical and intricate, but since the construction of the ring in the theorem is important for understanding the richness of the class of rings obtained as intersections of rings from the quadratic tree, we outline here the heuristic thinking that led to this example.

Recall that $Q_1(D)$ is an infinite collection of two-dimensional regular local rings that intersect to D. Let $\mathfrak{m} = (x, y)$ be the maximal ideal of D. We can place the rings in $Q_1(D)$ into three categories using natural choices for the parameters of their maximal ideals.

1. $\left(y, \frac{x}{y}\right)$
2. $\left(y, f\left(\frac{x}{y}\right)\right)$, where f is a non-monomial irreducible polynomial over $D/\mathfrak{m}$.
3. $\left(x, \frac{y}{x}\right)$

Clearly, categories 1 and 3 each contain only one ring, while category 2 contains infinitely many. The substance of the analysis we go into now involves valuations. However, we will approach things on a much more simplistic level:

- The local ring in category 1 contains $\frac{x}{y}$ as a non-unit. We consider this to be the arena in which x is strictly larger than y.
- All the rings in category 2 contain both $\frac{x}{y}$ and $\frac{y}{x}$. We consider this to be the arena in which x and y are the same size.
- The ring in category 3 contains $\frac{y}{x}$ as a non-unit. We consider this to be the arena in which y is strictly larger than x.

Let α be the ring in category 1. If we intersect just the rings in categories 2 and 3, omitting α, we get a ring which is not local. However, α can be represented as the intersection of the rings in $Q_1(\alpha)$. Is it possible that we can intersect all the rings in categories 2 and 3 with just one of the rings in $Q_1(\alpha)$ and obtain a local ring as a result? We, crudely, think of a local intersection resulting from the rings intersected as being tightly clustered together. Along that line of thinking, we geometrically characterize the rings in $Q_1(\alpha)$ via the parameters used to generate their maximal ideals just as we did with $Q_1(D)$.

1. $\left(y, \frac{x}{y^2}\right)$
2. $\left(y, f\left(\frac{x}{y^2}\right)\right)$ where f is a non-monomial irreducible polynomial over D/m.
3. $\left(\frac{x}{y}, \frac{y^2}{x}\right)$

Now we want to see if it is possible to intersect all of the rings in categories 2 and 3 of $Q_1(D)$ with just one of the rings in $Q_1(\alpha)$ and obtain a local ring as a result. To this end, we characterize the three categories above just as we did with the rings in $Q_1(D)$. We should keep in mind as we proceed that we are starting from the ring α, which means that we are assuming that we have x greater than y. What we can say about the three categories of rings in $Q_1(\alpha)$ then is:

- The ring in category 1 contains $\frac{x}{y^2}$ as a non-unit. We consider this to be the arena in which x is strictly larger than y and x is also strictly larger than y^2
- All the rings in category 2 contain both $\frac{x}{y^2}$ and $\frac{y^2}{x}$. We consider this to be the arena in which x and y^2 are the same size.

- The ring in category 3 contains $\frac{y^2}{x}$ as a non-unit. We consider this to be the arena in which x is strictly larger than y and y^2 is strictly larger than x.

Categories 2 and 3 of $Q_1(D)$ involve rings where y is greater than x or y and x are the same size. If we are going to include something from $Q_1(\alpha)$ and have the entire collection be tightly clustered, then we would want to allow x to be larger than y but only slightly larger. This would indicate that we should include the third of the choices immediately above.

The example cited in Theorem 5.6 is constructed by making exactly the type of substitution described above but for all of the rings in $Q_1(D)$ rather than for just one.

The rings in Theorems 5.2 and 5.6 are obtained by intersecting large collections of rings in $Q(D)$. For example, although the set X in Theorem 5.6 is a subset of $Q_2(D)$, there do not exist $\alpha_1, \ldots, \alpha_n \in Q_1(D)$ such that all the rings in the collection occur in $Q_1(\alpha_1) \cup \cdots \cup Q_1(\alpha_n)$. On the other hand, the set X in Theorem 5.2 is not contained in $Q_1(D) \cup \cdots \cup Q_n(D)$ for any $n \geq 0$. With regard to this last form of unboundedness, it would be interesting to have a comprehensive description of the rings obtained by intersecting rings from a subset of $Q(D)$ of bounded level. In lieu of such a description, we can at least give a representation theorem for such an intersection. Recall that an integral domain R is an *almost Dedekind domain* if each localization of R is a Dedekind domain, i.e., a DVR.

Theorem 5.7 ([12, Corollary 4.14]) *Let $n > 0$. An intersection of rings in a nonempty subset of $Q_1(D) \cup \cdots \cup Q_n(D)$ is the intersection of a flat PID and an almost Dedekind overring.*

The theorem relies on the fact that if X is a nonempty collection of prime divisors dominating D and there is $n > 0$ such that every prime divisor dominating D occurs at level at most n, then the ring $\bigcap_{V \in X} V$ is an almost Dedekind domain [12, Corollary 4.13]. (A version of this remains true if D is only assumed to be a two-dimensional Noetherian local ring; see [15].)

6 Projective Models and the Quadratic Tree

The quadratic tree can be used to give a different perspective on the classical topic of desingularization of surfaces. Let $x_0, x_1, \ldots, x_n$ be nonzero elements of D, and for each i let

$$D_i = D\left[\frac{x_0}{x_i}, \ldots, \frac{x_n}{x_i}\right].$$

The *projective model* over D defined by $x_0, \ldots, x_n$ is the set of local overrings:

$$X = \bigcup_{i=0}^{n} \{(D_i)_P : P \text{ is a prime ideal of } D_i\}.$$

Alternatively, the projective model is the *blowup of* D at the ideal $(x_0, \ldots, x_n)$ of D and can be identified with the projective scheme $\text{Proj}D[(x_0, \ldots, x_n)t]$, where t is an indeterminate. The Zariski topology on X has basic open sets

$$\mathcal{U}(t_1, \ldots, t_m) = \{R \in X : t_1, \ldots, t_m \in R\},$$

where $t_1, \ldots, t_m$ are in the quotient field of D. The closed points in the model are the *two*-dimensional rings in the model. For example, if x_0, x_1 generate the maximal ideal of D, then the projective model defined by x_0, x_1 has as its closed points the set $Q_1(D)$ of local quadratic transforms of D. In this case, the projective model is *nonsingular*, meaning that every local ring in the model is a regular local ring. However, a projective model over D need not be nonsingular nor even normal. (A projective model is *normal* if all its local rings are normal.) The closed points of a nonsingular projective model X over D that are not equal to D are contained in "small" subsets of the quadratic tree: There exist $\alpha_1, \ldots, \alpha_n \in Q(D)$ such that the closed points of X are in $Q_1(\alpha_1) \cup \cdots \cup Q_1(\alpha_n)$. For a more precise description of how the closed points of a nonsingular projective model sit in the quadratic tree, see [14, Theorem 4.6].

A *desingularization* of a projective model X is a nonsingular projective model Y over D such that each local ring in Y dominates a local ring in X. Every projective model over the two-dimensional regular local ring D has a desingularization; see [21]. Since a projective model is a blowup of $\text{Spec}D$ along an ideal of D, it is natural to ask precisely which ideals give the nonsingular blowups. This is answered in [11, Proposition 5.12], where it is shown that if X is a normal projective model that is the blowup of an integrally closed $\mathfrak{m}$-primary ideal I of D, then X is nonsingular if and only if I is "saturated." The definition of saturated ideal is somewhat technical and lengthy, so we omit it here and refer to [11] or [14] for details. In any case, using the concept of saturation, it is shown in [14, Theorem 5.2] that saturation corresponds to desingularization, in that if the ideal I is integrally closed, then the blowup of the saturation of I is the desingularization of the projective model obtained by blowing up I. It is in fact the *minimal desingularization* of X in the sense that every other desingularization of X dominates Y. (If there exists a desingularization, then there exists a unique minimal desingularization [20, Corollary 27.3].) Using this fact, desingularization has a simple interpretation in terms of the quadratic tree.

Theorem 6.1 ([14, Theorem 5.3]) *Let X be a normal projective model over D. The closed points of the minimal desingularization of X are the points in $Q(D)$ that are minimal with respect to dominating a closed point in X.*

Using similar ideas, it is shown in [14, Theorem 5.5(2)] that if $\alpha_1, \ldots, \alpha_n \in Q(D) \setminus \{D\}$ and these local rings are incomparable, then there is a unique nonsingular projective model X over D such that $\alpha_1, \ldots, \alpha_n \in X$ and every nonsingular projective model over D containing $\alpha_1, \ldots, \alpha_n$ dominates X.

We saw in Theorem 5.4 that every normal Noetherian overring R of D for which every maximal ideal has height 2 is the intersection of local rings from the quadratic tree. These rings can even be taken from a nonsingular projective model:

Theorem 6.2 ([14, Theorem 7.4]) *An overring R of D is a normal Noetherian domain whose maximal ideals have height 2 if and only if there is a nonsingular projective model X of D such that R is an intersection of rings in X.*

Acknowledgments We thank Dave Lantz for helpful comments on a previous draft of the article.

References

1. S. S. Abhyankar, On the valuations centered in a local domain, Amer. J. Math., 78 (1956) 321-348.
2. P.-J. Cahen and J.-L. Chabert, Integer-valued polynomials. Amer. Math. Soc. Surveys and Monographs, vol. 48, Providence, 1997.
3. P.-J. Cahen and J.-L. Chabert, What you should know about integer-valued polynomials, Amer. Math. Monthly, 123 (2016), 311–337.
4. C. Favre and M. Jonsson, The valuative tree, Springer-Verlag, 2004.
5. A. Granja, Valuations determined by quadratic transforms of a regular ring, J. Algebra, 280 (2004), no. 2, 699–718.
6. A. Granja, The valuative tree of a two-dimensional regular local ring, Math. Res. Lett., (2007), no. 1, 17–34.
7. A. Granja, M. C. Martinez and C. Rodriguez, Valuations dominating regular local rings and proximity relations, J. Pure Appl. Algebra, 209 (2007), no. 2, 371–382.
8. A. Granja and C. Rodriguez, Proximity relations for real rank-one valuations dominating a local regular ring. Proceedings of the International Conference on Algebraic Geometry and Singularities (Sevilla, 2001). Rev. Mat. Iberoamericana 19 (2003), no. 2, 393–412.
9. L. Guerrieri, W. Heinzer, B. Olberding and M. Toeniskoetter, Directed Unions of Local Quadratic Transforms of Regular Local Rings and Pullbacks, in Rings, Polynomials, and Modules, Springer, 2017, 257–280.
10. W. Heinzer, On Krull overrings of a Noetherian domain, Proc. Amer. Math. Soc., 22 (1969), 217–222.
11. W. Heinzer, B. Johnston, D. Lantz, and K. Shah, Coefficient ideal in and blowups of a commutative Noetherian domain, J. Algebra, 162 (1993), 355–391.
12. W. Heinzer and K. A. Loper and B. Olberding, The tree of quadratic transforms of a regular local ring of dimension two, J. Algebra, 560 (2020), 383–415.
13. W. Heinzer, K. A. Loper, B. Olberding, H. Schoutens and M. Toeniskoetter, Ideal theory of infinite directed unions of local quadratic transforms, J. Algebra, 474 (2017), 213–239.
14. W. Heinzer and B. Olberding, Noetherian intersections of regular local rings of dimension two, J. Algebra, 559 (2020), 320–345.
15. W. Heinzer and B. Olberding, The ideal theory of intersections of prime divisors dominating a normal Noetherian local domain of dimension two. Rend. Sem. Mat. Univ. Padova, 144 (2020), 145–158.
16. W. Heinzer, B. Olberding and M. Toeniskoetter, Asymptotic properties of infinite directed unions of local quadratic transforms, J. Algebra, 479 (2017), 216–243.

17. M. Hochster, Prime ideal structure in commutative rings, Trans. Amer. Math. Soc., 142 (1969), 43–60.
18. J. Hedstrom and E. Houston, Pseudo-valuation domains, Pacific J. Math., 75 (1978), 137–147.
19. W. Krull, Beiträge zur Arithmetik kommutativer Integritâtsbereiche II, Math. Zeit., 41 (1936), 665–679.
20. J. Lipman, Rational singularities, with applications to algebraic surfaces and unique factorization, Publ. Math. Inst. Haute Etudes Sci., 36 (1969), 195-279.
21. J. Lipman, Desingularization of two-dimensional schemes, Ann. Math., 107 (1978),151–207.
22. J. Lipman, On complete ideals in regular local rings, Algebraic Geometry and Commutative Algebra in Honor of Masayoshi Nagata, (1986) 203-231.
23. K.A. Loper and F. Tartarone, A classification of the integrally closed rings of polynomials containing $\mathbb{Z}[X]$, J. Commut. Algebra, 1 (2009), 91–157.
24. M. Nagata, On Krull's conjecture concerning valuation rings, Nagoya Math. J., 4 (1952), 29–33.
25. M. Nagata, Corrections to my paper "On Krull's conjecture concerning valuation rings.", Nagoya Math. J., 9 (1955), 209–212.
26. B. Olberding, Intersections of valuation overrings of two-dimensional Noetherian domains, in Commutative Algebra: Noetherian and non-Noetherian perspectives, Springer, 2010, 335–362.
27. B. Olberding, Affine schemes and topological closures in the Zariski-Riemann space of valuation rings, J. Pure Appl. Algebra, 219 (2015), 1720–1741.
28. B. Olberding and F. Tartarone, Integrally closed rings in birational extensions of two-dimensional regular local rings, Proc. Cambridge Phil. Soc., 155 (2013), 101–127.
29. P. Ribenboim, Sur une note de Nagata relative à un problème de Krull, Math. Z., 64 (1956), 159–168.
30. D. Shannon, Monoidal transforms of regular local rings, Amer. J. Math., **95** (1973), 294-320.
31. O. Zariski and P. Samuel, *Commutative Algebra, Vol. 2*, Van Nostrand, New York, 1960.

Reductions and Core of Ideals in Integral Domains: Some Recent Developments

Salah Kabbaj

Dedicated to the memory of Paul-Jean Cahen.

1 Introduction

Let R be a commutative ring, and let I be an ideal of R. A subideal J of I is called a *reduction* of I if $JI^n = I^{n+1}$, for some positive integer n. The ideal I is called *basic* if it has no proper reduction. The concept of (minimal) reduction was initially introduced and studied by Northcott and Rees in their work on the analytic theory of ideals in Noetherian local rings. In this class of rings, if I is nonbasic, then it has always a minimal reduction J, and, moreover, in the special case of infinite residue field, the minimal number of generators of J agrees with the analytic spread of I [50]. A comprehensive reference on this topic is Huneke and Swanson's book *Integral Closure of Ideals, Rings, and Modules* [33].

Hays investigated reductions in more general settings and proved that most results do not extend outside the class of Noetherian rings, including those results on the existence of minimal reductions [24, 25]. Song and Kim generalized some of Northcott-Rees' results on analytic spread and minimal reductions to Noetherian semi-local rings [56]. Heinzer, Ratliff, and Rush proved the existence of minimally generated reductions through an extension to finite free local unramified rings [30]. Fouli and Olberding studied the existence of proper reductions and their number of generators in the case of finite residue field [18].

The core of I, denoted core(I), is the intersection of all reductions of I. It was initially introduced by Judith Sally in the late 1980s [53] and then appeared in the context of Briancon and Skoda's theorem, which asserts that if R is regular with $\dim(R) = n$, then core(I) contains the integral closure of I^n. In 1995, Huneke and Swanson investigated the core for the special case of integrally closed ideals in the

S. Kabbaj (✉)
Department of Mathematics, KFUPM, Dhahran, Saudi Arabia
e-mail: kabbaj@kfupm.edu.sa

J.-L. Chabert et al. (eds.), *Algebraic, Number Theoretic, and Topological Aspects of Ring Theory*, https://doi.org/10.1007/978-3-031-28847-0_16

class of regular local rings of dimension 2 [32]. During 2001–2005, Corso, Polini, and Ulrich extended the results of [32] by providing an explicit description for the core of some classes of ideals in Cohen-Macaulay local rings [9, 10, 52]. In 2003, Hyry and Smith extended the results of [32] to commutative rings of arbitrary Krull dimension [35]. In 2005, Huneke and Trung resolved some open questions in the literature on core [34]. In 2008 and 2009, Fouli [16] and Fouli-Polini-Ulrich [19] studied the core in arbitrary characteristic. In 2012, Fouli and Morey probed the special case of edge ideals [17] .

The core appears difficult to compute, and most of the works were carried out in the class of Noetherian rings, particularly, Cohen-Macaulay rings. In 2016, we undertook the first study of the notion of core beyond Noetherian settings. This paper consists of four parts surveying four works which investigate reductions and core of ideals in various settings of integral domains such as Prüfer domains, Noetherian domains, and pullback constructions. The four papers involved in this survey are [41–44], co-authored with A. Mimouni. Results are presented and discussed without proofs, and numerous examples are provided (with full details from the original papers).

In Part 2, which covers [41], we appeal to techniques from multiplicative ideal theory to develop formulas for the core in some classes of integral domains, particularly valuation domains and Prüfer domains. In Part 3, which relates to [42], we study the core in the class of Noetherian rings. Here too, we consider objects from multiplicative ideal theory to prove formulas for the core in some classes of one-dimensional Noetherian domains. In Part 4, which relates to [43], we deal with reductions of ideals in diverse contexts of pullbacks. In Part 5, which covers [44], we investigate minimal reductions and core of ideals in various settings of pullback constructions. The main objective is to enrich the literature with new and original examples with explicit formulas for the core.

Throughout, for a domain R, $\mathrm{qf}(R)$ will denote its quotient field, $\overline{R}$ its integral closure, $\mathrm{Spec}(R)$ the set of its prime ideals, and $\mathrm{Max}(R)$ the set of its maximal ideals. If R/M is infinite for each $M \in \mathrm{Max}(R)$, R is said to have infinite residue fields. For a nonzero ideal I of R, let $(I : I) := \{x \in \mathrm{qf}(R) \mid xI \subseteq I\}$ and $I^{-1} = (R : I) := \{x \in \mathrm{qf}(R) \mid xI \subseteq R\}$. The ideal I is invertible in R if $II^{-1} = R$. Any unreferenced material on reduction theory, multiplicative ideal theory, or commutative ring theory is standard as in [14, 22, 33], or [45], respectively.

2 Core of Ideals in Integral Domains

This part covers [41], where techniques from multiplicative ideal theory are used to establish a formula for the core of ideals in some classes of integral domains, including valuation domains and Prüfer domains. Namely, in Sect. 2.1 we investigate the formula $\mathrm{core}(I) = I^2I^{-1}$ for nonzero ideals I and probe it, further, under the effect of stability conditions. We also study the important case of powers of prime ideals in Prüfer domains. Section 2.2 investigates the existence of minimal reductions for Noetherian domains and non-Noetherian domains. Minimal reductions were initially studied by Rees and Sally for local Noetherian rings [53]. Throughout this part, we give illustrative examples and answer some open questions on the core.

2.1 *Core of Ideals*

This section features formulas for the core of ideals in valuation domains and pseudo-valuation domains and the core of prime ideals in Prüfer domains and pseudo-valuation domains. We also investigate the core under stability condition in various classes of domains.

Recall that an ideal is said to be basic if it has no proper reduction. Observe that invertible ideals and idempotent ideals are basic. Indeed, suppose I is invertible, and let $J \subseteq I$ be a reduction of I; i.e., $JI^n = I^{n+1}$ for some positive integer n. Then, $JI^nI^{-1} = I^{n+1}I^{-1}$ so that $JI^{n-1} = I^n$. Reiterate the process n times to reach $J = I$. Now, if I is idempotent; i.e., $I^2 = I$, then $I = I^{n+1} = JI^n \subseteq J \subseteq I$, so that $J = I$.

A domain R has the trace property if $II^{-1} = R$ or $I(R : I) \in \text{Spec}(R)$, for every ideal I of R [13, 14, 47]. Typical examples of domains with the trace property are valuation domains [2] and, more generally, pseudo-valuation domains [29]. In this vein, it is worthwhile noting that in a domain with the trace property, every nonzero ideal I satisfies the inclusion $I^2I^{-1} \subseteq \text{core}(I)$. Indeed, let J be a reduction of I; i.e., $JI^n = I^{n+1}$, for some $n \geq 1$. Then, the trace property yields $II^{-1} = I^nI^{-n}$ [29, Remark 2.13 (b)]. Hence $I^2I^{-1} = I^{n+1}I^{-n} = JI^nI^{-n} \subseteq J$ and thus

$$I^2I^{-1} \subseteq \text{core}(I).$$

The first main result of this section asserts that equality holds in the class of valuation domains. As invertible ideals are basic, we restrict to non-invertible ideals. In the sequel, we denote by $Z(R, I)$ the set of all zero-divisors of R modulo I.

Theorem 2.1 *In a valuation domain V, every non-invertible ideal I satisfies the equality* $\text{core}(I) = I^2I^{-1} = IZ(V, I)$.

Notice that if I is an invertible ideal of V with maximal ideal $\mathfrak{m}$, then $Z(V, I) = \mathfrak{m}$ and so $IZ(V, I) = I\mathfrak{m} \subsetneqq I = \text{core}(I)$. Two open questions in the literature on core ask whether (1) $\text{core}(I)$ compares to $\text{core}(I^n)$ and (2) $\text{core}(I) \subseteq \text{core}(J)$ for any integrally closed ideals $I \subseteq J$. Huneke and Swanson proved that (2) always holds in the class of two-dimensional regular local rings with infinite residue field [32, Proposition 3.15] and $\text{core}(I^n) = I^{2n-2}\,\text{core}(I)$, provided I is an integrally closed ideal [32, Proposition 4.4]. Later, in 2008, Lee answered (2) in the negative in general [46]. Recall, for convenience, that the integral closure of I is the ideal $\overline{I}$ of all elements x of R which satisfies $x^n + a_1x^{n-1} + \cdots + a_n = 0$, where $a_i \in I^i$

for $i = 1, \cdots, n$. Then, I is called integrally closed if $I = \overline{I}$ [33]. Every ideal in a valuation domain is integrally closed [33, Proposition 6.8.1].

Theorem 2.1 offers complete answers to the above questions (1) and (2) in the class of valuation domains. Indeed, first observe that, for any nonzero ideal I and $n \geq 1$, $\text{core}(I^n) = I^{2n}I^{-n} = I^nI^nI^{-n} = I^nII^{-1} = I^{n-1}I^2I^{-1}$ and hence $\text{core}(I^n) = I^{n-1}\,\text{core}(I)$.

The next example shows that (2) does not hold in valuation domains (i.e., the notion of core is not stable under inclusion).

Example 2.2 Let k be a field and X, Y two indeterminates over k. Consider the valuation domain $V := k[[X]] + M$, where $M := Yk((X))[[Y]]$, and let $Q := YV = Yk[[X]] + M^2 \subsetneqq M$. By [31, Corollary 3.6 and Theorem 3.8], $MM^{-1} = M$, so that M is not invertible in V. By Theorem 2.1, $\text{core}(M) = MZ(V, M) = M^2 \subsetneqq Q = \text{core}(Q)$.

Theorem 2.1 does not extend to the class of Prüfer domains. If I is a nonzero ideal of a Prüfer domain, then each reduction J of I satisfies $JI = I^2$ [25, Proposition 1]. Hence $I^2I^{-1} = JII^{-1} \subseteq J$ and so $I^2I^{-1} \subseteq \text{core}(I)$. This inequality can be strict as shown by the next example.

Example 2.3 Let R be an non-Noetherian almost Dedekind domain [22, Example 42.6]. Then, R is a one-dimensional Prüfer domain with no idempotent maximal ideals [22, Theorem 36.5]. Since R is not Dedekind, it contains a non-invertible maximal ideal $\mathfrak{m}$. Since R is completely integrally closed, $\mathfrak{m}^{-1} = (\mathfrak{m} : \mathfrak{m}) = R$. Now, recall Hays' result that, for a domain R, every ideal has no proper reduction if and only if R is a one-dimensional Prüfer domain [24, Theorem 6.1] and [25, Theorem 10]. By this result, we get $\mathfrak{m}^2\,\mathfrak{m}^{-1} = \mathfrak{m}^2 \subsetneqq \mathfrak{m} = \text{core}(\mathfrak{m})$.

The next result establishes an explicit formula for the core of prime ideals and their powers in Prüfer domains.

Theorem 2.4 *Let R be a Prüfer domain and P a nonzero prime ideal of R. Then, for any integer $n \geq 1$, we have* $\text{core}(P^n) = \begin{cases} P^n, & \text{if } P \text{is maximal} \\ P^{n+1}, & \textit{if } P \textit{is not maximal} \end{cases}$

Next, we provide an example of an ideal I in a (Prüfer) domain R such that $I^2I^{-1} \subsetneqq \text{core}(I) \subsetneqq I$. It draws on [15, Example 8.4.1].

Example 2.5 Let B be the (Bézout) ring of entire functions, $\mathfrak{m}$ a maximal ideal of B of infinite height and residue field K, V a nontrivial valuation domain on K, and R the pullback determined by the following diagram of canonical homomorphisms:

$$\begin{array}{ccc} R := \varphi^{-1}(V) & \longrightarrow & V \\ \downarrow & & \downarrow \\ B & \xrightarrow{\varphi} & K = \dfrac{B}{\mathfrak{m}} \end{array}$$

Let $I := \mathfrak{m}^2$. Then, $I^2 I^{-1} \subsetneqq \mathrm{core}(I) \subsetneqq I$. Indeed, by [15, Example 8.4.1], R is Prüfer, $\mathfrak{m}$ is a non-maximal prime ideal of R with $\mathfrak{m}^2 \subsetneqq \mathfrak{m}$, $(R : \mathfrak{m}) = (B : \mathfrak{m}) = (\mathfrak{m} : \mathfrak{m}) = B$, and $P := \bigcap_{n\geq 1} \mathfrak{m}^n \in \mathrm{Spec}(R)$ with $0 \neq P \subsetneqq \mathfrak{m}$. By Theorem 2.4, $\mathrm{core}(I) = \mathrm{core}(\mathfrak{m}^2) = \mathfrak{m}^3$ and $I^2 I^{-1} = \mathfrak{m}^4(R : \mathfrak{m}^2) = \mathfrak{m}^4((R : \mathfrak{m}) : \mathfrak{m}) = \mathfrak{m}^4(B : \mathfrak{m}) = \mathfrak{m}^4 B = \mathfrak{m}^4$. Further, we have that $\mathfrak{m}^{n+1} \subsetneqq \mathfrak{m}^n$, for every positive integer n: Otherwise, assume $\mathfrak{m}^{n+1} = \mathfrak{m}^n$ for some n. Then, by induction on k, we have $\mathfrak{m}^n = \mathfrak{m}^k$ for all $k \geq n$. Hence $P = \mathfrak{m}^n$ and whence $P = \mathfrak{m}$, this is a contradiction. It follows that $\mathfrak{m}^4 \subsetneqq \mathfrak{m}^3 \subsetneqq \mathfrak{m}^2$.

At this point, recall that R is called a pseudo-valuation domain if it is local and shares its maximal ideal with a valuation overring V or, equivalently, if R is a pullback determined by the following diagram of canonical homomorphisms:

$$\begin{array}{ccc} R = \varphi^{-1}(k) & \longrightarrow & k \\ \downarrow & & \downarrow \\ V & \xrightarrow{\varphi} & K := \frac{V}{\mathfrak{m}} \end{array}$$

where $\mathfrak{m}$ is the maximal ideal of V, k is a subfield of K (cf. [26, 27], and $\mathfrak{m}$ is the maximal ideal of R with residue field k [3, Proposition 2.6].

The formula in Theorem 2.1 holds for the class of pseudo-valuation domains issued from algebraic field extensions, as stated below.

Theorem 2.6 *Let R be a pseudo-valuation domain, and let $(V, \mathfrak{m})$ be its associated valuation overring. Then,* $\mathrm{core}(I) = I^2 I^{-1}$, *for each nonzero ideal I of R, if and only if $V/\mathfrak{m}$ is an algebraic extension of $R/\mathfrak{m}$.*

For the prime ideals and their powers, we have the following result.

Proposition 2.7 *Let R be a nontrivial pseudo-valuation domain, P a nonzero prime ideal of R, and n a positive integer. Then,* $\mathrm{core}(P^n) = P^{n+1}$.

We close this section with a discussion of the reductions and core under stability conditions. A nonzero ideal I in a domain is called *strongly stable* (resp., *stable*) if I is principal (resp., invertible) in its endomorphism ring $(I : I)$ [2, 39, 54]. In [51], Olberding prepared the ground to address the correlation between stability and several concepts in multiplicative ideal theory [7, 23, 40]. The next result shed light on the effect of stability and strong stability on the core.

Theorem 2.8 *Let R be a domain and I a nonzero ideal of R.*

(1) Suppose I is stable. Then, J is a reduction of I if and only if $JI = I^2$ if and only if $JT = I$. Moreover, $I^2 I^{-1} \subseteq$ $\mathrm{core}(I)$.
(2) Suppose I is strongly stable. If either R is Prüfer or $(R : T)$ is maximal in R or $(I : I)$ is local, then $\mathrm{core}(I) = I^2 I^{-1}$.

In Sect. 2.2, Example 2.10 (Noetherian) and Example 2.15 (non-Noetherian) provide two illustrative examples for Theorem 2.8.

2.2 Minimal Reductions

Let $(R, \mathfrak{m})$ be a Noetherian local ring and let I be a nonbasic ideal of R (i.e., I admits a proper reduction). A reduction J of I is called *minimal* if no ideal strictly contained in J is a reduction of I. It is known that I admits at least one minimal reduction [33, 50], and it is not unique in general. In [53], Rees and Sally investigated the intersection of minimal reductions with the aim to counteract the lack of uniqueness of minimal reductions. If the residue field is infinite, then any l non-special elements of I generate a minimal reduction of I, where l denotes the analytic spread of I [50]. Consequently, in a Noetherian local ring with infinite residue field, we have

$$\text{“core}(I)\ is\ a\ reduction\ of I \Longrightarrow I is\ basic.\text{”} \tag{1}$$

This section investigates the existence of minimal reductions beyond the setting of Noetherian local rings. Precisely, we show that there are no minimal reductions in Prüfer domains and characterize the existence of minimal reductions in pseudo-valuation domains. We also examine the validity of (1) beyond the class of Noetherian local rings. Next, we appeal to Theorem 2.8 to compute the core and all minimal reductions for some maximal ideal of a Cohen-Macaulay domain. This example supports a conjecture by Corso, Polini, and Ulrich [10, Conjecture 5.1] (cf. [34, 52, 57]), which asserts that if R is a Cohen-Macaulay ring, I is an ideal of analytic spread ≥ 1, and J is a minimal reduction of I with reduction number r, then (under some additional assumptions) we have

$$\text{core}(I) = (J^{r+1} : I^r). \tag{2}$$

Throughout, let $\text{Red}(I)$ (resp., $\text{MinRed}(I)$) denote the set of all reductions (resp., minimal reductions) of I and $|S|$ denote the cardinality of a set S. Our example requires the following lemma. For this purpose, recall that an ideal I is said to be *divisorial* if $I = I_v := (I^{-1})^{-1}$ and a domain is divisorial if all its nonzero (fractional) ideals are divisorial. We will get back to this notion in Sect. 3.1.

Lemma 2.9 *$k[X^2, X^3]$ is a divisorial domain.*

Example 2.10 Let k be a field and X an indeterminate over k. Let $R := k[X^2, X^3]$ and $I := (X^2, X^3)$. Then:

(1) $\text{core}(I) = I^2 = X^4k[X]$.
(2) Every proper (minimal) reduction of I has reduction number equal to 1 and is of the form $J_x := X^2(1 + xX, X^2),\ x \in k$.
(3) $|\text{Red}(I)| = |\text{MinRed}(I)| + 1 = |k| + 1$.
(4) $\forall x \in k, \text{core}(I) = (J_x^2 : I)$.

Proof

(1) Notice that I is a maximal ideal in R. Set $T := (I : I) = I^{-1}$. Then $T = k[X]$, which is the (complete) integral closure of R. Hence $I = X^2T$, whence I is strongly stable. By Theorem 2.8, core$(I) = X^2(R\colon T) = X^2I = X^4k[X]$, as desired. Also, one can view R as $k[u, v]/(v^2 - u^3)$, which is a one-dimensional Cohen-Macaulay domain with a-invariant $a = 0$. By [20], core$(I) = I^{a+2} = I^2$ since I is the maximal ideal of R.

(2) Let $x \in k$. Clearly, $(1 + xX, X^2)T = T$ so $J_xT = X^2T = I$, that is, $J_xI = I^2$. Hence J_x is a reduction of I. Conversely, let J be a proper reduction of I. By Theorem 2.8, $JT = I$, hence $T = (R : I) = (R : JT) = ((R : T) : J) = (I : J)$ which yields $I = JT = J(I : J) \subseteq J(R : J) = JJ^{-1} \subseteq R$. If $JJ^{-1} \subsetneqq R$, then $I = JJ^{-1}$. Therefore $J^{-1} = (I : J) = T$. Hence, as R is divisorial (Lemma 2.9), $J = J_v = (R : J^{-1}) = (R : T) = I$, this is a contradiction. It follows that J is an invertible ideal of R. By [37, pages 27-42], J necessarily has the form $J = (f/g)(1 + xX, X^2)$ for some $f, g \in R$ and $x \in k$. We obtain $X^2T = I = JT = (f/g)(1 + xX, X^2)T = (f/g)T$, which yields $f/g = cX^2$ for some nonzero $c \in k$. So $J = cJ_x = J_x$ since c is a unit in R.

(3) Obvious by (2) and the facts that $J_x \subsetneqq I\ \forall\, x$ and $J_x \nsubseteq J_y\ \forall\, x \neq y$.

(4) Let $x \in k$ and let $J := J_x = X^2(1 + xX, X^2)$. Notice that the analytic spread of I is 1 since $J_0 = (X^2)$.

Claim 1: $I^2 \subseteq (J^2 : I) \subseteq I = X^2T$. Indeed, $J^2 = JI$ implies that $I^3 = JI^2 = J^2I \subseteq J^2$ so that $I^2 \subseteq (J^2 : I)$, as desired. Next, for $x \in k$, let $f \in (J^2 : I)$. Then $fI = fIT \subseteq J^2T = I^2$. So $f \in (I^2 : I) = (X^4T : X^2T) = X^2T = I$. Consequently, $(J^2 : I) \subseteq I$.

Claim 2: $X^5 \notin J^2$. Deny and let $X^5 = (X^4 + 2xX^5 + x^2X^6)f_1 + (X^6 + xX^7)f_2 + X^8f_3$, for some $f_1, f_2, f_3 \in R$. This yields $f_1(0) = 0$ and $2xf_1(0) = 1$, this is a contradiction.

Claim 3: $X^{-2}(J^2 : I)$ is a proper ideal of T. Indeed, $(J^2 : I)$ is an ideal of T and so is $X^{-2}(J^2 : I)$ by Claim 1. Assume by way of contradiction that $X^{-2}(J^2 : I) = T$; i.e., $(J^2 : I) = I$. Then $I^2 = I(J^2 : I) \subseteq J^2$, this is a contradiction by Claim 2 since $X^5 \in I^2 = X^4T$.

By Claim 3, there is $\mathfrak{m} \in \text{Max}(T)$ such that $X^{-2}(J^2 : I) \subseteq \mathfrak{m}$. However, by Claim 1, $X^2T = X^{-2}I^2 \subseteq X^{-2}(J^2 : I)$. So that $\mathfrak{m} = XT$ and thus $(J^2 : I) \subseteq X^3T$. By Claim 2, $(J^2 : I) \subsetneqq X^3T$. It follows that $\mathfrak{m} = X^{-3}I^2 \subseteq X^{-3}(J^2 : I) \subsetneqq T$. By maximality, $\mathfrak{m} = X^{-3}(J^2 : I)$. Consequently, $(J^2 : I) = X^3\mathfrak{m} = X^4T = \text{core}(I)$, as desired.

The first result of this section deals with the case of Prüfer domains. In particular, it proves that, in Prüfer domains, nonbasic ideals have no minimal reductions.

Theorem 2.11 *Let R be a Prüfer domain. Then, a nonzero ideal I has a minimal reduction if and only if* core(I) *is a reduction of I if and only if I is basic.*

Example 2.5 features Prüfer domains with nonzero ideals I such that $I^2I^{-1} \subsetneqq \text{core}(I) \subsetneqq I$. The next example features Prüfer domains with nontrivial basic ideals.

Example 2.12 Let R be a Prüfer domain with two maximal ideals M and N such that $M^{-1} = R$ and N is invertible. Then the ideal $I := MN$ is basic.

Indeed, for such an example, one can take R to be the ring of entire functions which possess infinite height maximal ideals and (height-one) invertible maximal ideals (cf. [15, Corollary 3.1.3, Proposition 8.1.1 (5) and Example 8.4.1]). Now notice that $I^{-1} = (R : MN) = ((R : M) : N) = (R : N) = N^{-1}$. So that $II^{-1} = M$. Suppose by way of contradiction that I is not basic. Let J be a proper reduction of I, that is, $JI = I^2$. Then, $JM = I^2I^{-1} = IM$. Since $JR_M \subsetneqq IR_M$, let $a \in IR_M \setminus JR_M$. Necessarily, we have $JR_M \subsetneqq aR_M$. Therefore $a^{-1}JR_M \subseteq MR_M$. We get $JR_M \subseteq aMR_M \subseteq IMR_M = JMR_M \subseteq JR_M$ and then $JR_M = aMR_M = IMR_M$.It follows that $I^2R_M = I^2MR_M = a^2MR_M$. Consequently, $a^2 \in a^2MR_M$ and thus $1 \in MR_M$; this is a contradiction.

The next two results deal with the case of pseudo-valuation domains. For the reader's convenience, we recall that if R is a pseudo-valuation domain issued from V, then $\mathrm{Spec}(R) = \mathrm{Spec}(V)$, and if P is a non-maximal prime ideal of R, then $R_P = V_P$. For ample details about spectra of pseudo-valuation domains and pullbacks, in general, we refer to [1, 3, 11, 36].

The first result addresses the problem of when the core is a reduction. Then we use it to establish our main result on the existence of minimal reductions in pseudo-valuation domains in which, unlike Prüfer domains and valuation domains, nonbasic ideals admit minimal reductions.

Proposition 2.13 *Let R be a pseudo-valuation domain, $(V, \mathfrak{m})$ its associated valuation overring, and I a nonzero ideal of R.*

(1) If I is an ideal of V, then $\mathrm{core}(I)$ *is a reduction of I if and only if I is basic.*
(2) If I is not an ideal of V, then $\mathrm{core}(I)$ *is a reduction of I only if either I is invertible or $I^2I^{-1} \subsetneqq \mathrm{core}(I)$.*

As a consequence of Theorem 2.6 and Proposition 2.13, if $V/\mathfrak{m}$ is algebraic over $R/\mathfrak{m}$, then $\mathrm{core}(I)$ is a reduction of I if and only if I is basic. A possible occurrence for (2) in the above proposition happens when I is a non-invertible basic ideal of R. For instance, if $V/\mathfrak{m}$ is not algebraic over $R/\mathfrak{m}$, x is a transcendental element of $V/\mathfrak{m}$ over $R/\mathfrak{m}$, and $0 \neq a \in \mathfrak{m}$. Then the ideal $I := a\varphi^{-1}(k + kx)$ of R satisfies $I^2I^{-1} \subsetneqq \mathrm{core}(I) = I$ as settled in the proof of Theorem 2.6.

The next result characterizes the ideals which possess minimal reductions in pseudo-valuation domains and describes these minimal reductions. We will break our findings into two separate and unrelated cases for a given ideal I of R, namely, when I is or is not an ideal of V. In the latter case, we restrict to those pseudo-valuation domains issued from finite extensions. This result also validates Corso-Polini-Ulrich's conjecture mentioned in (2) for pseudo-valuation domains.

Throughout, $U(A)$ denotes the set of all units of a ring A and $\mathrm{Frac}(A)$ denotes the set of all fractional ideals of A.

Theorem 2.14 *Let R be a pseudo-valuation domain and $(V, \mathfrak{m})$ its associated valuation overring with $R \subsetneqq V$. Let I be a nonzero ideal of R.*

(1) *Assume I is an ideal of V. Then I has a proper minimal reduction if and only if $I = aV$ for some $0 \neq a \in I$. Moreover,* $\text{MinRed}(I) = \{auR \mid u \in U(V)\}$.
(2) *Assume I is not an ideal of V and $[K : k] < \infty$. Then I has a proper minimal reduction if and only if $I = a\varphi^{-1}(W)$ for some $0 \neq a \in I$ and k-vector space W such that $k \subsetneqq W \subsetneqq K$ and $W^n = W^{n+1}$ for some (minimal) integer $n \geq 1$. Moreover,* $\text{MinRed}(I) = \{a\varphi^{-1}(kw) \mid w \in W \setminus \{0\}\}$.

Moreover, for both cases, $\text{core}(I) = (J^2 : I)$, *for each minimal reduction J of I.*

We close this section with an illustrative example for Theorem 2.8 (non-Noetherian context) and Theorem 2.14.

Example 2.15 Let $\mathbb{Q}$ denote the field of rational numbers and X an indeterminate over $\mathbb{Q}$. Consider the pseudo-valuation domain $R := \mathbb{Q} + \mathfrak{m}$ issued from $\mathbb{Q}(\sqrt{2}, \sqrt{3})[[X]]$, where $\mathfrak{m} := X\mathbb{Q}(\sqrt{2}, \sqrt{3})[[X]]$. Consider the ideal of R given by $I := X\big(\mathbb{Q}(\sqrt{2}) + \mathfrak{m}\big)$. Clearly, $T := (I : I) = \mathbb{Q}(\sqrt{2}) + \mathfrak{m}$. By Theorem 2.8 and Theorem 2.14, we get $\text{core}(I) = \mathfrak{m}^2$ and $\text{MinRed}(I) = \{wX\mathbb{Q} + \mathfrak{m}^2 \mid 0 \neq w \in \mathbb{Q}(\sqrt{2})\}$.

3 Core of Ideals in One-Dimensional Noetherian Domains

This part covers [42], which contributes to the study of the core in the class of Noetherian rings. Similarly to Part 2, we appeal to objects from multiplicative ideal theory to develop formulas for the core in some classes of one-dimensional Noetherian domains. A first result asserts that, in Noetherian settings, the class of domains satisfying $\text{core}(I) = I^2I^{-1}$ for all nonzero ideals lies strictly between the two classes of one-dimensional domains and TP domains. The equivalence holds in a large class of Noetherian domains. A second result sheds light on the core of ideals on a special class of one-dimensional Gorenstein local domains. The two results are backed with several illustrative examples, where we explicitly compute the core.

3.1 Two Results

In a domain R, if for each nonzero ideal I either $II^{-1} = R$ or $II^{-1} \in \text{Spec}(R)$, then R is said to have *the trace property*, or, for brevity, R is a TP domain [13, 14, 47]. Typical examples of TP domains are valuation domains, pseudo-valuation domains, and Dedekind domains [2, 14, 29].

In [14, Theorem 3.5], it is proved that a Noetherian domain with the trace property is either Dedekind or a one-dimensional domain with a unique non-invertible maximal ideal M such that M^{-1} is equal to its integral closure. Also, , for the reader's convenience, let us recall Hays' result that "in a Noetherian domain *R with infinite residue fields, each nonzero ideal of R has an invertible reduction* *if and only if* $\dim(R) \leq 1$" [24, Theorem 4.4]; and Huckaba-Papick's result that "given a Noetherian local domain (R, M), *the fractional ideal* M^{-1} *is a domain if and only if either* $\dim(R) \geq 2$ *or* $\dim(R) = 1$ *with* $R \subsetneq \overline{R}$ [31, Theorem 3.0].

The first main result (Theorem 4.1) of this section asserts that the class of Noetherian domains satisfying $\text{core}(I) = I^2 I^{-1}$ for all nonzero ideals lies strictly between the two classes of one-dimensional domains and TP domains and the equivalence holds in a special class of Noetherian domains. The proof of this theorem draws on the following lemma, which is interesting on its own.

Lemma 3.1 *Let R be a domain, I a finitely generated ideal, and $S := \bigcup_{n\geq 1}(I^n : I^n)$. Suppose I has an invertible reduction J_o, and let n_o be the smallest positive integer such that $J_o I^{n_o} = I^{n_o+1}$. Then:*

(1) $S = (I^n : I^n)$, *for any* $n \geq n_o$.
(2) $JS = J_o S = IS$, *for any reduction J of I.*
(3) $J_o(R : S) = I(R : S) \subseteq \text{core}(I)$.

Theorem 3.2 *Let R be a Noetherian domain with infinite residue fields. Consider the following conditions:*

(1) R is a TP domain,
(2) $\text{core}(I) = I^2 I^{-1}$, *for each nonzero ideal I of R,*
(3) $\dim(R) \leq 1$.

Then, (1) $\Longrightarrow$ (2) $\Longrightarrow$ (3), *and both implications are irreversible in general.*

Moreover, if (R, M) is local such that M^{-1} is a local domain with maximal ideal M, then the above three conditions are equivalent.

Observe that the assumption "infinite residue fields" is used only for the proof of (1) $\Longrightarrow$ (2). Further, the assumption "M^{-1} is local with maximal ideal M" is neither necessary nor superfluous. Indeed, Examples 3.5 and 3.6 feature one-dimensional Noetherian local domains (R, M) such that M^{-1} is local with maximal ideal $\neq M$ and Condition (2) above holds in Example 3.5 but not in Example 3.6.

Recall, once more, that a domain is divisorial if all its nonzero (fractional) ideals are divisorial. Divisorial domains were investigated by Bass [4] and Matlis [48] in the class of Noetherian rings, Heinzer [28] in the class of integrally closed rings, Bastida-Gilmer [5] for the $D + M$ rings, and Bazzoni [6] for more general settings. At this point, recall Matlis' result that *a nontrivial Noetherian local domain is divisorial if and only if* $\dim(R) = 1$ *and* M^{-1}/R *is a simple R-module* [6, Theorem A]. Also, two of Hays' results state that, *in a one-dimensional local Noetherian domain with infinite residue field, every nonzero ideal has a principal reduction*; and *a nonzero ideal is basic if and only if it is principal* [24, Theorem 4.4 & Corollary 4.5]. Moreover, a recent result, due to Fouli and Olberding, asserts that *if (R, M) is a one-dimensional local Noetherian domain, then* $\left|\text{Max}(\overline{R})\right| \leq \left|R/M\right|$ *if and only if every ideal of R has a principal reduction* [18, Corollary 3.3].

Throughout, for an ideal I, we denote by $\text{pcore}(I)$ the *principal core* of I, i.e., the intersection of all principal reductions of I. The second main result of this paper (Theorem 4.3) establishes correlation between the core and principal core in a special class of Noetherian local divisorial domains and will be used to compute the core in the next section. It is worthwhile noticing that, in general, $\operatorname{core}(I) \subsetneqq \operatorname{pcore}(I)$, as shown by Example 3.10.

The following lemma sheds light on principal reductions of an ideal and is used in the proof of Theorem 4.3 as well as in the examples of the next section.

Lemma 3.3 *Let R be a domain and I a finitely generated ideal. Suppose I has a principal reduction $a_o R$, and let $S := \bigcup_{n \geq 1}(I^n : I^n)$, $U_o := U(S) \cap (I : a_o R)$, and $P_o := \bigcap_{u \in U_o} uR$. Then, J is a principal reduction of I if and only if $J = a_o u R$ for some $u \in U_o$. Moreover,* $\operatorname{pcore}(I) = a_o P_o$.

Theorem 3.4 *Let $(R, \mathfrak{m})$ be a nontrivial Noetherian divisorial (i.e., one-dimensional Gorenstein) local domain such that $|\operatorname{Max}(\overline{R})| \leq |R/\mathfrak{m}|$ and $\mathfrak{m}^{-1}$ is a TP domain. Let I be a nonbasic ideal of R and aR any principal reduction of I.*

(1) Suppose $\mathfrak{m}^{-1}$ is local. Then, $\operatorname{core}(I) = \operatorname{pcore}(I)$.
(2) Suppose $\mathfrak{m}^{-1}$ is not local. Then, $\operatorname{core}(I) = a\,\mathfrak{m} \cap \operatorname{pcore}(I)$.

3.2 *Illustrative Examples*

In this section, all results obtained in the previous section are illustrated with explicit examples, where we compute the core. Examples are provided with full details from the original papers. The first example presents a Noetherian local non-TP domain in which $\operatorname{core}(I) = I^2 I^{-1}$ holds for each nonzero ideal I, showing that the implication (1) $\Longrightarrow$ (2) of Theorem 4.1 is irreversible in general. Recall that a nonzero ideal I is stable (resp., strongly stable) if it is invertible (resp., principal) in the ring $(I : I)$.

Example 3.5 Let k be a field and X an indeterminate over k. Let $R := k[[X^2, X^5]]$. Then, R is a one-dimensional Noetherian local domain with maximal ideal $M = (X^2, X^5)$. We claim that R is a strongly stable divisorial domain. Indeed, it is easy to check that $M^{-1} = k[[X^2, X^3]]$. Further, $k[[X^2, X^5]] \subsetneqq k[[X^2, X^3]]$ is a minimal extension; that is, M^{-1}/R is a simple R-module. Hence R is a divisorial domain by [6, Theorem A]. Next, let I be a non-invertible ideal of R and set $T := (I : I)$. Then $II^{-1} \subseteq M$ and so $k[[X^2, X^3]] = M^{-1} \subseteq (II^{-1})^{-1} = (I_v : I_v) = (I : I) = T \subseteq k[[X]]$. Since the extension $k[[X^2, X^3]] \subsetneqq k[[X]]$ is minimal, either $T = k[[X]]$ or $T = k[[X^2, X^3]]$. In the first case, $k[[X]]$ is a DVR and so I would be invertible in T. Next, assume $T = k[[X^2, X^3]]$ and $I(T : I) \subseteq (X^2, X^3)$. Then, we obtain $k[[X]] = (X^2, X^3)^{-1} \subseteq (T : (I(T : I))) = (I_{v_1} : I_{v_1})$. where v_1- denotes the v-operation with respect to T. Similar arguments as above ensure (via [6, Theorem A]) that T is a divisorial domain. So $(I_{v_1} : I_{v_1}) = (I : I) = T$ forcing $k[[X]] = T$, which is absurd. Necessarily, I is invertible in T. Consequently, in both cases, I is strongly stable. Since T is local, by Theorem 2.8, $\operatorname{core}(I) = I^2 I^{-1}$, as desired. But, R is not a TP-domain as $M^{-1} \subsetneqq \overline{R} = k[[X]]$.

Next, we show that (2) $\Longrightarrow$ (3) of Theorem 4.1 is irreversible in general.

Example 3.6 Let k be a field and X an indeterminate over k. Let $R := k[[X^3, X^4]]$. Then, R is a one-dimensional Noetherian local domain with maximal ideal $M := (X^3, X^4)$ and hence $T := (M : M) = M^{-1} = k[[X^3, X^4, X^5]]$. Since $R \subsetneqq T$ is a minimal extension, R is a divisorial domain [6, Theorem A]. Also, we have $S := R^M = (M^2 : M^2) = k[[X]]$ and so $(R : S) = M^2 = X^6k[[X]]$. Now, let $f \in \text{core}(M)$, say $f = X^3\left(a_o + a_3X^3 + a_4X^4 + a_6X^6 + \dots\right)$ and let $J_1 := X^3(X+1)R$ and $J_{-1} := X^3(X-1)R$. Both J_1 and J_{-1} are subideals of M with $J_1M^2 = J_{-1}M^2 = M^3$, that is, both are (principal) reductions of M in R. Hence $f \in J_1 \cap J_{-1}$ and a routine verification shows that $a_o = a_3 = a_4 = 0$. So $f \in X^9k[[X]] = M^3$; i.e., $\text{core}(M) \subseteq M^3$. On the other hand, let J be any reduction of M in R. By Lemma 3.1, $JS = MS$ and so $JM^2 = M^3 \subseteq J$. Hence $M^3 \subseteq \text{core}(M)$ and so $\text{core}(M) = M^3 \subsetneqq M^2 = M^2T = M^2M^{-1}$. □

Follow three illustrative examples for Theorem 4.3. The first two examples provide local Noetherian divisorial domains (R, M) such that M^{-1} is a *local* TP domain and R is a strongly stable domain in the first example, but not in the second example. The third example provides a local Noetherian divisorial domain (R, M) such that M^{-1} is a TP domain which is not local.

Example 3.7 Let k be an infinite field and X an indeterminate over k, and let $R := k[[X^2, X^5]]$. In Example 3.5, we saw that R is a local Noetherian strongly stable divisorial domain with maximal ideal $M = (X^2, X^5)$ and $M^{-1} = k[[X^2, X^3]]$. Since $\left(M^{-1} : (X^2, X^3)\right) = \overline{M^{-1}} = k[[X]]$, M^{-1} is a *local* TP-domain [14, Theorem 3.5]. Next, let $I := (X^4, X^5)R = X^4k[[X]]$. Now, consider S, U_o, and P_o from Lemma 3.3. Clearly, $X^4I = I^2$ and then Lemma 3.1 yields $S = (I : I) = k[[X]]$. Moreover, $U_o = U(S)$ and $P_o \subseteq M$. Let $f \in P_o$. Then, $f \in (1+X)R \cap M$, and, through polynomial identification, one can check that $f \in X^4k[[X]]$, that is, $P_o \subseteq X^4k[[X]]$. The reverse inclusion trivially holds since I is an ideal of S. It follows that $\text{pcore}(I) = X^4P_o = X^8k[[X]]$. By Theorem 4.3, $\text{core}(I) = \text{pcore}(I) = X^8k[[X]]$. □

Example 3.8 Let k be an infinite field, X an indeterminate over k, and let $R := k[[X^3, X^4]]$ and $I := (X^7, X^8) \subseteq M := (X^3, X^4)$. Then, (R, M) is divisorial [49, Theorem 2.2] and $T := M^{-1} = k[[X^3, X^4, X^5]]$ is *local* with maximal ideal $N := (X^3, X^4, X^5)$. Since $(T : N) = \overline{T} = k[[X]]$, T is a TP-domain [14, Theorem 3.5]. Notice that I (and a fortiori R) is not stable since $(I : I) = T$ and so $I(T : I) = I\big(X^{-4}k[[X]]\big) = N$. Next, consider S, U_o, and P_o from Lemma 3.3. Clearly $I^2 = (X^{14}, X^{15}, X^{16})$ and $I^3 = (X^{21}, X^{22}, X^{23})$, yielding $X^7I = (X^{14}, X^{15}) \subsetneqq I^2$ and $X^7I^2 = I^3$; that is, X^7R is a principal reduction of I and so $S = (I^2 : I^2) = k[[X]]$ by Lemma 3.1. Moreover, one can easily check that $U_o = \{a + bX + X^3g \mid a, b \in k \text{ and } g \in k[[X]]\}$. Let $f \in P_o$. Then, $f \in M \cap \left(\bigcap_p (1+pX)R\right)$, where p ranges over the positive prime integers. Through polynomial identification, we get $f \in X^6k[[X]]$; that is, $P_o \subseteq X^6k[[X]]$. On the other hand, for every $u \in U_o$, $u^{-1}X^6k[[X]] = X^6k[[X]] \subseteq R$ and so $X^6k[[X]] \subseteq uR$. Hence $X^6k[[X]] \subseteq P_o$

and therefore $P_o = X^6k[[X]]$. It follows that $\text{pcore}(I) = X^7P_o = X^{13}k[[X]]$. By Theorem 4.3, $\text{core}(I) = \text{pcore}(I) = X^{13}k[[X]]$. □

Example 3.9 Let k be an infinite field, X an indeterminate over k, and let $R := k + X(X-1)k[X]_{(X)\cup(X+1)}$. Then R is one-dimensional local Noetherian domain with maximal ideal $M := X(X-1)k[X]_{(X)\cup(X+1)}$. Since $M^{-1} = k[X]_{(X)\cup(X+1)}$ is a semi-local Dedekind domain with exactly two maximal ideals N_1 and N_2 with $M = N_1N_2$, M^{-1}/M is a 2-dimensional R/M-vector space. Hence R is a divisorial domain [6, Theorem A]. Also, since $M^{-1} = \bar{R}$ is Dedekind, R is a TP-domain [14, Theorem 3.5]. Next, let I be a nonbasic ideal of R with a principal reduction a_oR and consider S, U_o, and P_o from Lemma 3.3. Then, $II^{-1} = M$ and so $M^{-1} = (II^{-1})^{-1} = (II^{-1} : II^{-1}) = (I_v : I_v) = (I : I)$. Hence, $M^{-1} \subseteq S \subseteq \overline{R} = M^{-1}$; that is, $S = (I : I) = M^{-1}$. It follows that $U_o = U(S)$ and so, for every $u \in U(S)$ and $m \in M$, $mu^{-1} \in M \subseteq R$. Hence $m \in uR$, whence $M \subseteq P_o$. Thus $P_o = M$. Consequently, $\text{pcore}(I) = a_oP = a_oM$ and therefore $\text{core}(I) = a_oM \cap \text{pcore}(I) = a_oM$ by Theorem 4.3. □

Beyond the scope of Theorem 4.3, the next example features a *non-local* Noetherian divisorial domain R with a maximal ideal M such that M^{-1} is a *non-local* TP domain and with a non-basic ideal $I \subseteq M$ that has a principal reduction aR with $\text{core}(I) = aM \cap \text{pcore}(I) \subsetneqq \text{pcore}(I)$.

Example 3.10 Let k be an infinite field, X an indeterminate over k, $R := k[X^2, X^5]$ and $I := (X^4, X^5)R \subseteq M := (X^2, X^5)R$. Then, $M^{-1} = k[X^2, X^3]$ and, by [49, Theorem 2.2], R is divisorial. Moreover, $\left(M^{-1} : (X^2, X^3)\right) = \overline{M^{-1}} = k[X]$ and, for any $N \in \text{Max}(M^{-1})$, $N(M^{-1} : N) = N$ implies $(M^{-1} : N) = (N : N) \subseteq k[X]$ and so $N = (X^2, X^3)$. Hence, by [14, Theorem 3.5], M^{-1} is a TP domain. Further, $X^4I = I^2$, that is, X^4R is a principal reduction of I. By Lemma 3.1, $S := \bigcup_{n\geq 1}(I^n : I^n) = (I : I) = k[X]$ and $X^4k[X] = I$. Since $U(R) = U(S)$, by Lemma 3.3, X^4R is the unique principal reduction of I and so $\text{pcore}(I) = X^4R$. Moreover, it is easy to check that $I^{-1} = k[X]$ and hence $Q := (R : S) = (R : I^{-1}) = I_v = I$. By Lemma 3.1, $X^8k[X] = X^4Q \subseteq \text{core}(I)$. On the other hand, let $f := X^4g \in \text{core}(I) \subseteq \text{pcore}(I)$ with $g = a_o + a_2X^2 + X^4g'$, for some $a_o, a_2 \in k$ and $g' \in k[X]$, and let $J := X^4(1 + X, X^4)R \subset I$. The basic fact that $1 - X^4 \in (1 + X)S$ yields $(1 + X, X^4)S = S$ and so $JS = X^4S$. Hence $JI = I^2$; that is, J is a reduction of I. Therefore $f \in J$. It follows that $g = (1+X)h + X^4h'$, for some $h, h' \in R$, which forces $a_o = a_2 = 0$ and so $f \in X^8k[X]$. Consequently, we have $\text{core}(I) = X^8k[X] = X^4M \cap \text{pcore}(I) \subsetneqq X^4R = \text{pcore}(I)$. □

4 Reductions of Ideals in Pullbacks

This part covers [43], which deals with reductions of ideals in diverse contexts of pullbacks. We study reductions of several types of ideals in both classical and generic pullback constructions. Moreover, we characterize pullbacks in which reductions of ideals extend to reductions of their respective extended ideals. All results are backed with new examples, where we explicitly determine the reductions.

Let T be a domain, M a maximal ideal of T, K its residue field, $\varphi : T \longrightarrow K$ the canonical surjection, D a proper subring of K, and $k := \mathrm{qf}(D)$. Let R be the pullback issued from the following diagram of canonical homomorphisms:

$$\begin{array}{ccc} R & \longrightarrow & D \\ (\square)\ \downarrow & & \downarrow \\ T & \xrightarrow{\varphi} & K = T/M. \end{array}$$

So, $R := \varphi^{-1}(D) \subsetneqq T$. We shall refer to the diagram ($\square$) as *generic*, and we say R is a generic pullback issued from (T, M, D). If $T = V$ is a valuation domain, we say R is a classical pullback issued from (V, M, D), and in the special case when $D = k$, we say R is a pseudo-valuation domain (PVD, for short) issued from (V, M, k). Also, we will assume familiarity with the prime ideal structure along with basic ideal-theoretic properties of generic and classical pullbacks as in [1, 3, 5, 8, 11, 12, 21, 26, 36, 38].

Recall that, for a ring homomorphism $R \longrightarrow S$, if J is a reduction of I in R, then JS is a reduction of IS in S. The converse holds if the homomorphism is faithfully flat [33]. Also, in a Prüfer domain, if J is a reduction of I then $JI = I^2$ [25].

Let us recall some basic facts on the structure of ideals in a pullback R issued from (T, M, D). If I is an ideal of R with $M \subsetneqq I$, then $I = \varphi^{-1}(A)$ for some nonzero ideal A of D; and if T is local, then every ideal of R is comparable to M. Further, if T is a valuation domain and $I \subsetneqq M$, then either I is an ideal of T or $IT = aT$, for some nonzero $a \in M$; and in this case, if I is not an ideal of T, then $I = a\varphi^{-1}(W)$ for some D-submodule W of K with $D \subseteq W \subsetneqq K$ [5, 11].

Throughout, we denote by $\mathrm{PRed}_R(I)$ the set of all principal reductions of I in R; and recall that $\mathrm{Red}_R(I)$ denotes the set of all reductions of I in R. The first result of this section investigates reductions in generic pullbacks for three types of ideals: $I = M$, $I \supsetneqq M$, and $I + M = R$.

Theorem 4.1 *Let R be a pullback issued from (T, M, D) and I an ideal of R.*

(1) If $I = M$, then $\mathrm{Red}_R(M) = \mathrm{Red}_T(M) \Leftrightarrow M$ is basic in $R \Leftrightarrow M = M^2$.
(2) If $M \subsetneqq I := \varphi^{-1}(A)$, then $\mathrm{Red}_R(I) = \{\varphi^{-1}(B) \mid B \in \mathrm{Red}_D(A)\}$.
(3) If $I + M = R$, then: $\mathrm{Red}_R(I) = \{J \subseteq I \mid JT \in \mathrm{Red}_T(IT)$ with $J + M = R\} = \{H \cap R \mid H \in \mathrm{Red}_T(IT)\}$.

Here is an illustrative example for Theorem 4.1.

Example 4.2 Let $T := \mathbb{R}[X, Y] = \mathbb{R}+M$, where $M := (X, Y)T$, $R := \mathbb{Q}+M$, and $I := (X-1, Y)R$. Note that T is Noetherian with dimension 2, R is non-Noetherian with dimension 2 [8, Theorem 4], and M is a maximal ideal of T with $I + M = R$. Now, $IT = (X - 1, Y)T$ is a 2-generated height-two ideal in T. Hence IT is of the principal class and whence a basic ideal of T by [24, Theorem 2.3]. This argument cannot apply to I since R is not Noetherian. However, Theorem 4.1(3) yields $\text{Red}_R(I) = \{H \cap R \mid H \in \text{Red}_T(IT)\} = \{IT \cap R\} = \{I\}$. That is, I is a basic ideal of R.

The next result deals with the special case of classical pullbacks, for all four types of ideals; $I = M$, $M \subsetneqq I$, $0 \neq I \subsetneqq M$ with I an ideal of V, and $I \subsetneqq M$ with I not an ideal of V. The first two cases $I = M$ and $M \subsetneqq I$ are similar to Theorem 4.1(1)&(2). The last two cases are handled below.

Theorem 4.3 *Let R be a pullback issued from (V, M, D) and I an ideal of R.*

(1) If $0 \neq I \subsetneqq M$ and I is an ideal of V, then $\text{Red}_R(I) = \text{Red}_V(I) \Longleftrightarrow \text{PRed}_R(I) = \emptyset \Longleftrightarrow \text{PRed}_V(I) = \emptyset$.
(2) If $I \subsetneqq M$ and I is not an ideal of V, i.e., $I = a\varphi^{-1}(W)$, for some $0 \neq a \in M$ and D-submodule W with $D \subseteq W \subsetneqq K$, then $\text{Red}_R(I) = \{a\varphi^{-1}(H) \mid$ *H submodule of W with $HW^n = W^{n+1}$ for some $n \geq 0\}$.*

Here is an illustrative example for Theorem 4.3.

Example 4.4 Let $V := \mathbb{Q}\left(\sqrt{2}, \sqrt{3}\right)[[X]] = \mathbb{Q}\left(\sqrt{2}, \sqrt{3}\right) + M$, with $M := XV$, $R := \mathbb{Z} + M$, and $I := X(W + M)$, with $W := \mathbb{Q}\left(\sqrt{2}\right)$. Since W is a field, $W^n = W$ for every positive integer n. So, $\text{Red}_R(I) = \{X(H + M) \mid H\ \mathbb{Z}\text{-submodule of } W \text{ with } HW = W\}$ by Theorem 4.3 (e.g., $X(\mathbb{Q} + M)$ and $X(n\mathbb{Z} + M)$, for any integer $n \geq 1$, are proper reductions of I in R).

The next example shows that Theorem 4.3(1) does not carry up, in general, to generic pullbacks.

Example 4.5 Let $T_1 := \mathbb{Q}\left(\sqrt{2}, \sqrt{3}\right)[[X, Y]] = \mathbb{Q}\left(\sqrt{2}, \sqrt{3}\right) + M$, with $M := (X, Y)T_1$, $T := \mathbb{Q}\left(\sqrt{2}\right) + M$, $R := \mathbb{Q} + M$. Observe that T_1, T, and R are local Noetherian wit dimension 2 and share the same maximal ideal M. Let $0 \neq a \in M$ and $I := aM$. Clearly, M is not basic in T since $J = (X, Y)T$ is a proper reduction of M in T; and however M is basic in T_1 by the principal class property [24, Theorem 2.3]. Further, $a^{-1} \notin T_1 = (M : M)$ and so $I \subsetneqq M$. We claim that $\text{PRed}_R(I) = \emptyset$. Assume, for contradiction, that I has a principal reduction bR in R, for some $0 \neq b \in R$. Then, $a^{-1}bT_1 \subsetneqq M$ and $bI^n = I^{n+1}$ for some integer $n \geq 1$. Hence $a^{-1}bM^n = M^{n+1}$ and whence $a^{-1}bT_1$ is a proper reduction of M in T_1, this is a contradiction.

On the other hand, let $J := (aX, aY)R$. Since $JT_1 = aM = I$ and I is an ideal of T_1, $JI = I^2$ and so J is a reduction of I in R. But J is not an ideal of T and so $J \notin \text{Red}_T(I)$. Therefore, $\text{Red}_T(I) \subsetneqq \text{Red}_R(I)$.

The next result examines the extension of reductions in classical pullbacks, showing that this extension property characterizes a special class of PVDs.

Theorem 4.6 *Let R be a classical pullback issued from (V, M, D). Then, the following assertions are equivalent:*

(1) *For each ideal* $I \subseteq M$ *of* R, $J \subseteq I$ *is a reduction of* $I \Leftrightarrow JV$ *is a reduction of* IV.
(2) R *is a PVD issued from* (V, M, k), *where* K *is algebraic over* k *and, for every* k*-vector subspace* W *of* $K \supseteq k$, W^n *is a field for some positive integer* n.

Follow an illustrative example for Theorem 4.6 (and also Theorem 4.3).

Example 4.7 Let $V := \mathbb{Q}\left(\sqrt{2}, \sqrt{3}\right)[[X]] = \mathbb{Q}\left(\sqrt{2}, \sqrt{3}\right) + M$, with $M := XV$, $R := \mathbb{Q}+M$, and $I := X\left(\mathbb{Q}\left(\sqrt{2}\right) + M\right)$. Observe that $I \subsetneqq M$ and I is not an ideal of V. Further, since $W := \mathbb{Q}\left(\sqrt{2}\right)$ is a field, $W^n = W$ for every positive integer n. Then, $\text{Red}_R(I) = \left\{X(H + M) \mid H \ \mathbb{Q}\text{-subspace of } W \text{ with } HW = W\right\}$. by Theorem 4.3(2). Let $J := X(H + M) \in \text{Red}_R(I)$. If $\dim_{\mathbb{Q}} H = 1$, then $H = a\mathbb{Q}$, for some $0 \neq a \in W$ and so $J = aXR$. If $\dim_{\mathbb{Q}} H = 2$, then $H = W$ and so $J = I$. Moreover, by Theorem 4.3(2), $\text{Red}_R(I) = \left\{J \subseteq I \mid JV \in \text{Red}_V(IV)\right\}$. Now, $IV = M$ is basic in V and so $\text{Red}_R(I) = \left\{J \subseteq I \mid JV = M\right\} = \left\{I\right\} \cup \text{PRed}_R(I) = \left\{I\right\} \cup \left\{aXR \mid 0 \neq a \in \mathbb{Q}\left(\sqrt{2}\right)\right\}$. Consequently, a combination of both theorems points to the fact that the proper subideals of I which extend to M in V are exactly the principal reductions of I; i.e., they have the form aXR, where $0 \neq a$ ranges over $\mathbb{Q}\left(\sqrt{2}\right)$.

5 Minimal Reductions and Core of Ideals in Pullbacks

This part covers [44], which deals with minimal reductions and core of ideals in various setting of pullback constructions. The aim is to build new and original examples, where we explicitly compute the core. Once more, we use methods and objects from multiplicative ideal theory to probe the existence of minimal reductions in Sect. 5.1 and then develop explicit formulas for the core in Sect. 5.2, with illustrative examples and counterexamples to delimit the scoops of the main results.

5.1 *Minimal Reductions*

This section studies the existence of minimal reductions for some classes of ideals in pullbacks. Based on Part 4, the case when $M \subsetneqq I$ is simple, as shown below.

Proposition 5.1 *Let* R *be a generic pullback issued from* (T, M, D) *and* I *an ideal of* R. *Suppose* $M \subsetneqq I := \varphi^{-1}(I_o)$, *for some nonzero ideal* I_o *of* D. *Then, the minimal reductions of* I *have the form* $J := \varphi^{-1}(J_o)$, *where* J_o *is a minimal reduction of* I_o.

Notice that if I is an ideal of R with $IT = T$, then $M \subsetneq I$. Therefore, we restrict our investigation to the ideals of R that survive in T. We start with two results which link the existence of minimal reductions to the maximality of the conductor in pullbacks (and, in fact, in the general setting of extensions of domains).

Proposition 5.2 *Let $A \subseteq B$ an extension of domains such that the conductor ideal $(A : B)$ is maximal in B. Let I be a nonzero ideal of A and B that is finitely generated in $(I : I)$. If $(A : B)$ is maximal in A, then I has a minimal reduction in A.*

Note that Example 5.9 shows that Proposition 5.2 does not hold, necessarily, if I is not an ideal of both A and B. The second result establishes the converse for a special category of ideals.

Proposition 5.3 *Let $A \subseteq B$ an extension of domains such that the conductor ideal $(A : B)$ is maximal in B. Let I be an ideal of A with $I = aB$, for some $0 \neq a \in (A : B)$. If aA is a minimal reduction of I in A, then $(A : B)$ is maximal in A.*

As an application of Propositions 5.2 &5.3, the next result probes the existence of minimal reductions, in generic pullbacks, for a special category of stable ideals.

Proposition 5.4 *Let R be a generic pullback issued from (T, M, D) and $I \subseteq M$ be a strongly stable (resp., stable) ideal of R with $(I : I) = T$; i.e., $I = aT$ (resp., $IR_M = aT_M$), for some $0 \neq a \in I$. Then, aR (resp., $aR + IM$) is a proper minimal reduction of I in R if and only if I has a proper minimal reduction in R if and only if $D = k$.*

In particular, if R is a classical pullback issued from (V, M, D) such that M is not basic, then the above result implies that M has a minimal reduction in R if and only if R is a PVD. The first main result of this section characterizes the existence of minimal reductions, in classical pullbacks, for (strongly) stable ideals.

Theorem 5.5 *Let R be a classical pullback issued from (V, M, D). Then, every stable ideal of R has a minimal reduction if and only if every strongly stable ideal of R has a minimal reduction if and only if R is a one-dimensional PVD.*

The second main result of this section examines the existence of minimal reductions, in generic pullbacks, for the ideals which are incomparable to M.

Theorem 5.6 *Let R be a generic pullback issued from (T, M, D) and I an ideal of R with $I + M = R$. Then, $J \subseteq I$ is a minimal reduction of I if and only if JT is a minimal reduction of IT with $J + M = R$ if and only if $J = H \cap R$, for some minimal reduction H of IT.*

5.2 Core

This section investigates the core of ideals in pullback constructions. Recall, for convenience, that the core of an ideal I of R, denoted $\text{core}_R(I)$, is the intersection of all (minimal) reductions of I in R.

As an application of Theorem 5.6, the next result investigates the core and basic property, in generic pullbacks, for the ideals of R which are incomparable to M.

Corollary 5.7 *Let R be a generic pullback issued from (T, M, D), and let I be an ideal of R such that $I + M = R$. Then, $\text{core}_R(I) = \text{core}_T(IT) \cap R$. Moreover, I is basic in R if and only if IT is basic in T.*

Follow an illustrative example for Corollary 5.7.

Example 5.8 Let $T_1 = \mathbb{Q}(\sqrt{2})((Y))[[X]]$ and $T_2 = \mathbb{Q}(\sqrt{2})[[X]] + Y\mathbb{Q}(\sqrt{2})((X))[[Y]]$. Then, T_1 is a one-dimensional valuation domain with $\text{Spec}(T_1) = \{0 \subsetneqq M_1\}$, and T_2 is a two-dimensional valuation domain with $\text{Spec}(T_2) = \{0 \subsetneqq P_2 \subsetneqq M_2\}$. Then, $T := T_1 \cap T_2$ is a two-dimensional Prüfer domain with $\text{Spec}(T) = \{0 \subsetneqq P \subsetneqq M, 0 \subsetneqq N\}$. Here N is a one-height maximal ideal of T and T/N contains $\mathbb{Q}$. Now, consider the pullback R issued from $(T, N, \mathbb{Q})$. Then, $p := P \cap R$ is a prime ideal of R with $p + N = R$. So, by [41, Theorem 2.6], $\text{core}_T(P) = P^2$. Further, $pT_N = PT_N$ and $pT_M = PT_M$ so that $pT = P$. By Corollary 5.7, $\text{core}_R(p) = \text{core}_T(pT) \cap R = \text{core}_T(P) \cap R = P^2 \cap R = p^2$.

The next example shows that Corollary 5.7 does not hold, in general, if I is comparable to M in R; and also Proposition 5.2 is not valid if I is not an ideal of B.

Example 5.9 Let $T := \mathbb{Q}[X, Y] := \mathbb{Q} + M$, where $M := (X, Y)T$, $R := \mathbb{Z} + M$, and $I := (X, Y)R$. Then, $I \subsetneqq M$ with $IT = M$. As M is a 2-generated height-two ideal of the Noetherian domain T [8, Theorem 4], it is of the principal class and hence a basic ideal of T [24, Theorem 2.3]. Therefore, $\text{core}_R(I) \subsetneqq \text{core}_T(IT) \cap R$.

Next, let $I := (X - 1, Y)R$. Then, $I + M = R$ and similar arguments show that $IT = (X - 1, Y)T$ is a basic ideal of T. By Corollary 5.7, I is basic in R. Therefore, I is a minimal reduction of itself (in R), despite $(R : T) = M$ is not a maximal ideal of R.

The first main result of this section investigates, in generic pullbacks, the core for the category of ideals of both R and T.

Theorem 5.10 *Let R be a generic pullback issued from (T, M, D), and let I be a nonzero ideal of both R and T. Then*

(1) $\text{core}_R(I)$ *is an ideal of T with* $M\,\text{core}_T(I) \subseteq \text{core}_R(I) \subseteq \text{core}_T(I)$. *Moreover, If T_M is a valuation domain, then* $\text{core}_R(I) = \text{core}_T(I)$ *or* $M\,\text{core}_T(I)$.
(2) *If $k \subsetneqq K$, then* $M\,\text{core}_T(I) \subseteq \text{core}_R(I) \subseteq MI \cap \text{core}_T(I)$. *Moreover, if I is basic in T, then* $\text{core}_R(I) = MI$.

Follow an illustrative example for Theorem 5.10, which also shows that the '$k \subsetneqq K$' assumption is not necessary.

Example 5.11 Let $T := \mathbb{Q}(X)[Y] = \mathbb{Q}(X) + M$, where $M := YT$, $S := \mathbb{Q}[X] + M$, and $R := \mathbb{Z} + M$. Consider the ideal I of T, S, and R given by $I := Y(Y-1)T$. Then, $\text{core}_T(I) = I$ and, since $Y(Y-1)S$ is a reduction of I in S, $\text{core}_S(I) \subsetneqq I = \text{core}_T(I)$. Moreover, since T_M is a valuation domain, then Theorem 5.10(1)&(2) applied respectively to the pullbacks $S \subset T$ and $R \subset T$, yields $\text{core}_S(I) = M\,\text{core}_T(I) = MI = \text{core}_R(I) \subsetneqq I \subsetneqq M$.

The next two examples show that the inclusions of Theorem 5.10 can be strict.

Example 5.12 Let K be a field containing (strictly) a domain D and X an indeterminate over K. For the first inequality, let $T := K[[X^2, X^3]] = K + M$, where $M := (X^2, X^3)T$, and $R := D + M$. Then, $M^{-1} = (R : M) = (M : M) = (T : M) = K[[X]]$. Note that M is strongly stable, since $M = X^2K[[X]]$, and $(M : M)$ is local. By [41, Theorem 2.12] applied to R and T, we obtain $\text{core}_T(M) = M^2(T : M) = M^2 = M^2M^{-1} = \text{core}_R(M)$. It follows that $M\,\text{core}_T(M) = X^6K[[X]] \subsetneqq \text{core}_R(M) = X^4K[[X]]$. For the second inequality, let $T := K[[X]] = K + M$, where $M := XT$, and $R = D + M$. Then, $\text{core}_R(M) = M^2 \subsetneqq \text{core}_T(M) = M$ by Corollary 5.14

The next result investigates the core of ideals in the classical pullbacks for all four types of ideals; $I = M$, $M \subsetneqq I$, $0 \neq I \subseteq M$ with I an ideal of the valuation domain V, and $I \subsetneqq M$ with I not an ideal of V.

Theorem 5.13 *Let R be a pullback issued from (V, M, D) and I an ideal of R.*

(1) If $M \subsetneqq I := \varphi^{-1}(A)$, for some nonzero ideal A of D, then $\text{core}_R(I) = \varphi^{-1}(\text{core}_R(A))$. Moreover, I is basic in R if and only if A is basic in D.
(2) If $0 \neq I \subseteq M$ and I is an ideal of V, then $\text{core}_R(I) = I^2I^{-1}$. Moreover, I is basic in R if and only if I is not strongly stable in R.
(3) If $I \subsetneqq M$ and I is not an ideal of V, i.e., $I = a\varphi^{-1}(W)$, for some nonzero $a \in M$ and D-submodule W with $D \subseteq W \subsetneqq K$, then $\text{core}_R(I) = a\varphi^{-1}\left(\bigcap_{H\in\mathcal{H}} H\right)$, where $\mathcal{H} := \{$all D-submodules H of W with $HW^n = W^{n+1}$ for some $n \geq 0\}$. Moreover, I is basic in R if and only if $W = \bigcap_{H\in\mathcal{H}} H$.

The special case when $I = M$ is given below.

Corollary 5.14 *Let R be a classical pullback issued from (V, M, D). Then, for any positive integer $n \geq 1$, $\text{core}_R(M^n) = M^{n+1}$.*

Next, we recover two known results on PVDs.

Corollary 5.15 ([41, Proposition 2.10]) *Let R be a nontrivial PVD and $P \in \text{Spec}(R)$. Then, $\text{core}_R(P^n) = P^{n+1}$, for any integer $n \geq 1$.*

Corollary 5.16 ([41, Proposition 3.5(a)]) *Let R be a PVD issued from (V, M, k) and I a nonzero ideal of R and V. Then, $\text{core}(I)$ is a reduction of I if and only if I is basic.*

Follow three illustrative examples for Theorem 5.13(1-2-3).

Example 5.17 Let K be a field and let X, Y be two indeterminates over K.

(1) Let $V := K((X))[[Y]] = K((X)) + M$, where $M := YV$, $R := K[[X^2, X^3]] + M$, and $I := (X^2, X^3)K[[X^2, X^3]] + M$. Then, $\text{core}_R(I) = \text{core}_{K[[X^2,X^3]]}(X^2, X^3) + M$ by Theorem 5.13(1). Moreover, in Example 5.12, we saw that $\text{core}_{K[[X^2,X^3]]}(X^2, X^3) = X^4K[[X]]$. Therefore, $\text{core}_R(I) = X^4K[[X]] + YK((X))[[Y]]$.

(2) Let $V := K[[X]] + YK((X))[[Y]] = K + M$, where $M := XK[[X]] + YK((X))[[Y]]$, $R := D + M$, where D is any domain strictly contained in K. Let $I := Y^2K((X))[[Y]] = N^2$, where $N := YK((X))[[Y]]$. Clearly, $I \subsetneq M$ is an ideal of both R and V, and $I^{-1} = (R : I) = (R : N^2) = ((R : N) : N) = ((N : N) : N) = (V_1 : N) = Y^{-1}V_1$ where $V_1 := K((X))[[Y]]$. By Theorem 5.13(2), $\text{core}_R(I) = N^4Y^{-1}V_1 = N^3 = Y^3K((X))[[Y]]$.

(3) Let $V := K(X)[[Y]] = K(X) + M$, where $M := YV$, $R := K + M$, $W := K + KX$, and $I := Y(W + M)$. By Theorem 5.13(3), we get $\text{core}_R(I) = Y\left(\left(\bigcap_{H\in\mathcal{H}} H\right) + M\right)$, where $\mathcal{H} := \{\text{all } K\text{-subspaces } H \text{ of } W \text{ with } HW^n = W^{n+1} \text{ for some } n \geq 0\}$. Next, let $H \in \mathcal{H}$. If $\dim_K(H) = 2$, then $H = W$. Suppose $\dim_K(H) = 1$. Then, $H = (a + bX)K$, for some $a, b \in K$ such that $a + bX \neq 0$. If $a = 0$, then the assumption $HW^n = W^{n+1}$, for some $n \geq 1$, yields $KX + KX^2 + \cdots + KX^{n+1} = K + KX + KX^2 + \cdots + KX^{n+1}$, which is absurd. So, $a \neq 0$. Further, $b \neq 0$ since $K \notin \mathcal{H}$. It follows that $\bigcap_{H\in\mathcal{H}} H = \bigcap_{a,b\in K\setminus\{0\}} (a+bX)K \subseteq (1+X)K \cap (1-X)K$. Now, let $f \in \bigcap_{H\in\mathcal{H}} H$. Then, $f \in K[X]$ with degree equal to 1 and $f(1) = f(-1) = 0$. Consequently, for $K := \mathbb{Z}/2\mathbb{Z}$, we obtain $\text{core}_R(I) = Y\big((1 + X)K + M\big) = Y(1 + X)K + M^2 = Y(1 + X)K + Y^2K(X)[[Y]]$ and, for any $K \neq \mathbb{Z}/2\mathbb{Z}$, we have $\bigcap_{H\in\mathcal{H}} H = 0$. So $\text{core}_R(I) = YM = M^2 = Y^2K(X)[[Y]]$. Notice that I is not an ideal of V and not stable in R, and $\text{core}_R(I) = I^2I^{-1}$.

References

1. Anderson, D.F., Bouvier, A., Dobbs, D.E., Fontana, M., Kabbaj, S.: On Jaffard domains. Expo. Math. **6**, 145–175 (1988)
2. Anderson, D. D., Huckaba, J. A., Papick, I. J.: A note on stable domains. Houston J. Math. **13**, 13–17 (1987)
3. Anderson, D.F., Dobbs, D.E.: Pairs of rings with the same prime ideals. Canad. J. Math. **32**, 362–384 (1980)
4. Bass, H.: On the ubiquity of Gorenstein rings. Math. Z. **82**, 8–28 (1963)
5. Bastida, E., Gilmer, R.: Overrings and divisorial ideals of rings of the form $D + M$. Michigan Math. J. **20**, 79–95 (1973)
6. Bazzoni, S.: Divisorial domains. Forum Math. **12**, 397–419 (2000)
7. Bazzoni, S.: Clifford regular domains. J. Algebra **238**, 703–722 (2001)
8. Brewer, J.W., Rutter, E.A.: $D + M$ constructions with general overrings. Michigan Math. J. **23**, 33–42 (1976)
9. Corso, A., Polini, C., Ulrich, B.: The structure of the core of ideals. Math. Ann. **321**, 89–105 (2001)

10. Corso, A., Polini, C., Ulrich, B.: Core and residual intersections of ideals. Trans. Amer. Math. Soc. **357**, 2579–2594 (2002)
11. Fontana, M.: Topologically defined classes of commutative rings. Ann. Mat. Pura Appl. **123**, 331–355 (1980)
12. Fontana, M., Gabelli, S.: On the class group and the local class group of a pullback. J. Algebra **181**, 803–835 (1996)
13. Fontana, M., Houston, E., Lucas, T. G.: Factoring ideals in integral domains. Lecture Notes of the Unione Matematica Italiana, 14, Springer, Heidelberg (2013)
14. Fontana, M., Huckaba, J., Papick, I.: Domains satisfying the trace property. J. Algebra **107**, 169–182 (1987)
15. Fontana, M., Huckaba, J., Papick, I.: Prüfer domains. Monographs and Textbooks in Pure and Applied Mathematics **203** Marcel Dekker, Inc. New York (1997)
16. Fouli, L.: Computing the core of ideals in arbitrary characteristic. J. Algebra **319**, 2855–2867 (2008)
17. Fouli, L., Morey, S.: Minimal reductions and cores of edge ideals. J. Algebra **364**, 52–66 (2012)
18. Fouli, L., Olberding, B.: Generators of reductions of ideals in a local Noetherian ring with finite residue field. Proc. Amer. Math. Soc. **146**, 5051–5063 (2018)
19. Fouli, L., Polini, C., Ulrich, B.: The core of ideals in arbitrary characteristic. Special volume in honor of Melvin Hochster. Michigan Math. J. **57**, 305–319 (2008)
20. Fouli, L., Polini, C., Ulrich, B.: Annihilators of graded components of the canonical module, and the core of standard graded algebras. Trans. Amer. Math. Soc. **362**, 6183–6203 (2010)
21. Gabelli, S., Houston, E.: Coherent-like conditions in pullbacks. Michigan Math. J. **44**, 99–123 (1997)
22. Gilmer, R. .: Multiplicative Ideal Theory. Pure and Applied Mathematics, No. 12., Marcel Dekker, Inc., New York (1972)
23. Halter-Koch, F. : Clifford semigroups of ideals in monoids and domains. Forum Math. **21**, 1001–1020 (2009)
24. Hays, J.: Reductions of ideals in commutative rings. Trans. Amer. Math. Soc. **177**, 51–63 (1973)
25. Hays, J.: Reductions of ideals in Prüfer domains. Proc. Amer. Math. Soc. **52**, 81–84 (1975)
26. Hedstrom, J., Houston, E.: Pseudo-valuation domains. Pacific J. Math. **75**, 137–147 (1978)
27. Hedstrom, J., Houston, E.: Pseudo-valuation domains II. Houston J. Math. **4**, 199–207 (1978)
28. Heinzer, W.: Integral domains in which each nonzero ideal is divisorial. Mathematika **15**, 164–170 (1968)
29. Heinzer, W., Papick, I.: The radical trace property. J. Algebra **112**, 110–121 (1988)
30. Heinzer, W., Ratliff, L., Rush, D.: Reductions of ideals in local rings with finite residue fields. Proc. Amer. Math. Soc. **138**, 1569–1574 (2010)
31. Huckaba, J. A., Papick, I. J. : When the dual of an ideal is a ring. Manuscripta Math. **37**, 67–85 (1982)
32. Huneke, C., Swanson, I.: Cores of ideals in 2-dimensional regular local rings. Michigan Math. J. **42**, 193–208 (1995)
33. Huneke, C., Swanson, I.: Integral Closure of Ideals, Rings, and Modules. London Mathematical Society Lecture Note Series, 336. Cambridge University Press, Cambridge (2006)
34. Huneke, C., Trung, N. V.: On the core of ideals. Compos. Math. **141**, 1–18 (2005)
35. Hyry, E., Smith, K. E.: On a non-vanishing conjecture of Kawamata and the core of an ideal. Amer. J. Math. **125**, 1349–1410 (2003)
36. Kabbaj, S.: On the dimension theory of polynomial rings over pullbacks. In Multiplicative Ideal Theory in Commutative Algebra, pp. 263–277, Springer (2006)
37. Lam, T. Y.: Lectures on modules and rings. Graduate Texts in Mathematics, 189. Springer-Verlag, New York (1999)
38. Kabbaj, S., Lucas, T., Mimouni, A.: Trace properties and pullbacks. Comm. Algebra **31**, 1085–1111 (2003)
39. Kabbaj, S., Mimouni, A.: Class semigroups of integral domains. J. Algebra **264**, 620–640 (2003)

40. Kabbaj, S., Mimouni, A.: t-Class semigroups of integral domains. Reine Angew. Math. **612**, 213–229 (2007)
41. Kabbaj, S., Mimouni, A.: Core of ideals in integral domains. J. Algebra **445**, 327–351 (2016)
42. Kabbaj, S., Mimouni, A.: Core of ideals in one-dimensional Noetherian domains. J. Algebra **555**, 346–360 (2020)
43. Kabbaj, S., Mimouni, A.: Reductions of ideals in pullbacks. Algebra Colloq. **27**, 523–530 (2020)
44. Kabbaj, S., Mimouni, A.: Minimal reductions and core of ideals in pullbacks. Acta Math. Hungar. **163**, 512–529 (2021)
45. Kaplansky, I.: Commutative Rings. The University of Chicago Press, Chicago (1974)
46. Lee, K.: A short note on containment of cores. Comm. Algebra **36**, 3890–3892 (2008)
47. Lucas, T. G.: The radical trace property and primary ideals. J. Algebra **184**, 1093–1112 (1996)
48. Matlis, E.: Reflexive domains. J. Algebra **8**, 1–33 (1968)
49. Mimouni, A.: Note on the divisoriality of domains of the form $k[[X^p, X^q]]$, $k[X^p, X^q]$, $k[[X^p, X^q, X^r]]$, and $k[X^p, X^q, X^r]$.Turkish J. Math. **40**, 38–42 (2016)
50. Northcott, D.G., Rees, D.: Reductions of ideals in local rings. Proc. Cambridge Philos. Soc. **50**, 145–158 (1954)
51. Olberding, B. : On the classification of stable domains. J. Algebra **243**, 177–197 (2001)
52. Polini, C., Ulrich, B.: A formula for the Core of an ideal. Math. Ann. **331**, 487–503 (2005)
53. Rees, D., Sally, J.: General elements and joint reductions. Michigan Math. J. **35**, 241–254 (1988)
54. Sally, J., Vasconcelos, W. V. : Stable rings and a problem of Bass. Bull. Amer. Math. Soc. **79**, 574–576 (1973)
55. Smith, B.: A formula for the core of certain strongly stable ideals. J. Algebra **347**, 40–52 (2011)
56. Song, Y., Kim, S.: Reductions of ideals in commutative Noetherian semi-local rings. Commun. Korean Math. Soc. **11**, 539–546 (1996)
57. Wang, H.-J. : Core of ideals of Noetherian local rings. Proc. Amer. Math. Soc. **136**, 801–807 (2008)

Valuative Lattices and Spectra

Henri Lombardi and Assia Mahboubi

1 Introduction

This paper is written in the style of constructive mathematics à la Bishop, see [5, 6, 8, 38, 43, 53]. An updated French version of [38] is [39].

We use the terminology and notations of dynamical theories. See [3, 4, 11, 21, 32, 34, 35].

We continue the constructive approach to algebraic curves and valuation domains initiated in [12, 15, 17, 21] and [38, section XIII-8].

We hope to open the way for a simple constructive approach to the following topics.

- Understand (constructively) the theory of divisors of Weil-Kronecker when applied to a geometric ring[1] (continuing the study made in [15]).
- Understand (constructively) the theorem stating that the theory of the ring of algebraic integers is complete [45, 46, 50, 51].
- Understand (constructively) the Riemann-Roch's theorem in its general algebraic version [24, 25].
- Understand (constructively) the Grothendieck version of Riemann-Roch's theorem in [7].

[1] A finitely presented algebra over a discrete field.

H. Lombardi (✉)
Université de Franche-Comté, Besançon, France
e-mail: Henri.lombardi@univ-fcomte.fr

A. Mahboubi
Nantes Université, École Centrale Nantes, CNRS, INRIA, LS2N, UMR 6004, Nantes, France
e-mail: Assia.Mahboubi@inria.fr

J.-L. Chabert et al. (eds.), *Algebraic, Number Theoretic, and Topological Aspects of Ring Theory*, https://doi.org/10.1007/978-3-031-28847-0_17

The three first sections of the present article provide an overview of the constructive dynamical method. This method relies on dynamical theories and on dynamical algebraic structures. Its purpose is to unveil a computational content hidden in a number of abstract objects in classical mathematics. Dynamical methods provide effective counterparts even for objects seemingly inaccessible to a constructive treatment, e.g., the algebraic closure of a (discrete) field. They are typically able to turn a classical proof of a concrete result into an effective algorithm.

The second part of the article uses this method to study the theory of divisibility. In particular, we compare two notions of valuative spectrum present in the literature. The first one arises in the theory of valuation domains, where [27] introduced the valuative spectrum of an arbitrary commutative ring, akin to the Zariski spectrum and to the real spectrum. The second one is the constructive version of the valuative spectrum of an integral domain, introduced by Coquand [12].

Sections 2 to 4 give general facts about distributive lattices, spectral spaces, geometric theories, dynamical algebraic structures and relations between these objects. The central notion of dynamical algebraic structure intuitively corresponds to that of an incompletely specified algebraic structure. The existence of models for these structures is a pervasive tool in classical mathematics. Most of the time, the only constructive counterpart to the existence of such a purely idealistic existence is the fact that the corresponding dynamical algebraic structure does not collapse. But in general, this is in fact sufficient to establish constructively results obtained in classical mathematics via the purely ideal existence of these models.

Section 5 develops several dynamical theories for valuation domains, in relation with $\mathcal{val}$ and $\mathcal{Val}$, which correspond, respectively, to the approaches of [27] and of [12]. Theorem 5.5.6 establishes the isomorphism of distributive lattices (and hence that of spectral spaces) corresponding to the dynamical algebraic structures defined in theories $\mathcal{val}$ and $\mathcal{Val}$, for an algebra $\mathbf{k} \to \mathbf{K}$ when $\mathbf{k}$ is a sub-ring of a discrete field $\mathbf{K}$.

Section 6 addresses the valuative dimension of commutative rings. In particular, we prove that the valuative dimension of a commutative ring $\mathbf{A}$ is the Krull dimension of the associated distributive lattice with the dynamical algebraic structure $\mathcal{val}(\mathbf{A}, \mathbf{A})$. In the case of an integral domain, the isomorphism of distributive lattices of Theorem 5.5.6 establishes the correspondence with the valuative dimension defined constructively by Coquand [12]. We also show constructively the equivalence of several constructive approaches to the valuative dimension of a ring, or of an algebra. For this purpose, we introduce the minimal pp-closure of a ring $\mathbf{A}$, which replaces the too hypothetical "integral domain generated by $\mathbf{A}$."

Section 7 revisits the dynamical theory of valued discrete fields given in [21]. The objective of this section is akin to that of Sect. 5: showing that the dynamical algebraic structures associated with the theories $\mathcal{val}$, $\mathcal{Val}$, and $\mathcal{Vdf}$ are essentially the same. We rely on the formal Valuativstellensätze established for $\mathcal{Vdf}$ and we develop analogue formal Valuativstellensätze for $\mathcal{val}$ and $\mathcal{Val}$. This provides an interesting historical perspective on various results à la Valuativstellensatz. In particular, the formal Valuativstellensatz 7.4.8 explains the identity between the algebraic certificates associated with the formal theories, respectively, developed (implicitly or explicitly) in [21, 27] and [12].

2 Distributive Lattices and Spectral Spaces

References: [10, 14, 48] and [38, Chapters XI and XIII].

2.1 *The Seminal Paper by Stone*

In modern language, the main result of the seminal paper [48] can be stated as follows:
The category of distributive lattices is, in classical mathematics, antiequivalent to the category of spectral spaces.

Let us explain this with some details.

Ideals and Filters in a Distributive Lattice

If $\varphi : \mathbf{T} \to \mathbf{T}'$ is a distributive lattice homomorphism, $\varphi^{-1}(0)$ is called an *ideal of* $\mathbf{T}$. An ideal $\mathfrak{b}$ of $\mathbf{T}$ is a subset of $\mathbf{T}$ subjected to the following constraints

$$\left.\begin{array}{r} 0 \in \mathfrak{b} \\ x, y \in \mathfrak{b} \Longrightarrow x \vee y \in \mathfrak{b} \\ x \in \mathfrak{b},\ z \in \mathbf{T} \Longrightarrow x \wedge z \in \mathfrak{b} \end{array}\right\} \tag{1}$$

(the last is rewritten as $(x \in \mathfrak{b},\ y \leqslant x) \Rightarrow y \in \mathfrak{b}$). A *principal ideal* is an ideal generated by a single element a, it is equal to $\downarrow a := \{x \in \mathbf{T}; x \leqslant a\}$.

The ideal $\downarrow a$, equipped with the laws $\wedge$ and $\vee$ of $\mathbf{T}$, is a distributive lattice in which the maximum element is a. The canonical injection $\downarrow a \to \mathbf{T}$ is not a morphism of distributive lattices because the image of a is not equal to $1_{\mathbf{T}}$. However, the map $\mathbf{T} \to \downarrow a,\ x \mapsto x \wedge a$ is a surjective morphism, which therefore defines $\downarrow a$ as a quotient structure $\mathbf{T}/(a = 1)$.

The opposite notion to that of an ideal is the notion of a *filter*. The principal filter generated by a is equal to $\uparrow a$.

Let $\mathfrak{a}$ be an ideal and $\mathfrak{f}$ be a filter of $\mathbf{T}$, we say that $(\mathfrak{a}, \mathfrak{f})$ is a *saturated pair* in $\mathbf{T}$ if we have the following implications

$$(g \in \mathfrak{f},\ x \wedge g \in \mathfrak{a}) \Longrightarrow x \in \mathfrak{a} \text{ and } (a \in \mathfrak{a},\ x \vee a \in \mathfrak{f}) \Longrightarrow x \in \mathfrak{f}.$$

A saturated pair can also be defined as being $(\varphi^{-1}(0), \varphi^{-1}(1))$ for a morphism $\varphi : \mathbf{T} \to \mathbf{T}'$ of distributive lattices. When $(\mathfrak{a}, \mathfrak{f})$ is a saturated pair, we have the equivalences

$$1 \in \mathfrak{a} \iff 0 \in \mathfrak{f} \iff (\mathfrak{a}, \mathfrak{f}) = (\mathbf{T}, \mathbf{T}).$$

If A and B are two subsets of $\mathbf{T}$ we denote

$$A \vee B = \{a \vee b \mid a \in A,\ b \in B\} \text{ and } A \wedge B = \{a \wedge b \mid a \in A,\ b \in B\}. \tag{2}$$

Then the ideal generated by two ideals $\mathfrak{a}$ and $\mathfrak{b}$ is equal to

$$\mathcal{I}_{\mathbf{T}}(\mathfrak{a} \cup \mathfrak{b}) = \mathfrak{a} \vee \mathfrak{b}. \tag{3}$$

The set of ideals of $\mathbf{T}$ itself forms a distributive lattice[2] w.r.t. the inclusion and for greatest lower bound of $\mathfrak{a}$ and $\mathfrak{b}$, the ideal

$$\mathfrak{a} \cap \mathfrak{b} = \mathfrak{a} \wedge \mathfrak{b}. \tag{4}$$

Thus the operations $\vee$ and $\wedge$ defined in (2) correspond to the supremum and the infimum in the lattice of ideals.

When we consider the lattice of filters, we must pay attention to what the reversing of the order relation produces: $\mathfrak{f} \cap \mathfrak{g} = \mathfrak{f} \vee \mathfrak{g}$ is the infimum of the filters $\mathfrak{f}$ and $\mathfrak{g}$, whereas their supremum is equal to $\mathcal{F}_{\mathbf{T}}(\mathfrak{f} \cup \mathfrak{g}) = \mathfrak{f} \wedge \mathfrak{g}$.

The Spectrum of a Distributive Lattice

In classical mathematics, a *prime ideal* $\mathfrak{p}$ of a distributive lattice $\mathbf{T} \neq \mathbf{1}$ is an ideal whose complement $\mathfrak{f}$ is a filter (*a prime filter*). The quotient lattice $\mathbf{T}/(\mathfrak{p} = 0, \mathfrak{f} = 1)$ is isomorphic to $\mathbf{2}$. Giving a prime ideal of $\mathbf{T}$ is the same thing as giving a lattice morphism $\mathbf{T} \to \mathbf{2}$. We will write $\theta_{\mathfrak{p}} : \mathbf{T} \to \mathbf{2}$ the morphism corresponding to the prime ideal $\mathfrak{p}$.

If S is a system of generators for a distributive lattice $\mathbf{T}$, a prime ideal $\mathfrak{p}$ of $\mathbf{T}$ is characterized by its trace $\mathfrak{p} \cap S$ (see [10]).

Definition 2.1.1 The (Zariski) *spectrum of the distributive lattice* $\mathbf{T}$ is the set $\mathsf{Spec}\ \mathbf{T}$ whose elements are prime ideals of $\mathbf{T}$, with the following topology:

[2] Actually we need to introduce a restriction to truly obtain a set, in order to have a well-defined procedure to construct the ideals under consideration. For example, we can consider the set of ideals obtained from principal ideals via certain predefined operations, such as countable unions and intersections.

an open basis is provided by the subsets $\mathfrak{D}_{\mathbf{T}}(a) \overset{\text{def}}{=} \{\, \mathfrak{p} \in \mathsf{Spec}\,\mathbf{T}; a \notin \mathfrak{p} \,\} = \{\, \mathfrak{p}; \theta_{\mathfrak{p}}(a) = 1 \,\}$.

In classical mathematics we have

$$\left.\begin{array}{ll} \mathfrak{D}_{\mathbf{T}}(a \wedge b) = \mathfrak{D}_{\mathbf{T}}(a) \cap \mathfrak{D}_{\mathbf{T}}(b), & \mathfrak{D}_{\mathbf{T}}(0) = \emptyset, \\ \mathfrak{D}_{\mathbf{T}}(a \vee b) = \mathfrak{D}_{\mathbf{T}}(a) \cup \mathfrak{D}_{\mathbf{T}}(b), & \mathfrak{D}_{\mathbf{T}}(1) = \mathsf{Spec}\,\mathbf{T}. \end{array}\right\} \tag{5}$$

The complement of $\mathfrak{D}_{\mathbf{T}}(a)$ is a *basic closed set* denoted by $\mathfrak{V}_{\mathbf{T}}(a)$.

This notation is extended to $I \subseteq \mathbf{T}$: we let $\mathfrak{V}_{\mathbf{T}}(I) \overset{\text{def}}{=} \bigcap_{x \in I} \mathfrak{V}_{\mathbf{T}}(x)$. If $\mathfrak{I}$ is the ideal generated by I, one has $\mathfrak{V}_{\mathbf{T}}(I) = \mathfrak{V}_{\mathbf{T}}(\mathfrak{I})$. The closed set $\mathfrak{V}_{\mathbf{T}}(I)$ is also called *the subvariety of* $\mathsf{Spec}\,\mathbf{T}$ *defined by* I.

The closure of a point $\mathfrak{p} \in \mathsf{Spec}\,\mathbf{T}$ is provided by all $\mathfrak{q} \supseteq \mathfrak{p}$. Maximal ideals are the closed points of $\mathsf{Spec}\,\mathbf{T}$. The spectrum $\mathsf{Spec}\,\mathbf{T}$ is empty iff $0 =_{\mathbf{T}} 1$.

Definition 2.1.2 A topological space homeomorphic to a space $\mathsf{Spec}(\mathbf{T})$ is called a *spectral space*.

Spectral spaces come from [48]. Johnstone [28] calls them *coherent spaces*. Balbes and Dwinger [1] give them the name *Stone space*. Hochster [26] uses the name *spectral space* in a famous paper where he proves that all spectral spaces can be obtained as Zariski spectra of commutative rings.

With classical logic and choice axiom, the space $\mathsf{Spec}(\mathbf{T})$ has "enough points": the lattice $\mathbf{T}$ can be recovered from its spectrum.

An element (a point) x of a spectral space X is the *generic point of the closed subset* F if $F = \overline{\{x\}}$. This point (when it exists) is necessarily unique because spectral spaces are Kolmogoroff. In fact, closed subsets $\overline{\{x\}}$ are (in classical mathematics) all irreducible closed subsets of X. The order relation $y \in \overline{\{x\}}$ will be denoted as $x \leqslant_X y$, and we have equivalences

$$x \leqslant_X y \iff \overline{\{y\}} \subseteq \overline{\{x\}}. \tag{6}$$

When $X = \mathsf{Spec}(\mathbf{T})$ the order relation $\mathfrak{p} \leqslant_X \mathfrak{q}$ is merely the usual inclusion relation $\mathfrak{p} \subseteq \mathfrak{q}$ between prime ideals of $\mathbf{T}$.

Stone's Antiequivalence

First we have Krull's theorem.

Krull's theorem (In Classical Mathematics) *Let* $\mathfrak{a}$ *be an ideal and* $\mathfrak{v}$ *a filter of a distributive lattice* $\mathbf{T}$. *Suppose that* $\mathfrak{a} \cap \mathfrak{v} = \emptyset$. *Then there exists a prime ideal* $\mathfrak{p}$ *such that* $\mathfrak{a} \subseteq \mathfrak{p}$ *and* $\mathfrak{p} \cap \mathfrak{v} = \emptyset$.

One deduces the following.

- The map $a \in \mathbf{T} \mapsto \mathfrak{D}_{\mathbf{T}}(a) \in \mathcal{P}(\mathsf{Spec}\,\mathbf{T})$ is injective: it identifies $\mathbf{T}$ with a lattice of sets (*Birkhoff representation theorem*).
- If $\varphi : \mathbf{T} \to \mathbf{T}'$ is an injective homomorphism the dual map $\varphi^\star : \mathsf{Spec}\,\mathbf{T}' \to \mathsf{Spec}\,\mathbf{T}$ is onto.
- Any ideal of $\mathbf{T}$ is the intersection of prime ideals above it.
- The map $\mathfrak{J} \mapsto \mathfrak{V}_{\mathbf{T}}(\mathfrak{J})$, from ideals of $\mathbf{T}$ to closed subsets of $\mathsf{Spec}\,\mathbf{T}$, is an isomorphism of posets (w.r.t. to inclusion and reversed inclusion).

One proves also the following results.

- The quasi-compact open sets of $\mathsf{Spec}\,\mathbf{T}$ are exactly the $\mathfrak{D}_{\mathbf{T}}(a)$'s.
- The quasi-compact open sets of $\mathsf{Spec}\,\mathbf{T}$ form a distributive lattice of subsets of $\mathsf{Spec}\,\mathbf{T}$, isomorphic to $\mathbf{T}$ (equalities (5)).
- If X is a spectral space, its quasi-compact open sets form a distributive lattice, denoted as $\mathsf{Oqc}(X)$.
- For a distributive lattice $\mathbf{T}$, $\mathsf{Oqc}(\mathsf{Spec}(\mathbf{T}))$ is canonically isomorphic to $\mathbf{T}$ and for a spectral space X, $\mathsf{Spec}(\mathsf{Oqc}(X))$ is canonically homeomorphic to X.

Definition 2.1.3 A morphism $\varphi : \mathbf{T} \to \mathbf{T}'$ of distributive lattices gives *by duality* a continuous map

$$\varphi^\star : \mathsf{Spec}\,\mathbf{T}' \to \mathsf{Spec}\,\mathbf{T},\ \mathfrak{p} \mapsto \varphi^{-1}(\mathfrak{p})$$

which is called a *spectral map*.

A map between spectral spaces is spectral if and only if the preimage of any quasi-compact open set is a quasi-compact open set. So it is necessarily continuous.

The seminal paper by Stone gives the following characterization of spectral spaces. They are the topological spaces satisfying the following properties (see [28, 48, II-3.3, coherent locales]):

- The space is quasi-compact.[3]
- Every open set is a union of quasi-compact open sets.
- The intersection of two quasi-compact open sets is a quasi-compact open set.
- For two distinct points, there is an open set containing one of them but not the other.
- For any closed set F and any set S of quasi-compact open sets such that

  $$F \cap \bigcap_{U \in S'} U \neq \emptyset \text{ for any finite subset } S' \text{ of } S$$

  we have also $F \cap \bigcap_{U \in S} U \neq \emptyset$.

Hochster [26] shows that the last property can be replaced with:

[3] The nowadays standard terminology is quasi-compact, as in Bourbaki and Stacks, rather than compact.

- Every irreducible closed set is the closure of a point.

The precise modern formulation of Stone's antiequivalence is Theorem 2.1.4.

Theorem 2.1.4 *The contravariant functors* Spec *and* Oqc *between the categories of distributive lattices and spectral spaces define an antiequivalence.*

In constructive mathematics, spectral spaces may have no points, and we try to translate the discourse in classical mathematics on spectral spaces (very frequent in algebra) in a constructive discourse on corresponding distributive lattices.

A topological subspace Y of a spectral space X is called a *subspectral space* if the inclusion morphism $Y \to X$ is a spectral map. The notion of subspectral space is translated by the notion of quotient distributive lattice (Theorem 2.3.8). We have also good translations for the notions of Krull dimension, normal spectral space, lying over morphisms, going up and going down (see Sect. 2.4).

When we replace $\mathbf{T}$ with the opposite lattice $\mathbf{T}^{\mathrm{op}}$, by reversing the order, points of X remain the same ones in classical mathematics and the topology is replaced with the *opposite topology*, where quasi-compact open sets are the subsets $\mathfrak{V}_{\mathbf{T}}(a)$.

There is also the *constructible topology* (or patch topology) where quasi-compact open sets are Boolean combinations of $\mathfrak{O}_{\mathbf{T}}(a)$ and $\mathfrak{V}_{\mathbf{T}}(b)$. This spectral space is the dual of the Boolean algebra $\mathbb{B}\mathrm{o}(\mathbf{T})$ generated by the distributive lattice $\mathbf{T}$. This kind of spectral space is a *Stone's space* in the today terminology.

Finite Spectral Spaces

Finite distributive lattices correspond to finite spectral spaces. These ones are merely finite posets since it suffices to know the closure of points for defining the topology. Basic opens are the $\downarrow a$'s. In classical mathematics all open sets are quasi-compacts. These are the initial parts, and closed subsets are the final parts. Finally, a map between finite spectral spaces is spectral if and only if it is non-decreasing.

So, we see that the general notion of spectral space is a relevant generalization of the notion of finite poset. See [38, Theorem XI-5.6, duality between finite posets and finite distributive lattices].

In the finite case, if we identify the underlying sets of Spec $\mathbf{T}$ and Spec $\mathbf{T}^{\mathrm{op}}$, we get two opposite posets (reversing the order).

2.2 *Distributive Lattices and Entailment Relations*

A particularly important rule for distributive lattices, known as *cut*, is

$$\left(x \wedge a \leqslant b\right) \quad \& \quad \left(a \leqslant x \vee b\right) \quad \Longrightarrow \quad a \leqslant b. \tag{7}$$

For $A \in \mathrm{P_{fe}}(\mathbf{T})$ (finitely enumerated subsets of $\mathbf{T}$) we write

$$\bigvee A := \bigvee_{x \in A} x \qquad \text{and} \qquad \bigwedge A := \bigwedge_{x \in A} x.$$

We denote by $A \vdash B$ or $A \vdash_{\mathbf{T}} B$ the relation defined as follows over the set $\mathrm{P_{fe}}(\mathbf{T})$

$$A \vdash B \quad \overset{\text{def}}{\Longleftrightarrow} \quad \bigwedge A \leqslant \bigvee B.$$

This relation satisfies the following axioms, in which we write x for $\{x\}$ and A, B for $A \cup B$.

$$\begin{array}{rcll} & & a \ \vdash \ a & (R) \\ A \vdash B & \Longrightarrow & A, A' \ \vdash \ B, B' & (M) \\ (A, x \vdash B) \ \& \ (A \vdash B, x) & \Longrightarrow & A \ \vdash \ B & (T). \end{array}$$

We say that the relation is *reflexive*, *monotone*, and *transitive*. The third rule (transitivity) can be seen as a version of rule (7) and is also called the *cut* rule.

Definition 2.2.1 For an arbitrary set S, a relation over $\mathrm{P_{fe}}(S)$ which is reflexive, monotone, and transitive is called an *entailment relation*.

The following theorem is fundamental. It says that the three properties of entailment relations are exactly what is needed for the interpretation in the form of a distributive lattice to be adequate.

Theorem 2.2.2 (Fundamental Theorem of Entailment Relations) *See [10, Theorem 1], [38, XI-5.3], [40, Satz 7]. Let S be a set with an entailment relation $\vdash_S$ on* $\mathrm{P_{fe}}(S)$. *We consider the distributive lattice* $\mathbf{T}$ *defined by generators and relations as follows: the generators are the elements of S and the relations are the*

$$A \vdash_{\mathbf{T}} B$$

each time that $A \vdash_S B$. Then, for all A, B in $\mathrm{P_{fe}}(S)$, *we have*

$$A \vdash_{\mathbf{T}} B \Longrightarrow A \vdash_S B.$$

Remark The relation $x \vdash_S y$ is a priori a preorder, and not an order, on S. Let us denote by $\overline{x}$ the element x seen in the ordered set $\overline{S}$ defined by this preorder. For a subset A of S let us denote $\overline{A} = \{\overline{x}; x \in A\}$. In the theorem we consider a distributive lattice $\mathbf{T}$ which gives on S the same entailment relation as $\vdash_S$. Strictly speaking, we should have written $\overline{A} \vdash_{\mathbf{T}} \overline{B}$ instead of $A \vdash_{\mathbf{T}} B$ since the equality in $\mathbf{T}$ is coarser than in S. In particular, it is $\overline{S}$, and not S, which can be identified with a subset of $\mathbf{T}$. ∎

2.3 Gluing Distributive Lattices and Spectral Subspaces

Quotients, Covers, Gluing Procedures

A *quotient lattice* $\mathbf{T}'$ *of a lattice* $\mathbf{T}$ can also be given by a binary relation $\preccurlyeq$ over $\mathbf{T}$ satisfying the following properties

$$\left.\begin{array}{r} a \leqslant b \Longrightarrow a \preccurlyeq b \\ a \preccurlyeq b,\ b \preccurlyeq c \Longrightarrow a \preccurlyeq c \\ a \preccurlyeq b,\ a \preccurlyeq c \Longrightarrow a \preccurlyeq b \wedge c \\ b \preccurlyeq a,\ c \preccurlyeq a \Longrightarrow b \vee c \preccurlyeq a \end{array}\right\} \tag{8}$$

The relation $\preccurlyeq$ then induces a lattice structure over the quotient set $\mathbf{T}'$ obtained with the new equality

$$(a, b \in \mathbf{T}) \quad : \qquad a =_{\mathbf{T}'} b \overset{\text{def}}{\Longleftrightarrow} (a \preccurlyeq b \ \text{ and } \ b \preccurlyeq a)$$

Naturally if $\mathbf{T}$ is distributive, the same goes for $\mathbf{T}'$.

Proposition 2.3.1 *Let* $\mathbf{T}$ *be a distributive lattice and* (J, U) *be a pair of subsets of* $\mathbf{T}$. *Consider the quotient* $\mathbf{T}'$ *of* $\mathbf{T}$ *defined by the relations* $x = 0$ *for each* $x \in J$, *and* $y = 1$ *for each* $y \in U$. *Then the inequality* $a \leqslant_{\mathbf{T}'} b$ *is satisfied if and only if there exist* $J_0 \in \mathrm{P_{fe}}(J)$ *and* $U_0 \in \mathrm{P_{fe}}(U)$ *such that*

$$a \wedge \bigwedge U_0 \ \leqslant_{\mathbf{T}} \ b \vee \bigvee J_0. \tag{9}$$

We will denote by $\mathbf{T}/(J = 0, U = 1)$ *this quotient lattice* $\mathbf{T}'$.

In particular, for an ideal $\mathfrak{a}$, the natural morphism $\varphi : \mathbf{T} \to \mathbf{T}' = \mathbf{T}/(\mathfrak{a} = 0)$ satisfies $\varphi^{-1}(0_{\mathbf{T}'}) = \mathfrak{a}$. In the case of the quotient by a principal ideal $\downarrow a$ we obtain $\mathbf{T}/(\downarrow a = 0) \simeq \uparrow a$ with the morphism $y \mapsto y \vee a$ from $\mathbf{T}$ to $\uparrow a$.

We see in the example of totally ordered sets that a quotient structure of a distributive lattice is not generally characterized by the equivalence classes of 0 and 1.

In commutative algebra, when $\mathfrak{a}$ and $\mathfrak{b}$ are two ideals of a ring $\mathbf{A}$, there is an exact sequence of $\mathbf{A}$-modules

$$0 \to \mathbf{A}/(\mathfrak{a} \cap \mathfrak{b}) \xrightarrow{j} (\mathbf{A}/\mathfrak{a}) \times (\mathbf{A}/\mathfrak{b}) \xrightarrow{\delta} \mathbf{A}/(\mathfrak{a} + \mathfrak{b}) \to 0 \quad (\delta(x, y) = x - y \bmod \mathfrak{a} + \mathfrak{b})$$

otherwise said: the congruences system $x \equiv a \bmod \mathfrak{a}$, $x \equiv b \bmod \mathfrak{b}$ has a solution if and only if $a \equiv b \bmod \mathfrak{a} + \mathfrak{b}$, and in this case, the solution is unique modulo $\mathfrak{a} \cap \mathfrak{b}$. It is remarkable that this "Chinese remainder theorem" generalizes to an *arbitrary* system of congruences if and only if the ring is *arithmetic* [38, Theorem XII-1.6], i.e., if the lattice of ideals is distributive. By contrast, the usual Chinese remainder theorem concerns the special case of a family of pairwise comaximal ideals, and it works for an arbitrary ring.

Localizations provide other epimorphisms of the category of commutative rings. They enjoy a fruitful gluing principle, analogous to the Chinese remainder theorem, in this instance the basic local-global principle. This principle asserts that: *if* $(x_1, \ldots, x_n)$ *is a system of comaximal elements in a ring* $\mathbf{A}$*, the morphism* $\mathbf{A} \to \prod_{i \in [\![1..n]\!]} \mathbf{A}[1/x_i]$ *identifies* $\mathbf{A}$ *to a* subproduct[4] *of its localized rings* (see the concrete local-global principle XV-4.2 in [38], and also II-2.3, XV-2.1, XV-2.2, XV-2.3, XV-2.4 and XV-2.5). These principles are constructive versions of abstract local-global principles. They ensure that certain properties of an $\mathbf{A}$-module or an $\mathbf{A}$-algebra are satisfied if and only if they are satisfied after localization in any prime ideal (or, sometimes, in the neighborhood of any prime ideal).

Similarly, it is possible to reconstruct a distributive lattice from a finite number of quotients when the latter carry "enough" information. This can be seen either as a gluing procedure or as a Chinese remainder theorem for distributive lattices.

Covering a distributive lattice by quotient lattices is dual to covering a spectral space by spectral subspaces. Note, however, that a set-theoretical cover of the total space by a family of spectral subspaces does not suffice to reconstruct the spectral topology of the total space from those of the subspaces. Similarly, the three-by-three compatibility conditions on gluing isomorphisms do not suffice to glue spectral spaces: usually, gluing also requires some additional properties on their spectral subspaces.

Therefore, Theorems 2.3.3 and 2.3.6 consider very specific quotient distributive lattices. These results provide analogues, for the category of distributive lattices, to the similar results available in the category of ℓ-groups [38, Covering principles XI-2.10 and XI-2.21] and in the category of modules over a commutative ring [38, Covering principles XI-4.19 and XIII-3.3, Gluing principles XV-4.4, XV-4.4 bis and XV-4.6]. We can now describe the situation in detail.

Definition 2.3.2 Let $\mathbf{T}$ be a distributive lattice and $(\mathfrak{a}_i)_{i \in [\![1..n]\!]}$ (respectively, $(\mathfrak{f}_i)_{i \in [\![1..n]\!]}$) a finite family of ideals (respectively, of filters) of $\mathbf{T}$. We say that the ideals $\mathfrak{a}_i$ *cover* $\mathbf{T}$ if $\bigcap_i \mathfrak{a}_i = \{0\}$. Similarly we say that the filters $\mathfrak{f}_i$ *cover* $\mathbf{T}$ if $\bigcap_i \mathfrak{f}_i = \{1\}$.

Let $\mathfrak{b}$ be an ideal of $\mathbf{T}$; we write $x \equiv y \bmod \mathfrak{b}$ as meaning $x \equiv y \bmod (\mathfrak{b} = 0)$. Let us recall that for $s \in \mathbf{T}$ the quotient $\mathbf{T}/(s = 0)$ is isomorphic to the principal filter $\uparrow s$ (one sees this filter as a distributive lattice with s as 0 element).

Theorem 2.3.3 (Covering a Distributive Lattice by Suitable Quotients) *Let* $\mathbf{T}$ *be a distributive lattice,* $(\mathfrak{a}_i)_{i \in [\![1..n]\!]}$ *a finite family of principal ideals* ($\mathfrak{a}_i = \downarrow s_i$) *and* $\mathfrak{a} = \bigcap_i \mathfrak{a}_i$.

1. *If* (x_i) *is a family in* $\mathbf{T}$ *s.t. for each* i, j *one has* $x_i \equiv x_j \bmod \mathfrak{a}_i \vee \mathfrak{a}_j$*, then there exists a unique* x *modulo* $\mathfrak{a}$ *satisfying:* $x \equiv x_i \bmod \mathfrak{a}_i$ ($i \in [\![1..n]\!]$).

[4] A subobject of the considered finite product, in the category of commutative rings.

2. *Let us write* $\mathbf{T}_i = \mathbf{T}/(\mathfrak{a}_i = 0)$, $\mathbf{T}_{ij} = \mathbf{T}_{ji} = \mathbf{T}/(\mathfrak{a}_i \vee \mathfrak{a}_j = 0)$, $\pi_i : \mathbf{T} \to \mathbf{T}_i$ *and* $\pi_{ij} : \mathbf{T}_i \to \mathbf{T}_{ij}$ *the canonical maps. If the ideals* $\mathfrak{a}_i$ *cover* $\mathbf{T}$, *the system* $(\mathbf{T}, (\pi_i)_{i \in [\![1..n]\!]})$ *is the inverse limit of the diagram*

$$((\mathbf{T}_i)_{1 \leqslant i \leqslant n}, (\mathbf{T}_{ij})_{1 \leqslant i < j \leqslant n}; (\pi_{ij})_{1 \leqslant i \neq j \leqslant n}).$$

3. *The analogous result works with quotients by principal filters.*

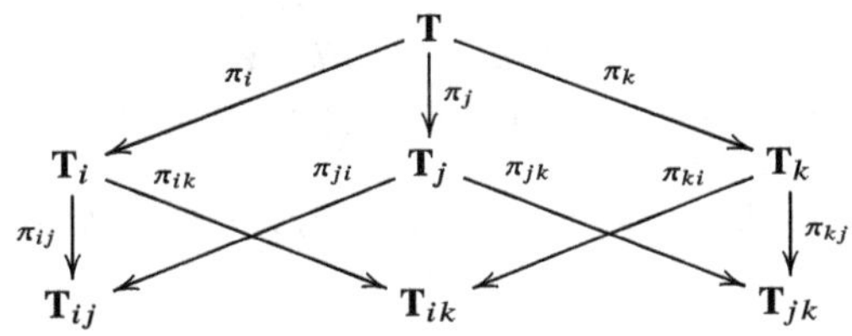

There is also a gluing procedure for quotient distributive lattices $\mathbf{T}_i \to \mathbf{T}_{ij}$ in certain particular cases.

Definition 2.3.4 (Morphism of Passage to Quotient) Let $\mathbf{T}$ be a distributive lattice and $u \in \mathbf{T}$. We identify $\uparrow u$ to the quotient $\mathbf{T}/(\downarrow u = 0)$ via the morphism $p_u : x \mapsto x \vee u$. More generally, a morphism of distributive lattices $\alpha : \mathbf{T} \to \mathbf{T}'$ is called a *morphism of passage to quotient by the ideal* $\downarrow u$ if there exists an isomorphism (necessarily unique) $\lambda : \mathbf{T}' \to \mathbf{T}/(u = 0)$ such that $\alpha = \lambda \circ p_u$.

Lemma 2.3.5 (In a Distributive Lattice, Principal Quotients Are "Split") *Let* $\pi : \mathbf{T} \to \mathbf{T}'$ *be a morphism of distributive lattices and* $s \in \mathbf{T}$. *T.F.A.E.*

1. π *is a morphism of passage to quotient of* $\mathbf{T}$ *by the principal ideal* $\mathfrak{a} = \downarrow s$.
2. *There exists a morphism* $\varphi : \mathbf{T}' \to \uparrow s$ *such that* $\pi \circ \varphi = \mathrm{Id}_{\mathbf{T}'}$.

In this case φ *is uniquely determined by* π *and* s.
Naturally, the "reversed" lemma is valid for a quotient by a principal filter.

Theorem 2.3.6 (Gluing Distributive Lattices) *Let* I *be a finite set, a diagram of distributive lattices*

$$\big((\mathbf{T}_i)_{i \in I}, (\mathbf{T}_{ij})_{i < j \in I}, (\mathbf{T}_{ijk})_{i < j < k \in I}; (\pi_{ij})_{i \neq j}, (\pi_{ijk})_{i < j, j \neq k \neq i}\big)$$

and a family of elements $(s_{ij})_{i \neq j \in I} \in \prod_{i \neq j \in I} \mathbf{T}_i$ *satisfying the following properties*

- *The diagram is commutative.*
- *If* $i \neq j$, π_{ij} *is a quotient morphism w.r.t. the ideal* $\downarrow s_{ij}$.
- *If* i, j, k *are distinct,* $\pi_{ij}(s_{ik}) = \pi_{ji}(s_{jk})$ *and* π_{ijk} *is a quotient morphism w.r.t. the ideal* $\downarrow \pi_{ij}(s_{ik})$.

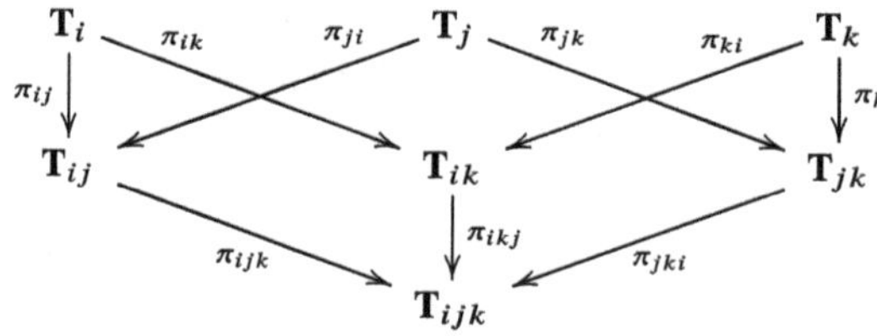

Let $\big(\mathbf{T}\,;\ (\pi_i)_{i\in I}\big)$ *be the limit of the diagram. Then there exist* s_i*'s in* $\mathbf{T}$ *such that the principal ideals* $\downarrow s_i$ *cover* $\mathbf{T}$ *and the diagram is isomorphic to the one in Theorem 2.3.3. More precisely each* π_i *is a quotient morphism w.r.t. the ideal* $\downarrow s_i$ *and* $\pi_i(s_j) = s_{ij}$ *for all* $i \neq j$.

The analogous result works with quotients by principal filters.

The Dual Viewpoint

Definition 2.3.7 A subset X' of a spectral space X is called a *subspectral space* when the topology on X' induced by X is spectral and $\mathsf{Oqc}(X') = \{\, U \cap X';\ U \in \mathsf{Oqc}(X) \,\}$ (in other words the canonical injection is a spectral morphism).

The following theorem explains that the notion of *spectral subspace* is translated by the notion of *quotient distributive lattice*. Some details are added. See also Theorem 2.4.1.

Theorem* 2.3.8 (Subspectral Spaces)

1. *Let* $\mathbf{T}'$ *be a quotient lattice of* $\mathbf{T}$ *and* $\pi : \mathbf{T} \to \mathbf{T}'$ *the quotient morphism. Let us write* $X' = \mathsf{Spec}\,\mathbf{T}'$, $X = \mathsf{Spec}\,\mathbf{T}$ *and* $\pi^\star : X' \to X$ *the dual map of* π. *Then* $\pi^\star$ *identifies* X' *with a subspectral space of* X.
2. *A subset* X' *of a spectral space* X *is a subspectral space if and only if it is closed for the patch topology.*
3. *If* Z *is an arbitrary subset of* $X = \mathsf{Spec}\,\mathbf{T}$, *its closure for the patch topology is given by* $X' = \mathsf{Spec}\,\mathbf{T}'$, *where* $\mathbf{T}'$ *is the quotient lattice of* $\mathbf{T}$ *defined by the following preorder* $\preccurlyeq$:

$$a \preccurlyeq b \quad \Longleftrightarrow \quad (\mathfrak{D}_{\mathbf{T}}(a) \cap Z) \subseteq (\mathfrak{D}_{\mathbf{T}}(b) \cap Z) \tag{10}$$

A gluing of distributive lattices as in Theorem 2.3.6 corresponds to a gluing of topological spaces along 2 by 2 intersections when they are compatible 3 by 3. In this way we glue spectral spaces along suitable quasi-compact open sets[5] when they are 3 by 3 compatible.[6]

---

[5] Variant: closed subsets complement of quasi-compact open sets. This corresponds to the opposite distributive lattices and the opposite topology on spectral spaces.

[6] In Theorem 2.3.6 the compatibility is described thanks to a distributive lattice $\mathbf{T}_{ijk}$. When gluing topological spaces we have a priori three distinct version of $U_1 \cap U_2 \cap U_3$, respectively, subspaces of U_1, U_2 and U_3. So it is necessary to first identify the two versions of $U_i \cap U_j$ inside U_i and U_j through homeomorphisms. And the three identifications have to give the same $U_1 \cap U_2 \cap U_3$.

Here, the gluing of spectral spaces corresponding to the figure is intuitively evident and easy to prove in classical mathematics. In contrast, Theorem 2.3.6 has a rather subtle proof (see [18]).

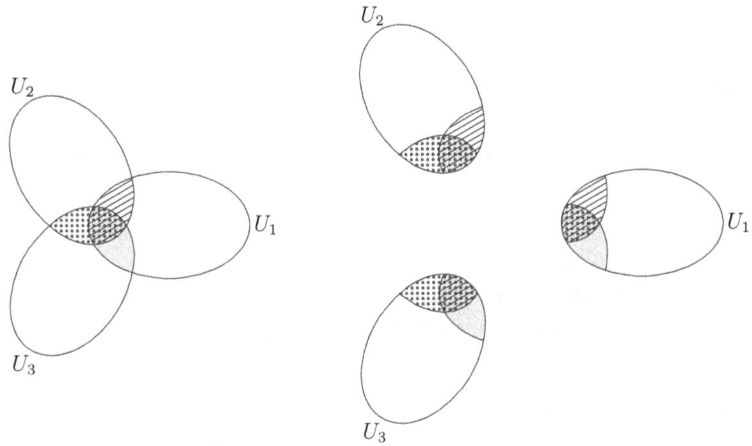

In the following subsection we give other relevant comparisons between distributive lattices and dual spectral spaces.

2.4 Short Dictionary of Stone's Antiequivalence

References: Krull's theorem page 279, [1, 10, 16, 36, Theorem IV-2.6].

We now recall a few results about Stone's antiequivalence of categories between distributive lattices and spectral spaces and provide references to the proofs available in the literature. We also prove Theorem 2.4.7, which finds applications in Sect. 6.

We are in the following context: $f : \mathbf{T} \to \mathbf{T}'$ is a morphism of distributive lattices and $\mathsf{Spec}(f)$, denoted as $f^\star$, is the dual morphism, from $X' = \mathsf{Spec}(\mathbf{T}')$ to $X = \mathsf{Spec}(\mathbf{T})$.

We first recall some usual definitions in classical mathematics.

- The morphism f is said to be *lying over* when $f^\star$ is onto: any prime ideal of $\mathbf{T}$ is the preimage of a prime ideal in $\mathbf{T}'$.
- The morphism f is said to be *going up* when one has: *if* $\mathfrak{q} \in X'$, $f^\star(\mathfrak{q}) = \mathfrak{p}$, *and* $\mathfrak{p} \subseteq \mathfrak{p}_2$ *in* X, *then there exists* $\mathfrak{q}_2 \in X'$ *such that* $\mathfrak{q} \subseteq \mathfrak{q}_2$ *and* $f^\star(\mathfrak{q}_2) = \mathfrak{p}_2$.
- In a similar way f is said to be *going down* when one has: *if* $\mathfrak{q} \in X'$, $f^\star(\mathfrak{q}) = \mathfrak{p}$, *and* $\mathfrak{p} \supseteq \mathfrak{p}_2$ *in* X, *then there exists* $\mathfrak{q}_2 \in X'$ *such that* $\mathfrak{q} \supseteq \mathfrak{q}_2$ *and* $f^\star(\mathfrak{q}_2) = \mathfrak{p}_2$.
- The morphism f *has the incomparability property* when one has: if $\mathfrak{q}_1 \subseteq \mathfrak{q}_2 \in X$ and $f^\star(\mathfrak{q}_1) = f^\star(\mathfrak{q}_2)$ in X' then $\mathfrak{q}_1 = \mathfrak{q}_2$.

- The spectral space X is said to be *normal* if for all x, the closure $\overline{\{z\}}$ contains a unique closed point.
- The spectral space $\mathsf{Spec}\,\mathbf{T}$ is said to be *completely normal* if for all x, y, z such that $x \in \overline{\{z\}}$ and $y \in \overline{\{z\}}$ one has $x \in \overline{\{y\}}$ or $y \in \overline{\{x\}}$.

Properties of Morphisms

Theorem 2.4.1 [1, Theorem IV-2.6] *In classical mathematics we have the following equivalences.*

1. *$f^\star$ is onto (f is lying over) $\iff$ f is injective $\iff$ f is a monomorphism $\iff$ $f^\star$ is an epimorphism.*
2. *f is an epimorphism $\iff$ $f^\star$ is a monomorphism $\iff$ $f^\star$ is injective.*
3. *f is onto*[7] *$\iff$ $f^\star$ is an isomorphism on its image, which is a subspectral space of X.*

There are bijective morphisms of spectral spaces that are not isomorphisms. For example, the morphism $\mathsf{Spec}(\mathbb{B}\mathrm{o}(\mathbf{T})) \to \mathsf{Spec}\,\mathbf{T}$ is rarely an isomorphism and the lattice morphism $\mathbf{T} \to \mathbb{B}\mathrm{o}(\mathbf{T})$ is an injective epimorphism which is rarely onto.

Theorem 2.4.2 ([16]) *In classical mathematics we have the following equivalences*

1. *f is going up $\iff$ for each $a, c \in \mathbf{T}$ and $y \in \mathbf{T}'$ we have*

$$f(a) \leqslant f(c) \vee y \;\Rightarrow\; \exists x \in \mathbf{T}\ (a \leqslant c \vee x \text{ and } f(x) \leqslant y).$$

2. *f is going down $\iff$ for each $a, c \in \mathbf{T}$ and $y \in \mathbf{T}'$ we have*

$$f(a) \geqslant f(c) \wedge y \;\Rightarrow\; \exists x \in \mathbf{T}\ (a \geqslant c \wedge x \text{ and } f(x) \geqslant y).$$

3. *f has the property of incomparability $\iff$ f is zero-dimensional.*[8]

Theorem 2.4.3 (Open Spectral Map [36]) *In classical mathematics t.f.a.e.*

1. *$Spec(f)$ is an open map.*
2. *There exists a map $\widetilde{f} : \mathbf{T}' \to \mathbf{T}$ with the following properties.*
   (a) *For $c \in \mathbf{T}$ and $b \in \mathbf{T}'$, one has $b \leqslant f(c) \Leftrightarrow \widetilde{f}(b) \leqslant c$.*
   In particular, $b \leqslant f(\widetilde{f}(b))$ and $\widetilde{f}(b_1 \vee b_2) = \widetilde{f}(b_1) \vee \widetilde{f}(b_2)$.
   (b) *For $a, c \in \mathbf{T}$ and $b \in \mathbf{T}'$, one has $f(a) \wedge b \leqslant f(c) \Leftrightarrow a \wedge \widetilde{f}(b) \leqslant c$.*
   (c) *For $a \in \mathbf{T}$ and $b \in \mathbf{T}'$, one has $\widetilde{f}(f(a) \wedge b) = a \wedge \widetilde{f}(b)$.*
   (d) *For $a \in \mathbf{T}$, one has $\widetilde{f}(f(a)) = \widetilde{f}(1) \wedge a$.*

[7] In other words, f is a quotient morphism.

[8] See Theorem 2.4.8.

3. *There exists a map* $\widetilde{f} : \mathbf{T}' \to \mathbf{T}$ *satisfying property 2.*
4. *For* $b \in \mathbf{T}$ *the g.l.b.* $\bigwedge_{b \leqslant f(c)} c$ *exists, and if we write it* $\widetilde{f}(b)$*, the property 2 holds.*

Dimension Properties

In classical mathematics, the dimension of a spectral space, based on chains of irreducible closed subsets, is called its *Krull dimension*. The dimension of the empty spectral space is -1. Theorem 2.4.4 explains why the definition of the Krull dimension of a distributive lattice in constructive mathematics, denoted $\mathsf{Kdim}\,\mathbf{T}$, is rather based on its Item *2*. The base case is for the trivial, singleton lattice, which has dimension -1. Since the equivalence between Item *2* and Item *3* is constructive, Item *3* also provides an appropriate constructive definition of the Krull dimension of a distributive lattice in non-trivial cases.

Coquand and Lombardi [13] explain the relation between Item *1* and Item *3* of the theorem, as well as the connection with the pioneering approach of [29]. Item *2* in the theorem goes back to [19].

Theorem 2.4.4 (Dimension of Spaces) *See [13, 19]. Let* $n \in \mathbb{N}$*, in classical mathematics t.f.a.e.*

1. *The spectral space* $\mathsf{Spec}(\mathbf{T})$ *has Krull dimension* $\leqslant n$ *(defined using chains of primes).*
2. *For any* $x \in \mathbf{T}$ *the quotient lattice* $\mathbf{T}/(x = 0, I_x = 0)$*, where* $I_x = \{\, y; x \wedge y = 0 \,\}$*, has dimension* $\leqslant n - 1$*.*
3. *For each sequence* $(x_0, \dots, x_n)$ *in* $\mathbf{T}$ *there exists a* complementary sequence $(y_0, \dots, y_n)$*, which means*

$$\left.\begin{array}{rcl} 1 & \vdash & y_n, x_n \\ y_n, x_n & \vdash & y_{n-1}, x_{n-1} \\ \vdots & \vdots & \vdots \\ y_1, x_1 & \vdash & y_0, x_0 \\ y_0, x_0 & \vdash & 0 \end{array}\right\} \tag{11}$$

For example, in dimension $n \leqslant 2$, the inequalities in (11) correspond to the following diagram in $\mathbf{T}$.

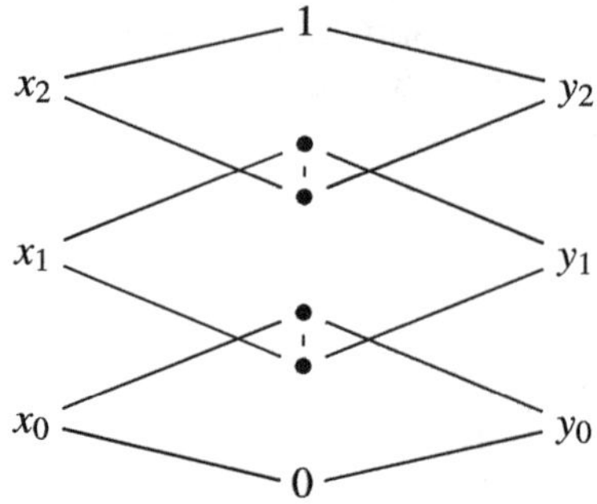

A zero-dimensional distributive lattice is a Boolean algebra.

Items *2* and *3* give a constructive meaning to the statement $\mathsf{Kdim}(\mathbf{T}) \leqslant n$ for $n \geqslant -1$. But in the general case, it seems impossible to understand constructively the statement $\mathsf{Kdim}(\mathbf{T}) = n$ (for $n \geqslant 0$).

Regarding the Krull dimension of commutative rings, see [13, 34] and [38, Chapter XIII]. This definition of the Krull dimension is close to (and probably more general than) the one given by Lurie [41, page 584], which was motivated by the case of certain non-Noetherian rings.

In addition, it is impossible to prove constructively $\mathsf{Kdim}(\mathbb{R}) \leqslant 0$. As $\mathbb{R}$ is a reduced local ring, $\mathsf{Kdim}(\mathbb{R}) \leqslant 0$ is equivalent to the assertion *any x is null or invertible*, i.e., **LPO**.

Lemma 2.4.5 *If $\mathbf{T}'$ is a quotient of $\mathbf{T}$, then $\mathsf{Kdim}(\mathbf{T}') \leqslant \mathsf{Kdim}(\mathbf{T})$.*

Proof Use Item *3* in Theorem 2.4.4. □

Lemma 2.4.6 *Let $\mathbf{T}_1$ and $\mathbf{T}_2$ be distributive lattices and $\mathbf{T} = \mathbf{T}_1 \times \mathbf{T}_2$. Then, for $n \geqslant -1$ we have*

$$\mathsf{Kdim}(\mathbf{T}) \leqslant n \text{ if and only if } \mathsf{Kdim}(\mathbf{T}_i) \leqslant n \text{ for } i = 1, 2.$$

In a shortened form: $\mathsf{Kdim}(\mathbf{T}_1 \times \mathbf{T}_2) = \sup(\mathsf{Kdim}\ \mathbf{T}_1, \mathsf{Kdim}\ \mathbf{T}_2)$.

Proof First, each $\mathbf{T}_i$ is a quotient of $\mathbf{T}$. In the other direction, assume each $\mathbf{T}_i$ is of dimension $\leqslant n$ and let $(x_0, \ldots, x_n)$ in $\mathbf{T}$, with $x_i = (a_i, b_i)$. If $(a_0, \ldots, a_n)$ has the complementary sequence $(y_0, \ldots, y_n)$ in $\mathbf{T}_1$ and $(b_0, \ldots, b_n)$ has the complementary sequence $(z_0, \ldots, z_n)$ in $\mathbf{T}_2$, then $(u_0, \ldots, u_n)$ (where $u_i = (y_i, z_i)$) is complementary of $(x_0, \ldots, x_n)$ in $\mathbf{T}$. □

Theorem 2.4.7 (Dimension of Distributive Lattices, Case of a Closed Cover) *Let $\mathbf{T}$ be a distributive lattice, $\mathfrak{a}$, $\mathfrak{b}$ two ideals such that $\mathfrak{a} \cap \mathfrak{b} = \{0\}$, $\mathbf{T}_\mathfrak{a} = \mathbf{T}/(\mathfrak{a} = 0)$ and $\mathbf{T}_\mathfrak{b} = \mathbf{T}/(\mathfrak{b} = 0)$. Then $\mathsf{Kdim}\ \mathbf{T} = \sup(\mathsf{Kdim}\ \mathbf{T}_\mathfrak{a}, \mathsf{Kdim}\ \mathbf{T}_\mathfrak{b})$.*

Proof We have a natural morphism $j_{\mathfrak{a},\mathfrak{b}} : \mathbf{T} \to \mathbf{T}_\mathfrak{a} \times \mathbf{T}_\mathfrak{b}$, $x \mapsto (\pi_\mathfrak{a}(x), \pi_\mathfrak{b}(x))$. Lemmas 2.4.5 and 2.4.6 imply we have $\sup(\mathsf{Kdim}\ \mathbf{T}_\mathfrak{a}, \mathsf{Kdim}\ \mathbf{T}_\mathfrak{b}) \leqslant \mathsf{Kdim}\ \mathbf{T}$.

Let us see the opposite inequality.

We begin by a proof in classical mathematics. Let $\mathfrak{p}$ be a prime ideal of $\mathbf{T}$. If $\mathfrak{p}$ disappears in $\mathbf{T}_{\mathfrak{a}}$ we have $\mathfrak{p} \vee \mathfrak{a} = \mathbf{T}$. In this case we have an $x \in \mathfrak{a}$ such that $x \notin \mathfrak{p}$, and for any $y \in \mathfrak{b}$, since $x \wedge y = 0$ and $\mathfrak{p}$ is prime, $y \in \mathfrak{p}$. So $\mathfrak{b} \subseteq \mathfrak{p}$ and $\mathfrak{p}$ remains a prime ideal in $\mathbf{T}_{\mathfrak{b}}$. Symmetrically, if $\mathfrak{p}$ disappears in $\mathbf{T}_{\mathfrak{b}}$, it persists in $\mathbf{T}_{\mathfrak{a}}$. Let us now consider a chain of prime ideals in $\mathbf{T}$ of maximal length. The minimal element in the chain contains $\mathfrak{a}$ or $\mathfrak{b}$, so all the chain persists in $\mathbf{T}_{\mathfrak{a}}$ or $\mathbf{T}_{\mathfrak{b}}$.

Finally let us give a proof in constructive mathematics.[9]

First we assume that $\mathfrak{a}$ and $\mathfrak{b}$ are principal ideals: $\mathfrak{a} = \downarrow a$ and $\mathfrak{b} = \downarrow b$ with $a \wedge b = 0$. We have a morphism of *unbounded distributive lattices*[10] $\mathbf{T}_a \to \mathbf{T}$, $\pi_a(x) \mapsto a \vee x$. Let us consider the morphism $r_{a,b} : \mathbf{T}_a \times \mathbf{T}_b \to \mathbf{T}$, $(x, y) \mapsto (a \vee x) \wedge (b \vee y)$. We see that it is a morphism of distributive lattices because $r_{a,b}(0, 0) = 0$. We have $r_{a,b} \circ j_{a,b} = \mathrm{Id}_{\mathbf{T}}$. Indeed, for $x \in \mathbf{T}$, one has $(a \vee x) \wedge (b \vee x) = (a \wedge b) \vee x = x$. So $r_{a,b}$ is a surjective morphism and $\mathbf{T}$ is a quotient of $\mathbf{T}_a \times \mathbf{T}_b$.

Let us see the general case, where $\mathfrak{a}$ and $\mathfrak{b}$ are not necessarily finitely generated. The quotient $\mathbf{T}_{\mathfrak{a}}$ is the filtered colimit of $\mathbf{T}_a$'s for $a \in \mathfrak{a}$ ($\mathfrak{a}$ is seen as a small filtered category). The lattice $\mathbf{T}_{\mathfrak{a}} \times \mathbf{T}_{\mathfrak{b}}$ is the filtered colimit of $\mathbf{T}_a \times \mathbf{T}_b$'s. Morphisms $r_{a,b}$ are compatible,[11] so they induce (by the universal property of a filtered colimit) a morphism $r_{\mathfrak{a},\mathfrak{b}} : \mathbf{T}_{\mathfrak{a}} \times \mathbf{T}_{\mathfrak{b}} \to \mathbf{T}$. For each (a, b) we have the filtered colimit morphism $\iota_{a,b} : \mathbf{T}_a \times \mathbf{T}_b \to \mathbf{T}_{\mathfrak{a}} \times \mathbf{T}_{\mathfrak{b}}$ and we have the commutative diagram:

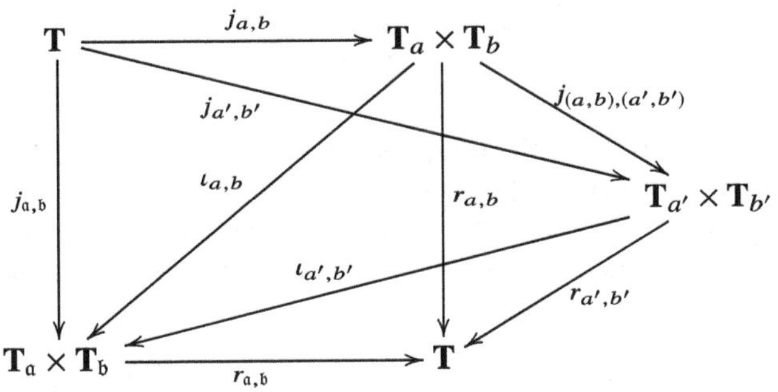

Hence $r_{\mathfrak{a},\mathfrak{b}} \circ j_{\mathfrak{a},\mathfrak{b}} = \mathrm{Id}_{\mathbf{T}}$. □

Theorem 2.4.8 (Dimension of Morphisms) See [16], [38, section XIII-7]. *Let $\mathbf{T} \subseteq \mathbf{T}'$ and f be the inclusion morphism. In classical mathematics t.f.a.e.*

1. *The morphism* $\mathsf{Spec}(f) : \mathsf{Spec}(\mathbf{T}') \to \mathsf{Spec}(\mathbf{T})$ *has Krull dimension* $\leqslant n$.
2. *For any sequence* $(x_0, \ldots, x_n)$ *in* $\mathbf{T}'$ *there exists an integer* $k \geqslant 0$ *and elements* $a_1, \ldots, a_k \in \mathbf{T}$ *such that for each partition* (H, H') *of* $\{1, \ldots, k\}$, *there exist* $y_0, \ldots, y_n \in \mathbf{T}'$ *such that*

---

[9] It should be interesting to check the constructive proof being a faithful translation of the classical one.

[10] The image of this morphism is $\uparrow a$. We have $\pi_a(1) = 1$, but $\pi_a(0) = a$.

[11] For all $a \leqslant a'$ and $b \leqslant b'$, the triangle constituted by $r_{a',b'}$, $r_{a,b}$ and the morphism of passage to quotient $\mathbf{T}_a \times \mathbf{T}_b \to \mathbf{T}_{a'} \times \mathbf{T}_{b'}$ is commutative.

$$\begin{array}{rcl} \bigwedge_{j \in H'} a_j & \vdash & y_n,\ x_n \\ y_n,\ x_n & \vdash & y_{n-1},\ x_{n-1} \\ \vdots\ \vdots & & \vdots \\ y_1,\ x_1 & \vdash & y_0,\ x_0 \\ y_0,\ x_0 & \vdash & \bigvee_{j \in H} a_j \end{array} \tag{12}$$

For example, for the relative dimension $n \leqslant 2$, the inequalities in (12) correspond to the following diagram in $\mathbf{T}$. with $u = \bigwedge_{j \in H'} a_j$ and $i = \bigvee_{j \in H} a_j$.

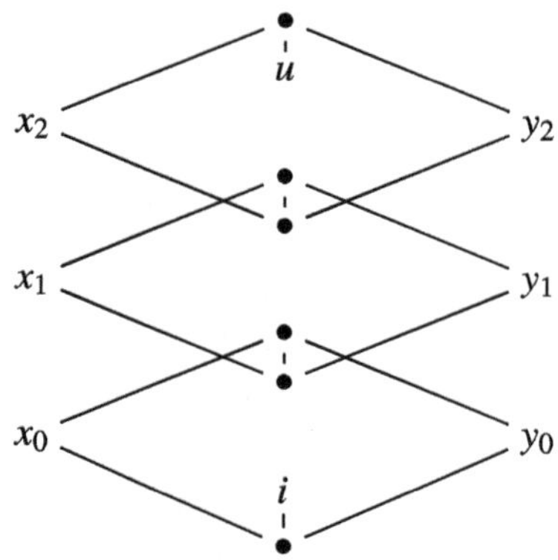

Note that the dimension of the morphism $\mathbf{T} \to \mathbf{T}'$ is bounded by the dimension of $\mathbf{T}'$: take the empty list ($k = 0$) in Item 2 of Theorem 2.4.8.

More generally we have a constructive proof of the fundamental inequality given in [47, A note on the dimension theory of rings] for the case of commutative rings: $1 + \dim \mathbf{T}' \leqslant (1 + \dim \mathbf{T})(1 + \dim f)$.

Properties of Spaces

A distributive lattice $\mathbf{T}$ is said to be *normal* if each time one has $a \vee b = 1$ in $\mathbf{T}$ there exist x, y such that $a \vee x = b \vee y = 1$ and $x \wedge y = 0$. See [23, 52]. Note that when replacing x and y with $x_1 = x \vee (a \wedge b)$ and $y_1 = y \vee (a \wedge b)$ we get $a \vee x_1 = b \vee y_1 = 1$ and $x_1 \wedge y_1 = a \wedge b$.

Theorem 2.4.9 *T.F.A.E.*

1. *The spectral space* $\mathsf{Spec}(\mathbf{T})$ *is normal.*
2. *The distributive lattice* $\mathbf{T}$ *is normal.*

Theorem 2.4.10 *T.F.A.E.*

1. *The spectral space* $\mathsf{Spec}(\mathbf{T})$ *is completely normal.*
2. *Each interval* $[a, b]$ *in* $\mathbf{T}$*, seen as a distributive lattice, is normal.*
3. *For all* $a, b \in \mathbf{T}$ *there exist* x, y *such that* $a \vee b = a \vee y = x \vee b$ *and* $x \wedge y = 0$.

Theorem 2.4.11 *T.F.A.E.*

1. *Any quasi-compact open set in* Spec(**T**) *is a finite union of irreducible quasi-compact open sets.*
2. *For all* $a_1, \ldots, a_n, b_1, \ldots, b_m$ *one has* $a_1, \ldots, a_n \vdash_{\mathbf{T}} b_1, \ldots, b_m$ *if and only if there is a* j *such that* $a_1, \ldots, a_n \vdash_{\mathbf{T}} b_j$.
3. *The distributive lattice* **T** *is constructed from a dynamical algebraic structure corresponding to a Horn theory.*

3 Finitary Dynamical Theories and Dynamical Algebraic Structures

References: [21, 32, 35]. A more detailed text is in preparation [37].

3.1 Finitary Dynamical Theories

Finitary dynamical theories have been introduced in [21]. They are a version "without logic, purely computational" of *coherent theories* (first order theories where all axioms are translations of dynamical rules).[12]

In this paper the authors introduce the notions of "dynamical theory" and of "dynamical proof." See also the paper [3] describing some advantages of this approach, and pioneering articles [42, 44, Sections 1.5 and 4.2] and [31].

Dynamical theories use only dynamical rules, i.e., deduction rules of the form

$$\Gamma \ \vdash \ \textbf{Introduce } \underline{y}^1 \textbf{ such that } \Delta_1 \textbf{ or } \cdots \textbf{ or Introduce } \underline{y}^m \textbf{ such that } \Delta_m \tag{13}$$

where Γ and the Δ_i's are lists of atomic formulae in the language $\mathcal{L}$ of the theory $\mathcal{T} = (\mathcal{L}, \mathcal{A})$.

A shortened form is

$$\Gamma \ \vdash \exists \underline{y}^1 \Delta_1 \textbf{ or } \cdots \textbf{ or } \exists \underline{y}^m \Delta_m \tag{14}$$

[12] A usual terminology in mathematical logic speaks about "formal first order theories" when quantifiers are only applied to usual variables, corresponding to elements of the reference set considered as a model of the theory. Second order is used for formal theories using quantifiers on variables in the powerset of the reference set. General geometric theory, with infinite **or** , is not exactly first order, but absolutely not second order. As a consequence general geometric theories are considered as first order theories in the corresponding literature. We use more or less this tradition in our use of geometric theories by omitting to speak of "first order geometric theories." We use instead "finitary geometric theories."

Axioms are dynamical rules, and theorems are valid dynamical rules (validity is described in a simple way and uses only a computational machinery).

If $\mathcal{T}$ is a coherent theory, the corresponding *(finitary) dynamical theory* uses limited methods of proof.

- First, only atomic formulae are used: no use of new formulae involving connectors or quantifiers. We manipulate only lists of atomic formulae in the language $\mathcal{L}$.
- Second, axioms are not seen as true formulae, but as *deduction rules*: an axiom such that (14) is used as a dynamical rule. Variables in the lists $\underline{y^j}$ are *bound variables*.
- Third, proofs are only proofs of *dynamical rules*.
- Forth, the unique way for proving a dynamical rule is a computational tree "without logic." At the root of the tree we find hypotheses of the theorem we want to prove. The tree is developed by applying axioms along a pure computational algebraic machinery in the structure described by $\mathcal{T}$. Precise formal definitions are given in [21].

  When using an axiom as (14), we substitute arbitrary terms t_i to free variables x_i in the rule. If these hypotheses are valid at a leaf of the proof tree, this leave becomes a node from which start branches of computation. In each branch one introduces fresh variables corresponding to bound variables $\underline{y^k}$ (these fresh variables must be distinct of free variables seen in the terms $\underline{t_i}$) and each conclusion of the list Δ_k is valid in its branch. Finally, a conclusion is *valid* when it is proved at each leave of a proof tree.

In a dynamical theory, each sort S has an equality predicate $\cdot =_S \cdot$. Axioms allow us to substitute a term t by a term t', when the rule $\vdash t = t'$ is valid, in any occurrence of an atomic formula in a valid dynamical rule.[13]

Simplest axioms for this scope are the following ones. First, equality has to be an equivalence relation

eq1 $\vdash x = x$ **eq2** $x = y \vdash y = x$

Eq3 $x = y, y = z \vdash x = z$

Second, for each function symbol f and each predicate P in the signature (we take them with arity one for simplicity) we have suitable compatibility axioms.

eq$_f$ $x = y \vdash f(x) = f(y)$ **Eq**$_P$ $x = y, P(x) \vdash P(y)$

[13] Naturally, it is not allowed that t or t' contains a variable x under the scope of an $\exists x$.

Collapsus

A dynamical rule is called a *collapsus rule* when the right member is "False," denoted by $\bot$. The symbol $\bot$ is necessarily in the language. It is an atomic formula, a predicate of arity 0. On can also see $\bot$ as the empty disjunction. When $\bot$ is proved, the universe of the discourse collapses, and all atomic formulae become "true," or at least "valid." This is the meaning of "ex falso quod libet," which is the relevant meaning of False in constructive mathematics. So, in a dynamical theory, the rules

$$\textbf{False}_P \quad \bot \ \vdash P$$

are valid for all atomic formulae.

The language has also the logical constant $\top$ meaning "True," with the following axiom.

$$\textbf{True} \vdash \top$$

The symbol $\top$ can also be seen as the empty conjunction.[14] Constants $\bot$ and $\top$ are the unique logical symbols used in dynamical theories.

Classification of Dynamical Rules

A dynamical theory is termed *propositional* when it does not involve any sort. In this case, constants are $\top$ and $\bot$, plus possibly other constants of arity zero, seen as propositional constants.

A dynamical rule with neither $\exists$ nor $\bot$, nor **or** on the right of $\vdash$ is called *Horn rule*. A dynamical theory is a *Horn theory* when all axioms are Horn rules. A Horn theory with a single, equality predicate, is called an *algebraic theory*.[15]

A Horn rule is *direct* when its hypothesis (on the left of $\vdash$) features a list of predicates over variables only, and when these variables are moreover pairwise distinct. Coste et al. [21] make use of direct rules for constructing formal Nullstellensätze and their variants. The latter are algebraic certificates of collapsus for certain dynamical theories.

A Horn rule which is not direct is a *simplification rule*.

We use the following typographical conventions: names of direct rules are in lower case, names of simplification rules start with a capital letter and names of other dynamical rules are in capital letters.

A dynamical theory is *weakly disjunctive* if in the axioms if its conclusion features no $\exists$. A dynamical rule is *existential simple* if its conclusion is of the form $\exists \underline{x}\ \Delta$, where Δ is a finite list of atomic formulae. A Horn rule can be considered

---

[14] When there is nothing to prove, prove nothing and all is OK. In a dynamical theory with at least one sort S, $\top$ is equivalent to $x =_S x$.

[15] In [21], Horn rules are called algebraic rules; Horn theories are called algebraic theories and algebraic theories are called purely equational. theories.

as a particular case of existential simple rule. A dynamical theory is *regular* if its axioms are Horn rules or existential simple rules . The theory of *Bézout rings* (each finitely generated ideal is principal) is clearly regular.

A theory is *existentially rigid* if its existential axioms are simple and correspond to provably unique existences. This is a slight generalization of weakly disjunctive theories. A regular theory which is existentially rigid is called *cartesian*. This is a slight generalization of Horn theories.

A theory is *rigid* (or disjunctive) when all its axioms are:

- Horn rules
- Disjunctive rules of type $\Gamma \vdash P$ **or** Q with the provable rule $\Gamma, P, Q \vdash \bot$
- Existentially rigid existential rules

The theory of discrete fields can be stated as a rigid dynamical theory using the invertibility predicate. The theory of discrete real closed fields can also be stated as a rigid dynamical theory, as opposed to the theory of discrete algebraically closed fields.

A Basic Example

The *theory Cr of commutative rings* is the paradigmatic example of a purely equational theory. The signature is $\Sigma_{Cr} = (\cdot = 0; \cdot + \cdot, \cdot \times \cdot, -\cdot, 0, 1)$ with only three axioms (direct rules):

cr1 $\vdash 0 = 0$ **cr2** $x = 0 \vdash x \times y = 0$
cr3 $x = 0, y = 0 \vdash x + y = 0$

The term "$x - y$" is an abbreviation of "$x + (-y)$" and the predicate "$\cdot = \cdot$" is *defined* by the convention: "$x = y$" is an abbreviation for "$x - y = 0$."

Explanation The computational machinery of polynomials with integer coefficients is added to the computational machinery of dynamical proofs. This machinery, external to the dynamical theory, rewrites any term as a polynomial with integer coefficients in a normal form. For example, the distributivity axiom $x(y + z) = xy + xz$ is replaced with the automatic computation which reduces to 0 the term $x(y+z) - (xy+xz)$. Similarly, transitivity of equality is obtained using axiom **cr3**.

3.2 Dynamic Algebraic Structures

Dynamical algebraic structures are explicit in [32, 35] and implicit in [21], where they are described through their presentations. They are also implicit in [34] and, last but not least, in [22, D5], which was a main source: it is possible to compute inside the algebraic closure of a discrete field, even if it is impossible to construct the structure. So it suffices to consider the algebraic closure as a dynamical algebraic structure à la D5 rather than a usual algebraic structure: *lazy evaluation à la D5 gives a constructive semantic for the algebraic closure of a discrete field.*

If $\mathcal{T} = (\mathcal{L}, \mathcal{A})$ is a dynamical theory, a *dynamical algebraic structure of type* $\mathcal{T}$ is given by a set G of generators and a set R of *relations*.

By definition a relation is a closed atomic formula $P(\underline{t})$ on the language $\mathcal{L} \cup G$ with closed terms t_i in this language. Such a relation gives the axiom "$\vdash P(\underline{t})$" in the dynamical algebraic structure.

From a constructive viewpoint, G can be seen as a set à la Bishop. So if two objects a, b give elements of G and if we have $a =_G b$, the relation $a = b$ is automatically present in the set R of relations that define the dynamical algebraic structure $\big((G, R), \mathcal{T}\big)$.

Notation 3.2.1 We shall indicate that the rule "$\Gamma \vdash \ldots$" is valid in the dynamical algebraic structure $\mathbf{S} = \big((G, R), \mathcal{T}\big)$ in the following abridged form: "$\Gamma \vdash_{\mathbf{S}} \ldots$".

Definition and notation 3.2.2 Let $\mathbf{S} = \big((G, R), \mathcal{T}\big)$ be a dynamical algebraic structure of type $\mathcal{T} = (\mathcal{L}, \mathcal{A})$. The set of closed terms of $\mathbf{S}$, i.e., terms constructed on $\mathcal{L} \cup G$, is denoted by Clt($\mathbf{S}$). The set of closed atomic formulae is denoted by Clat($\mathbf{S}$). A Horn rule $\vdash P$ with $P \in$ Clat($\mathbf{S}$) is called *a fact of* $\mathbf{S}$. The set of valid facts in $\mathbf{S}$ is denoted by VClat($\mathbf{S}$).

Intuitively, a dynamical algebraic structure is an incompletely specified usual algebraic structure.

Example 3.2.3 For instance, we obtain a dynamical algebraic structure of discrete field

$$\mathbf{K} = \big((G, R), \mathcal{D}f\big)$$

by taking $G = \{a, b\}$ and $R = \{105 = 0,\ a^2 + b^2 - 1 = 0\}$. This dynamical discrete field corresponds to an arbitrary field of characteristic 3 or 5 or 7 generated by two elements α and β such that $\alpha^2 + \beta^2 = 1$.

In addition to the dynamical rules valid in all discrete fields, we now also have all the ones obtained by extending the language with constants in G and by adding to the axioms the relations in R. ∎

Note that a valid dynamical rule in a dynamical algebraic structure uses for its proof a computational tree, without logic, with a finite number of generators, relations and axioms.

Notation 3.2.4 When $\mathbf{A}$ is a usual algebraic structure on the language of $\mathcal{T}$ we note $\mathcal{T}(\mathbf{A})$ the dynamical algebraic structure we get by taking the following presentation (G, R): G is the set of elements of $\mathbf{A}$ and R is the set of valid facts in $\mathbf{A}$. For example, if $\mathcal{T}$ is the theory of Bézout rings, and $\mathbf{A}$ is an arbitrary commutative ring, we put in R relations $a + b - c = 0$, $a'b' - c' = 0$ and $a + a'' = 0$ when $a, b \ldots, a''$ are elements of $\mathbf{A}$ and when the relations hold in $\mathbf{A}$. We say that we have added to the theory $\mathcal{T}$ the *positive diagram of* $\mathbf{A}$.

Constructive Models Versus Classical Models

Let us consider a dynamical algebraic structure $\mathbf{A} = \big((G, R), \mathcal{T}\big)$ of type $\mathcal{T}$. For simplifying notations we assume there is only one sort. A *model of* $\mathbf{A}$ is a usual (static) algebraic structure M on the language associated to $\mathbf{A}$ and satisfying axioms of $\mathbf{A}$ (the ones of $\mathcal{T}$ and those given by the presentation of $\mathbf{A}$).

When $\mathbf{A}$ is defined by the empty presentation, we have *models of* $\mathcal{T}$.

So, the notion of model is based on the intuitive notion of *algebraic structure* à la Bourbaki. We can say that these algebraic structures are "static" in contrast to general dynamical algebraic structures. Note that here the underlying set of the structure is a naive set (or several naive sets if there are several sorts) which is structured by giving predicates and functions (in the naive meaning) subject to certain axioms.

From a constructive viewpoint, axioms in models must hold with the constructive meaning of "or" and "there exists": in order to prove that a given algebraic structure satisfies the axioms, we have to use the intuitionist logic. Let us note also that the set theory we use in our (external) reasoning about dynamical algebraic structures is the informal set theory of Bishop.

3.3 Conservative Extensions

A dynamical theory $\mathcal{T}$ is defined by a pair of sets $(\mathcal{L}, \mathcal{A})$ where $\mathcal{L}$ is (the signature of) the formal language we use, and $\mathcal{A}$ is the set of axioms, which are dynamical rules on the language $\mathcal{L}$. From our constructive viewpoint, these two sets are intuitive sets à la Bishop. These sets are in all cases very different of sets which are usually considered in categorical logic, where the external framework is given as purely formal, in $\mathcal{ZFC}$ or in an extension of $\mathcal{ZFC}$. In a similar way, it should be possible to use a formalization of constructive mathematics compatible with Bishop's set theory for studying general properties of dynamical theories. In this paper we work within the informal set theory of Bishop.

A dynamical theory $(\mathcal{L}', \mathcal{A}')$ is a simple extension of $(\mathcal{L}, \mathcal{A})$ if $\mathcal{L}$ and $\mathcal{A}$ are subsets of $\mathcal{L}'$ and $\mathcal{A}'$ (with the categorical meaning in Bishop's book). In this case, the dynamical rules formulated in the language $\mathcal{L}$ and valid in $\mathcal{T}$ are valid in $\mathcal{T}'$.

Definition 3.3.1 Two dynamical theories *on the same language* are said to be *identical* if they prove the same dynamical rules, i.e., if axioms of each one are valid rules of the other one. In this case, models are the same in constructive mathematics as they are in classical mathematics.

Definition 3.3.2 We say that a dynamical theory $\mathcal{T}'$ is a *simple conservative extension of the theory* $\mathcal{T}$ if it is a simple extension of $\mathcal{T}$ and if dynamical rules of $\mathcal{T}$ which are valid in $\mathcal{T}'$ are valid in $\mathcal{T}$.

Informal definition Other extensions, more general than those of Definition 3.3.1, have to be considered as "changing nothing to the theory" on one hand because they are conservative and on the other hand because constructive models of any dynamical algebraic structure are the same. In this case we say that $\mathcal{T}'$ is an *intuitively equivalent* extension of the theory $\mathcal{T}$.

We are now looking at a number of these extensions.

Essentially Identical Extensions

- *Adding abbreviations in the language*

For example, we may introduce a function symbol "Som$(\cdot, \cdot, \cdot)$" as an abbreviation: $\mathrm{Som}(x, y, z)$ is an abridged notation for $(x + y) + z$.

Clearly this type of definitions "changes nothing" to the dynamical theory, it is an intuitively equivalent extension.

- *Adding predicates: conjunction, disjunction, existence*

Let $\Delta = (A_1, \ldots, A_n)$ be a list of predicates.

One may consider that the following introduction and elimination rules define the conjunction, as in natural deduction.

$$\textbf{Intro-}\wedge_\Delta \quad A_1, \ldots, A_n \ \vdash A_1 \wedge \cdots \wedge A_n$$
$$\textbf{Elim-}\wedge_\Delta \quad A_1 \wedge \cdots \wedge A_n \ \vdash A_1, \ldots, A_n$$

Last rule is equivalent to the conjunction of rules $A_1 \wedge \cdots \wedge A_n \ \vdash A_i$.

Disjunction is more complicated, because in dynamical proofs there is no **or** on the left of $\vdash$. One may introduce the disjunction connector $\vee$ in the language by using rules inspired by natural deduction.

$$\textbf{Elim-}\vee_\Delta \quad A_1 \vee \cdots \vee A_n \ \vdash A_1 \ \textbf{or} \ \ldots \ \textbf{or} \ A_n$$
$$\textbf{Intro-}\vee_{\Delta,1} \quad A_1 \ \vdash A_1 \vee \cdots \vee A_n$$
$$\vdots$$
$$\textbf{Intro-}\vee_{\Delta,n} \quad A_n \ \vdash A_1 \vee \cdots \vee A_n$$

We introduce the existential quantifier $\exists$ by the following introduction and elimination rules (note that it is not a joke).

$$\textbf{Intro-}\exists_{x,A} \quad A(x) \ \vdash \exists x\, A(x)$$
$$\textbf{Elim-}\exists_{x,A} \quad \exists x\, A(x) \ \vdash \exists x\, A(x)$$

- *Adding a function symbol in case of unique existence*

We consider a dynamical theory $\mathcal{T} = (\mathcal{L}, \mathcal{A})$. Assume that $\mathcal{L}$ has a predicate $P(u, x, y)$ of arity $k + 1$ (we give the example with $k = 2$) and that $\mathcal{T}$ prove the following dynamical rules

$$\textbf{Ex}_{P,u} \ \vdash \exists u\ P(u, x, y) \qquad\qquad \textbf{Uniq}_{P,u}\ P(u, x, y),\ P(v, x, y) \ \vdash u = v$$

Let $\mathcal{T}'$ be the dynamical theory we get by adding to $\mathcal{T}$ a new function symbol f and the following axiom

$$\textbf{df}_{P,u,f}\ \ P(u, x, y) \ \vdash u = f(x, y)$$

We say that *the rule* $\textbf{df}_{P,u,f}$ *defines the function symbol* f.

In this case the dynamical theory $\mathcal{T}'$ is a conservative extension of $\mathcal{T}$.

Lemma 3.3.3 *For a dynamical theory using previously defined extensions (abbreviations, conjunction predicate, disjunction predicate, existential quantifier, function symbols in case of unique existence) produces a conservative extension which does not change constructive models.*

Definition 3.3.4

1. Such an extension is called *essentially identical simple*.
2. Two dynamical theories $\mathcal{T}_1$ and $\mathcal{T}_2$ are said to be *essentially identical* if we have a dynamical theory which is, perhaps after renamings in the signatures, an essentially identical extension of $\mathcal{T}_1$ and $\mathcal{T}_2$.

Essentially Equivalent Extensions

References: [2, 49]. We now examine situations obtained by extending a dynamical theory by adding well defined new sorts. This corresponds to allowable set constructions in Bishop's set theory.

- *Introducing a subsort*

Let us consider a dynamical theory $\mathcal{T}$ and a sort S in this theory. Let us consider a unary predicate $P(\cdot)$ on objects of type S. We define a new dynamical theory by adding *the subsort* U *of* S *defined by the predicate* P in the following way.

1. We add U in the sorts.
2. We add a function symbol $j_P : U \to S$.
3. We add the following axioms.

$$\textbf{sso}_P\ \ \vdash_{u:U} P(j_P(u)) \qquad\qquad \textbf{SSO}_P\ \ P(a) \ \vdash_{a:S} \exists u\ j_P(u) = a.$$

4. We define the equality predicate $x =_U y$ on U as an abbreviation of $j_P(x) = j_P(y)$.

- *Introducing a sort for a finite product of sorts*

Let us consider a dynamical theory $\mathcal{T}$ and sorts $S_1, \ldots, S_n$ in this theory. We define a new dynamical theory by adding the *finite product* S *of sorts* S_i in the following way.

1. We add a sort S.
2. For each $i \in [\![1..n]\!]$ we add a function symbol π_i of type $S \to S_i$.
3. We define $x =_S y$ on S as an abbreviation of $\bigwedge_{i=1}^{n} \pi_i(x) = \pi_i(y)$.
4. We add a function symbol Pr of type $S_1 \times \cdots \times S_n \to S$.
5. We add axioms

$$\textsf{fps}_i \ \vdash_{x_1:S_1,\ldots,x_n:S_n} \ \pi_i\big(\mathrm{Pr}(x_1,\ldots,x_n)\big) = x_i$$

- *Introducing a quotient sort*

Let us consider a dynamical theory $\mathcal{T}$ and a sort S of this theory. Let us consider a binary predicate $E(x, y)$ on S. We assume that the rules saying that $E(x, y)$ is an equivalence relation on S are valid in $\mathcal{T}$. We define a new dynamical theory by adding *the quotient sort B of S w.r.t. the equivalence relation E* in the following way.

1. We add the sort B and an equality predicate $x =_B y$ on B.
2. We add a function symbol π_E of type $S \to B$.
3. We add axioms

$$\begin{array}{rl} \textsf{qs1}_E & E(a,b) \ \vdash_{a,b:S} \ \pi_E(a) =_B \pi_E(b) \\ \textsf{QS}_E & \vdash_{x:B} \ \exists a \ \pi_E(a) =_B x \\ \textsf{Qs2}_E & \pi_E(a) =_B \pi_E(b) \ \vdash_{a,b:S} \ E(a,b) \end{array}$$

- *Introducing a sort for a finite disjoint sum of sorts*

Let us consider a dynamical theory $\mathcal{T}$ and sorts $S_1, \ldots, S_n$ in this theory. We define a new dynamical theory by adding *the disjoint sum of sorts S_i* in the following way.

1. We add the sort S and an equality predicate $x =_S y$ on S.
2. For each $i \in [\![1..n]\!]$ we add a function symbol $j_{S_i,S}$ of type $S_i \to S$.
3. We add the following axioms (we use j_i as an abbreviation for $j_{S_i,S}$)

$$\begin{array}{rll} \textsf{Fds}_i & j_i(a) =_S j_i(b) \ \vdash_{a,b:S_i} \ a =_{S_i} b & \text{for } 1 \leqslant i \leqslant n \\ \textsf{FDS}_{i,k} & j_i(a) =_S j_k(b) \ \vdash_{a:S_i,b:S_k} \ \bot & \text{for } i \leqslant i < k \leqslant n \\ \textsf{FDS} & \vdash_{x:S} \ \exists a_1 \ j_1(a_1) =_S x \ \textsf{ or } \ \cdots \ \textsf{ or } \ \exists a_n \ j_n(a_n) =_S x & \end{array}$$

Lemma 3.3.5 *Consider a dynamical theory obtained by using previously defined extensions, as in Lemma 3.3.3 or by introducing a new sort as in the previous examples. Then the new theory is intuitively equivalent to the first one.*

Definition 3.3.6

1. Such an extension is called *essentially equivalent simple*.
2. Two dynamical theories $\mathcal{T}_1$ and $\mathcal{T}_2$ are said to be *essentially equivalent* if we have a dynamical theory which is, perhaps after renamings in the signatures, an essentially equivalent extension of $\mathcal{T}_1$ and $\mathcal{T}_2$.

3. One says that *the dynamical theory $\mathcal{T}'$ is an extension of the dynamical theory $\mathcal{T}$* if it is a simple extension of a theory which is essentially equivalent to $\mathcal{T}$.

Essentially equivalent extensions do not change constructive models of dynamical algebraic structures. Other conservative extensions may change constructive models. We are now looking at a number of these extensions.

Other Conservative Extensions

- *Adding classical logic*

For a dynamical theory, accepting classical logic is the same thing as allowing systematically beyond Definition 3.3.4, the introduction of a predicate negating a previously defined predicate P.

When introducing the opposite predicate Q (denoted as $\neg P$ or $\overline{P}$) we add the axioms of Boolean logic.

- $P, Q \vdash \bot$
- $\vdash P$ **or** Q

From a constructive viewpoint, adding classical logic means in models of dynamical algebraic structures that all predicates be decidable.

Theorem 3.3.7 (Cut Elimination) *When computing in a dynamical theory the use of classical logic produces a conservative extension.*

The use of classical logic is often seen by classical mathematicians as the possibility of transforming all proofs in proofs by contradiction. For example, the following rules are now equivalent.

- $A, B \vdash C$ **or** D
- $B, \overline{C}, \overline{D} \vdash \overline{A}$
- $A, B, \overline{C}, \overline{D} \vdash \bot$
- $A, \overline{C}, \overline{D} \vdash \overline{B}$

- *Skolemization*

In the following theorem, not only classical logic but also Skolemization is authorized. This can be viewed as a local form of Choice.

Theorem 3.3.8 (Skolemization [4]) *Let us consider a dynamical theory $\mathcal{T}$ and define $\mathcal{T}'$ to be the "Skolemized" theory: all existential axioms are transformed by replacing $\exists$ by the use of Skolem function symbols. Then $\mathcal{T}'$ is a conservative extension of $\mathcal{T}$.*

4 Distributive Lattices and Spectral Spaces Associated to a Dynamical Algebraic Structure

4.1 Zariski Spectrum and Zariski Lattice of a Commutative Ring

The Zariski lattice of a commutative ring can be defined using different extensions of the theory *Cr* of commutative rings.

We choose the theory of local rings since it plays a fundamental role in Grothendieck schemes.

We consider precisely the dynamical theory $\mathcal{L}r1$ of *local rings with units*, based on the signature $(\cdot = 0, \mathrm{Un}(\cdot); \cdot + \cdot, \cdot \times \cdot, -\cdot, 0, 1)$.

The predicate $\mathrm{Un}(x)$ is defined as the invertibility predicate with the suitable axioms. We add a collapsus axiom and Axiom **LR** of local rings.

$\mathbf{CL}_{Lr1}$ $\mathrm{Un}(0) \vdash \bot$ **LR** $\mathrm{Un}(x + y) \vdash \mathrm{Un}(x)$ **or** $\mathrm{Un}(y)$

Let $\mathbf{A}$ be a commutative ring. We consider the entailment relation $\vdash_{\mathbf{A},\mathrm{Zar}}$ on the underlying set of $\mathbf{A}$ which is defined by the following equivalence.

$$\begin{array}{c} a_1, \ldots, a_n \vdash_{\mathbf{A},\mathrm{Zar}} c_1, \ldots, c_m \qquad \overset{\mathrm{def}}{\Longleftrightarrow} \\ \mathrm{Un}(a_1), \ldots, \mathrm{Un}(a_n) \vdash_{\mathcal{A}l1(\mathbf{A})} \mathrm{Un}(c_1) \textbf{ or } \ldots \textbf{ or } \mathrm{Un}(c_m) \end{array} \tag{15}$$

We define the *Zariski lattice of* $\mathbf{A}$, denoted by $\mathsf{Zar}\,\mathbf{A}$ or $\mathsf{Zar}(\mathbf{A})$, as the one generated by the entailment relation $\vdash_{\mathbf{A},\mathrm{Zar}}$.

The corresponding map $\mathrm{D}_{\mathbf{A}} : \mathbf{A} \to \mathsf{Zar}\,\mathbf{A}$ is called the *Zariski support of* $\mathbf{A}$. When $\mathbf{A}$ is fixed by the context we merely note D.

The usual *Zariski spectrum* in classical mathematics is the dual spectral space of *Zar A*.

Note that since $\mathrm{D}(a_1) \wedge \cdots \wedge \mathrm{D}(a_n) = \mathrm{D}(a_1 \cdots a_n)$, elements of $\mathsf{Zar}\,\mathbf{A}$ can be written as $\mathrm{D}(c_1, \ldots, c_m) := \mathrm{D}(c_1) \vee \cdots \vee \mathrm{D}(c_m)$.

A more elementary theory is the theory $\mathcal{W}zdr$ of *without zerodivisor* nontrivial rings. It is obtained by adding to *Cr* a collapsus axiom and Axiom **WZD**

$\mathbf{CL}_R$ $1 = 0 \vdash \bot$ **WZD** $xy = 0 \vdash x = 0$ **or** $y = 0$.

One proves the equivalences in the following theorem. Item (4) is called a formal Nullstellensatz. Hilbert's Nullstellensatz is a more difficult topic.

Theorem 4.1.1 (Formal Nullstellensatz) *Let* $\mathbf{A}$ *be a commutative ring, and* $a_1, \ldots, a_n, c_1, \ldots c_m \in \mathbf{A}$. *T.F.A.E.*

$$\begin{array}{ll}(1) & \mathrm{D}(a_1), \ldots, \mathrm{D}(a_n) \vdash_{\mathsf{Zar}\mathbf{A}} \mathrm{D}(c_1), \ldots, \mathrm{D}(c_m) \\ (2) & \mathrm{Un}(a_1), \ldots, \mathrm{Un}(a_n) \vdash_{\mathcal{Al}(\mathbf{A})} \mathrm{Un}(c_1) \textbf{ or} \ldots \textbf{ or } \mathrm{Un}(c_m) \\ (3) & c_1 = 0, \ldots, c_m = 0 \vdash_{\mathcal{Asdz}(\mathbf{A})} a_1 = 0 \textbf{ or } \ldots \textbf{ or } a_n = 0 \\ (4) & \exists k > 0 \ (a_1 \cdots a_n)^k \in \langle c_1, \ldots, c_m \rangle \end{array}$$

As a consequence the element $\mathrm{D}(c_1, \ldots, c_m)$ of $\mathsf{Zar}\,\mathbf{A}$ can be identified with the ideal $\sqrt[\mathbf{A}]{\langle c_1, \ldots, c_m \rangle}$. Modulo this identification, the order relation is set inclusion.

Corollary 4.1.2 *The lattice* $\mathsf{Zar}\,\mathbf{A}$ *is generated by the least entailment relation on (the underlying set of)* $\mathbf{A}$ *such that*

- $0 \vdash 0$
- $1 \vdash 1$
- $ab \vdash a$
- $a, b \vdash ab$
- $a + b \vdash a, b$

In other words, the map $\mathrm{D} : \mathbf{A} \to \mathsf{Zar}\,\mathbf{A}$ *satisfies the relations*

$$\mathrm{D}(0) = 0,\ \mathrm{D}(1) = 1,\ \mathrm{D}(ab) = \mathrm{D}(a) \wedge \mathrm{D}(b),\ \mathrm{D}(a+b) \leqslant \mathrm{D}(a) \vee \mathrm{D}(b),$$

and any other map $\mathrm{D}' : \mathbf{A} \to T$ *satisfying these relations factorizes via* $\mathsf{Zar}\,\mathbf{A}$ *with a unique lattice morphism* $\mathsf{Zar}\,\mathbf{A} \to T$.

4.2 Real Lattice and Real Spectrum of a Commutative Ring

Let us consider a *dynamical theory of discrete ordered fields*, based on the signature $(\cdot = 0, \cdot \geqslant 0, \cdot > 0; \cdot + \cdot, - \cdot, 0, 1)$, for example, the theory $\mathcal{Dof}$ given in [21, Section 3].

The *real lattice* of a commutative ring $\mathbf{A}$, denoted by $\mathsf{Real}(\mathbf{A})$, is generated by the entailment relation $\vdash_{\mathbf{A},\mathrm{Real}}$ on $\mathbf{A}$ defined by the following equivalence.

$$\begin{array}{c} a_1, \ldots, a_n \vdash_{\mathbf{A},\mathrm{Real}} c_1, \ldots, c_m \qquad \overset{\mathrm{def}}{\Longleftrightarrow} \\ a_1 > 0, \ldots, a_n > 0 \vdash_{\mathcal{Dof}(\mathbf{A})} c_1 > 0 \textbf{ or } \ldots \textbf{ or } c_m > 0 \end{array} \tag{16}$$

We note $\mathrm{R} : \mathbf{A} \to \mathsf{Real}(\mathbf{A})$ the corresponding map. One proves the following equivalence (*formal Positivstellensatz*)

$$\mathrm{R}(a_1), \ldots, \mathrm{R}(a_n) \vdash_{\mathsf{Real}(\mathbf{A})} \mathrm{R}(c_1), \ldots, \mathrm{R}(c_m) \Longleftrightarrow \exists k \in \mathbb{N}\ \exists p \in C\ \ (a_1 \cdots a_n)^k + p = 0,$$

where C is the positive cone generated by the a_i's and $-c_j$'s. If $n = 0$ we replace $(a_1 \cdots a_n)^k$ with $1_{\mathbf{A}}$.

The usual real spectrum $\mathsf{Sper}(\mathbf{A})$ is the dual spectral space of $\mathsf{Real}(\mathbf{A})$. One can identify elements of $\mathsf{Sper}(\mathbf{A})$ to prime cones of $\mathbf{A}$. The quasi-compact open set corresponding to the element $\mathrm{R}(a) \in \mathsf{Real}\,\mathbf{A}$ is given by $\{\, \mathfrak{c} \in \mathsf{Sper}\,\mathbf{A};\ -a \notin \mathfrak{c} \,\}$.

One proves that the lattice $\mathsf{Real}(\mathbf{A})$ is generated by the least entailment relation on $\mathbf{A}$ satisfying the following relations

- $-x^2 \vdash$
- $\vdash 1$
- $x + y \vdash x, y$
- $x, y \vdash xy$
- $xy \vdash x, -y$

More generally we can define $\mathsf{Real}(\mathbf{A})$ and $\mathsf{Sper}(\mathbf{A})$ for any dynamical algebraic structure $\mathbf{A}$ of type $\mathcal{D}of$.

For more details see [10, 36].

4.3 Other Examples

- *First example.* Let us consider a dynamical algebraic structure $\mathbf{A} = \big((G, R), \mathcal{T}\big)$ for a dynamical theory $\mathcal{T} = (\mathcal{L}, \mathcal{A})$. If $P(x, y)$ is a binary predicate in the signature, and if $Clt = \mathrm{Clt}(\mathbf{A})$ is the set of closed terms of $\mathbf{A}$, we get an entailment relation $\vdash_{\mathbf{A},P}$ on $Clt \times Clt$ by letting

$$\begin{array}{c}(a_1, b_1), \ldots, (a_n, b_n) \vdash_{\mathbf{A},P} (c_1, d_1), \ldots, (c_m, d_m) \quad \overset{\text{def}}{\Longleftrightarrow} \\ P(a_1, b_1), \ldots, P(a_n, b_n) \vdash_{\mathbf{A}} P(c_1, d_1) \ \textsf{or} \ \ldots \ \textsf{or} \ P(c_m, d_m)\end{array} \tag{17}$$

Intuitively the distributive lattice generated by this entailment relation is the lattice of "truth values" of the predicate P in $\mathbf{A}$.

- *More generally.* Let us consider a dynamical algebraic structure $\mathbf{A} = \big((G, R), \mathcal{T}\big)$ for a dynamical theory $\mathcal{T} = (\mathcal{L}, \mathcal{A})$. Let S be a set of closed atomic formulae of $\mathbf{A}$. We define the entailment relation on S associated to $\mathbf{A}$ in the following way:

$$\begin{array}{c}A_1, \ldots, A_n \vdash_{\mathbf{A},S} B_1, \ldots, B_m \quad \overset{\text{def}}{\Longleftrightarrow} \\ A_1, \ldots, A_n \vdash_{\mathbf{A}} B_1 \ \textsf{or} \ \ldots \ \textsf{or} \ B_m\end{array} \tag{18}$$

We can note $\mathsf{Zar}(\mathbf{A}, S)$ the distributive lattice generated by this entailment relation.

- *A conservative extension $\mathcal{T}1$ of a dynamical theory $\mathcal{T}$* gives isomorphic Zariski lattices for dynamical algebraic structures $\mathcal{T}(\mathbf{A})$ and $\mathcal{T}1(\mathbf{A})$, with a same set S of closed terms. The lattice therefore gives a diminished image of the dynamical algebraic structure. For example, adding classical logic and Skolemizing a dynamical theory do not change associated lattices. Sometimes the study of the lattice is easier in the second theory. But to recover the wealth of dynamical theories seen from a constructive viewpoint, it becomes necessary in classical mathematics to use sheaves or toposes.

4.4 The Absolute Zariski Lattice of a Dynamical Algebraic Structure A

The (absolute) Zariski lattice of a dynamical algebraic structure $\mathbf{A} = \big((G, R), \mathcal{T}\big)$ is defined by taking S as the set Clat($\mathbf{A}$) of all closed atomic formulae of $\mathbf{A}$. We denote it $\mathsf{Zar}(\mathbf{A}, \mathcal{T})$ or with a suitable name corresponding to the theory $\mathcal{T}$. For example, $\mathsf{val}(\mathbf{A})$ for the theory $\mathcal{val}$ (see page 310). The spectral space dual is called the *Zariski spectrum of the dynamical algebraic structure* $\mathbf{A}$ (the valuative spectrum in the case of the theory $\mathcal{val}$).

When the theory $\mathcal{T}$ is weakly disjunctive, $\mathsf{Zar}(\mathbf{A})$ is (up to a canonical isomorphism) the distributive lattice defined by the entailment relation on Clat($\mathbf{A}$) generated by axioms in R and instantiations of axioms of $\mathcal{T}$ obtained by substituting variables by closed terms.

For example, for $\mathbf{A}$ a commutative ring, $(\mathsf{Zar}\,\mathbf{A})^{\mathrm{op}}$ can be seen as the absolute Zariski lattice of $\mathcal{Wzdr}(\mathbf{A})$.

4.5 Spectrum and Models in Classical Mathematics

Here models are generally seen from the viewpoint of classical mathematics.

One chooses for set S of closed atomic formulae few predicates in the language, such that other predicates can be defined in classical mathematics from those of S.

Giving a point of the spectrum $\mathsf{Spec}(\mathsf{Zar}(\mathbf{A}, S))$, i.e., a morphism $\alpha : \mathsf{Zar}(\mathbf{A}, S) \to \mathbf{2}$ means to give the truth value True or False to closed atomic formulae in S ($\alpha(B) = 1$ or 0). This implies giving also a truth value True or False to all closed atomic formulae in $\mathbf{A}$. This allows us to construct a model of $\mathbf{A}$ in the case of a weakly disjunctive theory. We get in this way a *minimal* model with the meaning that all elements are constructed from generators in G by using function symbols in the signature.

The choice of the set S is crucial for the topology of the dual spectral space. Two distinct choices for S can give the same points of the spectrum in classical mathematics but may define two distinct spectral spaces.

5 Valuative Lattice and Spectrum of a Commutative Ring

5.1 Valuative Divisibility Relation

Reference: [27].

Valuation Domains, Valuative Divisibility Relation

Remember that a subset P of a E is *detachable* when the property $x \in P$ is decidable for $x \in E$. For describing this situation in dynamical theories it is necessary to introduce a predicate $R(x)$ opposite to the predicate $Q(x)$ meaning $x \in P$. So the following rules hold: $\vdash Q(x)$ **or** $R(x)$, $Q(x), R(x) \vdash \bot$.

A ring is *integral* (or it is a *domain*) when each element is null or regular, and a ring is a *discrete field* when each element is null or invertible. These definitions do not exclude the trivial ring.

A ring is *without zerodivisor* when the nullity of a product ab of two elements induces an explicit alternative, $a = 0$ or $b = 0$. An integral domain is without zerodivisor. In classical mathematics the reciprocal is valid but not in constructive mathematics.

We say that an ideal is *prime* if the quotient ring is without zerodivisor. This definition does not exclude the ideal $\langle 1 \rangle$. The latter conventions are used in [38]. They allow the authors to avoid negation and certain case by case reasonings, non-legitimate from a constructive viewpoint.

Nevertheless, for corresponding dynamical theories, we come back to the usual tradition for which local, without zerodivisor, or integral rings, and discrete fields, have to be nontrivial: this is necessary for introducing collapsus axioms giving $\bot$ as a possible conclusion in a dynamical rule.

A *valuation domain* $\mathbf{V}$ is an integral domain whose divisibility relation (on the multiplicative monoid $\mathbf{V}/\mathbf{V}^{\times}$) is a total order: $\forall x, y\ (x \mid y \vee y \mid x)$.

If $\mathbf{K}$ is the fraction field of $\mathbf{V}$, $\mathbf{V}$ is called a *valuation ring of* $\mathbf{K}$ and $(\mathbf{K}, \mathbf{V})$ is called a *valued discrete field*. More generally, a subring $\mathbf{V}$ of a discrete field $\mathbf{K}$ is called a valuation ring of $\mathbf{K}$ if for all $x \in \mathbf{K}^{\times}$, x or $x^{-1} \in \mathbf{V}$.

From a constructive viewpoint, we define a *valued discrete field* by forcing the decidability of the relations $x \in \mathbf{V}$ and $x \in \mathbf{V}^{\times}$: this means that the divisibility in $\mathbf{V}$ has to be explicit.

A valuation domain can also be characterized a local residually discrete Bézout domain.[16]

In [21, Section 4] a dynamical theory $\mathcal{Vf}$ is introduced for valued discrete fields with this constructive viewpoint. The signature has three predicates $\mathrm{Vr}(x)$, $\mathrm{Un}(x)$

---

[16] A local ring $\mathbf{A}$ is said residually discrete when the residual field $\mathbf{A}/\mathrm{Rad}\mathbf{A}$ is discrete. If the ring is nontrivial this means that units are detachable.

and $\mathrm{Rn}(x)$, respectively, for $x \in \mathbf{V}$, $x \in \mathbf{V}^\times$ and $x \in \mathbf{V} \setminus \mathbf{V}^\times$. We come back to this theory in Sect. 7.

In a valued discrete field $(\mathbf{K}, \mathbf{V})$ we say that x divides y and we write $x \mid y$ if there exists a $z \in \mathbf{V}$ such that $xz = y$. We note $\Gamma = \Gamma(\mathbf{V})$ the group $\mathbf{K}^\times/\mathbf{V}^\times$ (in additive notation), with the order relation $\leqslant$ induced by the relation $\mid$ in $\mathbf{K}^\times$. We note $\Gamma_\infty = \Gamma \cup \{\infty\}$ (where ∞ is a purely formal maximum element). Thus, the natural map $v : \mathbf{K} \to \Gamma_\infty$ (with $v(0) = \infty$) is called the *valuation* of the valued discrete field. One has $v(xy) = v(x) + v(y)$ and $v(x + y) \geqslant \min(v(x), v(y))$ with equality if $v(x) \neq v(y)$. We have also $\boxed{\mathbf{V} = \{x \in \mathbf{K} v(x) \geqslant 0\}}$ and the unit group is characterized by $\boxed{\mathbf{V}^\times = \{x \in \mathbf{K} v(x) = 0\}}$.

In classical mathematics one defines a *valuative divisibility relation* $a \mid b$ on a commutative ring $\mathbf{A}$ as the reciprocal image of the divisibility relation on a valued discrete field $(\mathbf{K}, \mathbf{V})$ by a ring morphism $\varphi : \mathbf{A} \to \mathbf{K}$. In other words, we have $a \mid b$ in $\mathbf{A}$ if and only if $\varphi(a) \mid \varphi(b)$ in $(\mathbf{K}, \mathbf{V})$.

Points of the Valuative Spectrum in Classical Mathematics

Definition 5.1.1 In classical mathematics, elements of the *valuative spectrum* $\mathsf{Spev}(\mathbf{A})$ *of a commutative ring* $\mathbf{A}$ are defined in the following way (see [27]): a point of $\mathsf{Spev}\,\mathbf{A}$ is given by a pair $(\mathfrak{p}, \mathbf{V})$ where $\mathfrak{p}$ is a prime ideal of $\mathbf{A}$ and $\mathbf{V}$ a valuation ring of the fraction field $\mathbf{K} = \mathrm{Frac}(\mathbf{A}/\mathfrak{p})$.

This is analogous to the real spectrum: a point of $\mathsf{Sper}\,\mathbf{A}$ is given by a pair $(\mathfrak{p}, \mathbf{C})$ where $\mathfrak{p}$ is a prime ideal of $\mathbf{A}$ and $\mathbf{C}$ a positive cone of the fraction field $\mathbf{K} = \mathrm{Frac}(\mathbf{A}/\mathfrak{p})$ (i.e., $\mathbf{C} + \mathbf{C} \subseteq \mathbf{C}$, $\mathbf{C} \cdot \mathbf{C} \subseteq \mathbf{C}$, $\mathbf{C} \cup -\mathbf{C} = \mathbf{K}$, $\mathbf{C} \cap -\mathbf{C} = \{0\}$).

Distinct spectral topologies can be defined on the valuative spectrum, depending on the choice of basic quasi-compact open sets.

This corresponds to suitable distributive lattices (as those described in Sect. 4) which are defined w.r.t. dynamical theories that describe properties of a valuative divisibility relation.

We are now looking at a number of these theories.

5.2 *Weakly Disjunctive Theories for a Valuative Divisibility Relation*

The Theory *val0*

Definition 5.2.1 One chooses the signature

$$\Sigma_{\mathrm{val0}} = (\,\cdot \mid \cdot, \cdot = 0; \cdot + \cdot, \cdot \times \cdot, - \cdot, 0, 1\,).$$

Axioms for $x = 0$ and $a \mid b$ are the following ones.

vr0 $0 \mid x \vdash x = 0$	**CL**$_{\text{val}}$ $0 \mid 1 \vdash \bot$ (collapsus)
vr1 $\vdash 1 \mid -1$	**Vr2** $a \mid b, a \mid c \vdash a \mid b + c$
vr2 $a \mid b \vdash ac \mid bc$	**VR1** $\vdash a \mid b$ **or** $b \mid a$
Vr1 $a \mid b, b \mid c \vdash a \mid c$	**VR2** $ax \mid bx \vdash a \mid b$ **or** $0 \mid x$

We note *val0* this weakly disjunctive theory.

One proves easily $\vdash -1 \mid 1$, $\vdash 1 \mid 1$, $\vdash 1 \mid 0$, $\vdash x \mid x$, $\vdash x \mid 0$ and $x = 0 \vdash 0 \mid x$. Axioms **cr1**, **cr2**, and **cr3** for commutative rings are satisfied.

Here is a precise statement saying that our theory describes correctly a valuative divisibility relation.

Theorem* 5.2.2 (Models of the Theory *val0* in Classical Mathematics) *In classical mathematics a binary relation $a \mid b$ on a ring $\mathbf{A}$ satisfies axioms of val0 exactly in the following case.*

- *The set $\mathfrak{p} = \{ x \in \mathbf{A}; 0 \mid x \}$ is a prime ideal. We note $\overline{a}$ the element a of $\mathbf{A}$ seen in $\mathbf{A}/\mathfrak{p}$, and $\mathbf{K}$ the fraction field* $\mathrm{Frac}(\mathbf{A}/\mathfrak{p})$.
- *Fractions $\overline{a}/\overline{b} \in \mathbf{K}$ such that $\overline{b} \neq 0$ and $b \mid a$ make a valuation ring of $\mathbf{V}$ of $\mathbf{K}$.*

In other words, axioms for $\cdot \mid \cdot$ in val0 correspond to the definition of a valuative divisibility relation (in classical mathematics).

Proof First, one sees easily that the axioms hold if $\varphi : \mathbf{A} \to \mathbf{K}$ is a morphism from $\mathbf{A}$ to a field $\mathbf{K}$, if $\mathbf{V}$ is a valuation ring of $\mathbf{K}$ and if $a \mid b$ means: $\exists x \in \mathbf{V}\ x\varphi(a) = \varphi(b)$.

Let us now prove that the axioms describe correctly this situation (in classical mathematics).

Axioms **vr1** and **vr2** give valid rules $\vdash a \mid a$ and $\vdash a \mid 0$. So, recalling axiom **Vr1**, the relation $a \mid b$ defines a preorder. And if $0 \mid 1$ we get $0 \mid b$ and $a \mid b$ for all a, b without using Axiom **CL**$_{\text{val}}$.
A particular case of **vr2** is the following valid rule.

vr2' $1 \mid a \vdash b \mid ab$

Using **Vr2** and **vr2'**, one sees that for all $a \in \mathbf{A}$ such that $1 \mid a$, the set

$$\uparrow a := \{ b \in \mathbf{A}; a \mid b \}$$

is an ideal of $\mathbf{A}$. In particular $\mathfrak{p} = \uparrow 0$ is an ideal.

Let us prove that the relation $\cdot \mid \cdot$ passes to the quotient by $\mathfrak{p} = \uparrow 0$. Indeed, assume $0 \mid x$, it suffices to prove that $a \mid a + x$ and $a + x \mid a$. First, $a \mid 0 \mid x$ and $a \mid a$, hence $a \mid a+x$ using axiom **Vr2**. Now, since we have $a+x \mid a+x$ we get $a+x \mid a+x-x$, i.e., $a + x \mid a$.

Let us prove that $\mathbf{A}/\mathfrak{p}$ is without zerodivisor: if $0 \mid yx$ then $0x \mid yx$, and **VR2** gives $0 \mid xy \vdash 0 \mid x$ **or** $0 \mid y$. Using **vr2**, **Vr1**, and **Vr2** one sees that the fractions $\overline{a}/\overline{b}$ of $\mathbf{K}$ make a subring $\mathbf{V}$ of $\mathbf{K}$. Finally, two inverse elements in $\mathbf{K}$ are written $\overline{a}/\overline{b}$ and $\overline{b}/\overline{a}$. So, Axiom **VR1** implies $\mathbf{V}$ is a valuation ring of $\mathbf{K}$. □

Note that $\varphi(\mathbf{A})$ is not necessarily a subring of $\mathbf{V}$.

Theorem 5.2.2 legitimates the following definition in constructive mathematics.

Definition 5.2.3 A binary relation $a \mid b$ on a commutative ring is called a *valuative divisibility relation* if axioms of $\mathcal{val0}$ are satisfied.

Remark 5.2.4 The three axioms of commutative rings are valid in $\mathcal{val0}$.

cr1 $\vdash 0 = 0$ **cr2** $x = 0 \vdash xy = 0$
cr3 $\vdash x = 0, y = 0 \vdash x + y = 0$

So, we adopt the convention given in [21] that the purely computational part of commutative rings is treated outside the formal theory (see page 296).

If we consider a commutative ring **A** and the dynamical algebraic structure $\mathcal{val0}(\mathbf{A})$, all closed terms of $\mathcal{val0}(\mathbf{A})$ are equal to elements of **A** and there is no need of the three axioms for proving closed dynamical rules. ■

Remark 5.2.5 Without using the collapsus axiom, when the relation $0 \mid 1$ is valid, the ring becomes a singleton. Adding the axiom $0 \mid 1 \vdash \bot$ amounts to throw the trivial ring in the empty universe.[17] ■

Remark 5.2.6 In constructive mathematics if a ring **A** with a binary relation $\cdot \mid \cdot$ is a model of $\mathcal{val0}$ we get a prime ideal $\mathfrak{p}$ of **A** and a suitable subring **V** of the total fraction ring **K** of $\mathbf{A}/\mathfrak{p}$. But **K** is not a priori a discrete field. So we prefer to think about **A** as the dynamical algebraic structure $\mathcal{val0}(\mathbf{A})$, in which we can "do as if" we were in classical mathematics, where all fields are discrete. ■

The Theory $\mathcal{val}$ and Some Dynamical Rules Provable in It

Definition 5.2.7 The theory $\mathcal{val}$ is obtained from the theory $\mathcal{val0}$ by removing the predicate $\cdot = 0$, suppressing Axiom **vr0** and defining $x = 0$ as an abbreviation of $0 \mid x$.

We have the same conclusion as in Theorem 5.2.2: in classical mathematics the models of the theory $\mathcal{val}$ are valuation domains.

Without referring to models in classical mathematics, the dynamical theory $\mathcal{val}$ proves some usual properties of valued discrete fields as valid dynamical rules or, sometimes, as admissible dynamical rules.

First the fact that a valuation domain is local, normal and without zerodivisor correspond to the following valid rules. Proofs are without surprise directly translated from those of Theorem 5.2.2.

WZD $0 = bx \vdash 0 = b$ **or** $0 = x$.

Proof consider Axiom **VR2** with $a = 0 = 0x$. □

[17] Or, perhaps? In a black hole.

In particular we have $0 \mid x^2 \vdash 0 \mid x$.

LOC $1 \mid x, 1 \mid y, x + y \mid 1 \vdash x \mid 1$ **or** $y \mid 1$.

Using **VR1** we open two branches, the one where $x \mid y$, the other where $y \mid x$. In the first one, since $x \mid x$, **Vr2** gives $x \mid x + y$. So, by **Vr1**, $x \mid 1$. Symmetrically in the second branch $y \mid 1$.

The third rule is algebraic.

Nor $1 \mid a_1, \ldots 1 \mid a_n, y\big(\sum_{k=0}^{n} a_k x^k y^{n-k}\big) \mid x^{n+1} \vdash y \mid x$.

We open branches $x \mid y$ and $y \mid x$. We have to examine the first one; e.g., with $n = 2$. Since $x \mid y$, $1 \mid a_0$, $1 \mid a_1$ and $1 \mid a_2$, we get

$$x^2 \mid y^2 \mid a_0 y^2, \quad x^2 \mid xy \mid a_1 xy \quad \text{and} \quad x^2 \mid a_2 x^2.$$

So, $x^2 \mid a_2x^2 + a_1xy + a_0y^2$ by **Vr2**, and $yx^2 \mid x^3$. From $yx^2 \mid xx^2$ we deduce $\vdash 0 \mid x^2$ **or** $y \mid x$ by **VR2**. And in the branch where $0 \mid x^2$, we have $y \mid 0 \mid x$.

Now an admissible rule.

DIV $x \mid y \vdash \exists a\, (1 \mid a, ax = y)$

Lemma 5.2.8 *If a disjunctive rule is valid in the dynamical algebraic structure* $\mathbf{B} = \big((G, R), \mathit{val}\big)$ *when using* **DIV**, *it is also valid in* **B**.

The proof of this lemma seems rather difficult. We can catch it after having proved a formal Valuativstellensatz: see Remark 7.4.4.

Dynamical Algebraic Structures of Type *val*

Definition 5.2.9

1. Let **A** be a commutative ring, we define the dynamical algebraic structure *val*(**A**) as usual by taking the presentation given by the positive diagram of **A**.
2. Let $\mathbf{k} \subseteq \mathbf{A}$ be two rings,[18] or more generally let $\varphi : \mathbf{k} \to \mathbf{A}$ be an algebra. We note *val*(**A**, **k**) the dynamical algebraic structure whose presentation is given by
   - The positive diagram of **A** as commutative ring.
   - Axioms $\vdash 1 \mid \varphi(x)$ for elements x of **k**.

The two dynamical algebraic structures *val*(**A**) and *val*(**A**, **Z**), where **Z** is the least subring of **A**, are canonically isomorphic.

As a particular case of Rule **WZD**, if e is an idempotent of **A**, we get $\vdash e = 0$ **or** $e = 1$ in *val*(**A**). In particular, $\vdash 1 \mid e$ and $\vdash 1 \mid f$ (where $f = 1 - e$). We

[18] We use **k** for the first ring as an intuition given by the frequent context where **k** is a discrete field.

deduce for any $a, b \in \mathbf{A}$: $ae \mid be, af \mid bf \ \vdash a \mid b$. Indeed $a \mid ae \mid be$ and $a \mid af \mid bf$, so $a \mid be + bf = b$. Let us summarize the situation.

Lemma 5.2.10 *If e and $f \in \mathbf{A}$ are two complementary idempotents we have in $\mathcal{val}(\mathbf{A})$*

- $a \mid b \ \vdash (ae \mid be, af \mid bf)$
- $ae \mid be, af \mid bf \ \vdash a \mid b$

Now a consequence of the validity of **Nor** in $\mathcal{val}$.

Lemma 5.2.11 *Let $\mathbf{k} \subseteq \mathbf{A}$ be two rings and $x \in \mathbf{A}$, $y \in \mathbf{k}$. Then if x is integral*[19] *over the ideal $\langle y \rangle$ of $\mathbf{k}$, the dynamical algebraic structure $\mathcal{val}(\mathbf{A}, \mathbf{k})$ proves $y \mid x$.*

We shall see in Theorem 7.4.6 that this sufficient condition is also necessary.

5.3 Valuative Lattice and Spectrum of a Commutative Ring

Several Possible Spectral Topologies

Let $\Gamma = \Gamma(\mathbf{V})$, then the point $(\mathfrak{p}, \mathbf{V})$ of $\mathsf{Spev}(\mathbf{A})$ is characterized by the associated valuation

$$w : \mathbf{A} \to \Gamma_\infty, \ x \mapsto v(\pi_\mathfrak{p}(x)),$$

where $\pi_\mathfrak{p} : \mathbf{A} \to \mathbf{A}/\mathfrak{p}$ is the canonical surjection.

We use the notation $\mathsf{Spev}\ \mathbf{A}$ for the spectrum with the topology which seems most natural to us, where quasi-compact open sets generating the topology are the following $\mathfrak{O}(a, b)$'s:

$$\mathfrak{O}(a, b) := \{\, w;\ w(b) \leqslant w(a) \,\}, \quad a, b \in \mathbf{A}.$$

In [27] the topology is generated by the following $\mathfrak{U}(a, b)$'s:

$$\mathfrak{U}(a, b) := \{\, w;\ w(b) \leqslant w(a),\ w(b) \neq \infty \,\}, \quad a, b \in \mathbf{A}.$$

They note $\mathsf{Spv}(\mathbf{A})$ this spectral space.

Thus, we get (with $\overline{Y}$ denoting the complementary set of Y) the following equalities.

- $\mathfrak{U}(b, b) = \{\, w;\ w(b) \neq \infty \,\} = \overline{\mathfrak{O}(b, 0)}$
- $\mathfrak{U}(a, b) = \mathfrak{O}(a, b) \cap \overline{\mathfrak{O}(b, 0)}$
- $\overline{\mathfrak{U}(b, b)} = \{\, w;\ w(b) = \infty \,\} = \mathfrak{O}(b, 0)$

[19] The element x is said to be integral over the ideal $\mathfrak{a}$ of $\mathbf{k}$ if it is a zero of a polynomial $x^{n+1} + \sum_{i=0}^{n} a_i x^{n-i}$ with $a_i \in \mathfrak{a}^i$ for each i [38, Definition XII-2.1].

- $\mathfrak{O}(a,b) = \mathfrak{U}(a,b) \cup \left(\overline{\mathfrak{U}(b,b)} \cap \overline{\mathfrak{U}(a,a)}\right)$

Hence Spev **A** and Spv **A** define the same patch topology.

The topology of Spv **A** considers as essential the property "$b \mid a$ and $b \neq 0$." This definition seems unnatural. In the same paper [27], another spectral topology $\mathsf{Spv}'(\mathbf{A})$ is introduced with basic open sets $\mathfrak{F}(a,b) := \{ v; v(a) < v(b) \} = \overline{\mathfrak{O}(a,b)}$. This is the opposite spectral space of Spev **A**.

We define now distributive lattices corresponding to these spectral spaces.

The Lattice val(A) and Its Spectrum Spev A

Definition 5.3.1 Let **A** be a commutative ring. We consider the set (underlying) $\mathbf{A} \times \mathbf{A}$ and we define on it the entailment relation $\vdash_{\mathbf{A},\mathrm{val}}$ by the following equivalence.

$$\begin{array}{c} (a_1,b_1),\ldots,(a_n,b_n) \vdash_{\mathbf{A},\mathrm{val}} (c_1,d_1),\ldots,(c_m,d_m) \quad \overset{\mathrm{def}}{\Longleftrightarrow} \\ a_1 \mid b_1, \ldots, a_n \mid b_n \vdash_{\mathcal{val}(\mathbf{A})} c_1 \mid d_1 \text{ or } \ldots \text{ or } c_m \mid d_m . \end{array} \tag{19}$$

The lattice $\mathsf{val}(\mathbf{A})$ is defined as generated by the entailment relation $\vdash_{\mathbf{A},\mathrm{val}}$.

We note $\mathrm{Di}_{\mathbf{A}} : \mathbf{A} \times \mathbf{A} \to \mathsf{val}(\mathbf{A})$ (or simply Di) the corresponding map.

For example, with $a, b, c \in \mathbf{A}$

$$\mathrm{Di}(a,b) \wedge \mathrm{Di}(a,c) \leqslant \mathrm{Di}(a,b+c), \qquad \text{in } \mathsf{val}(\mathbf{A}).$$

In fact, since $\cdot \mid \cdot$ is the unique predicate of the dynamical theory $\mathcal{val}$, the lattice $\mathsf{val}(\mathbf{A})$ is the absolute Zariski lattice of the dynamical algebraic structure $\mathcal{val}(\mathbf{A})$.

Since the theory $\mathcal{val}$ satisfies the rule **WZD**, the natural morphism $\mathsf{val}(\mathbf{A}) \to \mathsf{val}(\mathbf{A}_{\mathrm{red}})$ is an isomorphism.

Theorem 5.2.2 says that the dynamical algebraic structure $\mathcal{val}(\mathbf{A})$ has in classical mathematics minimal models given by the points of the valuative spectrum $\mathsf{Spev}(\mathbf{A})$. After the explanations in paragraph *Spectrum and models in classical mathematics* page 306, and examining the definition of basic open sets $\mathfrak{O}(a,b)$ generating the spectral topology, we get the following theorem in classical mathematics.

Theorem* 5.3.2 *The spectral spaces* $\mathsf{Spev}(\mathbf{A})$ *and* $\mathsf{Spec}(\mathsf{val}(\mathbf{A}))$ *are canonically homeomorphic.*

In constructive mathematics, the *pointfree topology* given by the distributive lattice $\mathsf{val}(\mathbf{A})$ is generated by formal opens: the elements $\mathrm{Di}(a,b)$ of $\mathsf{val}(\mathbf{A})$.

In classical mathematics, where spectral spaces have enough points, the topology of $\mathsf{Spec}(\mathsf{val}(\mathbf{A}))$ is a usual one, On the other hand we have defined (on the same underlying set) a topology on $\mathsf{Spev}\mathbf{A}$ generated by open sets $\mathfrak{O}(a,b) := \{ w; w(b) \leqslant w(a) \}$. Clearly they correspond to formal $\mathrm{Di}(b,a)$'s: this explains the homeomorphism between Spev **A** and $\mathsf{Spec}(\mathsf{val}(\mathbf{A}))$.

The Lattice $\mathsf{val}^{\star}(\mathbf{A})$ and the Spectrum $\mathsf{Spv}\,\mathbf{A}$

Spaces $\mathsf{Spev}(\mathbf{A})$ and $\mathsf{Spv}(\mathbf{A})$ are different, but they have the same points given by suitable pairs $(\mathfrak{p}, \mathbf{V})$.

We consider the weakly disjunctive theory $\mathcal{val1}$ we get from $\mathcal{val}$ by adding the predicate $\cdot \neq 0$ opposite to $\cdot = 0$. The theory $\mathcal{val1}$ is a conservative extension of $\mathcal{val}$.

Let us consider now the predicate $\boxed{J(a,b) \overset{\text{def}}{=} (a \mid b \wedge a \neq 0)}$.

We say that the spectral space $\mathsf{Spv}(\mathbf{A})$ is homeomorphic to the spectrum of the following distributive lattice $\mathsf{val}^{\star}(\mathbf{A})$.

Definition 5.3.3 The distributive lattice $\mathsf{val}^{\star}(\mathbf{A})$ is generated by the entailment relation $\vdash_{\mathbf{A},\mathrm{val}^{\star}}$ on $\mathbf{A} \times \mathbf{A}$ defined by the equivalence

$$\begin{array}{c} (a_1, b_1), \ldots, (a_n, b_n) \vdash_{\mathbf{A},\mathrm{val}^{\star}} (c_1, d_1), \ldots, (c_m, d_m) \quad \overset{\text{def}}{\Longleftrightarrow} \\ J(a_1, b_1), \ldots, J(a_n, b_n) \vdash_{\mathcal{val1}(\mathbf{A})} J(c_1, d_1) \ \textsf{or} \ \ldots \ \textsf{or} \ J(c_m, d_m). \end{array} \tag{20}$$

We note $\mathrm{J}_{\mathbf{A}} : \mathbf{A} \times \mathbf{A} \to \mathsf{val}^{\star}(\mathbf{A})$ (or simply J) the corresponding map.

Thus, the element $\mathrm{J}(b, a)$ of $\mathsf{val}^{\star}(\mathbf{A})$ corresponds to the open set $\mathfrak{U}(a, b)$ of $\mathsf{Spv}(\mathbf{A})$.

5.4 Valuative Lattice and Spectrum of an Algebra

We consider in this section a $\mathbf{k}$-algebra $\mathbf{A}$, i.e., a morphism $\varphi : \mathbf{k} \to \mathbf{A}$ of commutative rings.

The Lattice $\mathsf{val}(\mathbf{A}, \mathbf{k})$ and the Spectrum $\mathsf{Spev}(\mathbf{A}, \mathbf{k})$

We define the distributive lattice $\mathsf{val}(\mathbf{A}, \mathbf{k})$ by using the dynamical algebraic structure $\mathcal{val}(\mathbf{A}, \mathbf{k})$ (see definition 5.2.9) in the same way as $\mathsf{val}(\mathbf{A})$ is defined by using the dynamical algebraic structure $\mathcal{val}(\mathbf{A})$.

Definition 5.4.1 Let $\mathbf{k}$ be a subring of a ring $\mathbf{A}$, or more generally let us consider a $\mathbf{k}$-algebra $\varphi : \mathbf{k} \to \mathbf{A}$. We define the distributive lattice $\mathsf{val}(\mathbf{A}, \mathbf{k})$ as generated by the entailment relation $\vdash_{\mathbf{k},\mathbf{A},\mathrm{val}}$ on the set $\mathbf{A} \times \mathbf{A}$, which is defined by the following equivalence.

$$\begin{array}{c} (a_1, b_1), \ldots, (a_n, b_n) \vdash_{\mathbf{k},\mathbf{A},\mathrm{val}} (c_1, d_1), \ldots, (c_m, d_m) \quad \overset{\text{def}}{\Longleftrightarrow} \\ a_1 \mid b_1, \ldots, a_n \mid b_n \vdash_{\mathcal{val}(\mathbf{A},\mathbf{k})} c_1 \mid d_1 \ \textsf{or} \ \ldots \ \textsf{or} \ c_m \mid d_m \end{array} \tag{21}$$

We note $\mathrm{Di}_{\mathbf{k},\mathbf{A}} : \mathbf{A} \times \mathbf{A} \to \mathsf{val}(\mathbf{A}, \mathbf{k})$ (or simply Di) the corresponding map.

Remark 5.4.2 Since the theory *val* satisfies the rule **Nor**, if $\overline{\mathbf{k}}$ is the integral closure of (the image of) $\mathbf{k}$ in $\mathbf{A}_{\text{red}}$, the dynamical algebraic structure $\mathit{val}(\mathbf{A}, \mathbf{k})$ proves $1 \mid x$ for $x \in \overline{\mathbf{k}}$. Thus, the natural morphism $\mathsf{val}(\mathbf{A}, \mathbf{k}) \to \mathsf{val}(\mathbf{A}_{\text{red}}, \overline{\mathbf{k}})$ is an isomorphism. ■

We note $\mathsf{Spev}(\mathbf{A}, \mathbf{k}) := \mathsf{Spec}(\mathsf{val}(\mathbf{A}, \mathbf{k}))$. Points of this spectrum correspond to points $(\mathfrak{p}, \mathbf{V})$ of $\mathsf{Spev}(\mathbf{A})$ such that $\mathbf{V}$ contains (the image of) $\mathbf{k}$. Since the lattice $\mathsf{val}(\mathbf{A}, \mathbf{k})$ is a quotient of $\mathsf{val}(\mathbf{A})$, $\mathsf{Spev}(\mathbf{A}, \mathbf{k})$ is a subspectral space of $\mathsf{Spev}\,\mathbf{A}$.

When $\mathbf{K}$ is a discrete field transcendent over a subfield $\mathbf{k}$, the spectral space $\mathsf{Spev}(\mathbf{K}, \mathbf{k})$ is often called the *Zariski-Riemann spectrum* of $(\mathbf{K}, \mathbf{k})$, or also, the *abstract Riemann surface* of $(\mathbf{K}, \mathbf{k})$. The points of this spectrum are the valuation rings of $\mathbf{K}$ containing $\mathbf{k}$.

In constructive mathematics we are mainly interested in the lattice $\mathsf{val}(\mathbf{K}, \mathbf{k})$. The abstract Riemann surface is more than a spectral space, a Grothendieck scheme. It has a good constructive description and the most important object is this scheme.

Another special case is given when $\mathbf{A}$ is the fraction field of a domain $\mathbf{k}$, for example, when $\mathbf{k}$ is the ring of all algebraic integers.

Lemma 5.4.3 *Let* $\mathbf{k}$ *be a domain with fraction field* $\mathbf{K}$*. The natural morphism* $\mathsf{val}(\mathbf{k}, \mathbf{k}) \to \mathsf{val}(\mathbf{K}, \mathbf{k})$ *is an isomorphism.*

Proof There are more atomic formulae in $\mathit{val}(\mathbf{K}, \mathbf{k})$ than in $\mathit{val}(\mathbf{k}, \mathbf{k})$. But if $x \mid y$ is an atomic formula of $\mathit{val}(\mathbf{K}, \mathbf{k})$ where $x = \frac{u}{v}$ and $y = \frac{s}{t}$ with $u, v, s, t \in \mathbf{k}$ and $v, t \neq 0$, formulae $x \mid y$ and $ut \mid sv$ are provably equivalent in $\mathit{val}(\mathbf{K}, \mathbf{k})$.
Thus it suffices to see that (for $a_i, b_i, c_j, d_j \in \mathbf{k}$)

$$a_1 \mid b_1, \ldots, a_n \mid b_n \vdash_{\mathit{val}(\mathbf{K},\mathbf{k})} c_1 \mid d_1 \textsf{ or } \ldots \textsf{ or } c_m \mid d_m$$

if and only if

$$a_1 \mid b_1, \ldots, a_n \mid b_n \vdash_{\mathit{val}(\mathbf{k},\mathbf{k})} c_1 \mid d_1 \textsf{ or } \ldots \textsf{ or } c_m \mid d_m.$$

Indeed, one may always avoid fractions in a proof by using the art of getting rid of denominators. □

The Center Map (1)

Proposition 5.4.4 (The Center Map for Distributive Lattices $\mathsf{val}(\mathbf{A}, \mathbf{A})$ and $\mathsf{Zar}\,\mathbf{A}$) *There exists a unique morphism of distributive lattice* $\gamma : \mathsf{Zar}\,\mathbf{A} \to \mathsf{val}(\mathbf{A}, \mathbf{A})$ *such that* $\gamma(\mathrm{D}_{\mathbf{A}}(a)) = \mathrm{Di}_{\mathbf{A}}(a, 1)$ *for all* $a \in \mathbf{A}$.

Proof Using Corollary 4.1.2, it is sufficient to see that the following rules are valid in $\mathit{val}(\mathbf{A}, \mathbf{A})$.

- $0 \mid 1 \vdash \bot$
- $\vdash 1 \mid 1$
- $ab \mid 1 \vdash a \mid 1$
- $a \mid 1, b \mid 1 \vdash ab \mid 1$

- $a+b \mid 1 \vdash a \mid 1$ **or** $b \mid 1$

And for this it is sufficient to see that the following rules are valid in $\mathit{val}(\mathbf{A})$.

- $0 \mid 1 \vdash \bot$
- $\vdash 1 \mid 1$
- $1 \mid b,\ ab \mid 1 \vdash a \mid 1$
- $a \mid 1,\ b \mid 1 \vdash ab \mid 1$
- $1 \mid a,\ 1 \mid b,\ a+b \mid 1 \vdash a \mid 1$ **or** $b \mid 1$

And this is easy. □

In classical mathematics, the dual viewpoint is given by the *center spectral map* $\mathsf{Spev}(\gamma) : \mathsf{Spev}(\mathbf{A}, \mathbf{A}) \to \mathsf{Spec}\,\mathbf{A}$: the image of the point $(\mathfrak{p}, \mathbf{V})$ of $\mathsf{Spev}(\mathbf{A}, \mathbf{A})$, is the prime ideal $\varphi^{-1}(\mathrm{Rad}(\mathbf{V}))$, where $\varphi : \mathbf{A} \to \mathrm{Frac}(\mathbf{A}/\mathfrak{p})$ is the canonical map (note that $\varphi(\mathbf{A}) \subseteq \mathbf{V}$ because we have $1 \mid a$ when $a \in \mathbf{A}$).

Remark 5.4.5 Theorem 7.4.7 proves that the center map is injective. A more direct proof should be interesting. ■

Theorem 5.4.6 *If* $\mathbf{k}$ *is an integral arithmetic ring (a Prüfer domain), the morphism* $\gamma : \mathsf{Zar}\,\mathbf{k} \to \mathsf{val}(\mathbf{k}, \mathbf{k})$ *in 5.4.4 is an isomorphism of distributive lattices.*

Proof Let us note $\mathbf{K}$ the fraction field of $\mathbf{k}$. Using Lemma 5.4.3, we replace $\mathsf{val}(\mathbf{k}, \mathbf{k})$ with $\mathsf{val}(\mathbf{K}, \mathbf{k})$ in the proof.

We assume now that γ is injective.

Let us prove that γ is onto. By definition, in an *arithmetic ring*, for any couple (a, b) we have s, t, u, v satisfying

$$sa = ub,\ \ tb = va \ \text{ and } \ s + t = 1.$$

Let us see the consequences of the hypothesis $a \mid b$ in $\mathit{val}(\mathbf{K}, \mathbf{k})$.

We have $ub = sa$ and $sa \mid sb$ (because $a \mid b$), thus $ub \mid sb$; and by **VR2** we get $a \mid b \vdash u \mid s$ **or** $b = 0$.

Since $s + t = 1$, the rule **LOC** gives $\vdash t \mid 1$ **or** $s \mid 1$.

If $s \mid 1$, we have $\vdash u \mid 1$ **or** $b = 0$. If $b \neq 0$, the branch $b = 0$ dies because b is invertible in $\mathbf{K}$. We get in $\mathit{val}(\mathbf{K}, \mathbf{k})$: $\vdash u \mid 1$.

Thus in the dynamical algebraic structure $\mathit{val}(\mathbf{K}, \mathbf{k})$ we have the valid following rules.

- $a \mid b \ \vdash t \mid 1$ **or** $u \mid 1$ if $b \neq 0$
- $u \mid 1 \ \vdash a \mid b$
- $t \mid 1 \ \vdash a \mid b$

So in the lattice $\mathsf{val}(\mathbf{K}, \mathbf{k})$ we have

$$\mathrm{Di}(a, b) = \mathrm{Di}(t, 1) \vee \mathrm{Di}(u, 1) \quad \text{if } b \neq 0.$$

As $\mathrm{Di}(a, 0) = 1$ this proves that $\gamma : \mathsf{Zar}\,\mathbf{k} \to \mathsf{val}(\mathbf{K}, \mathbf{k})$ is onto. □

5.5 *The Theory $\mathcal{Val}$ and the Lattice* **Val**(*K,k*)

The valuative lattice $\mathsf{Val}(\mathbf{K},\mathbf{k})$ is defined in the paper [12, Space of Valuations]. This lattice is very similar to $\mathsf{val}(\mathbf{K},\mathbf{k})$. But its definition is implicitly based on a weakly disjunctive theory $\mathcal{Val}$, which is distinct from $\mathcal{val}$.

The theory $\mathcal{Val}$ is an extension of the theory $\mathcal{Cr}$: one adds a predicate $\mathrm{Vr}(x)$ and axioms that are satisfied in a valuation ring of a discrete field. The predicate $x \mid y$ is not used, and axioms for Vr are minimalist.[20] The axioms we give are easily proved to be equivalent to those of [12, 20].

vf1	$x = 0, \mathrm{Vr}(y) \vdash \mathrm{Vr}(x + y)$	**vf2**	$\vdash \mathrm{Vr}(-1)$
vf3	$\mathrm{Vr}(x), \mathrm{Vr}(y) \vdash \mathrm{Vr}(xy)$	**vf4**	$\mathrm{Vr}(x), \mathrm{Vr}(y) \vdash \mathrm{Vr}(x + y)$
VF2	$xy = 1 \vdash \mathrm{Vr}(x)$ **or** $\mathrm{Vr}(y)$	$\mathbf{CL}_{\mathsf{Val}}$	$0 \mid 1 \;\vdash \bot$ (collapsus)

This theory is noted $\mathcal{Val}$. First an easy lemma.

Lemma 5.5.1 *In the theory* $\mathcal{val}$*, when reading* $\mathrm{Vr}(x)$ *as an abbreviation of* $1 \mid x$*, axioms of* $\mathcal{Val}$ *are valid rules.*

Proof The only point to be proved is the validity **VF2**. We prove the following disjunctive rule, which is a priori stronger.

- $\mathrm{Vr}(xy) \vdash \mathrm{Vr}(x)$ **or** $\mathrm{Vr}(y)$

As a special case of **VR1** we get

- $\vdash \mathrm{Vr}(x)$ **or** $x \mid 1$

If $x \mid 1$, then $x \mid 1 \mid xy$, thus $1.x \mid y.x$ and using **VR3**, $1 \mid y$ **or** $x \mid 0$; in the last case $1 \mid xy \mid 0$, collapsus! □

Notations in the article [12] are slightly different of ours. In [12], $\mathbf{k}$ is always a domain contained in a field $\mathbf{K}$. In this context our $\mathsf{val}(\mathbf{K},\mathbf{k})$ is the same as $\mathsf{Val}(\mathbf{K},\mathbf{k})$ in [12]. The notation $\mathsf{Val}(\mathbf{k})$ is an abbreviation of $\mathsf{Val}(\mathrm{Frac}(\mathbf{k}),\mathbf{k})$. This corresponds to our $\mathsf{val}(\mathrm{Frac}(\mathbf{k}),\mathbf{k})$, isomorphic to $\mathcal{val}(\mathbf{k},\mathbf{k})$. If we note $\mathbf{Z}$ the minimal subring of $\mathbf{k}$, our $\mathsf{val}(\mathbf{k}) \simeq \mathsf{val}(\mathbf{k},\mathbf{Z})$ corresponds to $\mathsf{Val}(\mathrm{Frac}(\mathbf{k}),\mathbf{Z})$,

By comparison with the theory $\mathcal{val}$, the only difficulty with the theory $\mathcal{Val}$ is it applies only for integral domains $\mathbf{k}$.

We prove now that $\mathsf{val}(\mathbf{K},\mathbf{k})$ and $\mathsf{Val}(\mathbf{K},\mathbf{k})$ are isomorphic distributive lattices in the context of the paper [12]. First we recall the definition of the lattice $\mathsf{Val}(\mathbf{K},\mathbf{k})$.

Definition 5.5.2 Let $\mathbf{K}$ be a discrete field and $\mathbf{k}$ a subring of $\mathbf{K}$.

1. The dynamical algebraic structure $\mathcal{Val}(\mathbf{K},\mathbf{k})$ is obtained by adding to the axioms of $\mathcal{Val}$ the positive diagram of $\mathbf{K}$ as a commutative ring and the rules $\vdash \mathrm{Vr}(x)$ for $x \in \mathbf{k}$.

[20] Names **vfi** and **VF2** come from the theory of valued discrete fields (see Sect. 7).

2. The lattice $\mathsf{Val}(\mathbf{K},\mathbf{k})$ defined in [12] is by definition the distributive lattice generated by the entailment relation $\vdash_{\mathbf{K},\mathbf{k},\mathrm{Val}}$ on $\mathbf{K}$ defined by the following equivalence.

$$\begin{array}{c} a_1,\ldots,a_n \vdash_{\mathbf{K},\mathbf{k},\mathrm{Val}} c_1,\ldots,c_m \qquad \overset{\text{def}}{\Longleftrightarrow} \\ \mathrm{Vr}(a_1),\ldots,\mathrm{Vr}(a_n) \vdash_{\mathcal{Val}(\mathbf{K},\mathbf{k})} \mathrm{Vr}(c_1) \textsf{ or } \ldots \textsf{ or } \mathrm{Vr}(c_m) \end{array} \tag{22}$$

We note $\mathrm{V}_{\mathbf{K},\mathbf{k}} : \mathbf{K} \to \mathsf{Val}(\mathbf{K},\mathbf{k})$ (or simply V) the corresponding map.

In the paper [12], the entailment relation is taken on $\mathbf{K}^*$ rather than on $\mathbf{K}$. This does not change the generated distributive lattice since $\mathrm{Vr}(0)$ may be replaced with $\mathrm{Vr}(1)$ (or T).

So the definition of $\mathsf{Val}(\mathbf{K},\mathbf{k})$ is similar to $\mathsf{val}(\mathbf{K},\mathbf{k})$ (definitions 5.2.9 and 5.4.1), and Lemma 5.5.1 gives a natural morphism

$$\theta : \mathsf{Val}(\mathbf{K},\mathbf{k}) \to \mathsf{val}(\mathbf{K},\mathbf{k}). \tag{23}$$

We have to prove that this is an isomorphism when $\mathbf{K}$ is a discrete field. A preliminary little job is necessary.

The Theory $\mathcal{Val}^+$

In order to prove that θ is an isomorphism, we first extend the theory $\mathcal{Val}$ by introducing the predicate $y \mid z$ with the following axioms (they prove that this predicate is equivalent to $\exists x\ (\mathrm{Vr}(x) \wedge z = xy)$).

Div $\mathrm{Vr}(x), z = xy \vdash y \mid z$ **DIV** $y \mid z \vdash \exists x\ (\mathrm{Vr}(x), z = xy)$

One sees easily that $\mathrm{Vr}(x)$ is equivalent to $1 \mid x$, and that $y \mid z$ is equivalent to $\exists x\ (\mathrm{Vr}(x) \wedge z = xy)$.

We note $\mathcal{Val}^+$ this new dynamical theory. It is an essentially identical extension of $\mathcal{Val}$ (see Lemma 3.3.3). In particular, disjunctive valid rules do not change, and the lattice defined from $\mathcal{Val}^+(\mathbf{K},\mathbf{k})$ is the same as the one defined from $\mathcal{Val}(\mathbf{K},\mathbf{k})$.

Note now that if we have a valid rule

$$a_1 \mid b_1,\ldots,a_n \mid b_n \vdash_{\mathcal{val}(\mathbf{K},\mathbf{k})} c_1 \mid d_1 \textsf{ or } \ldots \textsf{ or } c_m \mid d_m \quad (a_i, b_i, c_j, d_j \in \mathbf{K}), \tag{24}$$

a dynamical proof of this rule can use only terms in $\mathbf{K}$. Indeed, $\mathcal{val}$ is a weakly disjunctive theory and it should be necessary to have an existential axiom in order that new elements appear as fresh variables.

Lemma 5.5.3 *Let* $\mathbf{K}$ *be a discrete field and* $\mathbf{k}$ *a subring.*

1. *For dynamical algebraic structures $\mathcal{V}al^{+}(\mathbf{K}, \mathbf{k})$ and $\mathcal{V}al(\mathbf{K}, \mathbf{k})$ axioms of $\mathcal{V}al$ that are used for the definition of the lattice* $\mathsf{val}(\mathbf{K}, \mathbf{k})$*, i.e., for the proof of a rule* (24)*, are valid in $\mathcal{V}al^{+}$.*
2. *Consequently if a rule* (24) *is valid in $\mathcal{V}al(\mathbf{K}, \mathbf{k})$ it is also valid in $\mathcal{V}al^{+}(\mathbf{K}, \mathbf{k})$.*
3. *Consequently the morphism θ given in* (23) *is injective.*

Proof

1. For example let us see the rule **Vr2**. Assume $a \mid b$, one introduces a fresh variable z such that $az = b$ and $\mathrm{Vr}(z)$. So, $acz = bc$, and using **Div**, we get $a \mid b$. Other axioms have also direct proofs, except for disjunctive axioms **VR1** and **VR2**. In this case we shall use the fact that in a proof of a rule (24), elements are always in $\mathbf{K}$.

VR1 $\vdash a \mid b$ **or** $b \mid a$ **VR2** $ax \mid bx \vdash a \mid b$ **or** $0 \mid x$

Let us see **VR1**. In a proof, a and b are elements of $\mathbf{K}$. If $a = 0$, then $b \mid a$ in $\mathcal{V}al^{+}(\mathbf{K}, \mathbf{k})$ because $a = 0.b$ and $\mathrm{Vr}(0)$. In a similar way, if $b = 0$, then $a \mid b$ in $\mathcal{V}al^{+}(\mathbf{K}, \mathbf{k})$. When a and b are nonzero, one has in $\mathbf{K}$ an $x = ba^{-1}$ and a $y = ab^{-1}$, and $xy = 1$. So, in $\mathcal{V}al^{+}(\mathbf{K}, \mathbf{k})$, we have $\vdash \mathrm{Vr}(x)$ **or** $\mathrm{Vr}(y)$. If $\mathrm{Vr}(x)$ is valid, then $1 \mid x$ and $a \mid ax = b$ by **Vr2**. Similarly if $\mathrm{Vr}(y)$, then $b \mid a$.

Let us see **VR2**. Here a, b, x are elements of $\mathbf{K}$. If $x = 0$ then $0 \mid x$. If $x \neq 0$, we have an inverse x^{-1} in $\mathbf{K}$, and so $a = axx^{-1} \mid bxx^{-1} = b$ (we use **Vr2**). □

Remark 5.5.4 The proof of Lemma 5.5.3 uses the fact that $\mathbf{K}$ is a discrete field. It seems impossible to obtain an isomorphism $\mathsf{val}(\mathbf{K}, \mathbf{k}) \simeq \mathsf{Val}(\mathbf{K}, \mathbf{k})$ without assuming $\mathbf{K}$ to be zero-dimensional reduced. Extending further $\mathcal{V}al^{+}$ so as to prove all the axioms of $\mathcal{V}al$ would require adding additional axioms, but the latter would look too ad hoc. ■

Isomorphism of Lattices $\mathsf{val}(\mathbf{K}, \mathbf{k})$ and $\mathsf{Val}(\mathbf{K}, \mathbf{k})$

In order to prove that the morphism θ given in (23) is an isomorphism, it is now sufficient to prove the following lemma.

Lemma 5.5.5 *Let $\mathbf{K}$ be a discrete field and $\mathbf{k}$ a subring. In $\mathcal{V}al^{+}(\mathbf{K}, \mathbf{k})$, any rule*

$$a_1 \mid b_1, \ldots, a_n \mid b_n \vdash c_1 \mid d_1 \textbf{ or } \ldots \textbf{ or } c_m \mid d_m \qquad (a_i, b_i, c_j, d_j \in \mathbf{K})$$

is always equivalent to a rule

$$\mathrm{Vr}(x_1), \ldots, \mathrm{Vr}(x_k) \vdash \mathrm{Vr}(y_1) \textbf{ or } \ldots \textbf{ or } \mathrm{Vr}(y_\ell) \qquad (x_i, y_j \in \mathbf{K}^*).$$

Proof If $b_i = 0$ we cancel $a_i \mid b_i$ in the hypothesis (it is true). If we have a $c_j = 0$ with $d_j \neq 0$, we cancel $c_j \mid d_j$ in the conclusion (this branch collapses). If there is an $a_i = 0$ with $b_i \neq 0$, or if there is a $d_j = 0$, the rule is valid. It remains to be seen what happens when all elements are nonzero. In this case, we replace $a_i \mid b_i$ and $c_j \mid d_j$, respectively, with $\mathrm{Vr}(b_i a_i^{-1})$ and $\mathrm{Vr}(d_j c_j^{-1})$. □

We have proved the desired result.

Theorem 5.5.6 *Let* $\mathbf{k}$ *be a subring of a discrete field* $\mathbf{K}$. *The natural morphism* $\mathsf{Val}(\mathbf{K}, \mathbf{k}) \to \mathsf{val}(\mathbf{K}, \mathbf{k})$ *is an isomorphism.*

The following corollary applies to Theorem 5.5.6 the local-global elementary machinery n°1 that will be explained in Sect. 6.1.

Corollary 5.5.7 *Let* $\mathbf{k}$ *be a pp-ring and* $\mathbf{K} = \mathrm{Frac}(\mathbf{k})$. *The natural morphism* $\mathsf{Val}(\mathbf{K}, \mathbf{k}) \to \mathsf{val}(\mathbf{K}, \mathbf{k})$ *is an isomorphism.*

Note that the previous result does not work for an arbitrary ring $\mathbf{k}$.

6 Valuative Dimensions

In order to transfer results obtained in the case of integral domains to that of an arbitrary ring $\mathbf{A}$, one could hope for defining an idealistic "integral domain generated by $\mathbf{A}$." Unfortunately, such a generated integral domain does not exists in general as a usual object. A dynamical variant of this strategy would consists in defining a dynamical theory $\mathcal{Ai}$ of integral domains, and in considering the dynamical algebraic structure $\mathcal{Ai}(\mathbf{A})$ as a reasonable replacement for the desired ideal object.

In this section, devoted to the topic of valuative dimension, we propose instead to make use of the ring $\mathbf{A}_{\min}$, the "minimal pp-closure of $\mathbf{A}$." The latter was proposed by T. Coquand in an unpublished note about a substitute for the GCD algorithm in $\mathbf{A}[X]$ for the case of a commutative ring $\mathbf{A}$. Using $\mathbf{A}_{\min}$ has indeed proved to be efficient for studying the valuative dimension in [38].

For this purpose, we give in Sect. 6.1 a detailed constructive account of the classical theory of pp-rings, before constructing the ring $\mathbf{A}_{\min}$ in Sect. 6.2 and listing a few of its properties. We think that the zero-dimensional ring $\mathrm{Frac}(\mathbf{A}_{\min})$ is the best possible replacement for the fraction field of an integral domain. In particular, the natural morphism $\mathbf{A} \to \mathbf{A}_{\min}$ is an isomorphism when $\mathbf{A}$ is an integral domain.

In Sect. 6.3, we use the ring $\mathbf{A}_{\min}$ in the comparison of three possible constructive variants for the valuative dimension of a ring $\mathbf{A}$ and of an algebra $\mathbf{k} \to \mathbf{A}$.

6.1 pp-Rings

Definitions and Notations When the ring $\mathbf{A}$ is clear form the context, we use the abbreviated form $a^{\perp} := \mathrm{Ann}_{\mathbf{A}}(a)$. We note also $\mathfrak{a}^{\perp}$ the annihilator of the ideal $\mathfrak{a}$.

An element a such that $a^\perp = 0$ is called a *nonzerodivisor* or is said to be *regular*. A ring morphism $\mathbf{A} \to \mathbf{B}$ is said to be *regular* if it sends any regular element on a regular element.

We note $\mathbf{A}_{\text{red}} = \mathbf{A}/\sqrt[\mathbf{A}]{\langle 0\rangle}$ the reduced ring generated by $\mathbf{A}$.

We note $\mathcal{P}_n$ the set of finite subsets of $\{1, \ldots, n\}$.

Proposition and definition 6.1.1 *Let* $\mathbf{B}$ *be a commutative ring and* $a \in \mathbf{B}$.

1. *If there is an element* c *such that* $ca = a$ *and* $c^\perp = a^\perp$, *then* c *is an idempotent. This element is necessarily unique and we say it is an* attached idempotent to a in $\mathbf{B}$; *we note this* c *as* e_a *or* a°.
2. *An idempotent* $e \in \mathbf{B}$ *is an idempotent attached to* a *if and only if* $ea = a$ *and* $a + (1 - e)$ *is regular.*
3. *If* $\varphi : \mathbf{B} \to \mathbf{C}$ *is regular and if* $a \in \mathbf{B}$ *has an attached idempotent* e_a *in* $\mathbf{B}$, *then* $\varphi(e_a)$ *is an attached idempotent to* $\varphi(a)$ *in* $\mathbf{C}$.
4. *If* a *is an idempotent, it is an attached idempotent to itself.*
5. *The element* a *is regular if and only if* 1 *is an attached idempotent to* a.
6. *If* a° *and* b° *are attached idempotents to* a *and* b, *then* $a^\circ b^\circ$ *is an attached idempotent to* ab.

If each element of a ring $\mathbf{A}$ *has an attached idempotent we say that* $\mathbf{A}$ *is a* pp-ring *(principal ideals are projective). Otherwise said, the annihilator of any element is generated by an idempotent.*

Proof Left to the reader. □

In a pp-ring, for $a \in \mathbf{A}$, let e_a be the unique attached idempotent to a. We have $\mathbf{A} \simeq \mathbf{A}[1/e_a] \times \mathbf{A}/\langle e_a\rangle$. In the ring $\mathbf{A}[1/e_a]$, a is regular, and in $\mathbf{A}/\langle e_a\rangle$, a is null. We then have $e_{ab} = e_a e_b$, $e_a a = a$ and $e_0 = 0$.

Conversely, suppose that a commutative ring is equipped with a unary law $a \mapsto a^\circ$ which satisfies the following three axioms

$$a^\circ a = a, \quad (ab)^\circ = a^\circ b^\circ, \quad 0^\circ = 0. \tag{25}$$

Then, for all $a \in \mathbf{A}$, a° is an attached idempotent to a, thus the ring is a pp-ring.

Lemma 6.1.2 (Splitting Lemma for pp-Rings) *Let* $x_1, \ldots, x_n$ *be* n *elements in a pp-ring* $\mathbf{A}$. *There exists a fundamental system of orthogonal idempotents* (e_j) *of cardinality* 2^n *such that in each of the components* $\mathbf{A}[1/e_j]$, *each* x_i *is null or regular.*

Knowing how to systematically split a pp-ring into two components leads to the following general method. The essential difference with the previous splitting lemma is that we do not know a priori the finite family of elements which will provoke the splitting.

Local-Global Elementary Machinery n°1 ([38, Section IV-6]) *Most algorithms that work with nontrivial integral rings can be modified in order to work with pp-*

rings, by splitting the ring into two components each time that the algorithm written for the integral rings uses the "is this element null or regular ?" test. In the first component the element in question is null, in the second it is regular.

We state now a lemma analogous to Lemma XI-4.21 in [38], where we replace "quasi-inverse of a" with "attached idempotent to a."

Lemma 6.1.3 (The Ring Generated by an Attached Idempotent) *Let $\mathbf{A}$ be a reduced ring.*

1. *Let $a \in \mathbf{A} \subseteq \mathbf{C}$. Assume that $\mathbf{C}$ is reduced and a has an attached idempotent a° in $\mathbf{C}$. We note $\mathbf{B} = \mathbf{A}[a^\circ] \subseteq \mathbf{C}$.*
As an $\mathbf{A}$-module, we get $\mathbf{B} = a^\circ\mathbf{B} \oplus (1 - a^\circ)\mathbf{B}$, and as a ring
$$\mathbf{B} \simeq \mathbf{B}/\langle 1 - a^\circ\rangle \times \mathbf{B}/\langle a^\circ\rangle = \mathbf{A}_1 \times \mathbf{A}_2$$
with $\mathbf{A}_1 \simeq a^\circ\mathbf{B}$ and $\mathbf{A}_2 \simeq (1 - a^\circ)\mathbf{B}$ as $\mathbf{A}$-modules.
   (a) *The natural homomorphism $\mu_1 : \mathbf{A} \to \mathbf{A}_1$ (via $\mathbf{A} \to \mathbf{B} \to \mathbf{A}_1$) is onto. Its kernel is $a^\perp := \mathrm{Ann}_\mathbf{A}(a)$.*
   (b) *The natural homomorphism $\mu_2 : \mathbf{A} \to \mathbf{A}_2$ (via $\mathbf{A} \to \mathbf{B} \to \mathbf{A}_2$) is onto. Its kernel is the intersection $\mathfrak{a} = \mathbf{A} \cap a^\circ\mathbf{B}$ and satisfies the double inclusion*
$$\left(a^\perp\right)^\perp = \mathrm{Ann}_\mathbf{A}\big(\mathrm{Ann}_\mathbf{A}(a)\big) \supseteq \mathfrak{a} \supseteq \mathrm{D}_\mathbf{A}(a). \qquad (*)$$
In short $\mathbf{A}[a^\circ] = \mathbf{B} \simeq \mathbf{A}/a^\perp \times \mathbf{A}/\mathfrak{a}$.
2. *Conversely for $a \in A$, if an ideal $\mathfrak{a}$ of $\mathbf{A}$ satisfies inclusions $(*)$, the element $(1, 0)$ is an attached idempotent to (the image of) a in the ring*
$$\mathbf{B} := \mathbf{A}/a^\perp \times \mathbf{A}/\mathfrak{a} =: \mathbf{A}_1 \times \mathbf{A}_2$$
and the canonical homomorphism from $\mathbf{A}$ to $\mathbf{B}$ is injective.

Proof Let us note $\pi_i : \mathbf{B} \to \mathbf{A}_i$ the canonical morphisms. Since $\pi_i(a^\circ) = 0$ or 1, any element $\pi_i(x)$ is equal to a $\mu_i(y)$ for an $y \in \mathbf{A}$, so each μ_i is onto.

1a. The kernel of μ_1 is $\mathbf{A} \cap \mathrm{Ker}\pi_1 = \mathbf{A} \cap \mathrm{Ann}_\mathbf{B}(a) = \mathrm{Ann}_\mathbf{A}(a)$.

1b. The kernel of μ_2 is $\mathfrak{a} := \mathbf{A} \cap \mathrm{Ker}\pi_2 = \mathbf{A} \cap a^\circ\mathbf{B}$. Clearly $a \in \mathfrak{a}$, thus $\mathrm{D}_\mathbf{A}(a) \subseteq \mathfrak{a}$. Finally we prove that $\mathfrak{a}\,\mathrm{Ann}_\mathbf{A}(a) = 0$, which implies $\mathfrak{a} \subseteq \mathrm{Ann}_\mathbf{A}\big(\mathrm{Ann}_\mathbf{A}(a)\big)$. Indeed if $x \in \mathfrak{a}$ and $y \in \mathrm{Ann}_\mathbf{A}(a)$, we have $x = a^\circ z$ for a $z \in \mathbf{B}$ and $y \in \mathrm{Ann}_\mathbf{B}(a) = \mathrm{Ann}_\mathbf{B}(a^\circ)$, thus $xy = za^\circ y = 0$.

2. The image of a in $\mathbf{B}$ is $(a1_{\mathbf{A}_1}, 0_{\mathbf{A}_2})$, thus $(1_{\mathbf{A}_1}, 0_{\mathbf{A}_2})$ is an attached idempotent to a in $\mathbf{B}$. Consider now an $x \in \mathbf{A}$ whose image in $\mathbf{B}$ is 0. On one hand $x =_{\mathbf{A}_1} 0$, so $ax =_\mathbf{A} 0$. On the other hand $x\,\mathrm{Ann}_\mathbf{A}(a) = 0$, so $x^2 =_\mathbf{A} 0$, and $x =_\mathbf{A} 0$. □

One sees that the notation $\mathbf{A}[a^\circ]$ is ambiguous when $\mathrm{D}_\mathbf{A}(a) \neq \mathrm{Ann}_\mathbf{A}\big(\mathrm{Ann}_\mathbf{A}(a)\big)$. This leads in particular to two natural notions of pp-closure of a reduced ring, depending on whether one decides to systematically favor the side $D_{\mathbf{A}}(a)$ or the side $\mathrm{Ann}_{\mathbf{A}}\big(\mathrm{Ann}_{\mathbf{A}}(a)$.

Here is now a lemma which can simplify the construction of pp-rings. It is analogous to Lemma XI-4.22 in [38], by replacing "zero-dimensional reduced" with "pp-ring" and "quasi-inverse" with "attached idempotent."

Lemma 6.1.4

1. *When $a \in \mathbf{A} \subseteq \mathbf{C}$ where $\mathbf{C}$ is a pp-ring, we note a° the attached idempotent to a in $\mathbf{C}$. The smallest sub-pp-ring of $\mathbf{C}$ containing $\mathbf{A}$ is equal to $\mathbf{A}[(a^\circ)_{a\in\mathbf{A}}]$.*
2. *More generally, if $\mathbf{A} \subseteq \mathbf{B}$ where $\mathbf{B}$ is reduced and if each element a of $\mathbf{A}$ has an attached idempotent a° in $\mathbf{B}$, then the subring $\mathbf{A}[(a^\circ)_{a\in\mathbf{A}}]$ of $\mathbf{B}$ is a pp-ring. Moreover, each element of $\mathbf{A}[(a^\circ)_{a\in\mathbf{A}}]$ can be written* in a standard form, *i.e., as $\sum_j a_j e_j$ with a_j's in $\mathbf{A}$ and a list $(e_j)_j$ of orthogonal idempotents in the Boolean algebra generated by a°'s.*
   Finally, $\sum_j a_j e_j$ has an attached idempotent in $\mathbf{B}$ written in standard form: $\sum_j 1 f_j = \sum_j a_j^\circ e_j$.

Proof We prove Item 2, which implies clearly Item *1*.

Among elements of $\mathbf{B}$, let us consider the ones written as sums of products ab° with $a, b \in \mathbf{A}$. Clearly they form a subring of $\mathbf{B}$[21] which is equal to $\mathbf{A}[(a^\circ)_{a\in\mathbf{A}}]$. This ring contains the Boolean algebra generated by idempotents a°. So if an element is written in standard form, it is in $\mathbf{A}[(a^\circ)_{a\in\mathbf{A}}]$.

Let $x = \sum_j a_j e_j$ and $y = \sum_k b_k f_k$ be two elements written in standard form.

Then $xy = \sum_{j,k}(a_j b_k)e_j f_k$, which is standard. Let us see the sum $x + y$. One may assume $\sum_j e_j = 1$ by adding the idempotent $e = (1 - \sum_j e_j)$ to the list. Same thing for $\sum_k f_k$. Then $x + y = \sum_{j,k}(a_j + b_k)e_j f_k$, which is standard.

Finally, in a finite product of rings $\prod_\ell \mathbf{A}_\ell$, an element $x = (x_\ell)_\ell$ has an attached idempotent if and only if each x_ℓ has an attached idempotent $x_\ell{}^\circ$ in $\mathbf{A}_\ell$, and in this case $x^\circ = (x_\ell{}^\circ)_\ell$. This provides the last statement and proves that $\mathbf{A}[(a^\circ)_{a\in\mathbf{A}}]$ is a pp-ring. □

6.2 *The Ring* $\mathbf{A}_{\mathrm{min}}$

Référence: [38, section XIII-7].

Lemma 6.2.1 *Let $\mathbf{A}$ be a reduced ring and $a \in \mathbf{A}$. We define*

$$\mathbf{A}_{\{a\}} \stackrel{\text{def}}{=} \mathbf{A}\big/a^\perp \times \mathbf{A}\big/(a^\perp)^\perp$$

and we note $\psi_a : \mathbf{A} \to \mathbf{A}_{\{a\}}$ the canonical homomorphism.

1. *$\psi_a(a) = (\pi_1(a), 0)$ and $(1, 0)$ is an attached idempotent to $\psi_a(a)$ in $\mathbf{A}_{\{a\}}$.*

---

[21] Use Item *4* in Lemma 6.1.1.

2. *ψ_a is injective (we may consider* $\mathbf{A}$ *as a subring of* $\mathbf{A}_{\{a\}}$*).*
3. $\mathbf{A}_{\{a\}}$ *is a reduced ring.*
4. *If a has an attached idempotent, ψ_a is an isomorphism. For example, this is the case when* $\mathbf{A}$ *is a pp-ring.*
5. *Let $\mathfrak{b}$ be an ideal in* $\mathbf{A}_{\{a\}}$*, then the ideal* $\psi_a^{-1}(\mathfrak{b}^\perp) = \mathfrak{b}^\perp \cap \mathbf{A}$ *is an annihilator ideal in* $\mathbf{A}$*.*
6. *The morphism* $\psi_a : \mathbf{A} \to \mathbf{A}_{\{a\}}$ *is regular.*

For Item 6, see Exercice XIII-19 in [38].

Lemma 6.2.2 *Let* $\mathbf{A}$ *be a reduced ring and* $a, b \in \mathbf{A}$*. Then, with the notations of Lemma 6.2.1, the rings* $(\mathbf{A}_{\{a\}})_{\{b\}}$ *and* $(\mathbf{A}_{\{b\}})_{\{a\}}$ *are canonically isomorphic.*

Remark The case where $ab = 0$ is typical: when we meet it, we would like to split the ring into components where things are "clear." The previous construction then gives the three components

$$\mathbf{A}\big/(ab^\perp)^\perp\ ,\ \mathbf{A}\big/(a^\perp b)^\perp\ \text{ and }\ \mathbf{A}\big/(a^\perp b^\perp)^\perp\ .$$

In the first one, a is regular and $b = 0$, in the second one b is regular and $a = 0$, and in the third one $a = b = 0$. ■

Theorem and definition 6.2.3 (Minimal pp-Closure) *Let* $\mathbf{A}$ *be a reduced ring. We can define a ring* $\mathbf{A}_{\min}$ *as a filtering colimit by iterating the basic construction which consists in replacing* $\mathbf{E}$ *(the "current" ring, which contains* $\mathbf{A}$*) by*

$$\mathbf{E}_{\{a\}} \stackrel{\text{def}}{=} \mathbf{E}\big/a^\perp \times \mathbf{E}\big/(a^\perp)^\perp = \mathbf{E}/\mathrm{Ann}_{\mathbf{E}}(a) \times \mathbf{E}/\mathrm{Ann}_{\mathbf{E}}(\mathrm{Ann}_{\mathbf{E}}(a))\ ,$$

when a ranges over $\mathbf{A}$*.*

1. *This ring* $\mathbf{A}_{\min}$ *is a pp-ring, contains* $\mathbf{A}$*, and is integral over* $\mathbf{A}$*.*
2. *For all* $x \in \mathbf{A}_{\min}$*,* $x^\perp \cap \mathbf{A}$ *is an annihilator ideal in* $\mathbf{A}$*.*

This ring $\mathbf{A}_{\min}$ *is called the* minimal pp-closure of $\mathbf{A}$. *If* $\mathbf{A}$ *is a pp-ring, the natural morphism* $\mathbf{A} \to \mathbf{A}_{\min}$ *is an isomorphism.*
In the general case, we take $\mathbf{A}_{\min} \stackrel{\text{def}}{=} (\mathbf{A}_{\text{red}})_{\min}$ *(when we do not know whether* $\mathbf{A}$ *is reduced or not).*

We give now a description of each ring we get at a finite stage of the construction of $\mathbf{A}_{\min}$.

Lemma 6.2.4 *Let* $\mathbf{A}$ *be a reduced ring and* $(\underline{a}) = (a_1, \dots, a_n)$ *a list of n elements in* $\mathbf{A}$*. For* $I \in \mathcal{P}_n$*, we note* $\mathfrak{a}_I$ *the ideal*

$$\mathfrak{a}_I = \big(\textstyle\prod_{i\in I} \langle a_i\rangle^\perp \prod_{j\notin I} a_j\big)^\perp = \big(\langle a_i, i \in I\rangle^\perp \prod_{j\notin I} a_j\big)^\perp.$$

Then $\mathbf{A}_{\min}$ contains the following ring, product of 2^n quotient rings of $\mathbf{A}$ (some may be trivial):

$$\mathbf{A}_{\{\underline{a}\}} = \prod\nolimits_{I \in \mathcal{P}_n} \mathbf{A}/\mathfrak{a}_I \ .$$

6.3 Three Constructive Versions of Valuative Dimensions

The valuative dimension of a commutative ring has many possible definitions.

In classical mathematics the valuative dimension of a domain $\mathbf{k}$, noted as $\mathsf{Vdim}(\mathbf{k})$, is equal to the maximum height of the valuation group $\Gamma(\mathbf{V})$ of a valuation ring $\mathbf{V}$ satisfying $\mathbf{k} \subseteq \mathbf{V} \subseteq \mathrm{Frac}(\mathbf{k})$. The height of $\Gamma(\mathbf{V})$ is also the Krull dimension of $\mathbf{V}$. Another definition of $\mathsf{Vdim}(\mathbf{k})$ is the maximum length of a chain of valuation rings of $\mathrm{Frac}(\mathbf{k})$ containing $\mathbf{k}$.

Since the valuative dimension of a quotient domain of $\mathbf{k}$ is bounded by the valuative dimension of $\mathbf{k}$, one extends the definition to an arbitrary ring in the following way: the valuative dimension of a ring $\mathbf{A}$ is the l.u.b. of valuative dimensions of its quotients by prime ideals [9].

In the book [38, section XIII-8], in the case of a domain $\mathbf{k}$, authors use a classical characterization of the valuative dimension which is based on the Krull dimension of rings: the valuative dimension of a domain is the l.u.b. of Krull dimensions of overrings of $\mathbf{k}$ (the rings containing $\mathbf{k}$ and contained in its fraction field). We note it $\mathsf{vdim}(\mathbf{k})$. This definition of $\mathsf{vdim}(\mathbf{k})$ is easily extended to pp-rings. Finally, for an arbitrary ring, the authors use the definition $\mathsf{vdim}(\mathbf{A}) = \mathsf{vdim}(\mathbf{A}_{\min})$. They prove the following equivalence for their definition, for $n \geqslant 0$: $\boxed{\mathsf{vdim}(\mathbf{A}) \leqslant n \Leftrightarrow \mathsf{Kdim}(\mathbf{A}[X_1, \ldots, X_n]) \leqslant 2n}$. This characterization is known in classical mathematics. So the definition in [38] is an acceptable constructive version.

On the other hand, for a domain $\mathbf{k}$, it is clear in classical mathematics that the valuative dimension of $\mathbf{k}$ is equal to the dimension of the distributive lattice $\mathsf{val}(\mathrm{Frac}(\mathbf{k}), \mathbf{k})$. So it is also the Krull dimension of $\mathsf{Val}(\mathrm{Frac}(\mathbf{k}), \mathbf{k}) \simeq \mathsf{val}(\mathrm{Frac}(\mathbf{k}), \mathbf{k})$ (see Theorem 5.5.6). So this is $\mathsf{Vdim}(\mathbf{k})$ introduced in [12, Space of Valuations].

Since $\mathsf{val}(\mathrm{Frac}(\mathbf{k}), \mathbf{k}) \simeq \mathsf{val}(\mathbf{k}, \mathbf{k})$ (Lemma 5.4.3), it is reasonable to propose as a constructive definition for the valuative dimension of an arbitrary ring $\mathbf{A}$, the Krull dimension of the lattice $\mathsf{val}(\mathbf{A}, \mathbf{A})$. We note this dimension $\mathsf{Vdim}(\mathbf{A})$.[22]

Finally, in classical mathematics, [30] give a new constructive characterization of the valuative dimension of an arbitrary commutative ring.

So we have three constructive definitions: the one given in [38, section XIII-8], the Krull dimension of $\mathsf{val}(\mathbf{A}, \mathbf{A})$, and the characterization given by Kemper

[22] For the moment we are not sure that it coincides to the classical definition in classical mathematics.

and Yengui. We now need to prove constructively that these three constructive definitions are equivalent.

This proof will be given in an article by Yengui, Neuwirth, and Lombardi. The paper proves constructively that the first and the third definition are equivalent, and that for the case of domains they are equivalent to the second one (given in [12]).

In fact, we now need to prove the following lemma.

Lemma 6.3.1 *For an arbitrary commutative ring, the distributive lattices* $\mathsf{val}(\mathbf{A}, \mathbf{A})$ *and* $\mathsf{val}(\mathbf{A}_{\min}, \mathbf{A}_{\min})$ *have the same Krull dimension.*

Proof The ring $\mathbf{A}_{\min}$ is a filtered colimit of rings we construct, starting with $\mathbf{A}$, as described in Theorem 6.2.3. If $\mathbf{E}$ is obtained at a preceding stage and if $x \in \mathbf{E}$, the new ring is $\mathbf{E}/x^{\perp} \times \mathbf{E}/(x^{\perp})^{\perp}$. This is $\mathbf{E} \mapsto \mathbf{E}/\mathfrak{a}_1 \times \mathbf{E}/\mathfrak{a}_2$ with $\mathfrak{a}_1\mathfrak{a}_2 = 0$. If we show that this kind of construction does not change the dimension of the valuative lattice, we are done. In fact, since dimensions of $\mathbf{A}$ and $\mathbf{A}_{\text{red}}$ are the equal, we may assume our ring is reduced. □

Lemma 6.3.2 *Let* $\mathbf{A}$ *be a reduced ring* $\mathfrak{a}_1$, $\mathfrak{a}_2$ *two ideals s.t.* $\mathfrak{a}_1\mathfrak{a}_2 = 0$, $\mathbf{A}_1 = \mathbf{A}/\mathfrak{a}_1$ *and* $\mathbf{A}_2 = \mathbf{A}/\mathfrak{a}_2$. *Let us note* $\mathsf{av2}(\mathbf{A}) := \mathsf{val}(\mathbf{A}, \mathbf{A})$. *Then we have*

$$\mathsf{Kdim}(\mathsf{av2}(\mathbf{A})) = \sup(\mathsf{Kdim}(\mathsf{av2}(\mathbf{A}_1)), \mathsf{Kdim}(\mathsf{av2}(\mathbf{A}_2))) = \mathsf{Kdim}(\mathsf{av2}(\mathbf{A}_1 \times \mathbf{A}_2)).$$

Proof *First equality.* From $\mathbf{A}_i = \mathbf{A}/\mathfrak{a}_i$ we deduce that $\mathsf{av2}(\mathbf{A}_i) = (\mathsf{av2}\ \mathbf{A})/(\mathfrak{b}_i = 1)$ for a suitable filter $\mathfrak{b}_i$ of $\mathsf{av2}(\mathbf{A})$. Precisely, from the equivalence (21) of Definition 5.4.1, we get for example for $\mathbf{A}_1$ the following equivalences:[23]

$$(a_1, b_1), \ldots, (a_n, b_n) \vdash_{\mathsf{val}(\mathbf{A}_1,\mathbf{A}_1)} (c_1, d_1), \ldots, (c_m, d_m) \quad \stackrel{\text{def}}{\Longleftrightarrow}$$

$$a_1 \mid b_1, \ldots, a_n \mid b_n \vdash_{\mathit{val}(\mathbf{A}_1,\mathbf{A}_1)} c_1 \mid d_1 \ \textbf{or} \ \ldots \ \textbf{or} \ c_m \mid d_m \quad \stackrel{\text{def}}{\Longleftrightarrow}$$

$$0 \mid x_1, \ldots, 0 \mid x_r, a_1 \mid b_1, \ldots, a_n \mid b_n \vdash_{\mathit{val}(\mathbf{A},\mathbf{A})} c_1 \mid d_1 \ \textbf{or} \ \ldots \ \textbf{or} \ c_m \mid d_m \quad \stackrel{\text{def}}{\Longleftrightarrow}$$

$$(a_1, b_1), \ldots, (a_n, b_n) \vdash_{\mathsf{val}(\mathbf{A},\mathbf{A})/(\mathfrak{b}_1=1)} (c_1, d_1), \ldots, (c_m, d_m)$$

with some x_j's in $\mathfrak{a}_1$ and where $\mathfrak{b}_1$ is the filter of $\mathsf{val}(\mathbf{A}, \mathbf{A})$ generated by elements $(0, x)$ with $x \in \mathfrak{a}_1$.

Then let us prove that $\mathfrak{b}_1 \cap \mathfrak{b}_2 = \{1\}$. Indeed, the filter $\mathfrak{b}_1 \cap \mathfrak{b}_2$ is generated by elements $(0, x) \vee (0, y)$ with $x \in \mathfrak{a}_1$ and $y \in \mathfrak{a}_2$, and since $0 \mid xy$ we get by **WZD**, the valid rule $\vdash_{\mathit{val}(\mathbf{A},\mathbf{A})} 0 \mid x$ **or** $0 \mid y$.

We conclude with Theorem 2.4.7.

Second equality. First $\mathsf{av2}(\mathbf{A}_1 \times \mathbf{A}_2) \simeq \mathsf{av2}(\mathbf{A}_1) \times \mathsf{av2}(\mathbf{A}_2)$. Next, the dimension of the product of two lattices is the sup of the two dimensions. □

[23] Note the dynamical algebraic structure $\mathit{val}(\mathbf{A}_1, \mathbf{A}_1)$ is obtained by adding relations $0 \mid x$ when $x \in \mathfrak{a}_1$ to the dynamical algebraic structure $\mathit{val}(\mathbf{A}, \mathbf{A})$).

Note Since $\mathbf{A}_{\min}$ is a pp-ring, $\mathsf{val}(\mathbf{A}_{\min}, \mathbf{A}_{\min}) \simeq \mathsf{val}(\mathrm{Frac}(\mathbf{A}_{\min}), \mathbf{A}_{\min})$. So, $\mathsf{Vdim}(\mathbf{A})$ is equal to $\mathsf{Vdim}(\mathbf{A}_{\min})$ with the meaning in the article [12].

Definition 6.3.3 The dimension of the distributive lattice $\mathsf{val}\mathbf{A} \simeq \mathsf{val}(\mathbf{A}, \mathbf{Z})$ (where $\mathbf{Z}$ is the minimal subring of $\mathbf{A}$) is called the *absolute valuative dimension* of the ring $\mathbf{A}$ and is noted as $\mathsf{vdima}(\mathbf{A})$.

The dimension $\mathsf{vdima}(\mathbf{A}) := \mathsf{Kdim}(\mathsf{val}\mathbf{A})$ is a priori greater than or equal to $\mathsf{vdim}(\mathbf{A}) := \mathsf{Kdim}(\mathsf{val}(\mathbf{A}, \mathbf{A}))$ and it is often strictly greater, as for the ring $\mathbb{Q}$: $\mathsf{vdima}(\mathbb{Q}) = \mathsf{Kdim}(\mathsf{val}(\mathbb{Q}, \mathbb{Z})) = \mathsf{Kdim}(\mathsf{val}(\mathbb{Z}, \mathbb{Z})) = 1$, and $\mathsf{vdim}(\mathbb{Q}) = \mathsf{Kdim}(\mathsf{val}(\mathbb{Q}, \mathbb{Q})) = 0$. ■

Remark 6.3.4 Same proofs work for the dimension of distributive lattices $\mathsf{Zar}(\mathbf{A})$, $\mathsf{val}(\mathbf{A})$, $\mathsf{Real}(\mathbf{A})$ and $\mathsf{Heit}(\mathbf{A})$.[24] So we have the following analogous results.

$$
\begin{aligned}
\mathsf{Kdim}(\mathbf{A}) &= \mathsf{Kdim}(\mathbf{A}_{\min}) = \mathsf{Kdim}(\mathbf{A}_1 \times \mathbf{A}_2) \\
\mathsf{vdima}(\mathbf{A}) &= \mathsf{vdima}(\mathbf{A}_{\min}) = \mathsf{vdima}(\mathbf{A}_1 \times \mathbf{A}_2) \\
\mathsf{Rdim}(\mathbf{A}) &= \mathsf{Rdim}(\mathbf{A}_{\min}) = \mathsf{Rdim}(\mathbf{A}_1 \times \mathbf{A}_2) \\
\mathsf{Jdim}(\mathbf{A}) &= \mathsf{Jdim}(\mathbf{A}_{\min}) = \mathsf{Jdim}(\mathbf{A}_1 \times \mathbf{A}_2)
\end{aligned}
$$ ■

7 Comparisons with a Theory of Valued Discrete Fields

7.1 *Introduction*

In this section we prove that the weakly disjunctive theory $\mathcal{val}$ and the dynamical theory $\mathcal{Vdf}$ of valued discrete fields, which is introduced in [21, Section 4], are more or less identical. This is obtained through the formal Valuativstellensatz 7.3.1 for valued discrete fields. This implies various formal Valuativstellensätze for $\mathcal{val}$.

We compare also with the formal Valuativstellensatz for the theory $\mathcal{Val}$ given in [12].

Note that Theorem 5.5.6 is already an essential comparison result for the distributive lattices val and Val.

7.2 *The Theory $\mathcal{Vdf}$ of Valued Discrete Fields*

Section 4 in [21] gives a formal Positivstellensatz for the theory $\mathcal{Vdf}$ of valued discrete fields and a Positivstellensatz à la Hilbert for algebraically closed valued discrete fields. We prefer here the name of *Valuativstellensatz*. We are mainly

[24] See [18].

interested in formal Valuativstellensätze, they give algebraic certificates for the provability in diverse dynamical theories for some dynamical rules.

First we recall axioms given in [21] for the theory $\mathcal{Vdf}$ of valued discrete fields. The language is the one commutative rings to which we add predicates $\cdot \neq 0$, $\mathrm{Vr}(\cdot)$, $\mathrm{Rn}(\cdot)$, $\mathrm{Un}(\cdot)$. The first one, $\cdot \neq 0$, is seen as the invertibility predicate, the predicate $\mathrm{Vr}(\cdot)$ interprets belonging to the valuation ring, the predicate $\mathrm{Rn}(\cdot)$ interprets elements residually null and $\mathrm{Un}(\cdot)$ the units (invertible elements in the valuation ring). So we have the following signature

$$\Sigma_{\mathcal{Vdf}} = (\cdot = 0, \cdot \neq 0, \mathrm{Vr}(\cdot), \mathrm{Rn}(\cdot), \mathrm{Un}(\cdot); \cdot + \cdot, \cdot \times \cdot, - \cdot, 0, 1).$$

Axioms of the theory $\mathcal{Vdf}$ are the following dynamical rules. As consequence, the field is discrete and the divisibility relation is explicit.

- First we have direct rules.

cr1 $\vdash 0 = 0$	**cr2** $x = 0 \vdash xy = 0$
cr3 $x = 0, y = 0 \vdash x + y = 0$	
vf1 $x = 0, \mathrm{Vr}(y) \vdash \mathrm{Vr}(x + y)$	**vf2** $\vdash \mathrm{Vr}(-1)$
vf3 $\mathrm{Vr}(x), \mathrm{Vr}(y) \vdash \mathrm{Vr}(xy)$	**vf4** $\mathrm{Vr}(x), \mathrm{Vr}(y) \vdash \mathrm{Vr}(x + y)$
vf5 $x = 0, \mathrm{Rn}(y) \vdash \mathrm{Rn}(x + y)$	**vf6** $\vdash \mathrm{Rn}(0)$
vf7 $\mathrm{Rn}(x), \mathrm{Vr}(y) \vdash \mathrm{Rn}(xy)$	**vf8** $\mathrm{Rn}(x), \mathrm{Rn}(y) \vdash \mathrm{Rn}(x + y)$
vf9 $\mathrm{Rn}(x) \vdash \mathrm{Vr}(x)$	
vf10 $x = 0, \mathrm{Un}(y) \vdash \mathrm{Un}(x + y)$	**vf11** $\vdash \mathrm{Un}(1)$
vf12 $\mathrm{Un}(x), \mathrm{Un}(y) \vdash \mathrm{Un}(xy)$	**vf13** $\mathrm{Rn}(x), \mathrm{Un}(y) \vdash \mathrm{Un}(x + y)$
vf14 $\mathrm{Un}(x) \vdash \mathrm{Vr}(x)$	
vf15 $x = 0, y \neq 0 \vdash x + y \neq 0$	**vf16** $x \neq 0, y \neq 0 \vdash xy \neq 0$
vf17 $\mathrm{Un}(x) \vdash x \neq 0$	

- The collapsus is the same as for domains.

CL $0 \neq 0 \vdash \bot$

- Finally, we have simplification rules **Vf1** and **Vf2** followed by 4 dynamical rules.

Vf1 $xy = 1 \vdash x \neq 0$	**Vf2** $\mathrm{Vr}(xy), \mathrm{Un}(x) \vdash \mathrm{Vr}(y)$
VF1 $x \neq 0 \vdash \exists y\ xy = 1$	**VF2** $\vdash x = 0$ **or** $x \neq 0$
VF3 $xy = 1 \vdash \mathrm{Vr}(x)$ **or** $\mathrm{Vr}(y)$	**VF4** $\mathrm{Vr}(x) \vdash \mathrm{Un}(x)$ **or** $\mathrm{Rn}(x)$

If we want to force the valuation to be nontrivial, we introduce a constant c with the axiom

vf0 $\mathrm{Vr}(c) \vdash \bot$

There is only one existential axiom, **VF1**, and it corresponds to a unique existence: the theory is very near to a weakly disjunctive theory.

For the dynamical theory $\mathcal{V}\!\mathit{dfsc}$ of *separably closed valued discrete fields* we add to the theory $\mathcal{V}\!\mathit{df}$ axioms of separable algebraic closure.

$$\textbf{VF6}_n \quad \mathrm{disc}_Y(f) \neq 0 \ \vdash \exists y\ f(y) = 0, \qquad \text{where } f(Y) = Y^n + \textstyle\sum_{k=0}^{n-1} a_k Y^k.$$

For the dynamical theory $\mathcal{V}\!\mathit{dfac}$ of *algebraically closed valued discrete fields* we add to the theory $\mathcal{V}\!\mathit{df}$ axioms of algebraic closure.

$$\textbf{VF5}_n \quad \vdash \exists y\ y^n + a_{n-1}y^{n-1} + \cdots + a_1 y + a_0 = 0.$$

Lemma 7.2.1 *Let* **A** *be a commutative ring and* $\mathcal{V}\!\mathit{df}(\mathbf{A})$ *the dynamical algebraic structure of type* $\mathcal{V}\!\mathit{df}$ *constructed on (the positive diagram of)* **A**. *To give a minimal model of* $\mathcal{V}\!\mathit{df}(\mathbf{A})$ *is the same thing as giving a couple* $(\mathfrak{p}, \mathbf{V})$ *where* $\mathfrak{p}$ *is a detachable prime ideal of* **A** *and* **V** *is a valuation ring of* $\mathbf{K} = \mathrm{Frac}(\mathbf{A}/\mathfrak{p})$. *Moreover divisibility has to be decidable.*

Proof The prime ideal $\mathfrak{p}$ is the set of $x \in \mathbf{A}$ such that $x = 0$ is valid in the model. The quotient ring $\mathbf{B} = \mathbf{A}/\mathfrak{p}$ is integral, we note **K** its fraction field, present in the model thanks to Axiom **VF1**. The valuation ring **V** is then the set of fractions z in **K** such that $\mathrm{Vr}(z)$ is valid in the model.

The divisibility relation is decidable because **V** is a detachable subset of **K**: if $x \neq 0$, we have an inverse y by **VF1**. Then we have $\mathrm{Vr}(x)$ or $\mathrm{Vr}(y)$. In the second case, we have $\mathrm{Un}(y)$ or $\mathrm{Rn}(y)$, and $\mathrm{Vr}(x)$ if and only if $\mathrm{Un}(y)$. □

It seems that **VF1** is necessary at the end of the proof. It implies that the model contains only fractions of elements in **V**.

So, dynamical algebraic structures $\mathcal{V}\!\mathit{df}(\mathbf{A})$ and $\mathit{val1}(\mathbf{A})$ have the same minimal models in classical mathematics. But not necessarily in constructive mathematics since in $\mathit{val1}(\mathbf{A})$ the divisibility relation is not forced to be decidable.

We now define some theories closely related to the theory $\mathcal{V}\!\mathit{df}$, which share a number of theorems with the latter. We make use of the terminology of direct rules, simplification rules and dynamical rules, as explained in page 295.

Definitions 7.2.2 (Geometric Theories Closely Related to $\mathcal{V}\!\mathit{df}$)

1. Let us note $\mathcal{A}\mathit{pv}$[25] the direct theory we get from the theory $\mathcal{V}\!\mathit{df}$ when keeping only direct axioms, from **cr1** to **vf17**, and the collapsus.
2. The Horn theory $\mathcal{A}\mathit{qv}$[26] is intermediate between $\mathcal{A}\mathit{pv}$ and $\mathcal{V}\!\mathit{df}$: we replace in $\mathcal{V}\!\mathit{df}$ Axioms **VF1** to **VF4** with the following simplification axioms, which are valid rules in $\mathcal{V}\!\mathit{df}$:

Vf3 $\mathrm{Un}(xy), \mathrm{Vr}(x), \mathrm{Vr}(y) \ \vdash \mathrm{Un}(y)$	**Vf4** $\mathrm{Rn}(xy), \mathrm{Un}(x) \ \vdash \mathrm{Rn}(y)$
Vf5 $\mathrm{Rn}(x^2) \ \vdash \mathrm{Rn}(x)$	**Vf6** $xy \neq 0 \ \vdash x \neq 0$

25 Proto-valued rings in [21].

26 Quasi-valued rings in [21].

Vf7 $xy = 0, x \neq 0 \vdash y = 0$ **Vf8** $x^2 = 0 \vdash x = 0$
Vf9$_n$ $x^n = \sum_{k=0}^{n-1} a_k x^k, \mathrm{Vr}(a_{n-1}), \ldots, \mathrm{Vr}(a_0) \vdash \mathrm{Vr}(x)$

The last axiom scheme says that **V** is integrally closed.

3. The weakly disjunctive theory $\mathcal{Vdf}^-$ is obtained from $\mathcal{Aqv}$ by adding the three disjunctive axioms **VF2**, **VF3** and **VF4**.
4. The theory $\mathcal{Vdf}^+$ is the extension of $\mathcal{Vdf}$ we get when we add
   - The predicate $x \mid y$ as abbreviation of "$\exists z\, (\mathrm{Vr}(z) \wedge xz = y)$"
   - The predicate $x \nmid y$ as abbreviation of "$y \neq 0 \wedge \exists z\, (\mathrm{Rn}(z) \wedge yz = x)$"

We shall see that the theories $\mathcal{Apv}$ and $\mathcal{Vdf}$ collapse simultaneously (7.3.2) and that the theories $\mathcal{Aqv}$ and $\mathcal{Vdf}$ prove the same Horn rules (7.3.7). Theories $\mathcal{Vdf}^-$ and $\mathcal{Vdf}$ prove the same disjunctive rules, but we do not prove this result here.

The theory $\mathcal{Vdf}^+$ is by construction an essentially identical extension of $\mathcal{Vdf}$ (a fortiori a conservative extension).

Definition 7.2.3 Let $\mathbf{k} \subseteq \mathbf{A}$ be two rings, or more generally let $\varphi : \mathbf{k} \to \mathbf{A}$ be an algebra, we note $\mathcal{Vdf}(\mathbf{A}, \mathbf{k})$ the dynamical algebraic structure whose presentation is given by

- The positive diagram of **A** as commutative ring.
- Axioms $\vdash \mathrm{Vr}(\varphi(x))$ for x's in **k**.

The dynamical algebraic structure $\mathcal{Vdf}(\mathbf{A})$ is identical to $\mathcal{Vdf}(\mathbf{A}, \mathbf{Z})$ where **Z** is the minimal subring of **A**.

We define in a similar way dynamical algebraic structures $\mathcal{Apv}(\mathbf{A}, \mathbf{k})$, $\mathcal{Aqv}(\mathbf{A}, \mathbf{k})$, $\mathcal{Vdf}^-(\mathbf{A}, \mathbf{k})$ and $\mathcal{Vdf}^+(\mathbf{A}, \mathbf{k})$.

7.3 Formal Valuativstellensatz for $\mathcal{Vdf}$ and Consequences

Valuativstellensatz 7.3.1 (Formal Valuativstellensatz for the Theory $\mathcal{Apv}$**)** *Let* $GR = (G; R_{=0}, R_{\neq 0}, R_{\mathrm{Vr}}, R_{\mathrm{Rn}}, R_{\mathrm{Un}})$ *be a presentation over the signature* $\Sigma_{\mathcal{Vdf}}$[27] *for a dynamical algebraic structure. We note*

- $I_{=0}$ *the ideal of* $\mathbb{Z}[G]$ *generated by* $R_{=0}$
- $M_{\neq 0}$ *the monoid generated by* $R_{\neq 0}$
- V_{Vr} *the subring of* $\mathbb{Z}[G]$ *generated by* $R_{\mathrm{Vr}} \cup R_{\mathrm{Rn}} \cup R_{\mathrm{Un}}$
- I_{Rn} *the ideal of* V_{Vr} *generated by* R_{Rn}
- M_{Un} *the monoid generated by* R_{Un}

T.F.A.E.

[27] More precisely, $R_{=0}$ is the subset of $\mathbb{Z}[G]$ whose elements p are assumed to be $= 0$ in the dynamical algebraic structure. Same thing for $R_{\neq 0}$, R_{Vr}, R_{Rn}, R_{Un}.

1. *The dynamical algebraic structure* $(GR, \mathcal{A}pv)$ *collapses.*
2. *There is in* $\mathbb{Z}[G]$ *an equality*

$$s(u+t)+z=0$$

with $s \in M_{\neq 0}$, $u \in M_{\mathrm{Un}}$, $t \in I_{\mathrm{Rn}}$ *and* $z \in I_{=0}$.

The following theorem is a crucial result in [21]. It gives the condition of collapsus for dynamical algebraic structures of type $\mathcal{V}dfac$.

Theorem 7.3.2 (Simultaneous Collapsus) [21, Theorem 4.3] *A dynamical algebraic structure of type* $\mathcal{A}pv$ *collapses if and only if it collapses as a dynamical algebraic structure of type* $\mathcal{V}df$*, or of type* $\mathcal{V}dfac$*, or of any other theory between* $\mathcal{A}pv$ *and* $\mathcal{V}dfac$.

Remark 7.3.3 If (**K**, **V**) is a valued discrete field and if **L** is a field extension of **K** we deduce in classical mathematics that the valuation domain **V** can be extended in a valuation domain **W** of **L** such that $\mathbf{W} \cap \mathbf{K} = \mathbf{V}$ [21, Remark 4.6]. ■

Theorem 4.18 in [21] is a Valuativstellensatz à la Hilbert for algebraically closed valued discrete field. It is obtained from Theorem 7.3.2 by using the fact that the formal theory which corresponds to $\mathcal{V}dfac$ is complete (when the characteristic of **K** and the one of the residual field are fixed).

The following theorem describes provable facts in dynamical algebraic structures of type $\mathcal{V}df$.

Theorem 7.3.4 [21, Proposition 4.14] *Let* $GR = (G; R_{=0}, R_{\neq 0}, R_{\mathrm{Vr}}, R_{\mathrm{Rn}}, R_{\mathrm{Un}})$ *be a presentation over the signature* $\Sigma_{\mathcal{V}df}$. *Let* p *be an element of* $\mathbb{Z}[G]$. *We define* $I_{=0}$, $M_{\neq 0}$, V_{Vr}, I_{Rn} *and* M_{Un} *as in Theorem 7.3.1. We consider proofs in the dynamical algebraic structure* $(GR, \mathcal{V}df)$.

(a) A dynamical proof of $p = 0$ *gives an equality in* $\mathbb{Z}[G]$ *of type*

$$p^n m(u+j)+i=0$$

with $m \in M_{\neq 0}$, $u \in M_{\mathrm{Un}}$, $j \in I_{\mathrm{Rn}}$ *and* $i \in I_{=0}$.

(b) A dynamical proof of $p \neq 0$ *gives an equality in* $\mathbb{Z}[G]$ *of type*

$$m(u+j)+i+bp=0$$

with $m \in M_{\neq 0}$, $u \in M_{\mathrm{Un}}$, $j \in I_{\mathrm{Rn}}$, $i \in I_{=0}$ *and* $b \in \mathbb{Z}[G]$.

(c) A dynamical proof of $\mathrm{Vr}(p)$ *gives an equality in* $\mathbb{Z}[G]$ *of type*

$$m((u+j)p^{n+1}+a_n p^n+\cdots+a_1 p+a_0)+i=0$$

with $m \in M_{\neq 0}$, $u \in M_{\mathrm{Un}}$, $j \in I_{\mathrm{Rn}}$, *les* $a_k \in V_{\mathrm{Vr}}$ *and* $i \in I_{=0}$.

(d) *A dynamical proof of* Rn(p) *gives an equality in* $\mathbb{Z}[G]$ *of type*

$$m((u+j)p^{n+1}+j_n p^n+\cdots+j_1 p+j_0)+i=0$$

with $m \in M_{\neq 0}$, $u \in M_{\mathrm{Un}}$, j *and* $j_{k's} \in I_{\mathrm{Rn}}$ *and* $i \in I_{=0}$.

(e) *A dynamical proof of* Un(p) *gives an equality in* $\mathbb{Z}[G]$ *of type*

$$m((u+j)p^{n+1}+a_n p^n+\cdots+a_1 p+(u'+j'))+i=0$$

with $m \in M_{\neq 0}$, $u, u' \in M_{\mathrm{Un}}$, $j, j' \in I_{\mathrm{Rn}}$, $a_{k's}$ *in* V_{Vr} *and* $i \in I_{=0}$.

Corollary 7.3.5 *Let* $\mathbf{k} \subseteq \mathbf{A}$ *be two rings,* $x \in \mathbf{A}$ *and* $y_1, \ldots, y_n \in \mathbf{k}$.

1. *The rule* $\vdash x = 0$ *is valid in* $\mathcal{V}df(\mathbf{A}, \mathbf{k})$ *if and only if* x *is nilpotent.*
2. *The rule* $\vdash x \neq 0$ *is valid in* $\mathcal{V}df(\mathbf{A}, \mathbf{k})$ *if and only if* x *is invertible in* $\mathbf{A}$.
3. *The rule* $\vdash$ Vr(x) *is valid in* $\mathcal{V}df(\mathbf{A}, \mathbf{k})$ *if and only if* x *is integral over* $\mathbf{k}$.
4. *The rule* Rn(y_1), ..., Rn(y_n) $\vdash$ Rn(x) *is valid in* $\mathcal{V}df(\mathbf{A}, \mathbf{k})$ *if and only if* x *is weakly integral*[28] *over the ideal* $\langle y_1, \ldots, y_n \rangle$ *of* $\mathbf{k}$.

Remark 7.3.6 From Item *3* we deduce easily in classical mathematics that a domain **k** has for integral closure in its fraction field **K** the intersection of valuation rings of **K** containing **k** [21, Remark 4.15].

From Item *4*, we deduce in classical mathematics that if $(\mathbf{k}, \mathfrak{m})$ is a local domain, the ideal $\mathfrak{m}$ is the intersection of **k** with the maximal ideal of a valuation ring of the fraction field **K** [21, Corollary 4.7]. ■

A remarkable consequence of theorem 7.3.4 is theorem 7.3.7.

Theorem 7.3.7 [21, Theorem 4.17] *Theories* $\mathcal{A}qv$ *and* $\mathcal{V}dfac$ *prove the same Horn rules. Same thing for any intermediate theory.*

7.4 Formal Valuativstellensätze for $\mathcal{val}$ and $\mathcal{V}df^+$

In this section we prove a crucial result: a formal Valuativstellensatz for the theory $\mathcal{val}$, a consequence of Theorems 7.3.1 and 7.3.2.

We consider a ring **A**, we are searching for a formal Valuativstellensatz for the dynamical algebraic structure $\mathcal{val}(\mathbf{A})$.

Recall that the theory $\mathcal{V}df^+$ is defined at Item *4* of Definition 7.2.2 and that the theory $\mathcal{val}1$ is the conservative extension of $\mathcal{val}$ we get when we add the predicate $\cdot \neq 0$ opposite to $\cdot = 0$.

[28] This means that x is a zero of a monic polynomial whose nondominant coefficients are in the ideal.

Lemma 7.4.1

1. *The predicate $x \nmid y$ is the opposite predicate to $x \mid y$ in the theory $\mathcal{V}\!df^+$.*
2. *Axioms of val are valid in $\mathcal{V}\!df^+$.*
3. *The following Horn rule is valid in val1.*

VR2.5 $x \neq 0, ax \mid bx \vdash a \mid b$

Proof Left to the reader. □

A Formal Valuativstellensatz for val(A)

Valuativstellensatz 7.4.2 (A Formal Valuativstellensatz for val(A) and $\mathcal{V}\!df^+(\mathbf{A})$[29]) *Let* **A** *be a commutative ring and $a_i, b_i, c_j, d_j \in \mathbf{A}$. T.F.A.E.*

1.

$$(a_1, b_1), \ldots, (a_n, b_n) \vdash_{\mathsf{val}(\mathbf{A})} (c_1, d_1), \ldots, (c_m, d_m) \tag{26}$$

2.

$$a_1 \mid b_1, \ldots, a_n \mid b_n \vdash_{\mathit{val1}(\mathbf{A})} c_1 \mid d_1 \textsf{ or } \ldots \textsf{ or } c_m \mid d_m \tag{27}$$

3.

$$a_1 \mid b_1, \ldots, a_n \mid b_n \vdash_{\mathcal{V}df^+(\mathbf{A})} c_1 \mid d_1 \textsf{ or } \ldots \textsf{ or } c_m \mid d_m \tag{28}$$

4. *Introducing indeterminates X_i ($i \in [\![1..n]\!]$) and Y_j ($j \in [\![1..m]\!]$) we have in the ring $\mathbf{A}[\underline{X}, \underline{Y}]$ an equality in the following form*

$$d\left(1 + \sum\nolimits_{j=1}^{m} Y_j P_j(\underline{X}, \underline{Y})\right) \equiv 0 \bmod \left\langle (X_i a_i - b_i)_{i \in [\![1..n]\!]}, (Y_j d_j - c_j)_{j \in [\![1..m]\!]} \right\rangle \tag{29}$$

where d is in the monoid generated by d_j's, and $P_j(\underline{X}, \underline{Y})$'s are in $\mathbb{Z}[\underline{X}, \underline{Y}]$

5. *Let us note $y_j = \frac{c_j}{d_j}$ vied in $\mathbf{B} = \mathbf{A}[\frac{1}{d_1 \cdots d_m}]$. Let us introduce indeterminates X_i ($i \in [\![1..n]\!]$). One has in the ring $\mathbf{B}[\underline{X}]$ an equality in the following form*

$$1 + \sum\nolimits_{j=1}^{m} y_j P_j(\underline{X}, \underline{y}) \equiv 0 \bmod \left\langle (X_i a_i - b_i)_{i \in [\![1..n]\!]} \right\rangle \tag{30}$$

where $P_j(\underline{X}, \underline{Y})$'s are in $\mathbb{Z}[\underline{X}, \underline{Y}]$.

[29] Note that if $\mathbf{A} = \mathbb{Z}$ and if a_i, b_i, c_j and d_j are indeterminates we get a general formal Valuativstellensatz for the theory $\mathcal{V}\!df^+$ or for the theory val.

Proof Item *5* is a simple rewriting of Item *4*.
1 ⇔ *2*. By definition, Item *1* means that the following rule is valid:

$$a_1 \mid b_1, \ldots, a_n \mid b_n \vdash_{\mathit{val}(\mathbf{A})} c_1 \mid d_1 \ \textsf{or} \ \ldots \ \textsf{or} \ c_m \mid d_m \tag{31}$$

But *val1* is a conservative extension of *val*.
1 ⇒ *3*. By Item 2 of Lemma 7.4.1, the rule (28) is valid.
3 ⇒ *4*. The rule (28) is equivalent to the following collapsus

$$a_1 \mid b_1, \ldots, a_n \mid b_n, c_1 \nmid d_1, \ldots, c_m \nmid d_m \vdash_{\mathcal{V}\mathit{df}^+(\mathbf{A})} \bot. \tag{32}$$

Inside the theory $\mathcal{V}\mathit{df}^+$, $a_i \mid b_i$ is equivalent to the existence of an x_i such that $\mathrm{Vr}(x_i)$ and $x_i a_i = b_i$. This x_i can be represented by a fresh variable X_i. Similarly $c_j \nmid d_j$ is equivalent to the existence of an y_j such that $\mathrm{Rn}(y_j)$, $y_j d_j = c_j$ and $d_j \neq 0$. This y_j can be represented by a fresh variable Y_j. The validity of collapsus (32) is then equivalent to the collapsus of the family of conditions

$$(\mathrm{Vr}(X_i),\ X_i a_i - b_i = 0)_{i \in [\![1..n]\!]},\ (\mathrm{Rn}(Y_j),\ Y_j d_j - c_j = 0,\ d_j \neq 0)_{j \in [\![1..m]\!]}$$

in the dynamical algebraic structure constructed on the positive diagram of $\mathbf{A}$ by adding generators X_i and Y_j, i.e., in the ring $\mathbf{A}[\underline{X}, \underline{Y}]$. We conclude with Valuativstellensatz 7.3.1 (with $s = u = 1$ and $z = 0$).

4 ⇒ *2*. Let us assume an identity (29). We have to prove the validity of the rule (27). We make a case by case reasoning using $\vdash x = 0$ **or** $x \neq 0$ and $\vdash a \mid b$ **or** $b \mid a$.

In a branch where one d_j is supposed null the rule (27) is valid. So we can assume the d_j's are nonzero, hence regular (rule **VR2.5**).

In a branch where $a_i = 0$, the hypothesis of (27) implies $b_i = 0$ and we can replace a_i, b_i and X_i with 0 in (29).

Finally, we need to prove (27) in the case where a_i's and d_j's are nonzero, hence regular. Let us consider now $\mathbf{B}$, the total fraction ring of $\mathbf{A}$. We get in $\mathbf{B}$ an equality

$$0 =_{\mathbf{B}} 1 + \sum\nolimits_j \frac{c_j}{d_j} P_j \left(\frac{b_1}{a_1}, \ldots, \frac{b_n}{a_n}, \frac{c_1}{d_1}, \ldots, \frac{c_m}{d_m} \right), \tag{33}$$

where polynomials P_j's have coefficients in $\mathbb{Z}$. Let δ be a bound of degrees in Y_k's in P_j's and ϵ_i a bound of degrees in X_i's. We let $d := \prod_k d_k$, we multiply the preceding equality by $u = d^\delta \prod_i a_i^{\epsilon_i}$ and we get

$$u =_{\mathbf{B}} \sum\nolimits_j \frac{c_j}{d_j} Q_j(b_1, a_1, \ldots, b_n, a_n, c_1, d_1, \ldots, c_m, d_m), \tag{34}$$

where each Q_j is homogeneous of degree δ in each (c_k, d_k) and of degree ϵ_i in (a_i, b_i). We let $e_j = \prod_{k:k \neq j} d_k$ and we multiply the preceding equality by d. We get an equality in $\mathbf{A}$

$$ud = \sum_j c_j e_j q_j.$$

We now proceed by case analysis, and for the desired conclusion, we still need to deal with the case where we suppose $d_j \mid c_j$ for each j. By case analysis again, we assume that one $c_j e_j$, e.g., $c_1 e_1$, divides all the other ones. Then we get

- $u \mid q_j$ for each q_j (u divides each coefficient of q_j's)
- $uc_1 e_1 \mid c_j e_j q_j$ for each q_j
- so $c_1 e_1 u \mid \sum_j c_j e_j q_j = ud = d_1 e_1 u$

Finally, the rule **VR2.5** allows us to simplify by $e_1 u$: we get $c_1 \mid d_1$.
Phew! □

Remark 7.4.3 Perhaps a more direct proof of the implication *1* ⇒ *4* is possible, without using the formal Valuativstellensatz 7.3.1 for valued discrete fields. On one hand one should prove that property (29) defines an entailment relation over $\mathbf{A} \times \mathbf{A}$ (the cut rule seems difficult), on the other hand that this entailment relation satisfies axioms of *val*. For example the rule **Vr2**, $a \mid b, a \mid c \vdash a \mid b + c$ corresponds to the following equality

$$(b + c)(1 - y(x_1 + x_2)) \equiv 0 \bmod \langle x_1 a - b,\ x_2 a - c,\ y(b + c) - a\rangle .$$

■

Admissibility of the Rule DIV for the Weakly Disjunctive Theory *val*

Remark 7.4.4 A corollary of Valuativstellensatz 7.4.2 is Lemma 5.2.8 concerning the admissibility of the existential rule **DIV** in the theory *val*. Indeed, the formal Valuativstellensatz is established for the theory $\mathcal{V}df^+$ (see Item *4* in Definition 7.2.2), where we introduce $\cdot \mid \cdot$ with its definition. So, in this theory the rule **DIV** is valid. But the formal Valuativstellensatz works for the theory *val*. Thus, disjunctive rules that are valid for the predicate $x \mid y$ remain the same for the theory *val* when we add the axiom **DIV**. ■

A Formal Valuativstellensatz for *val*(A, k)

A slight variant of Valuativstellensatz 7.4.2.

Valuativstellensatz 7.4.5 (A Formal Valuativstellensatz for *val*(A, k) and $\mathcal{V}df^+$(A, k)) *Let* $\mathbf{k} \subseteq \mathbf{A}$ *be two commutative rings and* $a_i, b_i, c_j, d_j \in \mathbf{A}$. *T.F.A.E.*

1. One has

$$(a_1, b_1), \ldots, (a_n, b_n) \vdash_{\mathsf{val}(\mathbf{A},\mathbf{k})} (c_1, d_1), \ldots, (c_m, d_m). \tag{35}$$

2. *One has*

$$a_1 \mid b_1, \ldots, a_n \mid b_n \vdash_{val(\mathbf{A},\mathbf{k})} c_1 \mid d_1 \textbf{ or } \ldots \textbf{ or } c_m \mid d_m. \tag{36}$$

3. *One has*

$$a_1 \mid b_1, \ldots, a_n \mid b_n \vdash_{\mathcal{V}df^+(\mathbf{A},\mathbf{k})} c_1 \mid d_1 \textbf{ or } \ldots \textbf{ or } c_m \mid d_m. \tag{37}$$

4. *Let us note* $y_j = \frac{c_j}{d_j}$ *viewed in* $\mathbf{B} = \mathbf{A}[\frac{1}{d_1 \cdots d_m}]$. *Consider indeterminates* X_i $(i \in [\![1..n]\!])$. *One has in the ring* $\mathbf{B}[\underline{X}]$ *an equality in the following form*

$$1 + \sum_{j=1}^{m} y_j P_j(\underline{X}, \underline{y}) \equiv 0 \bmod \left\langle (X_i a_i - b_i)_{i \in [\![1..n]\!]}, \right\rangle \tag{38}$$

where $P_j(\underline{X}, \underline{Y})$*'s are in* $\mathbf{k}[\underline{X}, \underline{Y}]$.

Variant. *We are interested in understanding the equivalence between Items 2, 3 and 4 when some* a_i *and* c_j *are null, i.e., when we assume that some* b_i *or* d_j *are null. Let us note* e_i *and* f_j *the elements of* $\mathbf{A}$ *assumed to be null on the left and on the right of* $\vdash$. *T.F.A.E.*

5. *One has*

$$\begin{aligned} a_1 \mid b_1, \ldots, a_n \mid b_n, e_1 = 0, \ldots, e_k = 0 \vdash_{val(\mathbf{A},\mathbf{k})} \\ c_1 \mid d_1 \textbf{ or } \ldots \textbf{ or } c_m \mid d_m \textbf{ or } f_1 = 0 \textbf{ or } \ldots \textbf{ or } f_\ell = 0. \end{aligned}$$

6. *Let us note* $y_j = \frac{c_j}{d_j}$ *viewed in* $\mathbf{B} = \mathbf{A}[\frac{1}{d_1 \cdots d_m \cdot f_1 \cdots f_\ell}]$. *Consider indeterminates* X_i $(i \in [\![1..n]\!])$. *One has in the ring* $\mathbf{B}[\underline{X}]$ *an equality in the following form*

$$1 + \sum_{j=1}^{m} y_j P_j(\underline{X}, \underline{y}) \equiv 0 \bmod \left\langle (X_i a_i - b_i)_{i \in [\![1..n]\!]}, (e_i)_{i \in [\![1..k]\!]}, \right\rangle \tag{39}$$

where $P_j(\underline{X}, \underline{Y})$*'s are in* $\mathbf{k}[\underline{X}, \underline{Y}]$.

Proof Equivalence of Items *1*, *3* and *4* corresponds to the equivalence of Items *1*, *3* and *5* in Valuativstellensatz 7.4.2. Note that $\mathbb{Z}[\underline{X}, \underline{Y}]$ (in Item *5*) is now replaced with $\mathbf{k}[\underline{X}, \underline{Y}]$ (in Item *4*) since by hypothesis elements of $\mathbf{k}$ are now integral.
We now need to add a few words explaining the variant (a similar variant work also for Valuativstellensatz 7.4.2): Eq. (39) is simply the analogous of Eq. (38) for the ring $\mathbf{B}/\langle e_1, \ldots, e_\ell \rangle$. □

As corollary of Valuativstellensatz 7.4.5 we get the result which was announced after Lemma 5.2.11.

Theorem 7.4.6 *Let* $\mathbf{k} \subseteq \mathbf{A}$ *be two rings,* $x \in \mathbf{A}$ *and* $y \in \mathbf{k}$.

1. *A dynamical algebraic structure* $val(\mathbf{A}, \mathbf{k})$ *proves* $y \mid x$ *if and only if* x *is integral over the ideal* $\langle y \rangle$ *of* $\mathbf{k}$.

2. *Same thing for* $\mathcal{Vdf}^+(\mathbf{A},\mathbf{k})$.

Proof

1. This is a consequence of the equivalence of Items *2* and *4* in Valuativstellensatz 7.4.5 when $n = 0$ and $m = 1$.
2. Use Item *1* and the equivalence of Items *2* and *3* in Valuativstellensatz 7.4.5.

□

The Center Map (2)

We are now fulfilling the promise made in Remark 5.4.5. In particular, this finishes the proof of Theorem 5.4.6.

Theorem 7.4.7 (The Morphism γ : $\mathsf{Zar}\,\mathbf{A} \to \mathsf{val}(\mathbf{A},\mathbf{A})$ Is Injective) *Let* $\mathbf{A}$ *be a commutative ring and* $a_i, c_j \in \mathbf{A}$. *T.F.A.E.*

1. One has

$$(a_1, 1), \ldots, (a_n, 1) \vdash_{\mathsf{val}(\mathbf{A},\mathbf{A})} (c_1, 1), \ldots, (c_m, 1).$$

2. One has in the ring $\mathbf{A}$ *an equality*

$$\left(\prod_{i=1}^{n} a_i\right)^r + \sum_{j=1}^{m} c_j p_j = 0.$$

3. One has in the Zariski lattice of $\mathbf{A}$

$$\mathrm{D}(a_1), \ldots, \mathrm{D}(a_n) \vdash_{\mathsf{Zar}(\mathbf{A})} \mathrm{D}(c_1), \ldots, \mathrm{D}(c_m).$$

In particular the center map (definition 5.4.4) γ : $\mathsf{Zar}\,\mathbf{A} \to \mathsf{val}(\mathbf{A},\mathbf{A})$ *is injective.*

Proof Equivalence of Items *1* and *2* is a special case of the one given in Valuativstellensatz 7.4.5 (Items *1* and *4* equality (38)): computing modulo $Xa - 1$ in $\mathbf{A}[X]$ amounts to calculate in $\mathbf{A}[\frac{1}{a}]$). Equivalence of Items *2* and *3* is the formal Nullstellensatz 4.1.1. □

Another Formal Valuativstellensatz for $\mathit{val}(\mathbf{A},\mathbf{k})$

Another slight variant of Valuativstellensatz 7.4.2, which is particularly useful after Remark 7.4.9.

Valuativstellensatz 7.4.8 (Another Formal Valuativstellensatz for $\mathit{val}(\mathbf{A},\mathbf{k})$ and $\mathcal{Vdf}^+(\mathbf{A},\mathbf{k})$) *Let* $\mathbf{k} \subseteq \mathbf{A}$ *be two commutative rings and* $b_i, d_j \in \mathbf{A}$. *T.F.A.E.*

1. *One has*

$$(1, b_1), \ldots, (1, b_n) \vdash_{\mathsf{val}(\mathbf{A},\mathbf{k})} (1, d_1), \ldots, (1, d_m). \tag{40}$$

2. *One has*

$$\mathrm{Vr}(b_1), \ldots, \mathrm{Vr}(b_n) \vdash_{\mathcal{Vdf}(\mathbf{A},\mathbf{k})} \mathrm{Vr}(d_1) \textbf{ or } \ldots \textbf{ or } \mathrm{Vr}(d_m). \tag{41}$$

3. *One has*

$$\mathrm{vr}(b_1), \ldots, \mathrm{vr}(b_n) \vdash_{\mathcal{val2}(\mathbf{A},\mathbf{k})} \mathrm{vr}(d_1) \textbf{ or } \ldots \textbf{ or } \mathrm{vr}(d_m). \tag{42}$$

4. *Let us note* $\mathbf{B} = \mathbf{A}[\frac{1}{d_1 \cdots d_m}]$, *we have in the ring* $\mathbf{B}$ *an equality*

$$1 = \sum\nolimits_{j=1}^{m} d_j^{-1} P_j(b_1, \ldots, b_n, d_1^{-1}, \ldots, d_m^{-1}), \tag{43}$$

where $P_j(X_1, \ldots, X_n, Y_1, \ldots, Y_m)$*'s are in* $\mathbf{k}[\underline{X}, \underline{Y}]$.

5. *(Case when* $\mathbf{A} = \mathbf{K}$ *is a discrete field) One has*

$$\mathrm{Vr}(b_1), \ldots, \mathrm{Vr}(b_n) \vdash_{\mathcal{Val}(\mathbf{K},\mathbf{k})} \mathrm{Vr}(d_1) \textbf{ or } \ldots \textbf{ or } \mathrm{Vr}(d_m). \tag{44}$$

6. *(Case when* $\mathbf{A} = \mathbf{K}$ *is a discrete field) One has*

$$\mathrm{V}(b_1), \ldots, \mathrm{V}(b_n) \vdash_{\mathsf{Val}(\mathbf{K},\mathbf{k})} \mathrm{V}(d_1), \ldots, \mathrm{V}(d_m). \tag{45}$$

Proof First items are a special case of Valuativstellensatz 7.4.2. The equivalence with two last items (they are equivalent by definition) is a consequence of the lattice isomorphism between $\mathsf{val}(\mathbf{K}, \mathbf{k})$ and $\mathsf{Val}(\mathbf{K}, \mathbf{k})$ (Theorem 5.5.6). So we have $\mathbf{B} = \mathbf{K}$. □

Remark 7.4.9 As it was expected, we recover here the formal Valuativstellensatz for the predicate Vr given in the article [12] for the case where $\mathbf{A}$ is a discrete field and b_i's and d_j's are nonzero. See also [20] and [33]. ■

References

1. Balbes, R. and Dwinger, P. (1974). *Distributive lattices*. University of Missouri Press, Columbia, MO.
2. Barrett, T. W. and Halvorson, H. (2017). Quine's conjecture on many-sorted logic. *Synthese*, 194(9):3563–3582.
3. Bezem, M. and Coquand, T. (2005). Automating coherent logic. In *Logic for programming, artificial intelligence, and reasoning. 12th international conference, LPAR 2005, Montego Bay, Jamaica, December 2–6, 2005. Proceedings*, pages 246–260. Berlin: Springer.

4. Bezem, M. and Coquand, T. (2019). Skolem's theorem in coherent logic. *Fundam. Inform.*
5. Bishop, E. (1967). *Foundations of constructive analysis*. McGraw-Hill, New York.
6. Bishop, E. and Bridges, D. (1985). *Constructive analysis*, volume 279 of *Grundlehren der Mathematischen Wissenschaften [Fundamental Principles of Mathematical Sciences]*. Springer-Verlag, Berlin.
7. Borel, A. and Serre, J.-P. (1958). Le théorème de Riemann-Roch. *Bull. Soc. Math. Fr.*, 86:97–136.
8. Bridges, D. and Richman, F. (1987). *Varieties of constructive mathematics*, volume 97 of *London Mathematical Society Lecture Note Series*. Cambridge University Press, Cambridge.
9. Cahen, P.-J. (1990). Construction B, I, D et anneaux localement ou résiduellement de Jaffard. (B, I, D construction and locally or residually Jaffard rings). *Arch. Math.*, 54(2):125–141.
10. Cederquist, J. and Coquand, T. (2000). Entailment relations and distributive lattices. In *Logic Colloquium '98 (Prague)*, volume 13 of *Lect. Notes Log.*, pages 127–139. Assoc. Symbol. Logic, Urbana, IL.
11. Coquand, T. (2005). A completeness proof for geometrical logic. In *Logic, methodology and philosophy of science. Proceedings of the 12th international congress, Oviedo, Spain, August 2003*, pages 79–89. London: King's College Publications.
12. Coquand, T. (2009). Space of valuations. *Ann. Pure Appl. Logic*, 157(2–3):97–109.
13. Coquand, T. and Lombardi, H. (2003). Hidden constructions in abstract algebra: Krull dimension of distributive lattices and commutative rings. In *Commutative ring theory and applications (Fez, 2001)*, volume 231 of *Lecture Notes in Pure and Appl. Math.*, pages 477–499. Dekker, New York.
14. Coquand, T. and Lombardi, H. (2006). A logical approach to abstract algebra. *Math. Structures Comput. Sci.*, 16(5):885–900.
15. Coquand, T. and Lombardi, H. (2016). Anneaux à diviseurs et anneaux de Krull (une approche constructive). *Comm. Algebra*, 44:515–567.
16. Coquand, T. and Lombardi, H. (2018). Mise à jour en 2018 du texte de 2001: Constructions cachées en algèbre abstraite. Dimension de Krull, Going up, Going down. Technical report, Département de Mathématiques de l'Université de Franche-Comté.
17. Coquand, T., Lombardi, H., and Quitté, C. (2010). Curves and coherent Prüfer rings. *J. Symbolic Comput.*, 45(12):1378–1390.
18. Coquand, T., Lombardi, H., and Quitté, C. (2022). Dimension de Heitmann des treillis distributifs et des anneaux commutatifs. In *Publications Mathématiques de l'Université de Franche-Comté Besançon. Algèbre et théorie des nombres. Années 2003–2006*. Besançon: Laboratoire de Mathématiques de Besançon, 2006, p. 57–100, version corrigée.
19. Coquand, T., Lombardi, H., and Roy, M.-F. (2005). An elementary characterization of Krull dimension. In *From sets and types to topology and analysis*, volume 48 of *Oxford Logic Guides*, pages 239–244. Oxford Univ. Press, Oxford.
20. Coquand, T. and Persson, H. (2001). Valuations and Dedekind's Prague theorem. *J. Pure Appl. Algebra*, 155(2–3):121–129.
21. Coste, M., Lombardi, H., and Roy, M.-F. (2001). Dynamical method in algebra: effective Nullstellensätze. *Ann. Pure Appl. Logic*, 111(3):203–256.
22. Della Dora, J., Dicrescenzo, C., and Duval, D. (1985). About a new method for computing in algebraic number fields. In *EUROCAL '85. Lecture Notes in Computer Science no. 204, (Ed. Caviness B.F.)*, pages 289–290. Springer, Berlin.
23. Dickmann, M., Schwartz, N., and Tressl, M. (2019). *Spectral spaces*, volume 35. Cambridge: Cambridge University Press.
24. Edwards, H. M. (1990). *Divisor theory*. Birkhäuser Boston, Inc., Boston, MA.
25. Fulton, W. (1989). *Algebraic curves*. Advanced Book Classics. Addison-Wesley Publishing Company, Advanced Book Program, Redwood City, CA. An introduction to algebraic geometry, Notes written with the collaboration of Richard Weiss, Reprint of 1969 original.
26. Hochster, M. (1969). Prime ideal structure in commutative rings. *Trans. Am. Math. Soc.*, 142:43–60.

27. Huber, R. and Knebusch, M. (1994). *On valuation spectra. Recent advances in real algebraic geometry and quadratic forms*, volume 155 of *Contemp. Math.*, pages 167–206. Amer. Math. Soc., Providence, RI. MathSciNet review: 95f:13002.
28. Johnstone, P. T. (1986). *Stone spaces*, volume 3 of *Cambridge studies in advanced mathematics*. Cambridge university press, Cambridge. Reprint of the 1982 edition.
29. Joyal, A. (1976). Les théoremes de Chevalley-Tarski et remarques sur l'algèbre constructive. *Cah. Topologie Géom. Différ. Catégoriques*, 16:256–258.
30. Kemper, G. and Yengui, I. (2020). Valuative dimension and monomial orders. *J. Algebra*, 557:278–288.
31. Lifschitz, V. (1980). Semantical completeness theorems in logic and algebra. *Proc. Amer. Math. Soc.*, 79(1):89–96.
32. Lombardi, H. (1998). Relecture constructive de la théorie d'Artin-Schreier. *Ann. Pure Appl. Logic*, 91(1):59–92.
33. Lombardi, H. (2000). Une généralisation du Positivstellensatz pour les corps valués algébriquement clos. *C. R. Acad. Sci. Paris Sér. I Math.*, 331(5):345–348.
34. Lombardi, H. (2002). Dimension de Krull, Nullstellensätze et évaluation dynamique. *Math. Z.*, 242(1):23–46.
35. Lombardi, H. (2006). Structures algébriques dynamiques, espaces topologiques sans points et programme de Hilbert. *Ann. Pure Appl. Logic*, 137(1–3):256–290.
36. Lombardi, H. (2020). Spectral spaces versus distributive lattices: a dictionary. In *Advances in rings, modules and factorizations. Selected papers based on the presentations at the international conference on rings and factorizations, Graz, Austria, February 19–23, 2018*, pages 223–245. Cham: Springer.
37. Lombardi, H. (2022). Théories géométriques pour l'algèbre constructive. http://hlombardi.free.fr/Theories-geometriques.pdf.
38. Lombardi, H. and Quitté, C. (2015). *Commutative algebra: constructive methods. Finite projective modules*. Algebra and applications, 20. Springer, Dordrecht. Translated from the French (Calvage & Mounet, Paris, 2011, revised and extended by the authors) by Tania K. Roblot.
39. Lombardi, H. and Quitté, C. (2021). *Algèbre commutative. Méthodes constructives. Modules projectifs de type fini. Cours et exercices*. Paris: Calvage & Mounet, 2 edition. Second edition, revised and extended, of the 2011 book.
40. Lorenzen, P. (1951). Algebraische und logistische Untersuchungen über freie Verbände. *J. Symbolic Logic*, 16:81–106. Translation by Stefan Neuwirth: *Algebraic and logistic investigations on free lattices*, http://arxiv.org/abs/1710.08138.
41. Lurie, J. (2009). *Higher topos theory*, volume 170 of *Ann. Math. Stud.* Princeton, NJ: Princeton University Press.
42. Matijasevič, J. V. (1975). A metamathematical approach to proving theorems in discrete mathematics. *Zap. Naučn. Sem. Leningrad. Otdel. Mat. Inst. Steklov. (LOMI)*, 49:31–50, 177. Theoretical applications of the methods of mathematical logic, I.
43. Mines, R., Richman, F., and Ruitenburg, W. (1988). *A course in constructive algebra*. Universitext. Springer-Verlag, New York.
44. Prawitz, D. (1971). Ideas and results in proof theory. In *Proceedings of the Second Scandinavian Logic Symposium (Univ. Oslo, Oslo, 1970)*, pages 235–307. Studies in Logic and the Foundations of Mathematics, Vol. 63. North-Holland, Amsterdam.
45. Prestel, A. (1982). Decidable theories of preordered fields. *Math. Ann.*, 258:481–492.
46. Prestel, A. and Schmid, J. (1990). Existentially closed domains with radical relations. An axiomatization of the ring of algebraic integers. *J. Reine Angew. Math.*, 407:178–201.
47. Seidenberg, A. (1953). A note on the dimension theory of rings. *Pac. J. Math.*, 3:505–512.
48. Stone, M. H. (1937). Topological representations of distributive lattices and Brouwerian logics. *Cas. Mat. Fys.*, 67:1–25.
49. Tsementzis, D. (2017). A syntactic characterization of Morita equivalence. *J. Symb. Log.*, 82(4):1181–1198.

50. van den Dries, L. (1988). Elimination theory for the ring of algebraic integers. *J. Reine Angew. Math.*, 388:189–205.
51. van den Dries, L. and Macintyre, A. (1990). The logic of Rumely's local-global principle. *J. Reine Angew. Math.*, 407:33–56.
52. Wehrung, F. (2019). Real spectrum versus l-spectrum via Brumfiel spectrum. manuscript.
53. Yengui, I. (2015). *Constructive commutative algebra: projective modules over polynomial rings and dynamical Gröbner bases*. Lecture Notes in Mathematics, 2138. Springer, Cham.

Building Three-Variable Homogeneous Integer-Valued Polynomials Using Generalized Projective Planes

Marie MacDonald

1 Introduction

This paper will present a method of generating examples of homogeneous integer-valued polynomials (IVPs) in three variables. When constructing these polynomials, we are interested in finding examples of the form $f = \frac{g}{p^k}$ with $g \in \mathbb{Z}[x, y, z]$, p a prime, and k as large as possible. This method will only construct IVPs where the numerator is a product of linear polynomials. We will be working locally at a prime, usually $p = 2$ although our methods carry over to all primes. The two-variable case of this problem is covered in [7]; however, the methods used there do not extend to three or more variables.

One reason for interest in integer-valued polynomials and homogeneous ones in particular is their occurrence in stable homotopy theory. One of the first instances of this is from 1971 when Adams, Harris, and Switzer [1] explained some of the K-theory of BU through IVPs. Building on these results, Clarke [5] showed that the complex K-theory homology of the infinite complex projective space, $K_0(\mathbb{C}P^\infty)$ is isomorphic to $\text{Int}(\mathbb{Z}, \mathbb{Z})$, and this can be extended to $K_0(BT^n) \simeq \text{Int}(\mathbb{Z}^n, \mathbb{Z})$. The connection to homogeneous polynomials was made by Baker, Clarke, Ray, and Schwartz [4] who identified the primitive elements of $K_0(BU(n))$ as the symmetric homogeneous IVPs in n-variables. A survey of this application of integer-valued polynomials to topology can be found in [6].

M. MacDonald (✉)
Department of Mathematics, Cornell University, Ithaca, NY, USA
e-mail: mb2636@cornell.edu

J.-L. Chabert et al. (eds.), *Algebraic, Number Theoretic, and Topological Aspects of Ring Theory*, https://doi.org/10.1007/978-3-031-28847-0_18

2 Projective Planes over Finite Fields

Definitions 1 For $\mathbb{K}$ a field the projective n-space $\mathbb{K}\mathbf{P}^n$ is defined to be the quotient of $\mathbb{K}^{n+1} \setminus 0$ by the equivalence relation $(w_0, \ldots, w_n) \equiv (\lambda w_0, \ldots, \lambda w_n)$ for every $\lambda \in \mathbb{K} \setminus 0$ (thus a point in $\mathbb{K}P^n$ is an equivalence class of $n+1$ tuples in $\mathbb{K}^{n+1} \setminus 0$, which we will denote $[w_0, \ldots, w_n]$.

Definitions 2 A hyperplane in $\mathbb{K}\mathbf{P}^n$ is the zero set of a nontrivial homogeneous linear polynomial in $n+1$ variables, i.e., if the polynomial is $a_0x_0 + \cdots + a_nx_n$, then

$$L_{[a_0,\ldots,a_n]} = \{[w_0, \ldots, w_n] \in \mathbb{K}P^n : a_0w_0 + \cdots + a_nw_n = 0\}$$

Implicit in this notation in this definition is the observation that $n+1$ tuples $(a_0 \ldots, a_n)$ and $(\lambda a_0, \ldots, \lambda a_n)$ for $\lambda \in \mathbb{K} \setminus 0$ determine the same line in $\mathbb{K}\mathbf{P}^n$ and that conversely that two lines coincide only if their determining polynomials are related in this way.

Specializing to the case $n = 2$ we have

Lemma 3 Any two distinct points in $\mathbb{K}\mathbf{P}^2$ lie on a unique line and any two distinct lines in $\mathbb{K}P^2$ meet in a unique point.

This result shows that $\mathbb{K}\mathbf{P}^2$ is a model for the projective plane developed axiomaticly as described in [10] for example. It also shows the duality between points and lines that the axiomatic approach emphasizes.

Specializing further to the case $n = 2$, $\mathbb{K} = \mathbb{F}_2$, the finite field with two elements, we have the following from [2]:

Proposition 4 The finite projective plane $\mathbb{F}_2\mathbf{P}^2$ has the following properties:

(a) $\mathbb{F}_2\mathbf{P}^2$ has seven points.
(b) Each line in $\mathbb{F}_2\mathbf{P}^2$ contains three points.
(c) Each point of $\mathbb{F}_2\mathbf{P}^2$ lies on three lines.
(d) The three lines through a given point in $\mathbb{F}_2\mathbf{P}^2$ cover $\mathbb{F}_2\mathbf{P}^2$

The finite projective space $\mathbb{F}_2\mathbf{P}^2$ is sometimes called the Fano plane, and has the graphic representation (Fig. 1) [2]:

Parts (c) and (d) of this proposition will be used below to illustrate the connection between finite projective planes and homogeneous integer-valued polynomials.

Proposition 5

(a) If p_1, p_2, and p_3 are homogeneous linear polynomials in $\mathbb{Z}[x, y, z]$ whose reductions modulo 2 determine the three lines through a given point in $\mathbb{F}_2\mathbf{P}^2$, then $g = p_1p_2p_3/2$ is a homogeneous integer-valued polynomial.
(b) If $q_1, \ldots, q_7$ are seven homogeneous linear polynomials in $\mathbb{Z}[x, y, z]$ whose reductions modulo 2 determine the seven lines in $\mathbb{F}_2\mathbf{P}^2$, then $h = q_1 \ldots q_7/2^3$ is a homogeneous integer-valued polynomial.

Fig. 1 $\mathbb{F}_2\mathbf{P}^2$, the Fano plane

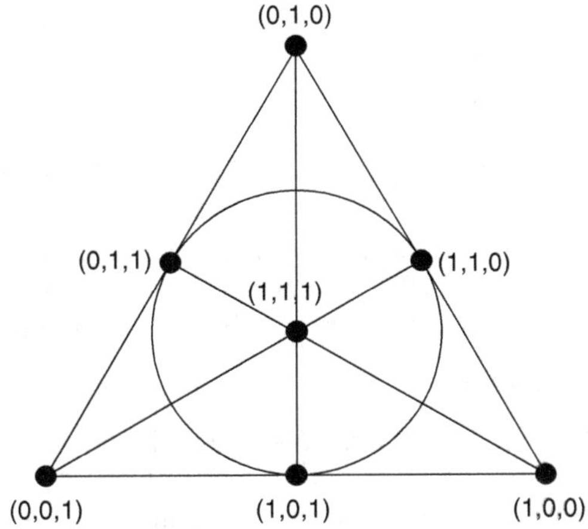

Proof If $(a, b, c) \in \mathbb{Z}^3$ has all entries divisible by 2, then any homogeneous polynomial in $\mathbb{Z}[x, y, z]$ will take an even value at this point and so g and h are integer-valued there. If not all of the entries are divisible by 2, then the reduction of (a, b, c) modulo 2 determines a point in $\mathbb{F}_2\mathbf{P}^2$ and so lies on one of the lines in part (a), by part (d) of the previous proposition. The polynomial p_i which determines that line therefore takes an even value at (a, b, c), and so g takes an integral value there. By part (c) of the previous proposition the reduction of (a, b, c) must lie on three of the lines in $\mathbb{F}_2\mathbf{P}^2$ hence three of the polynomials q_i must take even values at (a, b, c), and so h must be integer-valued there. □

To extend these results to obtain higher-degree homogeneous integer-valued polynomials with larger denominators, we need the extension of the finite projective plane in which $\mathbb{F}_2$ is replaced by the ring $\mathbb{Z}/2^k\mathbb{Z}$.

3 Finite Projective Hjelmslev Planes

The extension of the idea of finite projective spaces to spaces over rings rather than fields was originally investigated by the Danish mathematician Johannes Hjelmslev. Early accounts are in [9] and, in English, [8].

Definitions 6 For p a prime the projective Hjelmslev space, or projective H-space, $\mathbb{Z}/(p^k)\mathbf{P}^n$ over the ring $\mathbb{Z}/(p^k)$ is the quotient of the set of $n+1$ tuples from $(\mathbb{Z}/(p^k))^{n+1}$, such that not all entries in the $n+1$ tuple are divisible by p by the equivalence relation $(w_0, \ldots, w_n) \sim (\lambda w_0, \ldots, \lambda w_n)$ for all units λ in $\mathbb{Z}/(p^k)$ (thus a point in $\mathbb{Z}/(p^k)\mathbf{P}^n$ is an equivalence class of $n+1$ tuples in $(\mathbb{Z}/(p^k))^{n+1} \setminus p(\mathbb{Z}/(p^k))^{n+1}$, which we will denote $[w_0, \ldots, w_n]$.

Definitions 7 A line in $\mathbb{Z}/(p^k)\mathbf{P}^n$ is the zero set of a homogeneous linear polynomial in $n+1$ variables with at least one coefficient not divisible by p, i.e., if the polynomial is $a_0x_0 + \cdots + a_nx_n$ then

$$L_{[a_0,\ldots,a_n]} = \{[w_0, \ldots, w_n] \in \mathbb{Z}/(p^k)\mathbf{P}^n : a_0w_0 + \cdots + a_nw_n = 0\}$$

As for the finite projective spaces, the $n+1$ tuples $(a_0 \ldots, a_n)$ and $(\lambda a_0, \ldots, \lambda a_n)$ for λ a unit in $\mathbb{Z}/(p^k)$ determine the same line in $\mathbb{Z}/(p^k)\mathbf{P}^n$ and conversely two lines coincide in $\mathbb{Z}/(p^k)\mathbf{P}^n$ only if their determining polynomials are related in this way.

Proposition 8

(a) The projective H-space of dimension 1, $\mathbb{Z}/(p^k)\mathbf{P}^1$, contains $(p^{2k} - p^{2(k-1)})/(p^k - p^{k-1}) = p^{k-1}(p+1)$ points.
(b) The H-plane, $\mathbb{Z}/(p^k)\mathbf{P}^2$, contains $(p^{3k} - p^{3(k-1)})/(p^k - p^{k-1}) = p^{2(k-1)}(p^2 + p + 1)$ points.
(c) Each line in $\mathbb{Z}/(p^k)\mathbf{P}^2$ contains $p^{k-1}(p+1)$ points.
(d) Each point in $\mathbb{Z}/(p^k)\mathbf{P}^2$ lies on $p^{k-1}(p+1)$ lines.

Proof Since one of the coefficients in a pair (a_0, a_1) representing an element of $\mathbb{Z}/(p^k)\mathbf{P}^1$ is not divisible by p, if $(\lambda a_0, \lambda a_1) = (\lambda' a_0, \lambda' a_1)$, then $\lambda = \lambda'$, and so the equivalence class $[a_0, a_1]$ contains a distinct element for each unit of $\mathbb{Z}/(p^k)$, and so $p^k - p^{k-1}$ in all. Since $\mathbb{Z}/(p^k)^2 \setminus p(\mathbb{Z}/(p^k))^2)$ contains $p^{2k} - p^{2(k-1)}$ elements the size of $\mathbb{Z}/(p^k)\mathbf{P}^1$ is as stated. A similar argument applies to part (b).

Among the homogeneous linear polynomials, determining a given line in $\mathbb{Z}/(p^k)\mathbf{P}^2$ must be one for which one of the coefficients is 1. Without loss of generality, assume it is a_0 so that the line is

$$L_{[1,a_1,a_2]} = \{[w_0, w_1, w_2] : w_0 + a_0w_1 + a_2w_2 = 0\}$$

It follows that the map $\phi : [w_1, w_2] = [-a_1w_1 + a_2w_2, w_1, w_2]$ maps $\mathbb{Z}/(p^k)\mathbf{P}^1$ bijectively to $L_{[1,a_1,a_2]}$ and so the size of $L_{[1,a_1,a_2]}$ is as given in part (c). Part (d) follows by a similar arguement. □

Definitions 9 Let $r_\ell : \mathbb{Z}/(p^k) \to \mathbb{Z}/(p^\ell)$ be the reduction modulo p^ℓ map and for $k > \ell$ define

$$\pi_{k,\ell} : \mathbb{Z}/(p^k)\mathbf{P}^2 \to \mathbb{Z}/(p^\ell)\mathbf{P}^2$$

by $\pi_{k,\ell}([w_0, w_1, w_2]) = [r_\ell(w_0), r_\ell(w_1), r_\ell(w_2)]$.

Lemma 10 $\pi_{k,\ell}$ is surjective, $p^{k-\ell}$ to 1 and acts on lines by $\pi_{k,\ell}(L_{[a_0,a_1,a_2]}) = L_{\pi_{k,\ell}([a_0,a_1,a_2])}$.

These maps formalize the geometric idea of points or lines in $\mathbb{Z}/(p^k)\mathbf{P}^2$ sitting over, or projecting to, points or lines of $\mathbb{Z}/(p^\ell)\mathbf{P}^2$. The following two results assume that $k > \ell$.

Lemma 11 Two lines $L_{[a_0,a_1,a_2]}$ and $L_{[b_0,b_1,b_2]}$ in $\mathbb{Z}/(p^k)\mathbf{P}^2$ sit over the same line in $\mathbb{Z}/(p^\ell)\mathbf{P}^2$ if and only if all of the 2×2 minors of the matrix

$$\begin{bmatrix} a_0 & a_1 & a_2 \\ b_0 & b_1 & b_2 \end{bmatrix}$$

are all divisible by p^ℓ.

Proof If the lines coincide, then $\pi_{k,\ell}([a_0, a_1, a_2]) = \pi_{k,\ell}([b_0, b_1, b_2])$, and so there is a unit $\lambda \in \mathbb{Z}/(p^\ell)$ such that $r_\ell(b_i) = \lambda r_\ell(a_i)$ for $i = 0, 1, 2$ in which case the reduction of the matrix modulo p^ℓ is of rank 1 and all of the 2×2 minors vanish as required.

Conversely, one of the a_i's is a unit in $\mathbb{Z}/(p^k)$. Assume without loss of generality that it is a_0, and take $\lambda = r_\ell(b_0 a_0^{-1})$. The vanishing of the 2×2 minors modulo 2^ℓ implies that $r_\ell(b_i) = \lambda r_\ell(a_i)$ for $i = 1, 2$ and the fact that one of the b_i's is a unit implies that λ is a unit also. □

Proposition 12 If $L_{[a_0,a_1,a_2]}$ and $L_{[b_0,b_1,b_2]}$ are lines in $\mathbb{Z}/(p^k)\mathbf{P}^2$ which lie over the same line in $\mathbb{Z}/(p^\ell)\mathbf{P}^2$ but are distinct in $\mathbb{Z}/(p^{\ell+1})\mathbf{P}^2$ then $L_{[a_0,a_1,a_2]}$ and $L_{[b_0,b_1,b_2]}$ meet in p^ℓ points in $\mathbb{Z}/(p^k)\mathbf{P}^2$.

Proof The points of intersection of these lines are the equivalence classes of triples $(w_0, w_1, w_2) \in \mathbb{Z}/(p^k)^3 \setminus p(\mathbb{Z}/(p^k))^3)$ which are in the null space of the matrix:

$$\begin{bmatrix} a_0 & a_1 & a_2 \\ b_0 & b_1 & b_2 \end{bmatrix}$$

If $h = 0$, i.e., if the lines are distinct modulo p, then there is a unique equivalence class in the null space modulo p, and Hensel's lemma applies to show this lifts to a unique solution modulo p^k.

Now suppose $h > 0$. One of the entries in (a_0, a_1, a_2) is a unit in $\mathbb{Z}/(p^k)$. Assume without loss of generality that it is a_0. Since $h > 0$ we must also have that b_0 is a unit in $\mathbb{Z}/(p^h)$ and so in $\mathbb{Z}/(p^k)$ as well, and so we can pick representatives in which $a_0 = b_0 = 1$. Since the 2×2 minors of

$$\begin{bmatrix} 1 & a_1 & a_2 \\ 1 & b_1 & b_2 \end{bmatrix}$$

are divisible by p^h, and one of them is not divisible by p^{h+1}, we have $a_i - b_i$ is divisible by p^h for $i = 1, 2$ and one is not divisible by p^{h+1}. This linear system is equivalent to

$$\begin{bmatrix} 1 & a_1 & a_2 \\ 0 & b_1 - a_1 & b_2 - a_2 \end{bmatrix}$$

If $[w_0, w_1, w_2]$ is in the null space, then the first row implies w_0 is determined by w_1 and w_2 and that at least one of these is a unit. Choosing the representative with this unit equal to 1, the second row determines a congruence of the form $cx \equiv d \pmod{p^k}$ with c, d divisible by p^h and c not divisible by p^{h+1}. Such a linear congruence has exactly p^h solutions. □

Corollary 13 If $[a_0, a_1, a_2]$ and $[b_0, b_1, b_2]$ are points in $\mathbb{Z}/(p^k)\mathbf{P}^2$ which lie over the same point in $\mathbb{Z}/(p^\ell)\mathbf{P}^2$ but are distinct in $\mathbb{Z}/(p^{\ell+1})\mathbf{P}^2$, then $[a_0, a_1, a_2]$ and $[b_0, b_1, b_2]$ lie on exactly p^ℓ lines in $\mathbb{Z}/(p^k)\mathbf{P}^2$.

Specializing to the case $k = 2$, $p = 2$ we have

Lemma 14

(a) Any point of $\mathbb{Z}/(4)\mathbf{P}^2$ lies on *six* lines.
(b) The *six* lines through a given point in $\mathbb{Z}/(4)\mathbf{P}^2$ cover $\mathbb{Z}/(4)\mathbf{P}^2$.
(c) The *six* lines through a given point in $\mathbb{Z}/(4)\mathbf{P}^2$ divide into two sets of three lines in $\mathbb{Z}/(4)\mathbf{P}^2$ each of which project to the three lines through the projection of the given point in $\mathbb{Z}/(2)\mathbf{P}^2$.

Proposition 15 If $p_1, \ldots, p_6$ are homogeneous linear polynomials in $\mathbb{Z}[x, y, z]$ representing the *six* lines through a given point in $\mathbb{Z}/(4)\mathbf{P}^2$ and $f = p_1 \cdot \cdots \cdot p_6$, then $f/2^3$ is an integer-valued polynomial.

Proof If $(a, b, c) \in \mathbb{Z}$ has all entries divisible by 2, then, since f is homogeneous, $f(a, b, c)$ is divisible by 2^6. If not all entries are divisible by 2, then (a, b, c) represents a point of $\mathbb{Z}/(4)\mathbf{P}^2$, and so lies on a line determined by one of the p_i's; hence $p_i(a, b, c)$ is divisible by 4. Since the lines doubly cover $\mathbb{Z}/(2)\mathbf{P}^2$, one of the other p_j has $p_j(a, b, c)$ is divisible by 2. □

Proposition 16

(a) If $f = x \cdot y \cdot z \cdot (x+y) \cdot (x+z) \cdot (y+z) \cdot (x+y+z)$, then $f/2^3$ is a homogeneous integer-valued polynomial.
(b) If g is any one of the homogeneous linear polynomials $2x + y + 3z$, $y + z$, $y + 3z$, $2x + y + z$, $x + 2y + 3z$, $x + z$, $x + 3z$, $x + y + 2z$, $x + y$, $x + 3z$, $x + 3y + 2z$ and $x + 2y + z$, then $f \cdot g/2^4$ is a homogeneous integer-valued polynomial.

Proof Since the linear polynomials used to construct f represent the *seven* lines in $\mathbb{Z}/(2)\mathbf{P}^2$, part (a) follows from Proposition 5.

If the points on the lines in $\mathbb{Z}/(4)\mathbf{P}^2$ determined by the seven homogeneous linear polynomials used to construct f are enumerated (42 points in total), it will be discovered that every point of $\mathbb{Z}/(4)\mathbf{P}^2$ occurs with the exception of $[1, 1, 1]$. It follows that $f(a, b, c)$ is divisible by 2^4 for any $(a, b, c) \in \mathbb{Z}^3$ except if $[a, b, c] = [1, 1, 1]$ in $\mathbb{Z}/(4)\mathbf{P}^2$. The linear polynomials listed in part (b) are all ones taking an even value at $[1, 1, 1]$. □

Proposition 17

(a) If p_i for $i = 1, \dots, 28$ are the linear homogeneous polynomials in $\mathbb{Z}[x, y, z]$ whose coefficent triples are as in the following table:

$(0, 0, 1)$, $(4, 1, 0)$, $(0, 1, 5)$, $(1, 0, 1)$, $(1, 1, 4)$, $(1, 0, 4)$, $(1, 1, 1)$, $(0, 2, 1)$,

$(0, 1, 2)$, $(0, 1, 3)$, $(1, 4, 3)$, $(1, 1, 2)$, $(1, 0, 6)$, $(1, 1, 3)$, $(2, 0, 3)$, $(2, 1, 0)$,

$(2, 3, 3)$, $(1, 6, 0)$, $(1, 6, 1)$, $(1, 3, 0)$, $(1, 3, 5)$, $(2, 6, 1)$, $(2, 1, 2)$, $(2, 1, 7)$,

$(1, 2, 6)$, $(1, 2, 3)$, $(1, 3, 2)$, $(1, 7, 7)$

then the lines in $\mathbb{Z}/(8)\mathbf{P}^2$ determined by these polynomials cover all of the points of $\mathbb{Z}/(8)\mathbf{P}^2$ and their projections to $\mathbb{Z}/(4)\mathbf{P}^2$ cover all 28 lines of $\mathbb{Z}/(4)\mathbf{P}^2$.

(b) If f is the product of the 28 linear homogeneous polynomials in part (a) then $f/2^{19}$ is integer-valued.

Proof Part (a) is a simple but tedious computation. For part (b) let $(a, b, c) \in \mathbb{Z}^3$. If all three entries are even, then $f(a, b, c)$ is divisible by 2^{28}. If at least one entry is odd, then (a, b, c) represents a point in $\mathbb{Z}/(8)\mathbf{P}^2$, and so $p_i(a, b, c)$ is divisible by 2^3 for at least one i. Since the lines project to all of the lines in $\mathbb{Z}/(4)\mathbf{P}^2$, six of them $p_{j1}, \dots, p_{j6}$ must contain the point in $\mathbb{Z}/(4)\mathbf{P}^2$ represented by (a, b, c), i.e., $p_{jk}(a, b, c)$ is divisible by 2^2, for six values of k. At most one of those polynomials, p_{jk} is the previously identified p_i. The triple (a, b, c) will also represent a point in $\mathbb{Z}/(2)\mathbf{P}^2$ which lies on three lines in $\mathbb{Z}/(2)\mathbf{P}^2$ each of which is covered by *four* of the lines in $\mathbb{Z}/(4)\mathbf{P}^2$. Thus $p_{k\ell}(a, b, c)$ is divisible by 2 for at least 12 values of ℓ. At most *six* of these are the previously identified p_{jk}'s. Combining these we find that $f(a, b, c)$ is divisible by $2^3 \cdot (2^2)^{6-1} \cdot 2^{12-6} = 2^{19}$. □

Remark 18 By picking certain subsets of size 14 from the set of linear factors in the previous proposition it is possible to find degree 14 homogeneous 3 variable integer-valued polynomials with denominator 2^8. This is not optimal however. In [3] bases for the $\mathbb{Z}_{(2)}$-modules of homogeneous three-variable integer-valued polynomials are computed in degrees up to 20. Those calculations reveal that there is a degree 14 polynomial in 3 variables which is homogeneous integer-valued and has a denominator 2^9. It does not factor as a product of linear polynomials however.

The argument in the last proof can be generalized, although it depends on finding a collection of lines with a special property.

Proposition 19 If a collection of $7 \cdot 2^{2(k-1)}$ lines in $\mathbb{Z}/(2^{k+1})\mathbf{P}^2$ can be found that both cover all of $\mathbb{Z}/(2^{k+1})\mathbf{P}^2$ and reduce to the set of lines of $\mathbb{Z}/(2^k)\mathbf{P}^2$ then, taking the product of the homogeneous linear polynomials representing those lines will give a homogeneous integer-valued polynomial with denominator $(2^k)^{3\cdot 2^{k-1}} \cdot (2^{k-1})^{3\cdot 2^{k-1}} \cdots (2)^{3\cdot 2^{k-1}}$.

Acknowledgments The author would like to thank Keith Johnson at Dalhousie University for his incredible support and guidance throughout the project.

References

1. J. F. Adams, A. Harris, and R. Switzer, *Hopf algebras of cooperations for real and complex K-theory*, Proc. London Math. Soc. 23 (1971), 385–408.
2. A. Albert and R. Sandler, *An Introduction to Finite Projective Planes*, Holt, Rinehart and Winston, New York-Toronto, Ont.-London, 1968.
3. M. B.Langlois, *Homogeneous Integer-Valued Polynomials of Three Variables*, Ph.D. Thesis, 2018.
4. A. Baker, F. Clarke, N. Ray, and L. Schwartz, *On the Kummer congruences and the stable homotopy of BU*, Trans. Amer. Math. Soc. 316 (1989), 385–432.
5. F. Clarke, *Self maps of BU*, Math. Proc. Cambridge Philos. Soc. 89 (1981), 491–500.
6. K. Johnson, *Stable homotopy theory, formal group laws, and integer-valued polynomials*, Commutative Algebra, 2014, pp. 213–223.
7. K. Johnson and D. Patterson, *Projective p-orderings and homogeneous integer-valued polynomials*, Integers, 11 (2011), 597–604.
8. E. Kleinfeld, *Finite Hjelmslev Planes*, Illinois J. Math 3 (1959), 403–407.
9. W. Klingenberg, *Projektive und affine Ebenen mit Nachbarelementen*, Math. Z. 60 (1954), 384–406.
10. O.Veblen, *A system of axioms for geometry*, Trans. Amer. Math. Soc. 5 (1904), 343–384.

Around Prüfer Extensions of Rings

Gabriel Picavet and Martine Picavet-L'Hermitte

2010 Mathematics Subject Classification Primary 13B02, 13B22, 13B40; Secondary 13B30

1 Introduction and Notation

We consider the category of commutative and unital rings, whose flat epimorphisms will be strongly involved, like localizations with respect to a multiplicatively closed subset.

If $R \subseteq S$ is an (ring) extension, we denote by $[R, S]$ the set of all R-subalgebras of S and by $[R, S]_{fg}$ the set of all $T \in [R, S]$, such that T is of finite type over R. Any undefined material is explained in the next subsection and in the following sections.

1.1 An Overview of the Paper

We present some properties of Prüfer extensions of rings and derive from them new results, using the properties and definitions of Knebusch and Zhang [20]. It is well known that Prüfer extensions are nothing but normal pairs. Prüfer extensions are defined by flat epimorphisms, while normal pairs are defined by the integrally closed property. We will deal with the Prüfer aspect, except in Sect. 6.

In Sect. 2, we recall some facts about Prüfer extensions. We also give rules on flat epimorphisms and direct limits, in order to make our proofs easier.

If $R \subset S$ is a ring extension, an ideal I of R is called S-regular by [20] if $IS = S$. Such ideals are a useful concept in the next sections. Moreover, these

G. Picavet (✉) · M. Picavet-L'Hermitte
Mathématiques, Le Cendre, France
e-mail: picavet.mathu@orange.fr

J.-L. Chabert et al. (eds.), *Algebraic, Number Theoretic, and Topological Aspects of Ring Theory*, https://doi.org/10.1007/978-3-031-28847-0_19

ideals I permit us to factorize the extension through the ring of sections defined by the open subset associated to I. This is developed in Sect. 3, with some applications to Prüfer extensions. By the way, we give rules allowing us to calculate rings of sections.

In Sect. 4, we show that the integral closure $\overline{R}$ of a ring extension $R \subset S$ is the set intersection of all $T \in [R, S]$ such that $T \subseteq S$ is Prüfer. This statement generalizes a classical result on integral closures.

As an application, we show that an avoidance lemma with respect to finitely many integrally closed subrings holds. The proof is not easy and uses Manis valuations. We also show an avoidance lemma with respect to finitely many flat epimorphisms. This is explained in Sect. 5.

Section 6 deals with pullbacks results. Olivier proved that integrally closed extensions are characterized by pullbacks in which some morphism is of the form $V \to K$, where V is a semi-hereditary ring with total quotient ring K. We adapt this result to the Prüfer case and evidently reuse the normal pairs. Another result concerns a more classical situation.

In Sect. 7, we deal with extensions over local rings and introduce the strong divisors considered by Knebusch and Zhang [20]. A strong divisor is a regular element t of a ring R, such that Rt is comparable to each ideal of R. The maximal Prüfer extension of a local ring R is the localization of R with respect to the multiplicatively closed subset of all strong divisors of R. We develop a theory of strong divisors. The most striking results are that a regular element x of a local ring is a strong divisor if and only if $R \to R_x$ is Prüfer and that an extension of finite type of R is Prüfer if and only if it is of the form $R \to R_x$, where x is a strong divisor.

QR extensions $R \subset S$ are studied in Sect. 8: they are extensions such that each $T \in [R, S]$ is (isomorphic to) a localization. They are evidently Prüfer. We also look at the Bezout extensions of [20] and examine the Bezout and Prüfer hull of an extension. Over a local ring or a Nagata ring $R(X)$, the Prüfer, Bezout and QR properties are equivalent. To go further, we have introduced locally strong divisors. As locally strong divisors appear each time we are dealing with Prüfer extensions, we see that a ring R admits non-trivial Prüfer extensions if R has locally strong divisors that are non-units. This concept is more stable than that of strong divisors. An interesting result is that QR extensions are characterized by using locally strong divisors. Another one is that a QR extension $R \subset S$ satisfies that for each $s \in S$ there is a locally strong divisor ρ, such that $\rho s \in R$. The section ends on extensions whose supports are finite.

Section 9 is devoted to minimal or FCP extensions of a local ring that are either Prüfer or have the QR property. A special attention is paid to $\mathcal{B}$ extensions (extensions that are locally determined in some sense).

The paper ends by considering the set of all primitive elements in an extension, a study initiated by Dobbs and Houston. There is a link with quasi-Prüfer extensions.

1.2 Basics Concepts

As usual, $\operatorname{Spec}(R)$ and $\operatorname{Max}(R)$ are the set of prime and maximal ideals of a ring R and $\mathrm{U}(R)$ is the set of all its units.

We now give some notation for a ring morphism $f : R \to S$. We denote by ${}^a f$ the spectral map $\operatorname{Spec}(S) \to \operatorname{Spec}(R)$. Then $\mathcal{X}_R(S)$ (or $\mathcal{X}(S)$) is the image of the map ${}^a f$, and we say that f is an *i-morphism* if ${}^a f$ is injective. If Q is a prime ideal of S lying over P in R, the ring morphism $R_P \to S_Q$ is called the local morphism at Q of the morphism.

Then $(R : S)$ is the conductor of an extension $R \subseteq S$. The integral closure of R in S is denoted by $\overline{R}^S$ (or by $\overline{R}$ if no confusion can occur).

A *local* ring is here what is called elsewhere a quasi-local ring. For an extension $R \subseteq S$ and an ideal I of R, we write $\mathrm{V}_S(I) := \{P \in \operatorname{Spec}(S) \mid I \subseteq P\}$ and $\mathrm{D}_S(I)$ for its complement. If R is a ring, then $\mathrm{Z}(R)$ denotes the set of all its zero divisors. The support of an R-module E is $\operatorname{Supp}_R(E) := \{P \in \operatorname{Spec}(R) \mid E_P \neq 0\}$, and $\operatorname{MSupp}_R(E) := \operatorname{Supp}_R(E) \cap \operatorname{Max}(R)$. When $R \subseteq S$ is an extension, we will set $\operatorname{Supp}(T/R) := \operatorname{Supp}_R(T/R)$ and $\operatorname{Supp}(S/T) := \operatorname{Supp}_R(S/T)$ for each $T \in [R, S]$, unless otherwise specified.

If $R \subseteq S$ is a ring extension and Σ a *mcs* of R (i.e., a multiplicatively closed subset of R), then S_Σ is both the localization S_Σ as a ring and the localization at Σ of the R-module S, that is, $S \otimes_R R_\Sigma$.

Let Σ_1 and Σ_2 be two mcs of a ring R. We denote by $\Sigma_2/1$ the image of Σ_2 in R_{Σ_1}. We recall that $R_{\Sigma_1\Sigma_2} = (R_{\Sigma_1})_{\Sigma_2/1}$. It follows that if $x \in R$ and Σ is a mcs of R, then $(R_x)_\Sigma = (R_\Sigma)_{x/1}$.

Flat epimorphisms and their properties are the main tools in this chapter. We use the theory that was developed by D. Lazard [22, Chapter IV]. The reader may also use the scholium of our paper [31].

When $R \to S$ and $R \to T$ are ring morphisms, we will write $S \cong_R T$ (or $S \cong T$) if there is an isomorphism of R-algebras $S \to T$. It may happen that $\cong$ is replaced with $=$.

Let $R \subseteq S$ be an extension. A *chain* of R-subalgebras of S is a set of elements of $[R, S]$ that are pairwise comparable with respect to inclusion. We say that $R \subseteq S$ is *chained* if $[R, S]$ is a chain. We also say that the extension has FCP (or is an FCP extension) if each chain in $[R, S]$ is finite, or equivalently, the poset $[R, S]$ is Artinian and Noetherian. An extension is called FIP if $[R, S]$ has finitely many elements. An extension $R \subset S$ is called minimal if $[R, S] = \{R, S\}$. According to [14, Théorème 2.2], a minimal extension is either integral or a flat epimorphism. Finally, $|X|$ is the cardinality of a set X, and $\subset$ denotes proper inclusion (contrary to [20] where $\subset$ denotes the large inclusion). A compact topological space does not need to be separated. For a positive integer n, we set $\mathbb{N}_n := \{1, \ldots, n\}$.

2 Some Definitions, Notation, and Useful Results

An extension $R \subseteq S$ is called *Prüfer* if $R \subseteq T$ is a flat epimorphism for each $T \in [R, S]$ (or equivalently, if $R \subseteq S$ is a normal pair) [20, Theorem 5.2, p. 47]. A Prüfer integral extension is trivial.

We denote by $Q(R)$ the complete ring of quotients (Utumi–Lambeck) of a ring R.

Definition 2.1 ([20]) A ring extension $R \subseteq S$ has:

(1) A greatest flat epimorphic subextension $R \subseteq \widehat{R}^S$, called the *Morita hull* of R in S
(2) A greatest Prüfer subextension $R \subseteq \widetilde{R}^S$, called the *Prüfer hull* of R in S

We set $\widehat{R} := \widehat{R}^S$ and $\widetilde{R} := \widetilde{R}^S$, if no confusion can occur.

A ring R has:

(1) Lazard [22] a maximal flat epimorphic extension $R \subseteq \mathbb{M}(R) := \widehat{R}^{\mathrm{Q}(R)}$ (also termed the maximal flat epimorphic extension by some authors, like [22])
(2) Knebusch and Zhang [20] a maximal Prüfer extension $R \subseteq \mathbb{P}(R) := \widetilde{R}^{\mathrm{Q}(R)}$

Note that $\widetilde{R}^S$ is denoted by $\mathrm{P}(R, S)$ in [20] and $\widehat{R}^S$ coincides with the weakly surjective hull $\mathrm{M}(R, S)$ of [20]. Our terminology is justified because Morita's construction is earlier [23, Corollary 3.4]. The Morita hull can be computed by using an (transfinite) induction [23]. Let S' be the set of all $s \in S$, such that there is some ideal I of R, such that $IS = S$ and $Is \subseteq R$. Then $R \subseteq S'$ is a subextension of $R \subseteq S$. We set $S_1 := S'$ and $S_{i+1} := (S_i)' \subseteq S_i$. By [23, p.36], if $R \subset S$ is an FCP extension, then $\widehat{R} = S_n$ for some integer n.

We also note the following known consequence:

Proposition 2.2 *An extension $R \subseteq S$ is a flat epimorphism if and only if for each $s \in S$ there is some ideal I of R such that $IS = S$ and $Is \subseteq R$ (or equivalently $(R :_R s)S = S$).*

Corollary 2.3 *An extension $R \subseteq S$ is Prüfer if and only if $R[s] = (R :_R s)R[s]$ for each $s \in S$.*

Proof Use the definition of Prüfer extensions by flat epimorphisms. □

If an extension $R \subseteq S$ is Prüfer and Σ is a mcs of R, then $R_\Sigma \subseteq S_\Sigma$ is Prüfer. We have a converse.

Proposition 2.4 ([31, Proposition 1.1]) *An extension $R \subset S$ is Prüfer if and only if $R_M \subseteq S_M$ is Prüfer for each $M \in \mathrm{Max}(R)$ (resp., for each $M \in \mathrm{Spec}(R)$).*

Proposition 2.5 ([31, Corollary 3.15]) *Prüfer extensions are descended by faithfully flat morphisms.*

Proposition 2.6 *The Prüfer property of extensions $R \subseteq S$ is local on the spectrum; that is if* $\mathrm{Spec}(R) = \mathrm{D}(r_1) \cup \cdots \cup \mathrm{D}(r_n)$ *for some elements* $r_1, \ldots, r_n \in R$ *and* $R_{r_i} \subseteq S_{r_i}$ *is Prüfer for each* $i = 1, \ldots, n$, *then* $R \subseteq S$ *is Prüfer.*

Proof The extension $R_{r_1} \times \cdots \times R_{r_n} \subseteq S_{r_1} \times \cdots \times S_{r_n}$ is Prüfer [20, Proposition 5.20, p.56]. To conclude, use Proposition 2.5 since $R \to R_{r_1} \times \cdots \times R_{r_n}$ is faithfully flat. □

In [31], a minimal flat epimorphism is called a *Prüfer minimal* extension. An FCP Prüfer extension has FIP and is a tower of finitely many Prüfer minimal extensions [31, Proposition 1.3]. A minimal extension $R \subset S$ is such that there exists a maximal ideal M of R satisfying $\mathrm{Supp}(S/R) = \{M\}$. Such a prime ideal M is called the *crucial (maximal) ideal* $\mathcal{C}(R, S)$ of $R \subset S$ [9, Theorem 2.1].

In [31], we defined an extension $R \subseteq S$ to be *quasi-Prüfer* if it can be factored $R \subseteq R' \subseteq S$, where $R \subseteq R'$ is integral and $R' \subseteq S$ is Prüfer. In this case, R' is necessarily $\overline{R}$. An FCP extension is quasi-Prüfer [31, Corollary 3.4].

An extension $R \subseteq S$ is called *almost-Prüfer* if it can be factored $R \subseteq S' \subseteq S$, where the first extension is Prüfer and the second is integral. In this case, S' is necessarily $\widetilde{R}$. An almost-Prüfer extension is quasi-Prüfer [31].

We now give some rules on flat epimorphisms. The following result of Lazard is a key result. Let R be a ring. We denote by $\mathcal{FE}$ the collection of classes up to an isomorphism of flat epimorphisms whose domain is R and by $\mathcal{X}$ the set of subsets of $\mathrm{Spec}(R)$ that are affine schemes, when endowed with the induced sheaf. The elements of $\mathcal{X}$ are compact and stable under generalization.

Proposition 2.7 ([22, Proposition 2.5, p.112]) *The map* $\mathcal{FE} \to \mathrm{Spec}(R)$, *defined by* $T \mapsto \mathcal{X}(T)$, *is a bijection onto* $\mathcal{X}$. *The inverse map is as follows: an affine scheme* X *of* $\mathrm{Spec}(R)$ *gives* $R \to \Gamma(X)$, *the ring of sections over* X.

The next result, proved in [18, Proposition 3.4.10, p.242], will be useful in the sequel.

Proposition 2.8 *($\mathcal{L}$)-rule Let* $R \to E$ *be a ring morphism and* $E = \varinjlim E_i$ *where each* E_i *is an* R*-algebra, then* $\mathcal{X}(E) = \cap \mathcal{X}(E_i)$.

We will use Proposition 2.7 under the following form.

Proposition 2.9 *($\mathcal{X}$)-rule Let* $R \to E$ *be a flat epimorphism and* $R \to F$ *a ring morphism:*

(1) *There is a factorization* $R \to E \to F$ *if and only if* $\mathcal{X}(F) \subseteq \mathcal{X}(E)$.
(2) *If* $R \to F$ *is a flat epimorphism, then* $E \cong F$ *if and only if* $\mathcal{X}(F) = \mathcal{X}(E)$.

Proof (1) The ring morphism $\alpha : F \to F \otimes_R E$ is a flat epimorphism. If $\mathcal{X}(F) \subseteq \mathcal{X}(E)$, then the spectral morphism of α is surjective, because there is a surjective map $\mathrm{Spec}(F \otimes_R E) \to \mathrm{Spec}(E) \times_{\mathrm{Spec}(R)} \mathrm{Spec}(F)$ [18, Corollaire 3.2.7.1, p.235]. It follows that α is a faithfully flat epimorphism, whence an isomorphism by [22, Lemme 1.2, p.109] and one implication is proved. Its converse is obvious. Now (2) can be proved by using (1). But it is also a consequence of Proposition 2.7. □

Corollary 2.10 *($\mathcal{MCS}$)-rule Let $R \to E$ be a ring morphism:*

(1) *If $E = R_\Sigma$, where Σ is a mcs of R, then $\mathcal{X}(E) = \cap\{\mathrm{D}(s) | s \in \Sigma\}$.*
(2) *If $E = \varinjlim R_{s_i}$, where $\{s_i\}$ is a family of elements of R, then $E = R_\Sigma$, where Σ is the mcs of R generated by the family.*

Proof The proof is a consequence of the above rules. □

The reader may find some other information about Prüfer extensions in the recent paper [3] from Campanini and Finocchiaro.

3 S-Regular Ideals and Rings of Sections

If I is an ideal of a ring R, then $\Gamma(\mathrm{D}(I), R)$ (or $\Gamma(\mathrm{D}(I))$) denotes the ring of sections of the scheme $\mathrm{Spec}(R)$ over the open subset $\mathrm{D}(I)$. All that we need to know is that $\Gamma(\mathrm{D}(R)) = R$, $\Gamma(\emptyset) = 0$, and if $f : R \to S$ is a ring morphism, there is a commutative diagram, because ${}^a f^{-1}(\mathrm{D}(I)) = \mathrm{D}(IS)$:

$$\begin{array}{ccc} R & \longrightarrow & S \\ \downarrow & & \downarrow \\ \Gamma(\mathrm{D}(I)) & \longrightarrow & \Gamma(\mathrm{D}(IS)) \end{array}$$

We denote by $\mathrm{Ass}(R)$ the set of all (Bourbaki) prime ideals P associated to the R-module R; that is, $P \in \mathrm{Min}(\mathrm{V}(0 : r))$ for some $r \in R$. Recall that a ring morphism $f : R \to S$ is called *schematically dominant* if for each open subset U of $\mathrm{Spec}(R)$, the map $\Gamma(U, R) \to \Gamma({}^a f^{-1}(U), S)$ is injective [18, Proposition I.5.4.1]. The first author proved that a flat ring morphism $f : R \to S$ is schematically dominant if and only if $\mathrm{Ass}(R) \subseteq \mathcal{X}(S)$ [29, Proposition 52]. Clearly, if $\mathrm{Min}(R) = \mathrm{Ass}(R)$ (for example, if R is an integral domain) and f is injective and flat, then f is schematically dominant.

Lemma 3.1 *A flat extension $R \subseteq S$ is schematically dominant.*

Proof If $P \in \mathrm{Ass}(R)$, there is some $a \in R$, such that $P \in \mathrm{Min}(\mathrm{V}(0 : a))$. From $(0 :_S a) \cap R = 0 : a$, we deduce that $R/(0 : a) \to S/(0 :_S a)$ is injective and then $P/(0 : a)$ can be lifted up to a minimal prime ideal $Q/(0 :_S a)$. Hence, $Q \in \mathrm{Ass}(S)$ is above P. □

Let $R \subseteq S$ be an extension and an ideal I of R. Then I is called *S-regular* if $IS = S$ [20]. Note that S-regular ideals play a prominent role in [20]. They are involved in certain questions. For example, if $f : R \to S$ is a ring morphism, the fiber at a prime ideal P of R is ${}^a f^{-1}(P)$. This fiber is homeomorphic to the spectrum of the ring S_P/PS_P. Therefore, the fiber is empty if and only if $S_P = PS_P$, which means that PR_P is S_P-regular. If f is a flat epimorphism, the fiber at P is empty if and only if $S = PS$ [22, Proposition 2.4, p.111].

Remark 3.2 Let $f : R \hookrightarrow S$ be an extension:

(1) An ideal I of R is S-regular if and only if $\mathcal{X}(S) \subseteq \mathrm{D}(I)$ [28, Lemma 2.3]. Such an ideal I is dense; that is, $0 : I = 0$.

(1(a)) I is S-regular if and only if $\sqrt{I}$ is S-regular, because $\mathrm{D}(\sqrt{I}S) = {}^af^{-1}(\mathrm{D}(\sqrt{I})) = {}^af^{-1}(\mathrm{D}(I)) = \mathrm{D}(IS)$.

(1(b)) I is S-regular if and only if I_P is S_P-regular for each $P \in \mathrm{Spec}(R)$. We need only to show that if the local condition holds, then I is S-regular. Suppose that $IS \subset S$, then there is some prime ideal Q of S, such that $IS \subseteq Q$. If $P = Q \cap R$, then Q_P is a prime ideal of S_P, such that $I_PS_P \subseteq Q_P$, a contradiction.

(2) If I is S-regular, we have $\mathrm{Spec}(S) = \mathrm{D}(IS) = {}^af^{-1}(\mathrm{D}(I))$, so that there is a factorization $R \to \Gamma(\mathrm{D}(I)) \to S$. If, in addition, f is flat, then f is schematically dominant (Lemma 3.1), so that we can consider that there is a tower of extensions $R \subseteq \Gamma(\mathrm{D}(I)) \subseteq S$. Moreover, $\mathrm{D}(I)$ is an open subset that is (topologically) dense in $\mathrm{Spec}(R)$ because a schematically dominant morphism is dominant [18, Proposition I.5.4.3], i.e., its spectral image is dense. The density follows from $\mathcal{X}(S) \subseteq \mathrm{D}(I)$.
This result holds if the extension is Prüfer and then $R \to \Gamma(\mathrm{D}(I))$ is Prüfer.

(3) We will use the following consequence of Proposition 2.7. If I is an ideal of R, then $R \to \Gamma(\mathrm{D}(I))$ is a flat epimorphism and $\mathcal{X}(\Gamma(\mathrm{D}(I))) = \mathrm{D}(I)$ if and only if $\mathrm{D}(I)$ is an affine open subset of $\mathrm{Spec}(R)$ (for example if I is principal), in which case $\mathrm{D}(I) = \mathrm{D}(J)$ where J is a finitely generated ideal.

We can say more after looking at the following result adapted to ring morphisms (the reader is referred to [18, Definition I.4.2.1, p.260] for the definition of an open immersion of schemes). We will say that a ring morphism is an *open immersion* if the morphism of schemes associated is an open immersion.

Proposition 3.3 *Let $f : R \to S$ be a ring morphism:*

(1) *Grothendieck and Dieudonné [18, I.4.2.2] f is an open immersion if and only if* $\mathrm{Spec}(S) \to \mathcal{X}(S)$ *is a homeomorphism, $\mathcal{X}(S)$ is an open subset, and the local morphisms of f are isomorphisms.*

(2) *A flat epimorphism $R \to S$, such that $\mathcal{X}(S)$ is Zariski open, is an open immersion.*

(3) *Grothendieck [17, Théorème 17.9.1, p.79] f is an open immersion if and only if f is a flat epimorphism of finite presentation.*

(4) *Cox and Rush [4, Theorem 1.1] An injective flat epimorphism of finite type is of finite presentation, whence is an open immersion.*

Proof We need only to prove (2) and using (1). Since f is a flat epimorphism, its spectral map is an homeomorphism onto its image by [22, Corollaire 2.2, p.111] that is an open subset of the form $\mathrm{D}(I)$. Moreover, the local morphisms of the map are isomorphisms. □

Proposition 3.4 *Let $R \subset S$ be an injective flat epimorphism of finite type. Then $\mathcal{X}(S)$ is an open affine subset $\mathrm{D}(I)$, where I is a S-regular ideal, and there is an R-isomorphism $\Gamma(\mathrm{D}(I)) \cong S$, $R \subset S$ is of finite presentation, and I is a dense ideal.*

Conversely, if $\mathrm{D}(I)$ is an open affine subset, where I is a finitely generated dense ideal, then $R \to \Gamma(\mathrm{D}(I))$ is an injective flat epimorphism, of finite type (presentation), such that $\mathcal{X}(\Gamma(\mathrm{D}(I))) = \mathrm{D}(I)$.

Proof To apply Proposition 2.7, we need only to look at injective flat epimorphisms of finite type $R \to S$. We know that such a ring morphism $f : R \subset S$ is of finite presentation according to Proposition 3.3(4). By the Chevalley Theorem, $\mathcal{X}(S)$ is a Zariski quasi-compact open subset of $\mathrm{Spec}(R)$, therefore of the form $\mathrm{D}(I)$, where I is an ideal of R, of finite type. We have ${}^a f^{-1}(\mathrm{D}(I)) = \mathrm{D}(IS) = \mathrm{Spec}(S)$ because $\mathcal{X}(S) = \mathrm{D}(I)$, so that $IS = S$, and then I is dense because it is S-regular. Moreover, $\Gamma(\mathrm{D}(I)) \cong S$ by Proposition 2.9(2) because $\mathcal{X}(S) = \mathrm{D}(I) = \mathcal{X}(\Gamma(\mathrm{D}(I)))$.

Assume that the hypotheses of the converse hold. Since the morphism $R \to \Gamma(\mathrm{D}(I))$ is an open immersion by Proposition 3.3, we get that $R \to \Gamma(\mathrm{D}(I))$ is of finite presentation. Moreover, $0 : I = 0$ (which is equivalent to $\mathrm{Ass}(R) \subseteq \mathrm{D}(I)$ [22, Corollaire 1.14, p.93]), so that $R \to \Gamma(\mathrm{D}(I))$ is injective, by [22, Proposition 3.3, p.96]. □

We note the following result:

Proposition 3.5 ([20, Theorem 2.8, p.101, Theorem 2.6, p.100]) *Let $R \subseteq S$ be an extension that is a flat epimorphism. Then the extension is Prüfer if and only if for every finitely generated S-regular ideal I of R, the ring R/I is arithmetical (resp., I is locally principal).*

Proposition 3.6 *Let $R \subseteq S$ be a flat epimorphism. Then, $R \subseteq S$ is Prüfer if and only if for each $P \in \mathrm{Spec}(R)$, the set of S_P-regular ideals of R_P is a chain.*

Proof According to [31, Proposition 1.1(2)], the extension is Prüfer if and only if $R_P \subseteq S_P$ is Manis for each $P \in \mathrm{Spec}(R)$ and equivalently $R_P \subseteq S_P$ is Prüfer–Manis. The result follows from [20, Theorem 3.5, p.190]. □

We recall that the dominion of a ring morphism $f : R \to S$ is the subring $\mathrm{Dom}(f) = \{x \in S \mid x \otimes 1 = 1 \otimes x \text{ in } S \otimes_R S\}$ of S, which contains the subring $f(R)$. Actually, $\mathrm{Dom}(f)$ is the kernel of the morphism of R-modules $i_1 - i_2 : S \to S \otimes_R S$ where i_1, i_2 are the natural ring morphisms $S \to S \otimes_R S$.

Proposition 3.7 *If $f : R \to S$ is a flat morphism and I an ideal of R, such that $\mathcal{X}_R(S) = \mathrm{D}(I)$, then:*

(1) *$\Gamma(\mathrm{D}(I)) = \mathrm{Dom}(f)$ and $\Gamma(\mathrm{D}(I)) \to S$ is an injective flat morphism.*
(2) *If in addition f is a ring extension, $\widetilde{R} \subseteq \widehat{R} \subseteq \Gamma(\mathrm{D}(I))$, each of the extensions in S being flat. In particular, if $\mathrm{D}(I)$ is affine, then $\widehat{R} = \Gamma(\mathrm{D}(I))$.*
(3) *If $g : R \to B$ is a flat morphism, setting $C := S \otimes_R B$, then $\mathcal{X}_B(C) = \mathrm{D}(IB)$ and $\Gamma(\mathrm{D}(I)) \otimes_R B \cong \Gamma(\mathrm{D}(IB))$.*

(4) *If P is a prime ideal of R, then* $\Gamma(\mathrm{D}(I_P)) = (\Gamma(\mathrm{D}(I)))_P$. *In particular if* $P \in \mathrm{D}(I)$, *then* $(\Gamma(\mathrm{D}(I)))_P = R_P$.
(5) $\mathrm{D}(I) \subseteq \mathcal{X}(\Gamma(\mathrm{D}(I)))$.
(6) *If* $I\Gamma(\mathrm{D}(I)) = \Gamma(\mathrm{D}(I))$, *then* $\mathrm{D}(I) = \mathcal{X}(\Gamma(\mathrm{D}(I)))$, *so that* $\mathrm{D}(I)$ *is an open affine subset if in addition* $R \to \Gamma(\mathrm{D}(I))$ *is a flat epimorphism.*

Proof

(1) is a translation of [28, Theorem 2.7]. The flatness of $\Gamma(\mathrm{D}(I)) \to S$ follows from [22, Proposition 3.1 (2), p.112].
(2) If f is a ring extension, observe that $\widehat{R} \subseteq \mathrm{Dom}(f)$, because $R \to \widehat{R}$ is an epimorphism and then $y \otimes 1 = 1 \otimes y$ for each $y \in \widehat{R}$ [22, Lemme 1.0, p.108]. The flatness of the extensions $\widetilde{R}, \widehat{R} \subseteq S$ results from [22, Proposition 3.1(2), p.112]. At last, if $\mathrm{D}(I)$ is affine, then $R \to \Gamma(\mathrm{D}(I))$ is a flat epimorphism by Remark 3.2(2), so that $\widehat{R} = \Gamma(\mathrm{D}(I))$.
(3) Because $\mathrm{Spec}(C) \to \mathrm{Spec}(B) \times_{\mathrm{Spec}(R)} \mathrm{Spec}(S)$ is a surjective map [18, Corollaire 3.2.7.1, p.235], we have $\mathcal{X}_B(C) = {}^a g^{-1}(\mathrm{D}(I)) = \mathrm{D}(IB)$. To conclude use (1) and the fact that a kernel tensorized by B, which is flat over R, is the kernel of the tensorized map.
(4) is gotten by taking $B = R_P$ in (3).
(5) According to (4), an element P of $\mathrm{D}(I)$ is such that $(\Gamma(\mathrm{D}(I)))_P = R_P$. It follows that there is a prime ideal Q of $\Gamma(D(I))$ lying over P.
(6) holds because I is $\Gamma(\mathrm{D}(I))$-regular.

□

We can apply the above result in the following three contexts, when $I = (r_1, \dots, r_n)$ is an ideal of finite type of R (the hypothesis of this result entails that $\sqrt{I} = \sqrt{J}$, where J is an ideal of finite type). We can suppose that $\mathrm{D}(I) \neq \emptyset$ and that the set $\{\mathrm{D}(r_1), \dots, \mathrm{D}(r_n)\}$ is an antichain, so that the r_i's cannot be nilpotent. The first example is certainly the most interesting, because when $I = Rr$, we recover that $\Gamma(\mathrm{D}(r)) = R_r$.

(1) We can consider the flat ring morphism $\varphi : R \to R_{r_1} \times \cdots \times R_{r_n} := S_I$, which is such that $\mathcal{X}(S_I) = \mathrm{D}(I)$. Actually, φ is of finite presentation [18, Proposition 6.3.11, p.306] and its local morphisms are isomorphisms. But φ may not be a flat epimorphism, when it is not an i-morphism.
In case $\{\mathrm{D}(r_i)\}$ defines a partition on $\mathrm{D}(I)$, φ is an i-morphism, whence a flat epimorphism. In this case, $\Gamma(\mathrm{D}(I)) = \prod R_{r_i}$.
(2) Let $F_I := R[X_1, \dots, X_n]/(r_1X_1 + \cdots + r_nX_n - 1)$ be the forcing R-algebra with structural morphism f_I, associated to a finitely generated ideal $I = (r_1, \dots, r_n)$ (it would be more correct to write: associated to the sequence $\{r_1, \dots, r_n\}$). This ring is not zero, for otherwise $1 = (r_1X_1 + \cdots + r_nX_n - 1)P(X_1, \dots, X_n)$, for some $P(X_1, \dots, X_n) \in R[X_1, \dots, X_n]$, implies that $1 - (r_1X_1 + \cdots + r_nX_n) \in \mathrm{U}(R[X_1, \dots, X_n])$, so that $r_1X_1 + \cdots + r_nX_n$ would be nilpotent and then also the r_i's. Then I is F_I-regular and for every ring morphism $R \to S$ for which I is S-regular, there is a factorization $R \to F_I \to S$. But $F_I \to S$ does not need to be unique. According to [28, Theorem 2.7 and Remark 2.8(1)], the

ring morphism f_I is flat, $\mathcal{X}(F_I) = \mathrm{D}(I)$ and $\Gamma(\mathrm{D}(I)) = \mathrm{Dom}(f_I)$. Moreover, $\Gamma(\mathrm{D}(I)) \to F_I$ is an injective flat ring morphism.

(3) The first author introduced in [27] the following construction that we adapt to the present context. Let R be a ring and I an ideal of R. Denoting by $C(p(X))$ the content of a polynomial $p(X) \in R[X]$, we consider the mcs $\Sigma := \{p(X) \in R[X] \mid \mathrm{D}(I) \subseteq \mathrm{D}(C(p(X)))\}$. Setting $R(\mathrm{D}(I)) =: R[X]_\Sigma$, we get a flat morphism $R \to R(\mathrm{D}(I))$, such that $\mathcal{X}(R(\mathrm{D}(I))) = \mathrm{D}(I)$.

4 Integral Closures as Intersections

We start by giving some results that do not seem to have been observed. They are consequences of a paper by P. Samuel [34]. Let v be a valuation on a ring R. Following [20], we denote by A_v the valuation ring of v.

Lemma 4.1 ([34, Théorème 1(d)]) *An extension $R \subset S$, such that $S \setminus R$ is multiplicatively closed, is integrally closed. For example, $R \subseteq S$ is integrally closed if there is some valuation v on S such that $R = A_v$, the valuation ring of v.*

We will use the next result.

Lemma 4.2 ([34, Théorème 4]) *Let $R \subseteq S$ be an extension and P a prime ideal of R. Due to Zorn Lemma, there is a maximal pair (R', P') dominating (R, P) and $S \setminus R'$ is a mcs.*

A Manis valuation v on a ring S is a valuation such that $v : S \to \Gamma_v \cup \infty$ is surjective, where Γ_v is the value group of v.

A ring extension $R \subseteq S$ is called Manis if $R = A_v$ for some Manis valuation v on S. Prüfer–Manis extensions are defined as Prüfer extensions $R \subset S$ such that there is some Manis valuation v on S such that $A_v = R$ [20, Definition 1, p. 58].

By Knebusch and Zhang [20, Theorem 3.5, p.190], a flat epimorphism $R \subseteq S$ is Prüfer–Manis if and only if the set of all S-regular ideals of R is a chain.

Lemma 4.3 ([20, Theorem 3.3, p.187, Theorem 3.1, p.187, Proposition 5.1(iii), p. 46-47]) *The following statements are equivalent for an extension $R \subseteq S$:*

(1) *$R \subseteq S$ is Prüfer–Manis.*
(2) *$S \setminus T$ is a mcs for each $T \in [R, S[$.*
(3) *$R \subseteq S$ is integrally closed and chained.*
(4) *$R \subseteq S$ is Prüfer and $S \setminus R$ is a mcs.*

If the above condition (3) holds for an FCP extension, then $R \subseteq S$ has FIP.

By Knebusch and Zhang [20], we know that for a Prüfer extension $R \subset S$ and $P \in \mathrm{Supp}(S/R)$, the subset $S_P \setminus R_P$ is multiplicatively closed. Also [20, Proposition 5.1(iii), p. 46-47] shows that if $U \in [R, S]$ and $S \setminus U$ is a mcs, then $U \subset S$ is Prüfer–Manis.

Corollary 4.4 *A minimal extension $R \subset S$ is a flat epimorphism if and only if it is Prüfer and if and only if it is Prüfer–Manis.*

Proof The extension is a flat epimorphism if and only if it is integrally closed. To complete the proof, it is enough to use [14, Proposition 3.1] that states that $S \setminus R$ is a mcs when $R \subset S$ is a flat epimorphism. □

Corollary 4.5 *An FCP Prüfer extension has FIP and is a tower of finitely many Prüfer–Manis minimal extensions.*

We will need the two following results. They generalize known results about the integral closure of an integral domain, which is the intersection of valuation rings.

Lemma 4.6 *Let $R \subset S$ be an extension and $x \in S \setminus \overline{R}$. Then there is some $T \in [R, S]$, such that $T \subset S$ is Prüfer (respectively, Prüfer–Manis) and x is not integral over T, and then $x \notin T$.*

Proof It is enough to mimic the first part of the proof of [34, Théorème 8]. More precisely, let T be a maximal element of the $\cup$-inductive set $\{U \in [R, S] \mid x \notin \overline{U}^S\}$. We intend to show that $T \subset S$ is Prüfer–Manis. In view of the above results, we need only to show that any $V \in [T, S]$ with $V \neq S$ is such that $S \setminus V$ is multiplicatively closed and then such that $T \subset S$ is integrally closed. Now replace A with V in the second paragraph of the proof of [34, Théorème 8], and the result follows. □

Theorem 4.7 *Let $R \subset S$ be an extension, then $\overline{R}^S$ is the intersection of all $T \in [R, S]$ such that $T \subset S$ is Prüfer (resp., Prüfer–Manis) and also the intersection of all $U \in [R, S]$, such that $S \setminus U$ is a mcs.*

Proof The second result is [34, Théorème 8]. Now Lemma 4.6 shows that $\overline{R}^S$ contains the intersection of all $T \in [R, S]$ such that $T \subset S$ is Prüfer (resp., Prüfer–Manis).

For the reverse inclusion, consider an element $x \in S$, integral over R. Then $T \subseteq T[x]$ is integral and a flat epimorphism for any $T \in [R, S]$ such that $T \subset S$ is Prüfer. We deduce from [22, Lemme 1.2, p. 109] that $T = T[x]$ and x belongs to T. □

Remark 4.8 As a consequence of the above theorem, we get that an extension $R \subseteq S$ is quasi-Prüfer if and only if the set of all $T \in [R, S]$, such that $T \subseteq S$ is Prüfer, has a smallest element.

5 Avoidance Lemmata

Some of the following results are known in the context of integral domains and valuation domains. We will use the framework of their proofs but shorter different argumentations. Kostra proved the next Theorem [21, Lemma 2 and Theorem 2], in case S is a field. To prove it in our context, we follow the steps of his difficult proof by using Theorem 4.7.

If $V \subseteq S$ is Prüfer–Manis, S is endowed with a valuation $v : S \to \Gamma_v \cup \{\infty\}$, which is surjective. There is no need to consider invertible elements that may not exist but elements $x \in V$, such that $v(x) = 0$, that is $x \notin P_v$, the center of v. Moreover, if $v(x) > 0$, then $v(1 + x) = 0$.

Lemma 5.1 *Let $R, T, V, V_1, \dots, V_n$ be subrings of a ring S, where n is a positive integer and such that $V \subseteq S$ and $V_i \subseteq S$ are Prüfer–Manis for each $i \in \mathbb{N}_n$. Let v_i be the valuation associated to V_i. Assume that there is some $b \in [T \cap (\cap_{i\in\mathbb{N}_n} V_i)] \setminus V$. Then, there exists $c \in [T \cap (\cap_{i\in\mathbb{N}_n} V_i)] \setminus V$ such that $v_i(c) = 0$ for any $i \in \mathbb{N}_n$.*

Moreover, for any $W \in [R, S]$ such that $W \subseteq S$ is Prüfer–Manis with $b \notin W$, then $c \notin W$.

Proof We build by induction the sequence $\mathcal{S} := \{b_i\}_{i=0}^n$ in the following way: set $b_0 := b$ and $b_k := 1 + \prod_{i=0}^{k-1} b_i$ for any $k \in \mathbb{N}_n$. Then, $b_k \in T \cap (\cap_{i\in\mathbb{N}_n} V_i)$ for any $k \in \{0, \dots, n\}$, so that $v_i(b_k) \geq 0$ for any $i \in \mathbb{N}_n$ and any $k \in \{0, \dots, n\}$.

If $b_l = b_k$ for some $k \neq l$, assume that $k > l$. Then, $b_k = 1 + b_k \prod_{j=0, j\neq l}^{k-1} b_j$, giving that $b_k(1 - \prod_{j=0, j\neq l}^{k-1} b_j) = 1$, so that $v_i(b_k) + v_i(1 - \prod_{j=0, j\neq l}^{k-1} b_j) = v_i(1)$ $(*)$, with b_k and $1 - \prod_{j=0, j\neq l}^{k-1} b_j$ both in V_i for any $i \in \mathbb{N}_n$. It follows that $v_i(b_k) \geq 0$ and $v_i(1 - \prod_{j=0, j\neq l}^{k-1} b_j) \geq 0$. As $v_i(1) = 0$, $(*)$ leads to $v_i(b_k) = 0$ for any $i \in \mathbb{N}_n$ and the proof of the Lemma is gotten for b_k.

Assume now that $b_j \neq b_k$ for any $k, j \in \{0, \dots, n\}$, $k \neq j$, so that $|\mathcal{S}| = n + 1$.

We claim that for any $i \in \mathbb{N}_n$, there is at most one $b_{i_j} \in \mathcal{S}$ such that $v_i(b_{i_j}) > 0$ $(**)$.

If $v_i(b_k) = 0$ for any $k \in \{0, \dots, n\}$, then $(**)$ holds. Otherwise, let j_0 be the least integer of $\{0, \dots, n\}$ such that $v_i(b_{j_0}) \neq 0$, that is, $v_i(b_{j_0}) > 0$. It follows that for any $k \geq j_0$, we have $v_i(\prod_{j=0}^k b_j) > 0$, so that $v_i(b_{k+1}) = v_i(1 + \prod_{j=0}^k b_j) = v_i(1) = 0$. Since $v_i(b_k) = 0$ for any $k < j_0$, we get that $v_i(b_k) \neq 0$ if and only if $k = j_0$. Then $(**)$ holds. Hence, $|\{b_j \in \mathcal{S} \mid \exists i \in \mathbb{N}_n \text{ such that } v_i(b_j) \neq 0\}| \leq n < |\mathcal{S}|$. It follows that there exists some $c := b_k \in [T \cap (\cap_{i\in\mathbb{N}_n} V_i)]$ such that $v_i(c) = 0$ for any $i \in \mathbb{N}_n$.

It remains to show that $c \notin V$. We prove by induction on $j \in \{0, \dots, k\}$ that $b_j \notin V$ for any $j \in \{0, \dots, k\}$. This is satisfied for $j = 0$ since $b_0 = b$. Assume that $b_j \notin V$ for any $j \in \{0, \dots, l\}$ where $l < k$. But $S \setminus V$ is a mcs, so that $\prod_{j=0}^l b_j \notin V$, which implies that $b_{l+1} \notin V$ and then $c = b_k = 1 + \prod_{j=0}^{k-1} b_j \notin V$.

Now, let $W \in [R, S]$ be such that $W \subseteq S$ is Prüfer–Manis with $b \notin W$. We follow the proof of [21, Remark, page 173]. We consider the previous sequence $\{b_j\}$ with $b_0 := b$ and $b_j = 1 + \prod_{i=0,}^{j-1} b_i$. We still have $c \in T \cap (\cap_{i\in\mathbb{N}_n} V_i)$. Obviously, since $b \notin W$, so is any b_i, and then $b_k = c \notin W$ because $S \setminus W$ is a mcs. □

Theorem 5.2 *Let $R, B_1, \dots, B_n$ be subrings of a ring S, where n is a positive integer, $n > 1$. If the B_is are integrally closed in S, except at most two of them, and $R \subseteq B_1 \cup \dots \cup B_n$, then R is contained in some of the subrings B_i.*

Proof First, we may remark that $R \subseteq B_1 \cup B_2$ implies that R is contained in one of the subrings B_1, B_2 by an obvious property of additive subgroups. So, we may assume that $n \geq 3$ with B_i integrally closed in S for any $i \geq 3$. There is no harm to assume that n is the least integer such that $R \subseteq B_1 \cup \cdots \cup B_n$, that is $R \not\subseteq \cup_{i \in \mathbb{N}_n, i \neq j} B_i$ for each $j \in \mathbb{N}_n$ (*).

To prove the theorem, it is enough to show that if R is not contained in any of the subrings B_i, we get a contradiction, that is $R \not\subseteq B_1 \cup \cdots \cup B_n$, or equivalently, there exists some $x \in R \setminus (B_1 \cup \cdots \cup B_n)$. This x is gotten after five steps.

Step 1 Assume that $R \subseteq B_1 \cup \cdots \cup B_n$ with R not contained in any of the subrings B_i. According to (*), for any $j \in \mathbb{N}_n$, there exists $a_j \in (R \cap B_j) \setminus (\cup_{i \in \mathbb{N}_n, i \neq j} B_i)$.

Fix some $i \in \mathbb{N}_n$, $i \neq j$, $i > 2$. Since $B_i \subseteq S$ is integrally closed, by Theorem 4.7, there exists a family $\{V_{k,i}\} \subseteq [B_i, S]$ such that $V_{k,i} \subseteq S$ is Prüfer–Manis, with $B_i = \cap V_{k,i}$. Let $v_{k,i}$ be the Manis valuation associated to $V_{k,i}$. As $a_j \notin B_i$, there exists some $V_{j,i}$ such that $a_j \notin V_{j,i}$. Moreover, $a_j \in V_{k,j}$ for any k if $j \geq 2$.

Set $\mathcal{M} := \{V_{j,i} \mid i > 2, i \neq j\}$. Then, $B_3 \cup \cdots \cup B_n \subseteq \cup_{i>2, i \neq j} V_{j,i} = \cup_{V_{j,i} \in \mathcal{M}} V_{j,i}$. For each a_k, set $\mathcal{M}^{(k)} := \{V_{j,i} \in \mathcal{M} \mid a_k \in V_{j,i}\}$ so that $V_{k,j} \in \mathcal{M}^{(j)}$ for any k if $j \geq 2$.

Step 2. If $\mathcal{M}^{(k)} \neq \emptyset$, then $a_k \in [R \cap (\cap_{V_{j,i} \in \mathcal{M}^{(k)}} V_{j,i})] \setminus V_{k,i}$. It follows from Lemma 5.1 that there exists $c_k \in R$ such that $v_{j,i}(c_k) = 0$ for any $V_{j,i} \in \mathcal{M}^{(k)}$ and $c_k \notin V_{k,i}$. In particular, $c_k \in V_{j,i}$ for any $V_{j,i} \in \mathcal{M}^{(k)}$.

If $\mathcal{M}^{(k)} = \emptyset$, set $c_k := a_k \in R$. Since $a_k \notin V_{j,i}$ for any $V_{j,i} \in \mathcal{M}$, it follows that $c_k \notin V_{j,i}$ for any $V_{j,i} \in \mathcal{M}$.

Step 3 Set $d_0 := \prod_{k=1}^n c_k$. Then, $d_0 \in R$. We claim that $d_0 \notin V$, for any $V \in \mathcal{M}$. Let $V \in \mathcal{M}$. Then, there exist i_0, j_0, $i_0 > 2, i_0 \neq j_0$ such that $V = V_{j_0,i_0}$, so that $a_{j_0} \notin V$. Whatever is $\mathcal{M}^{(j_0)}$, we have that $c_{j_0} \notin V$. It is obvious if $\mathcal{M}^{(j_0)} \neq \emptyset$. If $\mathcal{M}^{(j_0)} = \emptyset$, then $c_{j_0} \notin V_{j,i}$ for any $V_{j,i} \in \mathcal{M}$. In particular, $c_{j_0} \notin V$. It follows that $v_{j_0,i_0}(c_{j_0}) < 0$.

Consider c_k for some $k \neq j_0$. If $c_k \notin V$, then $v_{j_0,i_0}(c_k) < 0$. If $c_k \in V$, we cannot have $\mathcal{M}^{(k)} = \emptyset$, so that $\mathcal{M}^{(k)} \neq \emptyset$. If $V \in \mathcal{M}^{(k)}$, then $v_{j_0,i_0}(c_k) = 0$ and $a_k \in V_{j_0,i_0}$. If $V \notin \mathcal{M}^{(k)}$, then $a_k \notin V_{j_0,i_0}$ and $c_k \notin V_{j_0,i_0}$ by Lemma 5.1, which leads to $v_{j_0,i_0}(c_k) < 0$. In any case, $v_{j_0,i_0}(c_k) \leq 0$.

To conclude $v_{j_0,i_0}(d_0) = \sum_{k=1}^n v_{j_0,i_0}(c_k) \leq v_{j_0,i_0}(c_{j_0}) < 0$. This implies that $d_0 \notin V$ for any $V \in \mathcal{M}$, and then $d_0 \notin B_3 \cup \cdots \cup B_n$.

Set $\mathcal{M}_0 := \{V_{1,i} \mid i > 2\} \cup \{V_{2,i} \mid i > 2\} \cup \{V_{3,i} \mid i > 3\}$, with $\{V_{3,i} \mid i > 3\} = \emptyset$ if $n = 3$. Obviously, $\mathcal{M}_0 \subseteq \mathcal{M}$, so that $d_0 \notin V_{j,i}$ for any $V_{j,i} \in \mathcal{M}_0$.

Let $t_1, t_2 \in \mathbb{N}$, $t_1 \neq t_2$. We claim that $v_{j,i}(d_0^{t_1}) \neq v_{j,i}(d_0^{t_2})$ for any $V_{j,i} \in \mathcal{M}_0$. Assume that $t_1 > t_2$ and set $t := t_1 - t_2$, that is, $t_1 = t + t_2$. It follows that $d_0^{t_1} = d_0^{t_2} d_0^t$, so that $v_{j,i}(d_0^{t_1}) = v_{j,i}(d_0^{t_2}) + v_{j,i}(d_0^t)$. Now, $v_{j,i}(d_0^{t_1}) = v_{j,i}(d_0^{t_2})$

implies $v_{j,i}(d_0^t) = 0$, that is, $d_0^t \in V_{j,i}$. But $V_{j,i} \subseteq S$ is Prüfer–Manis, and then integrally closed, so that $d_0 \in V_{j,i}$, a contradiction. Then, $v_{j,i}(d_0^{t_1}) \neq v_{j,i}(d_0^{t_2})$.

Let $l \in \{1, 2, 3\}$ and consider the corresponding a_l defined at the beginning of the proof. Then, there exists at most one $t_{j,i,l} \in \mathbb{N}$ such that $v_{j,i}(a_l) = v_{j,i}(d_0^{t_{j,i,l}})$. If there does not exist such $t_{j,i,l}$, we have $v_{j,i}(a_l) \neq v_{j,i}(d_0)$. In this case, set $t_{j,i,l} = 1$. It follows that in any case and for any $t > t_{j,i,l}$, we have $v_{j,i}(a_l) \neq v_{j,i}(d_0^t)$. Let

$$t_0 := \sup\{1 + t_{j,i,l} \mid j, l \in \{1, 2, 3\}, i \in \{3, \ldots, n\}, i > j\}.$$

Then, $v(d_0^{t_0}) \neq v(a_l)$ for any $V \in \mathcal{M}_0$ and any $l \in \{1, 2, 3\}$.

Step 4 Set $d := d_0^{t_0}$. Then, $v(d) \neq v(a_l)$ for any $V \in \mathcal{M}_0$ (**). Moreover, for any $V \in \mathcal{M}$, we have $d_0 \notin V$, which implies $d \notin V$ since $V \subset S$ is integrally closed. In particular, $d \notin B_3 \cup \cdots \cup B_n$, but $d_0 \in R$ implies $d \in R \subseteq B_1 \cup B_2 \cup B_3 \cup \cdots \cup B_n$, so that $d \in B_1 \cup B_2$. Now, $B_1 \cup B_2 = (B_1 \cap B_2) \cup (B_1 \setminus B_2) \cup (B_2 \setminus B_1)$.

Step 5 We are going to consider the three possible cases for d.

(1) $d \in B_1 \cap B_2$.

Set $x := a_3 + d \in R$. Since $a_3 \in (R \cap B_3) \setminus (\cup_{i \in \mathbb{N}_n, i \neq 3} B_i)$, we have $a_3 \notin B_1 \cup B_2$, so that $x \notin B_1 \cup B_2$. Moreover, $d \notin B_3$, which implies that $x \notin B_3$. Let $i > 3$.

If $v_{3,i}(d) < v_{3,i}(a_3)$, then $v_{3,i}(x) = v_{3,i}(a_3 + d) = v_{3,i}(d) < 0$ because $d \notin V_{3,i}$. Then, $x \notin B_i$.

If $v_{3,i}(d) \geq v_{3,i}(a_3)$, then $v_{3,i}(d) > v_{3,i}(a_3)$ by (**), so that $v_{3,i}(x) = v_{3,i}(a_3) < 0$ because $a_3 \notin V_{3,i}$. Then, $x \notin B_i$.

It follows that $x \notin B_1 \cup B_2 \cup B_3 \cup \cdots \cup B_n$, a contradiction.

(2) $d \in B_1 \setminus B_2$.

Set $x := a_2 + d \in R$. Since $a_2 \notin B_1$, we have $x \notin B_1$, and since $a_2 \in B_2$, this implies that $x \notin B_2$, so that $x \notin B_1 \cup B_2$. Let $i > 2$.

If $v_{2,i}(d) < v_{2,i}(a_2)$, then $v_{2,i}(x) = v_{2,i}(d) < 0$ because $d \notin V_{2,i}$. Then, $x \notin B_i$.

If $v_{2,i}(d) \geq v_{2,i}(a_2)$, then $v_{2,i}(d) > v_{2,i}(a_2)$ by (**), so that $v_{2,i}(x) = v_{2,i}(a_2) < 0$ because $a_2 \notin V_{2,i}$. Then, $x \notin B_i$.

It follows that $x \notin B_1 \cup B_2 \cup B_3 \cup \cdots \cup B_n$, a contradiction.

(3) $d \in B_2 \setminus B_1$.

The proof is similar as in (2) by changing B_1 and B_2.

To conclude, we get a contradiction in any case, so that there exists some i such that $R \subseteq B_i$. □

Proposition 5.3 *Let $R \subseteq S$ be a Prüfer extension and $U, B_1, \ldots, B_n \in [R, S]$ such that $B_1 \cap \cdots \cap B_n \subset U$ and $U \subseteq S$ is Prüfer–Manis. Then there is some i such that $B_i \subseteq U$.*

Proof Actually, this result is given under an equivalent form in [19, Theorem 1.4, p.4]. Let $B_1, \ldots, B_n \in [R, S]$ be such that $B_1 \cap \cdots \cap B_n \subset U$. Then, we have $U = U(B_1 \cap \cdots \cap B_n) = UB_1 \cap \cdots \cap UB_n$ by Knebusch and Zhang [20, Theorem 1.4(4), p.86-87]. Now $[U, S]$ is a chain [20, Theorem 3.1, p. 187], so that $U = UB_i$ for some i and then $B_i \subseteq U$. □

Note that if the extension $R \subseteq S$ is Prüfer–Manis, so is $U \subseteq S$ for any $U \in [R, S]$ [20, Corollary 3.2, P. 187].

Gotlieb proved the following result for a ring extension $R \subset K$, where R is an integral domain with quotient field K [16, Theorem 6].

Theorem 5.4 *Let $R \subset S$ be an extension and $T, T_1, \ldots, T_n \in [R, S]$, such that $T = R_\Sigma$, where Σ is a mcs of R and $R \to T_i$ is a flat epimorphism for $i = 1, \ldots, n$ such that $T \subseteq T_1 \cup \cdots \cup T_n$. Then T is contained in some T_i.*

Proof Assume that T is not contained in any T_i. By the ($\mathcal{X}$)-rule, there are prime ideals $P_i \in \mathcal{X}(T_i) \setminus \mathcal{X}(T)$ and $\mathcal{X}(T) = \{P \in \mathrm{Spec}(R) \mid P \cap \Sigma = \emptyset\}$. We set $I := P_1 \cap \cdots \cap P_n$. We deduce from $T \subseteq T_1 \cup \cdots \cup T_n$ that $I \cap \Sigma = \emptyset$. There exists some prime ideal P of R such that $I \subseteq P$ and $P \cap \Sigma = \emptyset$. Then some P_i is contained in P: so that $P_i \in \mathcal{X}(T)$. Hence, we get a contradiction. □

6 Pullback Results

Consider the following pullback diagram (D) in the category of commutative unital rings:

$$\begin{array}{ccc} R & \xrightarrow{i} & S \\ {\scriptstyle f}\downarrow & & \downarrow{\scriptstyle g} \\ V & \xrightarrow{j} & K \end{array}$$

where i and j are ring extensions. It can be considered as a composite of the two diagrams:

$$\mathrm{Ker}(D)\colon \begin{array}{ccc} R & \longrightarrow & S \\ \downarrow & & \downarrow \\ R/\mathrm{Ker}(f) & \longrightarrow & S/\mathrm{Ker}(g) \end{array} \quad \text{and } \mathrm{Im}(D)\colon \begin{array}{ccc} f(R) & \longrightarrow & g(S) \\ \downarrow & & \downarrow \\ V & \longrightarrow & K \end{array}$$

The first diagram is a pullback because $\mathrm{Ker}(f) = \mathrm{Ker}(g)$ thanks to the pullback diagram (D). It follows that $R = f(R) \times_{g(S)} S$. It is of the form

$$\begin{array}{ccc} A & \longrightarrow & B \\ \downarrow & & \downarrow \\ A/I & \longrightarrow & B/I \end{array}$$

where I is an ideal shared by the rings A and B. We recall that in this case, $A \subseteq B$ is Prüfer if and only if $A/I \subseteq B/I$ is Prüfer ([20, Proposition 5.8, p.52]).

It is easy to prove that the second diagram is a pullback and is such that $f(R)$ is isomorphic to $V \cap g(S)$.

Recall that a ring R is called *semi-hereditary* if each of its finitely generated ideals is a projective R-module.

Olivier proved that an extension of rings $R \subset S$ is integrally closed if and only if there is a pullback diagram (D), where V is a semi-hereditary ring with a (absolutely flat) total quotient ring K [25, Corollary p.56] or [24, Théorème de Ker Chalon (2.1)]. In this case, we call (DO) the diagram (D). Therefore, the Prüfer property is not descended in pullbacks, since $V \subset K$ is Prüfer [11, Theorem 2] and there are integrally closed extensions that are not Prüfer.

On the other hand, we have a pullback example provided by the following result.

Proposition 6.1 ([9, Theorem 6.8 and Theorem 6.10]) *If R is a local ring, an extension $R \subseteq S$ is Prüfer if and only if there exists $P \in \mathrm{Spec}(R)$ such that $S = R_P$, $P = SP$, and R/P is a valuation domain. Under these conditions, S/P is the quotient field of R/P and P is a divided prime ideal of R (i.e., comparable to each ideal of R). In particular, $[R, S]$ is a chain.*

Proof To complete the proof, observe that there is an order isomorphism $[R, S] \to [R/P, S/P]$ given by $T \mapsto T/P$ for $T \in [R, S]$. □

We now use Olivier's result to find a characterization of Prüfer extensions.

Theorem 6.2 *Let $R \subset S$ be an integrally closed extension and (DO) the pullback diagram where V is semi-hereditary with total quotient ring K. Then, $R \subset S$ is Prüfer if and only if $g(T)V \cap g(S) = g(T)$ for each $T \in [R, S]$ or equivalently, the following diagram (D_T) is a pullback, for each $T \in [R, S]$:*

$$\begin{array}{ccc} T & \longrightarrow & S \\ \downarrow & & \downarrow \\ g(T)V & \longrightarrow & K \end{array}$$

In that case, we have $R = V \times_{g(T)V} T$ and $g(T)V \cong V \otimes_R T$.

Proof We use the characterization of Prüfer extensions by normal pairs and flat epimorphisms. Suppose that (D_T) is a pullback. Since an overring of a semi-hereditary ring is semi-hereditary ([2, Corollary p.143]), Olivier's result implies that $T \subset S$ is integrally closed. Hence, $R \subset S$ is Prüfer. We now prove the converse. Suppose that $R \subset S$ is Prüfer. Then $R \subset T$ is a flat epimorphism. Tensoring the diagram (D) by $\otimes_R T$, we get another pullback diagram because the pullback R is a kernel of a morphism of R-modules and T is flat over R. We next identify the rings of the new pullback. We have clearly $T \cong R \otimes_R T$. Moreover, we also have $S \otimes_R T \cong S$. This is a consequence of [35, Satz 2.2 (d)] that states that if M is a T-module and $R \to T$ an epimorphism, then $M \otimes_R T \cong M$ (an isomorphism of T-modules). We next show that $V \otimes_R T \cong g(T)V$. Consider the natural map $V \otimes_R T \to K$; its image is $g(T)V$. Then $V \to V \otimes_R T$ is a flat epimorphism deduced from $R \to T$ by the base change $R \to V$ and $V \to V \otimes_R T \to g(T)V$ is injective. It follows that $V \otimes_R T \to g(T)V$ is an isomorphism because a flat epimorphism is essential by Lazard [22, Lemme 1.2, p.109]. Then we show that $K \otimes_R T \cong K$. We first observe that $K \to K \otimes_R T$ is a flat epimorphism whose domain is an absolutely flat ring. This map is surjective. To see this, if J is the kernel of the morphism, then $K/J \to K \otimes_R T$ is a faithfully flat epimorphism because K/J is absolutely flat whence is an isomorphism by Lazard [22, Lemme 1.2, p.109]. Moreover, $V \to V \otimes_R T$ identifies to $V \to g(T)V$, whence is injective. As $V \to K$ is flat, the map $K \to (V \otimes_R T) \otimes_V K \cong K \otimes_R T$ is injective, so that $K \cong K \otimes_R T$.

Therefore, we have proved that there is a pullback diagram (D_T). To complete the proof, it is enough to consider $\mathrm{Im}(D_T)$, in which case the pullback condition on T can be written $g(T)V \cap g(S) = g(T)$. □

Nevertheless, we give some examples of pullbacks where the ascent property holds.

Proposition 6.3 *Let I be an ideal of a ring S and set $S' = S/I$. Denote by φ the canonical map $S \to S/I$. Let R' be a subring of S' and R the pullback ring in the following diagram:*

$$\begin{array}{ccc} R & \longrightarrow & S \\ \downarrow & & \downarrow \\ R' & \longrightarrow & S' \end{array}.$$

Then $R \subset S$ is Prüfer if and only if $R' \subset S'$ is Prüfer.

Proof Clearly, I is an ideal shared by R and S. Now, observe that R' identifies to $(R+I)/I \cong R/(I \cap R) \cong R/I$. It is then enough to apply [20, Proposition 5.8, p.52]. □

7 The Case of a Local Base Ring

When the base ring R is local, we already gave a characterization of Prüfer extensions in Proposition 6.1.

Definition 7.1 An extension $R \subseteq S$ is called *module distributive* if $R \cap (X+Y) = (R \cap X) + (R \cap Y)$ for each pair of R-submodules (X, Y) of S (cf. [20, p.119]). We say that $R \subseteq S$ is distributive if the lattice $[R, S]$ endowed with compositum and intersection as laws is distributive.

Knebusch and Zhang [20, Theorem 5.4, p.121] show that an extension $R \subseteq S$ is module distributive if and only if it is Prüfer. As a consequence, we get that the set of

R_P-submodules of S_P/R_P is a chain [12, Corollary 2], when the extension is Prüfer. This gives a stronger result than that of Proposition 6.1. Moreover, we see that a Prüfer extension is both module distributive and distributive. For the distributivity, use [30, Proposition 5.18] since a Prüfer extension $R \subset S$ is arithmetical (that is $R_M \subset S_M$ is chained for any $M \in \mathrm{Max}(R)$).

In order to get more results, we introduce the following considerations.

In view of [20, Proposition 5.2, p.119], an ideal I of a ring R is called *distributive* if $I + (J \cap K) = (I + J) \cap (I + K)$ for all ideals J, K of R. When R is local, an ideal I is distributive if and only if I is comparable to each ideal (principal) of R. In this case (R is local), we will call I a *strong divisor* if in addition $0 : I = 0$. The following result will be useful.

Proposition 7.2 ([20, Example 5.1, p.119]) *Let $R \subset S$ be a Prüfer extension. An S-regular ideal I of R is distributive. In particular, such an ideal is a strong divisor if R is local.*

We can translate some results of [12, Lemma 1.1, Corollary] as follows. Let R be a ring and set $\Sigma := \{\sigma \in R \setminus \mathrm{Z}(R) \mid R\sigma \text{ is distributive}\}$. Then Σ is a saturated mcs of R. Moreover, if $\mathcal{T}$ is a mcs of R, such that $\mathcal{T} \subseteq R \setminus \mathrm{Z}(R)$ and $R \subseteq R_{\mathcal{T}}$ is Prüfer, then $\mathcal{T} \subseteq \Sigma$.

We next examine the local case. We may find in [20, p.123] the following definition and result.

Definition 7.3 A *strong divisor t* of a local ring R is an element t of R, such that the ideal Rt is a strong divisor. The set $\Delta(R)$ of all strong divisors of R is a saturated mcs of R and $\mathrm{U}(R) \subseteq \Delta(R)$.

We observe that for $t \in \Delta(R)$, the open subset $\mathrm{D}(t) = \{Q \in \mathrm{Spec}(R) \mid Q \subset Rt\}$.

Recall that a ring R has a maximal Prüfer extension $R \subseteq \mathbb{P}(R) := \widetilde{R}^{\mathrm{Q}(R)}$, where $\mathrm{Q}(R)$ is the complete ring of quotients of R (Utumi–Lambeck) [20]. Then $\mathbb{P}(R)$ is called the *Prüfer hull of R*.

It is known that a Prüfer extension $R \subset S$, where R is local, is a QR extension; that is, is such that each $T \in [R, S]$ satisfies $T \cong R_\Sigma$ (an isomorphism of R-algebras) for some mcs Σ of R. For more information, see Proposition 8.3. Next result refines this observation.

Proposition 7.4 *Let R be a local ring and $R \subset S$ an extension:*

(1) $\mathbb{P}(R) = R_{\Delta(R)}$.
(2) *An extension $R \subset S$ is Prüfer if and only if $S = R_\Sigma$, for some mcs $\Sigma \subseteq \Delta(R)$ and, if and only if $R \subseteq Rs$ (i.e., s^{-1} exists and belongs to R), for each $s \in S \setminus R$. In this case, $S \subseteq \mathrm{Tot}(R)$.*

Proof The proof is a consequence of the following facts: $R_{\Delta(R)}$ is the Prüfer hull of R. If $R \subset S$ is Prüfer, there is some mcs $\Sigma \subseteq \Delta(R)$ such that $S = R_\Sigma$ in which case $R \subseteq R_\Sigma \subseteq R_{\Delta(R)}$ [20, Remark 5.9, Proposition 5.10 p.123]. The last assertion is [6, Proposition 3.1]. □

Lemma 7.5 *Let $R \subset R_\Sigma := S$ be an extension of finite type, where Σ is a mcs of R. Then there is some $x \in \Sigma$ such that $S = R_x$.*

Proof It follows from [4, Theorem 1.1] that $R \subset R_\Sigma := S$ is of finite presentation because it is an injective flat epimorphism of finite type. Therefore, according to the $(\mathcal{MCS})$-rule, $\mathcal{X}(S) = \cap\{\mathrm{D}(r) | r \in \Sigma\}$ is an open subset of the patch topology (constructible topology) by the Chevalley Theorem and is even open because a flat morphism of finite presentation is open for the Zariski topology. As the patch topology is compact and the sets $\mathrm{D}(r)$ for $r \in \Sigma$ are closed in this topology, we get that $\mathcal{X}(S)$ is the intersection of finitely many $\mathrm{D}(r_i)$ for $i = 1, \ldots, n$ with $r_i \in \Sigma$. Setting $x = r_1 \cdots r_n$, we get that $\mathcal{X}(S) = \mathrm{D}(x)$ and then $S = R_x$ by the $(\mathcal{X})$-rule. □

The next result is now clear.

Proposition 7.6 *An extension $R \subset S$ of finite type over a local ring R is Prüfer if and only if there is some $s \in \Delta(R)$ such that $S = R_s$.*

Proof Suppose that $R \subset S$ is Prüfer, then $S = R_\Sigma$ for some mcs $\Sigma \subseteq \Delta(R)$ (Proposition 7.4). We deduce from Lemma 7.5 that $S = R_s$ for some $s \in \Delta(R)$. The converse is obvious. □

The following results will be useful.

Proposition 7.7 *Let R be a local ring and $x \in R$ a regular element. Then x is a strong divisor if and only if $R \subseteq R_x$ is Prüfer.*

Proof Proposition 7.6 gives one implication. Suppose that $R \subset R_x$ is Prüfer. From Proposition 7.6, we deduce that $R_x = R_s$ for some strong divisor $s \in R$. It follows that $\sqrt{Rx} = \sqrt{Rs}$ and then $s^n = yx$ for some $n \in \mathbb{N}$ and $y \in R$. Therefore, x is a strong divisor. □

Example 7.8 Let R be a local arithmetical ring. The set of all its ideals is a chain. It follows that each regular element x of R is a strong divisor and then $R \subseteq R_x$ is Prüfer.

Proposition 7.9 *Let $f : R \to R'$ be a faithfully flat ring morphism between local rings and $x \in R$. If $f(x)$ is a strong divisor, so is x.*

Proof Observe that x is regular in R. To conclude, use Proposition 2.5, because $R'_{f(x)} = R_x \otimes_R R'$. □

Let $R \subseteq S$ be an extension and Δ a mcs of R. The large quotient ring $R_{[\Delta]}$ of R (in S) with respect to Δ is the set of all $x \in S$ such that there is some $s \in \Delta$ with $sx \in R$. In case $\Delta = R \setminus P$, where P is a prime ideal of R, we set $R_{[P]} := R_\Delta$.

Proposition 7.10 *Let R be a local ring and $R \subset S$ a flat extension; then $\widetilde{R} = R_{[\Delta(R)]} = R_\Sigma$, where $\Sigma := \Delta(R) \cap \mathrm{U}(S)$.*

Proof By Proposition 7.4, there is some multiplicatively closed subset Θ of $\Delta(R)$ such that $\widetilde{R} = R_\Theta$. We have clearly $\Theta \subseteq \mathrm{U}(S)$, so that $\Theta \subseteq \Sigma$. It follows that $R_\Theta \subseteq R_\Sigma$, while $R \subseteq R_\Sigma$ is Prüfer and therefore $R_\Theta = R_\Sigma$.

Now let $z \in R_{[\Delta(R)]}$; there is some $t \in \Delta(R)$ such that (*): $x = tz \in R$. Since Rt is a strong divisor, Rt and Rx are comparable. Moreover, since $R \subset S$ is flat, t is also regular in S.

If $Rx \subseteq Rt$, then $x = at$, so that $z = a \in R$, because t is regular.

If $Rt \subseteq Rx$, then $t = bx$, and since $\Delta(R)$ is saturated, we get that $x \in \Delta(R)$ and x is regular. We deduce from (*) that $bz = 1$ in S. It follows that $z \in \mathrm{U}(S)$ and $z = b^{-1}$, with $b \in \Delta(R) \cap \mathrm{U}(S)$, so that $z \in R_\Sigma$.

To conclude, we have $R_{[\Delta(R)]} \subseteq R_\Sigma$. As the reverse inclusion is obvious, we get finally that $R_{[\Delta(R)]} = R_\Sigma$. □

If Q is a prime ideal of a ring R, we denote by $Q^\downarrow$ its generization i.e., $\{P \in \mathrm{Spec}(R) \mid P \subseteq Q\}$. The first author defined a *prime g-ideal* as a prime ideal Q such that $Q^\downarrow$ is an open subset of $\mathrm{Spec}(R)$ [26]. If Q is a g-ideal of R, then Q is a Goldman ideal of R; that is $R/P \subseteq \kappa(P)$ is of finite type as an algebra [26].

Proposition 7.11 *Let s be a non-unit strong divisor of a local ring R and $R \subset S := R_s$ the Prüfer extension associated. Then $P = \cap\{Rs^n \mid n \in \mathbb{N}\}$ is a prime g-ideal, $S = R_P$, $PS = P$ is a divided prime ideal of R and R/P is a valuation domain with quotient field S/P. We will denote by P_s the ideal P.*

Proof There exists $P \in \mathrm{Spec}(R)$ such that $S = R_P$, $PS = P$ is a divided prime ideal of R, and R/P is a valuation domain with quotient field S/P according to Proposition 6.1. Note $\langle s \rangle$ the saturated mcs generated by s and set $I := \cap\{Rs^n \mid n \in \mathbb{N}\}$. Remark that $\langle s \rangle = \{us^n \mid n \in \mathbb{N},\ u \in \mathrm{U}(R)\}$. Note that $P = R \setminus \langle s \rangle$. We are aiming to show that $I = P$. We have $I \subseteq P$, because if not, there is some $x \in I \cap \langle s \rangle$ and then $xy = s^p = bs^{p+1}$ for some $y, b \in R$ and $p \in \mathbb{N}$. Since s is regular, it follows that s is a unit, a contradiction. Now let $x \in P$ and suppose that $x \notin I$. Then $x \notin Rs^n$ for some positive integer n. Because s^n is a strong divisor, we get $Rs^n \subseteq Rx$, and then x belongs to $\langle s \rangle$, a contradiction. Now P is a prime g-ideal because $P^\downarrow = \mathrm{D}(s)$ is an open subset. □

If $R \subset S$ is a Prüfer extension of finite type over a local ring, there is some $s \in \Delta(R)$ such that $S = R_s$ by Proposition 7.6.

Remark 7.12 We use the notation of Proposition 7.11.

It is easy to prove that $P = RsP$ because s is regular and $P = \cap\{Rs^n \mid n \in \mathbb{N}\}$. Therefore, if (R, M) is Noetherian and local and s is not a unit, from $P = MP$, we deduce that $P = 0$ and R needs to be an integral domain, so that R is a Noetherian valuation domain, that is a discrete valuation domain, and S is the quotient field of R. Another consequence is that if R is not an integral domain, the only strong divisors of R are the units.

The next result is now clear.

Proposition 7.13 *Let $R \subset S$ be an extension over a local ring. The set of rings $\mathcal{F} := \{R_s \in [R, S] \mid s \in \Delta(R) \cap \mathrm{U}(S)\}$ is a chain and $\widetilde{R}$ is the set union of all elements of $\mathcal{F}$. It follows that $\widetilde{R} \hookrightarrow \mathrm{Tot}(R)$.*

Definition 7.14 We say that two ideals I and J of a ring R are *equivalent* if $\sqrt{I} = \sqrt{J}$ (equivalently $\mathrm{D}(I) = \mathrm{D}(J)$). We also say that two elements x, y of R are equivalent if $\mathrm{D}(x) = \mathrm{D}(y)$, and we write $x \simeq y$. This condition is equivalent to $R_x \cong R_y$ and also to $\sqrt{Rx} = \sqrt{Ry}$. Note that if x is a strong divisor and $x \simeq y$, then y is a strong divisor because $\sqrt{Rx} = \sqrt{Ry}$ and the set of all strong divisors is a saturated mcs.

Remark 7.15 We reconsider the context of Proposition 7.11, and we set $\delta(R) := \Delta(R) \setminus \mathrm{U}(R)$. There is a surjective map $\delta(R) \to \{P_s \mid s \in \delta(R)\}$, defined by $s \mapsto P_s$. Setting $\Delta\mathrm{Spec}(R) := \{P_s \mid s \in \delta(R)\}$, there is therefore a bijective map $(\delta(R)/\simeq) \to \Delta\mathrm{Spec}(R)$.

Then $\Delta\mathrm{Spec}(R)$ is a chain. It follows that the set intersection of all its elements is a prime ideal $\mathcal{R}$ that could be called the *strong radical* of the local ring R. Now, according to Proposition 7.13 and the ($\mathcal{MCS}$)-rule, $\mathcal{X}(\mathbb{P}(R)) = \cap\{\mathrm{D}(s) \mid s \in \Delta(R)\} = \cap\{P_s^{\downarrow} \mid s \in \delta(R)\} = \mathcal{R}^{\downarrow}$. It follows that $\mathbb{P}(R) = R_{\mathcal{R}}$. If $R \subset S$ is a ring extension, then $\widetilde{R} = \mathbb{P}(R) \cap S = R_{[\mathcal{R}]}$.

We think that the set $\Delta\mathrm{Spec}(R)$ deserves a deeper study, especially with respect to some classes of rings.

Proposition 7.16 *Let $R \subset S$ be a Prüfer extension, where R is local and I an ideal of finite type of R. Then I is S-regular if and only if $I = R\rho$ where ρ is a strong divisor of R, invertible in S.*

Proof Assume that I is S-regular. From $IS = S$, we deduce that I is a principal ideal $R\rho$ by Knebusch and Zhang [20, Theorem 1.13, p. 91 and Proposition 2.3, p.97], because $IS = S$ means that I is S-regular and, R being local, is S-invertible, whence principal of the form $I = R\rho$. An appeal to Proposition 7.2 yields that $R\rho$ is a strong divisor and $S\rho = S$ shows that ρ is invertible in S. The converse is obvious. □

8 QR Extensions

We first give some notation and definitions for an extension $R \subset S$. For $T \in [R, S]$, we set $\Sigma_T := \mathrm{U}(T) \cap R$, which is a mcs of R whose elements are regular and such that $R \subseteq R_{\Sigma_T} \subseteq T$.

A Prüfer extension $R \subseteq S$ is called *Bezout*, if each finitely generated S-regular ideal of R is principal [20, Definition 1; Theorem 10.2, p.145].

Let (R, M) be a local ring, then an extension $R \subseteq S$ is Bezout if and only if it is Prüfer, and if and only if (R, M) is Manis in S [20, Scholium 10.4 p.147].

We call a QR extension any extension $R \subseteq S$ such that each $T \in [R, S]$ is of the form $T \cong R_\Sigma$ (an isomorphism of R-algebras) for some mcs Σ of R, in which case the elements of Σ are regular, invertible in S, and $T = R_{\Sigma_T}$. It is easy to show that $R \subseteq S$ is a QR extension if and only if the defining property holds for each $T \in [R, S]_{fg}$. Moreover, an extension $R \subset S$ is a QR extension if and only if it is Prüfer, and each finitely generated S-regular ideal I of R satisfies $\sqrt{I} = \sqrt{Rx}$ for some $x \in R$ (which implies that $\mathrm{D}(I) = \mathrm{D}(x)$ is (special) affine) [20, Proposition 4.15, p.116].

A Prüfer extension does not need to be a QR extension: look at the example [15, Section 4, Examples].

Proposition 8.1 *A Bezout extension $R \subseteq S$ is a QR extension.*

Proof We first observe that a subextension $R \subseteq T$ is Bezout. Then [20, Proposition 10.16, p.152] shows that $T = R_\Sigma$, for some mcs Σ of R and therefore the extension is QR. □

Corollary 8.2 *Each extension $R \subseteq S$ has a unique Bezout subextension $R \subseteq T$ that contains any $T' \in [R, S]$, such that $R \subseteq T'$ is Bezout. Then T is called the Bezout hull of R and denoted here by $\beta(R)$.*

Proof It is enough to use [20, Theorem 10.14, p.151]. □

Moreover, we have the next result.

Proposition 8.3 *Let $R \subset S$ be an extension where projective R-modules of rank one are free. Then $R \subset S$ is Prüfer if and only if it is a QR extension, and if and only if $R \subset S$ is Bezout. If the above statements hold, then a finitely generated S-regular ideal I of R is of the form $I = R\rho$, where ρ is a locally strong divisor.*

Proof The first equivalence is [20, Proposition 4.16 p.116]. The second is a consequence of [20, Proposition 2.3, p.97] because under the hypotheses on R, a Prüfer extension is Bezout and the converse holds for an arbitrary ring R. The last statement is a consequence of Proposition 7.16. □

The conditions on projective modules that are involved in this paper are either R is semilocal or a Nagata ring $A(X)$ [13]. In particular, we recover Proposition 6.1 in case R is a local ring.

We will need an extension of the notion of strong divisors. A regular element of a ring R is called a *locally strong divisor* (shorten to *lsd*) if $R \subseteq R_x$ is Prüfer. In order to justify this definition, we recall that an extension $R \subseteq S$ is Prüfer if and only if all its localizations by a prime ideal of R are Prüfer. Hence if $x \in R$ is an lsd and P is a prime ideal of R, then $x/1 \in R_P$ is a strong divisor. For the converse, use that if $x \in R$ is regular in every ring R_P, where P is a prime ideal, so is x because $R \to \prod\{R_M \mid M \in \mathrm{Max}(R)\}$ is injective. The set of all locally strong divisors is a saturated mcs $\Lambda\Delta$. Clearly, a strong divisor of a local ring is an lsd. Now if $R \subseteq S$ is a ring extension, we denote by $\lambda\delta(R)$ the ring $R_{\Lambda\Delta\cap\mathrm{U}(S)}$.

Remark 8.4 Let $f : R \to S$ be a ring morphism:

(1) If f is a flat morphism and $x \in R$ is such that $f(x)$ is an lsd, then so is x. Indeed for $Q \in \mathrm{Spec}(S)$ lying over P, then $R_P \to S_Q$ is faithfully flat.
(2) If f is a flat epimorphism and $x \in R$ is an lsd so is $f(x)$, because for each $Q \in \mathrm{Spec}(S)$ and $P := f^{-1}(Q)$, the natural map $R_P \to S_Q$ is an isomorphism.

Theorem 8.5 *Let $R \subset S$ be an extension. Then $R \subset S$ is a QR extension if and only if each $T \in [R, S]$ of finite type over R is of the form $T = R_s$ for some lsd $s \in R$. In particular, if these conditions hold, each $T \in [R, S]$ is of the form $T = R_{\mathcal{T}}$, where $\mathcal{T} \subseteq \Lambda\Delta$ is a mcs.*

Proof One implication is clear. Suppose that the extension is QR. To see that the condition holds, it is enough to suppose that it is of finite type. According to Lemma 7.5, there is some $s \in R$ such that $S = R_s$. Then $R \subseteq S$ is Prüfer, whence so is $R_P \subseteq S_P$ for each prime ideal P of R and $S_P = (R_P)_{s/1}$. We also have $S_P = (R_P)_y$, where $y \in \Delta(R_P)$ by Proposition 7.6. It follows that $\mathrm{D}(s/1) = \mathrm{D}(y)$, and by Definition 7.14, $s/1$ is a strong divisor. The last statement follows from the $(\mathcal{MCS})$-rule applied to the flat epimorphism $R \subseteq T$, since T is a union of finitely generated QR extensions. □

Theorem 8.6 *Any extension $R \subset S$ has a QR hull; that is, there exists a largest QR extension $\chi(R) \in [R, S]$, contained in $\widetilde{R}$. As a consequence, $\chi(R)$ is the compositum of all QR extensions in $[R, S]$.*

Proof Let X be the set of all QR extensions in $[R, S]_{fg}$, which is directed upward: take $T, U \in X$. They are of the form R_x and R_y, where x and y are regular in R, because they are units in S. Then we have $R_x, R_y \subseteq R_{xy}$. We can now use the proof of [5, Theorem 5] that holds for an arbitrary extension $R \subset S$ and show that $R_{xy} \in X$.

Denote by $\chi(R)$ the set union of the elements of X. Since a QR extension in $[R, S]$ is a union of finitely generated QR extensions, it is contained in $\chi(R)$. To complete the proof, observe that an element of $[R, \chi(R)]_{fg}$ is contained in an element of X, whence is in X, from which we infer that $R \subseteq \chi(R)$ is a QR extension. □

Actually, the proof of Davis shows that the set of all elements $x \in R$ such that $R \to R_x$ is a QR extension is a mcs $\Omega(R)$ (also denoted Ω) contained in the mcs $\Lambda\Delta$. Moreover, in case R is either local or a Nagata ring, projective R-modules of rank one are free, so that an extension $R \subset S$ is Prüfer if and only if it is a QR extension by Proposition 8.3, giving $\Omega \cap \mathrm{U}(S) = \Lambda\Delta \cap \mathrm{U}(S)$. An application of the $(\mathcal{X})$-rule gives the following result.

Corollary 8.7 *If $R \subset S$ is an extension, then $\chi(R) = R_{\Omega \cap \mathrm{U}(S)}$. It follows that an extension $R \subset S$ is a QR extension if and only if for each $s \in S$ there is some $\rho \in \Omega \cap \mathrm{U}(S)$ such that $\rho s \in R$.*

We remark that $\beta(R) \subseteq \chi(R) \subseteq \lambda\delta(R) \subseteq \widetilde{R}$.

Proposition 8.8 *An extension $R \subset S$ is a QR extension if and only if each S-regular finitely generated ideal is equivalent to a principal ideal of R and there exists a mcs $\Sigma \subseteq \Lambda\Delta \cap \mathrm{U}(S)$ such that $S = R_\Sigma$. If these conditions hold, then $S = R_{\Lambda\Delta\cap \mathrm{U}(S)}$.*

Proof Assume first that $R \subseteq S$ is a QR extension. By the results mentioned before Proposition 8.1, each S-regular finitely generated ideal is equivalent to a principal ideal of R. According to Theorem 8.5, there exists a mcs Σ, whose elements are some lsd of R, and such that $S = R_\Sigma$.

Conversely, assume that each S-regular finitely generated ideal is equivalent to a principal ideal of R and there exists a mcs Σ whose elements are some lsd of R, and such that $S = R_\Sigma$. Let $M \in \mathrm{Max}(R)$. It follows that $S_M = (R_\Sigma)_M = (R_M)_{\Sigma'}$, where Σ' is a mcs whose elements are some lsd of R_M. Then, Proposition 7.4 implies that $R_M \subseteq S_M$ is Prüfer. Since this holds for any $M \in \mathrm{Max}(R)$, we get that $R \subset S$ is Prüfer, and then a QR extension by the recall before Proposition 8.1.

If these conditions hold, set $\Sigma' := \Lambda\Delta \cap \mathrm{U}(S) \subseteq S$. Since Σ' is a mcs whose elements are units of S, it follows that $R_{\Sigma'} \subseteq S$. But $\Sigma \subseteq \Sigma'$ implies $S = R_\Sigma \subseteq R_{\Sigma'} \subseteq S$, so that $S = R_{\Sigma'}$. □

We end this section by considering ring extensions $R \subset S$ that are flat epimorphisms, such that the support $\mathrm{Supp}(S/R)$ of the R-module (S/R) is finite. We recall that $R \subset S$ is a flat epimorphism $\Leftrightarrow$ for all $P \in \mathrm{Spec}(R)$, either $R_P = S_P$ is an isomorphism or $S = PS$, these two conditions being mutually exclusive [22, Proposition 2.4, p.112].

It is known that the support $\mathrm{Supp}(S/R)$ of the R-module S/R is the set of all $P \in \mathrm{Spec}(R)$, such that $PS = S$. Therefore, each element of the support is S-regular. Moreover, the support is closed because as any support, it is stable under specialization. Hence, the support equals to $\mathrm{V}(J)$, where J is the intersection of all elements $P_1, \ldots, P_n$ of the support. Now each P_i is the radical of an S-regular finitely generated ideal, as an examination of the proof of [1, Corollary 13] by Abbas and Ayache shows. Moreover, assume that $R \subset S$ is a QR extension. Using [20, Proposition 4.15, p.116], we get that P_i is of the form $\sqrt{Rx_i}$ for some $x_i \in R$. Then $J = \sqrt{Rx}$, where $x = x_1 \cdots x_n$. Now if I is an S-regular finitely generated ideal and Q is a prime ideal of R containing I, then Q is S-regular. Reasoning as above, we see that $\sqrt{I} = \sqrt{Ry}$, for some $y \in R$. Taking into account the characterization of QR extensions at the beginning of the section, we see that we have proved the following result:

Proposition 8.9 *Let $R \subset S$ be a Prüfer extension where $\mathrm{Supp}(S/R)$ is finite (in particular, if $R \subset S$ has FCP). We set $J := \cap\{P \mid P \in \mathrm{Supp}(S/R)\}$. The following statements are equivalent:*

(1) *$R \subset S$ is a QR extension.*
(2) *Each element of $\mathrm{Supp}(S/R)$ is equivalent to a principal ideal.*
(3) *Each S-regular finitely generated ideal of R is equivalent to a principal ideal.*

In case one of the above statements holds, J is a S-regular ideal equivalent to a principal ideal Rx and $\Gamma(\mathrm{D}(J)) = R_x$.

Proof We only need to prove the following. By the flatness of the extension, $JS = \cap\{PS \mid P \in \mathrm{Supp}(S/R)\} = S$. □

Remark 8.10 Proposition 20 of [1] states that if, in addition to the above hypotheses, S is an integral domain, then each $T \in [R, S]$ is of the form R_x for some $x \in R$. This proves that the extension is strongly affine. Actually, in the proof of [1, Proposition 20], we can replace the Kaplansky transform of an ideal by a ring of sections.

9 Prüfer FCP Extensions over a Local Ring

Clearly, a minimal extension is a flat epimorphism if and only if it is Prüfer. So we call such extensions Prüfer minimal. We note as a first result the following Proposition, which results from Proposition 8.9.

Proposition 9.1 *A Prüfer minimal extension with crucial maximal ideal M is a QR extension if and only if M is equivalent to a principal ideal.*

Proposition 6.1 takes the following form, observing that a Prüfer extension is integrally closed.

Proposition 9.2 ([9, Theorem 6.8 and Theorem 6.10]) *If R is a local ring, an extension $R \subseteq S$ is Prüfer FCP (resp., minimal) if and only if there exists $P \in \mathrm{Spec}(R)$ such that $S = R_P$, $P = SP$ and R/P is a finite-dimensional (resp., one-dimensional) valuation domain. Under these conditions, S/P is the quotient field of R/P and P is a divided prime ideal of R.*

The conductor of a ring extension $R \subset S$ is denoted by $(R : S)$. The following Corollary recalls, for a Prüfer minimal extension $R \subset S$, the link between the crucial maximal ideal $\mathcal{C}(R, S)$ and $(R : S)$.

Corollary 9.3 *If $R \subset S$ is a Prüfer minimal extension with crucial maximal ideal M, then $P := (R : S)$ is a prime ideal of R, $P \subset M$, and there is no prime ideal of R contained strictly between P and M.*

Proof First, $P := (R : S)$ is a prime ideal of R by Ferrand and Olivier [14, Lemme 3.2]. Moreover, $P_M = PR_M = (R_M : S_M)$ with $R_M \neq S_M$ shows that $P \subset M$. At last, $R_M \subset S_M$ is also a Prüfer minimal extension. Then, according to Proposition 9.2, R_M/P_M is a one-dimensional valuation domain, so that there is no prime ideal of R_M contained strictly between P_M and MR_M giving that there is no prime ideal of R contained strictly between P and M. □

Corollary 9.4 *Let $R \subset S$ be a Prüfer minimal extension over a local ring (R, M). Then, with the notation of Proposition 9.2, each element $t \in M \setminus P$ is a strong divisor of R, $S \cong R_t$, $P = \cap\{Rt^n | n \in \mathbb{N}\}$, and $M = \sqrt{Rt}$.*

Proof Because $t \notin P$, we have $t \in \mathrm{U}(S)$ and then a factorization $R \subseteq R_t \subseteq S$, so that $S = R_t$ by minimality. By Proposition 7.6, t is a strong divisor. The third statement follows from Corollary 7.11. Because R/P is one-dimensional, M is the only prime ideal containing P, so that $M = \sqrt{Rt}$. □

Proposition 9.5 *Let $R \subset S$ be a ring extension where (R, M) is a local ring. The following statements are equivalent:*

(1) *$R \subset S$ is a Prüfer minimal extension.*
(2) *There is a strong divisor $a \in R \setminus \mathrm{U}(R)$ such that $S = R_a$ and $\sqrt{Ra} \subseteq \sqrt{Rb} \Rightarrow \sqrt{Ra} = \sqrt{Rb}$ (or equivalently, $\mathrm{D}(a) \subseteq \mathrm{D}(b) \Rightarrow \mathrm{D}(a) = \mathrm{D}(b)$) for each $b \in R \setminus \mathrm{U}(R)$.*
(3) *There is a strong divisor $a \in R \setminus \mathrm{U}(R)$ such that $S = R_a$ and $M = \sqrt{Ra}$.*
(4) *There is a strong divisor $a \in R \setminus \mathrm{U}(R)$, such that $S = R_a$, and such that $\mathrm{D}(a)$ is an open affine subset, maximal in the set of proper open affine subsets.*

Proof We clearly have (2) $\Leftrightarrow$ (3) by Corollary 9.4, once (1) $\Leftrightarrow$ (2) is proved. We then prove that (1) is equivalent to (2). Suppose that $R \subset S$ is a minimal extension, that is a flat epimorphism. Then $R \subset S$ is clearly a Prüfer extension. By Proposition 7.10, there is a mcs Σ of R, whose elements are strong divisors and such that the extension identifies with $R \subset R_\Sigma$. Picking an arbitrary element $a \in \Sigma$, we get a factorization $R \to R_a \to R_\Sigma$. Its factors are injective because the flat epimorphism $R \to R_a$ satisfies [22, Lemme 3.4, p.114].

As a is not invertible, $R \neq R_a$ implies $S = R_a$, by minimality of $R \subset S$. In the same way, a factorization $R \subset R_b \subseteq R_a$ implies $R_b = R_a$, or equivalently $\mathrm{D}(a) \subseteq \mathrm{D}(b) \Rightarrow \mathrm{D}(a) = \mathrm{D}(b)$, which means that $\sqrt{Ra} \subseteq \sqrt{Rb} \Rightarrow \sqrt{Ra} = \sqrt{Rb}$.

We prove the converse. Observe that for any mcs Σ of R and $a \in R$, such that there is a factorization $R \to R_\Sigma \to R_a$, we have $\mathrm{D}(a) \subseteq \cap\{\mathrm{D}(\sigma) \mid \sigma \in \Sigma\}$. Suppose that $R \subset S$ satisfies the conditions of the proposition. Then $S = R_a$, where $a \in R \setminus \mathrm{U}(R)$ is a strong divisor, and then $R \subset S$ is Prüfer, so that any subextension $R \subset T \subseteq S$ is Prüfer. By Proposition 7.10, we get that $T = R_\Sigma$, for some mcs Σ of R. The above observation shows that $\mathrm{D}(a) \subseteq \mathrm{D}(\sigma)$ for any $\sigma \in \Sigma$. The last condition entails that $\Sigma = \langle a \rangle$, and then $T = R_a$. Therefore, $R \subset S$ is minimal and a flat epimorphism.

Clearly, (4) implies (2). The converse is a consequence of the following facts. If $\mathrm{D}(a) \subseteq \mathrm{D}(I)$, where $\mathrm{D}(I)$ is an open affine subset different from $\mathrm{Spec}(R)$, we have a factorization $R \subset \Gamma(\mathrm{D}(I)) \subseteq R_a$ and $R \to \Gamma(\mathrm{D}(I))$ is an injective flat epimorphism whose spectral image is $\mathrm{D}(I)$. □

Lemma 9.6 *Let R be a ring and $a \in R$. Then, there exists some $M \in \mathrm{Max}(R)$ such that $M = \sqrt{Ra}$ if and only if $\mathrm{Supp}(R_a/R) = \{M\}$.*

Proof Let $M \in \mathrm{Max}(R)$. Then $M = \sqrt{Ra} \Leftrightarrow M$ is the only $P \in \mathrm{Spec}(R)$ containing $a \Leftrightarrow$ for any $P \in \mathrm{Spec}(R) \setminus \{M\}$, $a \notin P$ and $a \in M \Leftrightarrow$ for any $P \in \mathrm{Spec}(R) \setminus \{M\}$, $a/1 \in \mathrm{U}(R_P)$ and $a/1 \notin \mathrm{U}(R_M) \Leftrightarrow$ for any $P \in \mathrm{Spec}(R) \setminus \{M\}$, $R_P = (R_a)_P$ and $R_M \neq (R_a)_M \Leftrightarrow \mathrm{Supp}(R_a/R) = \{M\}$. □

Proposition 9.7 *Let $R \subset S$ be a ring extension. The following statements are equivalent:*

(1) *$R \subset S$ is a minimal QR extension.*
(2) *There exists some $M \in \mathrm{Max}(R)$ such that $\mathrm{Supp}(S/R) = \{M\}$, and there is an lsd $a \in R \setminus \mathrm{U}(R)$ such that $S = R_a$.*
(3) *There is an lsd $a \in R \setminus \mathrm{U}(R)$ such that $S = R_a$, and there exists some $M \in \mathrm{Max}(R)$ such that $M = \sqrt{Ra}$.*

If these conditions are satisfied, then $M \not\subseteq \cup\{P \in \mathrm{Max}(R) \mid P \neq M\}$.

Proof (1) $\Rightarrow$ (2) Assume that $R \subset S$ is a minimal QR extension. Then, there exists some $M \in \mathrm{Max}(R)$ such that $\mathrm{Supp}(S/R) = \{M\}$. Since $R \subset S$ is a minimal QR extension, Theorem 8.5 asserts that there is an lsd $a \in R \setminus \mathrm{U}(R)$ such that $S = R_a$, since S is of finite type over R. It follows that $\mathrm{Supp}(R_a/R) = \{M\}$, which gives $M = \sqrt{Ra}$ by Lemma 9.6, so that $a \in M \setminus \cup\{P \in \mathrm{Max}(R) \mid P \neq M\}$, giving $M \not\subseteq \cup\{P \in \mathrm{Max}(R) \mid P \neq M\}$, proving the last assertion.

(2) $\Rightarrow$ (1) Assume that there exists some $M \in \mathrm{Max}(R)$ such that $\mathrm{Supp}(S/R) = \{M\}$, and there is an lsd $a \in R \setminus \mathrm{U}(R)$ such that $S = R_a$. These two conditions lead to $M = \sqrt{Ra}$ by Lemma 9.6. As $a/1$ is a strong divisor of R_M and $S_M = (R_a)_M = (R_M)_{(a/1)}$, Proposition 9.5 shows that $R_M \subset S_M$ is minimal Prüfer. Moreover, $R_P = S_P$ for any $P \in \mathrm{Max}(R)$, $P \neq M$ implies that $R \subset S$ is minimal Prüfer. At last, Theorem 8.5 shows that $R \subset S$ is a QR extension, since minimal.

(2) $\Leftrightarrow$ (3) by Lemma 9.6. □

Corollary 9.8 *A minimal Prüfer extension $R \subset S$ such that $S = R_s$ for some lsd s of R is a QR extension.*

Proof Since $R \subset S$ is minimal, there exists some $M \in \mathrm{Max}(R)$ such that $\mathrm{Supp}(S/R) = \{M\}$. Then Proposition 9.7 shows that $R \subset S$ is a QR extension. □

Example 9.9 Set $R := \mathbb{Z}$, $P := p\mathbb{Z}$, where p is a prime integer and $S := \mathbb{Z}_p$. Obviously, $\mathrm{Supp}(S/R) = \{P\}$, so that $S = R_p$ and $P = \sqrt{Rp}$. Then, $p/1$ is a strong divisor of R_P and $p/1 \in \mathrm{U}(R_M)$ for any $M \in \mathrm{Spec}(R) \setminus \{P\}$, showing that p is an lsd of R. We recover the fact that $\mathbb{Z} \subset \mathbb{Z}_p$ is a minimal Prüfer QR extension.

The next result shows that Prüfer FCP extensions can be described in a special manner.

Proposition 9.10 ([31, Proposition 1.3]) *Let $R \subset S$ be an FCP extension. Then $R \subset S$ is integrally closed $\Leftrightarrow$ $R \subset S$ is Prüfer $\Leftrightarrow$ $R \subset S$ is a tower of Prüfer minimal extensions.*

Theorem 9.11 *An FCP QR extension $R \subseteq S$ admits a tower of Prüfer minimal extensions $R \subset R_1 \subset \cdots \subset R_i \subset R_{i+1} \subset \cdots \subset R_n = S$, where $R_{i+1} = (R_i)_{a_i} = R_{a_i}$ for some lsd $a_i \in R$ and $S = R_{a_1 \cdots a_n} = R_{a_n}$. The integer n is independent of the sequence and is equal to $|\mathrm{Supp}(S/R)|$.*

Proof There is a tower of Prüfer minimal extensions $R_i \subset R_{i+1}$ by Proposition 9.10 because a QR extension is Prüfer. Therefore, each $T \in [R, S]$ is a localization R_a, for some lsd $a \in R$ by Theorem 8.5 and $R_i \subset R_{i+1}$ identifies to $R_i \subset R_{a_i}$ for some lsd $a_i \in R \setminus \mathrm{U}(R_i)$.

Then by minimality, we get that $R_{i+1} = R_{a_i} = (R_i)_{a_i}$ and the result follows. The last result is [9, Proposition 6.12]. □

The above result applies when $R \subseteq S$ is an FCP extension $A(X) \subseteq B(X)$, or equivalently, $A \subseteq B$ has FCP [10, Theorem 3.9]. We recall the definition of a Nagata ring [10, Section 3]. Let R be a ring and $R[X]$ the polynomial ring in the indeterminate X over R. We denote by $C(p)$ the content of any polynomial $p \in R[X]$. Then $\Sigma_R := \{p \in R[X] \mid C(p) = R\}$ is a saturated mcs of $R[X]$, each of whose elements is a non-zero-divisor of $R[X]$. The *Nagata ring of* R is defined to be $R(X) := R[X]_{\Sigma_R}$.

Proposition 9.12 *Let $R \subset S$ be an FCP Prüfer extension over a local ring (R, M):*

(1) *There is a sequence of Prüfer minimal extensions between local rings $R \subset R_1 \subset \cdots \subset R_i \subset R_{i+1} \subset \cdots \subset R_n = S$, where $R_{i+1} = (R_i)_{a_i} = R_{a_i}$ for some $a_i \in \Delta(R)$ and $S = R_{a_1 \cdots a_n} = R_{a_n}$. In fact, $[R, S] = \{R_i\}_{i=0}^n$. Moreover, $R \subset S$ is a QR extension.*
(2) *There is some subset $\{P_0, P_1, \ldots, P_n\}$ of* Spec(R), *with $P_0 = M$ and $P_i \subset P_{i-1}$ for $i = 1, \ldots, n$, such that $R_i = R_{P_i}$, $P_i R_i = P_i$, and R/P_i is a valuation domain whose dimension is i.*
(3) *For all $i = 1, \ldots, n$ and $t \in P_{i-1} \setminus P_i$, we have $R_i = R_t$. Moreover, each element of $M \setminus P_n$ is a strong divisor.*
(4) *Any finitely generated S-regular ideal is equivalent to a principal ideal.*

Proof

(1) We know that $R \subset S$ is chained because R is local [9, Theorem 6.10], and each R_i is local by Proposition 9.2. Moreover, this proposition shows that for each $i \in \mathbb{N}_n$, there exists $P_i \in \mathrm{Spec}(R)$ such that $R_i = R_{P_i}$, $P_i R_i = P_i$ and R/P_i is an i-dimensional valuation domain. Therefore, $R \subset S$ is a QR extension. Then apply Theorem 9.11.
(2) Since $R \subset R_1$ is Prüfer minimal, $P_0 := M = \mathcal{C}(R, R_1)$. As $S = R_n$, it follows that R/P_n is a n-dimensional valuation domain and $\{P_0, P_1, \ldots, P_{n-1}\} := \mathrm{Supp}(S/R)$ according to [9, Proposition 6.12] with $P_i \subset P_{i-1}$ for each $i \in \mathbb{N}_n$.
(3) Let $t \in P_{i-1} \setminus P_i$. Then, t is a unit of $R_{P_i} = R_i$ and $R_t \subseteq R_i$. As $t \in P_{i-1}$, this implies that t is a not a unit of $R_{P_{i-1}} = R_{i-1}$, so that $R_t \not\subseteq R_{i-1}$. But $[R, S]$ is a chain, which leads to $R_t = R_i$.
Let $x \in M \setminus P_n$. Since $\{P_0, P_1, \ldots, P_n\}$ is a chain, there exists some $i \in \mathbb{N}_n$ such that $x \in P_{i-1} \setminus P_i$. Then, $R_x = R_i$ by the first part of (3). To end, Proposition 7.6 and Definition 7.14 show that x is a strong divisor.
(4) Since $R \subset S$ is a QR extension, any finitely generated S-regular ideal is equivalent to a principal ideal according to the results at the beginning of Sect. 8. □

We end this section by a generalization of Proposition 9.12 to $\mathcal{B}$ extensions. We recall that an extension $R \subset S$ is a $\mathcal{B}$ *extension* if the map $\beta : [R, S] \to \prod\{[R_M, S_M] \mid M \in \mathrm{MSupp}(S/R)\}$ defined by $T \mapsto (T_M)_{M \in \mathrm{MSupp}(S/R)}$ is bijective. Actually, an FCP extension $R \subseteq S$ is a $\mathcal{B}$ extension if and only if R/P is a local ring for each $P \in \mathrm{Supp}(S/R)$ [32, Proposition 2.21]. The following lemma gives a first special case of $\mathcal{B}$ extensions. For any extension $R \subseteq S$, the *length* $\ell[R, S]$ of $[R, S]$ is the supremum of the lengths of chains of R-subalgebras of S. Notice that if $R \subseteq S$ has FCP, then there *does* exist some maximal chain of R-subalgebras of S with length $\ell[R, S]$ [10, Theorem 4.11].

Lemma 9.13 *Let $R \subset S$ be an FCP Prüfer extension such that* $|\mathrm{MSupp}(S/R)| = 1$. *Then, $R \subset S$ is a $\mathcal{B}$ extension where* $\mathrm{Supp}(S/R)$ *and* $[R, S]$ *are chains with* $n := |\mathrm{Supp}(S/R)| = |[R, S]| - 1$. *There is a tower of Prüfer minimal extensions $R \subset R_1 \subset \cdots \subset R_i \subset R_{i+1} \subset \cdots \subset R_n = S$, such that $[R, S] = \{R_i\}_{i=0}^n$. We define as follows a subset $\{P_0, P_1, \ldots, P_n\}$ of* $\mathrm{Spec}(R)$ *by* $\mathrm{Supp}(S/R) = \{P_0, P_1, \ldots, P_{n-1}\}$ *and $P_n := R \cap (R_{n-1} : S)$, where $P_i \subset P_{i-1}$ for each $i \in \mathbb{N}_n$. In particular,* $\mathrm{Supp}(R_i/R) = \{P_0, P_1, \ldots, P_{i-1}\}$.

Proof From $|\mathrm{MSupp}(S/R)| = 1$, we deduce that $R \subset S$ is a $\mathcal{B}$ extension by [32, Proposition 2.21]. If $\{M\} := \mathrm{MSupp}(S/R)$, the map $\beta : [R, S] \to [R_M, S_M]$ defined by $\beta(T) = T_M$ for any $T \in [R, S]$ is bijective. But, $R_M \subset S_M$ is chained by Proposition 9.12, whence $R \subset S$ is chained. According to [33, Theorem 3.10], $\mathrm{Supp}(S/R)$ has a least element P and $\mathrm{Supp}(S/R) = \mathrm{V}(P)$ is chained. Moreover, $|\mathrm{Supp}(S/R)| = |[R, S]| - 1 = \ell[R, S]$ by [9, Proposition 6.12]. If $n := |[R, S]| - 1$, there is a sequence of Prüfer minimal extensions $R \subset R_1 \subset \cdots \subset R_i \subset R_{i+1} \subset \cdots \subset R_n = S$ such that $[R, S] = \{R_i\}_{i=0}^n$ since $R \subset S$ is chained. Moreover, there is some subset $\{P_0, P_1, \ldots, P_n\}$ of $\mathrm{Spec}(R)$ such that $\mathrm{Supp}(S/R) = \{P_0, P_1, \ldots, P_{n-1}\}$ where $P_i \subset P_{i-1}$ for each $i \in \mathbb{N}_{n-1}$. In particular, $\mathrm{Supp}(R_i/R) = \{P_0, P_1, \ldots, P_{i-1}\}$ for each $i \in \mathbb{N}_n$. In fact, we have $P_i = R \cap \mathcal{C}(R_i, R_{i+1})$ for each $i = 0, \ldots, n-1$ by [9, Corollary 3.2] and also $P_i = R \cap (R_{i-1} : R_i)$ for each $i \in \mathbb{N}_{n-1}$ by Corollary 9.3. □

Proposition 9.14 *Let $R \subset S$ be an FCP Prüfer extension such that* $|\mathrm{MSupp}(S/R)| = 1$. *Set $[R, S] = \{R_i\}_{i=0}^n$, $P_n := R \cap (R_{n-1} : S)$ and* $\mathrm{Supp}(S/R) = \{P_0, P_1, \ldots, P_{n-1}\}$ *as defined in Lemma 9.13. The following conditions are equivalent:*

(1) *$R \subset S$ is a QR extension.*
(2) *For each $i \in \mathbb{N}_n$, there is some lsd $a_i \in R$ such that $R_i = R_{a_i}$.*
(3) *For each $i \in \mathbb{N}_n$, there is some $a_i \in R$ such that $P_{i-1} = \sqrt{Ra_i}$.*

If these conditions hold, then, for each $i \in \mathbb{N}_n$, we have $R_i = (R_{i-1})_{a_i}$ and $a_i \in P_{i-1} \setminus P_i$. Moreover, a_i satisfies (2) if and only if it satisfies (3).

Proof (1) $\Leftrightarrow$ (2) by Theorem 8.5 and (3) $\Leftrightarrow$ (1) by Proposition 8.9.

Assume that (2) holds with a_i an lsd such that $R_i = R_{a_i}$. Obviously, $R_i = (R_{i-1})_{a_i}$ (*).

Let $P \in \mathrm{Spec}(R)$. Then, $a_i \in P$ implies that $PR_i = R_i$, so that $P \in \mathrm{Supp}(R_i/R) = \{P_0, P_1, \ldots, P_{i-1}\}$. Then, there is some $j < i$ such that $P = P_j$ and $a_i \notin P_i$. To prove that $a_i \in P_{i-1}$, we localize the extension $R \subset S$ at M. Set

$R'_j := (R_j)_M$, $P'_j := P_jR_M$ for each $j = 0, \dots, n$ and $a'_i := a_i/1$. Then, $R' \subset S'$ is an FCP Prüfer extension over the local ring (R', M') with $n = \ell[R', S']$. Using Proposition 9.12, we get that $\{P'_0, P'_1, \dots, P'_n\}$ is the subset of $\mathrm{Spec}(R')$ such that $R'_i = R'_{P'_i}$, $P'_iR'_i = P'_i$. It follows that $(*)$ gives $R'_i = R'_{a'_i} = (R'_{i-1})_{a'_i}$, with R'_{i-1} a local ring with maximal ideal P'_{i-1}. Then $a'_i \in P'_{i-1}$ since $R'_i = (R'_{i-1})_{a'_i} \neq R'_{i-1}$, which gives $a_i \in P_{i-1} \setminus P_i$. To conclude, $P_{i-1} = \sqrt{Ra_i}$ is the least prime ideal of R containing a_i, and also the least element of $\mathrm{Supp}(R_i/R)$. It follows that a_i satisfies (3).

Conversely, if there is some b_i such that $P_{i-1} = \sqrt{Rb_i}$, then $\sqrt{Ra_i} = \sqrt{Rb_i}$ implies $R_i = R_{a_i} = R_{b_i}$ by Definition 7.14. □

In order to generalize Lemma 9.13 to an arbitrary FCP Prüfer $\mathcal{B}$ extension, we need the following definition introduced in [33]. Let $R \subseteq S$ be an FCP $\mathcal{B}$ extension and $M \in \mathrm{MSupp}(S/R)$. The elementary splitter $\sigma(M) := T$, associated to M, is defined by $\mathrm{MSupp}(T/R) = \{M\}$ and $\mathrm{MSupp}(S/T) = \mathrm{MSupp}(S/R) \setminus \{M\}$. Such a T always exists (see [33, Theorem 4.6 and the paragraph after Corollary 5.5]).

Proposition 9.15 *Let $R \subset S$ be an FCP Prüfer $\mathcal{B}$ extension and QR extension. Let $T \in [R, S]$, $T \neq R$ and set* $\mathrm{MSupp}(T/R) =: \{M_1, \dots, M_n\}$. *For each $i \in \mathbb{N}_n$, let P_i be the least element of* $\mathrm{Supp}(T/R)$ *contained in M_i. Then, there exists some lsd $t \in R$ such that $T = R_t$ and $\sqrt{Rt} = \cap\{P_i \mid i \in \mathbb{N}_n\}$.*

Proof Since $R \subset S$ is an FCP Prüfer $\mathcal{B}$ and QR extension, so is $R \subset T$ by Picavet and Picavet-L'Hermitte [33, Proposition 3.5] for the $\mathcal{B}$ extension property. Set $\mathrm{MSupp}(T/R) =: \{M_1, \dots, M_n\}$. For each $i \in \mathbb{N}_n$, according to [33, Theorem 3.10], there is a least element P_i of $\mathrm{Supp}(T/R)$ contained in M_i. The same reference gives that $\mathrm{V}(P_i)$ is a chain, whose greatest element is M_i and least element is P_i.

For each $i \in \mathbb{N}_n$, set $T_i := \sigma(M_i)$, so that $|\mathrm{MSupp}(T_i/R)| = 1$. Moreover, for any $P \in \mathrm{V}(P_i)$, we have $P \in \mathrm{Supp}(T_i/R)$. Then, $\mathrm{Supp}(T_i/R) = \mathrm{V}(P_i)$ because $\mathrm{MSupp}(T_i/R) = \{M_i\}$, and we can apply Proposition 9.14. For each $i \in \mathbb{N}_n$, we have $T_i = R_{a_i}$ for some lsd $a_i \in R$ and $P_i = \sqrt{Ra_i}$ for each $i \in \mathbb{N}_n$. Set $t := a_1 \cdots a_n$, which is still an lsd. Now, [33, Proposition 5.11] asserts that $T = \prod\{T_i \mid i \in \mathbb{N}_n\} = \prod\{R_{a_i} \mid i \in \mathbb{N}_n\} = R_t$. Moreover, $\sqrt{Rt} = \sqrt{\prod\{Ra_i \mid i \in \mathbb{N}_n\}} = \cap\{\sqrt{Ra_i} \mid i \in \mathbb{N}_n\} = \cap\{P_i \mid i \in \mathbb{N}_n\}$. □

10 The Set of All Primitive Elements

Let $R \subseteq S$ be an extension. An element $s \in S$ is called *primitive* (over R) if there exists a polynomial $p(X) \in R[X]$ whose content is R and such that $p(s) = 0$. An extension $R \subseteq S$ is called a $\mathcal{P}$ extension if all the elements of S are primitive over R. Important examples of $\mathcal{P}$ extensions are given by the Prüfer extensions of [20] (equivalently normal pairs [20, Theorem 5.2, p.47]). We recall that an element s of S is primitive if and only if $R \subseteq R[s]$ has the INC property [7, Theorem 2.3] and that an extension $R \subseteq S$ is a $\mathcal{P}$ extension if and only if $R \subseteq S$ is an INC pair [7, Corollary 2.4]. We proved that an INC pair is nothing but a quasi-Prüfer extension [31, Theorem 2.3].

It follows easily that an extension is quasi-Prüfer if and only if it is a $\mathcal{P}$ extension. Therefore, an FCP extension is a $\mathcal{P}$ extension [31, Corollary 3.4].

For an extension $R \subseteq S$, we denote by $\mathcal{P}_S(R)$ the set of all elements of S that are primitive over R, a set studied by Dobbs–Houston in [8].

We defined in [31, Theorem 3.18] the quasi-Prüfer closure (or hull) $\overset{\Longrightarrow}{R}$ of an extension $R \subseteq S$. This closure is the greatest quasi-Prüfer subextension $R \subseteq T$ of $R \subseteq S$ and is equal to $\widetilde{\overline{R}}$. It follows that $\overset{\Longrightarrow}{R} \subseteq \mathcal{P}_S(R)$. Obviously, $\overline{R} \subseteq \mathcal{P}_S(R)$.

Proposition 10.1 *Let $R \subseteq S$ be an extension. Then $\mathcal{P}_S(R)$ is a ring if and only if $\mathcal{P}_S(R) = \overset{\Longrightarrow}{R}$.*

Proof It is enough to show that if $\mathcal{P}_S(R)$ is a ring, it is contained in $\overset{\Longrightarrow}{R}$, that is $\overline{R} \subseteq \mathcal{P}_S(R)$ is Prüfer. But we may assume that $R \subset S$ is integrally closed because an element of S primitive over R is primitive over $\overline{R}$.

By Dobbs and Houston [8, Proposition 2.6], we can also assume that R is local. Let $s \in \mathcal{P}_S(R)$. Then, either $s \in R$ or s is a unit of S such that $s^{-1} \in R$ according to [8, Corollary 2.5]. If $s \in R$, then $R[s] = R[s^2]$. If s is a unit of S such that $s^{-1} \in R$, then $s^{-1} \in R[s^2]$ implies $s \in R[s^2]$, and we still have $R[s] = R[s^2]$. Therefore, for each $s \in \mathcal{P}_S(R)$, we have $R[s^2] = R[s]$. We deduce from [20, Chapter 1, Theorem 5.2] that the ring $\mathcal{P}_S(R)$ is Prüfer over R and then $\mathcal{P}_S(R) = \overset{\Longrightarrow}{R}$. □

It may happen that an extension $R \subseteq S$ is such that $\mathcal{P}_S(R) = S$. For example, if we denote by $\mathrm{Tot}(R)$ the total ring of quotients of a ring R, then $\mathcal{P}_{\mathrm{Tot}(R)}R$ is a ring if and only if $\mathcal{P}_{\mathrm{Tot}(R)}R = \mathrm{Tot}(R)$ [8, Corollary 2.9]. If $R \subseteq S$ defines a lying-over pair, then $\mathcal{P}_S(R) = \overset{\Longrightarrow}{R}$ [8, Proposition 3.11]. The next result generalizes some results of [8].

Corollary 10.2 *An extension $R \subseteq S$ is such that $\mathcal{P}_S(R) = S$ if and only if $R \subseteq S$ is quasi-Prüfer.*

References

1. Abbas, R., Ayache, A.: On questions related to normal pairs. Comm. Algebra **49** (3), 956–966 (2021)
2. Camillo, V.P.: A note on semi-hereditary rings. Arch. Math. **Vol. XXIV**, 142–143 (1973)
3. Campanini, F., Finocchiaro, C.A.: Some remarks on Prüfer rings with zero-divisors. J. Pure Appl. Algebra **225**, 106692 (2021)
4. Cox Jr, S.H., Rush, D.E.: Finiteness in flat modules and algebras. J. Algebra **32**, 44–50 (1974)
5. Davis, E.D.: Overrings of commutative rings. III: Normal pairs. Trans. Amer. Math. Soc. **182**, 175–185 (1973)

6. Davison, T.M.K.: Distributive homomorphisms of rings and modules. J. Reine Angew. Math. **271**, 28–34 (1974)
7. Dobbs, D.E.: Lying-over pairs of commutative rings. Can. J. Math., **(XXXIII)** (2), 454–475 (1981)
8. Dobbs, D.E., Houston, E.: On sums and products of primitive elements., Comm. Algebra **45**, 357–370 (2017)
9. Dobbs, D.E., Picavet, G., Picavet-L'Hermitte, M.: Characterizing the ring extensions that satisfy FIP or FCP. J. Algebra **371**, 391–429 (2012)
10. Dobbs, D.E., Picavet, G., Picavet-L'Hermitte, M.: Transfer results for the FIP and FCP properties of ring extensions. Comm. Algebra **43**, 1279–1316 (2015)
11. Endo, S.: On semi-hereditary rings. J. Math. Soc. Japan **13, No 2**, 109–119 (1961)
12. Erdoğdu, V.: The distributive hull of a ring. J. Algebra **132**, 263–269 (1990)
13. Ferrand, D.: Trivialisation des modules projectifs. La méthode de Kronecker. J. Pure Appl. Algebra **24**, 261–264 (1982)
14. Ferrand, D., Olivier, J.-P.: Homomorphismes minimaux d'anneaux. J. Algebra **16**, 461–471 (1970)
15. Gilmer, R., Ohm, J.: Integral domains with quotient overrings. Math. Ann. **153**, 813–818 (1964)
16. Gotlieb, C.: Finite unions of overrings of an integral domain. J. Commut. Algebra, DOI https://doi.org/10.1216/jca.2020.12.87.
17. Grothendieck, A.: Eléments de Géométrie Algébrique. IV. Etude locale des schémas et des morphismes de schémas IV. Inst. Hautes Etudes Sci. Publ. Math **32**, 361 pp. (1967)
18. Grothendieck, A., Dieudonné, J.: Eléments de Géométrie Algébrique I, Springer Verlag, Berlin, (1971).
19. Knebusch, M., Kaiser, T.: Manis Valuations and Prüfer Extensions II. Springer, Cham Heidelberg (2014)
20. Knebusch, M., Zhang, D.: Manis Valuations and Prüfer Extensions I. Springer, Berlin (2002)
21. Kostra, J.: The covering of rings by integrally closed rings. Mathematica Slovaca **34, no 2**, 171–176 (1984)
22. Lazard, D.: Autour de la platitude. Bull. Soc. Math. France **97**, 81–128 (1969),
23. Morita, K.: Flat modules, Injective modules and quotient rings. Math. Z. **120**, 25–40 (1971)
24. Olivier, J.-P.: Fermeture intégrale et changements de base absolument plats. Colloque d'Algèbre Commutative (Rennes 1972), Exp. No 9, 13 pp. Publ. Sém. Math. Univ. Rennes, Année 1972, Univ. Rennes, Rennes, (1972)
25. Olivier, J.-P.: Going up along absolutely flat morphisms. J. Pure Appl. Algebra **30**, 47–59 (1983)
26. Picavet, G.: Autour des idéaux premiers de Goldman d'un anneau commutatif. Ann. Sci. Univ. Clermont **57**, Math. No. 11, 73–90 (1975)
27. Picavet, G.: Propriétés et applications de la notion de contenu. Comm. Algebra **13**, 2231–2265 (1985)
28. Picavet, G.: Geometric subsets of a spectrum. Commutative ring theory and applications (Fez, 2001), 387–417, Lecture Notes in Pure and Appl. Math., 231, Dekker, New York, 2003.
29. Picavet, G.: Recent progress on submersions: a survey and new properties. Algebra, doi.org/10.1155/2013/128064, (2913). 2013, Article ID 128064, 14 p. (2013)
30. Picavet, G., Picavet-L'Hermitte, M.: Some more combinatorics results on Nagata extensions. Palest. J. Math. **5**, 49–62 (2016)
31. Picavet, G., Picavet-L'Hermitte, M.: Quasi-Prüfer extensions of rings. pp. 307–336, in: Rings, Polynomials and Modules, Springer, (2017)
32. Picavet, G., Picavet-L'Hermitte, M.: Boolean FIP ring extensions. Comm. Algebra **48**, 1821–1852 (2020)
33. Picavet, G., Picavet-L'Hermitte, M.: Splitting ring extensions. *Beitr. Algebra Geom.* https://doi.org/10.1007/s13366-022-00650-2 (2022).
34. Samuel, P.: La notion de place dans un anneau. Bull. Soc. Math. France **85**, 123–133 (1957)
35. Storrer, H.H.: Epimorphismen von kommutativen Ringen. Comment. Math. Helv. **43**, 378–401 (1968)

A Survey on Algebraic and Homological Properties of Amalgamated Algebras of Commutative Rings

Maryam Salimi, Elham Tavasoli, and Siamak Yassemi

1 Introduction

Throughout this paper all rings are considered commutative with identity. In [14], D'Anna and Fontana considered a construction obtained involving a ring A and an ideal $I \subset A$ that was denoted by $A \bowtie I$, called *amalgamated duplication*, and it was defined as the following subring of $A \times A$:

$$A \bowtie I = \{(r, r + i) \mid r \in A, i \in I\}.$$

This construction was studied from different points of view in [8, 13, 14, 27, 33, 34], and [37]. In [15], it is initiated a systematic study of a new ring construction called the *amalgamation of* A *with* B *along* J *with respect to* f, for a given homomorphism of rings $f : A \to B$ and ideal J of B. This construction finds its roots in a paper by J.L. Dorroh appeared in [18] and provides a general frame for studying the amalgamated duplication of a ring along an ideal. The amalgamation of A with B along J with respect to f is a subring of $A \times B$ which is defined as follows:

$$A \bowtie^f J = \{(r, f(r) + j) \mid r \in A, j \in J\}.$$

M. Salimi · E. Tavasoli
Department of Mathematics, East Tehran Branch, Islamic Azad University, Tehran, Iran
e-mail: maryamsalimi@ipm.ir; elhamtavasoli@ipm.ir

S. Yassemi (✉)
School of Mathematics, Statistics and Computer Science, College of Science, University of Tehran, Tehran, Iran
e-mail: yassemi@ut.ac.ir

J.-L. Chabert et al. (eds.), *Algebraic, Number Theoretic, and Topological Aspects of Ring Theory*, https://doi.org/10.1007/978-3-031-28847-0_20

This construction is a generalization of the amalgamated duplication of a ring along an ideal and several classical constructions (such as $A + XB[X]$, $A + XB[[X]]$ and $D + M$ constructions) can be studied as particular cases of the amalgamation (see [15, Examples 2.5 and 2.6]). Moreover, other classical constructions, such as the Nagata's idealization [15, Example 2.7 and Remark 2.8], and the CPI extensions (in the sense of Boisen and Sheldon [10]) are related to it (see [15, Example 2.7]). One of the key tools for studying $A \bowtie^f J$ is based on the fact that the amalgamation can be studied in the frame of pullback constructions [15]. This point of view allows to deepen the study initiated in [15] and continued in [17] and to provide an ample description of various properties of $A \bowtie^f J$, in connection with the properties of A, J, and f. In [15], it is provided necessary and sufficient conditions for $A \bowtie^f J$ to inherit the properties of Noetherian ring, integral domain, and reduced ring and characterized pullbacks that can be expressed as amalgamations. In [17], the authors provided a complete description of the prime spectrum of $A \bowtie^f J$ and gave bounds for its dimension. In [16], the prime spectrum of $A \bowtie^f J$, some of its invariants (like the embedding dimension), and relevant properties (like Cohen-Macaulayness and Gorensteiness) are studied. In particular, in [16, Remark 5.1], assuming A is a Cohen-Macaulay local ring, J is finitely generated as an A-module, and J is contained in the Jacobson radical of B, it is observed that $A \bowtie^f J$ is a Cohen-Macaulay ring if and only if it is a Cohen-Macaulay A-module if and only if J is a maximal Cohen-Macaulay A-module. Then in [32], the authors improved this observation as well as, in the case that J is not finitely generated, it provides conditions implying $A \bowtie^f J$ is Cohen-Macaulay. The above results lead to investigate further when the amalgamated algebra $A \bowtie^f J$ is (quasi-)Gorenstein, in [32]. Also, in [5] determined when the amalgamated algebra $A \bowtie^f J$ is Cohen-Macaulay (in the sense of Hamilton and Marley). Moreover, in [4] the property of Cohen-Macaulay in the sense of ideals, which is a general notion of the usual Cohen-Macaulay property (in the Noetherian case), is studied on the ring $A \bowtie^f J$.

It is known that the ring A is always embedded into the ring $A \bowtie^f J$, and the natural image of the ring A into $A \bowtie^f J$ is a retract of $A \bowtie^f J$ (see [15, Remark 4.6 or Proposition 4.7]). This leads to describe the transfer of Prüfer-like conditions of the amalgamations in [19]. Also in [11], necessary and sufficient conditions under which Prüfer-like properties transfer between a local ring A and the amalgamated duplication $A \bowtie I$ is investigated. Later, in [6], the authors attempted to generalize the results mentioned above to the case of the amalgamated algebra $A \bowtie^f J$. In [26], the stability of the divided ring, locally divided ring, going-down ring, and Gaussian ring properties in terms of amalgamation of rings are studied. In addition, the authors provided examples of new classes of commutative rings satisfying the above mentioned properties.

The study of algebraic structures by way of graph theory is an exciting research topics. There are many papers on assigning a graph to a ring. In this way, the zero-divisor graph of the amalgamated algebra $A \bowtie^f J$ is studied in [1] and [3]. In addition, in [35] the authors investigated some properties of the comaximal graph of the ring A which are transferred to the comaximal graph of $A \bowtie^f J$ and vice versa. In order to study of some graph-theoretic properties of the comaximal graph of $A \bowtie^f J$, certain algebraic properties of the ring $A \bowtie^f J$ were given by way of graph theory. The purpose of this paper is to survey briefly recent algebraic and homological properties of the amalgamated algebra $A \bowtie^f J$. The graph-theoretic properties of the ring $A \bowtie^f J$ are not mentioned in this paper.

2 Algebraic Properties

Let $f : A \to B$ be a ring homomorphism, and let J be an ideal of B. In [15] the following subring of $A \times B$

$$A \bowtie^f J = \{(a, f(a) + j) \mid a \in A, j \in J\}$$

called the amalgamation of A with B along J with respect to f is introduced. In this paper, we survey briefly recent algebraic and homological properties of the amalgamated algebra $A \bowtie^f J$.

Notation 2.1 Let $\mathrm{Reg}(A)$ be all regular elements of A. We also denote, respectively, by $\mathrm{Nilp}(A)$, $\mathrm{J}(A)$, and $\mathrm{Idem}(A)$ the ideal of all nilpotent elements of the ring A, Jacobson radical of A, and the set of all idempotent elements of A. □

Recall that, if $\alpha : A \to C$ and $\beta : B \to C$ are ring homomorphisms, the subring $D := \alpha \times_C \beta := \{(a, b) \in A \times B \mid \alpha(a) = \beta(b)\}$ of $A \times B$ is called the pullback (or fiber product) of α and β. In the following, we will denote by p_A (resp,. p_B) the restriction to $\alpha \times_C \beta$ of the projection of $A \times B$ onto A (resp,. B). The level of generality that we have choosen is due to the fact that the amalgamation can be studied in the frame of pullback constructions. This point of view allows us to provide easily an ample description of the properties of $A \bowtie^f J$, in connection with the properties of A, J, and f. The following result is a straightforward consequence of the definitions.

Proposition 2.2 ([15, Proposition 4.2]) *Let $\pi : B \to B/J$ be the canonical projection and $\breve{f} := \pi o f$. Then $A \bowtie^f J = \breve{f} \times_{B/J} \pi$.* □

In the following, the ideal-theoretic structure of the amalgamation $A \bowtie^f J$ is investigated, which are collected from [15, Proposition 5.1] and [16, Proposition 3.1].

Proposition 2.3 *The following statements hold.*

(i) *Let $\iota := \iota_{A,f,J} : A \to A \bowtie^f J$ be the natural ring homomorphism defined by $\iota(a) := (a, f(a))$, for all $a \in A$. Then ι is an embedding, making $A \bowtie^f J$ a ring extension of A with $\iota(A) = \Gamma(f)(:= \{(a, f(a)) \mid a \in A\}$ subring of $A \bowtie^f J)$.*

(ii) *Let I be an ideal of A and set $I \bowtie^f J := \{(i, f(i)+j) \mid i \in I, j \in J\}$. Then $I \bowtie^f J$ is an ideal of $A \bowtie^f J$, the composition of canonical homomorphisms*

$$A \hookrightarrow A \bowtie^f J \twoheadrightarrow \frac{A \bowtie^f J}{I \bowtie^f J},$$

is a surjective ring homomorphism and its kernel coincides with I. *Hence, we have the following canonical isomorphism:*

$$\frac{A \bowtie^f J}{I \bowtie^f J} \cong \frac{A}{I}.$$

(iii) *If* I *(resp,.* H*) is an ideal of* A *(resp,. of* $f(A)+J$*) such that* $f(I)J \subseteq H \subseteq J$, *then* $I \bowtie^f H := \{(i, f(i)+h) \mid i \in I, h \in H\}$ *is an ideal of* $A \bowtie^f J$.

(iv) *If* I *is an ideal of* A, *then the extension* $I(A \bowtie^f J)$ *of* I *to* $A \bowtie^f J$ *coincides with* $I \bowtie^f (f(I)B)J := \{(i, f(i)+\beta) \mid i \in I, \beta \in (f(I)B)J\}$.

(v) *If* I *is an ideal of* A *such that* $f(I)B = B$, *then* $I(A \bowtie^f J) = I'^f = \{(i, f(i)+j) \mid i \in I, j \in J\} = I \bowtie^f J$.

(vi) *Let* $p_A : A \bowtie^f J \to A$ *and* $p_B : A \bowtie^f J \to B$ *be the natural projections of* $A \bowtie^f J \subseteq A \times B$ *into* A *and* B, *respectively. Then* p_A *is surjective and* $\mathrm{Ker}(p_A) = \{0\} \times J$. *Moreover,* $p_B(A \bowtie^f J) = f(A) + J$ *and* $\mathrm{Ker}(p_B) = f^{-1}(J) \times \{0\}$. *Hence, the following canonical isomorphisms hold:*

$$\frac{A \bowtie^f J}{(\{0\} \times J)} \cong A \quad \text{and} \quad \frac{A \bowtie^f J}{f^{-1}(J) \times \{0\}} \cong f(A) + J.$$

(vii) *Let* $\gamma : A \bowtie^f J \to (f(A)+J)/J$ *be the natural ring homomorphism, defined by* $(a, f(a)+j) \mapsto f(a)+J$. *Then* γ *is surjective and* $\mathrm{Ker}(\gamma) = f^{-1}(J) \times J$. *Thus, there exists a natural isomorphism:*

$$\frac{A \bowtie^f J}{f^{-1}(J) \times J} \cong \frac{f(A)+J}{J}.$$

In particular, when f *is surjective we have*

$$\frac{A \bowtie^f J}{f^{-1}(J) \times J} \cong \frac{B}{J}.$$

□

Corollary 2.4 *Let* $(A, \mathfrak{m})$ *be a local ring. Then the following statements hold.*

(i) *[16, Corollary 3.2] Assume that* $f^{-1}(Q) \neq \mathfrak{m}$, *for each* $Q \in \mathrm{Spec}(B) \setminus \mathrm{V}(J)$. *Then the extension in* $A \bowtie^f J$ *of any* $\mathfrak{m}$*-primary ideal of* A *is* $(\mathfrak{m} \bowtie^f J)$*-primary.*

(ii) *[16, Remark 3.3] Assume that* J *is a finitely generated* A*-module and* $J \subseteq \mathrm{J}(B)$. *Then for every* $Q \in \mathrm{Spec}(B) \setminus \mathrm{V}(J)$, *we have* $f^{-1}(Q) \neq \mathfrak{m}$. *In particular, the extension in* $A \bowtie^f J$ *of any* $\mathfrak{m}$*-primary ideal of* A *is* $(\mathfrak{m} \bowtie^f J)$*-primary.*

□

2.1 Prime Ideals and Localization

In the following proposition, some results about the structure of the prime ideals of the ring $A \bowtie^f J$ are collected from [17, Proposition 2.6] and [16, Corollary 2.8 and Proposition 2.9]. The proof of the following proposition is based on well-known properties of rings arising from pullbacks.

Proposition 2.5 *Set* $X := \mathrm{Spec}(A)$, $Y := \mathrm{Spec}(B)$, $W := \mathrm{Spec}(A \bowtie^f J)$, $J_0 := \{0\} \times J (\subseteq A \bowtie^f J)$, $J_1 := f^{-1}(J) \times \{0\}$, *and* $\mathfrak{X} := \mathfrak{X}_{(f,J)} := \bigcup_{\mathfrak{q} \in Y \setminus \mathrm{V}(J)} \mathrm{V}(f^{-1}(\mathfrak{q} + J))$. *For all* $\mathfrak{p} \in X$ *and* $\mathfrak{q} \in Y$, *set*

$$\mathfrak{p}'^f := \mathfrak{p} \bowtie^f J := \{(p, f(p) + j) \mid p \in \mathfrak{p}, j \in J\},$$

$$\overline{\mathfrak{q}}^f := \{(a, f(a) + j) \mid a \in A, j \in J, f(a) + j \in \mathfrak{q}\}.$$

Then the following statements hold.

(i) *The prime ideals of* $A \bowtie^f J$ *are of the type* $\mathfrak{p}'^f$ *or* $\overline{\mathfrak{q}}^f$, *for* $\mathfrak{p} \in X$ *and* $\mathfrak{q} \in Y \setminus \mathrm{V}(J)$. *In particular*

$$\mathrm{Max}(A \bowtie^f J) = \{\mathfrak{p}'^f \mid \mathfrak{p} \in \mathrm{Max}(A)\} \cup \{\overline{\mathfrak{q}}^f \mid \mathfrak{q} \in \mathrm{Max}(B) \setminus \mathrm{V}(J)\}.$$

(ii) *For any prime* $\mathfrak{q} \in Y \setminus \mathrm{V}(J)$, *the ring* $(A \bowtie^f J)_{\overline{\mathfrak{q}}^f}$ *is canonically isomorphic to* $B_{\mathfrak{q}}$.

(iii) *For any prime* $\mathfrak{p} \in X \setminus \mathrm{V}(f^{-1}(J))$, *the ring* $(A \bowtie^f J)_{\mathfrak{p}'^f}$ *is canonically isomorphic to* $A_{\mathfrak{p}}$.

(iv) *Let* $\mathfrak{p} \in \mathrm{Spec}(A)$ *containing* $f^{-1}(J)$. *Consider the multiplicative subset* $S := S_{(f,\mathfrak{p},J)} := f(A \setminus \mathfrak{p}) + J$ *of* B, *and set* $B_S := S^{-1}(B)$ *and* $J_S := S^{-1}(J)$. *If* $f_{\mathfrak{p}} : A_{\mathfrak{p}} \to B_S$ *is the ring homomorphism induced by* f, *then the ring* $(A \bowtie^f J)_{\mathfrak{p}'^f}$ *is canonically isomorphic to* $A_{\mathfrak{p}} \bowtie^{f_{\mathfrak{p}}} J_S$.

(v) *The map* $\mathfrak{p} \mapsto \mathfrak{p}'^f$ *establishes a closed embedding of* X *into* W, *so its image, which coincides with* $\mathrm{V}(J_0)$, *is homeomorphic to* X. *Also, this map establishes a homomorphism of* $\mathrm{Min}(A) \setminus \mathfrak{X}$ *with* $\mathrm{Min}(A \bowtie^f J) \cap \mathrm{V}(J_0)$.

(vi) *The map* $\mathfrak{q} \mapsto \overline{\mathfrak{q}}^f$ *is a homeomorphism of* $Y \setminus \mathrm{V}(J)$ *onto* $W \setminus \mathrm{V}(J_0)$. *Also, this map establishes a homomorphism of* $\mathrm{Min}(B) \setminus \mathrm{V}(J)$ *with* $\mathrm{Min}(A \bowtie^f J) \setminus \mathrm{V}(J_0)$.

□

Proposition 2.6 *The following statements hold.*

(i) *[17, Proposition 4.1]* $\dim(A \bowtie^f J) = \max\{\dim(A), \dim(f(A) + J)\}$. *In particular, if* f *is surjective, then* $\dim(A \bowtie^f J) = \max\{\dim(A), \dim(B)\} = \dim(A)$.

(*ii*) *[17, Proposition 4.2] Let* $f_\diamond : A \to B_\diamond := f(A) + J$ *be the ring homomorphism induced from* f. *If* $f_\diamond$ *is integral (e.g.,* f *is integral), then* $\dim(A \bowtie^f J) = \dim(A)$.

□

2.2 Zero-Divisors

Let M be an A-module. An element $m \in M$ is called a *torsion element* if there exists $r \in \mathrm{Reg}(A)$ such that $rm = 0$. Also, M is called a *torsion module* if all its elements are torsion elements.

Proposition 2.7 *The following statements hold.*

(*i*) *[6, Lemma 2.1] The inclusion* $\mathrm{Z}(A \bowtie^f J) \subseteq \{(a, f(a)+j) \mid a \in \mathrm{Z}(A), j \in J\} \cup \{(a, f(a)+j) \mid a \in A, j \in J, \exists j' \in J \setminus \{0\} : j'(f(a)+j) = 0\}$ *always holds. In particular, the equality holds if at least one of the following conditions hold.*

(1) $f(\mathrm{Z}(A)) \subseteq J$ *and* $f^{-1}(J) \neq 0$.
(2) $f(\mathrm{Z}(A))J = 0$ *and* $f^{-1}(J) \neq 0$.
(3) $J \subseteq f(A)$.
(4) J *is a torsion A-module.*

(*ii*) *[25, Proposition 2.1] Let* $J \neq 0$ *and suppose that at least one of the following conditions hold:*

(1) $J \subseteq f(A)$.
(2) J *is a torsion A-module.*
(3) $J^2 = 0$.

Then $\mathrm{Z}(A \bowtie^f J) = \{\mathrm{Z}(A) \bowtie^f J\} \cup \{(a, 0) \mid f(a) \in J\} \cup \{(0, x) \mid x \in J\} \cup \{(a, f(a)+i) \mid a \in \mathrm{Reg}(A), \exists\, j \in J \setminus \{0\} : j(f(a)+i) = 0\}$.

(*iii*) *[24, Lemma 2.7] Let* $J \neq 0$ *be a proper ideal of* B *such that* $f^{-1}(J) \neq 0$, *and* $J^2 = 0$. *Then* $\mathrm{Z}(A \bowtie^f J) = \{\mathrm{Z}(A) \bowtie^f J\} \cup \{(a, f(a)+i) \mid \exists 0 \neq j \in J : j(f(a)+i) = 0\}$.

□

2.3 Integral Domain and Reduced Property

Note that, Proposition 2.3(i) implies that if $A \bowtie^f J$ is an integral domain, then A is also an integral domain. Also, we have the following result.

Proposition 2.8 ([15, Proposition 5.2]) *The following statements are equivalent.*

(*i*) $A \bowtie^f J$ *is an integral domain.*

(ii) $f(A) + J$ *is an integral domain and* $f^{-1}(J) = \{0\}$.

In particular, if B *is an integral domain and* $f^{-1}(J) = \{0\}$, *then* $A \bowtie^f J$ *is an integral domain.* □

The following proposition implies that the property of being reduced for $A \bowtie^f J$ is independent of the nature of f.

Proposition 2.9 ([15, Proposition 5.4]) *The following statements are equivalent.*

(i) $A \bowtie^f J$ *is a reduced ring.*
(ii) A *is a reduced ring and* $\mathrm{Nilp}(B) \cap J = \{0\}$.

In particular, if A *and* B *are reduced, then* $A \bowtie^f J$ *is reduced; conversely, if* J *is a radical ideal of* B *and* $A \bowtie^f J$ *is reduced, then* B *(and* A*) is reduced.* □

Remark 2.10 Note that Proposition 2.9 implies that if A and $f(A) + J$ are reduced rings, then $A \bowtie^f J$ is a reduced ring. But the converse is not true in general (see [15, Remark 5.5 (3)]). □

The ring $A \bowtie^f J$ satisfies the property $(*)$ if every ideal of $A \bowtie^f J$ has one of the following three forms:

(i) $I \times 0$, where $I \subseteq f^{-1}(J)$ is an ideal of A.
(ii) $0 \times K$, where $K \subseteq J$ is an ideal of $f(A) + J$.
(iii) $I \bowtie^f J$, where I is an ideal of A.

The following theorem provides necessary and sufficient conditions for the ring $A \bowtie^f J$ to satisfy the property $(*)$.

Theorem 2.11 ([26, Theorem 2.1]) *Let* $J \neq 0$ *be a proper ideal of* B. *Then the following statements hold.*

(1) *If* $A \bowtie^f J$ *satisfies the property* $(*)$, *then the following statements hold:*
    (i) $f(A)$ *is an integral domain.*
    (ii) $f(A) \cap J = 0$.
    (iii) $0 \times J \subseteq ((a, f(a)+j))$ *for all* $a \in A \setminus \{0\}$ *and* $j \in J$ *such that* $f(a)+j \neq 0$.
(2) *If* f *is injective and* $A \bowtie^f J$ *satisfies the property* $(*)$, *then* A *is an integral domain.*
(3) *If* f *is not injective and* A *is a ring with zero-divisors with* $A \bowtie^f J$ *satisfies the property* $(*)$, *then* $\mathrm{Ann}_{f(A)+J}(f(a) + j) \subseteq J$ *for all* $a \in A \setminus \{0\}$, *and* $j \in J$ *with* $f(a) \neq 0$. *Moreover, if* $f^{-1}(J) \nsubseteq \mathrm{Z}(A)$, *then* $f(a) + j \in \mathrm{Reg}(f(A) + J)$ *for all* $a \in \mathrm{Reg}(A)$, *and* $j \in J$ *with* $f(a) \neq 0$, *and* $\mathrm{Ann}_{f(A)+J}(j) \subseteq f(\mathrm{Z}(A) f^{-1}(J)) + J$ *for all* $j \in J$.
(4) *If* f *is not injective and* A *is an integral domain with* $A \bowtie^f J$ *satisfies the property* $(*)$, *then the following conditions hold:*
    (i) $f(A) + J$ *is an integral domain.*

(ii) *J is idempotent.*

(5) *If $\{0\} \times J \subseteq ((a, f(a)+j))$ for all $a \in A \setminus \{0\}$ and $j \in J$ such that $f(a)+j \neq 0$, then $A \bowtie^f J$ satisfies the property $(*)$.*

□

2.4 Noetherian and Coherent Property

In the following, some results about Noetherian property of the amalgamation are collected.

Proposition 2.12 ([15, Proposition 5.6]) *The following statements are equivalent:*

(i) $A \bowtie^f J$ is a Noetherian ring.
(ii) A and $f(A) + J$ are Noetherian rings.

□

In order to obtain more useful criteria for the Noetherianity of $A \bowtie^f J$, the following result specialize Proposition 2.12 in some relevant cases.

Proposition 2.13 *The following statements hold.*

(i) [15, Proposition 5.7] Assume that at least one of the following conditions holds:

(a) J is a finitely generated A-module.
(b) J is a Noetherian A-module.
(c) $f(A) + J$ is Noetherian as A-module.
(d) f is a finite homomorphism.

Then $A \bowtie^f J$ is Noetherian if and only if A is Noetherian. In particular, if A is a Noetherian ring and B is a Noetherian A-module (e.g., if f is a finite homomorphism), then $A \bowtie^f J$ is a Noetherian ring for all ideal J of B.

(ii) [15, Proposition 5.8] If B is a Noetherian ring and the ring homomorphism $\breve{f} : A \to B/J$ is finite, then $A \bowtie^f J$ is a Noetherian ring if and only if A is a Noetherian ring.

□

Proposition 2.14 ([19, Proposition 4.14]) *The following statements hold.*

(i) If $A \bowtie^f J$ is a coherent ring, then A is a coherent ring.
(ii) If A is a coherent ring, and J is a coherent A-module, then $A \bowtie^f J$ is a coherent ring.

□

2.5 Embedding Dimension

Let A be a ring and I be an ideal of A. If I is finitely generated, we denote, by $\nu(I)$, the minimum number of generators of the ideal I. Assume that $(A, \mathfrak{m})$ is a local ring and set $k = A/\mathfrak{m}$. If we suppose that $\mathfrak{m}$ is finitely generated, we call the embedding dimension of A the natural number

$$\mathrm{embdim}(A) := \nu(\mathfrak{m}) = \dim_k(\mathfrak{m}/\mathfrak{m}^2).$$

In the following, some bounds for the embedding dimension of $A \bowtie^f J$ is given, when A is a local ring with finitely generated maximal ideal.

Proposition 2.15 *Assume that $(A, \mathfrak{m})$ is a local ring and $J \subseteq \mathrm{J}(B)$. Then the following statements hold:*

(i) [16, Proposition 4.1] If $A \bowtie^f J$ has finitely generated maximal ideal, then A has also finitely generated maximal ideal and $\mathrm{embdim}(A) \leqslant \mathrm{embdim}(A \bowtie^f J)$.

(ii) [16, Proposition 4.1] If A has finitely generated maximal ideal and J is finitely generated, then $A \bowtie^f J$ has finitely generated maximal ideal and $\mathrm{embdim}(A \bowtie^f J) \leqslant \mathrm{embdim}(A) + \nu(J)$.

(iii) [16, Proposition 4.3] Let $f(\mathfrak{m})B = B$. Then, for every ideal J of B contained in the Jacobson radical of B, the amalgamation $A \bowtie^f J$ is a local ring with finitely generated maximal ideal, and $\mathrm{embdim}(A) \leqslant \mathrm{embdim}(A \bowtie^f J)$.

(iv) [16, Theorem 4.4] Suppose that J is a finitely generated ideal of B. If $f(\mathfrak{m})B \subseteq \mathrm{J}(B)$ and $J \subseteq \mathrm{J}(B)$, then $A \bowtie^f J$ is a local ring with finitely generated maximal ideal, and $\mathrm{embdim}(A) = \mathrm{embdim}(A \bowtie^f J) + \nu(J)$.

□

3 Cohen-Macaulay and Gorenstein Property

3.1 Cohen-Macaulay Property

In this subsection, assuming that $A \bowtie^f J$ is local and Noetherian, we investigate the problem of when $A \bowtie^f J$ is a Cohen-Macaulay ring.

Theorem 3.1 ([16, Remark 5.1]) *Let $(A, \mathfrak{m})$ be a Noetherian local ring, $J \subseteq \mathrm{J}(B)$, and let J be a finitely generated A-module. Then the ring $A \bowtie^f J$ is Cohen-Macaulay if and only if it is a Cohen-Macaulay A-module if and only if J is a maximal Cohen-Macaulay A-module.* □

In [16, Remark 5.2], the authors mentioned that, if J is not finitely generated as an A-module, it is more problematic to find conditions implying $A \bowtie^f J$ Cohen-Macaulay. One can get more information if the embedding $\iota : A \to A \bowtie^f J$ is flat (or, equivalently, if the A-module J is flat). In this case, $A \bowtie^f J$ is Cohen-Macaulay if and only if both A and $(A \bowtie^f J)/(\mathfrak{m}(A \bowtie^f J))$ are Cohen-Macaulay. In [32], the authors provided conditions implying $A \bowtie^f J$ is Cohen-Macaulay. Recall that a finitely generated module M over a Noetherian local ring A is called a *maximal Cohen-Macaulay* A-module if depth $M = \dim A$. An A-module N is said to be *big Cohen-Macaulay* if depth $N = \dim A$.

Theorem 3.2 ([32, Theorem 2.3]) *Let $(A, \mathfrak{m})$ be a local ring, and let $J \subseteq \mathrm{J}(B)$ such that $f^{-1}(\mathfrak{q}) \neq \mathfrak{m}$ for each $\mathfrak{q} \in \mathrm{Spec}(B) \setminus \mathrm{V}(J)$. Then the following statements hold:*

(i) If $A \bowtie^f J$ is Cohen-Macaulay, then so does A.
(ii) Further assume that $\mathrm{depth}_A(J) < \infty$. *Then $A \bowtie^f J$ is Cohen-Macaulay if and only if A is Cohen-Macaulay and J is a big Cohen-Macaulay module.*

□

Using the fact that a finitely generated big Cohen-Macaulay module is maximal Cohen-Macaulay, we have the following result.

Corollary 3.3 ([32, Corollary 2.5]) *Let $(A, \mathfrak{m})$ be a local ring, and let $J \subseteq \mathrm{J}(B)$ be an ideal of B such that J is a finitely generated A-module. Assume that $A \bowtie^f J$ is Noetherian. Then $A \bowtie^f J$ is Cohen-Macaulay if and only if A is Cohen-Macaulay and J is a maximal Cohen-Macaulay A-module.* □

Corollary 3.4 ([32, Corollary 2.7]) *Assume that $f^{-1}(\mathfrak{q}) \neq \mathfrak{m}$ for each $\mathfrak{q} \in \mathrm{Spec}(B) \setminus \mathrm{V}(J)$ and each $\mathfrak{m} \in \mathrm{Max}(A)$. If $A \bowtie^f J$ is Cohen-Macaulay, then so does A.* □

Theorem 3.5 ([2, Theorem 3.1]) *Let $(A, \mathfrak{m})$ be a local ring, and let $J \subseteq \mathrm{J}(B)$. Then the following statements hold.*

(1) *Let* $\dim(A \bowtie^f J) = \dim A$. *If A is Cohen-Maculay and* $\mathrm{depth}_A(J) \geqslant \dim A$, *then $A \bowtie^f J$ is Cohen-Maculay.*
(2) *Suppose that for every $\mathfrak{q} \in \mathrm{Spec}(B) \setminus \mathrm{V}(J)$ with property $f^{-1}(\mathfrak{q}) = \mathfrak{m}$, we have* $\mathrm{ht}(f^{-1}(\mathfrak{q})) \leqslant \mathrm{ht}(\mathfrak{q})$. *If $A \bowtie^f J$ is Cohen-Maculay, then A is Cohen-Maculay and* $\mathrm{depth}_A(J) \geqslant \dim A$.

□

3.2 Serre's Conditions

A finitely generated module M over a Noetherian ring A satisfies Serre's condition (S_n) if depth $M_\mathfrak{p} = \min\{n, \dim M_\mathfrak{p}\}$, for all $\mathfrak{p} \in \mathrm{Spec}(A)$. Note that if M is Cohen-Macaulay, then it satisfies Serre's condition (S_n) for any integer n. Also, when $\dim M = d$ and M satisfies Serre's condition (S_d), then M is Cohen-Macaulay. In the following results, the property (S_n) for the amalgamated algebra is investigated. Recall that an ideal is called a *nil ideal* if each of its elements is nilpotent.

Proposition 3.6 *Let $A \bowtie^f J$ be a Noetherian ring. Then the following statements hold:*

(i) [32, Corollary 2.9] Assume that for each $\mathfrak{p} \in \mathrm{V}(f^{-1}(J))$ and each $\mathfrak{q} \in \mathrm{Spec}(B) \setminus \mathrm{V}(J)$, $f^{-1}(\mathfrak{q}) \neq \mathfrak{p}$ (e.g., if f is surjective or J is a nil ideal of B). If $A \bowtie^f J$ satisfies (S_n), then so does A.
(ii) [32, Corollary 2.9] Assume that $J^2 = 0$ and that J is a finitely generated A-module. If $A \bowtie^f J$ satisfies (S_n), then so does J.
(iii) [32, Proposition 2.11] If A and B satisfy (S_n), and $J_{S_\mathfrak{p}}$ is a maximal Cohen-Macaulay $A_\mathfrak{p}$-module for each prime ideal $\mathfrak{p}$ of A, where $S_\mathfrak{p} = f(A \setminus \mathfrak{p}) + J$, then $A \bowtie^f J$ satisfies (S_n).

□

Recall that a finitely generated module M over a Noetherian local ring $(A, \mathfrak{m})$ is said to be a generalized Cohen-Macaulay A-module if $\mathrm{H}^i_\mathfrak{m}(M)$ is of finite length for all $i < \dim M$. A local ring is called generalized Cohen-Macaulay if it is a generalized Cohen-Macaulay module over itself. It is clear that every Cohen-Macaulay module is a generalized Cohen-Macaulay module.

Theorem 3.7 ([32, Theorem 2.13]) *Let $(A, \mathfrak{m})$ be a local ring, and $J \subseteq \mathrm{J}(B)$ be an ideal of B such that J is a finitely generated A-module. Assume that $A \bowtie^f J$ is Noetherian. Then $A \bowtie^f J$ is a generalized Cohen-Macaulay ring if and only if A and J are generalized Cohen-Macaulay and* $\dim J \in \{0, \dim A\}$. □

Theorem 3.8 *The following statements hold:*

(i) [16, Remark 5.4] If $(A, \mathfrak{m})$ is a local Cohen-Macaulay ring, having a canonical module isomorphic (as an A-module) to J, then $A \bowtie^f J$ is Gorenstein.
(ii) [16, Proposition 5.5] Assume that A is a local Cohen-Macaulay ring and that $\mathrm{Ann}_{f(A)+J}(J) = (0)$. If $A \bowtie^f J$ is Gorenstein, then A has a canonical module isomorphic to $f^{-1}(J)$.

□

3.3 Gorenstein and Quasi-Gorenstein Properties

In the sequel, we investigate Gorenstein and quasi-Gorenstein properties of the amalgamation collected from [16] and [32].

Proposition 3.9 ([16, Proposition 5.7]) *Assume that $(A, \mathfrak{m})$ is Noetherian local ring, $0 \neq J \subseteq \mathrm{J}(B)$, and J is a finitely generated A-module, and moreover, we assume that A is a Cohen-Macaulay ring, $f(A) + J$ satisfies (S_1) and* *equidimensional, and that $f^{-1}(J)$ is a regular ideal of A. Then, the following conditions are equivalent:*

(i) *$A \bowtie^f J$ is Gorenstein.*
(ii) *$f(A) + J$ is a Cohen-Macaulay ring, J is a canonical module of $f(A) + J$, and $f^{-1}(J)$ is a canonical module of A.*

□

Definition 3.10 Let R be a local ring with maximal ideal $\mathfrak{m}$ of dimension d. An R-module K is called a *canonical module* of R if

$$K \otimes_R \widehat{R} \cong \mathrm{Hom}_R(\mathrm{H}^n_{\mathfrak{m}}(R), \mathrm{E}_R(R/\mathfrak{m})).$$

□

The following notion of a quasi-Gorenstein ring is from [30].

Definition 3.11 A local ring R is said to be a *quasi-Gorenstein ring* if a canonical module of R exists and is a free R-module (of rank one). This is equivalent to saying that $\mathrm{H}^d_{\mathfrak{m}}(R) \cong \mathrm{E}_R(R/\mathfrak{m})$, where $d = \dim R$ and $\mathfrak{m}$ is the maximal ideal of R. □

It is known that a local ring R is quasi-Gorenstein if and only if so is $\widehat{R}$. Also R is Gorenstein if and only if it is a quasi-Gorenstein and Cohen-Macaulay ring.

Theorem 3.12 *Let A be a Noetherian local, and let $J \subseteq \mathrm{J}(B)$, which is finitely generated A-module. Then the following statements hold:*

(i) *[32, Theorem 3.2] If $\widehat{A}$ satisfies (S_2) and J is a canonical module of A, then $A \bowtie^f J$ is quasi-Gorenstein.*
(ii) *[32, Theorem 3.2] Let $J^2 = 0$ and $\mathrm{Ann}_A(J) = 0$. If $A \bowtie^f J$ is quasi-Gorenstein, then A satisfies (S_2), and J is canonical module of A.*
(iii) *[32, Theorem 3.2] Let $\mathrm{Ann}_{f(A)+J}(J) = 0$ and $A \bowtie^f J$ be quasi-Gorenstein. Then $f^{-1}(J)$ is a canonical module of A. Furthermore, if f is surjective, then $\widehat{A}$ satisfies (S_2).*
(iv) *[32, Theorem 3.2] Let J be a flat A-module and $A \bowtie^f J$ be quasi-Gorenstein. Then A is quasi-Gorenstein.*
(v) *[32, Corollary 3.3] Let A be Cohen-Macaulay and J be a canonical module of A. Then $A \bowtie^f J$ is Gorenstein.*
(vi) *[32, Corollary 3.4] Assume that at least one of the following conditions holds.*

  (1) *f is isomorphism and $\mathrm{Ann}_B(J) = 0$.*
  (2) *$J^2 = 0$ and $\mathrm{Ann}_A(J) = 0$.*

  Then $A \bowtie^f J$ is Gorenstein if and only if A is Cohen-Macaulay and J is a canonical module of A.
(vii) *[32, Corollary 3.5] Let $\mathrm{Ann}_{f(A)+J}(J) = 0$. If $A \bowtie^f J$ is Gorenstein, then A is Cohen-Macaulay and $f^{-1}(J)$ is a canonical ideal of A.* □

3.4 Cohen-Macaulay Property in the Sense of Ideals

In Theorems 3.1 and 3.2, the Cohen-Macaulay property of $A \bowtie^f J$ in the case that A is a Noetherian local ring is investigated. In [4], the authors investigated the property of Cohen-Macaulayness in the sense of ideals (respectively, maximal ideals, finitely generated ideals) on the amalgamation.

Definition 3.13 Let $\mathfrak{b}$ be an ideal of A such that is generated by the sequence $\mathbf{x} = x_1, \cdots, x_\ell$, and let M be an A-module. We denote the *Koszul complex* related to $\mathbf{x}$ by $\mathbb{K}_\bullet(\mathbf{x})$. The *Koszul grade* of $\mathfrak{b}$ on M is defined by

$$\mathrm{K.grade}_A(\mathfrak{b}, M) := \inf\{i \in \mathbb{N} \cup \{0\} \mid \mathrm{H}^i(\mathrm{Hom}_A(\mathbb{K}_\bullet(\mathbf{x}), M) = 0\},$$

and it is shown that this does not depend upon the choice of generating sets of $\mathfrak{b}$. Let $\mathfrak{a}$ be an arbitrary ideal of A. The Koszul grade of $\mathfrak{a}$ on M can then be defined $\mathrm{K.grade}_A(\mathfrak{a}, M) := \sup\{\mathrm{K.grade}_A(\mathfrak{b}, M) \mid \mathfrak{b}$ is a finitely generated subideal of $\mathfrak{a}\}$. This definition coincides with the original one for finitely generated ideals. Let $\mathcal{A}$ be a non-empty subclass of the class of all ideals of the ring A. We say that M is Cohen-Macaulay in the sense of $\mathcal{A}$ if $\mathrm{ht}_M(\mathfrak{a}) = \mathrm{K.grade}_A(\mathfrak{a}, M)$ for all ideals $\mathfrak{a}$ in $\mathcal{A}$. □

Theorem 3.14 ([4, Theorem 4.1]) *Let $(A, \mathfrak{m})$ be a local ring such that $\mathfrak{m}$ is finitely generated, and let $J \subseteq \mathrm{J}(B)$ be a finitely generated A-module. Then $A \bowtie^f J$ is Cohen-Macaulay (ring) in the sense of maximal ideals if and only if A is Cohen-Macaulay in the sense of maximal ideals and* $\mathrm{K.grade}_A(\mathfrak{m}, J) = \dim A$. □

Theorem 3.15 ([4, Theorem 4.6]) *Let $J \subseteq \mathrm{Nil}(B)$. Then $A \bowtie^f J$ is Cohen-Macaulay (ring) in the sense of ideals if and only if A is Cohen-Macaulay in the sense of ideals and* $\mathrm{K.grade}_A(\mathfrak{a}, J) \geqslant \mathrm{ht}\,\mathfrak{a}$ *for every ideal $\mathfrak{a}$ of A.* □

Corollary 3.16 ([4, Corollary 4.9]) *Let A be a Noetherian ring, and let $J \subseteq \mathrm{Nil}(B)$ be a finitely generated A-module. Then $A \bowtie^f J$ is Cohen-Macaulay if and only if A is Cohen-Macaulay and $J_\mathfrak{p}$ is maximal Cohen-Macaulay for every ideal* $\mathfrak{p} \in \mathrm{Supp}_A(J)$. □

Theorem 3.17 *With the notation of Proposition 2.3, the following statements hold:*

(i) [4, Theorem 4.11] Let $\mathcal{A}$ be a non-empty class of ideals of A. Assume that the homomorphism f satisfies the going-down property. If $A \bowtie^f J$ is Cohen-Macaulay (ring) in the sense of $\mathcal{A}^e := \{\mathfrak{a}^e \mid \mathfrak{a} \in \mathcal{A}\}$, then A is Cohen-Macaulay in the sense of $\mathcal{A}$, and $\mathrm{K.grade}_A(\mathfrak{a}, J) \geqslant \mathrm{ht}\,\mathfrak{a}$ *for every $\mathfrak{a} \in \mathcal{A}$.*

(ii) [4, Theorem 4.11] Assume that ι_A is an integral ring extension. If A is Cohen-Macaulay in the sense of ideals and $\mathrm{K.grade}_A(\mathfrak{a}, J) \geqslant \mathrm{ht}\,\mathfrak{a}$ *for every ideal $\mathfrak{a}$ of $\mathcal{A}$, then $A \bowtie^f J$ is Cohen-Macaulay (ring) in the sense of ideals.*

(iii) [4, Corollary 4.13] Assume that the homomorphism $f : A \to B$ satisfies the going-down property and that J is finitely generated as an A-module. Then,

$A \bowtie^f J$ is Cohen-Macaulay (ring) in the sense of ideals if and only if A is Cohen-Macaulay in the sense of ideals and $\mathrm{K.grade}_A(\mathfrak{a}, J) = \mathrm{ht}\,\mathfrak{a}$ *for every ideal $\mathfrak{a}$ of A.*

(iv) *[4, Corollary 4.14] Assume that $f : A \to B$ is a monomorphism of integral domains, A is integrally closed, and that B is integral over A. Then, $A \bowtie^f J$ is Cohen-Macaulay (ring) in the sense of ideals if and only if A is Cohen-Macaulay in the sense of ideals and* $\mathrm{K.grade}_A(\mathfrak{a}, J) \geqslant \mathrm{ht}\,\mathfrak{a}$ *for every ideal $\mathfrak{a}$ of A.*

(v) *[4, Corollary 4.15] Assume that $f : A \to B$ is flat and integral homomorphism. Then, $A \bowtie^f J$ is Cohen-Macaulay (ring) in the sense of ideals if and only if A is Cohen-Macaulay in the sense of ideals and* $\mathrm{K.grade}_A(\mathfrak{a}, J) \geqslant \mathrm{ht}\,\mathfrak{a}$ *for every ideal $\mathfrak{a}$ of A.*

□

3.5 Cohen-Macaulay Property in the Sense of Hamilton and Marley

One generalization of the usual concept of Cohen-Macaulayness to non-Noetherian rings has been given by Hamilton and Marley [21]. This notion then extended to modules in [23]. An A-module M is called Cohen-Macaulay if every strong parameter sequence on M is an M-regular sequence. This definition agrees with the usual definition of Cohen-Macaulay finitely generated modules over Noetherian rings [23].

Theorem 3.18 ([5, Theorem 3.3]) *Assume that $J \subseteq \mathrm{Nil}(B)$ and that every A-weakly proregular sequence is a J-weakly proregular sequence. Then $A \bowtie^f J$ is Cohen-Macaulay if and only if A is Cohen-Macaulay and every A-regular sequence is a weak J-regular sequence.* □

4 Clean Property and Prüfer-Like Conditions

4.1 Clean Property

The concept of clean rings was introduced by Nicholson [29]. Recall that a ring A is called (uniquely) clean if each element in A can be written (uniquely) as the sum of a unit and an idempotent. Examples of clean rings (uniquely clean rings) include all commutative von-Neumann regular rings (Boolean rings) and local rings (with residue field $\mathbb{Z}_2$). A basic property of clean rings is that any homomorphic image of a clean ring is again clean. This leads to the definition of a neat rings [28].

In the following, the transfer of clean property between A and its amalgamation is investigated.

Proposition 4.1 *The following statements hold:*

(i) *[12, Proposition 2.1] If $A \bowtie^f J$ is a clean (resp., uniquely clean) ring, then A is a clean (resp., uniquely clean) ring and $f(A) + J$ is a clean ring.*
(ii) *[12, Proposition 2.1] Let $(f(A) + J)/J$ be uniquely clean. Then $A \bowtie^f J$ is a clean ring if and only if A and $f(A) + J$ are clean rings.*
(iii) *[12, Theorem 2.4] Assume that $f(u) + j$ is invertible for each $u \in \mathrm{U}(A)$ and $j \in J$. Then $A \bowtie^f J$ is clean (resp., uniquely clean) if and only if A is clean (resp., uniquely clean).*

More generally, if $J \cap \mathrm{Idem}(B) = 0$ then, the following are equivalent:

(1) *$A \bowtie^f J$ is clean (resp., uniquely clean).*
(2) *A is clean (resp., uniquely clean) and $J \subseteq \mathrm{J}(B)$.*

(iv) *[12, Corollary 2.6] Let $J \subseteq \mathrm{J}(B)$. Then $A \bowtie^f J$ is clean (resp., uniquely clean) if and only if A is clean (resp., uniquely clean).*
(v) *[12, Corollary 2.11] Let $J \subseteq \mathrm{Nil}(B)$. Then $A \bowtie^f J$ is clean (resp. uniquely clean) if and only if A is clean (resp. uniquely clean).*
(vi) *[12, Corollary 2.16] The following statements are equivalent.*

(i) *$A \bowtie^f J$ is local and uniquely clean.*
(ii) *A is local and uniquely clean and $J \subseteq \mathrm{J}(B)$.*

(vii) *[12, Proposition 2.19] If $J \subseteq \mathrm{Idem}(B)$, then $A \bowtie^f J$ is clean if and only if A is clean.*

□

It is well-known that von-Neumann regular rings are particular cases of clean rings. In the following proposition, von-Neumann regular property of amalgamation is investigated.

Proposition 4.2 ([12, Proposition 2.21]) *The following statements hold.*

(i) *If A and $f(A) + J$ are von-Neumann regular rings, then so is $A \bowtie^f J$.*
(ii) *If $A \bowtie^f J$ is von-Neumann regular, then A is von-Neumann regular and $J \cap \mathrm{Nil}(B) = (0)$, and the converse holds if f is surjective.*

□

4.2 Prüfer-Like Conditions

Prüfer domains are introduced by H. Prüfer in [31]. This class was deeply studied by several authors, and many equivalent definitions of Prüfer domain were given.

Among the many equivalent conditions that make an integral domain A a Prüfer domain, we recall the following:

(1) Every finitely generated ideal of A is projective.
(2) $A_{\mathfrak{p}}$ is a valuation domain, for every $\mathfrak{p} \in \operatorname{Spec}(A)$.
(3) Every finitely generated ideal of A is locally principal.
(4) If T is an indeterminate over A, every polynomial $f \in A[T]$ is a Gauss polynomial over A.
(5) Every non-zero finitely generated ideal of A is invertible.

In [20], the notion of Prüfer domain was generalized to arbitrary commutative ring possibly with zero-divisors. In [9], the authors showed that none of the previous conditions is equivalent to the others, when A is a ring with zero-divisors. In general, the rings satisfying previous Prüfer-like conditions are distinct classes of rings leads us to recall the following definition.

Definition 4.3 Let A be a ring.

(P1) A is called a *semi-hereditary ring* if every finitely generated ideal of A is projective.
(P2) A has weak global dimension at most 1 if $A_{\mathfrak{p}}$ is a valuation domain, for every $\mathfrak{p} \in \operatorname{Spec}(A)$.
(P3) A is called an *arithmetical ring* if every finitely generated ideal of A is locally principal.
(P4) A is called a *Gauss ring* if every polynomial $f \in A[T]$ is a Gauss polynomial over A.
(P5) A is called a *Prüfer ring* if every regular and finitely generated ideal of A is invertible.

□

Theorem 4.4 ([19, Theorem 3.1 and Corollary 3.2]) *Let $n \in \{1, 2, 3, 4, 5\}$, and let $f^{-1}(J)$ and J be regular ideals. Then the following statements are equivalent:*

(i) $A \bowtie^f J$ satisfies Prüfer-like conditions (P_n).
(ii) A and B satisfy Prüfer-like conditions (P_n) and $J = B$.

□

Theorem 4.5 *The following statements hold:*

(i) [19, Proposition 4.2] Let $f(\operatorname{Reg}(A)) \subseteq \operatorname{Reg}(B)$, and let $A \bowtie^f J$ be a Prüfer ring. Then A is Prüfer.
(ii) [19, Corollary 4.5] Let $A \bowtie^f J$ be a Prüfer ring. If $\{0\} \times J$ is a regular ideal of $A \bowtie^f J$, then A is a Prüfer ring.
(iii) [19, Corollary 4.5] Let $A \bowtie^f J$ be a Prüfer ring. If $f^{-1}(J) \times \{0\}$ is a regular ideal of $A \bowtie^f J$, then $f(A) + J$ is a Prüfer ring.
(iv) [25, Theorem 2.2] Let $(A, \mathfrak{m})$ be a local ring, and let $J \subseteq \mathrm{J}(B)$. Assume that $J \subseteq f(A)$, and $f(\operatorname{Reg}(A)) = \operatorname{Reg}(B)$. Then $A \bowtie^f J$ is Prüfer if and only if so is A and $J = f(a)J$ for all $a \in \mathfrak{m} \setminus \mathrm{Z}(A)$. □

Total ring of quotients are important source of Prüfer rings. The next proposition studies when the amalgamation is total ring of quotient in case $J^2 = 0$.

Proposition 4.6 ([25, Proposition 2.10]) *Let $f^{-1}(J) \neq 0$, $f(\mathrm{Z}(A)) \subseteq J$, and $J^2 = 0$. Suppose that A is a total ring of quotient. Then the following statements hold:*

(i) $Q(A \bowtie^f J) = A \bowtie^f J$*. In particular,* $A \bowtie^f J$ *is Prüfer.*
(ii) $W.\dim(A \bowtie^f J) \geqslant 1$.

□

We say that the amalgamated ring $A \bowtie^f J$ has condition $\star$ if the equality $\mathrm{Z}(A \bowtie^f J) = \{(a, f(a) + j) \mid a \in \mathrm{Z}(A), j \in J\} \cup \{(a, f(a) + j) \mid a \in A, j \in J, \exists j' \in J \setminus \{0\} : j'(f(a) + j) = 0\}$ holds.

Theorem 4.7 *With the notation of Proposition 2.5, suppose that $f(\mathrm{Reg}(A)) \subseteq \mathrm{Reg}(B)$. Then the following statements hold:*

(i) *[6, Theorem 3.5] If $A \bowtie^f J$ is a Prüfer ring, then A is a Prüfer ring and $J_{S_\mathfrak{m}} = f(r)J_{S_\mathfrak{m}}$ for every $\mathfrak{m} \in \max(A)$ and $r \in \mathrm{Reg}(A)$.*
(ii) *[6, Theorem 3.5] Let $A \bowtie^f J$ has condition $\star$, and let $\mathrm{Z}(A \bowtie^f J) \subseteq \mathrm{J}(A \bowtie^f J)$. If A is a Prüfer ring and $J_{S_\mathfrak{m}} = f(r)J_{S_\mathfrak{m}}$ for every $\mathfrak{m} \in \max(A)$ and $r \in \mathrm{Reg}(A)$, then $A \bowtie^f J$ is a Prüfer ring.*
(iii) *[6, Corollary 3.6] Let $(A, \mathfrak{m})$ be a local ring, $J \subseteq \mathrm{J}(B)$. Then the following statements hold:*

  (1) *If $A \bowtie^f J$ is a Prüfer ring, then A is a Prüfer ring and $J = f(r)J$ for every $r \in \mathrm{Reg}(A)$.*
  (2) *Let $A \bowtie^f J$ has condition $\star$. If A is a Prüfer ring and $J = f(r)J$ for every $r \in \mathrm{Reg}(A)$, then $A \bowtie^f J$ is a Prüfer ring.*

□

Proposition 4.8 ([6, Proposition 3.15]) *The following statements hold:*

(i) *Let $J \subseteq \mathrm{J}(B)$, and let $A \bowtie^f J$ has condition $\star$. Then $A \bowtie^f J$ is a total ring of quotients, provided that A is a total ring of quotients.*
(ii) *Let $f(\mathrm{Reg}(A)) \subseteq \mathrm{Reg}(B)$. Then A is a total ring of quotients, provided that $A \bowtie^f J$ is a total ring of quotients.*

□

Theorem 4.9 ([6, Theorem 4.1]) *Let $(A, \mathfrak{m})$ be a local ring, and let $J \subseteq f(A) \cap \mathrm{J}(B)$ (e.g., f is surjective). Then $A \bowtie^f J$ is Gaussian if and only if A is Gaussian, $J^2 = 0$ and $f(r)J = f(r)^2 J$ for every $r \in \mathfrak{m}$.* □

Corollary 4.10 ([6, Corollary 4.2]) *Let $J \subseteq f(A) \cap \mathrm{J}(B)$. Then $A \bowtie^f J$ is Gaussian if and only if A is Gaussian, $J_\mathfrak{m}^2 = 0$ and $f(r)J_\mathfrak{m} = f(r)^2 J_\mathfrak{m}$ for every $\mathfrak{m} \in \max(A) \cap \mathrm{V}(f^{-1}(J))$ and $r \in \mathfrak{m}$.* □

Theorem 4.11 ([26, Theorem 5.1]) *Let $(A, \mathfrak{m})$ be a local ring, and let $J \subseteq \mathrm{J}(B)$. Then $A \bowtie^f J$ is Gaussian if and only if the following conditions hold:*

(1) *A is Gaussian.*
(2) $(f(a)+j)J = (f(a)+j)^2 J$ *for each* $a \in A$, *and each* $j \in J$.
(3) $J^2 = 0$.
or
(4) A *and* $f(A)+J$ *are Gaussian.*
(5) $x^2 = 0$ *for each* $x \in f^{-1}(J)$.
(6) $(f(a)+j)J = (f(a)+j)^2 J$ *for each* $a \in A$, *and each* $j \in J$.
(7) *If* $f(a^2) = 0$, *then* $a^2 = 0$.

□

Corollary 4.12 ([26, Corollary 5.2]) *With the notation of Proposition 2.5, the ring $A \bowtie^f J$ is Gaussian if and only if the following statements hold:*

(1) *A is Gaussian.*
(2) B_Q *is Gaussian for each* $Q \in \mathrm{Max}(B) \setminus \mathrm{V}(J)$.
(3) *For each* $\mathfrak{m} \in Max(A)$ *containing* $f^{-1}(J)$, *for each* $a \in A_{\mathfrak{m}}$, *and each* $j \in J_{S_{\mathfrak{m}}}$. $(f_{\mathfrak{m}}(a)+j)J_{S_{\mathfrak{m}}} = (f_{\mathfrak{m}}(a)+j)^2 J_{S_{\mathfrak{m}}}$,
(4) *For each* $\mathfrak{m} \in Max(A)$ *containing* $f^{-1}(J)$, $J^2_{S_{\mathfrak{m}}} = 0$.
or
(5) *A is Gaussian.*
(6) *For each* $\mathfrak{m} \in Max(A)$ *containing* $f^{-1}(J)$, $f_{\mathfrak{m}}(A_{\mathfrak{m}}) + J_{S_{\mathfrak{m}}}$ *is Gaussian.*
(7) B_Q *is Gaussian for each* $Q \in \mathrm{Max}(B) \setminus \mathrm{V}(J)$.
(8) $x^2 = 0$ *for each* $\mathfrak{m} \in Max(A)$ *containing* $f^{-1}(J)$, *and for each* $x \in f_{\mathfrak{m}}^{-1}(J_{S_{\mathfrak{m}}})$.
(9) *For each* $\mathfrak{m} \in Max(A)$ *containing* $f^{-1}(J)$. *for each* $a \in A_{\mathfrak{m}}$, *and each* $j \in J_{S_{\mathfrak{m}}}$.
(10) *If* $f_{\mathfrak{m}}(a^2) = 0$, *then* $a^2 = 0$, *for each* $\mathfrak{m} \in Max(A)$ *containing* $f^{-1}(J)$, *and for each* $a \in A_{\mathfrak{m}}$.

□

5 Other Properties

Note that Proposition 3.9 provides the necessary and sufficient conditions of self-injectivity of the ring $A \bowtie^f J$. As a nice generalization of injectivity for modules, Xu in [38] introduced the terminology of strongly cotorsion modules. Recall that an A-module M is called a strongly cotorsion module if $\mathrm{Ext}^1_A(F, M) = 0$ for all A-modules F with finite flat dimension. In [36], the authors investigated the strongly cotorsion property of $\mathrm{H}^{\dim A}_{\mathfrak{m}\bowtie^f J}(A \bowtie^f J)$, where $(A, \mathfrak{m})$ is a Noetherian local ring.

Theorem 5.1 ([36, Theorem 2.2]) *Let $(A, \mathfrak{m})$ be a Noetherian local ring with dimension d and $0 \neq J \subseteq \mathrm{J}(B)$, such that J is a finitely generated A-module.*

Then $\mathrm{H}^d_{\mathfrak{m}\bowtie^f J}(A \bowtie^f J)$ *is a strongly cotorsion A-module if and only if* $\mathrm{H}^d_{\mathfrak{m}}(A)$ *and* $\mathrm{H}^d_{\mathfrak{m}}(J)$ *are strongly cotorsion A-modules.* □

In the sequel, we investigate some homological properties of the amalgamation collected from [36].

Proposition 5.2 *Let J be a flat A-module. Then the following statements hold for any A-module M:*

(i) $\mathrm{fd}_A(M) = \mathrm{fd}_{A\bowtie^f J}(M \otimes_A (A \bowtie^f J)) = \mathrm{fd}_A(M \otimes_A (A \bowtie^f J))$.
(ii) $\mathrm{pd}_A(M) = \mathrm{pd}_{A\bowtie^f J}(M \otimes_A (A \bowtie^f J))$.
(iii) $\mathrm{id}_A(M) = \mathrm{id}_A(M \otimes_A (A \bowtie^f J))$.

□

Proposition 5.3 *The following statements hold for every A-module M:*

(i) If M is a (faithfully) injective A-module, then $Hom_A(A \bowtie^f J, M)$ *is a (faithfully) injective* $(A \bowtie^f J)$*-module.*
(ii) Every injective $(A \bowtie^f J)$*-module is a direct summand of the A-module* $Hom_A(A \bowtie^f J, M)$*, where M is an injective A-module.*

□

Recall that an A-module M is said to be uniserial if the set of its (cyclic) submodules is totally ordered by inclusion and A is a chained ring if it is uniserial as A-module.

Theorem 5.4 ([6, Theorem 5.1]) *Let* $J \neq 0$*, and let* $A \bowtie^f J$ *be a chained ring. Then A is a valuation domain and* $J = (f(a) + j)J$ *for every* $0 \neq a \in A$ *and* $j \in J$*. The converse holds provided that J is a uniserial A-module.* □

Corollary 5.5 ([6, Corollary 5.2]) *The following statements hold:*

(i) Let $J \neq 0$ *such that* $J^2 = 0$*. Then* $A \bowtie^f J$ *is a chained ring if and only if A is a valuation domain, J is a uniserial A-module, and* $J = f(a)J$ *for every* $0 \neq a \in A$*.*
(ii) Let $J \subseteq f(A)$*. Then* $A \bowtie^f J$ *is a chained ring if and only if A is a chained ring and* $J = 0$*.*

□

Recall that $\mathfrak{p} \in \mathrm{Spec}(A)$ is a *divided prime ideal* in A if $\mathfrak{p}$ is comparable under inclusion to each ideal of A and A is a *divided ring* if $\mathfrak{p} \in \mathrm{Spec}(A)$ is divided in A (see [7]). A ring A is called *locally divided* if $A_{\mathfrak{p}}$ is a divided ring for each $\mathfrak{p} \in \mathrm{Spec}(A)$. Each divided ring is locally divided [7, Proposition 4]. In the following theorem, the transfer of the properties of being a divided ring and a locally divided ring between rings A, B, and $A \bowtie^f J$ is studied.

Theorem 5.6 ([26, Theorem 3.1]) *With the notation of Proposition 2.5, the following statements hold:*

(1) *Let A be an integral domain. Then $A \bowtie^f J$ is a divided ring if and only if the following statements hold:*

(i) $f^{-1}(J) = 0$.
(ii) *A and $f(A) + J$ are divided rings.*
(iii) $\{0\} \times J \subseteq ((a, f(a) + j))$ *for all $a \in A - \{0\}$ and $j \in J$.*

(1′) *Let $A \bowtie^f J$ be a locally divided ring. Then A is locally divided and the following statements hold:*

(a) *For each $\mathfrak{p} \in \operatorname{Spec}(A)$ containing $f^{-1}(J)$ such that $A_\mathfrak{p}$ is an integral domain, the following statements hold:*

(i) $f_\mathfrak{p}^{-1}(J_{S_\mathfrak{p}}) = 0$, *where* $S_\mathfrak{p} = f(A \setminus \mathfrak{p}) + J$.
(ii) $f_\mathfrak{p}(A_\mathfrak{p}) + J_{S_\mathfrak{p}}$ *is a divided ring.*
(iii) $\{0\} \times J_{S_\mathfrak{p}} \subseteq ((a/s, f_\mathfrak{p}(a/s) + j/t))$ *for all* $a/s \in A_\mathfrak{p} - \{0\}$ *and* $j/t \in J_{S_\mathfrak{p}}$.

(b) *For each $\mathfrak{p} \in \operatorname{Spec}(A)$ containing $f^{-1}(J)$ such that $A_\mathfrak{p}$ is a ring with zero-divisors, the following statements hold:*

(i) $J_{S_\mathfrak{p}} \subseteq \operatorname{Nil}(B_{S_\mathfrak{p}})$.
(ii) $\{0\} \times J_{S_\mathfrak{p}} \subseteq ((a/s, f_\mathfrak{p}(a/s) + j/t))$ *for all* $\mathfrak{q} \in \operatorname{V}(\mathfrak{p})$, $a/s \in A_\mathfrak{p} \setminus \mathfrak{q}_\mathfrak{p}$ *and* $j/t \in J_{S_\mathfrak{p}}$.

(2) *Let $\operatorname{Z}(A) \neq \{0\}$. Then $A \bowtie^f J$ is a divided ring if and only if the following statements hold.*

(i) $J \subseteq \operatorname{Nil}(B)$.
(ii) *A is a divided ring.*
(iii) $\{0\} \times J \subseteq ((a, f(a) + j))$ *for all $\mathfrak{p} \in \operatorname{Spec}(A)$, for all $i \in A \setminus \mathfrak{p}$ and $j \in J$.*

(2′) *If moreover B is locally divided, then A is locally divided.*

□

Recall that a domain A is called a *going-down domain* if $A \subseteq T$ satisfies the going-down property GD for each domain T. The ring A is called *going-down ring* if $A/\mathfrak{p}$ is going-down domain for each $\mathfrak{p} \in \operatorname{Spec}(A)$.

In the following theorem, the transfer of the notion of going-down rings to their amalgamation is studied.

Theorem 5.7 ([26, Theorem 4.1]) *Let $J \neq 0$. Then the following statements hold:*

(i) *If $A \bowtie^f J$ is a going-down ring, then A is a going-down ring.*
(ii) *If $\mathrm{J} \subseteq \operatorname{Nil}(B)$, then A is a going-down ring if and only $A \bowtie^f J$ if is a going-down ring.*
(iii) *Suppose that $\mathfrak{q} \in \operatorname{Spec}(f(A) + J)$ for each $\mathfrak{q} \in \operatorname{Spec}(B) \setminus \operatorname{V}(J)$. Then $A \bowtie^f J$ if is a going-down ring provided that A and $f(A) + J$ are going-down rings.*

(iv) *If $f^{-1}(J) = 0$, then $A \bowtie^f J$ is a going-down ring if and only if $f(A) + J$ is a going-down ring.*

□

Recall that the ring A is called $(\mathcal{A})$-ring if for every finitely generated ideal I of A such that $I \subset \mathrm{Z}(A)$, we have $\mathrm{Ann}(I) \neq 0$, (see [22]).

Theorem 5.8 *The following statements hold:*

(i) *[24, Theorem 2.1] Let $f^{-1}(J)$ be a regular ideal of A, and let $f^{-1}(J)$ be a regular ideal of $f(A) + J$. Then $A \bowtie^f J$ is an $(\mathcal{A})$-ring if and only if so are A and B.*

(ii) *[24, Proposition 2.8] Let $J \neq 0$ be a proper ideal of B such that $f^{-1}(J) \neq 0$, and $J^2 = 0$. Then $A \bowtie^f J$ is an $(\mathcal{A})$-ring, provided that A is $(\mathcal{A})$-ring.*

(iii) *[24, Corollary 2.9] Let $J \neq 0$ be a proper ideal of B such that $f^{-1}(J) \neq 0$, and $J^2 = 0$. Suppose that A is a total ring of quotient which is $(\mathcal{A})$-ring and $f(\mathrm{Z}(A)) \subseteq J$. Then the following statements hold:*

(1) *$A \bowtie^f J$ is an $(\mathcal{A})$-ring.*
(2) *$Q(A \bowtie^f J) = A \bowtie^f J$ is not von-Neumann regular ring.*

□

References

1. Y. Azimi, *The diameter of the zero-divisor graph of an amalgamated algebra*, Collectanea Mathematica, **70**, (2019), 399–405.
2. Y. Azimi, *Linkage property under the amalgamated construction*, Communications in Algebra, **49**, (2021), Issue 9.
3. Y. Azimi, M. R. Doustimehr, *The zero-divisor graph of an amalgamated algebra*, Rendiconti del Circolo Matematico di Palermo, series 2, **70**, (2021), 1213–1225 (2021).
4. Y. Azimi, P. Sahandi, N. Shirmihammadi, *Cohen-Macaulay properties under the amalgamated construction*, J. Commut. Algebra **10**, (2018), 457–474.
5. Y. Azimi, P. Sahandi, N. Shirmihammadi, *Cohen-Macaulayness of Amalgamated Algebras*, Algebr Represent Theor. **23**, (2020), 275–280.
6. Y. Azimi, P. Sahandi, N. Shirmohammadi, *Prüfer conditions in amalgamated algebras*, Communications in Algebra, **47**, (2019), Issue 5, 2251–2261.
7. A. Badawi, *On Divided Commutative Rings*, Communications in Algebra, (1999), **27**, 1465–1474.
8. A. Bagheri, M. Salimi, E. Tavasoli, S. Yassemi, *construction of quasi- Gorenstein rings*, J. Algebra Appl. (2012), **11**(1):1250013 (9 pages).
9. S. Bazzzoni, S. Glaz, *Gaussian properties of total rings of quotients*, J. Algebra **310** (2007) 180–193.
10. M. B. Boisen, P. B. Sheldon, *CPI-extension: overrings of integral domains with special prime spectrum*, Canad. J. Math. **29** (1977), 722–737.
11. M. Chhiti, M. Jarrar, S. Kabbaj, N. Mahdou, *Prüfer conditions in amalgamated duplication of a ring along an ideal*, Communications in Algebra **43** (1), 249–261, (2015).
12. M. Chhiti, N. Mahdou, M. Tamekkante, *Clean property in amalgamated algebras along an ideal*, Hacettepe Journal of Mathematics and Statistics, Vol. 44 (1) (2015), 41–49.
13. M. D'Anna, *A construction of Gorenstein rings*, J. Algebra **306** (2006), 507–519.

14. M. D'Anna, M. Fontana, *An amalgamated duplication of a ring along an ideal: The basic properties.*, J. Algebra Appl. **6** (2007), no. **3**, 443–459.
15. M. D'Anna, C. A. Finocchiaro, M. Fontana, *Amalgamated algebras along an ideal*, in: Commutative Algebra and Applications, Proceedings of the Fifth International Fez Conference on Commutative Algebra and Applications, Fez, Morocco, 2008, W. de Gruyter Publisher, Berlin, 2009, pp. 155–172.
16. M. D'Anna, C. A. Finocchiaro, M. Fontana, *New algebraic properties of an amalgamated algebra along an ideal*, Communications in Algebra, **44**(5), (2016), 1836–1851.
17. M. D'Anna, C. A. Finocchiaro, M. Fontana, *Properties of chains of prime ideals in an amalgamated algebra along an ideal*, Journal of Pure and Applied Algebra, **214** (2010), 1633–1641.
18. J. L. Dorroh, *Concerning adjunctions to algebras*, Bull. Amer. Math. Soc. **38** (1932), 85–88.
19. C. Finocchiaro, *Prüfer-like conditions on amalgamated algebra along an ideal*, Houston J. math **40**(1), (2014), 63–79.
20. M. Griffin, *Prüfer rings with zero-divisors*, J. Reine Angew. Math **239/240** (1969) 55–67.
21. T. D. Hamilton, T. Marley, *Non-Noetherian Cohen-Macaulay rings*, J. Algebra **307**, (2007), 343–360.
22. J. A. Huckaba, *Commutative rings with zero-divisors*, Marcel Dkker, New York, 1988.
23. A. Mahdikhani, P. Sahandi, N. Shirmohammadi, *Cohen-Macaulayness of trivial extensions*, J. Algebra Appl. **17**, 1850008 (2018), (15 pages).
24. N. Mahdou, M. A. S. Moutui, *On (A)-rings and strong (A)-rings issued from amalgamations*, Studia Scientiarum Mathematicarum Hungarica, **55**(2) (2018), 270–279.
25. N. Mahdou, M. A. S. Moutui, *Prüfer Property in amalgamated algebras along an ideal*, Ricerche di Matematica, **69** (2020), 111–120.
26. N. Mahdou, S. Moussaoui, S. Yassemi, *The divided, going-down, and Gaussian properties of amalgamation of rings*, Communications in Algebra **49**, (2021), no.5, 1938–1949.
27. H. R. Maimani, S. Yassemi, *Zero-divisor graphs of amalgamated duplication of a ring along an ideal*, Journal of Pure an Applied Algebra **212**, no. 1, (2008), 168–174.
28. W. McGovern, *Neat rings*, J. Pure Appl. Algebra **205** (2006), 243–265.
29. W. K. Nicholson, *Lifting idempotents and exchange rings*, Trans. Amer. Math. Soc. 229 (1977), 278–279.
30. E. Platte and U. Storch, *Invariante regulare Differential formen auf Gorenstein-Algebren*, Math. Z., **157** (1997), 1–11.
31. H. Prüfer, *Untersuchungen über teilbarkeitseigenschaften in körpern*, J. Reine Angew. Math **168** (1932) 1–36.
32. P. Sahandi, N. Shirmohammadi, S. Sohrabi, *Cohen-Macaulay and Gorenstein Properties under the Amalgamated Construction*, Communications in Algebra, **44**, (2016), 1096–1109.
33. M. Salimi, E. Tavasoli, S. Yassemi, *The amalgamated duplication of a ring along a semidualizing ideal*, Rend. Sem. Mat. Univ. Padova **129** (2013), 115–127.
34. J. Shapiro, *On a construction of Gorenstein rings proposed by M. D'Anna*, J. Algebra **323**, (2010), 1155–1158.
35. H. Shoar, M. Salimi, A. Tehraninan, H. Rasouli, E. Tavasoli, *Comaximal gragh of amalgamated algebras along an ideal*, Journal of Algebra and its Application, **22**(1), (2023), 2350021 (11 Pages).
36. E. Tavasoli, *Some homological properties of amalgamation*, Mat. Vesn. **68**(4), (2016), 254–258.
37. E. Tavasoli, M. Salimi, A. Tehranian, *Amalgamated duplication of some special rings*, Bull. Korean. Math. Soc. **49** (2012), No. 5, pp. 989–996.
38. J. Xu, *Flat Covers of Modules*, Lecture Notes in Mathematics, Vol. 1634, Springer, New York, 1996.

The Ring of Integer-Valued Polynomials on 3 x 3 Matrices and Its Integral Closure

Asmita C. Sodhi

Mathematics Subject Classification: Primary: 13F20, Secondary: 11C08, 16S36

1 Introduction

Let $M_n(\mathbb{Z})$ be the ring of $n \times n$ matrices with integer entries. Then we can denote by

$$\text{Int}_{\mathbb{Q}}(M_n(\mathbb{Z})) = \{f \in \mathbb{Q}[x] : f(M) \in M_n(\mathbb{Z}) \text{ for all } M \in M_n(\mathbb{Z})\}$$

the set of rational polynomials mapping integer matrices to integer matrices. This is a $\mathbb{Z}$-module for which we have the inclusion:

$$\mathbb{Z}[x] \subseteq \text{Int}_{\mathbb{Q}}(M_n(\mathbb{Z})) \subseteq \text{Int}(\mathbb{Z}) \ . \tag{1}$$

We can see that $\text{Int}_{\mathbb{Q}}(M_n(\mathbb{Z}))$ has a regular basis—a basis containing exactly one polynomial for every degree k—via the following result of Cahen and Chabert:

Corollary 1 ([2, II.1.6]) *Let B be a domain such that $D[x] \subseteq B \subseteq$ Int(E, D) for some infinite fractional subset E of D. If D is a principal ideal domain, then B has a regular basis.*

Since $\mathbb{Z}$ is a principal ideal domain and is an infinite fractional subset of itself, we may conclude from this corollary and Eq. (1) that $\text{Int}_{\mathbb{Q}}(M_n(\mathbb{Z}))$ has a regular basis. Unlike for $\text{Int}(\mathbb{Z})$, however, it turns out that this regular basis is not easy to describe in closed form [3].

For our interests, the integral closure of $\text{Int}_{\mathbb{Q}}(M_n(\mathbb{Z}))$, the set of all polynomials $f(x) \in \mathbb{Q}[x]$ which are integral over $\text{Int}_{\mathbb{Q}}(M_n(\mathbb{Z}))$, is a very useful object (however,

A. C. Sodhi (✉)
Department of Mathematics and Statistics, University of Victoria, Victoria, BC, Canada
e-mail: acsodhi@uvic.ca

J.-L. Chabert et al. (eds.), *Algebraic, Number Theoretic, and Topological Aspects of Ring Theory*, https://doi.org/10.1007/978-3-031-28847-0_21

it should be noted that there are polynomials in the integral closure that are not in the ring of integer-valued polynomials itself; see [8] for a method of construction of such polynomials). We can learn information about computing regular bases for both of these rings by making use of the following two results of Frisch, and Loper and Werner, respectively.

Theorem 1 ([4, Lemma 3.4]) *Let $f(x) = \frac{g(x)}{c}$ with $g(x) \in \mathbb{Z}[x]$ and $c \in \mathbb{Z} \setminus \{0\}$. Then $f(x)$ maps $M_n(\mathbb{Z})$ to itself if and only if $g(x)$ is divisible modulo $c\mathbb{Z}[x]$ by all monic polynomials in $\mathbb{Z}[x]$ of degree n.*

Theorem 2 ([7, 3.8 and 4.6]) *Let $\mathfrak{O}_n$ denote the set of all algebraic integers in number fields $\mathbb{Q}(\theta)$ with $[\mathbb{Q}(\theta) : \mathbb{Q}] = n$. Then the integral closure of $Int_{\mathbb{Q}}(M_n(\mathbb{Z}))$ is equal to*

$$\bigcap_{\theta \in \mathfrak{O}_n} Int_{\mathbb{Q}}(\mathcal{O}_\theta) ,$$

where $\mathcal{O}_\theta$ denotes the ring of algebraic integers in $\mathbb{Q}(\theta)$ and $Int_{\mathbb{Q}}(\mathcal{O}_\theta)$ denotes the algebra of rational polynomials preserving $\mathcal{O}_\theta$.

To study the integral closure of $\text{Int}_{\mathbb{Q}}(M_n(\mathbb{Z}))$, we would like to describe its localizations at rational primes, which can be done using the localizations of the algebras $\text{Int}_{\mathbb{Q}}(\mathcal{O}_\theta)$ from Theorem 2. Loper and Werner [7] suggest that a basis for the integral closure of $\text{Int}(M_n(\mathbb{Z}))$ can be found by computing $\text{Int}_{\mathbb{Q}}(\mathcal{O}_\theta)_{(p)}$ for all possible $\mathcal{O}_\theta$ and a given rational prime p and then intersect, but computing the intersection becomes complicated.

Another way by which we can study the integral closure of $\text{Int}_{\mathbb{Q}}(M_n(\mathbb{Z}))$ is by using results about division algebras over local fields.

Theorem 3 (Embedding Theorem, in Appendix of [10]) *If D is a division algebra of degree n^2 over a local field K and F is a field extension of degree n of K, then F can be embedded as a maximal commutative subfield of D.*

From this theorem, it follows that if R_n is the maximal order of D, then by inclusion $\text{Int}_{\mathbb{Q}}(R_n)$ lies in the intersection of all the rings $\text{Int}_{\mathbb{Q}}(\mathcal{O}_\theta)$ (since each $\mathbb{Q}(\theta)$ can be embedded as a maximal commutative subfield of D). The rings $\text{Int}_{\mathbb{Q}}(R_n)$ and $\bigcap_{\theta \in \mathfrak{O}_n} \text{Int}_{\mathbb{Q}}(\mathcal{O}_\theta)$ are, in fact, equal, and so constructing an R_n-basis for $\text{Int}_{\mathbb{Q}}(R_n)$ via p-orderings will give the means to describe the integral closure of $\text{Int}_{\mathbb{Q}}(M_n(\mathbb{Z}))$.

Let p be a fixed prime, let D be a division algebra of degree n^2 over K a local field, and let R_n denote the maximal order in D. In all applications, we will take $K = \mathbb{Q}_p$ the p-adic numbers, equipped with the usual p-adic valuation.

Proposition 1 ([7, 4.6], [3, 2.1]) *The integral closure of $Int_{\mathbb{Q}}(M_n(\mathbb{Z})_{(p)})$ is $Int_{\mathbb{Q}}(R_n)$.*

Proposition 1 demonstrates that the problem of describing the integral closure of $\text{Int}_{\mathbb{Q}}(M_n(\mathbb{Z})_{(p)})$ is exactly that of describing $\text{Int}_{\mathbb{Q}}(R_n)$, so we can move our attention toward studying integer-valued polynomials over maximal orders. In so doing, we must recall the definition of a minimal polynomial of a finite set in a division ring:

Theorem 4 (Bray-Whaples, [6, 16.13]) *Let D be a division ring and $c_1, \dots, c_n$ be n pairwise nonconjugate elements of D. Then there exists a unique polynomial $g(x) \in D[x]$ with $\deg(g) = n$ that is monic and such that $g(c_1) = \cdots = g(c_n) = 0$. Moreover, $g(x)$ has the following properties:*

1. *$c_1, \dots, c_n$ are all the roots of $g \in D$.*
2. *If $h(x) \in D[x]$ vanishes on all c_i with $1 \le i \le n$, then $h(x) \in D[x] \cdot g(x)$.*

We can therefore refer to $g(x)$ as the minimal polynomial of the set $\{c_1, \dots, c_n\}$.

To describe $\mathrm{Int}_{\mathbb{Q}}(R_n)$, we require results analogous to Bhargava's p-ordering [1] for maximal orders of division algebras over a local field.

Definition 1 ([5, 1.1]) Let K be a local field with valuation ν, D be a division algebra over K to which ν extends, R the maximal order in D, and S a subset of R. Then a *ν-ordering* of S is a sequence $\{a_i : i = 0, 1, 2, \dots\} \subseteq S$ such that for each $k > 0$, the element a_k minimizes the quantity $\nu(f_k(a_0, \dots, a_{k-1})(a))$ over $a \in S$, where $f_k(a_0, \dots, a_{k-1})(x)$ is the minimal polynomial of the set $\{a_0, a_1, \dots, a_{k-1}\}$, with the convention that $f_0 = 1$. We call the sequence of valuations $\{\nu(f_k(a_0, \dots, a_{k-1})(a_k)) : k = 0, 1, \dots\}$ the *ν-sequence* of S.

Proposition 2 ([5, 1.2]) *As in Definition 1, let K be a local field with valuation ν, D be a division algebra over K to which ν extends, R the maximal order in D, and S a subset of R. Additionally, let $\pi \in R$ be a uniformizing element, meaning an element for which $(\pi^n) = (p)$, let $\{a_i : i = 0, 1, 2, \dots\} \subseteq S$ be a ν-ordering, and let $f_k(a_0, \dots, a_{k-1})$ be the minimal polynomial of $\{a_0, a_1, \dots, a_{k-1}\}$. Then the sequence $\{\alpha_S(k) = \nu(f_k(a_0, \dots, a_{k-1})(a_k)) : k = 0, 1, 2, \dots\}$ depends only on the set S, and not on the choice of ν-ordering. The sequence of polynomials*

$$\{\pi^{-\alpha_S(k)} f_k(a_0, \dots, a_{k-1})(x) : k = 0, 1, 2, \dots\}$$

forms a regular R-basis for the R-algebra of polynomials which are integer-valued on S.

To utilize Proposition 2, we first need to be able to construct a ν-ordering of our maximal order R_n. A recursive method for constructing ν-orderings for elements of a maximal order is based on two lemmas.

Lemma 1 (see [5, 6.2]) *Let $\{a_i : i = 0, 1, 2, \dots\}$ be a ν-ordering of a subset S of R with associated ν-sequence $\{\alpha_S(i) : i = 0, 1, 2, \dots\}$, and let b be an element in the center of R. Then:*

(i) *$\{a_i + b : i = 0, 1, 2, \dots\}$ is a ν-ordering of $S + b$, and the ν-sequence of $S + b$ is the same as that of S.*

(ii) *If p is the characteristic of the residue field of K (so that $(p) = (\pi)^n$ in R), then $\{pa_i : i = 0, 1, 2, \dots\}$ is a ν-ordering for pS and the ν-sequence of pS is $\{\alpha_S(i) + in : i = 0, 1, 2, \dots\}$.*

Definition 2 The *shuffle* of two nondecreasing sequences of integers is their disjoint union sorted into nondecreasing order. If the sequences are $\{b_i\}$ and $\{c_i\}$, their shuffle is denoted $\{b_i\} \wedge \{c_i\}$.

Lemma 2 ([5, 5.2]) *Let S_1 and S_2 be disjoint subsets of S with the property that there is a nonnegative integer k such that $\nu(s_1 - s_2) = k$ for any $s_1 \in S_1$ and $s_2 \in S_2$ and that S_1 and S_2 are each closed with respect to conjugation by elements of R, by which we mean $rsr^{-1} \in S_1$ for all $r \in R$ and $s \in S_1$, and, respectively, for S_2. If $\{a_i\}$ is a ν-ordering of $S_1 \cup S_2$ then the subsequence of this ordering consisting of those elements in S_1 is a ν-ordering of S_1 and similarly for S_2.*

Conversely, if $\{b_i\}$ and $\{c_i\}$ are ν-orderings of S_1 and S_2, respectively, with associated ν-sequence $\{\alpha_{S_1}(i)\}$ and $\{\alpha_{S_2}(i)\}$, then the ν-sequence of $S_1 \cup S_2$ is the sum of the linear sequence $\{ki : i = 0, 1, 2, \dots\}$ with the shuffle $\{\alpha_{S_1}(i) - ki\} \wedge \{\alpha_{S_2}(i) - ki\}$, and this shuffle applied to $\{b_i\}$ and $\{c_i\}$ gives a ν-ordering of $S_1 \cup S_2$.

As the linear sequence mentioned in the above Lemma will come up multiple times in this paper, we formalize its notation here:

Definition 3 The sequence (kn) denotes the linear sequence $\{kn : n = 0, 1, 2, \dots\}$, whose nth term is kn.

The case where $n = 2$, in which D is a division algebra of degree 4, has been described for the case where $p = 2$ in [5] and extended to the case where p is an odd prime in [3]. This paper will describe the case where $n = 3$ and $p = 2$. The author's PhD thesis [11], written under the supervision of Keith Johnson at Dalhousie University, examines the more general case where n is an odd prime and $p = 2$.

2 The Maximal Order Δ_3

When R is a complete discrete valuation ring with unique maximal ideal P, K is the quotient field of R, and D is a division ring whose center contains K and is such that $[D : K] = m$ is finite, then D contains a unique maximal R-order Δ. When the residue class field R/P is finite and $[D : K] = n^2$, the structures on the division ring D and maximal order Δ can be described explicitly and can be chosen to only depend on the index n.

The proof of Theorem 14.6 in [9] gives a construction for D. The proof shows that there exists a division ring D with any Hasse invariant r/n for any choice of $r \in \mathbb{Z}$ such that $1 \leq r \leq n$ and $\gcd(r, n) = 1$. Since we only care about the existence of *some* division ring of index n and not a specific one, we can take $r = 1$. Doing so provides a description of the division algebra D_3 and its maximal order

Δ_3, represented as subsets of the 3×3 matrices:

$$D_3 = \mathbb{Q}_2[\omega, \pi] \qquad\qquad \Delta_3 = \mathbb{Z}_2[\omega, \pi]$$

where $\mathbb{Q}_2, \mathbb{Z}_2$ denote the 2-adic numbers and integers, respectively, and

$$\omega = \begin{pmatrix} \zeta_7 & 0 & 0 \\ 0 & \zeta_7^2 & 0 \\ 0 & 0 & \zeta_7^4 \end{pmatrix} \qquad\qquad \pi = \begin{pmatrix} 0 & 1 & 0 \\ 0 & 0 & 1 \\ 2 & 0 & 0 \end{pmatrix}$$

with ζ_7 a primitive 7th root of unity. Note that we have the relations $\pi^3 = 2I_3$ and $\pi \cdot \omega \cdot \pi^{-1} = \omega^2$ and also a valuation ν in Δ_3 described by $\nu(z) = \nu_2(\det(z))$ for $z \in \Delta_3$ realized as a matrix, where ν_2 denotes the 2-adic valuation.

2.1 Conjugacy Classes of Δ_3 Modulo π

Each element in Δ_3 is expressible as a $\mathbb{Z}_2$-linear combination of the nine elements $\{\omega^i \cdot \pi^j : 0 \leq i, j \leq 2\}$. The quotient $\Delta_3/(\pi)$ is isomorphic to the finite field $\mathbb{F}_{2^3} = \mathbb{F}_8$ with nonzero residue classes modulo π represented by powers of ω. We would like to decompose Δ_3 using the conjugacy classes of Δ_3 modulo π and denote these classes as follows:

Definition 4 Define the sets

$$\begin{aligned}
T &= \{z \in \Delta_3 : z \equiv 0 \pmod{\pi}\} = \pi\Delta \\
T+1 &= \{z \in \Delta_3 : z \equiv I_3 \pmod{\pi}\} \\
S &= \{z \in \Delta_3 : z \equiv \omega \text{ or } \omega^2 \text{ or } \omega^4 \pmod{\pi}\} \\
S+1 &= \{z \in \Delta_3 : z \equiv \omega^3 \text{ or } \omega^6 \text{ or } \omega^5 \pmod{\pi}\} \\
&= \{z \in \Delta_3 : z \equiv \omega + I_3 \text{ or } \omega^2 + I_3 \text{ or } \omega^4 + I_3 \pmod{\pi}\}\,.
\end{aligned}$$

Lemma 3

(i) *If $z \in T$, then the characteristic polynomial of z is congruent to x^3 (mod 2).*
(ii) *If $z \in T + 1$, then the characteristic polynomial of z is congruent to $(x-1)^3$ (mod 2).*
(iii) *if $z \in S$, then the characteristic polynomial of z is congruent to $x^3 + x + 1$ (mod 2).*
(iv) *if $z \in S+1$, then the characteristic polynomial of z is congruent to x^3+x^2+1 (mod 2).*
(v) *Each of the sets $T, T + 1, S, S + 1$ is closed with respect to conjugation by elements of Δ_3, where "a conjugated by b" is the element bab^{-1}.*

(vi) *Each element of* Δ_3 *lies in exactly one of the sets* $T, T+1, S, S+1$, *so that their disjoint union is all of* Δ_3.
(vii) *If* $z, w \in \Delta_3$ *are not both simultaneously in one of* T, $T+1$, S, *or* $S+1$, *then* $\nu(z-w)=0$.

Proof

(i) If $z \in T$, then $z^3 \equiv 0 \pmod{\pi^3}$ so $z^3 \equiv 0 \pmod 2$, and hence z is a root of $x^3 \pmod 2$.
(ii) If $z \in T+1$, then $(z-I_3)^3 \equiv 0 \pmod{\pi^3}$ so $(z-I_3)^3 \equiv 0 \pmod 2$, and hence z is a root of $(x-1)^3 \pmod 2$.
(iii) When viewed as a matrix, ω has characteristic polynomial $x^3+x+1 \pmod 2$. As it is a diagonal matrix, it is easily seen that ω has the same eigenvalues as ω^2 and ω^4, so all three elements of Δ_3 have the same characteristic polynomial. If instead $z \equiv \omega, \omega^2$, or $\omega^4 \pmod{\pi}$ then z is still a root of the polynomial $x^3+x+1 \pmod 2$. Since this is an irreducible cubic polynomial, we can be certain that this is actually the characteristic polynomial of z.
(iv) When viewed as a matrix, $\omega+I_3$ has characteristic polynomial $x^3+x^2+1 \pmod 2$. This matrix is diagonal, and it is easily seen that ω^2+I_3 and ω^4+I_3 have the same entries as $\omega+I_3$, only permuted, and hence all three elements of Δ_3 have the same characteristic polynomial. If instead $z \equiv \omega+I_3, \omega^2+I_3$, or $\omega^4+I_3 \pmod{\pi}$ then z is still a root of the polynomial $x^3+x^2+1 \pmod 2$. Since this is an irreducible cubic polynomial, we can be certain that this is actually the characteristic polynomial of z.
(v) This follows from Dickson's Theorem [6, 16.8]. Since all nonconstant elements $z \in \Delta_3$ have irreducible characteristic polynomials (see Lemma 6), here characteristic and minimal polynomials coincide. Since by Dickson's Theorem conjugate elements share a characteristic polynomial, this implies that each of the subsets T, $T+1$, S, and $S+1$ is closed under conjugation by elements of Δ_3.
(vi) It is easy to see, since all nonzero residue classes of $\Delta_3 \pmod{\pi}$ are represented by powers of ω, that $T \cup (T+1) \cup S \cup (S+1) = \Delta_3$. The fact that each element of Δ_3 lies in exactly one of these four sets follows by the uniqueness of the characteristic polynomial.
(vii) If z, w are in different sets $T, T+1, S, S+1$, then z and w are by definition in different residue classes modulo π. Therefore $z - w \not\equiv 0 \pmod{\pi}$ and hence $\nu(z-w) = 0$ for all choices of $z, w \in \Delta_3$ such that z and w are not in the same subset of Δ_3 given in Definition 4.

□

Knowing this decomposition of Δ_3 into the union of disjoint sets, we can apply Lemma 2 to determine a recursive definition for the ν-ordering of Δ_3. Also by Lemma 1, we need only concern ourselves with the ν-sequences of the sets T and S, as $T+1$ and $S+1$ are simply translates, under which ν-sequences are invariant. However, we can further decompose Δ_3 by examining the subsets closed under conjugation modulo higher powers of π within the set T.

2.2 Decomposition of T

Definition 5 Let

$$T_1 = \{z \in \Delta_3 : z \equiv 0 \pmod{\pi^2}\} = \pi^2\Delta$$

$$T_2 = \{z \in \Delta_3 : z \equiv \omega^i\pi \pmod{\pi^2} \text{ for some } 0 \le i \le 6\}$$

Lemma 4

(i) *Every element in T is in exactly one of T_1 and T_2.*
(ii) *Each of T_1 and T_2 is closed with respect to conjugation by elements of Δ_3.*
(iii) *If $z \in T_1$ and $w \in T_2$, then $\nu(z - w) = 1$.*

Proof

(i) As all $z \in T$ are such that $z \equiv 0 \pmod{\pi}$, the fact that either $z \in T_1$ or $z \in T_2$ follows from the definition of these sets.
(ii) The fact that T_1 is closed under conjugation is clear from its definition. In the case of T_2, we can write any element of Δ_3 as a linear combination of the elements $\omega^k\pi^\ell$ with $0 \le k \le 6$, $0 \le \ell \le 2$. Using the known relations between ω and π, it follows that conjugating $\pi \in \Delta_3$ by an arbitrary element $\omega^k\pi^\ell$ of Δ_3 gives $\omega^k\pi^\ell \cdot \pi \cdot \pi^{-\ell}\omega^{-k} = \omega^{7-k}\pi$. Thus every element $\omega^i\pi$ is in the same orbit as π under the action of conjugation; hence T_2 is closed under conjugation by elements of Δ_3.
(iii) If $z \in T_1$ and $w \in T_2$, then $z - w \equiv \omega^i\pi \pmod{\pi^2}$ for some $0 \le i \le 6$. Therefore $\nu(z - w) = \nu(\omega^i\pi) = 1$ for any choice of $z \in T_1$ and $w \in T_2$.

□

We can, in fact, break the set T_1 into components even further.

Definition 6 Let

$$T_3 = \{z \in \Delta_3 : z \equiv 0 \pmod{\pi^3}\} = 2\Delta_3$$

$$T_4 = \{z \in \Delta_3 : z \equiv \omega^i\pi^2 \pmod{\pi^3} \text{ for some } 0 \le i \le 6\}$$

Lemma 5

(i) *Every element of T_1 is in exactly one of T_3 and T_4.*
(ii) *Each of T_3 and T_4 is closed with respect to conjugation by elements of Δ_3.*
(iii) *If $z \in T_3$ and $w \in T_4$, then $\nu(z - w) = 2$.*

Proof

(i) As all $z \in T_1$ are such that $z \equiv 0 \pmod{\pi^2}$, the fact that either $z \in T_3$ or $z \in T_4$ follows from the definition of these sets.
(ii) The fact that T_3 is closed under conjugation is clear from its definition. In the case of T_4, we can write any element of Δ_3 as a linear combination of the

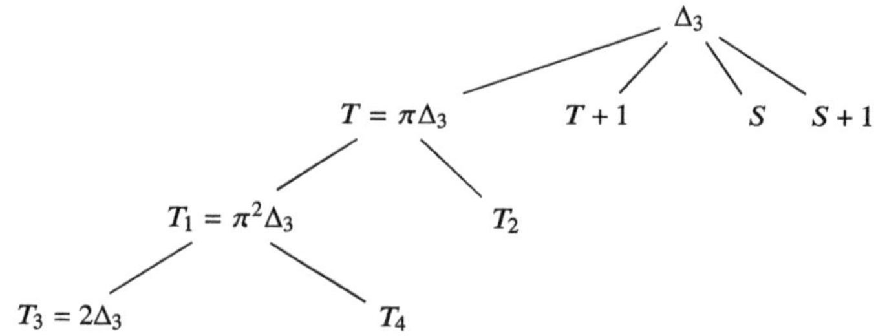

Fig. 1 Tree summarizing decomposition of Δ_3

elements $\omega^k\pi^\ell$ with $0 \le k \le 6, 0 \le \ell \le 2$. Using the known relations between ω and π, it follows that conjugating $\pi^2 \in \Delta_3$ by an arbitrary element $\omega^k\pi^\ell$ of Δ_3 gives $\omega^k\pi^\ell \cdot \pi^2 \cdot \pi^{-\ell}\omega^{-k} = \omega^{4k}\pi^2$. Since the equation $4k \equiv n \pmod 7$ has a solution for every $n \in \mathbb{Z}/(7)$, it follows that every element $\omega^i\pi^2$ is in the same orbit as π^2 under the action of conjugation; hence T_4 is closed under conjugation by elements of Δ_3.

(iii) If $z \in T_3$ and $w \in T_4$, then $z - w \equiv \omega^i\pi^2 \pmod{\pi^3}$ for some $0 \le i \le 6$. Therefore $\nu(z - w) = \nu(\omega^i\pi^2) = 2$ for any choice of $z \in T_3$ and $w \in T_4$.

□

From this analysis, it follows that

$$
\begin{aligned}
T &= T_1 \cup T_2 \\
&= (T_3 \cup T_4) \cup T_2 \\
&= 2\Delta_3 \cup T_4 \cup T_2
\end{aligned}
$$

with all unions disjoint and with all sets fulfilling the conditions of Lemma 2. The decomposition of Δ_3 into sets is demonstrated graphically in Fig. 1. By Lemma 1, the ν-sequence of $T_3 = 2\Delta_3$ can be written in terms of the ν-sequence for Δ_3, which provides the eventual recursive definition of α_{Δ_3} we seek, given in Proposition 3. Thus, to define the ν-sequence of T, it is sufficient to determine the ν-sequences of T_2 and T_4.

3 The ν-Sequence of Δ_3

From the description of the decomposition of Δ_3 into appropriate disjoint sets as in Sect. 2.2, coupled with the results of Lemmas 1 and 2, we obtain the following result.

Proposition 3 *The ν-sequence of Δ_3, denoted α_{Δ_3}, satisfies and is determined by the formula*

$$\alpha_{\Delta_3} = \left(\left[\left(\left[(\alpha_{\Delta_3} + (n)) \wedge (\alpha_{T_4} - (2n))\right] + (n)\right) \wedge (\alpha_{T_2} - (n))\right] + (n)\right)^{\wedge 2} \wedge (\alpha_S)^{\wedge 2},$$

where (kn) denotes the linear sequence whose nth term is kn.

Proof This formula follows from Lemmas 1, 2, 3, 4, and 5.

Since $T_3 = 2\Delta_3$, we have $\alpha_{T_3} = \alpha_{\Delta_3} + (3n)$. We then know that

$$\begin{aligned}\alpha_{T_1} &= \left[(\alpha_{\Delta_3} + (3n) - (2n)) \wedge (\alpha_{T_4} - (2n))\right] + (2n)\\ &= \left[(\alpha_{\Delta_3} + (n)) \wedge (\alpha_{T_4} - (2n))\right] + (2n)\end{aligned}$$

and therefore

$$\begin{aligned}\alpha_T &= \left[(\alpha_{T_1} - (n)) \wedge (\alpha_{T_2} - (n))\right] + (n)\\ &= \left[\left(\left[(\alpha_{\Delta_3} + (n)) \wedge (\alpha_{T_4} - (2n))\right] + (2n) - (n)\right) \wedge (\alpha_{T_2} - (n))\right] + (n)\\ &= \left[\left(\left[(\alpha_{\Delta_3} + (n)) \wedge (\alpha_{T_4} - (2n))\right] + (n)\right) \wedge (\alpha_{T_2} - (n))\right] + (n).\end{aligned}$$

We know that $T + 1$ is a translate of T and $S + 1$ of S, so that $\alpha_T = \alpha_{T+1}$ and $\alpha_S = \alpha_{S+1}$. Therefore

$$\begin{aligned}\alpha_{\Delta_3} &= \alpha_T \wedge \alpha_{T+1} \wedge \alpha_S \wedge \alpha_{S+1}\\ &= (\alpha_T^{\wedge 2}) \wedge (\alpha_S^{\wedge 2})\\ &= \left(\left[\left(\left[(\alpha_{\Delta_3} + (n)) \wedge (\alpha_{T_4} - (2n))\right] + (n)\right) \wedge (\alpha_{T_2} - (n))\right] + (n)\right)^{\wedge 2} \wedge (\alpha_S)^{\wedge 2}\end{aligned}$$

as claimed. □

Once we have determined the ν-sequences for S, T_2, and T_4, this formula will uniquely determine $\alpha_{\Delta_3}(i)$ for all i. For every $i > 0$, the ith term on the right-hand side consists of terms from α_S, α_{T_2}, and α_{T_4}, and also terms $\alpha_{\Delta_3}(j)$ for some $j < i$. As the first term of any ν-sequence is always 0, this formula gives an expression of α_{Δ_3} for all i.

3.1 Characteristic Polynomials of Subsets of Δ_3

We would like to be able to completely describe the subsets of Δ_3 in terms of the 2-adic valuation of coefficients in their characteristic polynomials in order to describe the appropriate ν-sequences.

Lemma 6 *Let $z \in \Delta_3$ be a nonconstant element. The characteristic polynomial of z is irreducible over $\mathbb{Q}_2$.*

The fact that characteristic polynomials of nonconstant elements of Δ_3 are irreducible allows us to make use of the following result, which has been restated for the degree 3 case with the convention of writing general polynomials as $f(x) = \sum_{i=0}^{n} a_i x^i$.

Lemma 7 ([9, 12.9 restated]) *Let $f(x) = a_0 + a_1 x + a_2 x^2 + x^3 \in K[x]$ be irreducible. Then*

$$\nu(a_j) \geq \frac{3-j}{3}\nu(a_0)\,, \qquad 0 \leq j \leq 2\,.$$

This lemma does not give us much information when it comes to the set S, as here $\nu(a_0) = 0$. In the case of this set, we do know definitively that the characteristic polynomial of $z \in S$ is equivalent to $x^3 + x + 1 \pmod 2$. This gives the result that if $f(x) = a_0 + a_1 x + a_2 x^2 + x^3$ the minimal polynomial of $z \in S$, then

$$\nu_2(a_0) = 0 \qquad \nu_2(a_1) = 0 \qquad \nu_2(a_2) \geq 1\,. \tag{2}$$

However, the aforementioned lemma does give us useful information for determining coefficients of the characteristic polynomial for elements in T_2 and T_4.

Proposition 4

(i) *Let $z \in T_2$, with $f(x) = a_0 + a_1 x + a_2 x^2 + x^3$ the minimal polynomial of $z \in \mathbb{Q}_2[x]$. Then*

$$\nu_2(a_0) = 1 \qquad \nu_2(a_1) \geq 1 \qquad \nu_2(a_2) \geq 1$$

(ii) *Let $z \in T_4$, with $f(x) = a_0 + a_1 x + a_2 x^2 + x^3$ the minimal polynomial of $z \in \mathbb{Q}_2[x]$. Then*

$$\nu_2(a_0) = 2 \qquad \nu_2(a_1) \geq 2 \qquad \nu_2(a_2) \geq 1$$

Proof

(i) We can write $T_2 = \pi\Delta_3 \setminus \pi^2\Delta_3$, so that every element $z \in T_2$ has $\nu(z) = 1$. Therefore $\nu(a_0) = \nu_2(\det(z)) = \nu(z) = 1$. Since $a_0 \equiv 0 \pmod 2$ but $a_0 \not\equiv 0 \pmod 4$, it must be the case that $a_0 \equiv 2 \pmod 4$. Lemma 7 gives the result.

(ii) We can write $T_4 = \pi^2\Delta_3 \setminus 2\Delta_3$, so that every element $z \in T_4$ has $\nu(z) = 2$. Therefore $\nu(a_0) = \nu_2(\det(z)) = \nu(z) = 2$. Since $a_0 \equiv 0 \pmod 4$ but $a_0 \not\equiv 0 \pmod 8$, it must be the case that $a_0 \equiv 4 \pmod 8$. Lemma 7 gives the result.

□

With this knowledge of the 2-adic valuations of coefficients of the characteristic polynomials, we can begin to construct elements that will feature in the integer-valued polynomials for these sets.

4 Toward Computing ν-Sequences

Given the expression of our sets in terms of characteristic polynomials given in Eq. (2) and Lemma 4, we can compute the ν-orderings and ν-sequences for S, T_2, and T_4. In this section, we establish some facts about the valuation of certain polynomials, with the goal of establishing these as the minimal polynomials of elements within their respective sets.

Given an element $z \in \Delta_3$, we can denote its characteristic polynomial by

$$x^3 - Tr(z)x^2 + \beta(z)x - \det(z)$$

where $Tr(z)$ and $\det(z)$ are the usual trace and determinant when z is viewed as a 3×3 matrix, and $\beta(z)$ is defined in terms of the 2×2 minors.

4.1 Characteristic Polynomials for Elements in S

For elements $z \in S$, we have

$$Tr(z) \equiv 0 \pmod 2 \qquad \beta(z) \equiv 1 \pmod 2 \qquad \det(z) \equiv 1 \pmod 2$$

Let us define the function

$$\phi = (\phi_2, \phi_1, \phi_0) : \mathbb{Z}_{\geq 0} \to 2\mathbb{Z}_{\geq 0} \times (1 + 2\mathbb{Z}_{\geq 0}) \times (1 + 2\mathbb{Z}_{\geq 0})$$

$$\phi(n) = \left(2\sum_{i \geq 0} n_{3i}2^i, 1 + 2\sum_{i \geq 0} n_{3i+1}2^i, 1 + 2\sum_{i \geq 0} n_{3i+2}2^i \right)$$

where $n = \sum_{i \geq 0} n_i 2^i$ is the expansion of n in base 2. Let

$$f_n(x) = \prod_{k=0}^{n-1} \left(x^3 - \phi_2(k)x^2 + \phi_1(k)x - \phi_0(k) \right) .$$

Lemma 8 *If $z \in S$ then*

$$\nu(f_n(z)) \geq 3n + 3\sum_{k>0} \left\lfloor \frac{n}{8^k} \right\rfloor$$

with equality if $Tr(z) = \phi_2(n)$, $\beta(z) = \phi_1(n)$, and $\det(z) = \phi_0(n)$.

Proof Let $z \in S$, and let $Tr(z) = 2\sum_{k \geq 0} a_k 2^k$ be the expansion of $Tr(z)$ in base 2. Similarly, let $\beta(z) = 1 + 2\sum_{k \geq 0} b_k 2^k$ and $\det(z) = 1 + 2\sum_{k \geq 0} c_k 2^k$ be the base

2 expansions of $\beta(z)$, $\det(z)$. Define $m := \sum_{k\geq 0} a_k 2^{3k} + b_k 2^{3k+1} + c_k 2^{3k+2}$, so that $\phi(m) = (Tr(z), \beta(z), \det(z))$.

For any $0 \leq k \leq n$,

$$\begin{aligned} z^3 - \phi_2(k)z^2 + \phi_1(k)z - \phi_0(k) \\ &= z^3 - \phi_2(k)z^2 + \phi_1(k)z - \phi_0(k) - (z^3 - Tr(z)z^2 + \beta(z)z - \det(z)) \\ &= (Tr(z) - \phi_2(k))z^2 + (\phi_1(k) - \beta(z))z + (\det(z) - \phi_0(k)) \\ &= (\phi_2(m) - \phi_2(k))z^2 + (\phi_1(k) - \phi_1(m))z + (\phi_0(m) - \phi_0(k)) \,. \end{aligned}$$

Since the characteristic polynomial for $z \in S$ is $x^3 + x + 1 \pmod 2$ and is irreducible over $\mathbb{F}_2$, it follows by Hensel's lemma that if $az^2 + bz + c \equiv 0 \pmod \pi$ in Δ_3 then $a \equiv b \equiv c \equiv 0 \pmod 2$.

Letting $a = 2^{t_1}\hat{a}$, $b = 2^{t_2}\hat{b}$, and $c = 2^{t_3}\hat{c}$ with $\hat{a}, \hat{b}, \hat{c}$ odd and $t_i > 0$, we obtain

$$\begin{aligned} az^2 + bz + c &= 2^{t_1}\hat{a}z^2 + 2^{t_2}\hat{b}z + 2^{t_3}\hat{c} \\ &= 2^{\min_i t_i}\left(2^{t_1 - \min_i t_i}\hat{a}z^2 + 2^{t_2 - \min_i t_i}\hat{b}z + 2^{t_3 - \min_i t_i}\hat{c}\right) \end{aligned}$$

with at least one of the $t_j - \min_i t_i = 0$. Therefore, the expression $2^{t_1 - \min_i t_i}\hat{a}z^2 + 2^{t_2 - \min_i t_i}\hat{b}z + 2^{t_3 - \min_i t_i}\hat{c}$ has at least one odd coefficient, and so it follows that

$$\nu\left(2^{t_1 - \min_i t_i}\hat{a}z^2 + 2^{t_2 - \min_i t_i}\hat{b}z + 2^{t_3 - \min_i t_i}\hat{c}\right) = 0 \,.$$

Therefore,

$$\begin{aligned} \nu(az^2 + bz + c) &= \nu\left(2^{\min_i t_i}\left(2^{t_1 - \min_i t_i}\hat{a}z^2 + 2^{t_2 - \min_i t_i}\hat{b}z + 2^{t_3 - \min_i t_i}\hat{c}\right)\right) \\ &= 3\nu_2(2^{\min_i t_i}) + \nu\left(2^{t_1 - \min_i t_i}\hat{a}z^2 + 2^{t_2 - \min_i t_i}\hat{b}z + 2^{t_3 - \min_i t_i}\hat{c}\right) \\ &= 3\nu_2(2^{\min_i t_i}) \\ &= 3\min_i t_i \\ &= 3\min(\nu_2(a), \nu_2(b), \nu_2(c)) \end{aligned}$$

We abuse notation and let $\nu_2(\phi_j) = \nu_2(\phi_j(m) - \phi_j(k))$ for $j = 0, 1, 2$ and so that

$$\nu(z^3 - \phi_2(k)z^2 + \phi_1(k)z - \phi_0(k)) = 3\min(\nu_2(\phi_2), \nu_2(\phi_1), \nu_2(\phi_0))$$

which gives

$$\nu(f_n(z)) = 3\sum_{k=0}^{n-1} \min(\nu_2(\phi_2), \nu_2(\phi_1), \nu_2(\phi_0)) .$$

If $k = \sum k_i 2^i$, $m = \sum m_i 2^i$ denote the expansions of k and m in base 2, then

$$\begin{aligned}
\nu_2(m-k) &= \min(i : k_i \neq m_i) \\
\nu_2(\phi_2) &= \min(i : k_{3i} \neq m_{3i}) + 1 \\
\nu_2(\phi_1) &= \min(i : k_{3i+1} \neq m_{3i+1}) + 1 \\
\nu_2(\phi_0) &= \min(i : k_{3i+2} \neq m_{3i+2}) + 1
\end{aligned}$$

Thus, we have

$$\min(\nu_2(\phi_2), \nu_2(\phi_1), \nu_2(\phi_0)) = \left\lfloor \frac{\nu_2(m-k)}{3} \right\rfloor + 1 .$$

Since $\left\lfloor \frac{\nu_2(m-k)}{3} \right\rfloor$ is the highest power of 8 dividing $m-k$, for simplicity let us denote $\nu_8(m-k) := \left\lfloor \frac{\nu_2(m-k)}{3} \right\rfloor$.

Using the fact that $\nu_p(n!) = \sum_{i=1}^{n} \nu_p(i) = \sum_{i>0} \left\lfloor \frac{n}{p^i} \right\rfloor = \frac{n-\sum n_i}{p-1}$ with $n = \sum n_i p^i$ for any prime p extends also to powers of primes [5, proof of Lemma 6.5], we obtain the result

$$\sum_{i=1}^{n} \nu_8(i) = \sum_{i>0} \left\lfloor \frac{n}{8^i} \right\rfloor = \frac{n - \sum n_i}{7}$$

where $n = \sum n_i 8^i$ is the expansion of n in base 8. Thus, we have

$$\begin{aligned}
\nu(f_n(z)) &= 3\sum_{k=0}^{n-1} \left(\left\lfloor \frac{\nu_2(m-k)}{3} \right\rfloor + 1 \right) \\
&= 3n + 3\left(\sum_{k=1}^{m} \left\lfloor \frac{\nu_2(k)}{3} \right\rfloor - \sum_{k=1}^{m-n} \left\lfloor \frac{\nu_2(k)}{3} \right\rfloor \right) \\
&= 3n + 3\left(\sum_{i>0} \left\lfloor \frac{m}{8^i} \right\rfloor - \sum_{i>0} \left\lfloor \frac{m-n}{8^i} \right\rfloor \right) \\
&= 3n + 3\left(\frac{m - \sum m_i}{7} - \frac{(m-n) - \sum (m-n)_i}{7} \right)
\end{aligned}$$

with $m = \sum m_i 8^i$, $m - n = \sum (m-n)_i 8^i$ as expansions base 8.

Noting that

$$\frac{m - \sum m_i}{7} - \frac{(m-n) - \sum (m-n)_i}{7} - \frac{n - \sum n_i}{7}$$
$$= \frac{\sum (m-n)_i + \sum n_i - \sum m_i}{7} \geq 0$$

since this is the number of carries in adding n and $m - n$ in base 8, and so is always nonnegative and equals zero only if $n = m$, we see that

$$\nu(f_n(z)) \geq 3n + 3 \sum_{k>0} \left\lfloor \frac{n}{8^k} \right\rfloor$$

for $z \in S$, with equality if $\phi(n) = (Tr(z), \beta(z), \det(z))$. □

Lemma 9 *Let a be a root of the polynomial $f(x) = x^3 - \phi_2(n)x^2 + \phi_1(n)x - \phi_0(n)$ in S, with θ the automorphism in Δ_3 given by $\theta(t) = \pi t \pi^{-1}$. The set of roots $a, \theta(a), \theta^2(a)$ are distinct modulo π, so that $\nu(\theta^i(a) - \theta^j(a)) = 0$ for $i \neq j$.*

Proof By Dickson's Theorem [6, 16.8], if a is a root of $f(x)$ then so too are $\theta(a)$ and $\theta^2(a)$. The element $a \equiv \omega^j \pmod{\pi}$ for some choice of $1 \leq j \leq 7$, and since $\theta(\omega^j) = \omega^{2j}$, it follows that the set of roots $\{a, \theta(a), \theta^2(a)\} \equiv \{\omega^j, \omega^{2j}, \omega^{4j}\} \pmod{\pi}$ and that these roots are distinct modulo π, as $\gcd(j, 7) = \gcd(2j, 7) = \gcd(4j, 7) = 1$. The result $\nu(\theta^i(a) - \theta^j(a)) = 0$ for $i \neq j$ follows. □

Lemma 10 *The ν-sequence α_S of $S \subseteq \Delta_3$ is given by*

$$\alpha_S(3n) = \alpha_S(3n+1) = \alpha_S(3n+2) = 3n + 3 \sum_{i>0} \left\lfloor \frac{n}{8^i} \right\rfloor$$

Proof The polynomial $x^3 - \phi_2(n)x^2 + \phi_1(n)x - \phi_0(n)$ is congruent to $x^3 + x + 1 \pmod{2}$, which is irreducible. Via Theorem 3, for any $n \in \mathbb{Z}_{\geq 0}$ there exists an element $a_n \in \Delta_3$ which is a root of the polynomial $x^3 - \phi_2(n)x^2 + \phi_1(n)x - \phi_0(n)$. Recalling that

$$f_n(x) = \prod_{k=0}^{n-1} \left(x^3 - \phi_2(k)x^2 + \phi_1(k)x - \phi_0(k) \right),$$

the inequality in Lemma 8 then implies that $f_n(x)$ is the minimal polynomial of the set

$$\{a_0, \theta(a_0), \theta^2(a_0), a_1, \theta(a_1), \theta^2(a_1), \ldots, a_{n-1}, \theta(a_{n-1}), \theta^2(a_{n-1})\}$$

where θ is a non-trivial automorphism in Δ_3. This shows that

$\{a_0, \theta(a_0), \theta^2(a_0), a_1, \theta(a_1), \theta^2(a_1), \ldots\}$ forms a ν-ordering for S, and that $f_n(x)$ is the minimal polynomial for the first $3n$ elements of this ν-ordering. Thus,

$$\alpha_S(3n) = \nu(f_n(a_n)) = 3n + 3\sum_{i>0} \left\lfloor \frac{n}{8^i} \right\rfloor ,$$

the minimum value of $\nu(f_n(z))$ attained for $z \in S$ in Lemma 8. By Lemma 9, the elements a_i, $\theta(a_i)$, and $\theta^2(a_i)$ will give rise to the same value in the ν-sequence for S, and so

$$\alpha_S(3n) = \alpha_S(3n+1) = \alpha_S(3n+2) = 3n + 3\sum_{i>0} \left\lfloor \frac{n}{8^i} \right\rfloor .$$

□

4.2 Characteristic Polynomials for Elements in T_2

For elements $z \in T_2$, we have

$$Tr(z) \equiv 0 \pmod 2 \qquad \beta(z) \equiv 0 \pmod 2 \qquad \det(z) \equiv 2 \pmod 4$$

Let us define the function

$$\psi = (\psi_2, \psi_1, \psi_0) : \mathbb{Z}_{\geq 0} \to 2\mathbb{Z}_{\geq 0} \times 2\mathbb{Z}_{\geq 0} \times (2 + 4\mathbb{Z}_{\geq 0})$$

$$\psi(n) = \left(2\sum_{i\geq 0} n_{3i+1}2^i, 2\sum_{i\geq 0} n_{3i}2^i, 2 + 4\sum_{i\geq 0} n_{3i+2}2^i \right)$$

where $n = \sum_{i\geq 0} n_i 2^i$ is the expansion of n in base 2. Let

$$g_n(x) = \prod_{k=0}^{n-1} \left(x^3 - \psi_2(k)x^2 + \psi_1(k)x - \psi_0(k) \right) .$$

Lemma 11 *If $z \in T_2$ then*

$$\nu(g_n(z)) \geq 4n + \sum_{i>0} \left\lfloor \frac{n}{2^i} \right\rfloor .$$

Proof Let $z \in T_2$, and let $Tr(z) = 2\sum_{k\geq 0} a_k 2^k$ be the expansion of $Tr(z)$ in base 2. Similarly, let $\beta(z) = 2\sum_{k\geq 0} b_k 2^k$ and $\det(z) = 2 + 4\sum_{k\geq 0} c_k 2^k$ be the base 2 expansions of $\beta(z)$, $\det(z)$. Define $m := \sum_{k\geq 0} a_k 2^{3k+1} + b_k 2^{3k} + c_k 2^{3k+2}$, so that $\psi(m) = (Tr(z), \beta(z), \det(z))$.

For any $0 \leq k \leq n$,

$$\begin{aligned} &z^3 - \psi_2(k)z^2 + \psi_1(k)z - \psi_0(k) \\ &\qquad = z^3 - \psi_2(k)z^2 + \psi_1(k)z - \psi_0(k) - (z^3 - Tr(z)z^2 + \beta(z)z - \det(z)) \\ &\qquad = (\psi_2(m) - \psi_2(k))z^2 + (\psi_1(k) - \psi_1(m))z + (\psi_0(m) - \psi_0(k)) \end{aligned}$$

Since $z \in T_2$ we have $\nu(z) = 1$, and therefore $\nu(az^2) = 2 + 3\nu_2(a)$, $\nu(bz) = 1 + 3\nu_2(b)$, and $\nu(c) = 3\nu_2(c)$. Because these have different residues modulo 3, we have

$$\nu(az^2 + bz + c) = \min(2 + 3\nu_2(a), 1 + 3\nu_2(b), 3\nu_2(c)) .$$

For the sake of simplicity, we abuse notation and let $\nu_2(\psi_j) = \nu_2(\psi_j(m) - \psi_j(k))$ for $j = 0, 1, 2$ and so

$$\nu(z^3 - \psi_2(k)z^2 + \psi_1(k)z - \psi_0(k)) = \min(2 + 3\nu_2(\psi_2), 1 + 3\nu_2(\psi_1), 3\nu_2(\psi_0)) ,$$

giving

$$\nu(g_n(z)) = \sum_{k=0}^{n-1} \min(2 + 3\nu_2(\psi_2), 1 + 3\nu_2(\psi_1), 3\nu_2(\psi_0)) .$$

If $k = \sum k_i 2^i$, $m = \sum m_i 2^i$ denote the expansions of k and m in base 2, then

$$\begin{aligned} \nu_2(m - k) &= \min(i : k_i \neq m_i) \\ \nu_2(\psi_2) &= \min(i : k_{3i+1} \neq m_{3i+1}) + 1 \\ \nu_2(\psi_1) &= \min(i : k_{3i} \neq m_{3i}) + 1 \\ \nu_2(\psi_0) &= \min(i : k_{3i+2} \neq m_{3i+2}) + 2 \end{aligned}$$

In this case, we find that the lower bounds on the $\nu(\psi_j)$ change depending on the residue of $\nu_2(m - k)$ (mod 3). We summarize the results in Table 1.

From the table, we see that

$$\min(2 + 3\nu_2(\psi_2), 1 + 3\nu_2(\psi_1), 3\nu_2(\psi_0)) = 4 + \nu_2(m - k) ,$$

giving

Table 1 Summary of lower bounds in T_2

$\nu_2(m-k) \pmod 3$:	0	1	2
$2+3\nu_2(\psi_2)$	$\geq 5+\nu_2(m-k)$	$= 4+\nu_2(m-k)$	$\geq 6+\nu_2(m-k)$
$1+3\nu_2(\psi_1)$	$= 4+\nu_2(m-k)$	$\geq 6+\nu_2(m-k)$	$\geq 5+\nu_2(m-k)$
$3\nu_2(\psi_0)$	$\geq 6+\nu_2(m-k)$	$\geq 5+\nu_2(m-k)$	$= 4+\nu_2(m-k)$

$$\begin{aligned}\nu(g_n(z)) &= \sum_{k=0}^{n-1}(4+\nu_2(m-k)) \\ &= 4n + \sum_{k=0}^{n-1} \nu_2(m-k) \\ &= 4n + \sum_{k=1}^{m} \nu_2(k) - \sum_{k=1}^{m-n} \nu_2(k) \\ &= 4n + \sum_{i>0} \left\lfloor \frac{m}{2^i} \right\rfloor - \left\lfloor \frac{m-n}{2^i} \right\rfloor \\ &\geq 4n + \sum_{i>0} \left\lfloor \frac{n}{2^i} \right\rfloor\end{aligned}$$

for $z \in T_2$, with equality if $\psi(n) = (Tr(z), \beta(z), \det(z))$. □

Lemma 12 *Let b be a root of the polynomial $g(x) = x^3 - \psi_2(n)x^2 + \psi_1(n)x - \psi_0(n)$ in T_2, with θ the automorphism in Δ_3 given by $\theta(t) = \pi t \pi^{-1}$. The set of roots $b, \theta(b), \theta^2(b)$ are distinct modulo π^2, so that $\nu(\theta^i(b) - \theta^j(b)) = 1$ for $i \neq j$.*

Proof If $b \equiv \pi \pmod{\pi^2}$, take instead $b \equiv \omega\pi \pmod{\pi^2}$—this choice can be made since π and $\omega\pi$ are conjugates: $\omega^{-1}\pi\omega = \omega\pi$. By Dickson's Theorem [6, 16.8], if b is a root of $g(x)$, then so too are $\theta(b)$ and $\theta^2(b)$. The element $b \equiv \overline{b}\pi \pmod{\pi^2}$ for some choice of $\overline{b} \not\equiv 0 \pmod{\pi}$. Applying our automorphism, we obtain modulo π^2

$$\theta(b) = \theta(\overline{b}\pi) = \theta(\overline{b})\theta(\pi) = \theta(\overline{b})\pi\ .$$

As in the proof of Lemma 9, the collection of elements $\{\overline{b}, \theta(\overline{b}), \theta^2(\overline{b})\}$ are distinct modulo π, and hence $\{b, \theta(b), \theta^2(b)\}$ are distinct modulo π^2. The result $\nu(\theta^i(b) - \theta^j(b)) = 1$ for $i \neq j$ follows. □

Lemma 13 *The ν-sequence α_{T_2} of $T_2 \subseteq \Delta_3$ is given by*

$$\alpha_{T_2}(3n) = \alpha_{T_2}(3n+1) - 1 = \alpha_{T_2}(3n+2) - 2 = 4n + \sum_{i>0} \left\lfloor \frac{n}{2^i} \right\rfloor$$

Proof Via Theorem 3, for any $n \in \mathbb{Z}_{\geq 0}$, there exists an element $b_n \in \Delta_3$ which is a root of the polynomial $x^3 - \psi_2(n)x^2 + \psi_1(n)x - \psi_0(n)$, which is irreducible by Eisenstein's criterion. Recalling that

$$g_n(x) = \prod_{k=0}^{n-1} \left(x^3 - \psi_2(k)x^2 + \psi_1(k)x - \psi_0(k)\right) ,$$

the inequality in Lemma 11 then implies that $g_n(x)$ is the minimal polynomial of the set

$$\{b_0, \theta(b_0), \theta^2(b_0), b_1, \theta(b_1), \theta^2(b_1), \dots, b_{n-1}, \theta(b_{n-1}), \theta^2(b_{n-1})\}$$

where θ is a non-trivial automorphism in Δ_3. This shows that $\{b_0, \theta(b_0), \theta^2(b_0), b_1, \theta(b_1), \theta^2(b_1), \dots\}$ forms a ν-ordering for T_2, and that $g_n(x)$ is the minimal polynomial for the first $3n$ elements of this ν-ordering. Thus,

$$\alpha_{T_2}(3n) = \nu(g_n(a_n)) = 4n + \sum_{i>0} \left\lfloor \frac{n}{2^i} \right\rfloor ,$$

the minimum value of $\nu(g_n(z))$ attained for $z \in T_2$ in Lemma 11. Since $\nu(b_n - \theta(b_n)) = 1$ by Lemma 12, we have

$$\alpha_{T_2}(3n) = \alpha_{T_2}(3n+1) - 1 = \alpha_{T_2}(3n+2) - 2 = 4n + \sum_{i>0} \left\lfloor \frac{n}{2^i} \right\rfloor .$$

□

4.3 Characteristic Polynomials for Elements in T_4

For elements $z \in T_4$, we have

$$Tr(z) \equiv 0 \pmod 2 \qquad \beta(z) \equiv 0 \pmod 4 \qquad \det(z) \equiv 4 \pmod 8$$

Let us define the function

$$\sigma = (\sigma_2, \sigma_1, \sigma_0) : \mathbb{Z}_{\geq 0} \to 2\mathbb{Z}_{\geq 0} \times 4\mathbb{Z}_{\geq 0} \times (4 + 8\mathbb{Z}_{\geq 0})$$

$$\sigma(n) = \left(2\sum_{i\geq 0} n_{3i}2^i, 4\sum_{i\geq 0} n_{3i+1}2^i, 4 + 8\sum_{i\geq 0} n_{3i+2}2^i\right)$$

where $n = \sum_{i\geq 0} n_i 2^i$ is the expansion of n in base 2. Let

$$h_n(x) = \prod_{k=0}^{n-1} \left(x^3 - \sigma_2(k)x^2 + \sigma_1(k)x - \sigma_0(k)\right) .$$

Lemma 14 *If $z \in T_4$ then*

$$\nu(h_n(z)) \geq 7n + \sum_{i>0} \left\lfloor \frac{n}{2^i} \right\rfloor$$

with equality if $Tr(z) = \sigma_2(n)$, $\beta(z) = \sigma_1(n)$, and $\det(z) = \sigma_0(n)$.

Lemma 15 *Let c be a root of the polynomial $h(x) = x^3 - \sigma_2(n)x^2 + \sigma_1(n)x - \sigma_0(n)$ in T_4, with θ the automorphism in Δ_3 given by $\theta(t) = \pi t \pi^{-1}$. The set of roots $c, \theta(c), \theta^2(c)$ are distinct modulo π^3, so that $\nu(\theta^i(c) - \theta^j(c)) = 2$ for $i \neq j$.*

Lemma 16 *The ν-sequence α_{T_4} of $T_4 \subseteq \Delta_3$ is given by*

$$\alpha_{T_4}(3n) = \alpha_{T_4}(3n+1) - 2 = \alpha_{T_4}(3n+2) - 4 = 7n + \sum_{i>0} \left\lfloor \frac{n}{2^i} \right\rfloor$$

The proofs of Lemmas 14, 15, and 16 are virtually identical to the proofs of Lemmas 11, 12, and 13, respectively, and so have been omitted from this paper.

5 A Regular Basis for Δ_3

Having written the ν-sequence for Δ_3 in terms of the ν-sequences for subsets in Proposition 3 and subsequently determined formulas for the ν-sequences of the subsets S, T_2, and T_4 in Sect. 4, we can now compute α_{Δ_3}.

Corollary 2 (To Proposition 3)
The first 200 terms of α_{Δ_3} are

0, 0, 0, 0, 0, 0, 0, 0, 1, 1, 2, 2, 3, 3, 3, 3, 3, 3, 3, 3, 5, 5, 6, 6, 6,
6, 6, 6, 6, 6, 7, 7, 8, 8, 9, 9, 9, 9, 9, 9, 10, 10, 12, 12, 12, 12, 12,
12, 12, 12, 13, 13, 14, 14, 15, 15, 15, 15, 15, 15, 15, 15, 17, 17,
18, 18, 18, 18, 18, 18, 18, 18, 19, 19, 21, 21, 21, 21, 21, 21, 21,
21, 23, 23, 25, 25, 26, 26, 27, 27, 27, 27, 27, 27, 27, 27, 28, 28,
30, 30, 30, 30, 30, 30, 30, 30, 31, 31, 32, 32, 33, 33, 33, 33, 33,
33, 33, 33, 36, 36, 36, 36, 36, 36, 36, 36, 37, 37, 38, 38, 39, 39,
39, 39, 39, 39, 39, 39, 41, 41, 42, 42, 42, 42, 42, 42, 42, 42, 43,

$$43, 45, 45, 45, 45, 45, 45, 45, 45, 47, 47, 48, 48, 48, 48, 48, 48,$$
$$49, 49, 51, 51, 52, 52, 53, 53, 54, 54, 54, 54, 54, 54, 54, 54, 56,$$
$$56, 57, 57, 57, 57, 57, 57, 57, 57, 58, 58, 60, 60, 60, 60, 60, 60.$$

The above results were generated using Mathematica. The code for the algorithm used to compute α_{Δ_3} can be found in Appendix A.1 of [11].

Following the results of Lemmas 10, 13, and 16, we obtain the following result as a corollary of Proposition 2.

Corollary 3 (c.f. [3, 2.14])

(i) *The sequence of polynomials*

$$\left\{\pi^{-\alpha_S(2n)} f_n(x), \pi^{-\alpha_S(3n+1)} x f_n(x), \pi^{-\alpha_S(3n+2)} x^2 f_n(x) : n = 0, 1, 2 \dots\right\}$$

forms a regular Δ_3-basis for Int(S, Δ_3).

(ii) *The sequence of polynomials*

$$\left\{\pi^{-\alpha_{T_2}(3n)} g_n(x), \pi^{-\alpha_{T_2}(3n+1)} x g_n(x), \pi^{-\alpha_{T_2}(3n+2)} x^2 g_n(x) : n = 0, 1, 2 \dots\right\}$$

forms a regular Δ_3-basis for Int(T_2, Δ_3).

(iii) *The sequence of polynomials*

$$\left\{\pi^{-\alpha_{T_4}(3n)} h_n(x), \pi^{-\alpha_{T_4}(3n+1)} x h_n(x), \pi^{-\alpha_{T_4}(3n+2)} x^2 h_n(x) : n = 0, 1, 2 \dots\right\}$$

forms a regular Δ_3-basis for Int(T_4, Δ_3).

The results in [3] regarding regular bases for subsets of Δ_2 in no way relies on the fact that the maximal order is of index 2 and can be extended without any trouble to Δ_3.

Lemma 17 (c.f. [3, 2.15]) *If two subsets of Δ_3 satisfying the hypotheses of Lemma 2 each have a regular basis whose elements are each quotients of polynomials in $\mathbb{Z}[x]$ by powers of π, then their union has a basis of this form also.*

Corollary 4 (c.f. [3, 2.16]) Int(Δ_3) *has a regular basis whose elements are each a quotient of a polynomial in $\mathbb{Z}[x]$ by a power of π.*

To conclude, it is noted that a very similar process to this extends to all n for which n is an odd prime over $\mathbb{Q}_2$ (as investigated in [11]). It seems there is no reason these results should not hold more generally over a different local field $\mathbb{Q}_p$, with p an odd prime. However, it is noted that this construction in this paper will not work precisely as given for n composite, and some adjustments will be needed in this case.

Acknowledgments The work presented in this paper was completed while the author was affiliated with Dalhousie University in Halifax, NS, Canada. It was supported in part by a Natural Sciences and Engineering Research Council of Canada (NSERC) Postgraduate Scholarship (PGS-D).

References

1. M. Bhargava. The factorial function and generalizations. *The American Mathematical Monthly*, 107(9):783–799, 2000.
2. P.-J. Cahen and J.-L. Chabert. *Integer-Valued Polynomials*, volume 48 of *Mathematical Surveys and Monographs*. American Mathematical Society, Providence, RI, USA, 1997.
3. S. Evrard and K. Johnson. The ring of integer valued polynomials on 2 × 2 matrices and its integral closure. *Journal of Algebra*, 441:660–677, 2015.
4. S. Frisch. Polynomial separation of points in algebras. *Arithmetical Properties of Commutative Rings and Monoids (Chapel Hill Conf. 2003), Lect. Notes in Pure and Appl. Math*, 241:253–259, 2005.
5. K. Johnson. p-orderings of noncommutative rings. *Proceedings of the American Mathematical Society*, 143(8):3265–3279, 2015.
6. T.Y. Lam. *A First Course in Noncommutative Rings*. Number 131 in Graduate Texts in Mathematics. Springer-Verlag, New York, 2nd edition, 2001.
7. K.A. Loper and N.J. Werner. Generalized rings of integer-valued polynomials. *Journal of Number Theory*, 132:2481–2490, 2012.
8. G. Peruginelli and N.J. Werner. Properly integral polynomials over the ring of integer-valued polynomials on a matrix ring. *Journal of Algebra*, 460:320–339, 2016.
9. I. Reiner. *Maximal Orders*. London Mathematical Society. Academic Press, London, 1975.
10. J-P. Serre. Local class field theory. In J.W.S. Cassels and A. Frohlich, editors, *Algebraic Number Theory*, chapter VI, pages 128–161. Thompson Book Company Inc., Washington, D.C., 1967.
11. A.C. Sodhi. *Polynomials integer-valued on maximal orders in division algebras*. PhD thesis, Dalhousie University, 2020.

Simultaneous p-Orderings and Equidistribution

Anna Szumowicz

1 Introduction

1.1 Integer-Valued Polynomials and Test Sets

Let D be a domain and let F be its field of fractions. We say that a polynomial $P(X) \in F[X]$ is **integer-valued** if $P(D) \subseteq D$. The sum, the product, and the difference of integer-valued polynomials are again integer-valued, so the set of integer-valued polynomials forms a ring:

$$\mathrm{Int}(D) = \{f(X) \in F[X] | \ f(D) \subseteq D\}.$$

More generally, for any subset $E \subseteq D$ we can consider

$$\mathrm{Int}(E, D) = \{f \in F[X] | \ f(E) \subseteq D\}.$$

Integer-valued polynomials do not necessarily have coefficients in D. For example, consider the case $D = \mathbb{Z}$, $F = \mathbb{Q}$. Then, any polynomial of the form

$$\binom{X}{n} = \frac{X(X-1)\dots(X-n+1)}{n!}$$

with $n \in \mathbb{N}$ is integer-valued. In fact, any integer-valued polynomial $P(X) \in \mathbb{Q}[X]$ of degree n can be uniquely written as a linear combination:

A. Szumowicz (✉)
Caltech, The Division of Physics, Mathematics and Astronomy, Pasadena, CA, USA
e-mail: anna.szumowicz@caltech.edu

J.-L. Chabert et al. (eds.), *Algebraic, Number Theoretic, and Topological Aspects of Ring Theory*, https://doi.org/10.1007/978-3-031-28847-0_22

$$P(X) = \sum_{i=0}^{n} \alpha_i \binom{X}{i},$$

where $\alpha_i \in \mathbb{Z}$.

This example shows that to check whether a degree n polynomial $P(X) \in \mathbb{Z}[X]$ is integer-valued, it is enough to check its values on the set $\{0, 1 \ldots, n\}$. Such testing sets can be defined in greater generality. Volkov and Petrov [13] introduced the notion of an n-universal set.

Definition 1 Let D be a domain and let F be its field of fractions. We say that a finite subset $S \subseteq D$ is n-**universal** if the following holds:

For every $P(X) \in F[X]$ of degree at most n, if $P(S) \subseteq D$ then $P(D) \subseteq D$, i.e., $P(X)$ is integer-valued polynomial.

We have the related notion of a Newton sequence.

Definition 2 Let D be a domain. A sequence $a_0, a_1, \ldots, a_n$ is called a Newton sequence if for every $0 \leq m \leq n$ the set $\{a_0, \ldots a_m\}$ is m-universal. The integer n is called the length of the Newton sequence.

Using the Lagrange interpolation it is easy to give a lower bound on the cardinality of an n-universal set.

Lemma 1 *Let D be a domain which is not a field. Then, every n-universal subset of D has at least $n + 1$ elements.*

Proof Indeed, for any fixed pairwise different elements $d_0, d_1, \ldots, d_n \in D$, we can construct a polynomial P that takes value 0 on $d_0, \ldots, d_{n-1}$ but a non-integer value α at d_n:

$$P(X) = \alpha \prod_{i=0}^{n-1} \frac{X - d_i}{d_n - d_i}.$$

Therefore P is not integer-valued, so $d_0, \ldots, d_{n-1}$ cannot be an n-universal set. □

On the other hand any Dedekind domain will contain an n-universal set with $n+2$ elements (see Theorem 3). Therefore, the case of n-universal sets of cardinality $n+1$ is particularly interesting. Some Dedekind domains will contain such sets, and as we shall see later, many do not.

Definition 3 Let D be a Dedekind domain. A subset $S \subseteq D$ with $|S| = n + 1$ is called n-**optimal** if it is n-universal.

Example

The set $\{x, x + 1, \ldots, x + n\}$ is n-optimal in $\mathbb{Z}$ for every $x \in \mathbb{Z}$.

The property of being n-optimal can be also understood as optimal equidistribution modulo all prime powers, hence the name. We expand more on that in the following section, after reviewing what is known on the minimal cardinality of an n-universal set. Petrov and Volkov [13] showed that there are no n-optimal sets in $\mathbb{Z}[i]$, for large enough n. Building on their method, together with Byszewski and Fraczyk, we generalized their result to the ring of integers in any quadratic imaginary number field.

Theorem 1 ([6]) *Let K be a quadratic imaginary number field and let $\mathcal{O}_K$ be its ring of integers. Then, there is no n-optimal sets in $\mathcal{O}_K$ for large enough n.*

For general quadratic number fields, Chabert and Cahen [8] proved that there are no 2-optimal sets except possibly in $\mathbb{Q}(\sqrt{d})$ with $d = -3, -1, 2, 3, 5$ and $d \equiv 1$ mod 8. The proof of Theorem 1 as well as the original method of Petrov and Volkov heavily relies on the fact that the norm of the field extension $K/\mathbb{Q}$ is convex. This is not the case in any number field beyond the imaginary quadratic extensions of $\mathbb{Q}$ and $\mathbb{Q}$ itself. Together with Fraczyk we used a new potential theoretic approach to extended the result to all number fields $K \neq \mathbb{Q}$.

Theorem 2 ([12]) *Let $K \neq \mathbb{Q}$ be a number field and let $\mathcal{O}_K$ be its ring of integers. There exists $n_0 \in \mathbb{N}$ dependent on K such that for any $n \geq n_0$ there is no n-optimal sets in $\mathcal{O}_K$.*

The picture becomes complete when we combine the above theorem with a very general upper bound on the minimal cardinality of an n-universal set.

Theorem 3 ([6]) *Let D be a Dedekind domain. Then for any $n \in \mathbb{N}$ there exists an n-universal set in D of size $n + 2$.*

The proof is an iterative construction using the Chinese Remainder Theorem. From Theorem 2 we can now deduce

Corollary 1 ([12]) *Let $K \neq \mathbb{Q}$ be a number field. For n large enough, the minimal cardinality of an n-universal set in $\mathcal{O}_K$ is $n + 2$.*

1.2 Equidistribution and Simultaneous $\mathfrak{p}$-Orderings

The methods of proving the nonexistence of large n-optimal sets are based on the almost equidistribution property of n-optimal sets.

Definition 4 Let A be a ring and let I be an ideal in A. A finite subset $E \subseteq A$ is called **almost uniformly equidistributed** modulo I if for any $a, b \in A$ we have

$$|\{x \in E|\ x - a \in I\}| - |\{x \in E|\ x - b \in I\}| \in \{-1, 0, 1\}.$$

Lemma 2 ([6, 13]) *Let K be a number field and let $\mathcal{O}_K$ be its ring of integers. Let $S \subseteq \mathcal{O}_K$ be a finite subset with $|S| = n + 1$. Then S is n-optimal if and only if S is* *almost uniformly equidistributed modulo* $\mathfrak{p}^l$ *for every prime ideal* $\mathfrak{p}$ *in* $\mathcal{O}_K$ *and all* $l \in \mathbb{N}$.

This characterization of n-optimal sets is reminiscent of the notion of a simultaneous $\mathfrak{p}$-ordering, introduced by Bhargava [4].

Definition 5 Let $E \subseteq \mathcal{O}_K$ and let $\mathfrak{p}$ be a non-zero proper prime ideal in $\mathcal{O}_K$. A sequence $(a_i)_{i\in\mathbb{N}} \subseteq E$ is called a $\mathfrak{p}$-**ordering** in E if for every $n \in \mathbb{N}$ we have

$$v_E(\mathfrak{p}, n) := v_\mathfrak{p}\left(\prod_{i=0}^{n-1}(a_i - a_n)\right) = \min_{x\in E} v_\mathfrak{p}\left(\prod_{i=0}^{n-1} a_i - x\right),$$

where $v_\mathfrak{p}$ denotes the additive $\mathfrak{p}$-adic valuation on K.

The value $v_E(\mathfrak{p}, n)$ does not depend on the choice of a $\mathfrak{p}$-ordering. Bhargava defined the generalized factorial as the ideal $n!_E := \prod_\mathfrak{p} \mathfrak{p}^{v_E(\mathfrak{p},n)}$ where $\mathfrak{p}$ runs over all prime ideals in $\mathcal{O}_K$. A sequence of elements in E is called a **simultaneous** $\mathfrak{p}$-**ordering** if it is a $\mathfrak{p}$-ordering for every prime ideal $\mathfrak{p}$ in $\mathcal{O}_K$ at the same time. One can show that $(a_i)_{i\in\mathbb{N}} \subseteq \mathcal{O}_K$ is a simultaneous $\mathfrak{p}$-ordering in $\mathcal{O}_K$ if and only if the set $\{a_0, \ldots, a_n\}$ is n-optimal (see Lemma 2). In [4, 5] Bhargava asked which subsets of Dedekind domains admit simultaneous $\mathfrak{p}$-orderings. In particular, he asked for which number fields K, the ring of integers $\mathcal{O}_K$ admits a simultaneous $\mathfrak{p}$-ordering. A partial progress was made by Wood in [15] where she showed that there are no simultaneous $\mathfrak{p}$-orderings in $\mathcal{O}_K$ when K is a quadratic imaginary number field. Adam and Cahen [1] extended this result to any quadratic number field $\mathbb{Q}(\sqrt{d})$, $d \in \mathbb{Z}$ square-free, except for possibly finitely many exceptional d's. Using a simultaneous $\mathfrak{p}$-ordering, one could construct n-optimal sets for every $n \in \mathbb{N}$. Therefore, Theorem 2 yields:

Corollary 2 ([12]) $\mathbb{Q}$ *is the only number field whose ring of integers admits a simultaneous* $\mathfrak{p}$*-ordering.*

This result answers the question of Bhargava [4, 5]. We remark that the methods used to prove Corollary 2 differ substantially from the methods used by Adam, Cahen and Wood.

1.3 Notation

By $|S|$ we denote the cardinality of the set S. For any $x \in \mathbb{R}$ denote by $\lfloor x \rfloor$ the largest integer less than or equal to x. Write $N_{K/\mathbb{Q}}$ for the norm of the extension $K/\mathbb{Q}$. We use the standard big-O and little-o notation. We write $B_\mathbb{R}(x, r)$ (resp. $B_\mathbb{C}(x, r)$) for a ball in $\mathbb{R}$ (resp. $\mathbb{C}$) of radius r around a point $x \in \mathbb{R}$ (resp. $x \in \mathbb{C}$). Denote by Leb the Lebesgue measure on $\mathbb{R}$, $\mathbb{C}$ and their products. We denote by Δ_K the discriminant of a field K.

1.4 Structure of the Paper

In Sect. 2 we give a sketch of the proof of nonexistence of large n-optimal sets in the ring of integers of quadratic imaginary number fields (Theorem 1). In Sect. 3, we estimate the energy of n-optimal sets. In Sect. 4 we describe the methods used to prove Theorem 2 in [12]. In Sect. 5 we state some questions and open problems.

2 n-Optimal Sets for Quadratic Imaginary Number Fields

In this section we give a sketch of the proof of nonexistence of large n-optimal sets in the ring of integers of a quadratic imaginary number field.

Theorem 4 ([6]) *Let K be a quadratic imaginary number field and let $\mathcal{O}_K$ be its ring of integers. There is no n-optimal sets for large enough n.*

The condition for a subset to be n-optimal can be expressed in terms of the energy ideal of a set.

Definition 6 Let $S = \{x_0, \ldots, x_n\}$ be a finite subset of $\mathcal{O}_K$. The principal ideal $E(S) := \prod_{i \neq j}(x_i - x_j)$ is called the energy of the set S.

The energy ideal was called the volume in [6, 13], but in the subsequent work, [12] it became clear that it is the arithmetic analogue of the energy functional in potential theory, hence the new name. The energy ideal of an n-optimal set is minimal possible in the sense that it should divide the energy of any other set of equal size. This can be made more precise using the factorial ideals:

Definition 7 ([4, 11]) Let K be a number field and let $\mathcal{O}_K$ be its ring of integers. The K-factorial of n is defined as the principal ideal:

$$n!_K = n!_{\mathcal{O}_K} = \prod_{\mathfrak{p} \in \mathrm{Spec}(\mathcal{O}_K)} \mathfrak{p}^{w_{\mathfrak{p}}(n)},$$

where $w_{\mathfrak{p}}(n) = \sum_{i=1}^{\infty} \lfloor \frac{n}{N(\mathfrak{p}^i)} \rfloor$.

Proposition 1 ([6, 13]) *Let $S \subseteq \mathcal{O}_K$ with $|S| = n + 1$. Then, the following conditions are equivalent:*

1. *S is n-optimal.*
2. *$E(S) = (\prod_{i=1}^{n} i!_K)^2$.*
3. *$E(S)$ divides $E(T)$ for any subset T of $\mathcal{O}_K$ with $n + 1$ elements.*

In other words, n-optimal sets are the sets with $n + 1$ elements which "minimize" the energy among all subsets of $\mathcal{O}_K$ with $n + 1$ elements.

Proof of Theorem 1 For the sake of contradiction, assume that for any n_0 there exists an n-optimal set with $n \geq n_0$. Identify $\mathcal{O}_K$ with its image via a fixed embedding $K \to \mathbb{C}$. The idea of the proof is as follows. Let S be an n-optimal set. We sketch the proof why S has to be contained in a polygon with $n + o(n)$ points from $\mathcal{O}_K$. With such a fine description of S, one can use the prime number theorem for number fields to show that there exists a prime power $\mathfrak{p}^l$ such that S fails to be almost uniformly equidistributed modulo $\mathfrak{p}^l$. Together with Proposition 1, this leads to a contradiction.

To show that an n-optimal set is contained in a suitable polygon with $n + o(n)$ points from $\mathcal{O}_K$, we use Proposition 1 and a procedure called discrete collapsing. Roughly speaking, the collapsing procedure takes a finite subset of $\mathcal{O}_K$ and makes it symmetric about a line ℓ by moving the points towards the line as close as possible. For the proof we will only need the formal definition of what it means to be collapsed with respect to a line.

Definition 8 ([6]) Let K be a quadratic imaginary number field. Let T be a finite subset of $\mathcal{O}_K$. Let l be a line in $\mathbb{C}$. The line ℓ divides the complex plane into two closed half-planes, say H_1 and H_2. Distinguish one of them, say H_1. The set T is **collapsed** along the pair (l, H_1) if the following conditions hold:

1. Let m be a line which is perpendicular to ℓ and contains at least one point from the set T. Let $x \in T \cap m$. Then every point in $\mathcal{O}_K$ which lies between $m \cap l$ and x is in T.
2. Let m be a line perpendicular to ℓ. Then $|H_1 \cap m \cap T| - |H_2 \cap m \cap T| \in \{0, 1\}$.

□

The discrete collapsing procedure produces collapsed sets, and one can show that if the set was not collapsed to begin with, it strictly decreases the norm of the energy [6]. One can show that an n-optimal set has to be collapsed in every direction. More precisely, from Proposition 1 we deduce the following.

Lemma 3 *Let K be a quadratic imaginary number field and let $\mathcal{O}_K$ be its ring of integers. Let T be an n-optimal set in $\mathcal{O}_K$ and let l be a line in the complex plane. Then there exists a line m parallel to l such that the set T is collapsed along the line m for some choice of the distinguished half-plane.* □

For the proof we refer to [6].

Definition 9 Let l be a line in the complex plane. A **strip** along l is a closed domain which is bounded by two lines parallel to l and symmetric with respect to l. A strip parallel to l is a strip along a line parallel to l. □

Let $K = \mathbb{Q}(\sqrt{d})$. The proof differs in the cases $d \not\equiv -1 \mod 4$ and $d \equiv -1 \mod 4$. We start with the case $d \not\equiv -1 \mod 4$. Then, $\mathcal{O}_K = \mathbb{Z} + \mathbb{Z}\sqrt{d}$. By Lemma 3, the set S has to be collapsed along some vertical line ℓ_1 and some horizontal line ℓ_2. By the Dirichlet's theorem on prime numbers in arithmetic progressions, one can find a rational prime number p_1 such that $p_1 = \sqrt{n} + o(\sqrt{n})$, $p_1 > n + 1$ and p_1 is prime in $\mathcal{O}_K$. Since S is supposed to be almost

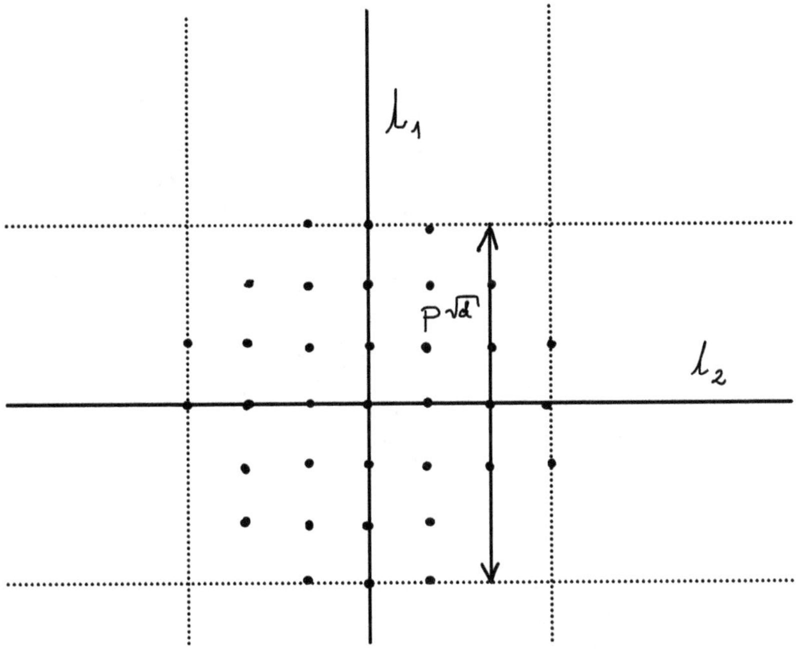

Fig. 1 $d \not\equiv -1 \mod 4$

uniformly equidistributed modulo p_1 by Lemma 2, the intersection of S with any horizontal or vertical line can contain at most p_1 consecutive points of $\mathcal{O}_K$. Therefore, S has to contained in the intersection of two stripes along ℓ_1, ℓ_2 of widths p_1 and $p_1\sqrt{d}$, respectively (see Fig. 1). This intersection is a rectangle containing roughly $n + o(n)$ lattice points. Since the set S has $n + 1$ elements, this means that it has to fill the rectangle perfectly, missing only $o(n)$ points.

In the case $d \equiv -1 \mod 4$, by Lemma 3, S is collapsed along the lines

$$k_0 \text{ parallel to } \{iy|\ \ y \in \mathbb{R}\},$$

$$k_1 \text{ parallel to } \left\{x + iy|\ \ y = -\frac{x}{\sqrt{d}}\right\},$$

$$k_2 \text{ parallel to } \left\{x + iy|\ \ y = \frac{x}{\sqrt{d}}\right\}.$$

Again, using Lemma 2, we deduce that in the case $d \equiv -1 \mod 4$ an n-optimal set has to be contained in a hexagon with $n + o(n)$ points from $\mathcal{O}_K$ (see Fig. 2). Since S has $n + 1$ points, this means that the set S fills the hexagon perfectly, missing only $o(n)$ points. Using Dirichlet's theorem on prime numbers in arithmetic progressions, we can find a rational prime p_1, non-split in $\mathcal{O}_K$, modulo which the set S fails to be almost uniformly equidistributed. The last part of the argument uses only the

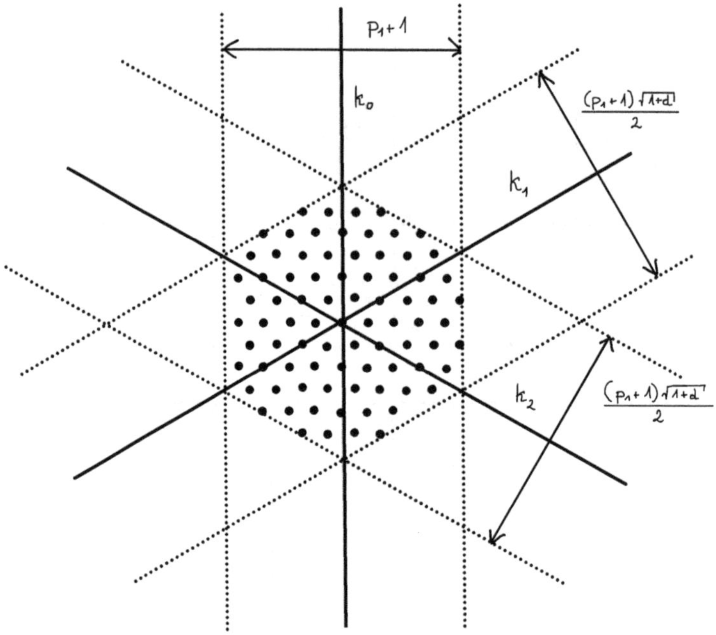

Fig. 2 $d \equiv -1 \mod 4$

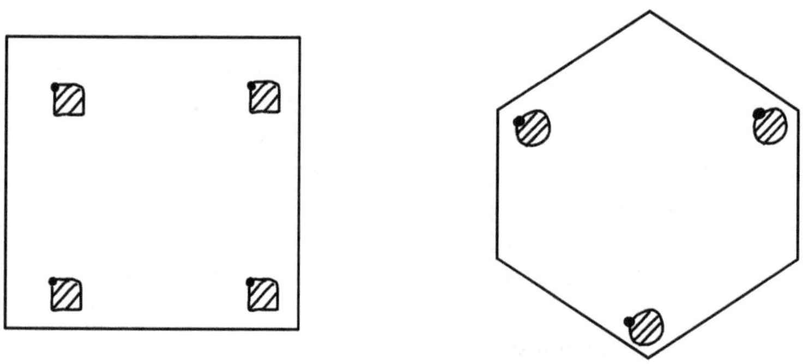

Fig. 3 Points congruent modulo p_1

geometry of the rectangle and the hexagon. For an appropriate size of p_1, both shapes contain too many points congruent mod p_1 (see Fig. 3). □

Remark 1 The key ingredient in the proof for the quadratic imaginary number fields is the fact that an n-optimal set must be collapsed in any direction. The proof of this property relies crucially on the fact that the norm $x \mapsto |N_{K/\mathbb{Q}}(x)|$ is a convex function. This is no longer the case in a general number field.

3 Estimate on the Energy of n-Optimal Sets

Let K be a number field, this time not necessarily quadratic imaginary. Using Proposition 1 and an estimate on the norm of the factorial ideals due to Lamoureux [11], it is possible to estimate the energy of n-optimal sets. The formulas will use the Euler-Kronecker constants, defined below.

Definition 10 ([9]) Let K be a number field and let $\zeta_K(z)$ be the Dedekind zeta function of K. Let

$$\zeta_K(z) = \frac{c_{-1}}{z-1} + c_0 + c_1(z-1) + \dots$$

be the Laurent expansion of ζ_K at $s = 1$. The quotient $\frac{c_0}{c_{-1}}$ is called the **Euler-Kronecker constant** of K and we denote it by γ_K.

In the case $K = \mathbb{Q}$, the Euler-Kronecker constant $\gamma_{\mathbb{Q}}$ is called the Euler-Mascheroni constant and is given by the following formula:

$$\gamma_{\mathbb{Q}} = \lim_{n\to\infty} \left(\sum_{i=1}^{n} \frac{1}{i} - \log n \right).$$

For more information on the Euler-Kronecker constants, see [9]. By Proposition 1, to estimate the energy of n-optimal sets, it is enough to estimate K-factorials. Thanks to Lamoureux [11] we have the following estimate.

Theorem 5 ([11]) *Let K be a number field. Then,*

$$\log N_{K/\mathbb{Q}}(n!_K) = n \log n - n(1 + \gamma_K - \gamma_{\mathbb{Q}}) + o(n).$$

Using Proposition 1 we deduce

Corollary 3 ([6]) *Let K be a number field and let $\mathcal{O}_K$ be its ring of integers. Let S be an n-optimal subset of $\mathcal{O}_K$. Then*

$$\log N_{K/\mathbb{Q}}(E(S)) = n^2 \log n - n^2(\frac{3}{2} + \gamma_K - \gamma_{\mathbb{Q}}) + o(n^2).$$

Moreover, for every subset $T \subseteq \mathcal{O}_K$ with $|T| = n + 1$, we have

$$\log N_{K/\mathbb{Q}}(E(T)) \geq n^2 \log n - n^2(\frac{3}{2} + \gamma_K - \gamma_{\mathbb{Q}}) + o(n^2).$$

4 n-Optimal Sets for an Arbitrary Number Field

As we already mentioned, the methods for quadratic imaginary number fields cannot be adapted to the case of a general number field, as in general the norm is not convex. In this section we sketch the proof for an arbitrary number field [12]. Fix a number field K of degree N and write $\mathcal{O}_K$ for the ring of integers.

We argue by contradiction. Assume there exists a sequence S_{n_i} of n_i-optimal sets where n_i tends to infinity. Let $V = K \otimes_{\mathbb{Q}} \mathbb{R} \cong \mathbb{R}^{r_1} \times \mathbb{C}^{r_2}$. Write $d = r_1 + r_2$. The absolute value of the norm $|N_{K/\mathbb{Q}}(\cdot)|$ extends to the map $\|\cdot\|: V \to \mathbb{R}$ defined by $\|v\| = \prod_{i=1}^{r_1} |v_i| \prod_{i=r_1+1}^{d} |v_i|^2$ for $v = (v_1, \ldots, v_d)$.

4.1 *Enclosure of n-Optimal Sets in Cylinders*

The first step is to show that we can enclose S_{n_i} in a cylinder of the volume $n_i + o(n_i)$.

Definition 11 A cylinder $\mathcal{C}$ in V is a coordinate-wise product of balls:

$$\mathcal{C} = \prod_{i=1}^{r_1} B_{\mathbb{R}}(x_i, r_i) \times \prod_{i=r_1+1}^{d} B_{\mathbb{C}}(x_i, r_i),$$

where $x_i \in \mathbb{R}$ for $i = 1, \ldots, r_1$, $x_i \in \mathbb{C}$ for $i = r_1 + 1, \ldots, d$ and $r_i \in \mathbb{R}_{\geq 0}$ for $i = 1, \ldots, d$.

The volume of a cylinder $\mathcal{C}$ is defined as its Lebesgue measure.

Theorem 6 ([12, Theorem 3.1]) *There exists a positive constant θ dependent only on K such that for every n-optimal set $S \subseteq \mathcal{O}_K$ there exists a cylinder $\mathcal{C}$ of volume θn with $S \subseteq \mathcal{C}$.*

And as a consequence we get:

Corollary 4 *There exists a positive constant $A > 0$ depending only on K such that the set $\Omega = B_{\mathbb{R}}(0, A)^{r_1} \times B_{\mathbb{C}}(0, A)^{r_2}$ has the following property. Let $S \subseteq \mathcal{O}_K$ be an n-optimal set. Then, there exist $s, t \in V$ such that $||s|| = n|\Delta_K|^{1/2}$ and $s^{-1}(S - t) \subseteq \Omega$.*

Theorem 6 was implicit in the proof of Theorem 2 for $K = \mathbb{Q}(i)$ [13] and for K quadratic imaginary [6]. Indeed, in Sect. 2 we remark that S is contained in a convex polygon of volume $|\Delta_K|^{1/2}n + o(n)$ which can be always enclosed in a cylinder of volume θn. It was proved using the collapsing procedure which heavily relies on the fact that the norm $N_{K/\mathbb{Q}}$ is convex for imaginary quadratic number field K. Proving Theorem 6 for a general case was one of the main difficulties in proving Theorem 2. One of the main ingredients is a result on counting the number of $x \in \mathcal{O}_K$ such that

$\|x(a-x)\| \leq X^2$ for some $X > 0$ and $a \in \mathcal{O}_K$ such that $\|a\| \geq Xe^{-B}$ where $B \in \mathbb{R}$ is fixed. This result, in a sense, substitutes for the role of the convexity of the norm. To give a precise statement of the result, we introduce the notion of a good fundamental domain.

Definition 12 A **good fundamental domain** of $\mathcal{O}_K^\times$ in $V^\times$ is a set $\mathcal{F}$ which is a finite union of convex closed cones in $V^\times$ such that $\mathcal{F}/\mathbb{R}^\times$ is compact in the projective space $\mathbb{P}(V)$, $V^\times = \bigcup_{\lambda \in \mathcal{O}_K^\times} \lambda\mathcal{F}$, $\mathrm{int}\mathcal{F} \cap \lambda(\mathrm{int}\mathcal{F}) = \emptyset$ for every $\lambda \in \mathcal{O}_K^\times$, $\lambda \neq 1$ and $\partial\mathcal{F}$ does not contain non-zero points from $\mathcal{O}_K$.

Fix a good fundamental domain $\mathcal{F}$. Let $a \in \mathcal{O}_K$, $a \neq 0$ and $X > 0$. Define

$$S(a, X) = \{(x, \lambda) \in (\mathcal{F} \cap \mathcal{O}_K) \times \mathcal{O}_K^\times | \ \|x(a - x\lambda^{-1})\| \leq X^2, \ \ \|x\| \leq X\}.$$

Denote $\log^\dagger x := \log x$ if $x > 1$ and $\log^\dagger x := 0$ otherwise.

Proposition 2 ([12, Proposition 2.5]) *Let K be a number field of degree N with d Archimedean places. Let $B \in \mathbb{R}$. Let $\kappa = \frac{1}{3}$ if $N = 1$ and $\kappa = \min\left\{\frac{1}{2N(N-1)}, \frac{1}{4N-1}\right\}$ otherwise. Fix a good fundamental domain $\mathcal{F}$. There exist constants $\theta_1, \theta_2, \theta_3, \theta_4$ dependent only on K, B and $\mathcal{F}$ such that for every $X > 0$ and $a \in \mathcal{O}_K$ such that $\|a\| \geq Xe^{-B}$ we have*

1. $|S(a, X)| \leq \theta_1 X^{1+\kappa}\|a\|^{-\kappa} + \theta_2(\log X)^{2d-2} + \theta_3\log^\dagger\log^\dagger\log^\dagger\log\|a\| + \theta_4.$
2. *Suppose $a \in \mathcal{F}$. Then, for every $\varepsilon > 0$ there exists M such that*

$$|\{(x, \lambda) \in S(a, X)| \ \|\lambda\|_\infty \geq M\}| \leq$$
$$\varepsilon X^{1+\kappa}\|a\|^{-\kappa} + \theta_2(\log X)^{2d-2} + \theta_3\log^\dagger\log^\dagger\log^\dagger\log\|a\| + \theta_4.$$

The proof of the proposition is based on Aramaki-Ikehara Tauberian theorem [2], Baker-Wüstholz's inequality on linear forms in logarithms [3, Theorem 7.1] and on some elementary estimates on the number of integer points in cylinders.

As a consequence we obtain the following result which may be of independent interest.

Theorem 7 *Let K be a number field of degree N with d Archimedean places. Let $B \in \mathbb{R}$ and let $\kappa = \frac{1}{3}$ if $N = 1$ and $\kappa = \min\left\{\frac{1}{2N(N-1)}, \frac{1}{4N-1}\right\}$ otherwise. There exist constants C_1, C_2, C_3, C_4 dependent only on K and B such that for every $X > 0$ and $a \in \mathcal{O}_K$ such that $\|a\| \geq Xe^{-B}$ we have*

$$|\{x \in \mathcal{O}_K| \ \|x(a-x)\| \leq X^2\}| \leq$$
$$C_1X^{1+\kappa}\|a\|^{-\kappa} + C_2(\log X)^{2d-2} + C_3\log^\dagger\log^\dagger\log^\dagger\log\|a\| + C_4.$$

4.2 Limit Measures

For the sake of contradiction, we assumed that there exists a sequence of n_i-optimal sets (S_{n_i}) in $\mathcal{O}_K$ where n_i tends to infinity. Theorem 6 shows that, up to translation and suitable rescaling, all of the sets S_{n_i} can be enclosed in one compact set. Using that fact we construct a tight family of measures associated to sets S_{n_i} and consider their weak-$*$ limits. In this section we study properties of such limits.

By Corollary 4, there exist sequences $(s_{n_i}), (t_{n_i}) \subseteq V$ with $\|s_{n_i}\| = n_i|\Delta_K|^{1/2}$ and a compact set Ω such that $s_{n_i}^{-1}(S_{n_i} - t_{n_i}) \subseteq \Omega$. Define the measures

$$\mu_{n_i} := \frac{1}{n_i} \sum_{x \in S_{n_i}} \delta_{s_{n_i}^{-1}(x - t_{n_i})}.$$

Since Ω is compact we can consider, passing to a subsequence if necessary, a weak-* limit of μ_{n_i}. Existence of such limits crucially uses the fact that Ω is compact. In this section we study the properties of such weak-* limits.

Definition 13 A probability measure μ on V is a **limit measure** if it is a weak-* limit measure of the measures μ_{n_i} defined above.

A limit measure is a probability measure supported on Ω, absolutely continuous with respect to the Lebesgue measure and of density at most one (see [12, Lemma 5.2]). The measure μ encodes some information about the large-scale geometry of the sets S_{n_i}. The idea for the rest of the proof is to use the properties of n-optimal sets to show that such limit measures cannot exist.

By analogy with the finite subsets of $\mathcal{O}_K$, one can define the energy of compactly supported, probability measures on V, absolutely continuous with respect to the Lebesgue measure and of bounded density.

Definition 14 Let ν be a compactly supported measure on V, absolutely continuous with respect the Lebesgue measure and of bounded density. The energy of ν is given by

$$I(\nu) = \int_V \int_V \log \|x - y\| d\nu(x) d\nu(y).$$

The estimate on the energy of n-optimal sets (Corollary 3) leads to a formula for the energy of limit measures.

Proposition 3 ([12, Proposition 5.3]) *Let μ be a limit measure. Then,*

$$I(\mu) = -\frac{1}{2} \log |\Delta_K| - \frac{3}{2} - \gamma_K + \gamma_{\mathbb{Q}}.$$

Using the estimate on the energy of n-optimal sets (Corollary 3) and the fact that n-optimal sets are energy minimizing, one can give a lower bound on the energy of a compactly supported probability measure on V with density at most 1.

Lemma 4 ([12, Lemma 5.4]) *Let ν be a compactly supported probability measure on V with density at most* 1. *Then*

$$I(\nu) \geq -\frac{1}{2}\log|\Delta_K| - \frac{3}{2} - \gamma_K + \gamma_{\mathbb{Q}}.$$

The idea of the proof of this estimate is to construct a sequence of sets E_n, $|E_n| = n+1$ such that the rescaled normalized counting measures $\frac{1}{n}\sum_{x\in E_n}\delta_{n^{-1/N}|\Delta_K|^{-1/2N}x}$ weakly-* converges to ν. Then, one can relate the asymptotic growth of the energies of E_n with the energy of ν and use Proposition 1.

Lemma 4 together with Proposition 3 implies that limit measures minimize the energy among all compactly supported probability measures on V of density at most 1. This gives a strong constrains on the structure of limit measures.

Proposition 4 ([12, Proposition 5.5]) *Let ν be a compactly supported probability measure on V of density at most 1 which is realizing the minimal energy among all such measures. Then, there exists an open set U and $v \in V$ such that*

1. $\nu = \mathrm{Leb}|_U$.
2. $\lambda(\overline{U} - v) \subseteq U - v$ *for every* $0 \leq \lambda < 1$.
3. $(\partial U - v) \cap V^\times$ *is a codimension 1 submanifold of* $V^\times$ *of class* C^1.

The proof uses a procedure of collapsing measures which is a continuous version of the discrete collapsing. As opposed to the quadratic imaginary case, here one can collapse the measure only along the hyperplanes parallel to a hyperplane contained in $V \setminus V^\times$.

Since any limit measure μ minimizes the energy among all compactly supported probability measures on V of density at most 1, μ satisfies the conclusion of Proposition 4. In particular, the set S_{n_i} is equal to $(s_{n_i}U + t_{n_i}) \cap \mathcal{O}_K$, modulo $o(n)$ points. In the remainder of the argument, the sets $(s_{n_i}U + t_{n_i})$ will play the role of the rectangles or the hexagons from the proof in the quadratic imaginary case.

4.3 Discrepancy

In the last step of the proof, one needs to find a prime power $\mathfrak{p}^l \subset \mathcal{O}_K$ modulo which the set S_{n_i} fails to be almost uniformly equidistributed. In the quadratic imaginary case, it was possible to describe the shape of S_{n_i} quite explicitly and find the prime power by hand. This is no longer possible, since one only knows that the shape is given by the set U from Proposition 4. The existence of a prime power with the desired properties can be shown using the discrepancy of the set U.

Definition 15 Let W be a bounded measurable subset of V. For $x \in V^{\times}$, $v \in V$ define $N_x(W, v) := |(xU + v) \cap \mathcal{O}_K|$. We define the discrepancy as

$$D_x(W, v) := N_x(W, v) - |\Delta_K|^{-\frac{1}{2}} \mathrm{Leb}(W)\|x\|$$

and the maximal discrepancy

$$D_x(W) := \operatorname*{ess\,sup}_{v \in V} |D_x(W, v)|.$$

Using the fact that n-optimal sets are almost uniformly equidistributed modulo every power of every prime ideal in $\mathcal{O}_K$ together with a version of the prime number theorem, one can give a uniform upper bound on $D_x(U)$ for all $x \in V^{\times}$.

Lemma 5 ([12, Lemma 6.3]) *Let μ be a limit measure on V. Let U be a non-empty open-bounded subset of V such that $\mu = \mathrm{Leb}|_U$ and ∂U is Jordan measurable of Jordan measure* 0. *Then $D_x(U) < 1$ for all $x \in V^{\times}$.*

On the other hand, using the smoothness of the boundary of U, we have:

Lemma 6 ([12, Lemma 6.4]) *Assume $V = \mathbb{R}^{r_1} \times \mathbb{C}^{r_2}$ with $r_1 + 2r_2 > 1$. Let W be an open-bounded subset of V such that $\partial W \cap V^{\times}$ is a submanifold of $V^{\times}$ of class C^1 and $\lambda \overline{U} \subseteq U$ for every $0 \le \lambda < 1$. Then, there exists $x \in V^{\times}$ such that $D_x(U) > 1$.*

4.4 *Proof of Theorem 2*

In this section we gather results described earlier to sketch the proof that large n-optimal sets do not exist.

Proof of Theorem 2 Let $V = K \otimes_{\mathbb{Q}} \mathbb{R} = \mathbb{R}^{r_1} \times \mathbb{C}^{r_2}$. For the sake of contradiction, let us assume that there exists a sequence of n_i-optimal sets $S_{n_i} \subseteq \mathcal{O}_K$ with $n_i \to \infty$. By Corollary 4, there exists a compact set $\Omega \subseteq V$ and sequences $(s_{n_i})_{i \in \mathbb{N}}, (t_{n_i})_{i \in \mathbb{N}} \subseteq V$ with $\|s_{n_i}\| = n_i |\Delta_K|^{1/2}$ such that $s_{n_i}^{-1}(S_{n_i} - t_{n_i}) \subseteq \Omega$. We define

$$\mu_{n_i} = \frac{1}{n_i} \sum_{x \in S_{n_i}} \delta_{s_{n_i}^{-1}(x - t_{n_i})}.$$

Since Ω is compact, after passing to a subsequence if necessary, these measures converge weak-* to a probability measure μ. This is a limit measure. By Sect. 4.2, μ is compactly supported, absolutely continuous with respect to the Lebesgue measure of density at most 1, so by Proposition 3 and Lemma 4, it minimizes the energy among all such measures. By Proposition 4, $\mu = \mathrm{Leb}|_U$ where U is an open set with piecewise C^1 boundary. Finally using Lemma 5 and Lemma 6, we get a contradiction, since the discrepancy would be at the same time bigger and smaller than 1. □

5 Open Problems

5.1 Function Fields

Let $\mathbb{F}_q$ be a finite field with q elements, $\mathbb{F}_q = \{a_0, \ldots, a_{q-1}\}$. Let $n \in \mathbb{N}$. Write $n = \sum_{i=0}^{k} d_i q^i$ with $d_i \in \{0, \ldots, q-1\}$. It was observed by Bhargava in [4] that the sequence $s_n := \sum_{i=0}^{k} a_{d_i} t^i \in \mathbb{F}_q[t]$ is a simultaneous p-ordering in $\mathbb{F}_q[t]$.

More generally one can consider a protective curve $\mathbf{C}$ over $\mathbb{F}_q$, select a finite set $\Sigma \subset \mathbf{C}(\mathbb{F}_q)$ and consider the ring $\Gamma(\mathbf{C} \setminus \Sigma, \mathcal{O})$ of the regular functions on $\mathbf{C} \setminus \Sigma$. The case of the ring $\mathbb{F}_q[t]$ is recovered by taking $\mathbf{C} = \mathbb{P}^1$ and Σ consisting of the point at infinity.

? Question

For which curves and sets Σ does the ring $\Gamma(\mathbf{C} \setminus \Sigma, \mathcal{O})$ admits a simultaneous $\mathfrak{p}$-orderings or arbitrarily large n-optimal sets?

It seems that the answer might depend of the genus of the curve, with high genus unlikely to contain large n-optimal sets.

5.2 Schinzel's Problem

The question on existence of simultaneous $\mathfrak{p}$-orderings seems a bit similar to an old problem called Schinzel's problem:

? Schinzel's Problem [10]

Let $K \neq \mathbb{Q}$ be a number field and let $\mathcal{O}_K$ be its field of fractions. Does there exist a sequence $(a_i)_{i \in \mathbb{N}} \subseteq \mathcal{O}_K$ such that for every ideal I in $\mathcal{O}_K$ with the norm $N(I)$, the sequence $a_0, \ldots, a_{N(I)-1}$ is a complete system of representatives of $\mathcal{O}_K/I$.

There have been partial results obtained: Wantula (1969) (unpublished) showed that K cannot be a quadratic number field and Wasén [14] showed that $\mathcal{O}_K$ has to be a principal ideal domain.

Definition 16 A sequence $(a_i)_{i \in \mathbb{N}}$ satisfying the condition from Question 5.2 is called a Schinzel sequence.

Frisch [7] showed that if $\mathcal{O}_K$ admits a Schinzel sequence, then it is Euclidean.

Proposition 5 ([7, p.102]) *Let K be a number field and let $\mathcal{O}_K$ be its ring of integers. Let $N(x)$ denote the norm of the principal ideal (x) in $\mathcal{O}_K$. If $\mathcal{O}_K$ admits a Schinzel sequence, then $\mathcal{O}_K$ is Euclidean with the norm N.*

Proof Assume there exists a Schinzel sequence $(a_i)_{i\in\mathbb{N}}$ in $\mathcal{O}_K$. We can assume $a_0 = 0$. By the definition of Schinzel sequence, $N(a_j) \leq j$ for every $j \in \mathbb{N} \setminus \{0\}$. Let $a, b \in \mathcal{O}_K$ with $a \neq 0$ and a not a unit. By the definition of a Schinzel sequence, there exists $m \in \{0, \ldots, N(a) - 1\}$ such that $b = ca + a_m$. As observed before, $N(a_m) \leq m < N(a)$. □

Could the methods used to prove Theorem 2 be used to address the Schinzel's problem? If one could construct limit measures attached to a Schinzel sequence, what would be their properties?

References

1. David Adam and Paul-Jean Cahen, *Newtonian and Schinzel quadratic fields*, Journal of Pure and Applied Algebra **215** (2011), no. 8, 1902–1918.
2. Junichi Aramaki, *On an extension of the Ikehara Tauberian theorem*, Pacific Journal of Mathematics **133** (1988), no. 1, 13–30.
3. Alan Baker and Gisbert Wüstholz, *Logarithmic Forms and Diophantine Geometry*, Cambridge University Press, 2008.
4. Manjul Bhargava, *P-orderings and polynomial functions on arbitrary subsets of Dedekind rings*, Journal für die reine und angewandte Mathematik **490** (1997), 101–128.
5. Manjul Bhargava, *The factorial function and generalizations*, The American Mathematical Monthly **107** (2000), no. 9, 783–799.
6. Jakub Byszewski, Mikolaj Fraczyk and Anna Szumowicz, *Simultaneous $\mathfrak{p}$-orderings and minimizing volumes in number fields*, Journal of Number Theory **173** (2017), 478–511.
7. Paul-Jean Cahen and Jean-Luc Chabert, *Old problems and new questions around integer-valued polynomials and factorial sequences* In James W. Brewer, Sarah Glaz, William Heinzer and Bruce Olberding, Multiplicative ideal theory in commutative algebra: A Tribute to the Work of Robert Gilmer. Springer (2006)
8. Paul-Jean Cahen and Jean-Luc Chabert, *Test sets for polynomials: n-universal subsets and Newton sequences*, Journal of Algebra **502** (2018), 277–314.
9. Yasutaka Ihara, *On the Euler-Kronecker constants of global fields and primes with small norms*, Algebraic geometry and number theory, 2006, pp. 407–451.
10. W. Narkiewicz, *Some unsolved problems*, Bull. Soc. Math France, Mémoire **25** (1971), 159–164.
11. Matthew Lamoureux, *Stirling's Formula in Number Fields*, Doctoral Dissertations, University of Connecticut (2014).
12. Anna Szumowicz and Mikolaj Fraczyk, *On the optimal rate of equidistribution in number fields*, available at: https://arxiv.org/pdf/1810.11110.pdf
13. Vladislav Volkov and Fedor Petrov, *On the interpolation of integer-valued polynomials*, Journal of Number Theory **133** (2013), no. 12, 4224–4232.
14. Rolf Wasén, *Remark on a problem of Schinzel*, Acta Arithmetica **29** (1976), no. 4, 425–426.
15. Melanie Wood, *P-orderings: a metric viewpoint and the non-existence of simultaneous orderings*, Journal of Number Theory **99** (2003), no. 1, 36–56.

A Survey on Flatness in Integer-Valued Polynomial Rings

Ali Tamoussit

1 Introduction

Historically, the notion of flatness was first introduced and studied in 1956 by J.-P. Serre [43] as follows: a module M over an integral domain D is said to be *flat* if whenever $N_1 \to N_2 \to N_3$ is an exact sequence of D-modules, the sequence $M \otimes_D N_1 \to M \otimes_D N_2 \to M \otimes_D N_3$ is exact as well (cf. [43, Définition 3, page 34]). Bourbaki treated extensively on this notion in their 1961 book *Algèbre Commutative* [3]. In 1969, D. Lazard, published his thesis titled "Autour de la platitude" completely concerned with the problem of flatness [33]. Earlier in 1965, F. Richman studied the notion of flatness in the context of overrings of an integral domain, and thus he was the first to characterize the flat overrings of integral domains. Particularly, he showed that an overring R of an integral domain D is a flat D-module if and only if $R_{\mathfrak{m}} = D_{\mathfrak{m} \cap D}$ for every maximal ideal $\mathfrak{m}$ of R [41, Theorem 2, page 795]. The theory of flatness plays an important role in commutative algebra, in particular; it provides a useful tool in the characterization of some classes of commutative rings, for instance, an integral domain D is Prüfer if and only if each overring of D is D-flat, and flat overrings of a Noetherian domain remain Noetherian. The notion of faithful flatness seems to have first appeared in Bourbaki's book *Algèbre Commutative* in the chapter on flatness and localization. Recently, a constructive approach to the study of flat modules and (faithfully) flat algebras has been developed [34, Chapter VIII].

A. Tamoussit (✉)
Department of Mathematics, The Regional Center for Education and Training Professions Souss-Massa, Inezgane, Morocco

Laboratory of Mathematics and Applications (LMA), Faculty of Sciences, Ibn Zohr University, Agadir, Morocco
e-mail: a.tamoussit@crmefsm.ac.ma

J.-L. Chabert et al. (eds.), *Algebraic, Number Theoretic, and Topological Aspects of Ring Theory*, https://doi.org/10.1007/978-3-031-28847-0_23

Let D be an integral domain with quotient field K and X an indeterminate over K. The set, denoted Int(D), of polynomials with coefficients in K that take values (on elements of D) in D, i.e., $\text{Int}(D) = \{f \in K[X] : f(D) \subseteq D\}$ forms a commutative D-algebra.

Originally, the theory of integer-valued polynomials goes back to 1919 publications of Pólya [40] and Ostrowski [38]. Their works focused on the study of the D-module structure of Int(D) if D is the ring of integers of a number field. Later, in 1936, T. Skolem [44] considered, for the first time, the structure of Int($\mathbb{Z}$) not only as a $\mathbb{Z}$-module but as a commutative ring also.

Major ring-theoretic studies of integer-valued polynomials began in the early 1970s independently in three different places and by various authors: Cahen and Chabert at the University of Paris, Gunji and McQuillan at the University of Wisconsin, and Brizolis at University of California, Los Angeles. They generalized the concept of "integer-valued polynomial ring" to the case of an arbitrary integral domain D. Since, a circle of investigations began about these constructions including their ring-theoretic properties, their module structure, their (prime) ideal structures, and the calculation of their Krull dimensions. Remarkable results were obtained over various classes of integral domains, and many questions about their ring-theoretic properties were formulated [9]. Note that this rich theory provides an excellent source of examples and counterexamples in various settings, particularly in the non-Noetherian commutative rings setting.

Since 2007, J. Elliott has published a series of articles involving a category-theoretic viewpoint in studying rings of integer-valued polynomials [18–22]. Much of his work, in this regard, is based on or dealt with the flatness of Int(D) as a D-module. Firstly, in [18, page 85], he pointed out the following open problem:

> Determine those Krull domains D, or more generally those integral domains D, for which Int(D) is flat as a D-module.

Two years later, in [19, Section 4], he wrote:

> The question of the flatness of Int(D) over D is motivated by considering that for any domain D and any set $\underline{X}$ there exists a canonical D-algebra homomorphism $\theta_{\underline{X}} : \bigotimes_{X \in \underline{X}} \text{Int}(D) \to \text{Int}(D^{\underline{X}})$ sending $X \in \text{Int}(D)$ to $X \in \text{Int}(D^{\underline{X}})$ for all $X \in \underline{X}$, where the (possibly infinite) tensor product is a tensor product of D-algebras. One might hope, if not expect, that this homomorphism be an isomorphism, at least if integer-valued polynomials are expected to behave anything like ordinary polynomials in this regard. After all, one does have $\text{Int}(\text{Int}(D^{\underline{X}})^{\underline{Y}}) = \text{Int}(D^{\underline{X} \amalg \underline{Y}})$ for any infinite domain D and any sets $\underline{X}$ and $\underline{Y}$, in perfect analogy with ordinary polynomial rings over anything but a finite field; and indeed for several large classes of domains it turns out that $\theta_{\underline{X}}$ is an isomorphism for all $\underline{X}$. However, it has not been proved that $\theta_{\underline{X}}$ is always an isomorphism, nor has there been found a counterexample. Proving injectivity is equivalent to showing that the given tensor product is D-torsion-free, and the easiest way to do that seems to be to prove that Int(D) is flat as a D-module.

Till now, there is no example in the literature of an integral domain D such that Int(D) is neither flat nor free as a D-module. In this vein, Elliott conjectured that if either $D = \mathbb{F}_2 + T\mathbb{F}_4[[T]]$ or $D = \mathbb{F}_2[[T^2, T^3]]$, then Int($D$) is not flat as a D-module [20, Conjectures 2.15 and 2.17], and this conjecture is still open.

All of these studies motivate the main open problems as stated in [10, Problem 19]:

[10, Problem 19a] *Is* Int(D) *a flat D-module for any domain D?*
[10, Problem 19b] *Is* Int(D) *a free D-module for any domain D?*

In 2020, under the supervision of Lahoucine Izelgue, the current author devoted a large part of his Ph.D. thesis [45] to studying the problems of flatness and local freeness of rings of integer-valued polynomials.

It is worth noting that, in 1919, Pólya [40] established that for any PID D, Int(D) is a free D-module. Cahen and Chabert, [8, conséquence du Corollaires (3), page 303] and [6, conséquence du Théorème 2, page 753], showed that for any Dedekind domain D, Int(D) is a (faithfully) flat D-module. Zantema in [50] asserted that Int(D) is D-free, for any ring of integers of a number field D.

Keeping in mind the containments $D \subseteq D[X] \subseteq \text{Int}(D)$ and the fact that $D[X]$ is always a flat D-algebra, one can easily see that the transitivity of the flatness (cf. [36, page 46]) guarantees that whenever the flatness of Int(D) holds over $D[X]$, it necessarily holds over D. So we are led to investigate the flatness in parallel to the following diagram of homomorphisms:

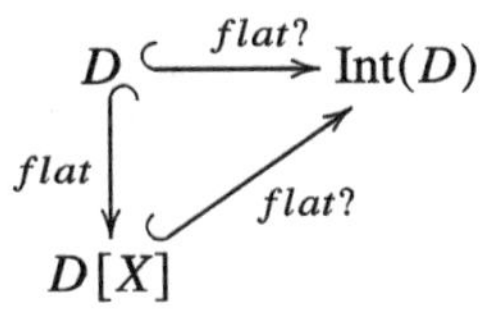

This fact motivated the authors of [30] and [29] to treat the problem of when Int(D) is flat over $D[X]$.

2 Flatness of Int(D), or More Generally Int(E, D), as a D-Module

We start with some well-known results concerning domains D such that Int(D) admits a regular basis, that is, it has a free D-module basis consisting of exactly one polynomial of each degree.

The following result is due to S. Frisch.

Proposition 1 ([24, Theorem 3.6 and Corollary 1]) *For any unique factorization domain D, the D-module* Int(D) *is free with a regular basis.*

A similar result holds for valuation domains.

Proposition 2 ([9, Remark II.2.14]) *For any valuation domain V, the V-module* Int(V) *is free with a regular basis.*

For Dedekind domains, we have:

Proposition 3 ([9, Remark II.3.7(iii)]) *For any Dedekind domain D,* $\text{Int}(D)$ *is a free D-module.*

We say that $\text{Int}(D)$ is a *locally free* D-module if $\text{Int}(D)_{\mathfrak{m}}$ is a free $D_{\mathfrak{m}}$-module for each maximal ideal $\mathfrak{m}$ of D. An integral domain D is said to be a *Prüfer v-multiplication domain*, in short a PvMD, if $D_{\mathfrak{m}}$ is a valuation domain for each t-maximal ideal $\mathfrak{m}$ of D (the notion of t-maximal ideal to be defined later).

Proposition 4 ([19, Proposition 3.5]) *Let D be a PvMD such that* $\text{Int}(D_{\mathfrak{p}}) = \text{Int}(D)_{\mathfrak{p}}$ *for every prime ideal $\mathfrak{p}$ of D. Then* $\text{Int}(D_{\mathfrak{p}})$ *has a regular basis for every prime ideal $\mathfrak{p}$ of D, and therefore the D-module* $\text{Int}(D)$ *is locally free, hence faithfully flat.*

From [11, Proposition 2.1] it follows that the ring $\text{Int}(D)$ has the property of good behavior under localization for Krull domains D, that is, $S^{-1}\text{Int}(D) = \text{Int}(S^{-1}D)$ for each multiplicative subset S of D, and then we derive the following:

Corollary 1 ([18, Corollary 3.6]) *For any Krull domain D, the D-module* $\text{Int}(D)$ *is locally free, hence faithfully flat.*

As the integral closure of any Noetherian domain is Krull (cf. [23, Theorem 4.3]), we deduce the following:

Corollary 2 *Let D be a Noetherian domain with integral closure D'. Then* $\text{Int}(D')$ *is a locally free D'-module.*

Further, it is worth noting that J. Elliott in [19] pointed out that the question of the flatness of the D-module $\text{Int}(D)$ can be reduced to the question of the flatness of Bhargava rings. So, let us recall the following:

For any element x of an integral domain D, the *Bhargava ring* over D at x is defined as follows:

$$\mathbb{B}_x(D) := \{f \in K[X] : \ \forall a \in D, \ f(xX + a) \in D[X]\},$$

where K is the quotient field of D.

Proposition 5 ([19, Proposition 6.4]) *Let D be an integral domain. If $\mathbb{B}_x(D)$ is flat as a D-module for every nonzero element x of D, then* $\text{Int}(D)$ *is also flat as a D-module*

Let $\text{Int}_n(D) := \{f \in \text{Int}(D) : \ \deg f \leqslant n\}$. Thus, as a particular case of [19, Proposition 7.1], we have:

Proposition 6 *Let D be an integral domain. Then* $\text{Int}(D)$ *is flat as a D-module whenever the D-module* $\text{Int}_n(D)$ *is flat for every nonnegative integer n.*

For a nonzero fractional ideal I of an integral domain D, we set $I^{-1} := \{x \in K \mid xI \subseteq D\}$ and $I_v := (I^{-1})^{-1}$, where K is the quotient field of D. On D the *t-operation* is defined by $I_t := \bigcup J_v$, where J ranges over the set of all nonzero finitely generated ideals contained in I. When $I_t = I$ we say that I is a *t-ideal*. A *t-maximal* ideal is a maximal ideal among all t-ideals of D.

Theorem 1 ([20, Theorem 1.2(a)]) *Let D be an integral domain such that $\mathfrak{m}D_{\mathfrak{m}}$ is principal and* $\mathrm{Int}(D_{\mathfrak{m}}) = \mathrm{Int}(D)_{\mathfrak{m}}$ *for every t-maximal ideal $\mathfrak{m}$ of D. Then* $\mathrm{Int}(D)$ *is a locally free, and hence faithfully flat, D-module.*

Recall that a *Krull-type domain* is a PvMD of finite t-character, that is, each nonzero non-unit is contained in only finitely many t-maximal ideals. Then, as a consequence of the previous result and [20, Theorem 1.2(b)], we have:

Corollary 3 *For any Krull-type domain D, the D-module* $\mathrm{Int}(D)$ *is locally free, hence faithfully flat.*

In the light of [12, Corollary 1.3], Theorem 1 can be viewed as a generalization of Proposition 4.

For polynomial rings, we always have $S^{-1}(D[X]) = S^{-1}D[X]$ for any integral domain D and any multiplicative subset S of D. However, in the case of $\mathrm{Int}(D)$, the inclusion $S^{-1}\mathrm{Int}(D) \subseteq \mathrm{Int}(S^{-1}D)$ always holds and when the reverse inclusion holds we say that $\mathrm{Int}(D)$ has the property of *good behavior under localization*. This last property is a key tool for studying both the local freeness and the flatness of $\mathrm{Int}(D)$ over D: particularly in Proposition 4 and Theorem 1. Remarkably, the good behavior under localization does not hold in general. First we recall that an integral domain D is said to be *almost Dedekind* if $D_{\mathfrak{m}}$ is a rank-one discrete valuation domain (a DVR for short) for all maximal ideals $\mathfrak{m}$ of D. We next notice that each of Examples 6.2 and 6.5 of [14] provides an almost Dedekind domain $\mathbf{D}$ with finite residue fields such that $\mathrm{Int}(\mathbf{D})$ does not behave well under localization, i.e., $\mathrm{Int}(\mathbf{D}_{\mathfrak{m}}) \neq \mathrm{Int}(\mathbf{D})_{\mathfrak{m}}$ for some maximal ideal $\mathfrak{m}$ of $\mathbf{D}$. In that case, $\mathrm{Int}(\mathbf{D})$ is a flat D-module because any almost Dedekind domain is Prüfer, and every torsion-free module over a Prüfer domain is always flat (see [42, Corollary 4.36, page 173] for the second statement).

In [29], Izelgue et al. continued the previous investigations and focused on the local freeness and flatness of rings of integer-valued polynomials over locally essential domains.

Given an integral domain D and a (non-empty) subset $\mathcal{P}$ of $\mathrm{Spec}(D)$, we say that D is an *essential domain* with defining family $\mathcal{P}$ if $D = \bigcap_{\mathfrak{p}\in\mathcal{P}} D_{\mathfrak{p}}$ and each $D_{\mathfrak{p}}$ is a valuation domain. As this notion does not carry up to localizations as established in [27], D is said to be a *locally essential domain* if $D_{\mathfrak{q}}$ is an essential domain for each $\mathfrak{q} \in \mathrm{Spec}(D)$. Recall also that a prime ideal $\mathfrak{p}$ of an integral domain D is called an *int prime* of D if $\mathrm{Int}(D)_{\mathfrak{p}} \neq D_{\mathfrak{p}}[X]$. Notice that int prime ideals are both maximal and t-maximal (cf. [9, Proposition I.3.4 and Remarks I.3.5(i)] and [19, Proposition 3.3]).

Theorem 2 ([29, Theorem 1]) *Let D be a locally essential domain. Then:*

1. $\mathrm{Int}(D)$ *is a flat D-module.*
2. *If* $\mathrm{Int}(D_{\mathfrak{m}}) = \mathrm{Int}(D)_{\mathfrak{m}}$ *for every int prime ideal $\mathfrak{m}$ of D of height at least two, then* $\mathrm{Int}(D)$ *is a locally free, and hence faithfully flat, D-module.*

Recall that an integral domain D is said to be *t-almost Dedekind* if $D_{\mathfrak{m}}$ is a DVR for short) for all t-maximal ideals $\mathfrak{m}$ of D. Recall also that an *almost Krull* domain is any integral domain D with $D_{\mathfrak{m}}$ is Krull for each maximal ideal $\mathfrak{m}$ of D.

Corollary 4 ([29, Corollary 1]) *Let D be an integral domain. If either D is almost Krull, t-almost Dedekind or* $\mathrm{Int}(D)$ *is a PvMD, then* $\mathrm{Int}(D)$ *is locally free as a D-module.*

From the previous two results, we deduce that the good behavior under localization hypothesis is no longer necessary when dealing with the flatness of $\mathrm{Int}(D)$ over a locally essential domain D and with the local freeness of $\mathrm{Int}(D)$ over domains D that are either $(t\text{-})$almost Dedekind or almost Krull.

Now, we turn our attention to two particular generalizations of the "classical" ring of integer-valued polynomials.

In 1991, Anderson et al. [1] treated the ring of *D-valued R-polynomials*:

$$I(R, D) := \{f \in R[X] : f(D) \subseteq D\},$$

where $D \subseteq R$ is an extension of integral domains, as a generalization of rings of integer-valued polynomials. In the sequel of this paper, we will use the notation $\mathrm{Int}_R(D)$ instead of $I(R, D)$. Thereafter, in 1993, Cahen [7] considered another generalization of $\mathrm{Int}(D)$, namely, the ring of *integer-valued polynomials over a subset E* of K, to be

$$\mathrm{Int}(E, D) := \{f \in K[X] : \ f(E) \subseteq D\},$$

where D is an integral domain with quotient field K.

In [22], the author showed that $\mathrm{Int}_R(D)$ is locally free as a D-module if D is a PvMD such that $\mathrm{Int}(D)_{\mathfrak{m}} = \mathrm{Int}(D_{\mathfrak{m}})$ for every maximal ideal $\mathfrak{m}$ of D and R is an overring of D (cf. [22, Theorem 7.11]). That includes the case of when D is a Krull-type domain. Recently, in [47], the current author generalized this last result as follows:

Theorem 3 ([47, Theorem 18]) *Let $D \subseteq R$ be an extension of integral domains with D a locally essential domain. Then:*

1. $\mathrm{Int}_R(D)$ *is a faithfully flat D-module.*
2. *Assume that each int prime ideal of D is of height-one (this holds, e.g., if either D is almost Krull or of dimension one). Then* $\mathrm{Int}_R(D)$ *is a locally free D-module.*
3. *Assume that R is an overring and for each int prime ideal $\mathfrak{m}$ of D of height at least two* $\mathrm{Int}(D)_{\mathfrak{m}} = \mathrm{Int}(D_{\mathfrak{m}})$. *Then* $\mathrm{Int}_R(D)$ *is a locally free D-module.*

Moreover, we investigate the faithful flatness of multivariable D-valued R-polynomial rings as a D-module. We first recall that, for any set $\mathbf{X}$:

- The ring of D-valued R-polynomials on $D^{\mathbf{X}}$:

$$\mathrm{Int}_R(D^{\mathbf{X}}) := \{f \in R[\mathbf{X}] : f(D^{\mathbf{X}}) \subseteq D\},$$

- The canonical D-algebra homomorphism

$$\theta_{\mathbf{X}} : \bigotimes_{X \in \mathbf{X}} \mathrm{Int}_R(D) \to \mathrm{Int}_R(D^{\mathbf{X}}),\ X \longmapsto X \text{ for all } X \in \mathbf{X},$$

where the tensor product is over D.

Proposition 7 ([47, Theorem 18]) *Let D be an integral domain and R be an overring of D. If $\theta_{\mathbf{X}}$ is an isomorphism for all finite sets $\mathbf{X}$ and $\mathrm{Int}_R(D)$ is a faithfully flat D-module, then $\mathrm{Int}_R(D^{\mathbf{X}})$ is a faithfully flat D-module for any set $\mathbf{X}$. In particular, for any Krull-type domain D and for any set $\mathbf{X}$, the D-module $\mathrm{Int}_R(D^{\mathbf{X}})$ is a faithfully flat.*

In the remainder of this section, we are interested in rings of integer-valued polynomials on a subset. Precisely, we pose to classify locally essential domains D and subsets E such that $\mathrm{Int}(E, D)$ is a locally free, or at least faithfully flat, D-module.

First notice that Chabert et al. [15] showed that $\mathrm{Int}(\mathbb{P}, \mathbb{Z})$ is a free $\mathbb{Z}$-module (with a regular basis), where $\mathbb{P}$ denotes the set of all prime integers. Then, Boulanger et al. [5] proved that, for an infinite subset E of a DVR V, $\mathrm{Int}(E, V)$ is a free V-module. More generally, if D is a principal ideal domain (in short PID) and E is an infinite fractional subset E of D, then $\mathrm{Int}(E, D)$ admits a regular basis and hence is a free D-module [9, Corollary II.1.6]. Once more, there is no example, in the literature, of an integral domain D such that $\mathrm{Int}(E, D)$ is not either flat or free as a D-module for some subset E of the quotient field of D.

Proposition 8 ([46, consequence of Lemma 1] and [31, Corollary 1]) *Let D be a Prüfer domain with quotient field K and let E be a fractional subset of K. We have:*

1. *$\mathrm{Int}(E, D)$ is a faithfully flat D-module.*
2. *Assume that D is almost Dedekind and E is an infinite subset of D. Then $\mathrm{Int}(E, D)$ is a locally free D-module.*

Example 1

The integral domain $D = \mathbb{Z}+T\mathbb{Q}[T]$, where T is an indeterminate over $\mathbb{Q}$, is known to be Prüfer. Then, by Proposition 8(1), $\mathrm{Int}(E, D)$ is a faithfully flat D-module for any subset E of $\mathbb{Q}(T)$.

Example 2

Let T be an indeterminate over $\mathbb{Q}$, and set $R = \cup_{n=0}^{\infty}\mathbb{Q}[T^{\frac{1}{2^n}}]$ and $S = \mathbb{Q}[T] \setminus (1 - T)\mathbb{Q}[T]$. According to [16, Section 3], $D := S^{-1}R$ is a non-Noetherian almost Dedekind domain. Thus, by Proposition 8(2), $\text{Int}(E, D)$ is locally free as a D-module for any infinite subset E of D.

Most of the results established in the case of $\text{Int}(D)$ are due to the fact that, for each $\mathfrak{p} \in \text{Spec}(D)$ with infinite residue field, either $\text{Int}(D_{\mathfrak{p}}) \subseteq D_{\mathfrak{p}}[X]$ or $\text{Int}(D)$ behaves well under localization. However, the inclusion $\text{Int}(E, D_{\mathfrak{p}}) \subseteq D_{\mathfrak{p}}[X]$, or at least $\text{Int}(E, D)_{\mathfrak{p}} \subseteq D_{\mathfrak{p}}[X]$, may not hold, which makes it harder to characterize the local freeness or the (faithful) flatness of $\text{Int}(E, D)$ as a D-module. Thus, under a restrictive condition on the subset E, we recover some well-known results stated in Sect. 2.

In [37], Mulay studied the ring $\text{Int}(E, D)$ under an extra condition on E. In fact, he introduced the concept of residual cofiniteness to extend some well-known results about $\text{Int}(D)$ to the case of $\text{Int}(E, D)$.

Recall that a non-empty subset E of an integral domain D is said to be *residually cofinite* with D [37] if E and D are either both finite or both infinite modulo any prime ideal of D. Obviously, D is residually cofinite with itself, and for more examples, we refer to [37, page 335].

In what follows, we give sufficient conditions for $\text{Int}(E, D)$ to be locally free, or at least (faithfully) flat, as a D-module.

Theorem 4 ([46, Theorem 1]) *Let D be a locally essential domain with defining family $\mathcal{P}$ and let $E \subseteq D$ be residually cofinite with D. We have:*

1. $\text{Int}(E, D)$ *is a faithfully flat D-module.*
2. *Assume that E is infinite and $D_{\mathfrak{p}}$ is a DVR, for each $\mathfrak{p} \in \mathcal{P}$ with $D/\mathfrak{p}$ finite. Then* $\text{Int}(E, D)$ *is a locally free D-module.*

As an application of Theorem 4, we have:

Corollary 5 ([46, Corollary 2]) *Let D be an integral domain and $E \subseteq D$ be infinite and residually cofinite with D. If D is either almost Krull or t-almost Dedekind, then* $\text{Int}(E, D)$ *is a locally free D-module.*

Example 3

Let $D = \mathbb{Z}\left[\{T/p_n\}_{n=1}^{\infty}\right]$, where $\{p_n\}_{n=1}^{\infty}$ is the set of all positive prime integers and T is an indeterminate over $\mathbb{Z}$. By [28, Example 166], D is a two-dimensional almost Krull domain which is neither Noetherian nor Krull. Then, by Corollary 5, for any infinite and residually cofinite subset E of D, the D-module $\text{Int}(E, D)$ is locally free.

When dealing with the integral domain D as a locally finite intersection of its localizations at some elements of $\text{Spec}(D)$, we have:

Proposition 9 *Let* $D = \bigcap_{\mathfrak{p}\in\mathcal{P}} D_{\mathfrak{p}}$*, where* $\mathcal{P} \subseteq \operatorname{Spec}(D)$*, be an integral domain and let* $E \subseteq D$ *be infinite and residually cofinite with* D*. Assume that the intersection* $\bigcap_{\mathfrak{p}\in\mathcal{P}} D_{\mathfrak{p}}$ *is locally finite and for each* $\mathfrak{m} \in \mathcal{P} \cap \operatorname{Max}(D)$ *with finite residue field,* $D_{\mathfrak{m}}$ *is a DVR. Then* $\operatorname{Int}(E, D)$ *is a locally free* D*-module.*

Proof First, we note that $\operatorname{Int}(E, D)_{\mathfrak{m}} = \operatorname{Int}(E, D_{\mathfrak{m}})$ for each prime ideal $\mathfrak{p}$ of D (cf. [30, Proposition 3.12]). Let consider the following partition of $\operatorname{Max}(D)$:

$$\mathcal{M}_0 := \{\mathfrak{m} \in \mathcal{P} \cap \operatorname{Max}(D),\ \ D/\mathfrak{m} \text{ is finite}\} \text{ and } \mathcal{M}_1 := \operatorname{Max}(D)\backslash\mathcal{M}_0.$$

Now, let $\mathfrak{m}$ be a maximal ideal of D. So, we need to discuss the following cases:

Case 1: $\mathfrak{m} \in \mathcal{M}_0$. Then, by assumption, $D_{\mathfrak{m}}$ is a DVR and hence, by [5, Proposition 2.2], $\operatorname{Int}(E, D)_{\mathfrak{m}} = \operatorname{Int}(E, D_{\mathfrak{m}})$ is free as a $D_{\mathfrak{m}}$-module.

Case 2: $\mathfrak{m} \in \mathcal{M}_1$. Then, by [30, Proposition 3.12], $\operatorname{Int}(E, D)_{\mathfrak{m}} = \operatorname{Int}(E, D_{\mathfrak{m}}) = D_{\mathfrak{m}}[X]$ is a free $D_{\mathfrak{m}}$-module.

Therefore, $\operatorname{Int}(E, D)$ is a locally free D-module, as desired.

We close this section with sufficient conditions for $\operatorname{Int}(E, D)$ to be flat as a D-module in terms of $\mathbb{B}_x(E, D)$ and $\operatorname{Int}_n(E, D)$, which are defined as follows:

For any element x of an integral domain D and any subset E of D, the *Bhargava ring* over E at x is defined as follows:

$$\mathbb{B}_x(E, D) := \{f \in K[X] :\ \forall a \in E,\ \ f(xX + a) \in D[X]\},$$

where K is the quotient field of D.

For any subset E of an integral domain D, let $\operatorname{Int}_n(E, D)$ be the D-module $\{f \in \operatorname{Int}(E, D) :\ \deg f \leqslant n\}$.

Proposition 10 ([19, Propositions 6.4 and 7.1]) *Let* D *be an integral domain and let* E *be a subset of* D*. Assume* ***one*** *of the following holds.*

1. $\mathbb{B}_x(E, D)$ *is flat as a* D*-module for every nonzero element* x *of* D*.*
2. $\operatorname{Int}_n(E, D)$ *is flat as a* D*-module for every nonnegative integer* n*.*

Then $\operatorname{Int}(E, D)$ *is flat as a* D*-module*

3 Flatness of Int(D) as an Overring of $D[X]$

Given a nonzero fractional ideal I of an integral domain D, we recall that on D the *w-operation* is defined by $I_w := \{x \in K \mid xJ \subseteq I$ for some nonzero finitely generated ideal J of D with $J^{-1} = D\}$, where K is the quotient field of D. When $I_w = I$ we say that I is a *w-ideal*. An integral domain D is said to be *Strong Mori* if it satisfies the ascending chain condition (a.c.c.) on integral w-ideals. Thus, the class of Strong Mori domains includes Noetherian domains and (infra-)Krull domains. We also recall that an integral domain D has *t-dimension one* if it is not a field and each t-maximal ideal of D has height one. Lastly, $\mathrm{Int}(D)$ is called *trivial* if $\mathrm{Int}(D) = D[X]$.

The following result characterizes the flatness of $\mathrm{Int}(D)$ over $D[X]$ when D is a Strong Mori domain.

Theorem 5 ([29, Theorem 2]) *Let D be a Strong Mori domain and consider the following statements:*

1. $\mathrm{Int}(D)$ *is flat over* $D[X]$.
2. $\mathrm{Int}(D)$ *is trivial.*
3. $\mathrm{Int}(D)$ *is Strong Mori.*
4. *Each height-one prime ideal of D has infinite residue field.*

Then (1) $\Leftrightarrow$ (2) $\Rightarrow$ (3) *and* (2) $\Rightarrow$ (4).

If, moreover, D is either integrally closed or of t-dimension one, then (3) $\Rightarrow$ (2) *and* (4) $\Rightarrow$ (2)*, and hence the four statements are equivalent.*

Notice that statements (2) and (3) are not, in general, equivalent as Example 11 shows.

Corollary 6 ([30, Proposition 2.1]) *Let D be a Noetherian domain and consider the following statements:*

1. $\mathrm{Int}(D)$ *is flat over* $D[X]$.
2. $\mathrm{Int}(D)$ *is trivial.*
3. $\mathrm{Int}(D)$ *is Noetherian.*
4. *Each height-one prime ideal of D has infinite residue field.*

Then (1) $\Leftrightarrow$ (2) $\Rightarrow$ (3) *and* (2) $\Rightarrow$ (4).

If, moreover, D is either integrally closed or of dimension one, then (3) $\Rightarrow$ (2) *and* (4) $\Rightarrow$ (2)*, and hence the four statements are equivalent.*

Let $I := (D : D') = \{x \in K : xD' \subseteq D\}$ be the *conductor* of D' in D.

Corollary 7 *Let D be a Noetherian domain such that $I := (D : D') \neq (0)$ and D/I is finite. Assume* ***one*** *of the following holds.*

1. *Each height-one prime ideal of D' has infinite residue field (this holds, e.g., if D' has no height-one maximal ideal).*
2. $(D' : I) \neq D'$.

Then $\mathrm{Int}(D)$ *is not flat over* $D[X]$.

Proof If (1) holds then, by [25, Corollary 3.4], $\mathrm{Int}(D)$ is not trivial (and Noetherian). Hence, by Corollary 6, $\mathrm{Int}(D)$ is not flat over $D[X]$.

If (2) holds then, by [25, Corollary 3.5], $\mathrm{Int}(D)$ is not Noetherian. Hence, by Corollary 6, $\mathrm{Int}(D)$ is not flat over $D[X]$.

Let $X^1(D)$ be the set of all height-one prime ideals of an integral domain D. Then D is called an *infra-Krull domain* [35]; if $D = \bigcap_{\mathfrak{p} \in X^1(D)} D_{\mathfrak{p}}$, the intersection is locally finite and $D_{\mathfrak{p}}$ is Noetherian for each $\mathfrak{p} \in X^1(D)$.

Proposition 11 ([29, Proposition 2]) *The following statements are equivalent for an infra-Krull domain D:*

1. $\text{Int}(D)$ *is flat over* $D[X]$.
2. $\text{Int}(D)$ *is trivial.*
3. $\text{Int}(D)$ *is infra-Krull.*
4. *Each height-one prime ideal of D has infinite residue field.*

Recall that an overring T of D is said to be t-*linked* over D if for each finitely generated fractional ideal I of D, $I^{-1} = D$ implies that $(IT)^{-1} = T$. Notice that any flat overring is a t-linked overring.

Corollary 8 *Let D be an infra-Krull domain. Then* $\text{Int}(D)$ *is t-linked over $D[X]$ if and only if* $\text{Int}(D)$ *is trivial.*

Proof As any t-linked overring of an infra-Krull domain is still infra-Krull (cf. [35, Lemma 2]), the desired conclusion follows immediately from Proposition 11.

An integral domain D is said to be t-*locally Strong Mori* if $D_{\mathfrak{m}}$ is Strong Mori for all t-maximal ideals $\mathfrak{m}$ of D. As any localization of a Strong Mori domain is also Strong Mori, then Strong Mori domains are t-locally Strong Mori, but the converse is not true in general. Indeed, any non-Noetherian almost Dedekind domain is t-locally Strong Mori but not Strong Mori because any almost Dedekind domain is Prüfer, and any Prüfer Strong Mori domain is Dedekind.

Recently, in [48], the current author extended Theorem 5 to t-locally Strong Mori with the condition of good behavior under localization as follows:

Theorem 6 ([48, Theorem 2.4]) *Let D be a t-locally Strong Mori domain such that* $\text{Int}(D_{\mathfrak{m}}) = \text{Int}(D)_{\mathfrak{m}}$, *for each int prime ideal $\mathfrak{m}$ of D. Consider the following statements:*

1. $\text{Int}(D)$ *is flat over* $D[X]$;
2. $\text{Int}(D)$ *is trivial;*
3. *Each height-one prime ideal of D has infinite residue field.*

Then (1) $\Leftrightarrow$ (2) $\Rightarrow$ (3). *If, in addition, D is either integrally closed or of t-dimension one, then* (3) $\Rightarrow$ (2), *and hence the three statements are equivalent.*

The next theorem shows that $\text{Int}(D)$ is flat over $D[X]$ only in the trivial case when D is an essential domain.

Theorem 7 ([29, Theorem 3 and Remark 1(2)]) *Let D be an essential domain with defining family $\mathcal{P}$ and consider the following statements:*

1. $\text{Int}(D)$ *is flat over* $D[X]$.
2. $\text{Int}(D)$ *is trivial.*

3. $D_{\mathfrak{p}}$ *has infinite residue field or nonprincipal maximal ideal, for each* $\mathfrak{p} \in \mathcal{P}$.

Then (1) $\Leftrightarrow$ (2) *and* (3) $\Rightarrow$ (2).

If, moreover, $\text{Int}(D)_{\mathfrak{p}} = \text{Int}(D_{\mathfrak{p}})$ *for each* $\mathfrak{p} \in \mathcal{P}$ *with finite residue field, then* (2) $\Rightarrow$ (3), *and hence the three statements are equivalent.*

Corollary 9 ([29, Corollary 4]) *Let* $D = \bigcap_{\mathfrak{p} \in \mathcal{P}} D_{\mathfrak{p}}$, *where* $\mathcal{P} \subseteq \text{Spec}(D)$, *be a Krull-type domain. The following statements are equivalent:*

1. $\text{Int}(D)$ *is flat over* $D[X]$.
2. $\text{Int}(D)$ *is trivial.*
3. $D_{\mathfrak{p}}$ *has infinite residue field or nonprincipal maximal ideal, for each* $\mathfrak{p} \in \mathcal{P}$.

The following result shows that $\text{Int}(D)$ may be locally free, and hence flat, over D without being so over $D[X]$.

Corollary 10 *Let D be an almost Dedekind domain which is not a field such that* $\text{Int}(D)$ *is a Prüfer domain (this holds, e.g., if* $\text{Int}(D_{\mathfrak{m}}) = \text{Int}(D)_{\mathfrak{m}}$ *for each maximal ideal* $\mathfrak{m}$ *of D with finite residue field). Then,* $\text{Int}(D)$ *is locally free over D but not flat over* $D[X]$.

Proof By Corollary 4, $\text{Int}(D)$ is locally free as a D-module. However, since $\text{Int}(D)$ is Prüfer and D is not a field, $\text{Int}(D) \neq D[X]$, and then it follows from Theorem 7 that $\text{Int}(D)$ is not flat over $D[X]$.

We recall that an integral domain D is *GCD*, if every pair of nonzero elements of D has a greatest common divisor, or equivalently, the intersection of any two principal ideals of D is principal. Clearly, UFDs and valuation domains are GCD and any GCD domain is a PvMD (cf. [49, page 95]).

The next result characterizes when $\text{Int}(D)$ is flat over $D[X]$ for a GCD domain D.

Proposition 12 ([29, Proposition 3]) *Let D be a GCD domain and consider the following statements:*

1. $\text{Int}(D)$ *is flat over* $D[X]$.
2. $\text{Int}(D)$ *is trivial.*
3. $\text{Int}(D)$ *is GCD.*
4. $D_{\mathfrak{p}}$ *has infinite residue field or nonprincipal maximal ideal, for each t-maximal ideal* $\mathfrak{p}$ *of D.*

Then (4) $\Rightarrow$ (1) $\Leftrightarrow$ (2) $\Rightarrow$ (3).

If, moreover, $\text{Int}(D_{\mathfrak{p}}) = \text{Int}(D)_{\mathfrak{p}}$ *for each t-maximal ideal* $\mathfrak{p}$ *of D with finite residue field, then* (3) $\Rightarrow$ (2) $\Rightarrow$ (4), *and the four statements are equivalent.*

As an immediate corollary of Proposition 12, we have the following characterization:

Corollary 11 ([29, Corollary 5]) *Let V be a valuation domain. The following statements are equivalent:*

1. $\text{Int}(V)$ *is flat over* $V[X]$.
2. $\text{Int}(V)$ *is trivial.*
3. V *has infinite residue field or nonprincipal maximal ideal.*

Since GCD domains have the property that every t-linked overring is a localization (cf. [17, Corollary 3.8]), we deduce from Proposition 12 the following corollary.

Corollary 12 *Let D be a GCD domain. Then* $\text{Int}(D)$ *is t-linked over* $D[X]$ *if and only if* $\text{Int}(D)$ *is trivial.*

The goal of the remaining part of this section is to investigate the (faithful) flatness of $\text{Int}(E, D)$ as a $D[X]$-module, when E is a subset of D (not necessary residually cofinite).

Proposition 13 ([30, Proposition 3.1]) *Let D be a Noetherian domain and let E be a subset of D. If* $\text{Int}(E, D)$ *is not Noetherian, then* $\text{Int}(E, D)$ *is not flat over* $D[X]$.

So, as a consequence of the previous proposition, we have:

Corollary 13 ([30, Corollaries 3.5 and 3.7]) *Let D be a Noetherian domain and let E be a non-empty subset of D. Assume* ***one*** *of the following two conditions holds:*

1. *There exists a height-one prime ideal of D' with finite residue field (this holds, e.g., if D is a Dedekind domain with finite residue fields).*
2. *The subset E is finite.*

Then $\text{Int}(E, D)$ *is not flat over* $D[X]$.

Corollary 14 *Let D be a Noetherian domain such that $I := (D : D') \neq (0)$ and let E be a subset of D. If* $\text{Int}(D)$ *is not Noetherian (this holds, e.g., if D/I is finite and $(D' : I) \neq D'$ [25, Corollary 3.5]), then* $\text{Int}(E, D)$ *is not flat over* $D[X]$.

Proof It follows from [25, Proposition 3.8] and Proposition 13.

The following example shows that it is possible to have $\text{Int}(E, D)$ free over D but not over $D[X]$.

Example 4

The ring $\mathbb{Z}$ of integers is Dedekind (in fact, it is a PID). For any subset E of $\mathbb{Q}$, $\text{Int}(E, \mathbb{Z})$ is a faithfully flat $\mathbb{Z}$-module (cf. Proposition 8(1)). In particular, since $\mathbb{P}$, the set of all prime integers, is an infinite subset of $\mathbb{Z}$, by [9, Corollary II.1.6], $\text{Int}(\mathbb{P}, \mathbb{Z})$ is a free $\mathbb{Z}$-module. However, by Corollary 13(1), the $\mathbb{Z}[X]$-module $\text{Int}(\mathbb{P}, \mathbb{Z})$ is not flat and hence not free.

The following proposition can be proved by adapting the arguments used in [29, Theorem 3].

Proposition 14 *Let D be an essential domain and let E be a subset of D. Then* $\mathrm{Int}(E, D)$ *is flat over* $D[X]$ *if and only if* $\mathrm{Int}(E, D) = D[X]$.

We end this section with a complete characterization of the faithful flatness of $\mathrm{Int}(E, D)$ over $D[X]$ for any integral domain D and any subset E of D. With some slight modifications in the proof of [29, Proposition 4], we get the following:

Proposition 15 *Let D be an integral domain and let E be a subset of D. Then* $\mathrm{Int}(E, D)$ *is faithfully flat over* $D[X]$ *if and only if* $\mathrm{Int}(E, D) = D[X]$.

4 Some Illustrating Examples

Next, we give a list of examples that illustrate the originality of our results which can be found in [29, 30, 45]. In fact, we distinguish three types of examples:

4.1 *Examples of Integral Domains D Such That* Int(D) *Is Either Locally Free or Faithfully Flat Over D*

Example 5

1. [14, Example 6.3] provides an almost Dedekind domain R such that $\mathrm{Int}(R)$ is not Prüfer. By Corollary 4, $\mathrm{Int}(R)$ is a locally free, and thus faithfully flat, R-module.
2. Example 6.5 of [14] provides an almost Dedekind, and thus an almost Krull, domain $\mathfrak{D}$ with finite residue fields, such that $\mathrm{Int}(\mathfrak{D})$ does not behave well under localization, i.e., $\mathrm{Int}(\mathfrak{D})_{\mathfrak{m}} \neq \mathrm{Int}(\mathfrak{D}_{\mathfrak{m}})$ for some maximal ideal $\mathfrak{m}$ of $\mathfrak{D}$. However, by Corollary 4, $\mathrm{Int}(\mathfrak{D})$ is a locally free, and thus faithfully flat, $\mathfrak{D}$-module.

Example 6

Set $D = \mathbb{Z}[\{T/p_n, U/p_n\}_{n=1}^{\infty}]$, where T and U are indeterminates over $\mathbb{Z}$ and $\{p_n\}_{n=1}^{\infty}$ is the set of all positive prime integers.

In [2, Example, page 52], the authors established that D is an almost Krull domain which is neither a Krull domain nor a PvMD. Then, by Corollary 4, $\mathrm{Int}(D)$ is a locally free, and hence faithfully flat, D-module.

Example 7

Let A be the domain of all algebraic integers and $\{p_n\}_{n=1}^{\infty}$ is the set of all positive prime integers. For each n choose a maximal ideal M_n of A lying over $p_n\mathbb{Z}$, and set $S = A \setminus \cup_{n=1}^{\infty} M_n$. In [26, Example 1, page 338], Gilmer proved that $D := S^{-1}A$ is a one-dimensional Prüfer domain which is not almost Dedekind (and hence not almost Krull since Prüfer almost Krull domains are exactly almost Dedekind domains). So, by taking $R = qf(D)$ in Theorem 3(2), we deduce that Int(D) is a locally free D-module.

Example 8

Let $\mathcal{E}$ be the ring of entire functions and set $D := \mathcal{E} + T\mathcal{E}_S[T]$, where T is an indeterminate over $\mathcal{E}$ and S is the set generated by the principal primes of $\mathcal{E}$.

According to [49, Example 2.6], D is a locally essential domain which is neither PvMD nor almost Krull. Then, as in the previous example, it follows from Theorem 3(1) that Int(D) is a faithfully flat D-module. Notice that this last conclusion can be obtained by taking $E = D$ in Theorem 4(1).

4.2 *Examples of Integral Domains D Such That* Int(D) *Is Not Flat Over* $D[X]$

Example 9

Let k be a finite field, T be an indeterminate over k. The domain $D = k[T^2, T^3]$ is a one-dimensional Noetherian domain. Set $f = \prod_{a \in k}(X - a)$ and $g = f^3/T^4$. Since g is lies in Int(D) but not in $D[X]$, Int(D) is not trivial and then, by Corollary 6, Int(D) is neither Noetherian nor flat over $D[X]$.

Example 10

Let k be a finite field, u, v two indeterminates over k and $B = k[[u, v]]$. We have B is a two-dimensional local Noetherian domain with maximal ideal $\mathfrak{M} = (u, v)$ such that each height-one prime ideal of B has infinite residue field.

Consider the following pullback diagram:

$$\begin{array}{ccc} D := k + \mathfrak{M}^2 & \longrightarrow & k \\ \downarrow & & \downarrow \\ B & \longrightarrow & B/\mathfrak{M}^2 \end{array}$$

Then D is a local Noetherian domain with maximal ideal $\mathfrak{M}^2$ such that $D' = B$, $I := (D : D') = \mathfrak{M}^2$ and $D/\mathfrak{M}^2 = k$ is finite. Hence, by Corollary 7(a), $\text{Int}(D)$ is not flat over $D[X]$. However, $\text{Int}(D)$ is Noetherian.

Example 11

Let R be a non-Noetherian Krull domain with a maximal ideal $\mathfrak{m}$, of finite residue field, such that $R_{\mathfrak{m}}$ is Noetherian. Assume that $R/\mathfrak{m}$ contains properly a finite field k and that the residue fields at height-one prime ideals of R are all infinite.

Park and Tartarone in [39, Example 3.5] assert the existence of such a domain and use it to construct a non-Noetherian Strong Mori domain D with $\text{Int}(D) \neq D[X]$, and $\text{Int}(D)$ is not Noetherian. In fact, $\text{Int}(D)$ is a Strong Mori domain which is neither Noetherian nor Krull. Moreover, it follows from Theorem 5 that $\text{Int}(D)$ is not flat over $D[X]$ because $\text{Int}(D) \neq D[X]$.

In [20], Elliott conjectured that if either $D = \mathbb{F}_2[[T^2, T^3]]$ or $D = \mathbb{F}_2 + T\mathbb{F}_4[[T]]$, then $D \hookrightarrow \text{Int}(D)$ is not flat. The conjecture is still open. The following example shows, in particular, that $\text{Int}(D)$ is not flat as a $D[X]$-module.

Example 12

If either $D = \mathbb{F}_2[[T^2, T^3]]$ or $D = \mathbb{F}_2 + T\mathbb{F}_4[[T]]$, we have D is one-dimensional Noetherian domain such that $\text{Int}(D)$ is not trivial because $f(X) = (X^2 + X)/T^2 \in \text{Int}(D) \setminus D[X]$ for $D = \mathbb{F}_2[[T^2, T^3]]$, and $g(X) = (X^2 + X)/T \in \text{Int}(D) \setminus D[X]$ for $D = \mathbb{F}_2 + T\mathbb{F}_4[[T]]$. Then, by Corollary 6, $\text{Int}(D)$ is not flat over $D[X]$.

4.3 Examples of Integral Domains D Such That Int(D) *Is Flat Over D But Not Over* $D[X]$

The following examples show that $\text{Int}(D)$ can be locally free and hence (faithfully) flat, over D, without being flat over $D[X]$.

Example 13

1. It is well-known that $\mathbb{Z}$ is one-dimensional Noetherian domain such that $\text{Int}(\mathbb{Z}) \neq \mathbb{Z}[X]$. Then, by Corollary 6, $\text{Int}(\mathbb{Z})$ is not flat as a $\mathbb{Z}[X]$-module. However, $\text{Int}(\mathbb{Z})$ is free, and hence flat, as a $\mathbb{Z}$-module.
2. In [14, Example 6.4], Chabert constructed an almost Dedekind domain $\mathbf{D}$ that is not Dedekind, such that $\text{Int}(\mathbf{D})$ is a Prüfer domain. In fact, he established that $\mathbf{D}$ has finite residue fields and $\text{Int}(\mathbf{D})_{\mathfrak{m}} = \text{Int}(\mathbf{D}_{\mathfrak{m}})$ for each maximal ideal $\mathfrak{m}$ of $\mathbf{D}$. By Corollary 4, $\text{Int}(\mathbf{D})$ is a locally free, and then faithfully flat, $\mathbf{D}$-module. Furthermore, it follows from Corollary 10 that $\text{Int}(\mathbf{D})$ is not flat over $\mathbf{D}[X]$.
3. The domain D in [12, Example 5.1], is an almost Dedekind domain such that $\text{Int}(D)$ is not a PvMD. So, by Corollary 4, $\text{Int}(D)$ is locally free, and hence faithfully flat, as a D-module. Moreover, $\text{Int}(D)$ is not flat over $D[X]$. Deny, then it follows from Theorem 7 and [32, Theorem 3.7] that $\text{Int}(D) = D[X]$ is a PvMD, a contradiction.

Example 14

The domain $D = \mathbb{Z} + T\mathbb{Q}[T]$, where T is an indeterminate over $\mathbb{Q}$, is a two-dimensional Prüfer domain that is not of finite t-character (and then D is not a ring of Krull-type). Then, if we take $R = qf(D)$ in Theorem 3(1), we infer that $\text{Int}(D)$ is a faithfully flat D-module.

On the other hand, since each maximal ideal $\mathfrak{m}$ of D is principal, $\text{Int}(D)_{\mathfrak{m}} = \text{Int}(D_{\mathfrak{m}})$ [13, Proposition 1.8(ii)]. Hence, by Theorem 3(3), $\text{Int}(D)$ is a locally free D-module. However, there exists a principal maximal ideal of D with finite residue field, then it follows from Proposition 12 that $\text{Int}(D)$ is neither flat over $D[X]$ nor GCD. Moreover, by Corollary 12, $\text{Int}(D)$ is not t-linked over $D[X]$.

Notice also that D is an H-domain, that is, every t-maximal ideal is divisorial, which is neither almost Krull nor TV, that is, $t < v$. Then, by [19, Proposition 3.3], int prime ideals of D are exactly the maximal ideals of D with finite residue fields.

Acknowledgments The author would like to thank the referee for carefully reading of the manuscript and several valuable suggestions and corrections.

References

1. Anderson, D.D., Anderson, D.F., Zafrullah, M.: Rings between $D[X]$ and $K[X]$. Houston J. Math. **17**, 109–129 (1991)

2. Arnold, J.T., Matsuda, R.: An almost Krull domain with divisorial height one primes. Canad. Math. Bull. **29**(1), 50–53 (1986)
3. Bourbaki, N.: Algèbre Commutative, chapitres 1 à 4. 1ère édition Hermann, Paris, (1961)
4. Bourbaki, N.: Commutative Algebra, chapters 1–4. Herman Translated from the French Reprint of the 1989 English translation. Elements of Mathematics (Berlin). Springer-Verlag, Berlin, (1998)
5. Boulanger, J., Chabert, J.-L, Evrard, S.E., Gerboud, G.: The characteristic sequence of integer-valued polynomials on a subset. In: Advances in Commutative Ring Theory, (Fez 1997), Lecture Notes in Pure and Applied Mathematics, vol. 205, pp. 161–174. Dekker, New York (1999)
6. Cahen, P.-J.: Polynômes à valeurs entières. Canad. J. Math. **24**, 747–754 (1972)
7. Cahen, P.-J.: Integer-valued polynomials on a subset. Proc. Amer. Math. Soc. **117**, 919–929 (1993)
8. Cahen, P.-J., Chabert, J.-L.: Coefficients et valeurs d'un polynôme. Bull. SC. Math. Série 2, **95**, 295–304 (1971)
9. Cahen, P.-J., Chabert, J.-L.: Integer-Valued Polynomials, Mathematical Surveys and Monographs, vol. 48, American Mathematical Society, Providence (1997)
10. Cahen, P.-J., Fontana, M., Frisch, S., Glaz, S.: Open problems in commutative ring theory. In: Fontana, M., Frisch, S., Glaz, S. (eds.) Commutative Algebra: Recent Advances in Commutative Rings, Integer-Valued Polynomials, and Polynomial Functions, pp. 293–305. Springer, Berlin (2014)
11. Cahen, P.-J., Gabelli, S., Houston, E.G.: Mori domains of integer-valued polynomials. J. Pure Appl. Algebra **153**, 1–15 (2000)
12. Cahen, P.-J., Loper, A.K, Tartarone, F.: Integer-valued polynomials and Prüfer v-multiplication domains. J. Algebra **226**(2), 765–787 (2000)
13. Chabert, J.-L.: Les idéaux premiers de l'anneau des polynômes à valeurs entières. J. Reine Angew. Math. 293/294, 275–283 (1977)
14. Chabert, J.-L: Integer-valued polynomials, Prüfer domains and localization. Proc. Amer. Math. Soc. **118**(4), 1061–1073 (1993)
15. Chabert, J.-L, Chapman, S.T., Smith, W.: A Basis of the ring of polynomials integer-valued on prime numbers. In: Factorization in Integral Domains, (Iowa 1996), Lecture Notes in Pure and Applied Mathematics, vol. 189, pp. 271–284. Dekker, New York (1997)
16. Chang, G.W., Kang, B.G., Toan, P.T.: The Krull dimension of power series rings over almost Dedekind domains. J. Algebra **438**, 170–187 (2015)
17. El Baghdadi, S.: Semistar GCD domains. Comm. Algebra **38**, 3029–3044 (2010)
18. Elliott, J.: Universal properties of integer-valued polynomial rings. J. Algebra **318**, 68–92 (2007)
19. Elliott, J.: Some new approaches to integer-valued polynomial rings. In: Fontana, K., Olberding, S. (eds.) Proceedings of the Fifth International Fez Conference on Commutative Algebra and Applications, pp. 223–237. de Gruyter, New York (2009)
20. Elliott, J.: Integer-valued polynomial rings, t-closure, and associated primes. Comm. Algebra **39**(11), 4128–4147 (2011)
21. Elliott, J.: Presentations and module bases of integer-valued polynomial rings. J. Algebra Appl. **12**(1), 1–25 (2013)
22. Elliott, J.: Idempotent plethories. J. Algebra **463**, 33–79 (2016)
23. Fossum, R.: The Divisor Class Group of a Krull Domain. Springer, New York, (1973)
24. Frisch, S.: Integer-valued polynomials on Krull rings. Proc. Amer. Math. Soc. **124**, 3595–3604 (1996)
25. Gabelli, S., Tartarone, F.: Complete integral closure and the Noetherian property for the integer-valued polynomial rings. In: Commutative Ring Theory and Applications, Proceedings of the Fourth International Fez Conference on Commutative Ring Theory and Applications, (Fez 2001), Lect. Notes in Pure Appl. Math. vol. 231, pp. 173–189. Dekker, New York (2003)
26. Gilmer, R.: Overrings of Prüfer domains. J. Algebra **4**, 331–340 (1966)

27. Heinzer, W.: An essential integral domain with a nonessential localization. Canad. J. Math. **33**, 400–403 (1981)
28. Hutchins, H.C.: Examples of Commutative Rings. Polygonal Publishing House, Washington, (1981)
29. Izelgue, L., Mimouni, A., Tamoussit, A.: On the module structure of the integer-valued polynomial rings. Bull. Malays. Math. Sci. Soc. **43**, 2687–2699 (2020)
30. Izelgue, L., Tamoussit, A.: On the flatness of Int(D) as a $D[X]$-module. Gulf J. Math. **4**(4), 39–47 (2016)
31. Izelgue, L., Tamoussit, A.: When is Int(E, D) a locally free D-module. In: Badawi, A., Vedadi, M., Yassemi, S., Yousefian Darani, A. (eds.) Homological and Combinatorial Methods in Algebra, SAA 2016, Springer Proceedings in Mathematics & Statistics, vol. 228, pp. 105–110. Springer, Cham (2018)
32. Kang, B.G.: Prüfer v-multiplication domains and the ring $R[X]_{N_v}$. J. Algebra **123**, 151–170 (1989)
33. Lazard, D.: Autour de la platitude. Bulletin de la S. M. F., tome **97**, 81–128 (1969)
34. Lombardi, H., Quitté, C.: Commutative Algebra: Constructive Methods. Finite Projective Modules. Springer, New York, (2015)
35. Martin, M.B., Zafrullah, M.: t-Linked overrings of Noetherian weakly factorial domains. Proc. Amer. Math. Soc. **115**(3), 601–601 (1992)
36. Matsumura, H.: Commutative Ring Theory, Cambridge University Press, Cambridge, (1986)
37. Mulay, S.B.: On integer-valued polynomials. In: Zero-Dimensional Commutative Rings, Pure Applied Mathematics, Dekker, New York, vol. 171, pp. 331–345 (1995)
38. Ostrowski, A.: Über ganzwertige Polynome in algebraischen Zahlkörpern. J. Reine Angew. Math. **149**, 117–124 (1919)
39. Park, M.H., Tartarone, F.: Strong Mori and Noetherian properties of integer-valued polynomial rings. J. Pure Appl. Algebra **186**(3), 297–309 (2004)
40. Pólya, G.: Über ganzwertige Polynome in algebraischen Zahlkörpern. J. Reine Angew. Math. **149**, 79–116 (1919)
41. Richman, F.: Generalized quotient rings. Proc. Amer. Math. Soc. **16**, 794–799 (1965)
42. Rotman, J.: An Introduction to Homological Algebra. (Second Edition), Universitex, Springer, New York, (2009)
43. Serre, J.-P.: Géométrie algébrique et géométrie analytique. Ann. Inst. Fourier **6**, 1–42 (1956)
44. Skolem, T.: Ein Satz über ganzwertige Polynome. Det Kongelige Norske Videnskabers Selskab (Trondheim) **9**, 111–113 (1936)
45. Tamoussit, A.: Rings of Integer-Valued Polynomials as Flat Modules, Almost Krull Domains and Star-Operations. Ph.D. Thesis, Cadi Ayyad University, Marrakech, (2020)
46. Tamoussit, A.: Note on integer-valued polynomials on a residually cofinite subset. Beitr. Algebra Geom. **62**, 599–604 (2021)
47. Tamoussit, A.: On the ring of D-valued R-polynomials over E. J. Algebra Appl. **21**(5), 2250087 (2022). https://doi.org/10.1142/S0219498822500876
48. Tamoussit, A.: Some results on t-locally Strong Mori domains and their rings of integer-valued polynomials. Rend. Circ. Mat. Palermo, II. Ser **71**, 349–360 (2021)
49. Zafrullah, M.: The $D + XD_S[X]$ construction from GCD-domains. J. Pure Appl. Algebra **50**, 93–107 (1988)
50. Zantema, H.: Integer-valued polynomials over a number field. Manuscripta Math. **40**, 155–240 (1982)

Equivalent Characterizations of Non-Archimedean Uniform Spaces

Daniel Windisch

1 Introduction

Throughout this manuscript, let X be a set. A *uniform space* is a pair (X, M), where M is a system of pseudo-metrics on X, that is,

(i) The elements of M are generalizations of metrics for which two distinct points of X may have distance 0.
(ii) M is closed under the operation $\max(\cdot, \cdot)$, where $d = \max(d_1, d_2)$ is the pseudo-metric defined by $d(x, y) = \max(d_1(x, y), d_2(x, y))$ for $d_1, d_2 \in M$ and $x, y \in X$.

Equivalently, one can define a uniform space by a diagonal uniformity or by a covering uniformity, respectively, and the three concepts easily translate back and forth, see Sect. 2.1. For a general introduction to uniform spaces, see [5].

Following Monna [4], a uniform space (X, M) is called *non-Archimedean* if the pseudo-metrics $d \in M$ satisfy the non-Archimedean triangle inequality

$$d(x, z) \leq \max(d(x, y), d(y, z))$$

for all $x, y, z \in X$. In many settings, especially in commutative algebra and Abelian group theory, non-Archimedean uniformities are used extensively. For instance, the completion of a commutative ring R with respect to a valuation or a directed system of ideals is a special case of the more general concept of completions of uniform spaces, and the original uniformity defined on R is non-Archimedean, cf. [1, Chapter 10]. In this context, one often freely uses the correspondence of

D. Windisch (✉)
Institut für Analysis und Zahlentheorie, Technische Universität Graz, Graz, Austria
e-mail: dwindisch@math.tugraz.at

J.-L. Chabert et al. (eds.), *Algebraic, Number Theoretic, and Topological Aspects of Ring Theory*, https://doi.org/10.1007/978-3-031-28847-0_24

diagonal uniformities, covering uniformities and systems of pseudo-metrics. So the following question arises naturally:

Question Is it possible to characterize those diagonal uniformities and covering uniformities, respectively, that correspond to systems of pseudo-metrics satisfying the non-Archimedean triangle inequality?

We give a positive answer to this question in Theorem 1: the analogs of systems of pseudo-metrics satisfying the non-Archimedean triangle inequality are diagonal uniformities admitting a basis of equivalence relations and covering uniformities with a basis of partitions, respectively.

It is well-known that uniformizability for a topological space is equivalent to $T_{3\frac{1}{2}}$ (Willard [5] uses the term *completely regular* for $T_{3\frac{1}{2}}$). We introduce the following stronger separation axiom to characterize in Theorem 2 those topological spaces X stemming from a uniform space of non-Archimedean pseudo-metrics:

(T_A) For every closed subset $A \subseteq X$ and every $x \in X \setminus A$, there are open sets $U_1, U_2 \subseteq X$ such that

(i) $U_1 \cap U_2 = \emptyset, U_1 \cup U_2 = X$.
(ii) $A \subseteq U_1, x \in U_2$.

Moreover, we are able to characterize when a non-Archimedean uniform space is pseudo-metrizable by a single non-Archimedean pseudo-metric. Results closely related to ours were given by Monna [4, Théorème 13], Banaschewski [2, Satz 6] and de Groot [3, Theorem II]. These authors assume uniformities to be separating (i.e., the induced topology is Hausdorff), and therefore they prove results dealing with metrizability. Our Theorem 3 is similar to the general result that a uniform space is pseudo-metrizable if and only if there is a countable system of pseudo-metrics inducing the uniformity (cf. [5, Theorem 38.3]).

Overview of results

	General uniform space	Non-Arch. uniform space
Diagonal uniformity	Basis of relations	Basis of equivalence relations
Covering uniformity	Basis of covers	Basis of partitions
Pseudo-metrics	System of pseudo-metrics	System of non-Arch. pseudo-metrics
Uniformizability	$T_{3\frac{1}{2}}$	T_A
Pseudo-metrizability	Countable system of pseudo-metrics	Countable system of non-Arch. pseudo-metrics

2 Equivalent Approaches to Non-Archimedean Uniform Spaces

2.1 The General Case

We first want to recall the various approaches to uniformities in the general case. We follow the text book by Willard [5]. It is worth noting that the definitions of uniformities and their bases, which might seem a bit clumsy and unnatural in the general case, simplify a lot when passing over to non-Archimedean uniformities. In general, for all of the three types of uniformities introduced, a set X together with a uniformity of the respective type is called a *uniform space*.

We begin with the definition of diagonal uniformities. For relations $E, F \subseteq X \times X$, we denote

$$E \circ F = \{(x, z) \in X \times X \mid \exists y \in X\ ((x, y) \in E \wedge (y, z) \in F)\}$$

and

$$E^{-1} = \{(y, x) \in X \times X \mid (x, y) \in E\}.$$

Definition 1 ([5, Definition 35.2]) A *diagonal uniformity* $\mathfrak{D}$ on the set X is a set of relations on X satisfying:

(a) $(x, x) \in D$ for all $x \in X$ and $D \in \mathfrak{D}$.
(b) $D_1 \cap D_2 \in \mathfrak{D}$ whenever $D_1, D_2 \in \mathfrak{D}$.
(c) For each $D \in \mathfrak{D}$ there is $E \in \mathfrak{D}$ such that $E \circ E \subseteq D$.
(d) For each $D \in \mathfrak{D}$ there is $E \in \mathfrak{D}$ such that $E^{-1} \subseteq D$.
(e) $D \in \mathfrak{D}$ and $D \subseteq E \subseteq X \times X$ implies $E \in \mathfrak{D}$.

Note that, assuming (a), the assertions (c) and (d) may be seen as a generalization of asking that all relations in $\mathfrak{D}$ be equivalence relations.

Definition 2 ([5, Definition 35.2]) A *basis* of a diagonal uniformity $\mathfrak{D}$ on X is a subset $\mathcal{B} \subseteq \mathfrak{D}$ such that each $D \in \mathfrak{D}$ contains some $E \in \mathcal{B}$.

It is easy to see that a basis $\mathcal{B}$ of a diagonal uniformity satisfies (a), (c), and (d) of Definition 1 (substituting $\mathfrak{D}$ with $\mathcal{B}$). Moreover, the following modified form of (b) holds:

(b') For all $D_1, D_2 \in \mathcal{B}$ there exists some $E \in \mathcal{B}$ such that $E \subseteq D_1 \cap D_2$.

Conversely, given a set $\mathcal{B}$ of relations on X satisfying (a), (b'), (c) and (d), $\mathcal{B}$ is a basis of the diagonal uniformity:

$$\mathfrak{D} = \{D \subseteq X \times X \mid \exists E \in \mathcal{B}\ E \subseteq D\}.$$

We now recall the concept of covering uniformities. It is an equivalent description of diagonal uniformities that uses covers of X rather than relations on X. A *cover* $\mathcal{U}$ of a set X is a collection of subsets of X such that $\bigcup_{U\in\mathcal{U}} U = X$. A cover $\mathcal{U}$ is a *refinement* of a cover $\mathcal{V}$ if each $U \in \mathcal{U}$ is contained in some $V \in \mathcal{V}$. In this case, we write $\mathcal{U} < \mathcal{V}$.

For a cover $\mathcal{U}$ of X and $A \subseteq X$ define the *star* of A with respect to $\mathcal{U}$ as

$$star(A,\mathcal{U}) = \bigcup_{\substack{U \in \mathcal{U} \\ A\cap U \neq \emptyset}} U.$$

$\mathcal{U}$ is a star-refinement of a cover $\mathcal{V}$ if for each $U \in \mathcal{U}$ there exists $V \in \mathcal{V}$ such that $star(U,\mathcal{U}) \subseteq V$. In this case, we write $\mathcal{U} <^* \mathcal{V}$.

Definition 3 ([5, page 244 ff.]) A *covering uniformity* on a set X is a collection μ of covers of X satisfying:

(a) For all $\mathcal{U}_1, \mathcal{U}_2 \in \mu$ there exists $\mathcal{U}_3 \in \mu$ such that $\mathcal{U}_3 <^* \mathcal{U}_1$ and $\mathcal{U}_3 <^* \mathcal{U}_2$.
(b) For all covers $\mathcal{V}$ of X, if $\mathcal{U} < \mathcal{V}$ for some $\mathcal{U} \in \mu$ then $\mathcal{V} \in \mu$.

Definition 4 ([5, Definition 36.3]) A *basis* of a covering uniformity μ on X is a subset β of μ such that for every $\mathcal{V} \in \mu$ there exists $\mathcal{U} \in \beta$ such that $\mathcal{U} < \mathcal{V}$.

Like for diagonal uniformities, every basis of a covering uniformity satisfies (a) of Definition 3. Conversely, any collection β of covers of X satisfying (a) is a basis of the covering uniformity:

$$\mu = \{\mathcal{V} \text{ covering of } X \mid \exists \mathcal{U} \in \beta\, \mathcal{U} < \mathcal{V}\}.$$

Finally, one can approach uniformities from a more analytic point of view, namely, via pseudo-metrics. A *pseudo-metric* on a set X is a function $d : X \times X \to \mathbb{R}_{\geq 0}$ satisfying for all $x, y, z \in X$:

(a) $d(x,x) = 0$.
(b) $d(x,y) = d(y,x)$.
(c) $d(x,z) \leq d(x,y) + d(y,z)$.

$d(x,y)$ can be seen as a distance between two points $x, y \in X$. In contrast to metrics, it can happen that two distinct points have distance 0 with respect to a pseudo-metric.

Let $d_1, d_2 : X \times X \to \mathbb{R}_{\geq 0}$ be two pseudo-metrics on X. It is easy to see that

$$\begin{aligned} \max(d_1, d_2) : X \times X &\to \mathbb{R}_{\geq 0} \\ (x,y) &\mapsto \max(d_1(x,y), d_2(x,y)) \end{aligned}$$

is again a pseudo-metric on X.

Definition 5 A *system of pseudo-metrics* on X is a set M whose elements are pseudo-metrics on X such that $\max(d_1, d_2) \in M$ for all $d_1, d_2 \in M$.

The following standard observations on uniformities follow immediately from the definitions.

Proposition 1

(1) *Let $\mathfrak{D}$ be a diagonal uniformity on X, and let $\mathcal{B}$ be a basis of $\mathfrak{D}$. For $x \in X$ and $D \in \mathcal{B}$, let $D[x] := \{y \in X \mid (x, y) \in D\}$ and $\mathcal{U}_D := \{D[x] \mid x \in X\}$. Then $\beta_{\mathfrak{D}} := \{\mathcal{U}_D \mid D \in \mathcal{B}\}$ is a basis of a covering uniformity $\mu_{\mathfrak{D}}$ on X, and $\mu_{\mathfrak{D}}$ is independent of the choice of $\mathcal{B}$.*

(2) *Let β be a basis of a covering uniformity μ on X. For $\mathcal{U} \in \beta$, we define $D_{\mathcal{U}} := \{(x, y) \in X \times X \mid \exists U \in \mathcal{U}\ x, y \in U\}$. Then $\mathcal{B}_\mu := \{D_{\mathcal{U}} \mid \mathcal{U} \in \beta\}$ is a basis of a diagonal uniformity $\mathfrak{D}_\mu$ on X, and $\mathfrak{D}_\mu$ is independent of the choice of β.*

(3) *[5, Theorem 36.4] The maps $\mathfrak{D} \mapsto \mu_{\mathfrak{D}}$, $\mu \mapsto \mathfrak{D}_\mu$ are bijections of the set of all diagonal uniformities on X and the set of all covering uniformities on X which are inverse to each other.*

(4) *Let M be a system of pseudo-metrics on X. For every $d \in M$ and every $\varepsilon \in \mathbb{R}_{>0}$ we consider the binary relation $D_\varepsilon^d := \{(x, y) \in X \times X \mid d(x, y) < \varepsilon\}$. The set $\mathcal{B}_M := \{D_\varepsilon^d \mid d \in M,\ \varepsilon \in \mathbb{R}_{>0}\}$ is a basis of a diagonal uniformity on X, denoted by $\mathfrak{D}_M$.*

Proposition 1(3) shows that the covering uniformities and the diagonal uniformities on a set X are in one-to-one correspondence. This is not true in general for systems of pseudo-metrics and diagonal uniformities (respectively covering uniformities). Thus, it makes sense to consider systems of pseudo-metrics up to equivalence. Two systems M and N of pseudo-metrics on X are said to be *equivalent* if $\mathfrak{D}_M = \mathfrak{D}_N$.

2.2 The Non-Archimedean Case

It is probably most natural to motivate non-Archimedean uniformities from the view-point of pseudo-metrics. A pseudo-metric d on X is called *non-Archimedean* if it satisfies the non-Archimedean triangle inequality, i.e.,

$$d(x, y) \leq \max(d(x, z), d(z, y))$$

for all $x, y, z \in X$. Pseudo-metrics of this type appear in several different areas of mathematics, most prominently as absolute values induced by prime ideals (or more generally by rank-one valuations) in algebraic number theory.

If d_1 and d_2 are non-Archimedean pseudo-metrics on X, then so is $d = \max(d_1, d_2)$. Indeed, if $x, y, z \in X$ then

$$
\begin{aligned}
d(x, y) &= \max(d_1(x, y), d_2(x, y)) \\
&\leq \max(\max(d_1(x, z), d_1(z, y)), \max(d_2(x, z), d_2(z, y))) \\
&= \max(d_1(x, z), d_2(x, z), d_1(z, y), d_2(z, y)) \\
&= \max(\max(d_1(x, z), d_2(x, z)), \max(d_1(z, y), d_2(z, y))) \\
&= \max(d(x, z), d(z, y)).
\end{aligned}
$$

Definition 6 A *non-Archimedean uniform space* is a pair (X, M), where X is a set and M is a system of pseudo-metrics on X whose elements are non-Archimedean.

In what follows, we see that the sensible analog of non-Archimedean uniform spaces is that of diagonal uniformities admitting a basis whose elements are equivalence relations. Note that a collection of equivalence relations on X automatically satisfies (a), (c), and (d) from Definition 1 and is therefore always a basis of a uniformity provided it has property (b') from Definition 2. Also, it is rather convenient that in the special case of partitions, star-refinements and refinements are the same and thus, every partition is a star-refinement of itself.

Proposition 2 *A diagonal uniformity $\mathfrak{D}$ on a set X admits a basis of equivalence relations if and only if $\mathfrak{D} = \mathfrak{D}_M$ for a system M of non-Archimedean pseudo-metrics on X.*

Proof First, we show that a diagonal uniformity induced by a system of non-Archimedean pseudo-metrics M on X admits a basis of equivalence relations. Let $\varepsilon \in \mathbb{R}_{>0}$ and $d \in M$. Clearly, D^d_ε is reflexive and symmetric. So let $(x, z), (z, y) \in D^d_\varepsilon$. Then $d(x, y) \leq \max(d(x, z), d(z, y)) < \varepsilon$, which means that $(x, y) \in D^d_\varepsilon$.

Now, assume that $\mathfrak{D}$ admits a basis $\mathcal{B}$ whose elements are equivalence relations. For $E \in \mathcal{B}$, define a pseudo-metric d_E on X by

$$
d_E(x, y) = \begin{cases} 0 & \text{if } (x, y) \in E, \\ 1 & \text{else.} \end{cases}
$$

Let M be the system of pseudo-metrics generated by the pseudo-metrics of the form d_E for $E \in \mathcal{B}$. In other words, M is the set of all pseudo-metrics $d = \max(d_{E_1}, \ldots, d_{E_n})$, where n varies over all positive integers, $E_1, \ldots, E_n \in \mathcal{B}$ and $\max(d_{E_1}, \ldots, d_{E_n}) = \max(\max(d_{E_1}, \ldots, d_{E_{n-1}}), d_{E_n})$ recursively. By induction, the elements of M are non-Archimedean pseudo-metrics.

M induces a diagonal uniformity on X which admits a basis $\mathcal{B}'$ formed by the relations

$$
\{(x, y) \in X \times X \mid d(x, y) < \varepsilon\}
$$

where $d \in M$ and $\varepsilon > 0$ is a real number. This diagonal uniformity coincides with $\mathfrak{D}$. Indeed, $d_E \in M$ for each $E \in \mathcal{B}$, and hence $\mathcal{B} \subseteq \mathcal{B}'$. Conversely, let $d \in M$, $\varepsilon >$

0 and $E = \{(x, y) \in X \times X \mid d(x, y) < \varepsilon\} \in \mathcal{B}'$. Then $d = \max(d_{E_1}, \ldots, d_{E_n})$ for some $E_i \in \mathcal{B}$ and $E = \bigcap_{i=1}^{n}\{(x, y) \mid d_{E_i}(x, y) < \varepsilon\}$. If $\varepsilon \leq 1$ then $E = \bigcap_{i=1}^{n} E_i$ contains an element of $\mathcal{B}$ by definition of a basis. If $\varepsilon > 1$ then $E = X \times X$ also contains an element of $\mathcal{B}$.

Next, we see that the covering uniformities analogous to non-Archimedean uniform spaces are those admitting a basis whose elements are partitions.

Proposition 3

(1) If a diagonal uniformity $\mathfrak{D}$ has a basis of equivalence relations, then the covering uniformity $\mu_{\mathfrak{D}}$ has a basis whose elements are partitions.
(2) If a covering uniformity μ has a basis consisting of partitions then the diagonal uniformity D_μ admits a basis of equivalence relations.
(3) The maps in Proposition 1(3) restrict to bijections of diagonal uniformities having a basis of equivalence relations and covering uniformities having a basis of partitions.

Proof

(1) Let $\mathcal{B} \subseteq \mathfrak{D}$ be a basis consisting of equivalence relations. Then, for all $D \in \mathcal{B}$, we have that $\mathcal{U}_D = \{D[x] \mid x \in X\}$ is a partition of X.
(2) Let $\beta \subseteq \mu$ be a basis consisting of partitions of X. For $\mathcal{U} \in \beta$, $D_{\mathcal{U}} = \{(x, y) \in X \times X \mid \exists U \in \mathcal{U} : x, y \in U\}$ is an equivalence relation.
(3) is an immediate consequence of (1) and (2).

Theorem 1 *For a diagonal uniformity $\mathfrak{D}$ on X, the following are equivalent:*

(A) $\mathfrak{D}$ admits a basis of equivalence relations.
(B) There exists a system M of non-Archimedean pseudo-metrics on X that induces the uniformity $\mathfrak{D}$.
(C) The corresponding covering uniformity $\mu_{\mathfrak{D}}$ of $\mathfrak{D}$ has a basis consisting of partitions of X.

3 An Equivalent Separation Axiom

3.1 The General Case

A diagonal uniformity $\mathfrak{D}$ on a set X canonically induces a topology on X in the following way: let $\mathcal{B} \subseteq \mathfrak{D}$ be any basis and define $E[x] = \{y \in X \mid (x, y) \in E\}$ for $x \in X$ and $E \in \mathcal{B}$. For $x \in X$, the collection $\mathcal{U}_x = \{E[x] \mid E \in \mathcal{B}\}$ forms a neighborhood basis of a topology on X; see [5, Theorem 35.6]. This topology is called the *topology induced by* $\mathfrak{D}$ and is independent of the choice of $\mathcal{B}$.

Definition 7 ([5, Definition 35.7]) Let (X, τ) be a topological space. X is called *uniformizable* if there exists a uniformity $\mathfrak{D}$ such that τ is the topology induced by $\mathfrak{D}$.

In the general case, the uniformizable topological spaces can be characterized by the following separation axiom: a topological space X is $\mathrm{T}_{3\frac{1}{2}}$ (or, in the terminology of Willard [5], *completely regular*) if for every closed set A in X and every $x \in X \setminus A$ there exists a continuous function $f : X \to [0, 1]$ such that $f(x) = 0$ and $f(A) = \{1\}$, where $[0, 1]$ denotes the unit interval with the canonical topology of real numbers.

Proposition 4 ([5, Theorem 38.2]) *A topological space is uniformizable if and only if it is* $T_{3\frac{1}{2}}$.

3.2 *The Non-Archimedean Case*

Considering a stronger version of $\mathrm{T}_{3\frac{1}{2}}$, we are able to give a characterization of topologies induced by non-Archimedean uniformities analogous to Proposition 4.

Definition 8 Let (X, τ) be a topological space.

(1) X is called *uniformizable by a non-Archimedean uniformity* if there exists a diagonal uniformity on X that admits a basis of equivalence relations and induces τ in the sense of Sect. 3.1.
(2) X is said to satisfy T_{A} if for every closed subset $A \subseteq X$ and every $x \in X \setminus A$, there are open sets $U_1, U_2 \subseteq X$ such that

   (i) $U_1 \cap U_2 = \emptyset$, $U_1 \cup U_2 = X$.
   (ii) $A \subseteq U_1$, $x \in U_2$.

(3) X is said to be *zero-dimensional* (*with respect to the small inductive dimension*) if it has a basis of clopen sets.

The equivalence of (A) and (C) in the following theorem was shown by Banaschewski [2] for Hausdorff spaces.

Theorem 2 *For a topological space* (X, τ), *the following are equivalent:*

(A) X is uniformizable by a non-Archimedean uniformity.
(B) X satisfies T_A.
(C) X is zero-dimensional.
(D) For every closed set A in X and every $x \in X \setminus A$ *there exists a continuous function* $f : X \to \{0, 1\}$ *such that* $f(x) = 0$ *and* $f(A) = \{1\}$, *where* $\{0, 1\}$ *carries the discrete topology.*

Proof The equivalence of (B) and (D) is obvious.

"(B) $\Rightarrow$ (A)" Assume that X satisfies $\mathrm{T_A}$. We consider the set of continuous functions $S = \{f : X \to \{0,1\} \mid f \text{ is continuous}\}$, where $\{0,1\}$ carries the discrete topology. For every $f \in S$, we define an equivalence relation:

$$D_f = \{(x, y) \in X \times X \mid f(x) = f(y)\}.$$

It is immediate that $\mathcal{B} := \{D_{f_1} \cap ... \cap D_{f_n} \mid f_i \in S\}$ is a basis consisting of equivalence relations for a non-Archimedean uniformity $\mathfrak{D}$ on X. We denote by σ the topology on X induced by $\mathfrak{D}$ and assert that $\tau = \sigma$.

Let $A \subseteq X$ be closed with respect to τ and $x \in X \setminus A$. Let $U_1, U_2 \subseteq X$ open with respect to τ such that $U_1 \cap U_2 = \emptyset$, $U_1 \cup U_2 = X$, $x \in U_2$, $A \subseteq U_1$ and let $f_x : X \to \{0,1\}$ be such that f_x takes the value 0 on U_2 and the value 1 on U_1. Then clearly f_x is continuous with respect to τ. Per definition,

$$D_{f_x}[x] = \{y \in X \mid f_x(y) = f_x(x) = 0\} = U_2$$

is open with respect to σ and disjoint from A. Since $x \in X \setminus A$ was arbitrarily chosen, we conclude that

$$A = X \setminus \bigcup_{x \in X \setminus A} D_{f_x}[x]$$

is closed with respect to σ. So σ is finer than τ.

For the reverse direction, it suffices to show that $D[x]$ is open with respect to τ for all $x \in X$ and $D \in \mathcal{B}$. But since $D = D_{f_1} \cap ... \cap D_{f_n}$ for some $f_1, ..., f_n \in S$ and $D[x] = D_{f_1}[x] \cap ... \cap D_{f_n}[x]$ for all $x \in X$, we just have to show this for D_f with $f \in S$. For every $x \in X$, we have $D_f[x] = \{y \in X \mid f(x) = f(y)\} = f^{-1}(\{f(x)\})$. Since f is continuous and $\{f(x)\}$ is open, it follows that $D_f[x] \in \tau$.

So X is uniformizable by a non-Archimedean uniformity.

"(A) $\Rightarrow$ (C)" Let $\mathfrak{D}$ be a non-Archimedean diagonal uniformity inducing τ and let $\mathcal{B} \subseteq \mathfrak{D}$ be a basis consisting of equivalence relations. For every $x \in X$, the sets $D[x]$ where D runs through $\mathcal{B}$ form a neighborhood basis for x. Since every $D[x]$ is a neighborhood of each of its points, it is an open set. Conversely, given an open set $U \in \tau$, for every $x \in U$ there exists $D_x \in \mathcal{B}$ such that $D_x[x] \subseteq U$, since these sets form neighborhood bases as said before. Now clearly, $U = \bigcup_{x \in U} D_x[x]$. Therefore

$$\{D_1[x_1] \cap ... \cap D_n[x_n] \mid D_i \in \mathcal{B}, x_i \in X\}$$

is a basis of clopen sets for τ, since $D[x]$ is clopen for every $D \in \mathcal{B}$ and every $x \in X$.

"(C) $\Rightarrow$ (B)" Let $A \subseteq X$ be closed and $x \in X \setminus A$. Since $X \setminus A$ is open and therefore a union of clopen sets, there exists some $U_2 \subseteq X \setminus A$ that is clopen and contains x. Set $U_1 = X \setminus U_2$.

On the one hand, T_A implies $T_{3\frac{1}{2}}$ by assertion (D) in Theorem 2. On the other hand, T_A does not fit into the classical hierarchy of separation axioms, as normal spaces are not necessarily T_A. Consider for instance $\mathbb{R}$ carrying the canonical topology. This space is normal, but it is connected, hence does not satisfy T_A.

4 Pseudo-Metrizability

Finally, we give a characterization of uniform spaces that are pseudo-metrizable by a single non-Archimedean pseudo-metric. We first recall the situation for arbitrary uniform spaces.

4.1 The General Case

Definition 9 A diagonal uniformity $\mathfrak{D}$ on a set X is *pseudo-metrizable* if there exists a single pseudo-metric d on X such that $\mathfrak{D} = \mathfrak{D}_{\{d\}}$ in the sense of Proposition 1(4).

Proposition 5 ([5, Theorem 38.3]) *Let $\mathfrak{D}$ be a diagonal uniformity on a set X. Then the following are equivalent:*

(1) $\mathfrak{D}$ pseudo-metrizable.
(2) $\mathfrak{D}$ is induced by a countable system of pseudo-metrics in the sense of Proposition 1(4).
(3) $\mathfrak{D}$ possesses a countable basis.

4.2 The Non-Archimedean Case

Inspired by the general case, we give the following definition.

Definition 10 A diagonal uniformity $\mathfrak{D}$ on a set X is *pseudo-metrizable by a non-Archimedean pseudo-metric* if there exists a single non-Archimedean pseudo-metric d on X such that $\mathfrak{D} = \mathfrak{D}_{\{d\}}$ in the sense of Proposition 1(4).

Theorem 3 *Let $\mathfrak{D}$ be a diagonal uniformity on a set X. Then the following are equivalent:*

(A) $\mathfrak{D}$ is pseudo-metrizable by a non-Archimedean pseudo-metric.
(B) $\mathfrak{D}$ is induced by a countable system of non-Archimedean pseudo-metrics in the sense of Proposition 1(4).
(C) $\mathfrak{D}$ possesses a countable basis consisting of equivalence relations.

Proof (A)⇒(B) is trivial.

For (B)⇒(C), let $(d_n)_{n\in\mathbb{N}}$ be a countable system of non-Archimedean pseudo-metrics inducing $\mathfrak{D}$. Then $(D^{d_n}_{1/m})_{n,m\in\mathbb{N}}$ is a countable basis of $\mathfrak{D}$ consisting of equivalence relations.

To see (C)⇒(A), let $(E_n)_{n\in\mathbb{N}}$ be a basis for $\mathfrak{D}$ consisting of equivalence relations and assume without restriction that $E_1 = X \times X$. We get another countable basis $(D_n)_{n\in\mathbb{N}}$ (of equivalence relations) by applying the properties of a basis of a uniformity and iterating the following recursion:

$$D_1 := E_1$$

$$\text{Choose } D_{i+1} \in \{E_n \mid n \in \mathbb{N}\} \text{ such that } D_{i+1} \subseteq D_i \cap E_{i+1},$$

$$\text{for } i \in \mathbb{N}.$$

Note that $(D_n)_{n\in\mathbb{N}}$ is a decreasing chain, i.e., $X \times X = D_1 \supseteq D_2 \supseteq D_3 \supseteq \ldots$, which allows us to define a non-Archimedean pseudo-metric as follows:

$$d : X \times X \to \mathbb{R}_{\geq 0}$$

$$(x, y) \mapsto \begin{cases} 0 \text{ if } \forall n \in \mathbb{N} : (x, y) \in D_n \\ \frac{1}{n} \text{ if } n = \max\{m \in \mathbb{N} \mid (x, y) \in D_m\}. \end{cases}$$

To see that d induces $\mathfrak{D}$, let $\varepsilon > 0$ and $n \in \mathbb{N}$ such that $\frac{1}{n} < \varepsilon$. For $(x, y) \in D_n$, we have $d(x, y) \leq \frac{1}{n} < \varepsilon$, which implies $(x, y) \in D^d_\varepsilon$.
On the other hand, let $n \in \mathbb{N}$. For $(x, y) \in D^d_{\frac{1}{n+1}}$, it follows that $d(x, y) \leq \frac{1}{n+1} < \frac{1}{n}$, hence $(x, y) \in D_n$. This completes the proof.

We want to illustrate the concepts and results of this manuscript by an easy and well-known example that is frequently used in algebraic number theory and commutative algebra.

Let R be a commutative ring and let M be a maximal ideal of R. The map

$$d : R \times R \to \mathbb{R}_{\geq 0}$$

$$(x, y) \mapsto 2^{-\mathrm{ord}(x-y)},$$

where $\mathrm{ord}(z) = \max\{n \in \mathbb{N}_0 \mid z \in M^n\}$ for $z \in R$ and $2^{-\infty} = 0$, is a pseudo-metric on R. Its induced uniformity has as a basis the equivalence relations $\{(x, y) \in R \times R \mid x - y \in M^n\}$ for $n \in \mathbb{N}_0$. The topology on R induced by this uniformity admits as a neighborhood basis for $x \in R$ the residue classes $x + M^n$ for $n \in \mathbb{N}_0$.

This topology is Hausdorff if and only if d is a metric which is the case if and only if $\bigcap_{n\in\mathbb{N}_0} M^n = 0$. This is true, for instance, for a Noetherian integral domain R by Krull's Intersection Theorem. It follows from [5, Theorem 39.9] that the completion of R with respect to d is compact if and only if the residue rings R/M^n are finite for all positive integers n.

Acknowledgments I wish to thank the anonymous referee for his/her suggestions which improved the structure of the present manuscript a lot and made possible a short and elegant proof of Proposition 2.

I am also very grateful to the Austrian Science Fund (FWF) for supporting me within the project P 30934.

References

1. M.F. Atiyah, I.G. MacDonald, *Introduction to Commutative Algebra*. Addison-Wesley Publishing Company, Reading, MA (1969).
2. B. Banaschewski, *Über nulldimensionale Räume*. Math. Nachr. **13** (1955), 129-140.
3. J. de Groot, *Non-Archimedean metrics in topology*. Proc. Am. Math. Soc. **7:5** (1956), 948–953.
4. A.F. Monna, *Rémarques sur les métriques non-archimédiennes* I - II. Indag. Math. **12** (1950), 122-133, 179-191.
5. St. Willard, *General Topology*. Addison-Wesley Publishing Company, Reading, MA (1970).

Printed in the USA
CPSIA information can be obtained
at www.ICGtesting.com
LVHW020557061123
763146LV00005B/72